The Great Ordovician Biodiversification Event

Critical Moments and Perspectives in Earth History and Paleobiology

Critical Moments and Perspectives in Earth History and Paleobiology
David J. Bottjer, Richard K. Bambach, and Hans-Dieter Sues, Editors

The Emergence of Animals:
The Cambrian Breakthrough
Mark A. S. McMenamin and
Dianna I. S. McMenamin

Phanerozoic Sea-Level Changes
Anthony Hallam

The Great Paleozoic Crisis:
Life and Death in the Permian
Douglas H. Erwin

Tracing the History of Eukaryotic Cells:
The Enigmatic Smile
Betsy Drexler and Robert Allan Obar

The Eocene-Oligocene Transition: Paradise Lost
Donald R. Prothero

The Late Devonian Mass Extinction:
The Frasnian/Famennian Crisis
George R. McGhee Jr.

Dinosaur Extinction and the End of an Era:
What the Fossils Say
J. David Archibald

One Long Experiment:
Scale and Process in Earth History
Ronald E. Martin

Interpreting Pre-Quaternary Climate
from the Geologic Record
Judith Totman Parrish

Theoretical Morphology:
The Concept and Its Applications
George R. McGhee Jr.

Principles of Paleoclimatology
Thomas M. Cronin

The Ecology of the Cambrian Radiation
Andrey Yu. Zhuravlev and Robert Riding, eds.

Plants Invade the Land:
Evolutionary and Environmental Perspectives
Patricia G. Gensel and Dianne Edwards, eds.

Exceptional Fossil Preservation:
A Unique View on the Evolution of Marine Life
David J. Bottjer, Walter Etter, James W. Hagadorn,
and Carol M. Tang, eds.

The Great Ordovician Biodiversification Event

EDITED BY

Barry D. Webby, Florentin Paris,
Mary L. Droser, and Ian G. Percival

Columbia University Press
NEW YORK

Columbia University Press
Publishers Since 1893
New York   Chichester, West Sussex

Copyright © 2004 Columbia University Press
All rights reserved

Library of Congress Cataloging-in-Publication Data
The great Ordovician biodiversification event / Barry D. Webby,
Florentin Paris, Mary L. Droser and Ian G. Percival, editors.
   p. cm.—(Critical moments and perspectives in Earth history and paleobiology)
Includes bibliographical references and index.
ISBN (cloth): 0231-12678-6
1. Biodiversity—Ordovician. 2. Paleontology—Ordovician. 3. Earth History—Ordovician. I. Webby, Barry D., 1934– II. International Geological Correlation Program, Project No. 410 (Ordovician biodiversity).   III. Series

Columbia University Press books are printed on permanent and durable acid-free paper.
Printed in the United States of America

c 10 9 8 7 6 5 4 3 2 1

This book is dedicated to all paleontologists who, over the past 200 years, have described the Ordovician biotas of the world. Their contributions have provided the knowledge base that our team of specialists have drawn on for this comprehensive stocktaking of arguably the greatest-ever interval of sustained diversification of life on earth.

CONTENTS

Acknowledgments ix

1. Introduction 1
   Barry D. Webby

PART I

Scaling of Ordovician Time and Measures for Assessing Biodiversity Change

2. Stratigraphic Framework and Time Slices 41
   Barry D. Webby, Roger A. Cooper, Stig M. Bergström, and Florentin Paris

3. Calibration of the Ordovician Timescale 48
   Peter M. Sadler and Roger A. Cooper

4. Measures of Diversity 52
   Roger A. Cooper

PART II

Conspectus of the Ordovician World

5. Major Terranes in the Ordovician 61
   L. Robin M. Cocks and Trond H. Torsvik

6. Isotopic Signatures 68
   Graham A. Shields and Ján Veizer

7. Ordovician Oceans and Climate 72
   Christopher R. Barnes

8. Was There an Ordovician Superplume Event? 77
   Christopher R. Barnes

9. End Ordovician Glaciation 81
   Patrick J. Brenchley

10. Ordovician Sea Level Changes: A Baltoscandian Perspective 84
    Arne Thorshøj Nielsen

PART III

Taxonomic Groups

11. Radiolarians 97
    Paula J. Noble and Taniel Danelian

12. Sponges 102
    Marcelo G. Carrera and J. Keith Rigby

13. Stromatoporoids 112
    Barry D. Webby

14. Conulariids 119
    Heyo Van Iten and Zdenka Vyhlasová

15. Corals 124
    Barry D. Webby, Robert J. Elias, Graham A. Young, Björn E. E. Neuman, and Dimitri Kaljo

16. Bryozoans 147
    Paul D. Taylor and Andrej Ernst

Contents

17　Brachiopods　157
　　David A. T. Harper, L. Robin M. Cocks, Leonid E. Popov,
　　Peter M. Sheehan, Michael G. Bassett, Paul Copper,
　　Lars E. Holmer, Jisuo Jin, and Rong Jia-yu

18　Polyplacophoran and Symmetrical
　　Univalve Mollusks　179
　　Lesley Cherns, David M. Rohr, and Jiří Frýda

19　Gastropods　184
　　Jiří Frýda and David M. Rohr

20　Bivalve and Rostroconch Mollusks　196
　　John C. W. Cope

21　Nautiloid Cephalopods　209
　　Robert C. Frey, Matilde S. Beresi, David H. Evans,
　　Alan H. King, and Ian G. Percival

22　Tube-Shaped Incertae Sedis　214
　　John M. Malinky, Mark A. Wilson, Lars E. Holmer,
　　and Hubert Lardeux

23　Worms, Wormlike and
　　Sclerite-Bearing Taxa　223
　　Olle Hints, Mats Eriksson, Anette E. S. Högström,
　　Petr Kraft, and Oliver Lehnert

24　Trilobites　231
　　Jonathan M. Adrain, Gregory D. Edgecombe,
　　Richard A. Fortey, Øyvind Hammer, John R. Laurie,
　　Timothy McCormick, Alan W. Owen, Beatriz G. Waisfeld,
　　Barry D. Webby, Stephen R. Westrop, and Zhou Zhi-yi

25　Eurypterids, Phyllocarids,
　　and Ostracodes　255
　　Simon J. Braddy, Victor P. Tollerton Jr.,
　　Patrick R. Racheboeuf, and Roger Schallreuter

26　Crinozoan, Blastozoan,
　　Echinozoan, Asterozoan, and
　　Homalozoan Echinoderms　266
　　James Sprinkle and Thomas E. Guensburg

27　Graptolites: Patterns of Diversity
　　Across Paleolatitudes　281
　　Roger A. Cooper, Jörg Maletz, Lindsey Taylor,
　　and Jan A. Zalasiewicz

28　Chitinozoans　294
　　Florentin Paris, Aïcha Achab, Esther Asselin,
　　Chen Xiao-hong, Yngve Grahn, Jaak Nõlvak,
　　Olga Obut, Joakim Samuelsson, Nikolai Sennikov,
　　Marco Vecoli, Jacques Verniers, Wang Xiao-feng,
　　and Theresa Winchester-Seeto

29　Conodonts: Lower to Middle
　　Ordovician Record　312
　　Guillermo L. Albanesi and Stig M. Bergström

30　Vertebrates (Agnathans
　　and Gnathostomes)　327
　　Susan Turner, Alain Blieck, and Godfrey S. Nowlan

31　Receptaculitids and Algae　336
　　Matthew H. Nitecki, Barry D. Webby,
　　Nils Spjeldnaes, and Zhen Yong-yi

32　Acritarchs　348
　　Thomas Servais, Jun Li, Ludovic Stricanne,
　　Marco Vecoli, and Reed Wicander

33　Miospores and the Emergence of
　　Land Plants　361
　　Philippe Steemans and Charles H. Wellman

PART IV

Aspects of the Ordovician Radiation

34　The Ichnologic Record of the
　　Ordovician Radiation　369
　　M. Gabriela Mángano and Mary L. Droser

35　The Ordovician Radiation:
　　Toward a New Global Synthesis　380
　　Arnold I. Miller

List of Figures and Tables　389
References　395
List of Contributors　467
Index　473

ACKNOWLEDGMENTS

This book derives directly from a significant part of the work of UNESCO-IUGS–supported IGCP Project no. 410: *The Great Ordovician Biodiversification Event: Implication for Global Correlation and Resources* (1997–2002). As the first globally directed IGCP program to highlight Ordovician rocks and fossils, IGCP 410 established (1) closely coordinated "regional teams" for data collection and analysis of biodiversity; (2) a wholly integrated and well-calibrated "timescale" for global and regional correlation, and (3), from 1998, a complementary "clade team" work program to assess global biodiversity patterns in time and space Also a Web-based relational "database" managed by Arnold Miller (University of Cincinnati) was available for input of relevant biotal data.

We acknowledge particularly UNESCO and IUGS and the members of the IGCP Scientific Board, who gave IGCP 410 six years of continuing encouragement and support. They, together with a number of regional organizations, helped us sponsor highly successful meetings each year in different parts of the world.

The following individuals and their organizations were especially supportive in arranging our globally focused meetings: Tatjana N Koren' (A. P. Karpinsky All-Russian Geological Research Institute—VSEGEI, St. Petersburg, Russia), Duck K. Choi (Institute of Geological and Environmental Sciences, Seoul National University, Korea), Rong Jia-yu, Zhou Zhi-yi, and Chen Xu (Laboratory of Palaeobiology and Stratigraphy, Nanjing Institute of Geology and Palaeontology, Academia Sinica, China), Wang Xiao-feng (Yichang Institute of Geology and Mineral Resources, MGMR, China), Petr Kraft and Oldřich Fatka (Institute of Geology–Palaeontology, Faculty of Science, Charles University, Prague, Czech Republic), John Talent, Ruth Mawson, and Glen Brock (Centre of Ecostratigraphy and Palaeobiology, Macquarie University, North Ryde, Sydney, Australia), Mary Droser (Department of Earth Sciences, University of California at Riverside, United States), Evgeny Yolkin (Institute of Petroleum Geology, Russian Academy of Sciences, Novosibisk, Russia), and Minjin Chuulin (School of Geology, Mongolian Technical University, Ulaanbaatar, Mongolia).

We maintained important linkages in our cooperative work programs with the IUGS/ICS Subcommission on Ordovician Stratigraphy (Chairman Stan Finney) and with IGCP project no. 421—North Gondwana Mid-Palaeozoic Biodynamics (co-leaders Raymund Feist and John Talent). A former chairman of the Ordovician Subcommission, Reuben J. Ross Jr., encouraged the publication of a worldwide series of 12 Ordovician correlation charts between 1980 and 1995 (IUGS Publications nos. 1–2, 6, 8, 11–12, 21–22, 26, 28–29, 31), and these were invaluable in the work to establish precise stratigraphic ranges for many of the taxa used in this survey.

From 1998 it was planned that our "clade team" work would include the compilation of an edited volume covering the biodiversity profiles of as many taxonomic groups as possible. The main "clade team" meeting of IGCP 410 was held at the University of California (Riverside) in June 2001. Organized by Mary Droser and her Riverside colleagues, it focused on a wide range of global and regional biodiversity topics across the main Ordovician taxonomic groups and the Ordovician timescale. Many chapter authors attended this most productive meeting, but a number of active, mainly overseas participants were unable to attend. All of these have been encouraged to submit contributions to the volume. In addition, a few other specialist contributors were solicited after the meeting to fill obvious gaps so that a wider global coverage could be presented in the book.

During the preparation of this book, Barry Webby benefited from being able to use the facilities of the Centre of Ecostratigraphy and Palaeobiology, the Department of Earth and Planetary Sciences, and the University Library of Macquarie University in Sydney.

In terms of the editorial processing, Alan Owen (trilobites) and Dave Harper (brachiopods) assisted by coordinating their large chapters. Roger Cooper's advice and his contribution on diversity measures are also much appreciated. Collective thanks are extended, also, to all the other authors who contributed manuscripts to the volume; many, additionally, helped by critically reading drafts submitted by other authors in the project.

The editorial process could not have been undertaken in the available time frame without the substantial support of numerous referees worldwide. Special thanks to them all for their insightful reviews. They include Jonathan Aitchison, Frits Akterberg, Bob Anstey, Claude Babin, Arthur Boucot, Simon Conway Morris, Kent Condie, James Crampton, Dave Elliott, Ray Ethington, Dianne Edwards, Stan Finney, Dan Fisher, Michael Foote, Dan Goldman, Roland Goldring, Joe Hannibal, Charles Holland, Steven Holland, Richard Hoare, Warren Huff, Nigel Hughes, Peter Jell, Ron Johns, Anita Löfgren, Merrell Miller, Heldur Nestor, Paddy Orr, Robert Owens, Mark Patzkowsky, Geoffrey Playford, John Pojeta Jr., John Pickett, John Peel, June Philips Ross, Charles Ross, Enrico Serpagli, Paul Selden, John Shergold, Colin Stearn, Desmond Strusz, Julie Trotter, Fons VandenBerg, Jean Vannier, Pat Wilde, Henry Williams, A. (Tony) D. Wright, Gavin Young, Ellis Yochelson, and three reviewers who wish to retain their anonymity.

The Great Ordovician Biodiversification Event

1 Introduction

Barry D. Webby

> *The historical scientist focuses on detailed particulars—one funny thing after another—because their coordination and comparison permits us, by consilience of induction, to explain the past with ... confidence (if the evidence is good).*
> STEPHEN JAY GOULD, *Wonderful Life*

This introductory chapter presents the scope, aims, and organization of the volume; outlines previous work on Ordovician biodiversity topics; and gives an overview of the chapters in the volume, from those briefly appraising the Ordovician World to those more comprehensively surveying the diversification patterns of the main Ordovician taxonomic groups. The chapter ends with some closing remarks on the Ordovician Radiation and future directions.

■ Scope and Aims

Two of the greatest evolutionary events in the history of life on earth occurred during Early Paleozoic time. The first was the Cambrian explosion of skeletonized marine animals—what Wilson (1992: 188) called the "big bang of animal evolution"—about 540 million years (m.y.) ago. The second was the "Great Ordovician Biodiversification Event," the focus of this book. During the 46-m.y. span of Ordovician time (489–443 m.y. ago), an extraordinarily varied range of evolutionary radiations of "Cambrian-, Paleozoic- and Modern-type" biotas appeared, the most diversified occurring in marine continental platform to open ocean habitats. Animals (arthropods) also first walked on land, and based on their cryptospore record, the first nonvascular bryophytelike plants colonized damp areas on land.

The Cambrian explosion of skeletonized animals is now comparatively well documented, though the timing of the initial event still needs to be reassessed (Bowring and Erwin 1998). Books have been published on a number of aspects of this explosion, for example, Glaessner's *The Dawn of Animal Life: A Biohistorical Study* (1984), Lipps and Signor's *Origin and Early Evolution of the Metazoa* (1992), and Zhuravlev and Riding's *The Ecology of the Cambrian Radiation* (2000). Also well documented is the history of the extraordinary Mid Cambrian Burgess Shale faunas, about 505 m.y. ago. The most notable publications are Gould's *Wonderful Life* (1989), Briggs, Erwin, and Collier's *The Fossils of the Burgess Shale* (1994), and Conway Morris's *The Crucible of Creation* (1998b).

In contrast, until very recently, there was no Ordovician biodiversity volume, or at least none that focused significantly on aspects of Ordovician biodiversity.

However, nearly half the contributions in Crame and Owen's *Palaeobiogeography and Biodiversity Change: The Ordovician and Mesozoic-Cenozoic Radiations* (2002) have addressed issues relating to Ordovician diversity change. The important role of plate tectonics during the Ordovician was singled out—examples, such as the fragmentation of the Gondwanan margin, the drift of Avalonia, and the development of an array of marginal to oceanic terranes (including island chains) that effectively partitioned ocean circulation patterns within the Iapetus Ocean—as prominent in engendering the patterns of diversity change.

Ordovician biodiversity has typically been studied in three different ways: (1) *taxonomic diversity*, which involves a focus on the taxonomic richness and turnover (originations and extinctions) within fossil groups; (2) *ecologic diversity*, which examines how organisms (or their communities) adapt to fill niche spaces in order to exploit available food resources more successfully; and (3) *morphological diversity* (usually termed *disparity*), which traces the patterns of morphological (design) change in various fossil groups.

This book is devoted primarily to documenting the taxonomic diversification of Ordovician biotas, in both global and regional contexts, and to firming up the timing of the most important diversification events. The volume has a broad coverage, but inevitably there are gaps. A few aspects of ecologic diversity have been treated, but the coverage is limited, and aspects of morphological diversity (disparity) are barely touched on. Moreover, only limited discussion of possible causes for the radiation events is included.

All these Ordovician diversification events form part of the "Great Ordovician Diversification Event," also called the Ordovician Radiation (Droser et al. 1996). Although the most intensive part of the Ordovician Radiation was during the Mid to Late Ordovician epochs, an interval of 28 m.y. (until the second extinction pulse of the end Ordovician mass extinctions—Sheehan 2001b; chapter 9 in this volume), some taxonomic groups (e.g., trilobites, inarticulated brachiopods, graptolites, conodonts, and rostroconch mollusks) also diversified significantly during the Early Ordovician. Consequently, all these evolutionary events from the beginning to virtually the end of the Ordovician Period—through nearly 46 m.y. of earth history—should be treated as belonging to the Ordovician Radiation (figure 1.1).

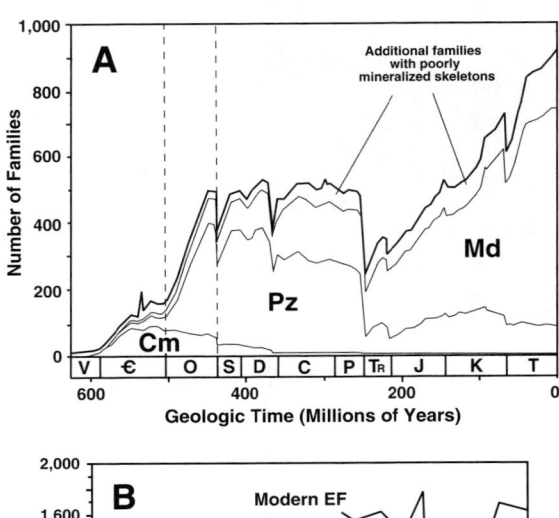

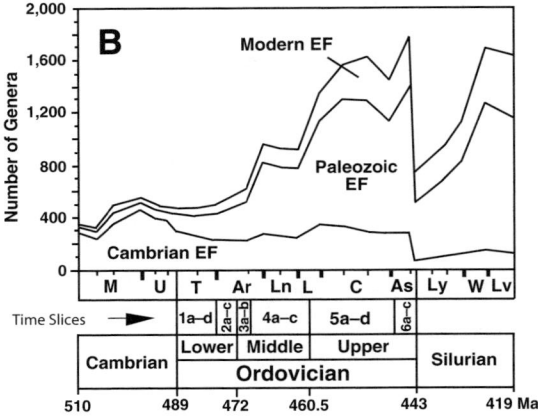

FIGURE 1.1. (A) Phanerozoic taxonomic diversity of marine animal families (slightly modified from Sepkoski 1984: figure 1). The fields Cm, Pz, and Md represent the Cambrian, Paleozoic, and Modern "evolutionary faunas" (EFs), respectively. Note also the generalized field of additional, poorly preserved families at the top of the diagram. Sepkoski's timescale follows Harland et al. (1982) and includes the following abbreviations: V = Vendian; € = Cambrian; O = Ordovician; S = Silurian; D = Devonian; C = Carboniferous; P = Permian; Tr = Triassic; J = Jurassic; K = Cretaceous; T = Tertiary. (B) The Middle to Upper Cambrian, Ordovician, and Silurian (except Pridoli) taxonomic diversity of marine animal genera (modified from Sepkoski 1995: figure 1). The main field of Cambrian, Paleozoic, and Modern EFs is shown, as well as Sepkoski's time units, including British series for his Ordovician and Silurian subdivisions. His abbreviations are as follows: M = Middle Cambrian; U = Upper Cambrian; T = Tremadocian; Ar = Arenig; Ln = Llanvirn; L = "Llandeilo"; C = Caradoc; As = Ashgill; Ly = Llandovery; W = Wenlock; Lv = Ludlow. The "Llandeilo" series has now been abandoned in favor of an enlarged Llanvirn, the name Llandeilian retained for its upper stage (figure 2.1), and the overlying Caradoc expanded downward to fill the gap (Fortey et al. 1995, 2000). Note also the added scale bar at the bottom of the diagram that comprises the tripartite global series subdivisions for the Ordovician System, radiometric dates in millions of years (Ma), and the time-slice subdivisions used in this volume.

No comparable, well-defined major extinction episode has been recognized at the beginning of the Ordovician Period. According to Sepkoski (1981b, 1995, 1997), after the dramatic Early Cambrian radiations there was a phase of apparent "stagnation" or quiescence that persisted through the Mid Cambrian to the Early Ordovician, with a comparatively lower overall diversity thoughout (figure 1.1B). Yet it remained an interval of high turnover—with high extinction rates limiting the overall buildup of diversity, including the well-defined successive pulses of mainly trilobite-based Mid to latest Cambrian "biomere" extinctions (Palmer 1979; Ludvigsen 1982; Zhuravlev 2000). The last of these more or less coincides with the base of the North American Ibexian (Ross et al. 1997), that is, three conodont zones below the base of the Ordovician (Cooper et al. 2001). A few extinction horizons have been identified within the Ordovician, but they all appear to be relatively minor pulses that have yet to be demonstrated as being truly global events. For example, Ji and Barnes (1993, 1996) documented an Early Ordovician (mid Tremadocian) conodont extinction event in Laurentia; Sepkoski (1992b, 1995, 1996) reported extinction horizons in the late Mid Ordovician (near the bottom and top of the Darriwilian, respectively); and Patzkowsky et al. (1997) identified a climatic event with associated Late Ordovician (mid Caradoc) brachiopod-dominated extinction in Laurentia. The last has wider, probably even global, importance, correlating with climatic and other changes—including the "late Keila" extinctions of chitinozoans, acritarchs, and ostracodes—in Baltoscandia (Kaljo et al. 1996; Ainsaar et al. 1999).

Biotas

This book was compiled by a large team of Ordovician specialists from around the world, under the aegis of the joint UNESCO and International Union of Geological Sciences (IUGS) geosciences initiative—the International Geological Correlation Programme (IGCP)—which supported IGCP project no. 410. The aim in establishing this first globally oriented, internationally sponsored IGCP project was to have the "Great Ordovician Biodiversification Event" comprehensively evaluated in a collective effort. When IGCP 410 became formally established in 1997, data collection and analysis of biotas were coordinated mainly on a regional basis, but in 1998 a separate study program was developed with a more constrained global approach on the major fossil groups. This became an evaluation of how each taxonomic group diversified through Ordovician time, with assessments of patterns of diversity change, and rates of origination and extinction, based on the assembly of genus- and species-level taxonomic diversity data of the various fossil groups. Leading specialists and their colleagues were invited to participate in the three-year work program and to attend a major meeting organized by Mary Droser in June 2001 at the University of California in Riverside to discuss the results of the compilations. The book derives in part from these contributions, with a number of additional studies from other specialists added to the project after the Riverside meeting to widen the coverage to include nearly all fossil groups.

The leader of each fossil group chose his or her own team of co-workers to tackle the compilation of data and assembly of the manuscript for the particular chapter. The groups varied in size from large (with a coordinating author and many coauthors) to small, single-author presentations. We encouraged each team to establish its diversity surveys so that they highlighted patterns of diversity change, originations, and extinctions, where possible down to species level at least for the pelagic groups, and to genus, and where possible to species level for benthic groups. Most authors assembled their primary taxonomic data using their own databases (spreadsheets or census lists).

The pelagic groups (e.g., graptolites, chitinozoans, radiolarians) and some benthic groups (e.g., bryozoans, sponges, stromatoporoids, echinoderms) are presented as species-level, or combined species- and genus-level, diversity surveys. Larger benthic groups (e.g., trilobites, brachiopods, gastropods, bivalves, nautiloids) have been surveyed mainly at the genus level at this stage. Although we attempted to provide the widest possible coverage of Ordovician biotas, it was inevitable that some unevenness in the levels of documentation and analysis would occur from chapter to chapter. For a variety of reasons, a number of groups are presented with rather incomplete global diversity analyses. This may be because (1) only part of the group's Ordovician record has been treated; (2) reliable data of well-preserved and diagnostic material are available only from one or two regions in the world, so the survey does not have a global focus; (3) genus- and species-level assignments of a group are unreliable or at least need further revision before a worldwide analysis can be attempted; or (4) present-day expertise on a particular group is lacking. Consequently, for a few groups, only preliminary statements could be included, and in some cases these assessments remain predominantly regionally based.

Although we have attempted to include most Ordovician groups in this genus- and species-level coverage (table 1.1), there are some gaps in the documentation—for instance, the Mid to Late Ordovician conodont diversity record and a few small groups of comparatively limited Ordovician occurrence, for example, foraminiferans (Lipps 1992a, 1992b), hydroids (Stanley 1986; Foster et al. 1999), and hyolithelminths. In addition, a few groups of microorganisms—such as two unicellular planktic green algal prasinophyte groups (leiospheres, tasmanitids) and the prolific, organic-walled, benthic cyanobacterium *Gloeocapsomorpha prisca,* a microorganism of considerable economic interest because it forms significant matlike accumulations of late Mid to Late Ordovician "kukersite" oil shales in the intracratonic basins of Estonia (Körts 1992), North America (Jacobson et al. 1988), and Australia (Foster et al. 1990)—have not been included in the survey.

An estimate of total numbers of Ordovician genera and species for each fossil group treated in the book (table 1.1) is provided by authors for general guidance only. It is not intended as a comprehensive listing of totals for all Ordovician groups. The list comprises about 4,605 genera (excluding the trace fossils). The total number of animal genera (less the "plants"—algae, acritarchs, miospores) is 4,254 genera. Earlier, Sepkoski (1995) employed a database of 4,367 animal genera (i.e., 12 percent of all known animal genera in the Phanerozoic) to prepare his outline of Ordovician diversity history. Only the few small groups mentioned earlier, the late Mid to Late Ordovician conodont record and about 70 genera of nautiloids from China and Russia, have been excluded from our present survey.

With recent publication of Sepkoski's (2002) comprehensive listing of all known fossil marine animal genera, it is now possible to make more-up-to-date comparisons between the two lists of Ordovician animal generic diversity records (table 1.2). However, it is difficult to compare the two lists in detail because they represent completely different types of fossil data assemblies. The present study was not intended to be comprehensive but rather to involve as many available Ordovician specialists as possible in a stocktaking of their fossil group. In some cases these workers have

TABLE 1.1. A Preliminary Listing of Genus- and Species-Level Totals for the Ordovician Fossil Groups

Fossil Groups	No. of Genera	No. of Species
Radiolarians	24	98
Sponges	135	280
Stromatoporoids	29	155
Conulariids	12	51 (97)
Corals	128	n.d.
Bryozoans	169	1,120
Inarticulated brachiopods	190	n.d.
Articulated brachiopods	540	n.d.
Polyplacophoran mollusks	15	31
Symmetrical univalves	75	c. 700
Gastropods	140	c. 2,500
Bivalves	136	530–680
Rostroconchs	21	107
Nautiloid cephalopods	305 (375)	n.d.
Hyoliths	32	n.d.
Cornulitids	3	n.d.
Coleoloids	3	n.d.
Sphenothallids	1	n.d.
Byroniids	1	n.d.
Tentaculitids	1	n.d.
Scolecodonts (polychaete jaws)	50	500–1,000
Chaetognaths	4	7
Palaeoscolecidans	8	21
Machaeridians	7	65
Trilobites	842	n.d.
Eurypterids	6	9
Ostracodes	559	+2,600
Phyllocarid crustaceans	9,+4?	22,+9?
Echinoderms	431	1,216
Chitinozoans	35	397
Graptolites	192	800
Conodonts (Early to Mid Ordovician)	106	430
Vertebrates	35	37
Receptaculitids	9	48
Algae (including cyclocrinitids)	88	+164
Acritarchs	250	c. 1,300
Miospores: cryptospores	12	35
: trilete spores	1	1
Land plants (no megafossil record)	n.d.	n.d.
Trace fossils (ichnogenera)	49	n.d.

Source: Compiled from data supplied by individual authors.
Note: This compilaton of animals, plants, and trace fossils is provided for general guidance only because some genus-level and especially the species-level taxonomy of fossil groups remains in a state of flux. For many groups a large backlog of taxonomic revision work still needs to be done.
Bivalve and scolecodont species data are shown as ranges, namely, between 530 and 680 and between 500 and 1,000 species, respectively. For entries of nautiloid genera and conulariid species, two numbers are included (one in parentheses): the lower number is the actual count (a mainly regional summation), and the higher number in parentheses is the global estimate. For the phyllocarids, the valid genera and species numbers are shown, plus the numbers of doubtful taxa (with a question mark). For trace fossils, only numbers of form genera (ichnogenera) could be supplied.
Abbreviations: c. = about; n.d. = data not determined (or not supplied).

TABLE 1.2. Comparative Generic Lists of Ordovician Animal Diversity Data

Animal (and Animal-like Protist) Fossil Groups	Genera: This Study	Genera: Sepkoski (2002) List
Protozoans: radiolarians	24	25
Foraminiferans	-	19
Spirotrichs (ciliates)	-	1
Sponges	135	126
Stromatoporoids[a]	29	25
Problematic receptaculitids[b]	9	-
Cnidaria–Medusa/Melanosclerites	-	3
Conulariids	12	16
Hydrozoans	-	13
Corals	128	157
Bryozoans	169	229
Inarticulated brachiopods	190	154
Articulated brachiopods	540	601
Polyplacophoran mollusks	15	18
Symmetrical univalves[c]	75	34
Gastropods[c]	140	139
Bivalves	136	146
Rostroconchs	21	20
Scaphopods	-	1
Nautiloid cephalopods	375	436
Tubelike taxa: hyoliths	32	28
Cornulitids and coleoloids	6	6
Sphenothallids[d]	1	-
Byroniids[e]	1	-
Tentaculitids	1	1
Hyolithelminthes	-	6
Incertae sedis	-	13
Scolecodonts (polychaete jaws)	50	36
Other Polychaete worms	-	14
Palaeoscolecidan worms	8	7
Myzostomian "worms"	-	1
Pentastome "worms"	-	1
Problematic machaeridians	7	9
Arthropods: eurypterids	6	5
Other merostomes	-	6
Mimetasterids	-	1
Trilobites	842	833
Phyllocarid crustaceans	13	11
Ostracodes	559	462
Class incertae sedis	-	6
Echinoderms	431	407
Chitinozoans	35	33
Graptolites-Graptoloids	192	119
Other hemichordates	-	78
Conodonts	106 (*partim*)	133
Chaetognaths (arrow worms)	4	1
Vertebrates	35	11
Total	4,327	4,391

Source: Compiled from data supplied by individual authors (this study) and from Sepkoski (2002).

[a] Stromatoporoids are here listed (in this study and in Sepkoski's list) as separate from other sponges.

[b] The problematic receptaculitids is the only group omitted from Sepkoski's list. They are included in this study as a possible metazoan group.

[c] The bellerophontids have been included with the symmetrical univalves of this study, whereas they are included in Sepkoski's list with the gastropods.

[d] The genus (*Sphenothallus*) is included in Sepkoski's list as a conulariid.

[e] The genus (*Byronia*) is included in Sepkoski's lists as a hyolithelminth.

adopted revised taxonomic categories from those listed in Sepkoski's *Compendium of Fossil Marine Animal Genera* (2002). The *Compendium* represented a mammoth compilation of Phanerozoic animal genera, assembled from the data coverage of animal groups in Moore et al., *Treatise on Invertebrate Paleontology* (1953–1992), and many hundreds of other literature sources published up to 1998. Nevertheless, there is a remarkable degree of similarity between the two lists. The comprehensive Sepkoski (2002) compilation also provides a means to appreciate the full range of Ordovician taxonomic groups, including those biodiversity records that have not been documented here.

Only a comparatively few groups have good-quality assignments of species data (table 1.1). These include mainly pelagic groups such as the graptolites, conodonts, chitinozoans, and radiolarians, but correct species assignments have also been established for a few benthic groups (e.g., stromatoporoids, conulariids, bryozoans, rostroconchs, eurypterids, echinoderms, and receptaculitids).

Wilson (1992) has an estimate of 1.4 million organisms (plants, animals, and microorganisms) currently known to be living on earth, though millions of insects, microbes, and other organisms remain undescribed across habitats from rain forests to the ocean deeps (Wilson 1992; Thorne-Miller 1999). The problems of estimating fossil species numbers is a more daunting task. Paul (1998) reviewed the available approaches, namely, to establish what proportion of living animal species is likely to be preserved in modern settings, as well as to establish the relationships between soft-bodied and skeletonized organisms in fossil *Lagerstättern* such as the Middle Cambrian Burgess Shale (Conway Morris 1986). These two approaches produce rather similar results, suggesting to Paul that only about 10 percent of all Phanerozoic species are likely to have been preserved. In assessing the Ordovician global record, we currently have only reasonably complete numbers for the genera—some 4,600 known Ordovician genera. From the generic and specific data in the 26 listed fossil groups (table 1.1), there are on average about five species for each genus. Multiplying this value by the generic total gives the very approximate total of 23,000 Ordovician skeletonized species. If this estimate represents only 10 percent of species likely to be preserved (Paul 1998), then the overall total (including the soft-bodied organisms) may be of the order of 200,000 to 250,000 Ordovician species.

Ordovician Time and Time Slices

A specific aim of IGCP 410 was to develop a more highly integrated, well-calibrated Ordovician timescale capable of providing a more reliable global and regional basis for correlating range data and establishing age relationships of the diverse biotas. It has long been recognized (Jaanusson 1960, 1979) that establishing biostratigraphic correlations using Ordovician biotas, especially on a global scale, is a difficult task because of the extensive biogeographic and ecologic differentiation of the faunas. Furthermore, the comparative lack of good reliable radiometric ages has limited the degree to which age relationships could be determined (Webby 1998). The priority to establish a stabilized global stratigraphic framework for the Ordovician System has been the main responsibility of the International Subcommission on Ordovician Stratigraphy (ISOS). It is an independent subcommission of the International Commission on Stratigraphy (ICS), and both ISOS and ICS exist under the umbrella of IUGS. Since the mid-1980s, ISOS has made good progress toward the establishment of a single set of globally based Series and Stage divisions, using the most highly resolved biologic indices (graptolite, conodont, and chitinozoan zones) coupled with available physical and chemical tools (table 1.3).

Stratigraphers, especially those closely associated with global correlation work of international stratigraphic bodies like ICS and ISOS, have long been aware of the need to draw clear distinctions between "chronostratigraphic units" based on rock sequences and "geochronologic units" that reflect intervals of geologic time (Salvador 1994). The terminology is straightforward for the bulk of conventional hierarchical usages, such as "system," "series," and "stage" with their geochronologic counterparts "period," "epoch," and "age." However, most "series" have tripartite subdivisions, with names derived from their position in the "system," for example, "Lower Ordovician Series," "Middle Ordovician Series," and "Upper Ordovician Series." The nomenclature "Early Ordovician Epoch" and "Late Ordovician Epoch" is also acceptable, but the terminology for the intervening epoch has remained a problem because the same name "Middle Ordovician" has been applied to both series (for the rocks) and epoch (for time) usages. Only British stratigraphers (see Holland et al. 1978) have drawn a distinction in their regional work between "middle" (for rocks) and "mid-" (for time). According to S. C. Finney, who is the current chair of ISOS and second vice-chair of ICS, the terms "Middle Ordovician" and "Mid Ordovician" are now "used for the series and its correlative epoch, respectively" (Finney et al. 2003:351). Following such recent moves at the international level, we have adopted the formalized global names "Middle Ordovician Series" for the rock sequences and "Mid Ordovician Epoch" for time throughout this book. The categories of lower, middle, upper (for rocks) and early, mid, late (for time) have similarly been recognized for ratified, tripartite, global stage subdivisions and for informal, mainly lower-rank, usages, as depicted in table 1.3.

Of critical importance, especially for biodiversity studies, was the level of resolution of the timescale (Erwin 1993; Conway Morris 1998a). Clearly it was essential to establish the most refined timescale with the shortest-term subdivisions for the widest applicability in studies of global biodiversity change. Consequently, two new, highly integrated and resolved correlation charts were prepared by a small group of specialists in IGCP 410, in cooperation with members of ISOS. These present the global and regional stratigraphic frameworks, main biostratigraphic (graptolite, conodont, and chitinozoan) schemes for establishing cross ties, key radiometric dates, and a set of 19 time-slice divisions by Webby et al. (see chapter 2: figures 2.1, 2.2).

Sadler and Cooper's (chapter 3: figure 3.1) new calibration of the Ordovician timescale based on computer-assisted constrained optimization (CONOP) procedures allowed the subdivisions in the correlation charts to be calibrated with respect to geologic time. CONOP calibrations involved processing a huge data set of first- and last-appearance events based on 1,100 or so Ordovician and Silurian graptolite species from nearly 200 measured sections. A "best fit" sequence of events was derived in the CONOP process to establish a relative timescale from the thousands of event levels. The scale became numerically calibrated when radiometric dates and ranges of key graptolites

TABLE 1.3. Summary of the Global Subdivisions of the Ordovician System/Period

System/Period	Series/Epoch	Stage/Age	Substage/Subage	Time Slices (TS)
O R D O V I C I A N [443Ma] [489Ma]	**Upper/Late**	Ashgill	upper/late[a] middle/mid lower/early	6c 6b 6a
		Caradoc	upper/late middle/mid lower/early	5d 5b–c 5a
	Middle/Mid	**Darriwilian**[b]	upper/late middle/mid lower/early	4c 4b 4a
		lower Middle/early Mid[c]	N/A	3a 3b
	Lower/Early	upper Lower/late Early[c]	N/A	2c 2b 2a
		Tremadocian	upper/late middle/mid lower/early	1d 1b–c 1a

Ratified global subdivisions are shown in bold and are capitalized; informal stage/age and substage/subage usages are not capitalized; estimated radiometric ages (Ma) for the base and top of the Ordovician are shown in square brackets; N/A = subdivisions not applicable.
  Note that a distinction is drawn between the "System," or body of rocks, and the "Period" representing geologic time. The 19 time slices adopted for use in this book are shown in the right column.
  [a] The British Hirnantian stage is equivalent to the upper/late Ashgill interval.
  [b] The British Llanvirn is correlative with the middle/mid to upper/late parts of the Darriwilian Stage.
  [c] The British lower/early Arenig and the middle/mid Arenig are more or less correlative with the unnamed upper Lower/late Early Ordovician and lower Middle/early Mid Ordovician global stages/ages, respectively.

(and conodonts) were interpolated along a regression line.

In a more recent development, Cooper and Sadler (2002) have shown that the CONOP application can be used directly, without the "time unit bias," to establish a global diversity curve. Again taking the well-studied and widely distributed graptolites, they have depicted a "running diversity curve" of 2,272 estimates of "interval free" graptolite species standing diversity through Ordovician and Silurian time.

The CONOP calibrations (chapter 3) for this biodiversity project have added significantly to establishing the Ordovician timescale as an integrated, high-resolution dating and correlation tool, to a point where it is now arguably one of the best-resolved intervals of time within the Paleozoic Era. This indicates, despite Jaanusson's (1960, 1979) earlier misgivings, that the extreme levels of biogeographic and ecologic differentiation of the Ordovician biotas were not limiting factors. The 19 slices employed for the Ordovician Period (46 m.y. duration; table 1.2), with time-slice intervals of relatively equal units of time (intervals that each span between 1.6 and 3.0 m.y.), provide a most practical, best-resolved, and reliable basis for use in this present biodiversity survey. These standardized time slices, as emphasized by Cooper (1999c:441), are "as fine as is practical" to allow precision in global studies of diversity change of the fossil groups. Two types of abbreviations are employed through this work to distinguish between time, as a duration or interval in millions of years (m.y.), and time as a specific age assignment including radiometric ages in millions of years (Ma). Authors of the chapters in this volume were provided with the correlation charts (chapter 2: figures 2.1, 2.2) so that time equivalences could be maintained in data compilations across the different groups. In addition, abbreviated left- and bottom-margin versions of the key elements of these Ordovician stratigraphic charts were supplied so that they could be incorporated at the margins of diversity plots when compiling diversity data and printed with a consistent time frame for comparative purposes throughout the chapters of the book. The majority of authors have, in compiling their diversity data, successfully employed the standardized global time-slice scheme across a wide range of environmentally and provincially differentiated profiles.

Diversity Measures

Another important aspect of this biodiversity study was to have some consistency in the kinds of measures used in the volume for summarizing the data on diversity patterns and turnover rates. A short chapter on diversity measures is contributed by Cooper (chapter 4). The guidelines distributed to contributors provided general recommendations for establishing the best estimate of the "true mean standing diversity," irrespective of the length of the time unit and taxonomic rank (Cooper 1999c), and the means of determining origination and extinction rates, the best approach being to estimate the rates in million-year intervals, which, of course, is dependent on the reliability of the calibration of the timescale (see earlier in this chapter). The measures have been adopted by about half the contributors, giving overall a relatively uniform set of presentations without limiting flexibility for those authors who wished to use other measures to emphasize particular features in their databases. Some additional measures that were circulated in a final update of the guidelines could not be adopted by some authors because they had already compiled their data and submitted their manuscripts.

Cooper's standardized symbols (d, o, e, and subscript i, for diversity, originations, extinctions, and interval in m.y.) have been used widely throughout the book. On the other hand, no symbols were recommended to denote the taxonomic level of the compilation (genus or species), as it was thought preferable to explain the hierarchical level of the compilation in the text or in an appropriate caption. Cooper's (chapter 4) contribution is important in identifying "normalized diversity" (d_{norm}) as consistently providing the best estimate of "true mean standing diversity"; hence it is the most appropriate measure for determining diversity change through geologic time.

With the highly resolved and calibrated global and regional time frame, the contributors of the various fossil groups were able to plot their data (genera and/or species) accurately as global and/or regional stratigraphic ranges and then, using the diversity measures provided, derive the patterns of diversity change and turnover rates from their plotted range-chart data.

The diversity data have been, in some cases, plotted in two sets of diversity curves, one based on actual "sampled" diversity through time slices and the other as "range-through" diversity, whereby a taxon is shown as having a continuous range, though absent from a time slice (possibly owing to sampling failure or biofacies shift) between occurences in the immediately preceding and succeeding time slices (for examples, see Regional Patterns in Australia and New Zealand, the Anglo-Welsh Sector of Avalonia and in South China, chapter 24).

■ Organization

The present book derives from the contributions of 96 Ordovician paleontologists, stratigraphers, geochemists, and other geologists from 17 countries throughout the world. The book comprises this introduction and thirty-four other chapters, which have been divided into four main sections: part I: Scaling of Ordovician Time and Measures for Assessing Biodiversity Change; part II: Conspectus of the Ordovician World; part III: Taxonomic Groups; and part IV: Aspects of the Ordovician Radiation.

The three chapters of part I deal with the stratigraphic framework, time slices, and calibration of the Ordovician timescale (chapters 2 and 3) and the measures of diversity (chapter 4) that formed the basis of this biodiversity study. Outlines for these chapters have been presented earlier.

The remainder of this introductory chapter comprises (1) a brief history of earlier work on Ordovician biodiversity topics and (2) an overview of all the remaining chapters, divided into two sections, for part II (chapters 5–10) under the subheading "Ordovician World in Brief" and for parts III and IV (chapters 11–35) under the subheading "Synopses of Fossil Groups."

Part II (Conspectus of the Ordovician World; chapters 5–10) includes a broad overview of a number of significant physical and chemical features of the Ordovician world, presented in order to (1) illustrate what marked contrasts exist between the Ordovician and the present-day climates and in the dispositions of continents and oceans and (2) provide background to some of the features that may have been influential in shaping this greatest diversification of marine life.

Part III (Taxonomic Groups; chapters 11–33) comprises a nearly complete coverage of taxonomic groups. For reasons discussed earlier in this introduction, the chapters vary considerably, from comprehensive, integrated surveys of global and regional genus- and/or species-level diversity data to far more preliminary compilations, focusing mainly on a few aspects of regional genus- or species-level biodiversity. The two concluding chapters of part IV (Aspects of the Ordovician Radiation; chapters 34 and 35) deal with the ichnofossil record and a global synthesis of the Ordovician Radiation.

■ Ordovician Biodiversity Perspectives: Earlier Work

Evolutionary Faunas (EFs) and Floras: Some Global Considerations

Faunal Patterns

Sepkoski's (1978, 1979, 1981a, 1984, 1991a, 1997) numerous contributions on the patterns of the Phanerozoic global marine animal diversity change have given much attention to the development of the three "Great Evolutionary Faunas" (Cambrian, Paleozoic, and Modern EFs) and the pivotal role played by the Ordovician Radiation. The representatives of these EFs were recognized by Sepkoski as somewhat "fuzzy" sets of unrelated higher taxa with similar histories of diversification through extended intervals of time. At each level of taxonomic hierarchy (order, family, or genus), there were marked differences in the relative intensities of the Cambrian and Ordovician radiations. In comparison with the Cambrian increases, approximately twice as much ordinal biodiversity was added to the Ordovician marine system, some three times more familial diversity (figure 1.1A), and nearly four times more genus-level diversity (Sepkoski 1978, 1988; Sepkoski and Sheehan 1983). Each successive EF exhibited a greater diversity and ecologic complexity (Sheehan 2001a).

The Cambrian EF was an explosive evolutionary event involving appearances of the first well-skeletonized metazoans—groups such as trilobites, inarticulated brachiopods, hyoliths, "monoplacophorans," and eocrinoids. The fauna expanded exponentially during the Early Cambrian, an interval of 20 m.y. (Sepkoski 1979, 1997: figure 1; Zhuravlev 2000); leveled off during the Mid to Late Cambrian, at a peak diversity of about 100 families; and then declined gradually through the Ordovician as the Paleozoic EF expanded. The Cambrian EF was reduced still further to about 30 families by the short-lived, glacially induced end Ordovician mass extinction. The fauna included appearances of a larger number of higher-level categories—most of the skeletonized metazoan phyla and nearly two-thirds of the classes (Sepkoski 1981a). On the other hand, lower-level categories were sometimes depauperate—"rather plain, even grubby" species, according to Valentine (1973: 451). The communities were mainly benthic, rather generalized, and intergrading, composed of surface deposit feeders, grazers, carnivores, and suspension feeders with low epifaunal or infaunal tiering. Overall the fauna exploited little of the potential ecospace (Bambach 1983).

This contrasts markedly with the patterns of global diversity change during the Ordovician Period. The main contributors in this new wave of major diversification were elements of Sepkoski's (1979, 1984) Paleozoic EF. The fauna included the articulated brachiopods, cephalopods, crinoids, ostracodes, stenolaemate bryozoans, and corals. Components of the Paleozoic EF diversified very slowly in the Cambrian, with only slight increase during the Late Cambrian. In the Ordovician an exponential rise of diversity occurred, which was at a more rapid rate than at any other time during the Phanerozoic (Sepkoski 1995). The communities of attached epifaunal suspension feeders, deep burrowers, and carnivores greatly expanded (Bambach 1983; Droser and Sheehan 1997a), as did more specialized reef and hardground communities. Compared with those of the Cambrian EF, the taxa of the Paleozoic EF became more specialized, with the development of narrower ecologic requirements, use of resources, and competitive abilities. Suspension feeders of all types began to appear, along with the scavengers and carnivores that preyed on them from the earliest Ordovician (Signor and Vermeij 1994).

More than 350 new "Paleozoic" families were added in what Sepkoski (1981b:204) called the "largest turnover in composition of marine faunas seen in the history of the oceans" (figure 1.1A). The familial diversity curve, then, exhibits a sudden flattening after its steep upward climb, and this is interpreted to mean that the ecospace had finally been filled to capacity, to its equilibrium level, as shown by the nearly horizontal, Paleozoic-wide diversity plateau (Sheehan 2001a). However, two extinction/rebound perturbations disrupted the continuity of this diversity plateau; the first was the end Ordovician mass extinction, and the second was represented by the pulses of Late Devonian extinction. Despite these, the Paleozoic EF remained dominant for nearly 240 m.y. until it was far more severely disrupted in the end Permian mass extinction (Erwin 1993).

Zhuravlev (2000) noted that, with the exception of the bryozoans, all the phyla that participated actively in the major Ordovician diversifications were present in the Cambrian (e.g., precursors to chordates, conodonts and graptolites, the trilobites, brachiopods, sponges, echinoderms, and the mollusk groups, such as rostroconchs, cephalopods, gastropods, and polyplacophorans). Most of these groups (apart from trilobites and inarticulated brachiopods and, to a lesser extent, sponges and echinoderms) remained unimportant in Cambrian communities, but they gradually emerged from the shadows as opportunities arose following the repeated extinctions of late Early, Mid, and Late Cambrian time (Zhuravlev 2000). By the latest Cambrian (Sunwaptan) they had already taken over in the sense that the last diversity peak was produced by them.

Sepkoski's (1979, 1984) Modern EF had a comparatively limited development in the Ordovician, though its two dominant mollusk members (gastropods and bivalves) did expand markedly during Ordovician time (Sepkoski and Sheehan 1983; Sepkoski 1991a). Crustaceans, gymnolaemate bryozoans, foraminiferans, echinoids, fishes, reptiles, and mammals were other typical Modern EF components. The fauna exhibited a greater variety of predators and more highly complex infaunal communities (Bambach 1983, 1985; Thayer 1983). Such changes allowed ecospace to be divided more finely and still higher diversities to be achieved. The changes through the Ordovician Period mainly involved a very slow and steady increase in importance of members of the Modern EF (mainly bivalves and gastropods) relative to the overwhelmingly dominant members of the Paleozoic EF. This pattern continued through the Paleozoic (figure 1.1A), until the end Permian mass extinction, when differential survival of the faunal components occurred—the Paleozoic EF losing 79 percent of its familial diversity, compared with the Modern EF's 27 percent loss (Sepkoski 1984). The Modern EF became the dominant fauna, and the expansion continues to the present with no sign of the appearance of a diversity plateau (Sepkoski 1997).

Overview of the Ordovician Diversity Record

Sepkoski (1995) presented a very important summary of the diversification history of Ordovician biotas that focused not only on all the marine animal genera (see figure 1.1B) but also on typical representatives of individual fossil groups, with separate diversity curves for each group. Figure 1.1B is basically a reproduction of Sepkoski's original (1995) figure 1—that is, it includes his primary data compiled in Ordovician generic-level diversity curves of marine animal genera against 12 Ordovician "time units" derived from British series subdivisions, with a scale bar added at the bottom of his diagram, for comparative purposes, to illustrate relevant new global Lower, Middle, and Upper Ordovician series, radiometric dates, and time-slice data as used throughout this volume. Sepkoski (1995) established his 12 "time unit" subdivisions by dividing each British series into two, except for the "Llandeilo," which he left undivided, and the Caradoc, which he divided into three. In addition, his lower Arenig and upper Ashgill subdivisions were allocated longer and shorter durations, respectively, in comparison with the other time units. Similar stratigraphic subdivisions have been used in Sepkoski's (1992a, 2002) family- and generic-level compendia. Stratigraphic relationships between Sepkoski's subdivisions and the global series and time-slice (*TS*) subdivisions used in this volume are shown in figure 1.1B, in particular, the base of the Middle Ordovician with a position toward the middle Arenig, and the base of the Upper Ordovician with a level close to the middle "Llandeilo."

Sepkoski's (1995) figure 1 (also reproduced here in figure 1.1B) for all the marine animal genera illustrates that the global diversity levels during the Early Ordovician were maintained at levels broadly similar to those of the preceding late Mid to Late Cambrian. Both of these Cambrian and Early Ordovician intervals were interpreted by Sepkoski (1995) as intervals of high turnover (i.e., including high rates of extinction, as discussed earlier) with several "Paleozoic" EF groups expanding slowly at the expense of Cambrian EF members through Late Cambrian to Early Ordovician time. This pattern contrasts with representations of the Ordovician part of the diversity trajectory shown in Sepkoski's Phanerozoic-wide compilations. In both generic-level plots (see Sepkoski 1997: figure 1-1, 1998: figure 2) and familial-level plots (see Sepkoski 1979: figure 7, 1984: figure 1, 1997: figure 1-2), the dramatic, exponential rise of diversity is shown commencing near the beginning of the Ordovician and continuing more or less unabated—the slope of the trajectory remaining steeply inclined upward—until leveling off in the late Caradoc–early Ashgill (see Sepkoski 1984: figure 1, reproduced here as figure 1.1A).

The first major diversification pulse of "Paleozoic" groups commenced early in the Mid Ordovician (mid Arenig—*TS*.3a–b) and continued to an initial early Darriwilian (late Arenig—*TS*.4a) peak; then there was a short-lived mid Darriwilian (or Llanvirn—*TS*.4b) lag that Sepkoski linked to the North American "Middle Ordovician" Sauk-Tippecanoe sequence boundary (an interval, or intervals, of lowered sea level highstands; see Ross and Ross 1995; Golonka and Kiessling 2002). The second major diversification started during the late Darriwilian (formerly latest Llanvirn–early "Llandeilo"—*TS*.4c) and reached its culmination in a mid Caradoc (*TS*.5b–c) maximum. A limited decline followed in the late Caradoc (*TS*. 5d), before rapid increase resumed to the sharp, third and highest diversity peak during the early to mid Ashgill (*TS*.6a–b). The late Ashgill mass extinctions then caused the dramatic decline of the diversity, virtually back to levels previously attained in the early Mid Ordovician, near the start of the first of the three great diversification pulses of the Ordovician Radiation, some 25 m.y. earlier.

The second feature of Sepkoski's (1995) documentation is the presentation of individual plots for some of the major taxonomic groups, the Cambrian EF represented by two groups (trilobites and inarticulated brachiopods), the Paleozoic EF by six groups, (articulated brachiopods, echinoderms, corals, bryozoans, nautiloid cephalopods, and rostroconch mollusks), the Modern EF by two groups (gastropods and bivalves), and two other mainly pelagic groups (graptolites and conodonts) that Sepkoski assigned to the Paleozoic EF. Some features of these patterns are outlined here, but detailed comparisons of his curves with counterparts documented in chapters of this book are not attempted because of the differing fossil data assemblies, taxonomic categories, and methods of presentation employed by many authors.

For the Early Ordovician record of more than 550 genera, the trilobite and inarticulated brachiopod components of the Cambrian EF make up about 49 percent of the total; the articulated brachiopod, nautiloid, rostroconch, bryozoan, graptolite*, and conodont* components of the Paleozoic fauna constitute about 42 percent of the total (14 percent of them pelagics—the groups with asterisks); and the gastropod and bivalve components of the Modern fauna make up about 8 percent of the total. Of these various groups, the rostroconchs have their main diversification through the Late Cambrian to Tremadocian interval, so this is one small "Paleozoic" group that started to diversify in the Late Cambrian, reaching its maximum in the Early Ordovician. Among other "Paleozoic" groups, at least one order (Ellesmerocerida) of nautiloids diversified in successive pulses across the Late Cambrian to Tremadocian boundary, and the "pelagic" graptolites and conodonts also radiated markedly during the Early Ordovician. This latter fact suggested to Sepkoski that, in the early history of the Paleozoic EF, some decoupling of the pelagic realm from the benthic realm may have occurred.

The marine generic diversity doubled to about 1,200 genera during the Mid Ordovician, and there were significant changes to the proportions of the three EFs. The Cambrian EF (trilobites and inarticulated brachiopods) was now represented by only about 30 percent of the total, whereas the Paleozoic EF (articulated brachiopods, echinoderms, corals, bryozoans, cephalopods, rostroconchs, graptolites*, and conodonts*) had expanded to about 62 percent of the total (nearly 12 percent of them pelagics). The Modern EF (gastropods and bivalves) remained a minor component, as in the Early Ordovician, constituting close to 8 percent of the total. A significant feature of the Paleozoic EF is that benthic elements (especially articulated brachiopods, cephalopods, echinoderms, and bryozoans) expanded more rapidly than the pelagic components through the Mid Ordovician. On the other hand, of all the groups, only the graptolites and cephalopods are shown attaining their maximum generic diversities during the Mid Ordovician. Frey et al. (chapter 21), however, show the cephalopods as reaching a relatively slightly higher maximum diversity in the Late Ordovician.

The Late Ordovician record comprises two greatest diversity increases, the first as shown in the broad, gently arching diversity curve of increase to a mid Caradoc maximum of about 1,600 genera, and the second in a higher, sharper, mid Ashgill peak of nearly 1,800 genera (figure 1.1B). It seems that all the groups surveyed by Sepkoski (1995), except apparently the rostroconchs and graptolites, attained their highest Ordovician generic diversities during this epoch. The Cambrian EF (trilobites and inarticulated brachiopods) had now declined further to about 23 percent of the total, while the Paleozoic EF (articulates, echinoderms, corals, bryozoans, cephalopods, rostroconchs, graptolites*, and conodonts*) had expanded further to about 64 percent of the total (of which nearly 8 percent were pelagics). In addition, the Modern EF (gastropods and bivalves) had now increased slightly in importance, to 12 percent of the total. All groups exhibit a sharp major decline associated with the end Ordovician extinctions.

The Late Ordovician diversity peaks of individual groups show some interesting patterns, with all the maxima associated with the early Caradoc (*TS*.5a), mid Caradoc (*TS*.5b–c), and/or the mid Ashgill (*TS*.6b) intervals. For the Cambrian EF groups, the trilobites have a diversity peak in the early Caradoc, while the inarticulated brachiopods have a mid Caradoc peak, though both groups exhibit a small, secondary mid Ashgill peak. Three of the Paleozoic EF groups (articulates, echinoderms, and bryozoans) show rather similar-sized mid Caradoc and mid Ashgill diversity peaks; the conodonts also show these same peaks, though they are less well defined. The nautiloids and graptolites, on the other hand, have their two prominent diversity peaks spaced farther apart, the first in the early Caradoc and the second in the mid Ashgill (*TS*.6b). The rostroconchs have only one minor early Caradoc peak, while the corals show only one prominent, major mid Ashgill peak. Of the components of the Modern EF, the bivalves show all three diversity peaks (early and mid Caradoc and mid Ashgill), and the gastropods show a steadily rising diversity curve to a single mid Ashgill peak.

Floral Patterns

The Ordovician floral groups have received comparatively little attention. The calcified benthic marine Ordovician "algae" were originally grouped by Chuvashov and Riding (1984) in their "Ordovician Flora." This was differentiated as one of the three major Paleozoic "evolutionary" floras based on characteristic Cambrian, Ordovician, and Carboniferous assemblages associated with reefs, stromatolites, oncoids, or "debris" in the shallow marine carbonates. The Cambrian flora was dominated by cyanophytes (blue-green "algae") that appeared during a brief span (5 m.y.) of the earliest Cambrian. The "blue greens" are now commonly referred to cyanobacteria because they are prokaryotic and therefore related to bacteria (Riding 1991) but differ in having chlorophyll *a* and in developing a typical thallus without roots, stems, or leaves (for further discussion, see chapter 31). Riding (2000) outlined the cyanophyte radiation of the Cambrian flora, from its dramatic appearance in the earliest Cambrian through its progressive decline during the Mid to Late Cambrian into the Early Ordovician, when it finally more or less disappeared.

The Ordovician flora—represented by mineralized (calcified) thallophytes—was dominated by dasyclads, codiaceans/udotaceans, and solenoporans (green and red algae). Chuvashov and Riding (1984) indicated that this flora appeared through the first two-thirds of the Ordovician Period. However, the major radiations of these groups took place later, mainly through Darriwilian to Caradoc time (see chapter 31). The great expansions of metazoans and calcified algae ("Paleozoic" fauna and Ordovician flora) had a profound impact on the "Cambrian" cyanophytes. Their decline is well documented in the reef habitat (Rowland and Shapiro 2002: figure 8; Webby 2002), but it is not yet known whether the disappearances relate to real extinctions or merely represent the loss of a preserved record when cyanophyte calcification processes ceased in unsuitable environmental conditions, possibly related to temperature decline (Riding 1992).

The other floral groups are represented by nonmineralized, microscopic thallophytes, the acritarchs, and the dispersed spores of the first bryophytelike land plants. The thallophytes are unicellular green algae (e.g., prasinophytes) and form a small component of the organic-walled phytoplankton in the Ordovician oceans. The acritarchs represent a major, organic-walled phytoplankton component of the oceans, but their identification as cysts of unicellular thallophytes remains uncertain (chapter 32). The cryptospores are predominantly dispersed spores and probably derived from small nonvascular bryophytelike plants (chapter 33). The acritarchs had an important early record in Proterozoic-Cambrian oceans—in the Early Cambrian up to 100 species (Vidal and Moczydłowska-Vidal 1997)—and then became significantly more diversified and abundant during the Ordovician (some 250 genera and about 1,300 species; table 1.1). Servais et al. (chapter 32) indicate that well-defined warmer water and cooler "provincial" assemblages were developed during the Early to Mid Ordovician. Based on British species data, there was apparently a progressive rise through the Early Ordovician to a Darriwilian diversity high, then continuous decline during the Late Ordovician. This contrasts with the Late Ordovician record in Baltoscandia, where Kaljo et al. (1996) have shown fluctuating, relatively high levels of diversity through the same interval, including localized peaks of diversity—two during the mid Caradoc and another in the early Ashgill. These somewhat divergent regional Mid to Late Ordovician acritarch pelagic diversity results do not closely match the record of mainly Darriwilian to Caradoc radiation events for the benthic calcified algae.

Two separate diversification phases seem to have been responsible for the earliest land-based floras. The first is indicated by the appearances of moderately abundant and cosmopolitan cryptospore records from the Darriwilian onward, and the second is suggested by the record of trilete spores in the latest Ordovician, which possibly signals the emergence of the earliest vascular plants (chapter 33). From this timing, only the first of these events can be related to the main radiation events for the benthic, calcified algae. However, it seems more likely that all these different Ordovician floral radiation events occurred completely independently of one another in their separate benthic, pelagic, and terrestrial realms.

Evolutionary Faunas at the Ecologic Level

North American and Other Continental Platforms

Sepkoski (1981b, 1991a), Sepkoski and Sheehan (1983), and Sepkoski and Miller (1985) highlighted the close relationships that exist between global diversity change within the successive "Evolutionary Faunas" (EFs) and localized to regional community-based diversity change. The sequential diversifications of the three EFs were recognized in the onshore-offshore patterns of expansion of the three basic trilobite-, brachiopod-, and mollusk-dominated community types. These were seen to be environmentally controlled and governed by successive onshore originations and offshore expansion, with replacement, or displacement, through time. Sepkoski and Sheehan (1983), Sepkoski and Miller (1985), and Sepkoski (1991a) employed about 500 Paleozoic (100 of them Ordovician) level-bottom "communities" across the North American platform for their compilations. They first converted the faunal lists to ordinal counts of generic (or species) numbers for each community and then used cluster and factor analyses to analyze the assembled data, with assembly of the results in a series of time-environmental "maps." These "maps" depict the major changes through Ordovician time: (1) the more complexly structured Paleozoic benthic communities displacing preexisting, less-structured Cambrian communities to the outer shelf and slope; and (2) the Modern mollusk-rich communities originating onshore and, in turn, displacing the Paleozoic-dominated communities to more offshore sites.

Consequently, the Ordovician Radiation, with its significant components of the Paleozoic and Modern EFs in the brachiopod-rich and mollusk-rich communities, exerted a profound impact on the ecologic structure, with the older (Cambrian-type), and to a lesser extent the Paleozoic-type, communities being displaced diachronously offshore (Sepkoski and Sheehan 1983; Sepkoski and Miller 1985; Sepkoski 1991a). These events triggered great benthic community restructurings in the low latitudes of the North American continental shelf to slope during the Ordovician, but we are still some way from settling whether they were part of an overall global restructuring across the full range of geotectonic settings, facies profiles, and paleolatitudes or strictly regional (North American) events. Only by establishing similarly intensive, wide-ranging, and rigorous programs of Ordovician biodiversity study in other major platform regions such as Baltoscandia and South China can we necessarily expect to determine convincingly whether global processes were ultimately responsible for triggering these restructurings.

On the other hand, Miller (1997a) has shown that promising results are attained if the global patterns are dissected—that is, if diversification patterns are compared at different geographic and environmental scales. His comparative database survey of more than 6,570 genus-level Ordovician occurrences of major faunal representatives of each of Sepkoski's (1981a) three EFs—trilobite-, brachiopod-, and mollusk-dominated—across six different continental blocks (Laurentia, Baltoscandia, North China, South China, Bohemia, and East Avalonia) revealed distinctive continent-to-continent differences in generic richness and compositional changes between the faunas through Ordovician time. Both raw data and rarefaction analyses were used, the latter to compensate for the uneven sample sizes—each bin equivalent to one very unequal British series subdivision of time. Further comparative analysis of the Laurentian and South Chinese data was provided by Miller and Mao (1998) outlining the generic diversity patterns: (1) generally higher in Laurentia than in South China; (2) levels rather static in the Early to Mid Ordovician, then becoming higher in the Caradoc for Laurentia; (3) levels in the Mid and Late Ordovician probably somewhat higher than for the Early Ordovician of South China; and (4) increases of benthic mollusks only in the Late Ordovician of Laurentia. The major contrast in the patterns of diversification of the two continental areas relates to the presence or absence of siliciclastic-rich environments. Carbonate sedimentation was predominant in South China, but there were influxes of terrigenous sediments across eastern and central Laurentia during the Late Ordovician, associated with orogenic activity (Taconic Orogeny), and this facilitated the radiation of benthic mollusks, especially the bivalves (chapter 20). Miller and Mao (1998) recognized that the orogenic activity had importance as an abiotic factor in this continental-scale Late Ordovician radiation of benthic mollusks. In a more recent survey, Connolly and Miller (2002) have sought to explain global origination and extinction patterns by using diversity- and productivity-dependent models to analyze their standard data set of trilobite, brachiopod, gastropod, and bivalve occurrences. The bivalves were recognized as the only group to show productivity-dependent origination, as a consequence of the increased orogenic activity in the Late Ordovician, but this overall had a rather limited impact on origination patterns.

Another community-based approach for studying Ordovician biodiversity has involved the assessment of taxonomic richness using two main types of measurement: (1) "inventory" diversity, which is usually represented by *alpha* (within habitat) diversity, or the numbers of taxa per unit area, and basically records taxonomic richness, or packing, in the habitat; and (2) "differentiation" diversity, which is a measure of the dissimilarity (or similarity) between components of the inventory diversity. Following Sepkoski (1988), this comprises *beta* (between habitat) diversity, or a measure of variation in taxonomic composition between areas of alpha diversity, and *gamma* (between province) diversity, a measure of taxonomic differentiation between geographic regions (a reflection of provinciality or endemicity). Sepkoski (1988) took the 500 or so Paleozoic assemblages from the North American platform—much the same database as used in the studies of faunal change mentioned earlier in this chapter—though he used a more precisely defined onshore-offshore framework encompassing six environmental "zones" for this compilation. The survey focused on the patterns of generic alpha and beta diversity, particularly the contributions during the Ordovician radiations. Averages of alpha diversity were computed across each environmental "zone" and for each geologic period. Each period exhibited a trend of alpha diversity that increased offshore to a high about midshelf then declined farther offshore. Sepkoski was only able to demonstrate an overall generic alpha diversity increase of 70 percent from the Cambrian to later in the Paleozoic (i.e., of diversity increase that could be related to Ordovician radiations). This value falls far below the 300 percent global generic diversity increase derived from the compilation of the marine animal genera based on the *Treatise* (Moore et al. 1953–1992) and other sources (Sepkoski 1986, 2002).

Sepkoski (1988) sought a number of explanations for this missing diversity. Was it incorporated in the beta diversity or derived from other sources? The trends of beta diversity between the Cambrian and Ordovician suggest some increased habitat specialization, as well as addition of soft-bottom communities and appearances of new community types (reef and hardground communities, bryozoan thickets, and crinoid gardens), but still the amounts are insufficient to relate to the missing diversity. Although Sepkoski (1988) discounted gamma diversity as having a role, it is apparent that the significant levels of provinciality that have been recognized in many groups of organisms through Ordovician time were overlooked (Jaanusson 1979; Patzkowsky 1995c; Webby et al. 2000). Moreover, Sepkoski's assessments of alpha and beta diversity were based on regional North American community studies rather than global data. Miller (1997c) and Miller and Mao (1998) also noted that Sepkoski (1988) did not measure patterns throughout the Ordovician. Miller and Mao's (1998) comparative analyses of data show beta diversity in South China declining markedly through late Mid to Late Ordovician time, whereas in Laurentia the beta diversity exhibits only a slight overall decline. The alpha diversity patterns also change through time, with both continental platform areas exhibiting modest increases through the latest Darriwilian to Caradoc interval. The global summation of community-scale diversification has yet to be determined because it requires assessments across all continental and oceanic terranes.

Other Geotectonic Settings and the Oceanic Sampling Biases Caused by Subduction

Ordovician community-based diversity change has been recognized in other geotectonic settings (especially the small continental blocks and terranes with oceanic and island-arc remnants), for example: (1) in the smaller, faunally distinctive, high-latitude peri-Gondwanan terrane of Perunica–Bohemia (see Kraft and Fatka 1999); (2) in the low-latitude microcontinental block of the Argentine Precordillera (Waisfeld et al. 1999); (3) in the intraoceanic to marginal island sites of Newfoundland in the Iapetus Ocean (Harper and Mac Niocaill 2002); (4) in the Kazakhstanian terranes (Popov et al. 1997); (5) and in the peri-Gondwanan Macquarie terrane with associated volcanic islands in eastern Australia (Webby 1992c; Percival and Webby 1996, 1999). All these geotectonically distinctive sites need to be well documented, and ecologically, and in a few cases provincially, distinct diversity profiles need to be fully developed.

Depending on their size, age, isolation, topography, and climate, islands and archipelagoes are potential sites for evolution and dispersal and are most important contributors to the biodiversity record (MacArthur and Wilson 1967; Soja 1992). Yet the patterns of ancient marine island biotas have been largely ignored. The greater part of the pre-Mesozoic oceanic crustal record (with its sedimentary and volcanic island records) has been destroyed by subduction (Menard 1986). Much of the preserved Ordovician record of the great Panthalassic Ocean (figure 5.4) that presumably straddled the Northern Hemisphere, for example, seems to have been lost. Nevertheless, some important oceanic and island-arc remnants have been incorporated into the terranes of Paleozoic fold belts. The oceanic deposits that survived the various accretionary processes are commonly metamorphosed and/or structurally disrupted, but some of these "windows" may exhibit a reasonably well-preserved fossil record of oceanic (pelagic and deep benthic) life and shallower benthic faunas from the fringes of volcanic island sites.

The importance of Ordovician islands as sites for evolution and dispersal was first emphasized by Neuman (1972). He noted the islands in the Iapetus Ocean as providing (1) a wide range of shallow-water habitats, (2) pathways for migration, (3) centers for evolution of new taxa, and (4) potential areas for development of biogeographically distinct island populations. R. B. Neuman (1984) recognized the mid-oceanic and marginal islands (some arc-related) as having played important parts in the development of the two biogeographically distinct Iapetus faunas, and these patterns have more recently been confirmed by Harper et al. (1996). First, the Early to early Mid Ordovician Celtic faunas developed in the intermediate to high latitudes of (1) island and microcontinental sites of Avalonia (to the south side of Iapetus) after it had split off from Gondwana in the Late Cambrian or Early Ordovician and (2) arc settings in marginal and intraoceanic sites (between the southern margins and the middle of the Iapetus Ocean), but these became accreted to Laurentia later in the Ordovician. Second, a contrasting, low-latitude Early Ordovician fauna (Toquima–Table Head assemblage) developed on the northern side of Iapetus, in Laurentian marginal sites (including west-central Ireland and Norway).

Harper and Mac Niocaill (2002) described these various marginal and intraoceanic Iapetus sites as acting, alternately, as "cradles" that provided sources for radiations onto adjacent platforms and "museums" (or refugia) for the otherwise relict taxa.

Island sites in the Late Ordovician of eastern Australia (Macquarie Arc) have also acted as cradles for the dramatic radiations of siliceous sponges in the upper slope (Rigby and Webby 1988; Percival and Webby 1996) and the heliolitine corals in level-bottom areas of the midshelf (Webby and Kruse 1984). There was also ecologic differentiation of trimerellid brachiopod shell beds in the inner shelf (Webby and Percival 1983) and tetradiid coral biostromes and shoals in the midshelf (Webby et al. 1997).

The wholesale recycling of Paleozoic oceanic crust in subduction-related plate tectonics has left a patchy record of preserved ocean-sourced fossiliferous rocks, the sequences commonly disrupted, frequently limited in continuity across time and space, and sometimes metamorphosed. In contrast, the stable continental platforms typically have accessible, temporally and spatially continuous fossiliferous successions that may exhibit well-preserved communities across a wide range of habitats and a fossil record that has largely survived intact. Consequently, the fossil record as reflected by our global biodiversity databases does not provide a balanced view of the diversification of Ordovician marine life. We need to find some way to correct for the preservational bias that exists between the more complete biotal records in continental platform deposits and the incomplete and fragmentary records of biotas from oceanic sites.

Other Preservation and Sampling Biases

Considerable efforts have also been made in this biodiversity study to minimize potential biases, especially in the areas of evaluating taxa, establishing a refined timescale, and standardizing the diversity measures, as already stated. Other sampling biases have significantly important implications for Ordovician diversity studies across global and regional scales (Miller and Foote 1996; Holland 1997; Patzkowsky and Holland 1999; Alroy et al. 2001). Miller and Foote (1996) raised the possibility that the great proliferation of familial (and generic) diversity through Ordovician time, as illustrated, for example, by Sepkoski (1981a, 1995; see figure 1.1) using raw data, might be affected by sampling bias. They depicted analyses showing raw and rarefied data (the latter to allow for the uneven sizes of the coarse, British-based series sampling intervals used by them) for their nearly global-scale sample of more than 6,570 Ordovician occurrences (trilobite, brachiopod, and benthic mollusk genera). The data are from Laurentia, Avalonia, Bohemia, Baltoscandia, North China, South China, and, to a limited extent, Australia. The diversity curve based on the raw data showed an upward rise during the Late Ordovician (Caradoc), in contrast to the "sample standardized" rarefied plot, which exhibits a flattened, plateaulike trajectory of stabilized diversity through the Late Ordovician.

At the regional level, sequence stratigraphy architecture has been shown to exert a primary control on the fossil record (Holland 1995, 2000; Holland and Patzkowsky 1999). It not only has a significant role for use in establishing global correlations and age relationships but also has important implications for biodiversity studies of at least a regional, basinwide scale. Holland (2000) predicted from sequence stratigraphic modeling (with verification of some examples from the field) that four types of biases occur within the sequence packages: sampling biases (rare, short range forms especially likely to be missing), facies biases (concentrations of first and last appearances at abrupt facies changes such as flooding surfaces), unconformity biases (concentrations of first and last appearances at sequence boundaries), and condensation biases (clustering of occurrences during slow sedimentation).

Some of these biases have been recognized by Patzkowsky and Holland (1996, 1997, 1999) and Holland (1997) in their full faunal analyses of the biotas of the Upper Ordovician (Mohawkian-Cincinnatian) depositional sequences in the eastern United States. These studies demonstrate how sequence stratigraphic architectures bias the fossil record, especially the representation of fossil range data. This in turn affects interpretations of patterns of biodiversity change in such deposits. All such successions therefore need full stratigraphic analyses to establish whether the biodiversity results reflect true biologic change or merely relate to apparent diversity change, driven or controlled by sequence stratigraphic architectures. Smith (2001) also emphasized the important role of sequence stratigraphic architectures, and also major transgressive-regressive cycles, in controlling Paleozoic faunal diversification. Holland and Patzkosky's insights demonstrate that, in future biodiversity studies, paleontologists should, if possible, work in close cooperation with sequence stratigraphers.

Alroy et al. (2001) adopted a new database program of compilation and analysis with a variety of sampling and analytical protocols to minimize sampling bias for translating the sampled fossil occurrence data into global Phanerozoic marine diversity curves. It aims to standardize sampling levels and to define counts of taxa. Eventually this major database initiative is intended to cover the entire Phanerozoic record of terrestrial and marine fossils across all geographic regions, but currently the sampling focuses only on well-studied regions through two halves of the Phanerozoic: interval 1 from the base of the Upper Ordovician ("Llandeilo") through Carboniferous (apparently mainly across the Paleozoic diversity plateau), and interval 2 from middle Jurassic through Oligocene. The core taxonomic groups for interval 1 analysis include anthozoans, brachiopods, echinoderms, mollusks, and trilobites, with the current compilation dominated by inclusion of data from North American and European localities. Sampling of bryozoans, conodonts, and graptolites is as yet inadequate. Interval 1 has 15 roughly equal bins of 10.7 m.y., so the Late Ordovician Epoch spans only about two or three sampling bins. From an Ordovician perspective, not only is this a rather limited span of genus-level sampling (only depicting the last part of the Ordovician Radiation), but the primary data have significant biases—the core groups are not completely representative, given that all the groups are benthic to the exclusion of pelagics and they are geographically restricted in North American and European sites that, during the Late Ordovician, predominantly occupied low paleolatitude positions. The establishment of standardized analytical protocols to minimize sampling biases for calculating diversity estimates of previously collected data is indeed a laudable development but no real substitute for maintaining a wide array of intensive field- and laboratory-based biodiversity studies—especially regional and global investigations that fully document and sample biotas from all habitats (where applicable, making allowances for sequence stratigraphy architectures) and across all paleolatitudes, tied to highly calibrated and resolved timescales.

Timing of Major Radiations of Level-Bottom, Reef, and Infaunal Communities

Level-Bottom Communities. The timing of the major Ordovician radiations of the level-bottom and the reef communities proves to be markedly different (Webby 2002). The best records of these faunal changes are preserved on the North American (Laurentian) platform, and they seem to be decoupled from one another (cf. Sheehan 1985). The significant radiation event that relates to the appearance of the level-bottom community is recorded from the base, into the lowest part, of the Middle Ordovician (lower Whiterockian, i.e., *TS*.3a–b; see table 1.3). The base of the Whiterockian represents the boundary between Boucot's (1983) level-bottom community groups, Ecologic-Evolutionary Units (EEUs) III and IV—typically units lasting tens of millions of years with community stability (stasis) bounded by pulses of extinction (Brett 1995). These Ordovician units were renumbered EEUs P1 and P2 by Sheehan (1996, 2001a) on the basis that they were equivalent to the first two subdivisions of the Paleozoic EF. Droser and Sheehan (1997) noted that, because EEUs P1 and P2 are associated with the Ordovician radiations, they are not typical. The community complexity increased during EEU P1 because members of the Paleozoic EF were being added while representatives of the Cambrian EF were still in place. By the beginning of EEU P2, however, the community ecology of the Paleozoic EF was "fully in place" (Sheehan 2001a:244).

The early Mid Ordovician radiation is characterized by both taxonomic and ecologic diversification changes that are especially well documented in the Great Basin successions of the western United States (Droser and Sheehan 1995, 1997b; Droser et al. 1996). The features include major turnover of faunas, changes in shell-bed composition, and hardground evolution. The shell-bed concentrations were dominated by brachiopods and ostracodes, instead of the previously important trilobite-dominated shell beds (Droser et al. 1996). The trilobites (representatives of the Whiterock Fauna; Adrain et al. 1998; see also chapter 24) diversified significantly, but the group no longer had a prominent role in shell-bed development

—that is, the taxonomic and ecologic diversifications had effectively become decoupled. Bottjer et al. (2001) have regarded the early Mid Ordovician radiation event as exhibiting signals of second-, third-, and fourth-level ecologic changes, which reflect shifts ranging from major structural change in the ecosystem to minor community-level changes.

Kanygin (2001) outlined the Ordovician diversification of biotas in the Siberian Platform and the Verkhoyansk-Chukchi fold-belt region of northeastern Siberia, highlighting a dramatic diversity change in the level-bottom communities of these successions across virtually all biotal components, apparently close to the base of the Llanvirn (i.e., mid Darriwilian—TS.4b). Both taxonomic and ecologic changes were involved. The Early to early Mid Ordovician (TS.1a–4a) biotas were reported to be closer in aspect to Cambrian assemblages than to the markedly diverse mid Darriwilian to Late Ordovician record (Kanygin 2001). The Early Ordovician platform succession is dolomitic and dominated by occurrences of stromatolites. Trilobites are the most common faunal components, with brachiopods, cephalopods, conodonts, "monoplacophorans," and sponges less common, and ecologic relationships remained generalized. In contrast, the record above the base of the Llanvirn has sudden appearances of abundant bryozoans and ostracodes, some crinoids, acritarchs, and chitinozoans, as well as continuation of moderately common trilobites, brachiopods, and conodonts. Most of the benthic species were endemics, and a host of ecologic innovations was also reported—for example, development of short-ranging benthic and "subpelagic" ostracodes (the latter adapted to passive hovering and active swimming) that densely occupied both shallow-shelf and deep-shelf habitats. This may prove to be a fundamentally wide-ranging, coupled taxonomic and ecologic diversification event that occurred long after the ecologic diversifications in "level-bottom" communities of the early Whiterockian (TS.3a–b) in North America. However, the fact that this dramatic diversity change is associated with a regional deepening event, involving a sudden shift from restricted conditions (presence of dolomitic and stromatolitic deposits) to more normal open marine conditions, suggests that these Siberian successions require fuller stratigraphic analyses to determine whether the fossil records represent true or apparent biodiversity change.

Reef Communities. Modern coral reefs have the greatest levels of known diversity and productivity in marine ecosystems, and there seems little reason to doubt that Ordovician reefs, at least the metazoan/calcified algal-dominated structures, relative to other Ordovician communities, were equally richly diverse and productive. Webby (2002) established that the distribution of Ordovician reefs, prior to the Mid to Late Ordovician boundary and the first appearances of well-skeletonized metazoan- and algal-dominated reef communities, included mainly microbial-dominated reef structures. Prokaryotic, photosynthetic microorganisms (mainly blue-green "algae" or cyanobacteria) were ultimately responsible for establishing the microbial communities that precipitated the various types of stromatolites, thrombolites, mats, and films and contributed the calcified microbes, such as *Girvanella, Renalcis,* and *Epiphyton,* to the initial Ordovician reef ecosystem. Sometimes these microbial constructions were associated in consortia with subordinate receptaculitids (*Calathium*), lithistid sponges (*Archaeoscyphia*), stromatoporoids (*Pulchrilamina*), bryozoans (*Batostoma*), and even rarer, tabulate corals (*Lichenaria*). Some structures, especially in the late Tremadocian, developed into huge, kilometer-scale barrier reef complexes along parts of the Laurentian shelf margin, and other large (near 100 m high) stromatactis-bearing carbonate mud mounds formed in parts of the outer platform and slope. These patterns were maintained through the Early Ordovician to the mid Darriwilian, until the dramatic change in the latest Darriwilian when the new, well-calcified metazoan- and algal-dominated reef ecosystem emerged.

The earliest stage of the great metazoan/algal reef expansion is best exhibited in on-shelf sites through the Chazyan to early Mohawkian interval (TS.4c–5b) of eastern North America (Webby 2002: figure 6). Stromatoporoids, tabulate corals, bryozoans, lithistid sponges, pelmatozoan echinoderms, and solenoporan algae established themselves during the late Darriwilian (Chazyan) in an array of new and complex community interrelationships, dependent on their differing frame building, encrusting, and sediment-producing roles. Taxonomic diversification involved the bryozoans, lithistid sponges, a few stromatoporoids, and a few tabulate corals. Framework cavities became colonized by three types of cryptic niche dwellers: a first indubitable colonial organism (bryozoan *Batostoma*), as well as taxa representing bioeroders and bioturbators. Microbial components such as localized stromatolites and *Girvanella* crusts continued to be associated, but they were now subordinate in the reefs. The second stage of reef development during the Caradoc (early Mohawkian) involved expansion of bryozoan-dominated reefs into the offshore—shelf-edge to downslope—habitats. The record is again most complete in eastern North American sites. Some of the more massive, composite structures were up to 50 km across and 80 m high near the shelf edge. By the mid Caradoc (*TS*.5c), the reefs had spread circumglobally to Greenland, Baltoscandia, Arctic Russia and the Urals, Kazakhstan, China, and Australia, to on-shelf sites near continental margins, and to island arcs.

Another significant event is Wilson and Palmer's (2001a) Ordovician "bioerosion revolution" (see discussion in chapter 34), involving a marked increase in the numbers of bioeroding organisms actively colonizing reefs and hardgrounds. This bioerosion event seems to be strictly correlative with the initial, late Darriwilian expansion of metazoan and calcified algal reefs.

Infaunal Communities. A major increase in the intensity and depth of bioturbation has been recognized by Droser and Bottjer (1989) in the infaunal records of shallow-water carbonate successions of the Great Basin, western United States. This change marks a major infaunal biodiversity event, signifying an increased utilization of infaunal ecospace. It presumably reflects a restructuring of the infaunal ecosystem (at least in shallow-water carbonate platform habitats) and probably represents the evolutionary development of new and markedly different soft-bottom burrowers. It is highlighted by the morphological changes of one particularly common trace fossil (a complex branching burrow) called *Thalassinoides*, which in its Lower to Middle Ordovician record forms mazelike structures, which are replaced in the Upper Ordovician record by more complex boxwork structures to a depth of 0.3 to 1 m. The precise timing of this intensification of bioturbation, with the accompanying morphological change from the mazelike to boxwork structures, has not been determined. Unfortunately, much of the upper Darriwilian to middle Caradoc succession is not in the typical "carbonate inner shelf" facies that contains the key ichnofossil—either it is in a coarse nearshore quartzite and sandstone sequence, or the record is missing. Consequently, it is impossible at the present time to determine precisely when, within that undiagnostic interval representing about 15 m.y., the bioturbation change occurred and whether it was sudden or gradual. It may represent an infaunal radiation event of late Darriwilian (*TS*.4c) age—more or less contemporaneous with the initial expansion of metazoan and calcified algal reefs—or later, at some time during the Late Ordovician (Caradoc—*TS*.5a–d). Sections with a more complete Middle to Upper Ordovician stratigraphic record in the "carbonate inner shelf" facies need to found elsewhere in North America and/or in other continental platform successions so that this infaunal event can be more precisely defined and correlated.

Ordovician Radiation in the Pelagic Realm

Pelagic organisms—broadly divided into planktic (suspended, floating, and/or drifting) and nektic (independently mobile, swimming) types—lived in waters of the open ocean, as well as the waters overlying the continental shelves (and platforms). Generalized aspects of the origins and early diversification of pelagic faunas have been outlined by Rickards (1990), Signor and Vermeij (1994), and Rigby (1997). Rigby (1997), for example, has argued that all the main planktic groups had benthic origins—indeed, the plankton and benthos were inextricably linked (Signor and Vermeij 1994). Apart from the graptolites, few preserved, predominantly Ordovician pelagic groups have been studied intensively enough to understand how they became ecologically differentiated (Rickards 1975; Cooper et al. 1991; Underwood 1993; Finney and Berry 1997; Cooper 1999c). Furthermore, rather limited detailed attention has yet been given to studying the associated deep-water sediments, for example, as represented by biogenic siliceous sediments (radiolarian oozes) and organic-rich black shales (graptolite- and/or unicellular algal-dominated). Preservational biases were probably greater for pelagic forms than for benthos, given their exposure to predation, dissolution, or decay in their long descents through the water column before accumulation in sediments on the seafloor (Butterfield 1997). However, to some extent this would have been offset by a more continuous record of deep-water sedimentation and by the larger population sizes, especially in plankton blooms, which would have increased the chances of a few skeletons becoming entombed and preserved in bottom sediments.

The preserved Ordovician record includes the planktic groups (unicellular green algae, acritarchs, chitinozoans, graptolites, radiolarians) and the mainly nektic groups (some conodonts, some nautiloid cephalopods, a few trilobites, and a few phyllocarid crustaceans). A few representatives of the conulariids and the polychaetes (scolecodonts) also may have had pelagic life habits, but there is insufficient evidence to review them meaningfully at this stage. Probably only a very small percentage (perhaps less than 10 percent) of the diversity of organisms that inhabited the open oceans in the Ordovician is likely to have been preserved—the bulk were soft-bodied organisms that left no trace. In addition, the overwhelming majority of dispersive planktic larval stages of Ordovician metazoan phyla, including most of the skeletonized benthic marine animals, are unlikely to have been preserved. These would have formed a very large component of the plankton in the water column—a very high diversity of soft-bodied larvae (Signor and Vermeij 1994)—that would have been most common closer to coastal areas.

Acritarchs, Unicellular Green Algae, and Chitinozoans

Acritarchs and unicellular green algae were the main preserved microplanktic groups of primary producers in the Ordovician oceans. The acritarchs exhibit a broad twofold "provincial" differentiation across paleolatitudes (chapter 32), and high levels of diversification were maintained (Tappan and Loeblich 1973) in one region or another through all intervals of Ordovician time. In addition, levels of productivity were high, with samples of more than 100,000 specimens per gram of rock recorded in certain offshore sites (Dorning 1999). The highest acritarch productivity is usually in the upper offshore, much reduced in the lower offshore, and mainly replaced by leiospheres (featureless sphaeromorph members of unicellular green algae) in more distal sediments.

The chitinozoans are presumed to be eggs of unknown soft-bodied metazoans and probably also a part of the plankton (Paris and Nõlvak 1999; chapter 28). A similar broadly twofold provincial differentiation of higher-paleolatitude and lower-paleolatitude assemblages was maintained (Achab 1991) and abundances of up to a few thousand specimens per gram of rock in the higher paleolatitude samples, compared with a few tens from equivalent low paleolatitude samples (chapter 28). From their initial appearances in the Early Ordovician (early Tremadocian), the chitinozoans expanded steadily to a first diversity peak in the Darriwilian and then, after a short-lived lowering, to a second, comparable diversity pulse during the early Late Ordovician (mid to late Caradoc).

Graptolites

Rickards (1975, 1990) considered the graptolites to be probably the first abundant macroplankton in marine habitats, with their food supply mainly minute phytoplankton and zooplankton. According to Cooper (1999c), the planktic graptolites probably arose from a sessile dendroid-type ancestor that occupied a deeper-water, outer-shelf site close to the shelf-slope margin in the earliest Ordovician (early Tremadocian). The major evolutionary event is thought by Cooper to have involved the larval stage acquiring the ability to initiate its skeletal growth without first attaching to the substrate. What followed was the major Tremadocian expansion of these earliest planktic graptolites into onshore (shelf) and offshore (oceanic) habitats, with vertical differentiation into the shallow-water epipelagic depth "zone" and the deeper-water mesopelagic and bathypelagic depth "zones." The boundaries between epipelagic-mesopelagic and mesopelagic-bathypelagic "zones" were placed at depths of 200 m and 1,000 m, respectively (Cooper et al. 1991). The vertical depth-related differentiation of graptolite assemblages was followed during the late Early Ordovician to Darriwilian (Arenig-Llanvirn) interval by the equally important lateral, "provincial" differentiation across latitudes and the greatest radiation of graptolite species—with a spectacular diversity spike in the early Arenig representing dichograptid appearances (Cooper 1999c: figure 1).

Radiolarians

The radiolarians utilized seawater saturated with silica for their skeleton formation and were consequently major contributors to the formation of Ordovician cherts. They therefore had an important role in establishing sinks of oceanic silica in some deeper basinal habitats (Malvina et al. 1990). Ordovician radiolarian chert localities are widely distributed in Kazakhstan, New South Wales (Australia), Newfoundland, Nevada, Scotland, Spitsbergen, and Gansu (North Central China). Although Nazarov and Ormiston (1985) originally claimed that spherical-type zooplanktic radiolarians appeared in abundance for the first time during the Early Ordovician, recent reports of assemblages in the mid to late Late Cambrian of Newfoundland (Won and Iams 2002), the Siberian Altai Mountains, and central Kazakhstan (Tolmacheva et al. 2001a) now suggest that their initial major diversification into continental slope and deep basinal habitats may have commenced slightly earlier. Similar assemblages are reported from the early Tremadocian (chapter 11) but are then replaced in successive pulses of early Arenig and Mid Ordovician diversity increase by more typical, Ordovician-type associations.

Conodonts

Sweet (1988b) interpreted conodonts as pelagic, probably nektic, marine animals—a swimming predator that occupied a broad range of geographic and environmental habitats. Two main ecologic models were originally proposed to explain the distribution patterns: (1) Seddon and Sweet (1971) favored a depth-stratified model that was more or less based on the assumption that conodonts were pelagic organisms; and (2) Barnes and Fåhraeus (1975) preferred a biofacies-controlled model, related to the lateral segregation of predominantly nektobenthic organisms and some benthics—only simple coniform taxa being interpreted as pelagics. Many refinements and modifications to the two basic "pelagic" and "nektobenthic" conodont schemes have since been proposed (see review in Pohler and Barnes 1990), and more generally applicable fossil distributional modeling applications have also been outlined by Tipper (1980). Some of the taxa recorded as from "deep/cold environments" in Albanesi and Bergström's analysis (chapter 29) are pelagic forms.

Many detailed studies of conodont biofacies have been undertaken over the past two decades, especially in North America (e.g., Sweet and Bergström 1984; Ji and Barnes 1994a; Pohler 1994; Johnston and Barnes 1999), but Rasmussen's (1998) detailed, computer-based cluster analysis of mid Darriwilian conodont faunas was the first to demonstrate that the deep-water *Protopanderodus-Periodon* biofacies of the North Atlantic provincial realm differed from other biofacies in exhibiting exactly the same associations of cosmopolitan forms (including coniform taxa) in both Baltica and Laurentia (specifically eastern Laurentia). This was despite the fact that these continental blocks occupied different sides of the Iapetus Ocean and lay in different paleolatitudes during Mid Ordovician time. When these cosmopolitan taxa were removed from the biogeographic analysis of the eastern Laurentia fauna, the region became much more provincially similar to the North American Midcontinent provincial realm than was previously shown. His study established the deep-water *Protopanderodus-Periodon* association as an oceanic biofacies, implying that the conodonts were pelagic forms.

Armstrong and Owen (2002a, 2002b) provided further insights on how the conodont biofacies (and biodiversity) became differentiated, with the shelf biofacies composed of nektobenthic forms, and the oceanic biofacies of wide-ranging pelagic taxa, recognizing that they lived in shallower, stratified oceanic water masses. Their presence-absence data for conodont genera were compiled from sections across onshore-offshore profiles in Laurentia and Avalonia (to either side of the Iapetus Ocean) for three Mid to Late Ordovician "time-slice" intervals—late Darriwilian–early Caradoc (*TS*.4c–5a), late Caradoc (*TS*.5d), and mid to late Ashgill (Rawtheyan, *TS*.6b)—with all genera coded to a particular biofacies. Three depth-related oceanic biofacies and three (usually two) laterally segrated shelf biofacies were recognized. The shelf biofacies were regionally distributed, ranging from onshore to offshore and terminating offshore. The oceanic biofacies included faunal elements usually regarded as "North Atlantic" provincial components and many coniform genera. It had a spread from the inner to outer shelf that was independent of benthic biofacies. Higher diversities were associated with the inner shelf and the upper water column in both Laurentia and Avalonia.

Armstrong and Owen (2002a) stressed the importance of defining the biofacies in order to reconstruct the biofacies architecture of the regions and as a basis for establishing meaningful analysis of paleogeographic relationships. Clearly many previous studies of provinciality are now outdated, as they employ admixtures of shelf and oceanic taxa. The oceanic biofacies reflected water-mass structure, and this was likely to have caused different ecologic constraints and dispersal mechanisms for oceanic taxa, as compared with time-correlative benthic taxa. Other complications may apply where changes in temperature and/or density cause vertical movements of the thermally stratified water masses, such as in upwelling zones at the continental margins. Armstrong and Owen's biofacies insights have important implications for biodiversity studies. In the future the biofacies records must be dissected into their benthic and pelagic components so that more meaningful patterns of biodiversity change can be determined for each very different ecologic realm.

Zhen and Percival (2002), working in the Australasian region, have independently recognized the difficulties of utilizing the biogeographic subdivisions based on the Midcontinent and North Atlantic provincial realms (or provinces). They have adopted a global approach that to some extent overlaps with the new initiatives of Rasmussen, Armstrong, and Owen, arguing that most of the organisms occupying the shelf regions were nektobenthic or benthic (representatives of their "Shallow-Sea Realm" [SSR]) and recognizing a separate, important component of pelagic forms that have cosmopolitan faunal relationships and include coniform taxa (their "Open-Sea Realm" [OSR]). Temperature, depth, and salinity are regarded as controlling influences, though endemic forms show more restricted habitat preferences in the SSR. The SSR is subdivided into six provinces: Australia, Laurentia, North China (Tropical Domain), South China and Argentine Precordillera (Temperate Domain), and Baltoscandia (Temperate and Cold domains). The OSR (apart from a tropical domain) remains undivided. Zhen and Percival (2002) have recommended that usage of the traditional North American Midcontinent conodont "province" now be strictly equated with the low-latitude domain (Laurentia) of their SSR. They further argue that the North Atlantic conodont "province," which represents a composite of temperate and cold domains (Baltoscandia) in the SSR and an as yet undefined domain of the OSR, should now be abandoned. In terms of Ordovician biodiversity relationships, the assemblages in Zhen and Percival's OSR are significantly pelagic biotas.

Armstrong (1997) recognized a deep or cool, predominantly shelf and slope assemblage of probable nektic taxa from the Southern Uplands of Scotland that he referred to the "*Periodon-Pygodus* Restricted Species Association" (sensu Bergström and Carnes 1976). This association of pandemic species, from the *Pygodus anserinus* Zone (latest Darriwilian–earliest Caradoc), is identical to correlatives in eastern North America (Newfoundland and Tennessee) and Baltoscandia (i.e., clearly a part of the oceanic biofacies). Armstrong (1996) had previously interpreted the patterns of distribution of conodonts across the Ordovician-Silurian boundary in terms of the changing ocean state, presenting a broad outline of biofacies profiles from the late Caradoc onward. This included reference to Sweet and Bergström's (1984) biofacies data and to the deepest association from basinal cherts, called the *Dapsilodus-Periodon* biofacies (see further discussion of the chert association later in this chapter). Sweet and Bergström (1984) recognized this biofacies as the deepest they encountered in their survey of the late Caradoc–earliest Ashgill interval. It is preserved with graptolites and radiolarians in the Ouachita trough—a rifted reentrant into the Laurentian continental interior of Oklahoma that lay in low paleolatitudes.

Tolmacheva et al. (2001a) described a remarkable 35-m-thick biogenic accumulation of ribbon-banded radiolarian cherts and shales with a nearly continuous zonal record of conodont assemblages in a composite section of the Burubaital Formation, central Kazakhstan. These range from Late Cambrian (with continuity across the Cambrian-Ordovician boundary), through the Tremadocian, to the early Arenig. The assemblages were considered to represent pelagic components of the oceanic biofacies, possibly associated with an area of equatorial upwelling. Earlier, Dubinina (1991) and Popov and Tolmacheva (1995) outlined the conodont succession in sections in central Kazakhstan, at Sarykum and Burubaital (about 240 km apart), respectively, through essentially the same interval of chert-dominated sediments and revealing essentially similar temporal patterns of faunal change. In the lower part of the succession, that is, through the Late Cambrian (*Eoconodontus notchpeakensis* Zone) to Tremadocian (lower *Drepanoisodus deltifer* Zone) interval, the associations were dominated by two intermixed faunal components, the mainly undescribed, very simple, conelike protoconodonts and paraconodonts and the true conodonts (euconodonts)—including many diagnostic zonal species of *Cordylodus, Eoconodontus, Hirsutodontus, Cambrooistodus,* and *Teridontus*. The intermixture of components was explained by Dubinina (1991) and Popov and Tolmacheva (1995) as due to thermally stratified water masses—the protoconodont and paraconodont elements representing the cooler association that occupied waters below the thermocline, while the component of euconodonts lived in warmer, shallower waters above the thermocline.

During the *D. deltifer* Zone (early to mid Tremadocian), however, a significant faunal change occurred in the Kazakhstan sections, interpreted by Popov and Tolmacheva as indicating the disappearance of the well-defined, two-layered stratification. The warmer association disappeared, and the protocondont and paraconodont assemblages were replaced by a single, low-diversity euconodont assemblage. This association continued to be represented in Kazakhstan from the *D. deltifer* Zone, through *Paraoistodus proteus* and *Prioniodus elegans,* to *Oepikodus evae* zones (mid Tremadocian–mid Arenig). This major oceanic event in the *D. deltifer* Zone also seems to equate with emergence of well-marked Ordovician conodont provincialism in the Ordovician, especially in the SSR. Two less-pronounced Late Cambrian phases of provincialism have been identified by Miller (1984), the first with the appearance of differentiated pelagic euconodonts in warmer waters of the late Franconian (mid Late Cambrian), and the second when the species of the mainly pelagic *Cordylodus* and the questionably nektobenthic genera of the *Teridontus* lineage invaded shallow, low-latitude areas in the latest Cambrian (base of Ibexian).

Trilobites

Most Ordovician trilobites lived as benthic or nektobenthic organisms in reasonably close contact with the seafloor, though, of course, like most metazoan phyla, they probably had dispersive planktic larval stages. A few small trilobite groups, however, developed active swimming roles in a pelagic adult life (Fortey 1985; McCormick and Fortey 1998, 1999; see also Fortey in chapter 24). Fortey (1985), using three main criteria—their functional design, nature of geologic association, and analogy with living, large-eyed, ocean-dwelling isopod and amphipod crustaceans—determined that two main groups of pelagic trilobites were initially differentiated early in Ordovician time. The first was mainly based on the family Telephinidae, with characteristic *Carolinites, Opipeuter,* and four other genera—forms that were regarded by Fortey (1985) as having large bulbous eyes and comparatively poorly streamlined, smaller body forms. They have been interpreted as becoming well adapted to active surface swimming in near equatorial waters, constrained between 30 degrees north and south paleolatitude and in the upper part (epipelagic "zone") of the water column, being a widespread community across all biofacies (McCormick and Fortey 1998, 1999). Elements of this epipelagic community maintained a pan-tropical span from early Arenig through Mid Ordovician time, but *Telephina* and *Phorocephala* dispersed to higher paleolatitudes during the Late Ordovician, prior to their end Ordovician demise.

The second group was characterized by representatives of the family Cyclopygidae—mainly large-eyed genera such as *Pricyclopyge, Cyclopyge, Degamella,* and perhaps up to seven others—as well as other, rather aberrant forms having possible remopleuridioid affinities (*Bohemilla* and *Cremastoglottos*). Some of these, such as *Cyclopyge* and *Bohemilla,* were smaller, poorly streamlined, sluggish swimmers and others, such as *Degamella,* larger and streamlined, more efficient swimmers. Given these differences, Fortey and Owens (1999) suggested that the two groups may have fed on different varieties of plankton or that the larger forms may have preyed on smaller plankton feeders such as the phyllocarid crustaceans (see later in this chapter). Their overall distribution was markedly different from that of the telephinids, characterized by the cyclopygid biofacies that occupied the mesopelagic "zone" (probably below 200 m water depth), offshore from marginal areas across moderately high paleolatitude regions of peri-Gondwana, mainly between western and central Europe (Britain,

France), to Kazakhstan and China (McCormick and Fortey 1998). This pelagic cyclopygid biofacies was long ranging and conservative, first developing in the Tremadocian of Argentina, becoming more widespread through late Early Ordovician (Arenig) to Caradoc time, then being more restricted within low paleolatitudes during the Ashgill (Scotland, Quebec), before its extinction at the end of the Ordovician (Fortey in chapter 24).

Phyllocarid Crustaceans

Although many phyllocarid crustaceans appear to have maintained benthic or nektobenthic habits (Rolfe 1969), a few distinctive Ordovician representatives of the group, in particular the widely distributed genus *Caryocaris* (Racheboeuff in chapter 25), adopted a truly pelagic mode of life, with preservation in deep-water graptolitic shale deposits of offshore, mainly slope to basinal, sites. The morphology of *Caryocaris* was characterized by thin carapaces and expanded, leaflike furcal rami that would have allowed it to adopt a free-swimming habit in the water column. In places, large concentrations of specimens may be preserved on bedding planes that perhaps suggest that they also lived in schools at a particular depth within the water column. They have a wide distribution across high to low paleolatitudes and a long Ordovician record of occurrences, from the early Tremadocian to Caradoc. Fortey (1985) has noted that phyllocarids were common in the deeper, pelagic cyclopygid trilobite biofacies and that they may have been preyed on by larger, streamlined trilobite species of that biofacies. In comparing the Ordovician species of *Caryocaris* and related forms with certain free-swimming modern crustaceans, Vannier et al. (2003) recognized that in all probability the Ordovician "caryocaridids" lived in epipelagic to mesopelagic depth zones of the water column, especially "marginal" outer-shelf to slope settings.

Nautiloid Cephalopods

Late Cambrian nautiloid cephalopods (members of the order Ellesmereocerida) exhibit the initial development of a siphuncle (Yochelson et al. 1973), which allowed intercommunication through a strand of soft tissue for supply of blood and gaseous exchange to the closed off, septate, apical part of the shell. This unique evolutionary innovation paved the way for all cephalopods to achieve very effective buoyancy control during their Ordovician and later development. The paleoecology of the Late Cambrian precursors is not well understood. According to Flower (1957), the known shells are small, have closely spaced septa, and are associated in limestones with benthic trilobites. Hence, they probably lived near or on the bottom. While the close spacing of septa in these early nautiloids may have been a simple device to reduce buoyancy in the shells (Teichert 1967), it is not known whether the intervening spaces (camerae) could yet be used for gaseous exchange.

Holland (1987: figure 4) depicted how the two basic Cambrian types of short, nearly vertically oriented, slightly curved shells gave rise to the two great lines of nautiloid descent—coiled forms deriving from shells with an exogastric curve (ventral to the outer convex side) and the longicones evolving by straightening of shells with an endogastric curve (ventral to the inner concave side). The apical part of the longicone of later forms became partially filled by cameral and siphuncular deposits, enabling a horizontal life mode to be maintained.

The greatest ever diversification of nautiloids occurred during the Ordovician, with all the major higher taxon groups appearing by the early Mid Ordovician (Flower 1976; Crick 1981; House 1988). The various representatives have been interpreted broadly as exhibiting a full range of benthic, nektobenthic, and pelagic life habits, but at the present time only a few can be assigned indubitably to the pelagic realm. No detailed systematic survey of Ordovician nautiloids has yet been attempted, comparable to that undertaken by Fortey (1985) to elucidate the life habits of trilobites. It seems that pelagic forms have been derived from the two main body plans mentioned earlier. In terms of the major Ordovician groups, the order Tarphycerida was dominantly coiled, with the discoidal forms becoming nimble and very maneuverable swimmers, while the orders Michelinocerida, Actinocerida, and Endocerida had mainly horizontally disposed longicones, though only the first two groups may have had roles as active swimmers.

Probably the majority of the larger Ordovician longicones were carnivores (predators and scavengers) that occupied positions toward the top of the Ordovician food chain. With the ability to maintain precise buoyancy controls and an efficient means of jet propulsion, both forward and backward, dependent on the way they oriented their hyponome, they would have been capable of maintaining an active swimming mode. The largest longiconic forms may, however, have had problems because of their size and weighting (Holland 1987), restricting them necessarily to a mainly nektobenthic existence near the seafloor. Examples include the giants referred to by Holland (1987) with estimated lengths of nearly 9 m (including a body chamber), based on estimates from a number of mainly incomplete endocerid longicones. The specimen from Watertown, New York (Flower 1957), was probably of early Late Ordovician age.

Crick (1990:147–148) considered that Paleozoic nautiloids were representatives of the "shallow-shelf vagrant benthos," not strictly a part of the nekton capable of oceanic dispersal. Consequently, they were capable of dispersing only "over shallow stretches of open ocean." Stait et al. (1985), Webby (1985, 1987), Stait and Burrett (1987), and Percival in Webby et al. (2000) documented the biogeographic distribution of Ordovician nautiloids in the Australasian region, recognizing that nautiloids preferred the extensive shallow waters of carbonate platform areas. There they adopted nektobenthic habits, for example, as represented by the diverse and predominantly endemic Tasmanian faunas described by Stait (1988). However, in central New South Wales—originally a part of an island arc–related terrane, at least 1,000 km away from the cratonic East Gondwanan margin, including Tasmania (Webby and Percival in Webby et al. 2000)—the few nautiloids in the island shelf carbonates were longiconic orthoceratids and discoidally coiled tarphycerids, almost exclusively free-swimming, long-ranging, cosmopolitan forms. An additional occurrence of the wide-ranging, cosmopolitan genus *Bactroceras,* representing a specialized, surviving group of ellesmerocerids (chapter 21), is associated in deep-water graptolitic shales. It also appears to be a nektic component. Hewitt and Stait (1985) argued that the species, *B. latisiphonatum,* with its tubelike thickening of connecting rings that strengthened the septa, was able to withstand higher hydrostatic water pressures before imploding, compared with the extant *Nautilus.* Data from Westermann and Ward (1980) indicate that the depth of implosion of *Nautilus* shells is 600 m. Applying strength indices to the *Bactroceras* data, Hewitt and Stait (1985) estimated a slightly greater implosion depth of about 800 m for the New South Wales species. Consequently, there seem to be two likely scenarios, given that the offshore New South Wales island arc was separated from other parts of the Gondwanan margin by the Wagga Sea (a marginal sea). If the nautiloids tried to cross the Wagga Sea as benthic or nektobenthic elements, maintaining close contact with the seafloor, they would have exceeded implosion depths of 800 m (using comparable depths across the present-day counterpart, the Tasman Sea). The alternative, and preferred, view is that the New South Wales nautiloid assemblages crossed the oceanic barrier by swimming (perhaps also in part drifting in ocean currents) but remaining in the epipelagic to upper mesopelagic zones of the water column during this oceanic dispersal. Hence the Late Ordovician Wagga Sea was a barrier to the Tasmanian benthic and nektobenthic nautiloids but not to wider-ranging nektic forms.

Crick (1980) presented a global assessment of late Early Ordovician to early Darriwilian (Arenig) nautiloid biogeography and differentiated a number of groups that had higher mobility and a more cosmopolitan distribution, recognized as common in the oceanic environment (his "geosynclinal facies"). Longicones of the ellesmerocerid and protocycloceratid families (order Ellesmerocerida), the actinocerid family (order Actinocerida), and proterocamerocertatid and endocerid families (order Endocerida) are well represented, as well as coiled-shelled tarphyceratid and estonioceratid families (order Tarphycerida). It may be inferred, then, that by the early Arenig at least some representatives of these forms had adopted pelagic life habits, possibly mainly within the epipelagic zone. On the other hand, Teichert (1967) considered that all the Early Ordovician cephalopods with longicones probably remained as bottom dwellers, given that calcareous endosiphuncular deposits—weighting structures so important for buoyancy control—had not developed significantly prior to the beginning of the Mid Ordovician, though he regarded the coiled shells of tarphycerids as already having partially colonized the pelagic realm.

In the major diversification of the Mid Ordovician nautiloids, three groups—the actinocerids, orthocerids, and ascocerids—seem to have adapted at least in part to living in the pelagic realm. Possibly, components of twofold differentiation of distinctive faunal provinces across higher and lower paleolatitudes during the Mid Ordovician, outlined by Frey et al. (chapter 21), also extended into the pelagic realm. The longiconic actinocerids and orthocerids are medium- to large-sized groups that developed a variety of weighting structures (especially endosiphuncular and cameral deposits) toward the apical end, helping to stabilize their orientation in a horizontal position, in their role as active swimmers. The orthocerids have shells characterized by well-developed cameral deposits, subcentrally placed, tubular siphuncles that may either be empty or sometimes associated with endosiphuncular deposits, as well as orthochoanitic septal necks, thin connecting rings, and pointed apices. The secretion of calcareous cameral deposits added weight to the apical (posterior) end as the shell grew, helping to maintain the horizontal life position in an equilibrium, with centers of gravity and buoyancy near one another and also acting to stabilize the shell with respect to its ventral and dorsal sides. Meanwhile, buoyancy control was maintained by gaseous exchange into the anterior camerae, effectively lightening the shell for active swimming (Flower 1976). The actinocerids, in contrast, feature a siphuncle with mainly cyrtochoanitic-type septal necks, an elaborate development of annulosiphonate, endosiphuncular deposits, and an associated vascular system that includes radial canals extending from the central tube. Larger siphuncles are usually placed toward the ventral side, though smaller siphuncles tend to be more subventral to subcentral. Cameral deposits also occur in many genera, and the apices are short and blunt. Both these longicone groups have long histories through the Paleozoic.

The ascocerids were a small, comparatively poorly preserved, slightly cyrtoconic group, with thin, rather fragile shells, specialized septa that almost directly overlay the main body chamber, without cameral or siphuncular deposits, and an overall streamlined body design—well adapted to swimming in a horizontal orientation, presumably within the water column (Flower 1957; Furnish and Glenister 1964; see also Holland 1984: figure 1). They did not appear until the late Darriwilian (Chazyan) and then survived to the Late Silurian (Flower 1976).

Nautiloid assemblages, as previously mentioned, sometimes have close associations with deeper-water graptolite shales. Marek (1999) recorded a rich nautiloid assemblage from the Šárka Formation of the Prague Basin, Czech Republic (Bohemia), from a deeper-water black shale succession with rich accompanying biotas. Havlíček (1998:53), in outlining the stratigraphy of the Šárka Formation, gave special attention to the marked transgression and facies changes that accompanied the start of black shale deposition and the "sudden influx of new benthic and planktic biotas," comprising graptolites, acritarchs, chitinozoans, conulariids, cystoids, crinoids, carpoids and other echinoderms, brachiopods, trilobites, gastropods, univalves, phyllocarids, and ostracods. The succession ranges through two graptolite zones: the *Corymbograptus retroflexus* and succeeding *Didymograptus clavulus* zones of mid to late Darriwilian (Llanvirn) age. Mergl (1999) outlined the Early to Mid Ordovician brachiopod community distribution for the Prague Basin, depicting the onshore-offshore community profile in time and place. As depicted (see Mergl's figure reproduced in this volume—figure 17.10), the rich biotas mentioned here come mainly from the deeper-water environments of the Šárka Formation. Bohemia (part of the Perunican terrane; chapter 5: figure 5.1) lay off Gondwana in high paleolatitudes during Mid Ordovician time, forming part of the cooler "Mediterranean Province." The trilobites comprise representatives of the mesopelagic community (cyclopygid biofacies), as well as some elements of the "atheloptic" assemblage—deep benthic forms with blind or much reduced eyes (Fortey in chapter 24). The nautiloid assemblage includes 20 species, all longicones, mainly orthocerids, and a few ellesmerocerids, endocerids, and an actinocerid as well (Marek 1999). It seems likely that this community was largely composed of nektic forms. Marek (1999) referred to the unusual richness of the nautiloid fauna as probably caused by a warming event (and related sea level rise mentioned by Havlíček 1998) and suggested that the nautiloids were deposited in depths no greater than 150 m. But the presence of the mesopelagic trilobite community and accompanying atheloptic elements suggests that depths may have been a little greater. Bogolepova (1999) briefly reviewed the Mid Ordovician nautiloid geographic distributions of the Mediterranean Province (Iberian, Armorican, and Bohemian regions), suggesting a strong invasion of elements, especially of orthocerids and endocerids, from the Baltic (which lay in intermediate, ?warmer, paleolatitudes) to the "Mediterranean Province." That mid Darriwilian invasion of more mobile swimming orthocerids and endocerids seems to be related to sea level rise and climatic amelioration, which accords with the patterns in the Prague Basin. It may represent the initial major colonization of higher paleolatitude parts of the pelagic realm.

A few orthocerids from graptolitic shales (Utica Shale and equivalents) of Caradoc (late Mohawkian to early Cincinnatian) age across eastern and midcontinental North America have been reported as being relatively small and apparently thinner shelled, though larger specimens also occur (Flower 1957: 835). These may be pelagic nautiloids, but investigation of the mode of life of nautiloids preserved in black shales remains a challenge for future workers.

■ Ordovician World in Brief

The six short introductory essays outline the disposition of major terranes (chapter 5), the isotope patterns (chapter 6), the oceans and climate (chapter 7), the evidence for a possible Mid Ordovician superplume event (chapter 8), the end Ordovician glaciation (chapter 9), and the sea level changes (chapter 10).

Cocks and Torsvik (chapter 5) provide one global and three South Polar Ordovician map reconstructions based on paleomagnetic, faunal, and facies data that show the positions of terranes (Gondwana, Laurentia, Baltica, and Siberia and many other blocks), the Panthalassic, Iapetus, Rheic, and Tornquist oceans, the equator, midlatitudes, and South Pole. The successive Ordovician map reconstructions (figures 5.1–5.4) illustrate the changing patterns of displacement of continental blocks such as Baltica and Avalonia and the progressive closure of the Iapetus Ocean through Ordovician time. Island arcs are also identified.

Shields and Veizer (chapter 6) look at the fluctuations of strontium, neodymium, carbon, oxygen, and sulfur isotope compositions in Ordovician seawater and how these records help in interpreting global dynamics and environmental change. Of particular significance is the sharp excursion in the strontium values across the Middle to Upper Ordovician boundary, mentioned by Barnes (chapter 8) as an indicator of enhanced interaction by a possible mantle superplume with the hydrosphere.

Barnes presents broad overviews of oceanographic and climatic events in the Ordovician world (chapter 7) and evidence for a mantle superplume event during the Mid Ordovician (chapter 8). The highlights are a world in a greenhouse state, most land located in the Southern Hemisphere, a complex spread of island arcs and microcontinents in the Iapetus Ocean, major phases of orogenesis and associated volcanism, a probable mantle superplume event, high sea levels, widespread epeiric seas, and end Ordovician glaciation. Barnes also identifies a number of events that may have been particularly important in shaping the overall patterns of Ordovician biodiversification.

Brenchley (chapter 9) provides an outline of the main features of the short-lived end Ordovician (Hirnantian) glaciation in near polar areas of Gondwana and the timing of the associated climatic, isotopic, and biologic events. The rapid rate of environmental change is seen as being especially important in triggering the two phases of mass extinction, the first related to global cooling and the second associated with global warming with accompanying spread of anoxia.

Nielsen (chapter 10) offers a new Ordovician sea level curve based on the Norwegian, Swedish, Estonian, and Latvian successions of Baltica, establishing close ties with parts of the Ross and Ross (1992, 1995) sea level curve of the North American (Laurentian) platform. Correlation of some 30 different sea level changes across the two regions suggests that they represent substantially eustatic fluctuations of a consistent, wide-ranging (possibly global) sequence stratigraphic framework. Relationships between the sea level changes in Baltoscandia and the global time slices (chapter 2) are shown in figure 10.3. Potentially, if the pattern of 30 short-term sea level changes can be matched on a truly global scale, then it will, especially if used in close conjunction with key biostratigraphic indicators and radiometric dates, provide a very good complementary basis for studying aspects of biodiversity change, especially species-level analyses that require the most finely resolved subdivision of Ordovician time.

■ Synopses of Fossil Groups

The individual topics in parts III (Taxonomic Groups) and IV (Aspects of the Ordovician Radiation) are presented in the following order: part III—the siliceous unicellular radiolarians (chapter 11), the main animal groups—sponges (chapter 12) to vertebrates (chapter 30), the problematic receptaculitids (of possible metazoan affinity) and the algae (chapter 31), the organic-walled, unicellular acritarchs (chapter 32), the miospore record and evidence for land plants (chapter 33); and part IV—the ichnofossil (trace fossil) record (chapter 34) and a new global synthesis (chapter 35).

Noble and Danelian present the diversification history of radiolarians in chapter 11. This major siliceous, zooplanktic component of the Ordovician oceans shows a more or less progressive rise in genus-level diversity through Ordovician time, to the mid Ashgill (table 1.3). The first assemblages in the Late Cambrian and Tremadocian differ markedly from typical Ordovician radiolarians. Ordovician-type groups with new body plans appear initially in the late Early Ordovician (early Arenig), and then successive family- and generic-level diversification pulses follow, during the Darriwilian and the Late Ordovician (mainly late Caradoc—*TS*.5d), leading to a diversity peak of 17 genera in the mid Ashgill (*TS*.6b). On the other hand, the species exhibit diversity maxima in the Darriwilian (*TS*.4b) and the early Caradoc (*TS*.5a).

Carrera and Rigby (chapter 12), in their outline of species-level sponge diversification, depict a diversity curve with overall rise through Ordovician time but punctuated by increases to lesser peaks in the late Tremadocian and latest Early Ordovician (mid Arenig) and a more sharply peaked maximum in the mid Darriwilian due mainly to the radiation of orthocladine demosponges. Several higher-level sponge groups appeared next during the early Caradoc, contributing significantly to the overall Late Ordovician diversity rise, with culmination in a major early Ashgill (*TS*.6a) spike. This peak was followed by a dramatic mid Ashgill (*TS*.6b) decline. This decline occurs about one time slice earlier than the loss of diversity directly associated with the end Ordovician (*TS*.6c) mass extinctions, as seen in most other groups of organisms. Biogeographic and paleoecologic factors are shown to be influential in shaping the sponge diversification.

In chapter 13, Webby records the genus- and species-level diversity patterns of the stromatoporoids, a group of calcified sponges that commonly contribute to the growth of reefs. Apart from a short-lived development of the problematic Early to early Mid Ordovician pulchrilaminids, the group has a mainly Darriwilian to Late Ordovician record. The species-level diversity curve appears steplike, with each diversity peak (successively the late Darriwilian, mid Caradoc, and mid Ashgill peaks) being higher than the preceding peak. Then a marked late Ashgill decline followed in the interval associated with the end Ordovician glaciation. Rates of genus-level origination were highest in the late Darriwilian and mid Caradoc, whereas the rates of origination at the species level were highest in the early Ashgill. Extinction rates at both the genus and species levels were the highest in the mid to late Ashgill.

Van Iten and Vyhlasová (chapter 14) report that the conulariids had the highest ever generic-level diversities during the Darriwilian to mid Caradoc. That interval also exhibits the highest Ordovician species-level diversities. The conulariid records are from the low-latitude North American platform and the higher-latitude Prague Basin (Czech Republic). A relatively high proportion of genera are confined to either one region or the other. Only about one-third of the genera are common to the two regions. The authors also recognize a latitudinal diversity gradient with generic diversity levels declining toward the higher latitudes.

Webby, Elias, Young, Neuman, and Kaljo (chapter 15) document a number of aspects of coral diversity in a general overview of coral groups, a global analysis of the tetradiid corals, and three regional analyses—on North American Late Ordovician (Cincinnatian) corals, Baltoscandian rugose corals, and Australasian corals, respectively. Of the two major groups, the tabulate corals appeared first in the earliest Ordovician (Tremadocian) but remained at a background level (one genus only) for nearly 25 m.y. (more than half the length of the Ordovician). Only then did they start to diversify through the late Darriwilian to Late Ordovician, becoming the dominant Ordovician coral group. In the Cincinnatian of Laurentia, for example, tabulates approximately outnumber the rugose corals by 2:1. In terms of the main higher-level diversification of the two groups, the tabulates diversified mainly between the late Darriwilian and the mid Caradoc, whereas the rugose corals first appeared during the early Caradoc, and by mid Caradoc time their main higher-level diversification had already occurred.

The tetradiid corals are a small, exclusively Ordovician group that are here separated from the tabulates. They first appeared and remained at background levels in the Darriwilian but then diversified rapidly during the early Caradoc, reaching a species-level peak of 17 species by the mid Caradoc. A small diversity decline followed, and then a second diversity peak was attained in the early to mid Ashgill, prior to their dramatic late Ashgill decline to end Ordovician extinction. The Baltoscandian rugose corals show a comparatively similar distribution pattern with more rapid diversification during the mid Caradoc and mid to late Ashgill against a background of generally rising diversity.

Taylor and Ernst (chapter 16) provide a comprehensive analysis of the major radiations of bryozoans through Ordovician time, during which all Phanerozoic marine orders make their first appearances except one. The database comprises a total of 169 genera and 1,120 species. Patterns of diversity change and turnover show an exponential rise for both genus- and species-level diversity from the Early Ordovician (mid Arenig) to Late Ordovician (mid Caradoc), with the most dramatic increase commencing in the Darriwilian. Then two sharp, steplike declines of species occur to the end Ordovician, whereas the genus-level diversity drop commences later with a more dramatic, mid to late Ashgill fall. This loss of genus-level diversity is considered to be related to Lazarus taxa and a poor fossil record rather than to extinction. All eight major groups diversified most markedly through the Darriwilian to Caradoc interval, though two showed significant increase earlier, and others attained their diversity peaks only later, in the early to mid Ashgill. Provinciality was probably greatest in the Arenig, declining progressively to the end of the Caradoc and then increasing in the latest Ordovician.

A genus-level diversity survey of the brachiopods is presented by Harper, Cocks, Popov, Sheehan, Bassett, Copper, Holmer, Jin, and Rong (chapter 17). Formerly recognized as comprising only two broad divisions ("inarticulates" and "articulates"), the phylum is now grouped in three subphyla: the linguliformeans, craniiformeans, and rhynchonelliformeans. Each group shows markedly different taxonomic diversification histories. Most of the evolutionary features of linguliformeans appeared long before the Early Ordovician, including their burrowing adaptation. After significant Late Cambrian loss of diversity, the generic diversity was restored during two pulses of Early Ordovician diversification, in the late Tremadocian and the late Early Ordovician (mid Arenig). By the Darriwilian, however, the linguliformeans had declined to subordinate elements in most environments, for example, displaced to marginal (shallower or deeper basin) environments.

The craniiformeans lived free or cemented/encrusted on the substrate with no trace of a pedicle and remained a comparatively minor group. They first appeared in the late Early Ordovician (early Arenig). Then their main radiation commenced in the late Darriwilian and continued into the Late Ordovician, with maximum diversity through the late Caradoc and early Ashgill. This included especially the development of the large, thick-shelled trimerellides, initially centered in island arcs of Kazakhstan and eastern Australia.

The rhynchonelliformeans were arguably the most highly successful and completely diversified benthic group to inhabit Ordovician marine ecosystems. The group proliferated from an initial 4 Cambrian to 19 superfamilies through Ordovician time. In addition, the majority of the morphological adaptations exhibited by brachiopods developed during Ordovician time. The Early Ordovician record included the surviving Cambrian groups and two new higher-level groups, but, except for the orthidines and pentamerides, diversification levels remained at comparatively low levels. However, during the Mid Ordovician another five higher-level groups were added, and diversification levels increased dramatically. The Mid to Late Ordovician genus-level trajectories of 10 higher-level groups shows a dramatic, two-step, exponential rise of diversity, with early Mid Ordovician and latest Darriwilian to early Caradoc pulses to the first major diversification peak during the mid Caradoc. A sharp late Caradoc decline followed, then another rise to a higher, sharper early Ashgill peak. The large decline of diversity in the late Ashgill is attributed to the end Ordovician mass extinction. Individual diversification histories of a number of the higher-level taxonomic groups are also depicted.

In the Ordovician Radiation the brachiopods played a significant role in establishing new adaptive strategies, life modes, and feeding habits. The marked diversity increases were accompanied by a general trend to greater specialization and decreased niche sizes to utilize the existing ecospace more completely. With low food requirements, brachiopods probably had advantages over competitors. Morphological innovations that developed include (1) the adoption of planar shell profiles allowing a free-lying, recumbent lifestyle by the strophomenides; (2) the appearances of more robust ball-and-socket articulations (cyrtomatodont teeth) for the hinge that clearly benefited the rhynchonellides, atrypides, athyridides, and spiriferides; and (3) appearance of spiral calcified ribbons to support complex lophophores in atrypides and earliest spiriferides.

The diversity records of polyplacophorans (chitons) and symmetrical univalves (the tryblidiids—formerly "monoplacophorans"—and the bellerophontids) are reviewed by Cherns, Rohr, and Frýda (chapter 18). The polyplacophorans diversified mainly in the Early Ordovician (especially Tremadocian to early Arenig). Occurrences are mainly associated with carbonates of the low-latitude Laurentian margins. Otherwise, their Ordovician record is temporally and spatially restricted. The tryblidiids similarly have highest diversities in the Early Ordovician and then decline through Mid and Late Ordovician time. In contrast, the bellerophontids have a low Early to early Mid Ordovician diversity, gradually increasing to a peak in the late Caradoc, and then steadily decline to the end Ordovician.

Genus-level gastropod diversity, as a whole, is shown by Frýda and Rohr (chapter 19) rising steadily from a low at the beginning of the Ordovician to a maximum in the late Caradoc–early Ashgill. But when the diversity records of some higher-level groups are dissected, two very distinctly different patterns emerge. The Archaeogastropoda (Selenimorpha), Mimospirina, and Archaeogastropoda (Trochomorpha)—though the diversification of the last group commenced later—show a similar pattern of rising diversity change. This contrasts markedly with the diversity records of two other major groups, the Euomphalomorpha and Macluritoidea. Again, their initial diversifications do not coincide, but both groups exhibit a remarkably similar Mid to Late Ordovician record—of increase to a peak in the early Mid Ordovician, then a dramatic drop to a plateaulike diversity low through Darriwilian to Late Ordovician time. Overall the gastropods exhibit four peaks of origination (late Early Ordovician, early Mid Ordovician, early Caradoc, and early Ashgill) and two extinction peaks. The first extinction in the early Darriwilian is less intense (it mainly affects euomphalomorphs and macluritoids), and the second is the largest, involving loss of nearly half the genera across all groups, and coincides with the end Ordovician mass extinction.

In chapter 20 Cope documents the contrasting diversification patterns of the bivalve and rostroconch mollusks. A set of range charts illustrates the genus-level diversity patterns, and a summary of the diversity measures is also depicted. The first major radiation of bivalves in the Early Ordovician was limited geographically to the Gondwanan margins, across a wide range of latitudes, and to Avalonia, involving significant family-level diversification and a diversity peak in the early Arenig. Bivalves are known from many more localities in the Mid Ordovician, but the diversity did not increase markedly until later in the epoch and remained largely confined to Gondwana and surrounding areas. Then, in the early Caradoc (*TS*.5a), the second (and greatest) wave of expansion commenced after the migration and early diversification of bivalves in the low-latitude carbonate platforms regions of Baltica and Laurentia. The rostroconch mollusks, on the other hand, are a much smaller, infaunal group that exhibits a relatively high level of genus-level diversity in the Tremadocian, declined in extra-Gondwanan areas during the early Mid Ordovician, and then became reduced to near extinction at the end Ordovician.

The nautiloid cephalopods were roving, commonly large predators and scavengers that occupied nektic to nektobenthic habitats in the Ordovician oceans and participated in their greatest ever expansion during Ordovician time. From one order at the beginning of the Ordovician, the group diversified to nine orders by the early Late Ordovician. The progressive nature of this radiation is outlined by Frey, Beresi, Evans, King, and Percival (chapter 21). The initial radiation during the Tremadocian involved the diversification of ellesmerocerids and first appearances of two other groups (endocerids, tarphycerids). These latter were the next to diversify, producing the first major diversification peak in the late Early Ordovician, and again were accompanied by first appearances of new groups (orthocerids, actinocerids, dissidocerids). These latter groups became actively involved in the next rise in diversity during the Mid Ordovician, when again a number of new groups appeared. The second major diversification peak was in the Darriwilian and, unlike earlier radiation events, involved a significant biogeographic differentiation of the faunas into equatorial carbonate platform and high-latitude, mixed clastic and carbonate settings. Some of the biogeographically differentiated genera, however, suffered early extinction in the latest Darriwilian. The third major radiation in the Late Ordovician included appearances of further new orders and families. The faunas were diverse and more cosmopolitan, especially in the carbonate platform facies. A peak of diversification occurred in the early Ashgill, followed by rapid decline and extinction of families and genera in the end Ordovician mass extinction.

Malinky, Wilson, Holmer, and Lardeux (chapter 22) document a number of problematic tube-shelled organisms of uncertain affinities. They include groups with an earlier (Cambrian) record, such as hyoliths, coleoloids, sphenothallids, and bryoniids, and two others, the cornulitids and tentaculitids, that are first confirmed in the Late Ordovician. Only the largest group, the hyoliths, has a significant biodiversity record, though, compared with its Cambrian record, the diversities and abundances are lower. The Ordovician members of the group are provincially well differentiated and diversified into higher-latitude assemblages, as well as forming a low diversity component in more diverse shelly assemblages. The provincial differentiation includes higher-latitude "Mediterranean" and intermediate-latitude "Baltic" provinces. Genus-level diversity increased from Early to Mid Ordovician time but declined in the Late Ordovician as the hyoliths of the Baltic province disappeared, probably as a result of the drift of Baltica into low latitudes. Faunas at the species level are profoundly endemic. Hyolith species diversity is low in the Tremadocian but rises toward the Early to Mid Ordovician boundary. Subsequently, in the Darriwilian, the diversity increased dramatically owing to the biogeographic differentiation of the faunas. The diversity continues to a maximum in the Caradoc and then loss of half the species in the Ashgill.

Hints, Eriksson, Högström, Kraft, and Lehnert discuss aspects of biodiversity of four unrelated worm-like and sclerite-bearing groups in chapter 23. The groups include the scolecodonts, machaeridians, palaeoscolecidans, and chaetognaths. The organic-walled polychaete jaw apparatuses of scolecodonts provide a distinctive Ordovician diversity record. In the Early to early Mid Ordovician they have a low diversity; they then increase sharply in the late Darriwilian, as the most common genera diversified with a sharp peak of originations and accompanying increase of species diversity and abundances. After this rapid diversification event there was comparatively little diversity change—a slow Late Ordovician rise to a "high" in the latest Caradoc to early Ashgill and then decline toward the Ordovician-Silurian boundary. Sclerites are probably better known from the Cambrian (from groups such as the tommotiids and halkieriids) than in the Ordovician, but nevertheless the sclerite-bearing Ordovician machaeridians are widely distributed, with a relatively continuous record: (1) confirmed sclerites in the Tremadocian; (2) indications from the record in the latest Early Ordovician (early to mid Arenig) that a first radiation of the group had occurred; (3) further widening of distribution and increase in the number of taxa in the Mid Ordovician; and (4) evidence that suggests a continuation of the general trend toward a substantially higher diversity in the Late Ordovician. The other groups comprise the palaeoscolecidans of probably annelid or priapulid affinities and the chaetognaths (or arrow worms). They include comparatively few genera, with mainly long ranges, from the Cambrian through Ordovician to the Silurian (or younger). No marked Ordovician diversification or turnover events are recorded. Only two new paleoscolecidan genera and a new chaetognath genus originate in the Ordovician.

Trilobite diversity patterns are addressed by Adrian, Edgecombe, Fortey, Hammer, Laurie, McCormick, Owen, Waisfeld, Webby, Westrop, and Zhou (chapter 24) in a presentation divided into three parts: (1) a survey of global (and Laurentian) taxonomic genus-level diversity patterns; (2) regional reports of taxonomic genus-level (and some species-level) diversity patterns for Australasia, South America, Avalonia, Baltica, and South China; and (3) a review of patterns of biofacies differentiation. The global data are presented using a nine-unit subdivision of Ordovician time and recognition of distinctive EFs, represented by family-level clusters—the Ibex Fauna I, Ibex Fauna II, and Whiterock Fauna. Each shows markedly different diversity trajectories. The Ibex Fauna I diversity was high in the Tremadocian and then declined; Ibex Fauna II diversified in the late Early Ordovician to early Darriwilian (Arenig) and then rapidly declined in the latest Darriwilian (both faunas failing to survive the end Ordovician mass extinction). Expansion of the Whiterock Fauna commenced in the Arenig, but the radiation was most intensive from the Mid Ordovician to the early Caradoc diversity peak. A slow initial decline followed, then more rapid decline toward the end Ordovician, but with more than two-thirds of these families surviving into the Silurian. The major diversification is recognized across three of the four provincial realms: that is, it is developed in the low latitudes of Laurentia, the midlatitudes of Baltica, and the higher latitudes of Gondwana (North Africa) and peri-Gondwana (Avalonia, Perunica), but the timing of the comparable radiation event is difficult to establish in Australia (part of low-latitude East Gondwana) because of a less than complete Mid Ordovician to mid Caradoc record. In terms of the environmental patterns during the radiation in Laurentia, the Whiterock Fauna dominated only the deep subtidal and reef habitats initially, in the early Mid Ordovician, but by the Late Ordovician, elements of this fauna had spread both onshore and offshore to virtually dominate all habitats across the environmental spectrum.

Regional surveys exhibit some significantly different patterns. In Avalonia the Whiterock Fauna became dominant earlier, in the late Early Ordovician (early Arenig) across a wider range of shelf to slope habitats than in most other regions. In South America elements of the Whiterock Fauna appeared mainly in the mid Darriwilian and early Caradoc radiation pulses. In South China the main increase in the Whiterock Fauna occurred in the early Mid Ordovician, but the radiation was relatively less intensive from the Darriwilian to a mid Caradoc diversification peak than that shown on the global (and Laurentian) scale. In Australia the Whiterock Fauna first appeared in the Darriwilian, but its expansion is not recorded until much later, during the mid to late Caradoc.

Trilobite faunas are shown to be widely distributed and highly adapted to benthic and pelagic ecosystems. The group adopted a wider range of feeding and habitat adaptations in the Ordovician than at any other time—from the more richly populated shallower epipelagic and deeper mesopelagic biotopes, to the widest possible array of benthic biofacies developed across shelf to slope habitats, and across latitudinal zones. The benthic adaptation included (1) the olenid biofacies that occupied oxygen-deficient waters, with taxa apparently adapted to symbiotic living with sulfur bacteria; (2) the blind "atheloptic" trilobites that invaded deeper, poorly lit sites; (3) the suspension feeding trilobites, such as trinucleids and raphiophorids, that were more common in quieter-water, outer-shelf sites; (4) the small to medium-sized deposit feeders, such as the hystricurines and proetids, that occupied shallow-water carbonates, muds, and silts; and (5) the predators and scavengers, with their attached hypostomes, large size (e.g., asaphids), and well-developed eyes (e.g., phacopids, encrinurids), that largely dominated shelf environments.

In chapter 25, Braddy, Tollerton, Racheboeuf, and Schallreuter outline aspects of the diversity of three other arthropod groups (eurypterids, phyllocarids, and ostracodes). The eurypterids are a small group of vagile (swimming and walking), mainly large predatory arthropods (water scorpions) that in Ordovician time occupied marine and probably some marginal marine settings. They have a poor fossil record, mainly fragmentary, though a few well-articulated skeletons are known; only about 20 sites have confirmed or doubtful records of eurypterids worldwide. The occurrences are mainly Late Ordovician, suggesting that the main diversification occurred during the Caradoc and Ashgill, though eurypterid trackways have been recorded from deposits without body fossils from an as yet unverified Early Ordovician age in South Africa. Phyllocarid crustaceans are widely distributed in time and space, especially in deeper-water black shale facies where the highest diversities and abundances occur. However, the taxonomy of the group needs to be reassessed. Preliminary work in South America suggests that rapid diversification occurred in the late Early Ordovician (early Arenig) and again during the early Caradoc.

The ostracodes are a much larger group—again, not evenly or adequately documented in all parts of the world. The small, bivalved benthic forms (swimmers and crawlers) lived in marine waters of normal salinities on shallow shelves and shelf-slope habitats across a wide range of latitudes. Best-known records of ostracode faunas are preserved in the carbonate facies of Baltoscandia, and these to some extent mirror the worldwide diversity patterns. The region is recognized as occupying higher to intermediate latitudes in the Early to Mid Ordovician but moved into the tropics during the Late Ordovician (chapter 5). Confirmed Early Ordovician biodiversity records are poor—only one early Tremadocian species but more diverse faunas in the late Early Ordovician (five families). More rapid diversification followed in the Mid Ordovician, especially in the mid Darriwilian. Significant diversity maxima occur in the Late Ordovician, first in the early Caradoc and then, after loss of about one-third of the species owing to the great Kinnekulle K-bentonite ash fall, higher peaks in the late Caradoc and mid Ashgill, prior to the dramatic decline to end Ordovician extinction of all species and loss of about a third of the families. In higher latitudes of Bohemia (Prague Basin), the ostracodes were preserved in clastics. No diversity record exists in the Early Ordovician, but the early Mid Ordovician has a low diversity; then there are three successive diversification pulses—in the late Darriwilian, mid Caradoc, and late Caradoc—before the Ashgill decline.

Typically echinoderms occur in interbedded thin limestones and shales of shallow cratonic seas or locally in reefs, hardgrounds, or, less commonly, deltaic, slope, and deeper-water deposits across a wide range of latitudes. Echinoderm diversity of the five subphyla is fully described by Sprinkle and Guensburg (chapter 26). The class-level diversity more than doubled from levels in the Cambrian to the highest ever, class-level diversification of echinoderms by the early Late Ordovician. The diversity change of the five subphyla—that is, the four subphyla of Cambrian survivors (crinozoans, blastozoans, echinozoans, and homalozoans) and the new subphylum (asterozoans)—and the class- to order-level clades is depicted in a series of spindle diagrams. New clades were progressively added, from the 21 clades at the beginning of the Ordovician to the 30 clades in the Caradoc. Only a few groups exhibit significant species-level diversification during the Early to Mid Ordovician. The most remarkable diversification event for the echinoderms is during the mid Caradoc—(*TS*.5b–c), when the majority of groups, at almost all hierarchical levels (four out of five subphyla and all classes, genera, and species), radiated dramatically over a relatively short interval of time. This diversification extended, as well, to an almost complete range of lifestyle modes. In terms of the two major Ordovician echinoderm groups, (1) the blastozoans were more common before the mid Caradoc expansion; (2) the crinoids achieved comparatively larger expansion during the mid Caradoc event; (3) both groups attained approximately equal importance in the latter part of the Ordovician; and (4) the crinoids became the dominant group after the latest Ordovician (Hirnantian), then remaining the larger group for the next 200 m.y. There was only a comparatively small decline in the numbers of higher-level groups affected by the end Ordovician mass extinction, given that, of the 27 groups present in Ashgill, 23 survived into the Silurian. Outlines of the lifestyle modes and the biogeographic relationships are also presented.

The graptolites are a major planktic group with a worldwide distribution in sequences mainly representing outer-shelf, slope, and ocean basin habitats and geographic regions that range from low to high paleolatitudes. The species-level diversity analysis of the graptolites presented by Cooper, Maletz, Taylor, and Zalasiewicz (chapter 27) is based on a data set of species lists and zonal ranges from deeper shales of more or less continuous sections in three main regions: low paleolatitudes of Australasia, high to intermediate paleolatitudes of Avalonia (southern Britain), and the mainly intermediate paleolatitudes of Baltica. A set of standardized measures of diversity and turnover is employed in compiling the data. Patterns of diversity change, evolutionary rates, and environmental event significance are assessed. The initial diversification of these planktic organisms at the beginning of the Ordovician was slow, but by the late Early Ordovician (early Arenig) they had expanded spectacularly, especially in low-latitude Australasia, in a first of several successive waves of diversification, apparently each associated with global transgression and sea level highstand. Later waves of diversification were less intensive, the second occurring in the mid Darriwilian and a third in the mid Caradoc. Extinction events, in contrast, are recorded in the early Mid Ordovician (late Arenig), late Caradoc, and possibly late Darriwilian, and these are associated with regressive events, though they did not occur in all three regions simultaneously. In Australasia the graptolites came close to extinction in the late Ashgill, whereas in middle-high latitude Avalonia the decline started much earlier, during the mid Caradoc. The near extinction of graptolites in the late Ashgill is related to major regression and the end Ordovician (Hirnantian) glaciation.

Paris, Achab, Asselin, Chen, Grahn, Nõlvak, Obut, Samuelsson, Sennikov, Vecoli, Verniers, Wang, and Winchester-Seeto analyze the global and regional species-level diversity patterns for chitinozoans in chapter 28. These organic-walled, flask-shaped microfossils are found only in Ordovician to Devonian sequences. The preferred view is that they represent egg capsules of an unknown, exclusively marine softbodied animal. The microfossils appear to have a pelagic distribution and are widely distributed in shallower to deeper environments as well as across paleolatitudes. The group includes 35 genera, which diversified more rapidly through the Early and Mid Ordovician than in the Late Ordovician. Only a few genera became extinct in the Early and Mid Ordovician, in marked contrast to the larger number (19 genera) that suffered extinction at the end of the Ordovician.

The global chitinozoan species-level database shows a record of continuous diversification from the early Tremadocian to a first major peak in the late Darriwilian, then a short-lived interval of lowered diversity in the early Caradoc, followed by a rise to a second major diversity maximum in the mid Caradoc. For chitinozoans the initial major Early Ordovician (Tremadocian) radiation is the most significant (first-order) origination event. The most extreme (first-order) extinction event is at the end of the Ordovician, though there are less significant extinction pulses in the late Early Ordovician (early Arenig) and late Caradoc. Species diversity data for North Gondwana, Laurentia, Baltica, Avalonia, and South China were also analyzed to assess the nature of diversity change, turnover, and origination/extinction rates on a region-by-region basis employing rigorous sampling and counting methods. These results show some important regional differences.

Albanesi and Bergström (chapter 29) discuss the diversity patterns of Lower and Middle Ordovician conodonts based on an extensive global database. These phosphatic microfossil elements are presumed to be from apparatuses in the cephalic part of a living eel-shaped pelagic animal that may have been distantly related to chordates. The elements are found in a wide range of marine environments, including the two broadly differentiated major biogeographic "Midcontinent" or "Atlantic" realms. A genus- and species-level biodiversity analysis of the Lower to Middle Ordovician conodonts is presented using diversity measures that display changing patterns and turnover rates. Overall, a progressive increase in diversity is shown for genera and species through six biostratigraphically distinct "intervals" of Early to Mid Ordovician time. New species arise at a rate of from 7.0 to 9.4 species appearances per m.y. Extinctions continue at a high rate with values of between 7.7 and 8.0 disappearances per m.y., though in one late Tremadocian "interval" the rate dropped to 2.5 per m.y. Assessments of holdover and carryover percentages were also calculated for genera and species. The results were comparatively high (more than 40 percent) for both holdover and carryover values, suggesting rapid diversity change throughout the entire Early to Mid Ordovician interval. Two significant diversification pulses are recognizable from the overall results, the first spanning from the late Tremadocian to near the end of the Early Ordovician and the second from the early Mid Ordovician to the mid Darriwilian.

The Ordovician agnathan (jawless fish) and gnathostome (jawed fish) vertebrate diversity record is outlined by Turner, Blieck, and Nowlan (chapter 30). Two temporally and geographically distinct assemblages are recognized: in the Mid Ordovician of Gondwana, and in the Late Ordovician of Laurentia, Baltica, and Siberia. Only a few problematic taxa are recorded from the Early Ordovician (Tremadocian to late Early Ordovician), and all have doubtful affinities except for the confirmed small "bony" plate impressions of genus *Porophoraspis* from localities of early Arenig age in Australia. The more complete diversification of this Gondwanan vertebrate assemblage occurred in the Darriwilian of Australia and in the Darriwilian to early Caradoc interval of South America (Bolivia, Argentina). The greatest diversification, however, occurred in the mid Caradoc. Initially this was recorded only in Laurentia, with the dramatic appearance of at least 12 different types of vertebrate taxa in the Harding Sandstone and correlatives across North America. Taxa such as agnathans (including the first thelodonts) and the first stem-group gnathostomes were involved, and these components spread also to Baltica and

Siberia. Most elements of these Caradoc faunas were short lived. Then in the latest Ashgill (Hirnantian), during the interval of the end Ordovician Gondwanan glaciation, a further diversification of new vertebrate taxa occurred, apparently only in Baltica and Siberia. The timing of this event possibly coincides with the climatic amelioration immediately following the glaciation (chapter 9).

In chapter 31, Nitecki, Webby, Spjeldnaes, and Zhen outline the species diversity patterns of two unrelated Paleozoic groups, the problematic receptaculitids and the algal cyclocrinitids, and review selected algal groups belonging to Chuvashov and Riding's (1984) "Ordovician Flora" in order to clarify the timing of their initial diversifications. The receptaculitids were widely distributed in near-equatorial platform carbonates including reefs, especially in the Early to early Mid Ordovician. Their affinities are uncertain but probably linked to animals. They have a relatively continuous, low diversity record, but with weakly discernible peaks in the late Tremadocian to late Early Ordovician (early Arenig) and Mid Ordovician intervals and a higher peak within the late Caradoc. The cyclocrinitids had a wide distribution in carbonate and siliciclastic facies and are probably related to dascyclad algae. Their diversification commenced in the late Darriwilian, increased to a sharp diversity peak in the mid Caradoc, and then progressively lost species diversity to the end Ordovician. Like the cyclocrinitids, most of the other main Ordovician components of calcified green and red algae (e.g., dascyclads, udotaceans/codiaceans, solenoporans) exhibit diversity patterns with the greatest diversity change across the Mid to Late Ordovician boundary—initial appearances in the Darriwilian and significant diversification of these groups extending well into the Caradoc. In terms of timing, these diversification patterns in Ordovician calcified algae match in a rather similar way the major diversifications of a number of animal groups (e.g., bryozoans, corals), and they also played a part in the great late Darriwilian to Caradoc expansion of metazoan- and algal-dominated reefs (Webby 2002). The uncalcified brown algae also diversify significantly through the Late Ordovician (chiefly Caradoc). In contrast, the "blue-green algae," which are mainly Cambrian cyanobacterial components, show a major decline in their importance through Ordovician time.

In chapter 32, Servais, Li, Stricanne, Vecoli, and Wicander review the acritarchs—their definition, biologic affinities, history of previous study, distribution, record of available literature, and data sets—and provide some new species-level data for South China and North Africa. This large group of organic-walled microfossils formed an important phytoplankton component of Ordovician oceans. The group has a markedly higher species diversity than in the preceding Cambrian. A global acritarch species database was compiled but remains difficult to use in establishing diversity curves because of the many doubtful taxonomic assignments and uncertain stratigraphic ranges reliant on earlier documentation. Similar problems exist in using the regional data sets (e.g., Baltica, North America), and in some sections the sampling remained inadequate. Species-level diversity data are therefore presented from only two regions—South China and North Africa. In South China, the diversity trend derived from eight biostratigraphically well-contrained acritarch assemblages shows a progressive increase from early Tremadocian to an early Mid Ordovician peak, then a slight decline during the early to mid Darriwilian. The North African record is mainly derived from borehole data, but significant stratigraphic breaks limit the continuity of the record to three stratigraphic levels: (1) a continuous Late Cambrian to mid Tremadocian record with diversity peak in the early Tremadocian; (2) a significant diversification record from the early Mid Ordovician (mid Arenig) to early Caradoc with the maximum diversity in the late Darriwilian; and (3) a shorter-ranging latest Ordovician diversity pulse, with first appearances of "Silurian"-type floral elements. These floral elements presumably reflect the climatic amelioration following the Ordovician glaciation (chapter 9), especially given the proximity of these occurrences to the south paleopole.

Steemans and Wellman (chapter 33) record the progressive increase in the diversity of miospore (mainly cryptospore) taxa, from the Darriwilian through the Late Ordovician to a maximum across the Ordovican/Silurian boundary. The end Ordovician climatic and glacial events that caused the mass extinctions in other fossil groups had no apparent adverse effects on the spore diversity. The abundant and cosmopolitan mid Darriwilian to Late Ordovician cryptospore assemblages are suggested to indicate the presence of a widely dispersed flora of small, nonvascular, bryophytelike plants on land, at least in the areas where damp conditions prevailed. In addition, an isolated occurrence of trilete spores in the latest Ordovician points to the initial development of vascular plants, possibly the antecedents to Siluro-Devonian rhyniophytes. This event may be linked with the postglacial amelioration of climate immediately following the glaciation (chapter 9). Evidence of actual Ordovician land plant megafossils, however, remains elusive.

Mángano and Droser (chapter 34) review the trace fossil (ichnofossil) record in terms of their development, colonization trends, and ecologic diversity patterns. The development of macroboring activity and aspects of trace fossil utility in biostratigraphy are also outlined. A steady rise of ichnofossil diversity is recognized through Ordovician time, with the highest ichnodiversity developed during the Ashgill. The rise reflects the changes toward more diverse and complex behavioral patterns as new areas of ecospace became occupied and the development of more complicated deep tiering and pre- and postdepositional structures. The changing ichnodiversity patterns are assessed across a wide range of infaunal habitats in shallow marine (both siliciclastic and carbonate), deep marine, marginal marine, continental, and volcanic settings. During Ordovician time the shallow siliciclastic marine habitats developed more varied behavioral patterns, especially those formed by trilobites, and more distinctive tiering patterns, whereas in the shallow carbonate settings markedly increased bioturbation and tiering developed as a consequence of greater infaunal ecospace utilization. The deep marine environments also exhibit trends toward higher diversity and complexity, with the patterns changing from mainly feeding traces in the earlier Ordovician to more complicated, patterned grazing activity later in the Ordovician. In volcanic areas the diversity remained low, and the assemblages were of limited complexity, while in marginal marine settings there is evidence to suggest that some trilobites entered areas with brackish waters. Continental environments exhibit trackways that suggest that arthropods occupied coastal dunes, perhaps even early in the Ordovician.

Miller (chapter 35) presents the final overview of the Ordovician Radiation, drawing on aspects of contributions in this volume and other relevant research published since the late 1960s. He notes that compilations in the volume include some significant advances in the documentation of global diversity trajectories across a nearly complete suite of Ordovician taxonomic groups. The contributions provide new data from previously undersampled areas, a better understanding of phylogenetic relationships in some clades, and an improved global timescale—"a small sample of what has been—or could be—accomplished by the integration of new kinds of data and analyses . . . of Ordovician diversity." The late Jack J. Sepkoski Jr. initially raised our awareness of the magnitude of this dramatic, extended radiation in his seminal papers of the 1970s, 1980s, and 1990s of compilations and analyses of global Phanerozoic marine diversity and showed how pivotal the Ordovician Radiation was in shaping the future of marine diversification worldwide.

Miller's chapter focuses mainly on two fundamental questions about the major radiation: why this "extensive, worldwide increase in total biodiversity" occurred during Ordovician time, and why "certain taxa radiated more appreciably than others." These matters are discussed, using examples from chapters in the volume, in the context of (1) whether a global agent was principally responsible for shaping the diversity trajectory or whether local and regional factors also played an important part and (2) whether environmental and geographic limitations were overriding influences as well.

Miller's remarks conclude with five important key recommendations for future work. Briefly stated they are to (1) standardize analytical procedures; (2) dissect diversity patterns at finer geographic scales; (3) involve definitive integration of physical and chemical data with the biodiversity data; (4) undertake field-based studies across important sections to focus on biotic transitions; and (5) further investigate relationships between taxonomic diversification and morphological differentiation (disparity).

■ Future Directions

A great many things remain to be done in the future to better understand the nature, and likely causes, of the great Ordovician biodiversification event. Miller (chapter 35) has made suggestions to a similar effect, recommending a number of topics that need to be highlighted. Work initially should be directed toward completing, in a fully integrated and comprehensive way, the analyses that were started here: of taxonomic diversity patterns for all groups in time and space and across all taxonomic hierarchies from class to species. Other more exhaustive studies of ecologic diversity and morphological disparity should also be given priority.

Then there are a host of other topics of relevance to Ordovician biodiversity studies that remain poorly understood, such as (1) the plankton ecology and productivity of the Ordovician oceans; (2) the overall impact of microbial life in all environmental settings of the Ordovician; (3) the reconstruction of trophic webs and assessment of nutrient levels; (4) fully integrated analyses of all the main Ordovician ecosystems, from terrestrial to the open ocean, and across latitudes; (5) fullest possible documentation of biodiversity data in at least two other major continental platform areas with complete and well-preserved Ordovician biodiversity records (e.g., Baltoscandia, South China) to rigorously test against the known patterns of onshore-offshore diversity change recognized across the North American platform; and (6) intensive documentation of biotas from the comparatively few well-exposed and well-preserved "windows" of island, island-arc, and oceanic segments.

In a wider geologic context, we also need a much greater focus on the following topics: (1) global syntheses of Ordovician volcanic and orogenic histories; (2) a fully integrated and rigorously tested global Ordovician sea level curve; and (3) an improved understanding of global continental and oceanic configurations through Ordovician time—the recently published, cooperative quantitative biogeographic, paleomagnetic, and plate tectonic modeling approach (Lees et al. 2002) is a promising start. The only possible way to establish the real global significance of volcanic, orogenic, and plate tectonic effects on Ordovician biodiversity is to have such fully comprehensive integrated worldwide analyses of these Ordovician volcanic, orogenic, and plate tectonic histories available to assess their respective roles.

Finally, it is essential that more cooperative approaches are developed with other geologists, geochemists, and geophysicists, especially the scientists actively involved in past climatic and oceanographic modeling work. Future programs of Ordovician biodiversity studies should be more closely linked to the multidisciplinary groups working on earth-system processes and modeling of such past climatic and oceanographic change. The interactive roles of and responses to changing atmospheric compositions, ocean chemistry, and patterns of ocean circulation would have been critically important for the Ordovician biosphere and therefore are vitally important and relevant matters for our future understanding of the Ordovician radiations. As the ocean and climate states become better understood using the earth-system approach (see chapters 7 and 8), the extrinsic factors that may have been influential in shaping the major diversification events may be expected to begin to emerge more clearly.

ACKNOWLEDGMENTS

I thank Alan W. Owen, Arnold. I. Miller, and my editorial colleagues, Florentin Paris and Ian G. Percival, for their comments on the manuscript. Florentin Paris assisted also in the compilation of generic and specific data from authors, in the preparation of table 1.1, and provided advice on patterns of microplankton productivity. The chapter, like the many others in this volume, is a contribution to IGCP 410, "The Great Ordovician Biodiversification Event."

PART I

Scaling of Ordovician Time and Measures for Assessing Biodiversity Change

2 Stratigraphic Framework and Time Slices

Barry D. Webby, Roger A. Cooper,
Stig M. Bergström, and Florentin Paris

In compiling data for this volume, it was regarded as crucial for all participants to have access to a well-integrated Ordovician timescale based on the global and regional subdivisions, the key graptolite, conodont, and chitinozoan zonations (those with the greatest utility for wide-ranging correlation, though they are provincially distinct), and a set of close-spaced global time lines. This provides the highest-resolution correlation for the Ordovician biodiversity data collected and for global analysis of the data (for example, in determining the patterns of biologic and other significant marker events). In establishing a precise stratigraphic framework, it was vitally important to utilize as many cross ties between the provincially distinct zonations as possible and to integrate all the available, reliable radiometric dates to achieve the most finely calibrated scale possible.

The main calibration involved a computer-based optimization approach (CONOP9 software) and is outlined by Sadler and Cooper in the following chapter. They compiled first and last appearances of more than 1,100 (mainly graptolite) taxa from almost 200 Ordovician and Silurian stratigraphic sections worldwide. This resulted in an ordered and scaled sequence of 2,306 biostratigraphic events within which stages and zone boundaries can be located. Twenty-two radiometrically dated samples were included in this sequence and serve to test the linearity of the scale and to calibrate it. Sadler and Cooper's calibration results in a significantly more highly resolved timescale than previously was possible and allows a particularly refined subdivision of 19 time slices to be employed. The time-slice intervals (*TS*) represent durations of between 1.6 and 3.0 million years (m.y.) (on average ~2.5 m.y.).

With this level of refinement participants have been able to undertake comprehensive assessments of biodiversity change through Ordovician time, surveying a variety of different patterns of diversity, such as taxonomic richness, rates of speciation, turnover, species-genus ratios, and extinction rates in specific clades. These results allow more precise (temporally constrained) comparisons between the diversification patterns of different clade groups, and at different levels of the taxonomic hierarchy, than were previously possible.

Based on the Sadler and Cooper calibration of the following chapter, the Ordovician is taken to range from 489 to 443 m.y., that is, it represents a duration of 46 m.y.

■ The Ordovician System and Its Systemic Boundaries

Lapworth's (1879:11) original proposal and naming of the "Ordovician System" were influenced greatly by his recognition that the intermediate division of Lower Paleozoic rocks contained a fauna of "grand

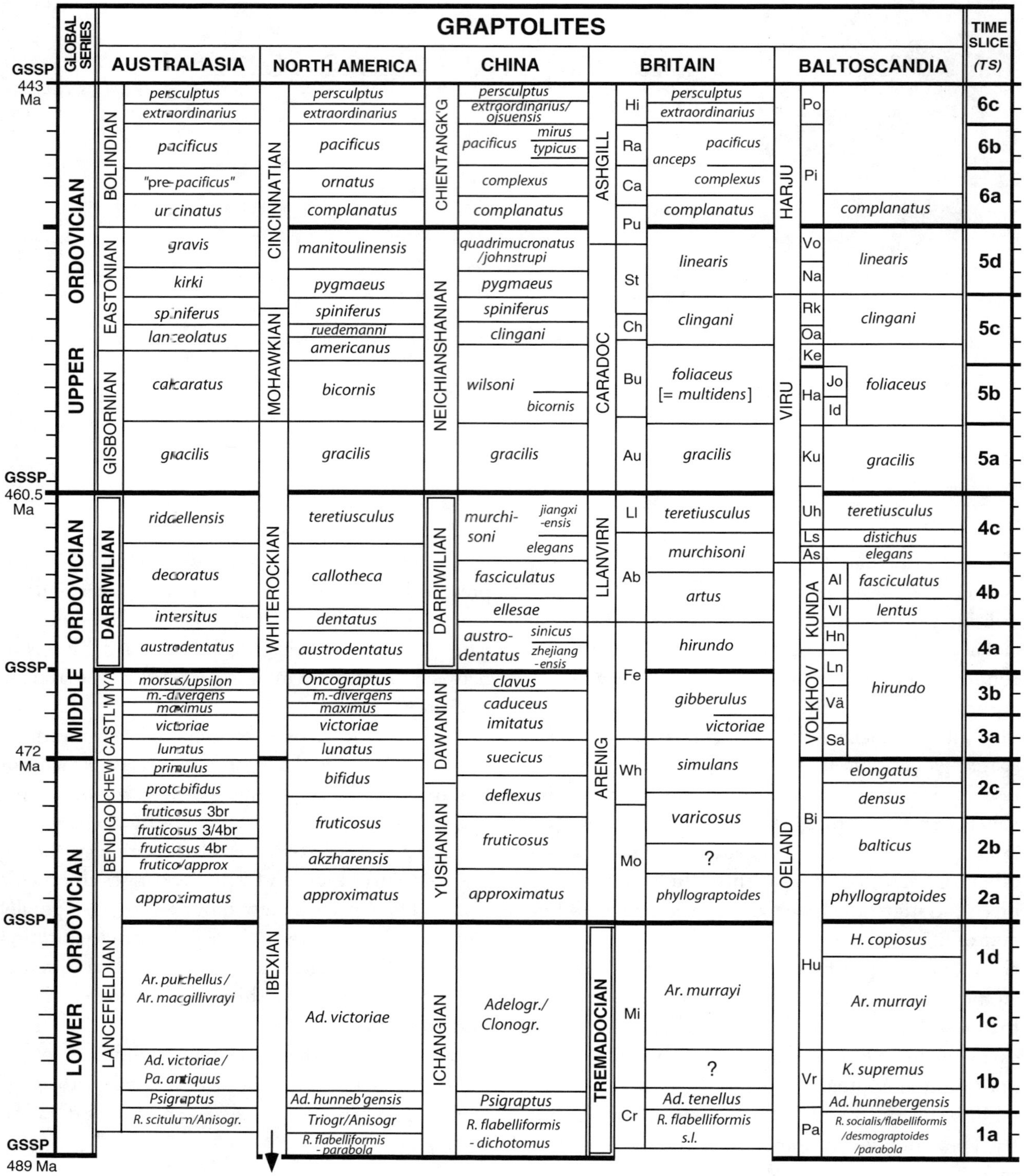

FIGURE 2.1. Ordovician stratigraphic chart showing the correlations between the main graptolite zonal sequences of Australasia, North America, China, Britain, and Baltoscandia, the global Series, the global Tremadocian and Darriwilian Stages (see boxes surrounded by double lines), the mainly regional series and stages, and the 19 numbered and lettered time slices (*TS*) (last column); at the left margin, the million-year divisions, specific radiometric ages (Ma), and IUGS-ratified global boundary stratotype sections and points (GSSPs) are shown. The formalized zonal graptolite taxa are cited here mainly using key species (or subspecies) only, except for the Tremadocian graptolite taxa, which are shown as customary with both generic and species names. Explanation of the abbreviated generic names: *Ad.* = *Adelograptus*, *Ar.* = *Araneograptus*, *H.* = *Hunnegraptus*, *K.* = *Kiaerograptus*, *Pa.* = *Paradelograptus*, and *R.* = *Rhabdinopora*.

distinctiveness" from older (Cambrian) and younger (Silurian) faunas. However, his proposal failed to win widespread early support because the Lower Paleozoic rocks in the British Isles had long been subdivided into two systems—Sedgwick's Cambrian and Murchison's Silurian—and the adherents of these opposed schools of thought preferred to maintain the bipartite division while continuing to argue about in which of the two systems Lapworth's Ordovician rocks belonged. However, as the influence of Sedgwick and Murchison waned toward the end of the nineteenth century, Lapworth's simple and practical compromise to use a tripartite subdivision of Lower Palaeozoic rocks gradually became more generally accepted and used by geologists worldwide. The name Ordovician was eventually adopted for global use by the International Geological Congress at Copenhagen in 1960 (Holland 1976).

In terms of the two systemic boundaries, the first is the Cambrian/Ordovician boundary, and it is defined at the base of the conodont-based *Iapetognathus fluctivagus* Zone at Green Point, western Newfoundland (Cooper et al. 2001). This level is also the base of the lowest Ordovician global Tremadocian Stage (figures 2.1, 2.2). The second is the Ordovician/Silurian boundary, and it is defined at the base of the graptolite-based *Akidograptus acuminatus* Zone (sensu lato) at Dob's Linn, Scotland (Cocks 1988). This is one zone higher than the *Glyptograptus persculptus* Zone (figure 2.1).

■ Status of Global Series and Stages, and the Regional Subdivisions

Since 1974 the International Subcommission on Ordovician Stratigraphy (ISOS), under the aegis of the International Union of Geological Sciences (IUGS) and affiliated International Commission on Stratigraphy (ICS), has been actively focusing on how best to establish a single set of global series and stage divisions. In order to achieve these objectives it has been necessary to (1) identify significant levels (biohorizons) with wide-ranging global correlation potential using graptolites or conodonts, or both; (2) document best available sections as a basis for selecting a single global boundary stratotype section and point (GSSP) for global definition and correlation purposes; and (3) assign appropriate formal names for the chronostratigraphic intervals (Series and Stages) between the GSSPs (Webby 1995). The task has been a difficult one because of the highly provincial and ecologically differentiated Ordovician biotas (Jaanusson 1979), leading to a paucity of good stratigraphic sections containing associated graptolites, conodonts, and chitinozoans and the uneven distribution of reliable radiometric dates.

In 1995, a 90 percent majority of voting members of the ISOS agreed on the use of a tripartite global (Series) subdivision of the Ordovician System into Lower, Middle, and Upper Ordovician Series. It was further accepted that each of the series would be further subdivided into two, giving a sixfold global Stage division for the Ordovician as a whole. A level near the base of the North American Whiterockian series was approved for defining the base of the Middle Ordovician (see figures 2.1, 2.2), and the base of the *Nemagraptus gracilis* Zone (now equated with the base of the British Caradoc) has been accepted (Bergström

---

FIGURE 2.1. (*Continued*)
Key to abbreviated regional names: **Australasia:** Chew = Chewtonian, Castl'm = Castlemainian, Ya = Yapeenian; **China:** Chientangk'g = Chientangkiangian; **Britain:** Cr = Cressagian, Mi = Migneintian, Mo = Moridunian, Wh = Whitlandian, Fe = Fennian, Ab = Abereiddian, Ll = Llandeilian, Au = Aurelucian, Bu = Burrellian, Ch = Cheneyan, St = Streffordian, Pu = Pusgillian, Ca = Cautleyan, Ra = Rawtheyan, Hi = Hirnantian; **Baltoscandia:** Pa = Pakerort, Vr = Varangu, Hu = Hunneberg, Bi = Billingen, Sa = Saka*, Vä = Väänä*, Ln = Langevoja*, Hn = Hunderum*, Vl = Valaste*, Al = Aluoja*, As = Aseri, Ls = Lasnamägi, Uh = Uhaku, Ku = Kukruse, Ha = Haljala, Id = Idavere*, Jo = Jõhvi*, Ke = Keila, Oa = Oandu, Rk = Rakvere, Na = Nabala, Vo = Vormsi, Pi = Pirgu, Po = Porkuni. The units shown with asterisks and compiled in a third, discontinuous Baltoscandian column are regarded as regional substages; the two other columns show regional series and stages. The former Baltoscandian Latorp stage name (not shown) is represented by its former lower and upper substages, now raised to full stage rank as the Hunneberg and Billingen stages (Meidla 1997).

Note that in China a threefold series and a sixfold stage subdivision is now employed in relatively close conformity to international usages. However, some alternative stage names (and spellings) remain; for example, the Lower Ordovician Tremadocian has alternative names (either Ichangian or Xinchangian), and the succeeding Yushanian is alternatively the Dobaowanian. The Middle Ordovician includes most of the Dawanian and the Darriwilian (formerly Zhejiangian). The Upper Ordovician is subdivided into the Neichiashanian (alternatively, Aijiashanian), and the Chientangkiangian (alternatively, Qiangtangjiangian).

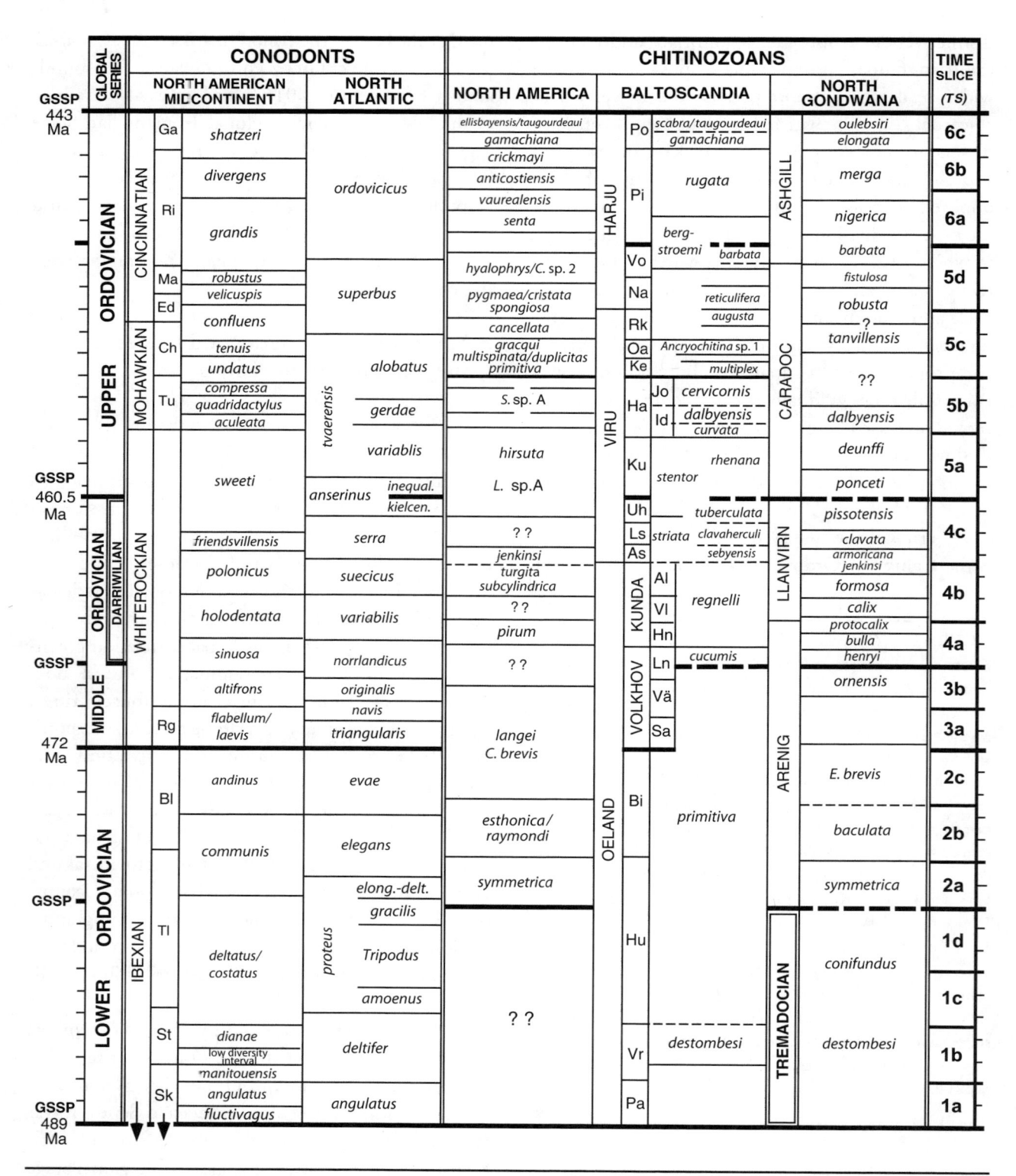

FIGURE 2.2. Ordovician stratigraphic chart illustrating correlations between the main conodont zonal sequences of the North American Midcontinent and North Atlantic provinces, the chitinozoan zonal sequences of North America, Baltoscandia, and North Gondwana, the global Series, the Tremadocian and Darriwilian Stages, the mainly regional series and stages, the 19 numbered and lettered time slices (*TS*) (last column), and the million-year divisions (at left margin). The formalized zonal conodont and chitinozoan taxa are cited here mainly using key species. In the chitinozoan sequences a few zonal intervals are defined informally using abbreviated generic name and informal (numbered or lettered) species, for example, *C.* sp. 2 = *Conochitina* sp. 2, *L.* sp. A = *Lagenochitina* sp. A, and *S.* sp. A = *Spinachitina* sp. A. In addition, two key taxa have been assigned the same species name; these are distinguished by having different generic names, namely, *C. brevis* = *Conochitina brevis,* and *E. brevis* = *Eremochitina brevis.*

et al. 2000) and ratified as the base of the Upper Ordovician (figure 2.1). The Sadler and Cooper calibration presented in the following chapter indicates that the global Series have the following durations: Lower Ordovician (17 m.y.), Middle Ordovician (11.5 m.y.), and Upper Ordovician (17.5 m.y.).

Two global Stages have been ratified. The lowest division, the Tremadocian Stage, was adopted recently (Cooper et al. 2001), and the Darriwilian Stage, which is the upper division of the Middle Ordovician, was ratified earlier (Mitchell et al. 1997). The other four global Stage divisions remain unnamed, although GSSPs for the second and fifth Stages have recently been formally approved by the ICS and ratified by the IUGS (figures 2.1, 2.2). British regional series units may continue to be used by some Ordovician stratigraphers in a kind of de facto global nomenclature, but these names will be replaced gradually as the unnamed units become defined and ratified as global Stages.

The formalized tripartite global Series (Epoch) usages of Lower (Early) Ordovician, Middle (Mid) Ordovician, and Upper (Late) Ordovician should now be maintained. These global Series usages differ from earlier, more traditional regional concepts of "lower," "middle," and "upper" as employed particularly in North America and Baltoscandia (as well as the former Soviet Union). In North America the Ibexian has been assigned to the lower Ordovician but now includes an interval of topmost Cambrian beds (equivalent to three conodont zones) in its lowermost part (Cooper et al. 2001). The Whiterockian and Mohawkian were combined as representing the middle Ordovician, and the Cincinnatian was retained as upper Ordovician (figure 2.1), but Sweet (1995) advocated that the Mohawkian and Cincinnatian be combined as "upper Ordovician," which is more consistent with the scope of the global Upper Ordovician Series as recently ratified by the ICS.

FIGURE 2.2. (*Continued*)
Key to abbreviated regional names: **North American Midcontinent**: Sk = Skullrockian, St = Stairsian, Tl = Tulean, Bl = Blackhillsian, Rg = Rangerian, Tu = Turinian, Ch = Chatfieldian, Ed = Edenian, Ma = Maysvillian, Ri = Richmondian, Ga = Gamachian. For key to other abbreviated names and other symbols, see figure 2.1.

In the British Isles five regional series are recognized, instead of six as previously, with the "Llandeilo" now abandoned in favor of incorporating the lower part in an expanded Llanvirn, and including the upper part in the overlying Caradoc (Fortey et al. 1995, 2000). The "Llandeilo" encompassed the interval of the graptolite-based *Husteograptus teretiusculus* and *Nemagraptus gracilis* zones combined, consequently straddling the newly formalized Middle-Upper Ordovician boundary. Therefore the series usage of "Llandeilo" should be discontinued. A more restricted conception has been retained by Fortey et al. (1995, 2000), named the "Llandeilian stage" and comprising the upper part of the Llanvirn series (figure 2.1). In this volume we have drawn a clear distinction between the Llandeilian stage, and the former, expanded, series usage of "Llandeilo." The latter term is quoted in parentheses in this volume because a few workers have continued to recognize the old name.

In Baltoscandia a tripartite subdivision into the Oeland (lower Ordovician), Viru (middle Ordovician), and Harju (upper Ordovician) series has been used. The Oeland series includes the Pakerort to Kunda interval (figure 2.1). A similar tripartite series subdivision of lower, middle, and upper Ordovician was applied across wide tracts of Russia and parts of Asia (Sokolov et al. 1960). Further information about the regional series and stage subdivisions shown in figures 2.1 and 2.2 is presented in Webby (1998), Fortey et al. (2000), and Chen et al. (2001).

■ Toward a Biostratigraphically Integrated Zonal Framework

The existing global and regional stratigraphic frameworks with their accompanying successions of biozones that represent unidirectional, nonreversible evolutionary change comprise, for all practical purposes, chronostratigraphic scales (Harland et al. 1990). The biostratigraphically most useful schemes for wide-ranging correlation are the graptolite, conodont, and chitinozoan zonal indices, but long-distance correlations are complicated because of the marked biogeographical and ecological differentiation even among these groups of organisms. Consequently, each representative group exhibits at least two biogeographically distinct zonal successions (figures 2.1, 2.2).

Pacific-type provincial graptolites are present in the Australasian and North American columns, representing low paleolatitude regions, and European-type ("Atlantic") provincial graptolite zones in the British columns of figure 2.1, representing high paleolatitudes (Skevington 1973; Berry 1979). On the other hand, the Baltoscandian and Chinese columns have mixed Pacific and European faunal affinity (Cooper et al. 1991).

Cooper et al. (1991) found ecological depth stratification to have a marked effect on graptolite distribution, while Finney and Berry (1997) recognized graptolite abundance in Nevada to be concentrated along the continental margins. The conodont zonal successions include both the North Atlantic type and the North American Midcontinent type, based on data from Bergström (1986), Sweet (1995), and Sweet and Tolbert (1997). Sweet (1995) and Sweet and Tolbert (1997) additionally employed a graphic analysis approach for correlation of the Midcontinent province conodonts. The chitinozoan zonal successions also include distinctly different provincial assemblages from North America, Baltoscandia, and the higher paleolatitude regions of North Gondwana (Paris 1996).

Bergström (1986) has presented the most comprehensive documentation of cross ties between the different representative zonal successions. His survey extended to the graptolite and conodont successions of northwestern Europe and North America. A total of 44 separate zonal ties were recognized between the European graptolite and North Atlantic conodont successions through the Ordovician of northwestern Europe (mainly in Britain and Baltoscandia), and a further 34 zonal ties were identified between Pacific-type graptolite and North Atlantic conodont successions across North America. These cross-zonal ties served to establish a much more firmly integrated biostratigraphically based correlation framework for the chronostratigraphic Ordovician timescale.

The much revised timescale compilation illustrated in figures 2.1 and 2.2 adds many further cross-tie adjustments to establish the present correlation framework. It uses the same Bergström cross-tie approach, not only for the graptolites and conodont data but also for available data from the three most important, provincially distinct chitinozoan successsions (figure 2.2).

■ Definition of Time Slices (*TS*)

The six primary time slices, *TS*.1–6, more or less coincide with the six approved global Stages (figures 2.1, 2.2). Formalized Tremadocian and Darriwilian Stages coincide with *TS*.1 and *TS*.4, respectively; the other four Stages remain unnamed.

The time slices are further subdivided into a total of 19 secondary divisions using the supplementary lowercase lettering a, b, c, and d. In most cases the boundaries dividing these smaller units are tied to important graptolite or conodont zonal boundaries, or both.

The Tremadocian Stage is subdivided into four time slices (*TS*.1a–d), with *TS*.1a broadly equating with the Lower (Early) Tremadocian (and exhibiting the classic *Rhabdinopora flabelliformis* series), and *TS*.1b–d correlative with the Upper (Late) Tremadocian of Cooper (1999a). In terms of the Baltoscandian regional subdivisions, *TS*.1a equates with the Estonian Pakerort stage, *TS*.1b is correlative with the Estonian Varangu stage, and *TS*.1c–d is equivalent to the Swedish lower-middle Hunnebergian (Cooper and Lindholm 1990).

The unnamed upper Lower Ordovician Stage is subdivided into *TS*.2a–c, which are broadly correlative with the lower-middle parts of the British Arenig. The *TS*.2a is equivalent to the wide-ranging, distinctive *Tetragraptus approximatus* Zone (and partly correlative *T. phyllograptoides* Zone of Baltoscandia), and *TS*.2b–c more or less span the Swedish Billingen and the Australasian Bendigonian to Chewtonian stages.

The unnamed lower Middle Ordovician Stage (equivalent to the Australian Castlemainian and Yapeenian stages) is represented by *TS*.3a–b, and the base of *TS*.3a more or less equates with the bases of the North American Whiterockian series (the Rangerian stage), the Australian Castlemainian stage, and the Baltoscandian Volkhov stage.

Fortey et al. (2000) commented on the difficulties of correlating the base of the Middle Ordovician into parts of northern Gondwana and associated terranes, where the British regional Arenig series subdivision is applied. The base is equated with a level in the mid Arenig (probably within the Whitlandian stage), but this correlation is regarded by Fortey et al. as speculative. In South Wales where the regional stages of the Arenig are defined (Fortey and Owens 1987), in particular, through the Whitlandian interval, the continuity of the faunal succession is incomplete, and there is a lack of diagnostic ties to the graptolite zonation recorded elsewhere.

Consequently, in the higher paleolatitude regions of North Gondwana and associated terranes it may not always be easy to differentiate confidently between the biotas of *TS.2* and *TS.3*. The North Gondwanan chitinozoan zonation presented in figure 2.2 (and the acritarch counterparts) may, however, provide more precise stratigraphic control for use in the clade group analyses of these biotas across North Gondwanan regions, and the differentiation between the upper Lower Ordovician (*TS.2*) and the lower Middle Ordovician (*TS.3*) is less ambiguous.

The Darriwilian Stage is defined as including *TS.4a–c*. The *TS.4a* interval coincides with the *Undulograptus austrodentatus* Zone (and its Subzones in China; Chen and Bergström 1995), and the *TS.4a–b* boundary coincides with the base of the British Llanvirn series (figure 2.1). Consequently, the uppermost part of the British Arenig extends into the lowest part of the Middle Ordovician (*TS.4a*). The *TS.4b–c* boundary is equivalent to the Baltoscandian series boundary between the Oeland (top of Kunda stage) and the Viru series.

The unnamed lower Upper Ordovician stage is represented by *TS.5a–d* and spans slightly more than the entire British Caradoc series. The *TS.5a* coincides with the remarkably widespread *Nemagraptus gracilis* Zone. The base of the North American Mohawkian series (and base of the Turinian stage) correlates with the base of *TS.5b*. The base of the Baltoscandian Harju series (and Nabala stage) coincides with the base of *TS.5d*.

The *TS.5b–c* boundary has been equated with the Kinnekulle K-bentonite (volcanic ash bed) at the base of the Estonian Keila stage (Männil and Meidla 1994) in Baltoscandia, and its possible correlative, the Millbrig K-bentonite bed in North America, that defines the base of the Chatfieldian stage, near the middle of the Mohawkian (figure 2.2). This supposedly correlative, trans-Iapetus ash marker has been dated at 454.6 Ma (Huff et al. 1992; Leslie and Bergström 1995). However, Min et al. (2001) have recently cast doubt on this intercontinental correlation, citing significant differences in the geochemical data of primary mineral phases and a wide discrepancy (about 7 m.y.) between their apparent ^{40}Ar/^{39}Ar ages, as evidence against the Kinnekulle and Millbrig K-bentonites being a part of a single eruptive event. Though only a few apparent radiometric ages have been determined based on U/Pb data (e.g., Tucker and McKerrow 1995), they may be more reliable and do not exhibit such markedly dissimilar apparent ages.

The unnamed upper Upper Ordovician stage is represented by *TS.6a–c* and spans most of the British Ashgill series (except the lower part). The lower part of *TS.6a* is recognized by the wide-ranging *Dicellograptus complanatus* Zone. The *TS.6b* more or less coincides with the *Paraorthograptus pacificus* Zone. The *TS.6c* correlates with the British Hirnantian stage and is approximately equivalent to the North American Gamachian stage and the Estonian Porkuni stage.

The British column shows the correlation of the mainly shelly faunal-based Ashgill stages (figure 2.1) with the standard graptolite zonation, following Fortey et al. (2000). In the Fortey et al. scheme the Caradoc-Ashgill boundary is placed in the upper half of the *Pleurograptus linearis* Zone (e.g., upper *TS.5d*). However, Rickards (2002) recently identified beds in the middle part of the Rawtheyan of the type Cautley area in northern England as having a *P. linearis* Zone age, which suggests that the accepted Fortey et al. correlation may need to be revised.

The Hirnantian incorporates the end Ordovician glaciation in its lower two-thirds (chapter 9). There are two recognizable phases of extinction, the first at the base of the Hirnantian (base of *TS.6c*) and the second within the *Glyptograptus persculptus* Zone in the mid–late Hirnantian (mid–late *TS.6c*).

3 Calibration of the Ordovician Timescale

Peter M. Sadler and Roger A. Cooper

Ordovician deep-water shales contain the prerequisites for a high-resolution timescale: rich successions of graptolite faunas, datable ash-fall K-bentonites, and minimally interrupted accumulation. Traditionally, the first appearances of selected graptolite taxa define provincial sets of zones, into which radiometrically dated bentonites are subsequently arrayed to achieve a numerical timescale. Provincial differences and the modest numbers of zones impose the primary limits on resolution. We have taken a different approach that avoids the constraint of zones.

■ The New Approach

To achieve a unifying timescale for this book, we applied computer-assisted optimization to combine graptolite range charts from all provinces directly, without using zones. The optimization process searches for a model sequence of all range-end events that best fits all the locally observed taxon ranges. Dated ashfall events are included with the range-end events from the outset of the search; thus, they too receive optimal placements in the model sequence and permit numerical ages for biostratigraphic events to be estimated by interpolation. By this method, we interpolated ages for the Australasian graptolite zone boundaries (VandenBerg and Cooper 1992). Ages for the boundaries in other subdivision schemes were derived, using all available criteria to correlate into the Australasian graptolite zonation (chapter 2).

■ The Raw Data

The timescale is based on an ordered and scaled sequence of 2,306 events: 22 dated bentonites that are associated with graptolites or other fossils for which contemporary graptolites are known (table 3.1); 12 undated bentonite beds that help tie together short sections from the Mohawk Valley (C. E. Mitchell pers. comm.); and the first and last appearances of 1,136 taxa (1,119 graptolites, plus 17 trilobites and conodonts) as reported from almost 200 stratigraphic range charts worldwide, from the basal Ordovician to early Devonian. The range charts represent Arctic (31), Cordilleran (10), Midwestern (6), and Northeastern (29) North America, as well as South America (9), Great Britain (24), Iberia (9), Germany (10), Scandinavia (16), eastern Europe (14), the Middle East to Central Asia (14), Siberia (1), China (16), and Australasia (9). These are all the range charts we could find that meet modern standards of graptolite systematics and depict collections from shaley facies. Inclusion of Silurian range charts and radiometric dates permits a more robust calibration. Cambro-Ordovician trilobite and conodont taxa were added to improve the constraints near the base of the Or-

TABLE 3.1. Radiometric Control Points

Age (Ma)		Unit	Stratigraphic Constraint	Reference	
491 ± 1	[U-Pb]	Latest Cambrian	*Peltura scarabaeoides scarabaeoides*	Davidek et al. 1998	
489 ± 0.6	[U-Pb]	Cambrian/Ordovician	*Peltura scarabaeoides scarabaeoides*	Landing et al. 2000	
483 ± 1	[U-Pb]	Late Tremadocian	*Hunnegraptus* sp.	Landing et al. 1997	
469 +5 -3	[U-Pb]	Arenig/Llanvirn	*Undulograptus austrodentatus*	Tucker & McKerrow 1995	(12)
465.7 ± 2.1	[U-Pb]	Llanvirn	above *Didymograptus artus*	"	(13)
464 ± 2	[U-Pb]	Llanvirn	[Cerro Viejo section—Argentina]	Huff et al. 1997; Mitchell et al. 1998	
464.6 ± 1.8	[U-Pb]	Llanvirn	*Holmograptus spinosus* *Didymograptus murchisoni*	Tucker & McKerrow 1995	(14)
460.4 ± 2.2	[U-Pb]	Llanvirn	*Pterograptus elegans* *Hustedograptus teretiusculus*	"	(15)
455 ± 3	[Ar-Ar]	"Kinnekulle" K-bentonite	*Climacograptus bicornis* *Climacograptus wilsoni*	"	(18)
456.9 ± 1.8	[U-Pb]	"Kinnekulle" K-bentonite	*Climacograptus bicornis* *Climacograptus wilsoni*	"	(18)
454.8 ± 1.7	[U-Pb]	"Pont-y-ceunant ash"	*Dicranograptus clingani* *Diplograptus foliaceus*	"	(19)
457.4 ± 2.2	[U-Pb]	"Pont-y-ceunant ash"	*Dicranograptus clingani* *Diplograptus foliaceus*	"	(19)
453.1 ± 1.3	[U-Pb]	"Millbrig" K-bentonite	*Climacograptus bicornis* *Ensigraptus caudatus*	"	(20a)
454.1 ± 2.1	[Ar-Ar]	"Millbrig" K-bentonite	*Climacograptus bicornis* *Ensigraptus caudatus*	Kunk et al. 1985; Tucker & McKerrow 1995	(20a)
454.5 ± 0.5	[U-Pb]	"Deicke" K-bentonite	*Climacograptus bicornis* *Ensigraptus caudatus*	Tucker & McKerrow 1995	(20b)
445.7 ± 2.4	[U-Pb]	Ashgill	[Dobs Linn section—UK]	"	(21)
438.7 ± 2.1	[U-Pb]	Llandovery	[Dobs Linn section—UK]	"	(22)
436.2 ± 5.0	[Ar-Ar]	Llandovery	*Coronograptus cyphus*	"	(23)
430.1 ± 2.4	[U-Pb]	Llandovery/Wenlock	*Oktavites spiralis*	"	(24)
423.7 ± 1.7	[Ar-Ar]	Ludlow	*Lobograptus scanicus*	Kunk et al. 1985	
421.0 ± 2	[K-Ar]	Ludlow	*Neodiversograptus nilssoni*	Tucker & McKerrow 1995	(25)
417.6 ± 1.0	[U-Pb]	Lochkovian	*Monograptus uniformis*	Tucker et al. 1998	

Note: The "stratigraphic constraint" column names the local range chart in which the dated bed can be placed or the taxon used to constrain an isolated bed in the optimal sequence. *P. s. scarabaeoides* is a trilobite; all other listed taxa are graptolites. "Sensitive High Resolution Ion Microprobe" (SHRIMP) dates (Compston and Williams 1992) were avoided pending resolution of systematic differences from the isotope dilution dates and questions about analytical standards. The entries "U-Pb" and "Ar-Ar" distinguish dates determined by the uranium-lead and argon-argon methods, respectively.

dovician, where graptolite species richness is very low. Because habitats and fossil preservation are patchy, the local range charts disagree in detail concerning the sequence of first and last appearances of taxa. The rules for resolving these discrepancies are straightforward, but the data set is so large that computer assistance is essential.

■ Computer-Assisted Calibration

Optimization algorithms in the CONOP9 software (Kemple et al. 1995; Sadler 2001) proceed by iterative improvement from an initially randomized sequence of events toward a "best-fit" sequence. They arrive at a model sequence for all 2,306 events with the best-known fit to the field observations in the sense that a minimum of stratigraphic range extensions is required to fit every local range chart to the model. The means of measuring the length of range extensions is critical to the outcome of the optimization. For this calibration task, we minimized the number of other range ends overlapped by the extensions, not the stratigraphic thickness of the extensions as minimized in graphic correlation techniques. This change eliminates bias due to variations in accumulation rate and favors those sequences preserved in the most richly fossiliferous sections. Such a misfit measure can solve ordinary correlation problems in which all sections span approximately the same time interval. It needs augmentation here because the local sections span only small fractions of the total time interval. We ensure that local sections are "stacked" in the correct order by simultaneously minimizing a second component of misfit borrowed from unitary association techniques (Guex 1991; Alroy 1992). The additional term counts the number of coexistences of pairs of taxa that are implied by the model sequence but not observed anywhere. The outcome of searching on these two criteria is an optimally ordered sequence of events.

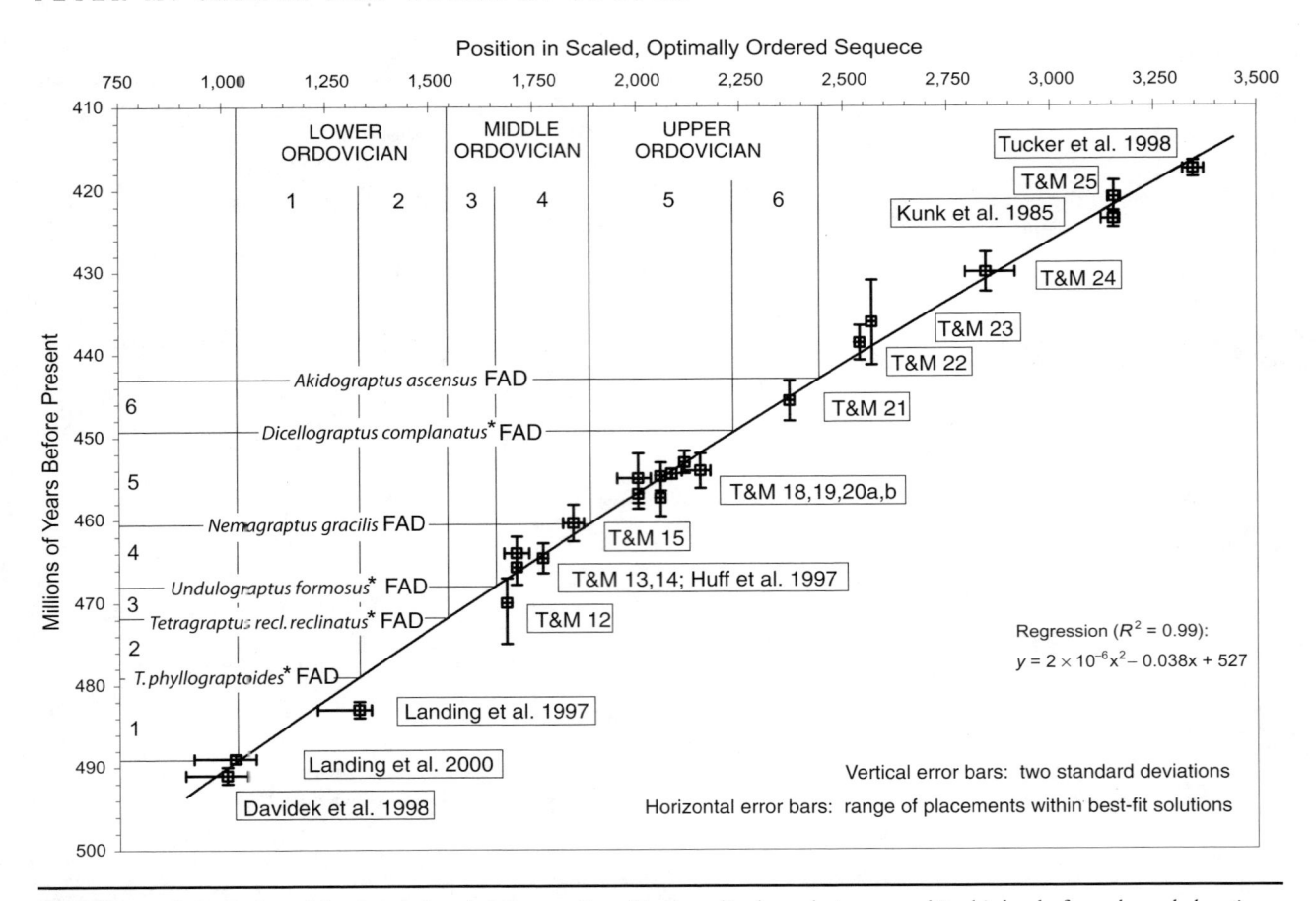

FIGURE 3.1. A projection of the six numbered chronostratigraphic time-slice boundaries, as used in this book, from the scaled optimally ordered sequence into the numerical timescale, by regression on the optimal placements of dated ash-fall events. We use, somewhat arbitrarily, a very small second-order polynomial term and force the line through the high-quality Cambro-Ordovician date. Similar results can be achieved with cubic splines (Frits Agterberg pers. comm.) or the locally estimated sum of squares. T&M refers to numbered dated items in Tucker and McKerrow (1995), many of them from Tucker et al. (1990).

Biostratigraphic field observations alone support the optimal *sequence,* but severe simplifying assumptions are unavoidable for scaling the time *intervals* between events. Two explicit simplifications were chosen as the least objectionable: in the long term, net taxonomic change (numbers of first- and last-appearance events) is a guide to the relative duration of whole sections; in the short term, stratigraphic thickness is a better guide to relative duration, especially in deep-water shales. The scaling assumptions are applied as follows. First, all local range charts are adjusted to fit the optimal sequence: observed taxon ranges are extended, if necessary, and missing taxa are inserted. Second, the total thickness of each section is rescaled according to the fraction of the composite sequence that it spans. Third, the spacing of adjacent events in the best-fit composite sequence is set equal to the average of all the local rescaled spacings. Zero values are included in the average to allow for mass extinctions and rapid radiations. The result at this stage is a relative timescale—a scaled, optimally ordered sequence in which events are spaced in proportion to one possible approximation of their separation in time.

■ Tests of the Calibration

The ultimate test of the scaling process plots the position of the dated events in the putative relative timescale against their radiometric ages on a regular numeric timescale. The near-linear regression in figure 3.1 justifies the scaled, optimally ordered sequence as a *plausible* proxy for a timescale and therefore suitable for interpolating the age of traditional zone boundaries. This regression test was actually applied in three stages. The first stage omitted the Australasian composite section (VandenBerg and Cooper 1992) and used only the 607 taxa observed in more than one non-Australasian section; the second stage added Australasian ranges; the third stage (figure 3.1) added those taxa known from only a single section. The fit to a linear regression improved slightly with each step. Because of provincial differences, presumably aggravated by the difficulty of achieving perfect optimization for huge data sets, the ordinal regression between our optimally ordered sequence and the Australasian sequence is not perfect, but it has a high correlation coefficient. Australasian zone boundaries were placed in the spaced, optimally ordered sequence at the first appearance of a cluster of zonal taxa (usually three to four). If this level did not coincide with the appearance of the customary defining species, a substitute species was selected from the cluster (asterisks in figure 3.1).

ACKNOWLEDGMENTS

This work was supported in part by National Science Foundation grant EAR 9980372 to Sadler. Fritz Agterberg, Barry Webby, and Henry Williams suggested improvements on an earlier draft.

4 Measures of Diversity

Roger A. Cooper

Measuring diversity change through geologic time faces difficulties from several sources. The obvious ones include variable preservation quality and collection completeness. Incomplete collecting will tend to shorten the stratigraphic ranges of taxa and reduce diversity. It will also be unlikely to detect the rare species, affecting diversity estimates as well as origination and extinction rates. A less obvious bias stems from the nature of most biostratigraphic data sets and is discussed in this chapter.

The basic data for the study of diversity change through geologic time are generally in the form of stratigraphic range charts. They are compiled from either individual sections or isolated collecting localities. When one is dealing with single sections, the stratigraphic ranges of taxa can be plotted in terms of the stratigraphic thickness of strata through which the taxon ranges. But when one is dealing with several sections, or with broad regions, the ranges of taxa are generally recorded in terms of biostratigraphic units such as zones or chronostratigraphic units such as stages. Fossil data collected from isolated localities must be assigned to stages or zones and compiled into composite zonal (or stadial) range charts. For large-scale, and global, studies such as the Great Ordovician Biodiversification Project (IGCP 410), ranges expressed in terms of zones, stages, or larger units are the norm.

The problem of converting stratigraphic ranges of taxa into measures of diversity and taxonomic rates of evolutionary change has been extensively discussed in the literature (Harper 1975, 1996; Sepkoski 1975; Raup 1985; Sepkoski and Koch 1996; Foote 1997a, 1997b, 2000a, 2000b with numerous references; Alroy 2000). A range of measures has been suggested that attempt to avoid introduced bias. However, it appears that no one measure of diversity, or of species origination or extinction intensity, avoids all problems (Jablonski 1995; Foote 2000a). The purpose of this chapter is to examine some of the more commonly used measures and assess which is the most appropriate for deriving diversity patterns from range charts. Commonly used measures of rates of origination, extinction, and faunal turnover are also outlined.

■ Mean Standing Diversity (MSD)

Ideally, the aim of paleontological diversity measures is to estimate the standing taxic diversity of a broad region through geological time (Sepkoski 1975; Sepkoski et al. 1981), equivalent to gamma diversity or total species richness (R, Rosenweig 1995). Standing taxic diversity is the total species diversity at a given instant in time. Because of the difficulty in directly measuring the standing diversity at a given instant in geologic time, we try instead to estimate the MSD over a specified time interval. This involves counting species present in strata that span the time interval. But there are several ways of expressing the number

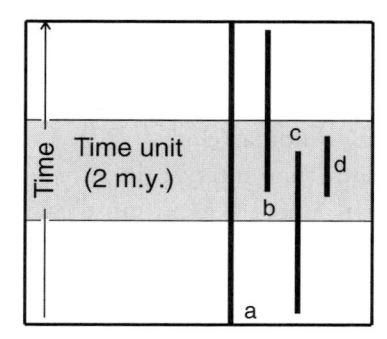

Total diversity (d_{tot}) = 4
Species/m.y. (d_i) = 2
Normalized diversity
(d_{norm}) = 2.5

FIGURE 4.1. The four ways in which a species can be present in a time interval; a, range through; b, originate within the interval and range beyond it; c, range into the interval and terminate within it; d, confined to the time interval. The three measures of diversity for the time interval are shown.

of species present and of allowing for introduced biases. The different methods can produce very different diversity curves for the same data set.

There are four ways in which a taxon can be present in a time interval, illustrated in figure 4.1: (a) range from the preceding interval to the following interval (range-through taxa); (b) originate within the interval and range into the following interval; (c) range from the preceding interval and become extinct within the interval; (d) confined to the time interval. For the purposes of the present exercise, range ends in categories b–d that coincide with an interval boundary are regarded as lying within the interval.

■ Effects of Time Interval Duration

A principal source of bias derives from unevenness in time interval duration (Sepkoski and Koch 1996; Foote 2000a). Sepkoski and Koch recommend minimizing this problem by combining zones of short duration and subdividing zones of long duration in order to minimize their differences. Foote (2000a) has modeled the effects on measures of diversity, and origination and extinction rates, of time interval duration, preservation quality, "edge effects," and the taxonomic rates themselves. He concluded that single-interval taxa produce many undesirable distortions, and he recommended that diversity and rate measures be adopted that do not count single-interval taxa (taxa confined to a single time interval). Rather than counting the taxa within a time interval, he recommended counting only those taxa whose stratigraphic ranges cross interval boundaries. However, this procedure would preclude a significant proportion of paleontological data for some fossil groups (chapter 27) and is unlikely to be acceptable to most paleontologists.

Some of the concerns raised by Foote may be minimized by using a precise timescale with subequal time intervals, as recommended by Sepkoski and Koch (1996). As discussed later in this chapter, by using model data sets that are comparable to those found in the fossil record, it is possible to test directly the performance of a range of diversity measures in estimating MSD through time intervals that range widely in duration.

■ Estimators of MSD

Three commonly used estimators of MSD are total taxic diversity, taxa per million years (m.y.), and normalized total taxic diversity here referred to as normalized diversity. In the following discussion the species is taken as the taxonomic unit, but genera and other taxonomic categories can be substituted as appropriate.

Total Diversity (d_{tot})

The simplest and most commonly used measure for estimating MSD is *total diversity* (d_{tot})—the total number of species (or other taxa) that are recorded from the time interval.

Species per m.y. (d_i)

This measure is simply total diversity divided by the duration of the time interval (i) in m.y. and allows for the probability that longer time intervals will capture more species. This is a rate measure and is therefore not strictly comparable with the other two measures, which are counts. The absolute value yielded by the measure is less important than its trend through time.

Normalized Diversity (d_{norm})

Species ranges in zonal range charts will overestimate true ranges because few species ranges will completely span the zones in which they first appear or last appear or to which they are confined. To compensate for this, Sepkoski (1975) devised a diversity measure here referred to as the normalized diversity measure. It is the sum of species that range from the interval below to the interval above, plus half the number of species that range beyond the time interval but originate or become extinct within it, plus half those that are confined to the time interval itself. The measure also normalizes for variability in time interval duration to the extent that the longer a time interval is, the more species will begin or end within it or are confined to it.

■ Model Data Sets

For the purposes of testing and comparing the three diversity measures, six model data sets have been constructed. The durations of time intervals and of species life spans in the data sets are comparable with those found in the graptolite data sets for Australasia, Baltica, and Avalonia (chapter 27). In the model data sets, both time interval durations and species longevities are uneven, as in most paleontological data sets. For simplicity, it is assumed that species ranges are complete and that all species that can be present in a bed *are* present. In addition, it is assumed for the purposes of the exercise that the observed ranges are the true, global ranges.

The six model data sets give a total of 40 time intervals. The smallest set contains five time intervals, which range from 1 to 2.5 m.y. in duration. This set is shown in figure 4.2, which also shows how the zonal range (thin lines), as recorded in range charts, considerably overestimates the true range (bold lines) for most taxa. Twenty-nine species are present and average 2.49 m.y. in duration. The largest set (set 6) contains 10 time intervals, which range in duration from 1.5 m.y. to 5 m.y. Eighty-three species are present and range in duration from 0.3 to 3 m.y., averag-

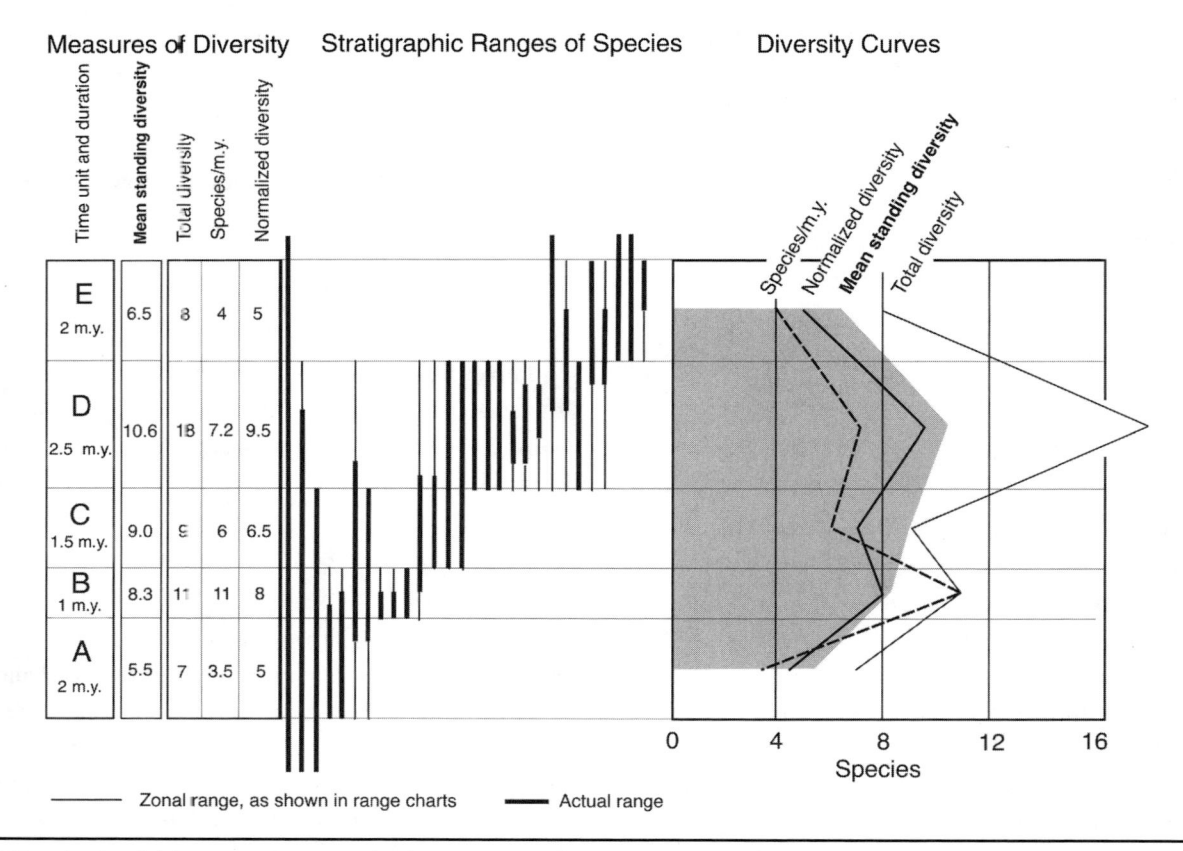

FIGURE 4.2. Model data set 1 comparing the three measures of diversity—total diversity (d_{tot}), species/m.y. (d_i), and normalized diversity (d_{norm})—with mean standing diversity (MSD; shaded area).

TABLE 4.1. Properties of the Six Trial Data Sets and Comparison of Trial Diversity Measures with Mean Standing Diversity (MSD)

Data Sets	1	2	3	4	5	6
Properties of model data sets						
Number of time units	5	5	6	8	10	10
Longest time unit	2.5	6	7	4	4	5
Shortest time unit	1	2	1	1	1	1.5
Mean duration of time units	1.80	3.80	3.30	2.20	2.10	1.80
Number of species (total diversity)	29	29	37	62	76	83
Mean duration of species	2.49	4.50	5.40	3.90	6.20	2.60
Number of range through species	15	16	37	67	115	62
Number of confined species	11	11	17	22	33	30
Number of originations	14	14	18	35	41	41
Number of extinctions	13	13	15	37	36	47
Net % difference from MSD (allows for +ve and −ve)						Mean
Total diversity (d_{tot})	13.1	17.1	20.0	36.8	47.5	39.5	29.0
Species/m.y. (d_i)	−8.2	−17.7	−29.3	−18.0	−41.8	31.9	−13.9
Normalized diversity (d_{norm})	−5.9	−2.0	−4.0	−9.2	−10.5	−21.5	−8.8
Mean % difference from MSD (regardless of +ve or −ve)						Mean
Total diversity (d_{tot})	15.3	23.6	20.3	28.6	29.1	30.1	24.5
Species/m.y. (d_i)	17.3	26.4	29.7	20.1	29.9	52.8	29.4
Normalized diversity (d_{norm})	7.4	9.0	5.6	9.8	14.1	13.0	9.8

ing 2.6 m.y. The basic properties of the data sets are given in table 4.1.

Because the exact stratigraphic ranges of species in the model data sets are known, the values for MSD can be precisely calculated for each of the time intervals. In the first data set (figure 4.2) it is highest in time interval D, where it reaches 10.6. At any given time in interval D there are, on average, 10.6 species in existence. Time interval B has more species than interval C yet scores a lower MSD. This is because more of the species in interval B have ranges that do not span the entire interval. Generally, this level of precision is not available in the fossil record, and species are recorded only as present or absent in each time interval (shown as the zonal range in figure 4.2).

The performance of the three measures in estimating MSD can thus be directly tested against its "true" value for each time interval in each data set (figure 4.2 and table 4.1). Two measures of comparison with MSD are given (table 4.1). The first is the mean net difference from MSD, expressed as the cumulative difference in value from MSD for each measure in each time interval. The lowest value indicates the closest comparison. However, positive and negative differences cancel each other out, so that it is possible for a curve to be the mirror image of MSD yet score a low value. The second measure of comparison with MSD is the mean percentage difference from MSD, expressed as the cumulative difference, regardless of whether it is positive or negative. The second measure requires the two curves to have similar shapes and similar magnitude in order to score a low value and is most meaningful in the present context.

■ Results of Trials

In model data set 1 (figure 4.2) it can be seen that no one measure closely approximates true MSD in all time intervals. Total diversity (d_{tot}) overestimates MSD in all time intervals except interval C. In interval D it overestimates MSD by 80 percent. The species per m.y. (d_i) curve generally underestimates MSD. The normalized diversity (d_{norm}) measure also underestimates MSD but is the best estimator. All three measures indicate a diversity peak in interval B where none exists.

The performance of the three measures through 40 time intervals in the six trial data sets is shown in table 4.1 and figure 4.3. The total diversity (d_{tot}) measure consistently and sometimes substantially overestimates true MSD, deviating from it by 25 percent on average, and the species per m.y. (d_i) measure generally underestimates it, deviating by 29 percent on average. The species per m.y. measure is the most erratic of the three. Normalized diversity (d_{norm}) is

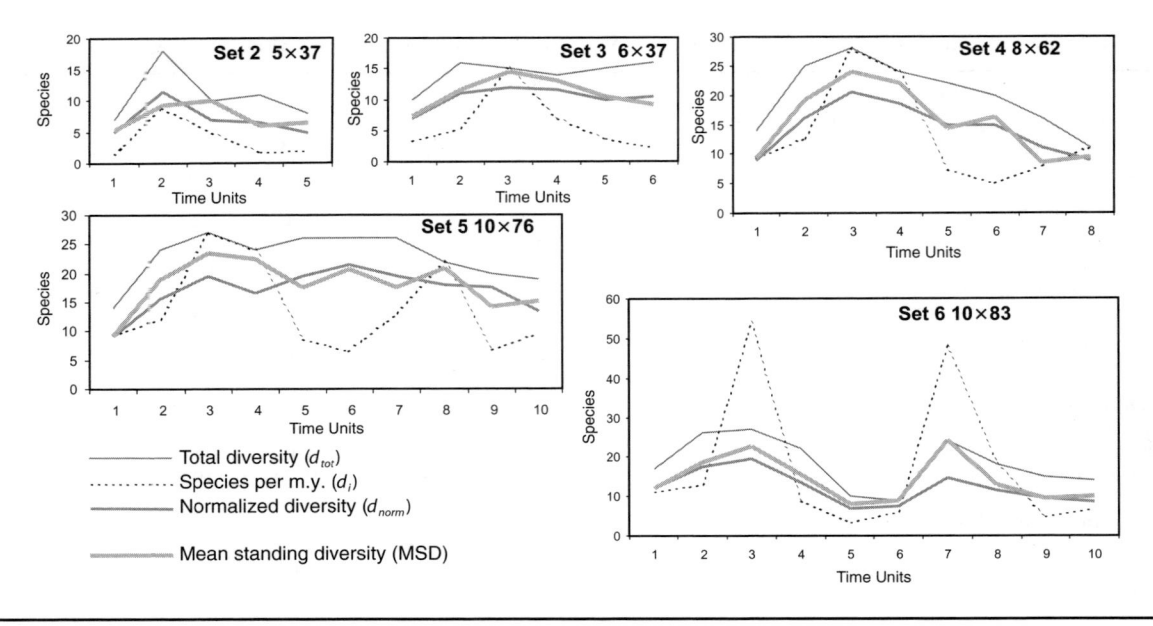

FIGURE 4.3. Comparison of alternative diversity measures in model data sets 1–5. The normalized diversity measure (d_{norm}) consistently gives the closest estimate.

consistently the closest approximation of true MSD, deviating by 10 percent on average.

The MSD curve in all model data sets is a smoother curve than the species per m.y. curve, which inserts, or exaggerates, peaks and troughs, sometimes in mirror image of the MSD curve. The normalized diversity curve copies the shape of the MSD curve more faithfully than either total diversity or species per m.y.

The difference between total diversity and MSD diminishes as the number of taxa that begin, end, or are confined to the time interval diminishes. This happens when larger taxonomic categories with long stratigraphic ranges (genera, families, orders) are used or when very fine time intervals are used. If all taxa present in a time interval range from the interval below to the interval above, then total diversity will equal both MSD and normalized diversity. Conversely, when low-level taxonomic categories such as species are used, a higher proportion of them will begin, end, or be confined to the time interval. Total diversity becomes a poor estimator of MSD in these circumstances. Using time intervals that are relatively long in comparison with the ranges of the taxa produces similar results.

Two further diversity measures were also tested. The first (Sepkoski et al. 1981) uses the normalized diversity measure discussed earlier but further normalizes for the difference between the mean duration of a species in the data set and the mean duration of a time interval. The second diversity measure was a composite derived from all four measures, the most appropriate being chosen for each time interval as judged by a range of criteria. However, in neither case was there a significant improvement on the normalized diversity curve in estimating MSD.

■ **Measures of Evolutionary Change**

Apart from the effects of regional migration, diversity within a region is a function of origination (o), here taken to mean a cladogenetic event, and extinction (e). The combination of origination and extinction within any time interval gives the faunal turnover ($o + e$). Similarly, net increase or decrease in diversity is given by originations minus extinctions ($o - e$).

The raw data for these four parameters of evolutionary change will be influenced by two main factors. The first is total diversity (d) in the time interval; high-diversity time intervals are more likely to have higher counts for each measure. The second factor is duration of the time interval (i); longer time intervals are likely to have higher counts for each measure. For these reasons Sepkoski and Koch (1996) recommend that a variety of measures be used for these evolutionary parameters. For extinctions they recommend (1) the total number of extinction events in the time interval (the "raw count"), (2) the percentage of extinction, which is the number of extinction events divided by the total number of taxa present, expressed as a percentage, and (3) the mean per capita extinction rate, which is the percentage of extinction divided by the duration of the time interval in m.y. Following are the three measures along with their equivalents for origination, faunal turnover, and net diversity increase/decrease.

Number of extinctions	$= e$
Percentage of extinction	$= e_d \times 100$
Per capita rate of extinction	$= e_{di}$
Number of originations	$= o$
Percentage of origination	$= o_d \times 100$
Per capita rate of origination	$= o_{di}$
Faunal turnover	$= (o + e)$
Percentage of faunal turnover	$= (o + e)_{2d} \times 100$
Per capita rate of faunal turnover (in m.y.)	$= (o + e)_{2di}$
Net increase/decrease	$= (o - e)$
Percentage of net increase/decrease	$= (o - e)_{2d} \times 100$
Per capita rate of net increase/decrease	$= (o - e)_{2di}$

■ Conclusions

The conversion of stratigraphic range data to diversity estimates involves inescapable biases. None of the three measures tested faithfully estimates MSD for data sets in which both species longevities and time interval durations are nonuniform, that is, for most paleontological data sets. Of the measures tested, that most commonly used—total diversity count (d_{tot})—generally overestimates MSD. Species per m.y. (d_i) produces diversity curves that fluctuate more strongly than MSD and can have inflections that mirror those of MSD. It is the most erratic of the three measures tested. The normalized diversity measure (d_{norm}) consistently gives the closest estimate of true MSD and is recommended for use in studies of diversity change through geologic time.

Because origination, extinction, faunal turnover, and net diversity increase/decrease in a time interval are influenced by the interval's duration and total diversity, the measures that normalize for these biases (percentages and per capita rates) should be given in addition to the raw counts.

ACKNOWLEDGMENTS

I thank Drs. Arnie Miller and James Crampton for their comments. The project was supported in part by the Marsden Fund, Royal Society of New Zealand.

PART II

Conspectus of the Ordovician World

5 Major Terranes in the Ordovician

L. Robin M. Cocks and Trond H. Torsvik

From Jurassic times onward, magnetic stripes from spreading centers are preserved on the modern ocean floors, and unraveling these are good guides to the migration of terranes through time. However, the old ocean floors present in the Ordovician have been either lost through subduction or distorted and displaced through obduction, and thus the identification and positioning of the terranes present at that time must rely on less direct methods. Chief among these are paleomagnetism, but that can indicate only paleolatitude, not paleolongitude, and faunal studies, which can indicate terrane separation and to a general extent paleolatitude, but in a more subjective way than paleomagnetism. In addition, the distribution of distinctive facies can be helpful for some periods, particularly, for example, at the time of the latest Ordovician glacial event. We have summarized our methods elsewhere (Cocks and Torsvik 2002) and collaborate to place the Ordovician terranes in positions that satisfy the current constraints from paleomagnetism, faunas, and sediments. The faunal evidence to support the reconstructions is reviewed in Fortey and Cocks (2003). However, the degree of confidence varies greatly with both place and time. In general, the present-day North Atlantic area, which in the Ordovician included Laurentia, Baltica, and West Gondwana and their adjacent terranes and island arcs, is well constrained. However, the many terranes that today make up most of Asia are less securely defined and their Ordovician locations much less certain.

In Ordovician times there was one supercontinent, Gondwana, occupying paleolatitudes from the South Pole to north of the equator, and three other major terranes, Laurentia (largely North America), Baltica (northern Europe), and Siberia. There were also many other smaller but still substantial terranes, mostly surrounding Gondwana and known as peri-Gondwanan; chief among them were Avalonia, Sibumasu (Shan-Thai), Annamia (Indochina), South China, and North China. Smaller but faunally distinctive terranes included Perunica (Bohemia), the Taurides and Pontides of Turkey, Tarim, the Kazakh terranes, the Himalayan terranes, and the Argentine Precordillera. In addition to these there were several island arcs in the various oceans, many of which carried faunas that have survived. These terranes are reviewed in turn, with appropriate reference to their changing positions as the Ordovician progressed. In addition to the terranes discussed in this chapter, Apulia and the Hellenic Terrane (both in southern Europe) and various Mexican terranes are shown on our maps, but since there is little evidence for their existence and detailed positions in the Ordovician, they are not discussed further here.

In this short summary the various major terranes are listed and their positions presented in four maps (figures 5.1 to 5.4) depicting Early Ordovician (480 Ma,

FIGURE 5.1. Southern Hemisphere Ordovician terrane disposition in latest Tremadocian and base Arenig (about 480 Ma) time (from Cocks and Torsvik 2002: figure 4). 13a, Apulia; 13b, Hellenic; 14a, Taurides; 14b, Pontides; 15a, Lhasa; 15b, Qiantang; Kar., Karakum; Kaz, Kazakh terranes (shown arbitrarily). The three small areas with asterisks between Laurentia and Baltica denote the only three places on the Iapetus island arcs from which reliable paleomagnetic data are known. North China is absent because it was entirely within the Northern Hemisphere. The thick dashed and dotted line surrounds core Gondwana. Figures 5.1 to 5.3 are Schmidt's Equal Area projection, with projection center at the South Pole.

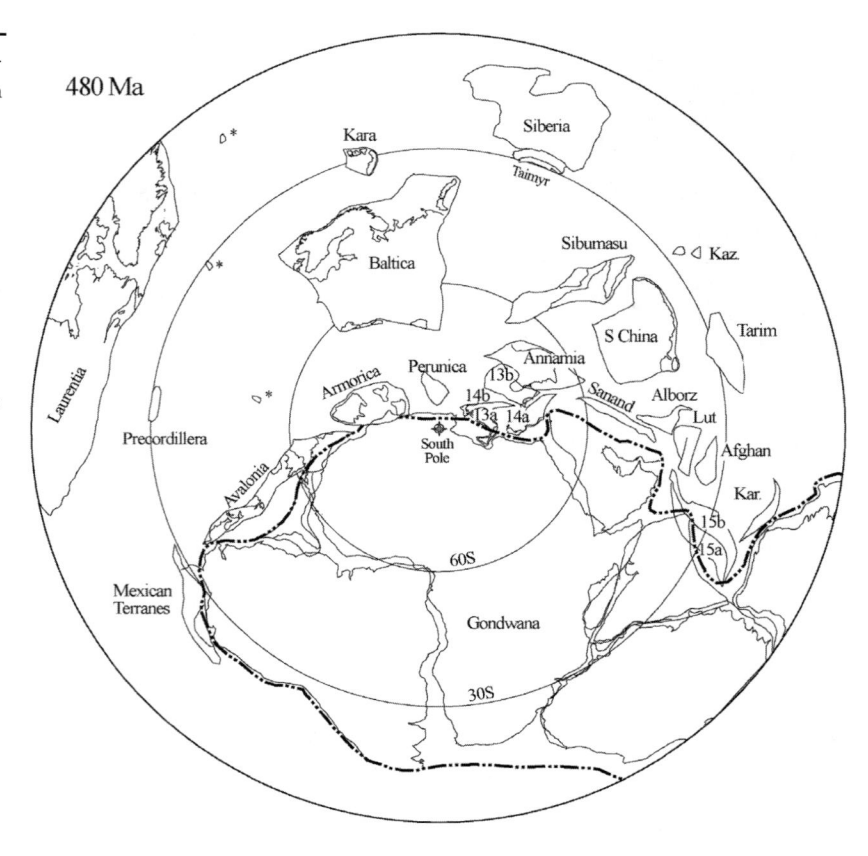

late Tremadocian and earliest Arenig), earliest Late Ordovician (about 460 Ma, early Caradoc), and latest Ordovician–earliest Silurian (approximately 440 Ma, Ashgill-Llandovery) times, respectively. The first three figures show only the Southern Hemisphere, where most of the Ordovician terranes lay, but the fourth shows the whole earth. The latter emphasizes how much of the Northern Hemisphere was occupied by the vast Panthalassic Ocean.

■ Gondwana

Core Gondwana consisted of a very large area, divided for descriptive reasons into two, West and East Gondwana. West Gondwana included Africa, Madagascar, Florida, South America, and Arabia. East Gondwana included Greater India, Antarctica, New Guinea, and most of Australia. Core Gondwana is shown by a distinctive dashed line in our figures.

Although Armorica (including Iberia) is shown outside that line as a peri-Gondwanan terrane, it is now thought that it did not split off from West Gondwana until well after the end of the Ordovician. This is because (a) the Ordovician faunas of Brittany, the Montagne Noire, and the Iberian Peninsula are essentially the same as those of North Africa (Morocco, Algeria, and Libya) and (b) the position of Gondwana is now better constrained by both paleomagnetic and faunal data than it was until recently, all of which negates published models showing a substantial seaway between Armorica and Gondwana.

In the earliest Ordovician, West Gondwana also included Avalonia and Perunica, but they became separate from it during the period (see later in this chapter). Arabia has been thought by some authors to have been formed from several terranes (termed Lut, Sanand, and others), but there seems little evidence that it formed anything other than part of core Gondwana in Ordovician times.

Gondwana formed at about 550 Ma (Meert and Van der Voo 1997), and its early dispersal history commenced with the rifting off of Avalonia from the northern South America–northwestern Africa part of West Gondwana in Arenig time. The whole continent occupied more than 100 degrees of paleolatitude in the Ordovician and thus spanned the area from the

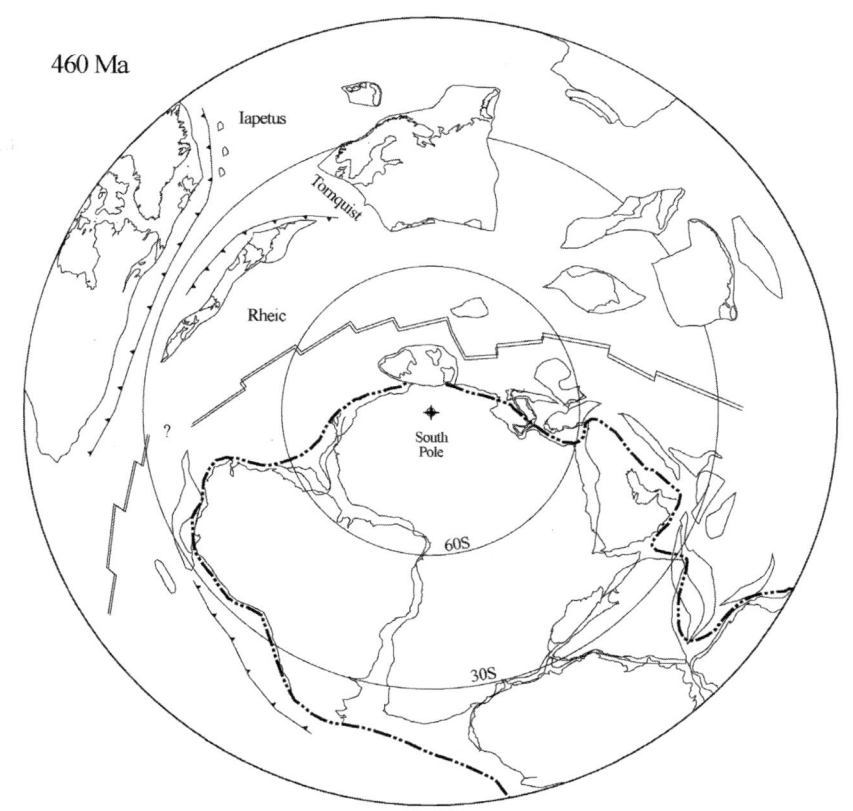

FIGURE 5.2. Southern Hemisphere Ordovician terrane disposition in early Caradoc (about 460 Ma) time (base map derived from Cocks and Torsvik 2002: figure 5). The thick dashed line surrounds core Gondwana, and some putative spreading centers and subduction zones are shown. The terrane names are shown on figure 5.1. North China was entirely within the Northern Hemisphere.

South Pole to north of the equator. In the Early Ordovician the South Pole lay under North Africa, perhaps Libya, but as time and drift proceeded, by the end of the Ordovician the pole lay under central West Africa or perhaps slightly farther to the west under adjacent Brazil (Cocks and Fortey 1988).

■ Baltica and Kara

Baltica follows the traditional outline as used by many authors (references in Cocks and Fortey 1998) but includes the Malopolska and Lysogory areas of the Holy Cross Mountains, Poland (Cocks 2002), and the hidden southward extension of the Urals as far south as the northern Caspian Sea (Cocks 2000: figure 6). It excludes the highest nappes in the Trondheim area, Norway (which were part of Laurentia), and also the Taimyr region of Siberia, which had previously been considered as part of Baltica (Cocks and Fortey 1998). North Taimyr is now considered to be part of the Kara Terrane and central and southern Taimyr as parts of the main Siberian continent. We have restored Novaya Zemlya to its pre-Triassic position as a direct extension of the Urals (Torsvik and Andersen 2002).

The whole terrane rotated by more than 100 degrees between the Vendian and the Middle Ordovician, and in particular 55 degrees of that rotation took place in the Late Cambrian and Early Ordovician (references in Torsvik and Rehnström 2001), and thus today's southwestern Tornquist margin in central Europe faced toward Laurentia in the Early Ordovician. Baltica moved from fairly high temperate paleolatitudes in the earliest Ordovician northward to more equable climates by the Late Ordovician. The movement of the Kara Terrane, which includes northern Taimyr and Severnaya Zemlya, shown in figures 5.1 to 5.4, is well constrained paleomagnetically (Torsvik and Rehnström 2001); the terrane includes Ordovician faunas, but these are not yet well evaluated.

■ Siberia

Although persisting with the traditional use of this name for an Ordovician terrane its margins were very different from those of Siberia today. We follow the outline of Rundqvist and Mitrofanov (1993), which includes most of the Baikalides, which accreted on to the main Angaran craton in the center of Siberia in the Late Precambrian. We also include south and central Taimyr as an integral part of the terrane. The paleomagnetic data are good (Smethurst et al. 1998), and from them we can see that the paleocontinent moved from south to north across the paleoequator and into northern low to temperate latitudes during the Ordovician, making it, North China, and parts of Laurentia and East Gondwana (Australia and New Zealand) the only lands of consequence yet identified from the then Northern Hemisphere. More important, Siberia underwent rotation after the Ordovician, so that in our maps today's north faces southward.

■ Laurentia

Ordovician Laurentia included most of today's North America north of the Ouachita Front, in addition to Greenland, Spitzbergen, and all the British Isles north of the Iapetus suture (following the fit of Bullard et al. 1965). It also included the eastern part of today's Siberia, the Chukhot Peninsula. From both paleomagnetic and faunal evidence Laurentia appears to have occupied a transequatorial position with little movement during the whole Ordovician, and we have no evidence of any substantial rotation during the period.

■ Avalonia

This region (Cocks et al. 1997) includes the eastern seaboard of North America north from Cape Cod, Massachusetts, through the Maritime Provinces of Canada, eastern Newfoundland, southeastern Ireland, England, Wales, Belgium, and Holland, and eastward to the Trans-European Suture Zone. There is no convincing evidence of it being split between "East" and "West" Avalonia in the Ordovician—the use of these terms should be discontinued. To the south Avalonia is bordered today by the various terranes of Gondwana and peri-Gondwana, notably Armorica and Perunica (Bohemia). In the earliest Ordovician Avalonia formed an integral part of West Gondwana but separated from it, probably in Arenig times, and drifted northward across the narrowing Iapetus Ocean, leaving a widening Rheic Ocean to its south. Avalonia docked, probably softly, with Baltica at about 443 Ma, the close of the Ordovician. The movement of the terrane is well constrained paleomagnetically (Torsvik et al. 1993).

■ Sibumasu

The Sibumasu, sometimes termed the Shan-Thai, Terrane stretches today from Burma (Myanmawr) in the north, through western Thailand and western Malaysia, to Sumatra in the south. The paleomagnetic data are scanty, but the faunas place the terrane close to South China in the peri-Gondwanan collage (Fortey and Cocks 1998).

■ Annamia

This terrane occupies nearly all of Indochina. There are few paleomagnetic data, and the faunal descriptions date from the early twentieth century or before. Nevertheless, enough is known to understand that it had little in common with Sibumasu, with which it did not unite until the late Jurassic, but that it too was also a member of the peri-Gondwanan collage in the Ordovician. The Early Ordovician faunas reviewed by Fortey and Cocks (2003) suggest the relatively high paleolatitude shown in figure 5.1.

■ South China

The margins shown in figures 5.1 to 5.4 follow the outline of Rong et al. (1995), and we show the modern coastline only to aid recognition in the diagrams. There are some paleomagnetic data for the Ordovician, and the terrane appears to have moved from temperate to tropical paleolatitudes during the course of the period. There is much faunal evidence (discussed in Fortey and Cocks 2003) to place the terrane near East Gondwana.

■ North China

This terrane, which also includes the Korean Peninsula, also follows the outlines of Rong et al. (1995). From both the fauna and the paleomagnetic data we

Major Terranes in the Ordovician 65

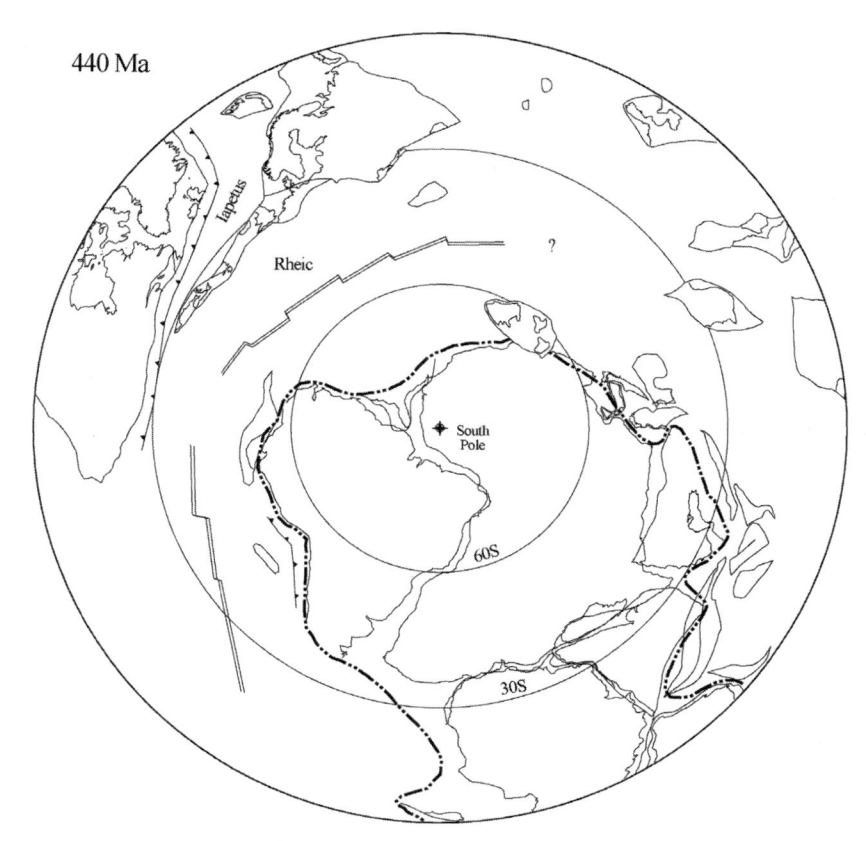

FIGURE 5.3. Southern Hemisphere Ordovician terrane disposition in latest Ordovician–earliest Silurian (about 440 Ma) time (base map derived from Cocks and Torsvik 2002: figure 6). Siberia and Tarim are absent because they were entirely within the Northern Hemisphere. Part of North China (with today's outline of the Korean peninsula) is to be seen at the top right. The other terrane names are shown on figure 5.1. The thick dashed line surrounds core Gondwana, and some putative spreading centers and subduction zones are shown.

can see that it was not too far from Laurentia in the Early Ordovician but moved from north to south across the paleoequator, and by the latest Ordovician it appears to have approached, or even formed part of, the peri-Gondwanan terrane collage. It eventually docked with South China in the Permian.

■ Perunica

The Bohemian Massif of central Europe, which forms the core of the old terrane of Perunica, was an integral part of West Gondwana until some time in the Early Ordovician. However, certainly by early

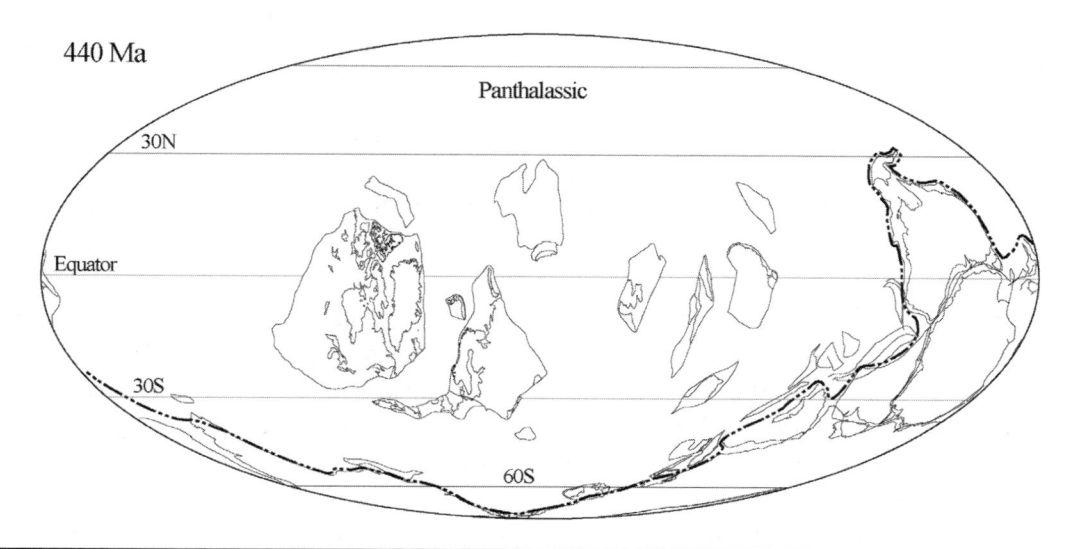

FIGURE 5.4. Global reconstruction for latest Ordovician–earliest Silurian time (about 440 Ma), from Cocks and Torsvik (2002: figure 7). Mollweide projection, showing all the major areas absent from figure 5.3 and the extensive Panthalassic Ocean.

Caradoc times (figure 5.2) it was adrift in the Rheic Ocean on both faunal (Havlíček et al. 1994) and paleomagnetic (Tait et al. 1997) grounds. It remained as a separate small terrane until Early Devonian time and during the period appears to have undergone some rotation.

■ Taurides and Pontides

The Pontides, which are developed north of the east-west Anatolian Fault, and the Taurides, which are developed to the south of that fault and north of the main Arabian plate, form the greater part of modern Turkey. There are few paleomagnetic constraints, but the faunas (Dean et al. 2000) indicate separation of the two terranes, with the Pontides at a higher latitude off West Gondwana than the Taurides in the Ordovician. Early Ordovician brachiopods from the Taurides (Cocks 2000) confirm this intermediate paleolatitude.

■ Tarim and the Kazakh Terranes

The many terranes that make up central Asia between the Urals (to the west of which was Baltica) and the Siberian craton are the least well constrained in figures 5.1 to 5.4: little reliable paleomagnetism is known from them. Şengör and Natal'in (1996) have identified more than 20 terranes in the area, which they postulated as forming an enormous Kipchak Arc between Baltica and Siberia in the Lower Paleozoic, which subsequently collapsed and coalesced to form the central Asian collage of today. Many of these structural units have Precambrian cores, and a large number of differing Ordovician benthic and planktonic faunas are known from this large area. Nikitin et al. (1991) show the modern extent of some of these units. However, recent analysis of the faunas (Fortey and Cocks 2003) has demonstrated that at least some of these areas—for example, the Chingiz, Tien Shan, and Chu-Ili terranes, as well as Tarim—have affinities with Gondwana and peri-Gondwana rather than with Baltica or Siberia. Thus these areas are shown only diagrammatically as two small units in figures 5.1 to 5.4, apart from Tarim, which is shown near South China—this is based on a combination of faunal similarity and some paleomagnetic data from the Devonian.

■ The Himalayan Terranes

Figures 5.1 to 5.4 show the Afghan Terrane, the Lhasa Terrane, South Tibet (15a in figure 5.1), and the Qiangtang Terrane, north Tibet (15b in figure 5.1), and their boundaries are well delineated from Upper Paleozoic and later faunal and tectonic studies. There are Ordovician localities, many with faunas, scattered at intervals through these Himalayan regions. However, there are no reliable Lower Paleozoic paleomagnetic data from these areas, and the faunal data have not yet been intensively evaluated for the Ordovician; thus we show these terranes as tied to Gondwana in our reconstructions.

■ Precordillera

The Precordillera terrane, which today lies in an area in the northwest of Argentina, carries Cambrian and Early Ordovician faunas of Laurentian affinity. Whether the Precordillera actually formed an integral part of Laurentia or lay close offshore as a separate terrane is uncertain. However, it clearly became separate from Laurentia soon afterward, since faunal endemicity there was at its highest in Caradoc times (Benedetto 1998). The terrane docked with Gondwana sometime during the Silurian, sandwiching the Famatina island arc between it and the supercontinent.

■ Island Arcs

Figures 5.1 to 5.4 omit island arcs, apart from three small areas in the Iapetus Ocean representing paleomagnetic data from intra-arc localities in eastern North America published by Mac Niocaill et al. (1997). However, there were at least two island arcs in the Iapetus Ocean in the Ordovician (Mac Niocaill et al. 1997) that amalgamated with each other and with the neighboring continents as the Iapetus narrowed during the Ordovician and closed in the Silurian. There were further island arcs in the Chukhot area to the north of Laurentia (Natal'in et al. 1999), the Famatina range between the Precordillera and

Gondwana in South America (Benedetto 1998), and off Gondwana in what is today southeastern Australia (Percival and Webby 1996), from all of which Ordovician rocks and various faunas are known (Cocks 2001). In addition, the substantial area preserved within central Asia today (see Kazakh terranes and Tarim earlier in this chapter) certainly carried island-arc faunas in the Ordovican, but the identity, extent, disposition, and tectonic evolution of the various arcs there are not yet certain.

6 Isotopic Signatures

Graham A. Shields and Ján Veizer

Here we review current understanding of the isotope chemistry of Ordovician seawater as determined from published analyses of the isotopic compositions (strontium, neodymium, sulfur, carbon, and oxygen) of marine authigenic precipitates. Isotopic data have implications for our understanding of global dynamics and paleoenvironmental evolution; however, many published interpretations are still controversial. In general, the past 20 years have witnessed a laudable departure from using bulk samples to carefully selected calcite components, such as microstructurally pristine brachiopods and early marine cements, as subjects of geochemical study.

■ Strontium Isotopes

Because of the long, 2–5 m.y., residence time of strontium in the oceans, the $^{87}Sr/^{86}Sr$ ratio of seawater is the same worldwide and can be used for both global stratigraphic correlation and interpretation of major trends in planetary dynamics (Veizer 1989). Seawater $^{87}Sr/^{86}Sr$ ratios decrease worldwide across the Cambrian-Ordovician boundary from >0.7091 to <0.7090 (figure 6.1). Minor shifts in $^{87}Sr/^{86}Sr$ around the Cambrian-Ordovician transition mark breaks in sedimentation that coincide with biozone boundaries and regressive events (Ebneth et al. 2001). Subsequently, $^{87}Sr/^{86}Sr$ decreased gradually from ~0.7090 to ~0.7087 through the Early Ordovician, attaining a plateau ~0.7087 during most of the Mid Ordovician (figure 6.1). From the late Mid Ordovician to the early Late Ordovician, an estimated period of 6–12 m.y., $^{87}Sr/^{86}Sr$ decreased rapidly from ~0.7087 to ~0.7078 (Denison et al. 1998; Qing et al. 1998; Shields et al. 2003). There is no evidence for significant seawater $^{87}Sr/^{86}Sr$ fluctuation during the rest of the Late Ordovician–earliest Silurian (figure 6.1).

The high $^{87}Sr/^{86}Sr$ ratios at the beginning of the Ordovician Period are indicative of a relative dominance of continental weathering processes on ocean chemistry, possibly related to widespread orogeny. The generally decreasing trend in $^{87}Sr/^{86}Sr$ during the Ordovician is interpreted to be a response to the waning Pan-African orogeny (Qing et al. 1998). The much sharper drop in seawater $^{87}Sr/^{86}Sr$ across the Mid to Late Ordovician transition, which coincides with a major transgression, possibly the largest in the Phanerozoic (Ross and Ross 1992), is more likely to be related to an acceleration of ocean spreading rates (and of hydrothermal exchange rates) combined with low overall continental weathering rates and widespread volcanicity around this time (Shields et al. 2003). Higher sea levels would also have favored low seawater $^{87}Sr/^{86}Sr$ by decreasing continental runoff and submerging low-lying juvenile extrusives.

■ Neodymium Isotopes

Neodymium, unlike strontium, is removed efficiently from seawater and so has a shorter ocean residence time (~10^2 years) than the turnover rate of water in the oceans (~10^3 years). Consequently, the ratio $^{143}Nd/^{144}Nd$ will tend to vary between and even within individual ocean basins, thus providing a powerful tracer of water masses and sediment provenance. Only a few studies have tried to estimate the Nd isotopic composition of Ordovician seawater (e.g., Keto and Jacobsen 1987; Fanton et al. 2002; Wright et al. 2002) using mostly biogenic apatite. Wright et al. (2002) show that Laurentian seaways were isotopically distinct during the Early and Mid Ordovician but that this distinction disappeared by the Late Ordovician, when the global ocean appears to have adopted a more juvenile volcanic signature (chapter 7). Detailed Nd isotope stratigraphy of North American carbonates (Fanton et al. 2002) suggests that this Mid to Late Ordovician change and other higher-order changes in εNd may be linked to the submergence and reexposure to weathering of Precambrian basement during marine transgressions and regressions, respectively. The study by Fanton et al. demonstrates that εNd profiles can reflect regional sea level change and so be used for interbasinal correlation. Wright et al. (2002) postulate that Nd isotope data will prove useful in generating more accurate paleogeographic reconstructions, but this potential has yet to be realized in pre-Mesozoic studies.

■ Carbon and Oxygen Isotopes

Carbonate stable isotopes form the mainstay of much isotope stratigraphy because of the usefulness of $\delta^{13}C$ excursions in stratigraphic correlation and $\delta^{18}O$ in paleoenvironmental interpretation (temperature, salinity). However, there is still much we do not understand about both isotope systems. Ripperdan and co-workers established a global chronostratigraphic framework for the Cambrian-Ordovician boundary, incorporating biozones, magnetic reversals, regressive events, and $\delta^{13}C$ excursions (Ripperdan et al. 1992, 1993). Although there is a broad, long-term trend toward more negative $\delta^{13}C$ values from the Late Cambrian to the Early Ordovician (figure 6.1), $\delta^{13}C$ excursions are rather muted, which limits their usefulness for correlation. A positive $\delta^{13}C$ excursion of 1–2‰ (= per mil, meaning "per thousand"), peaking at ~+1‰, occurs close to the boundary globally and at the incoming of the conodont *Iapetognathus* at Lawson Cove, United States (Ripperdan and Miller 1995).

Oxygen isotope studies of the Early and Mid Ordovician have attracted controversy because of anomalously low $\delta^{18}O$ values (figure 6.1), which have been variously interpreted to indicate diagenetic alteration (e.g., Land 1995), very high seawater temperatures (e.g., Karhu and Epstein 1986), salinity stratification (Railsback et al. 1990), high seawater pH (Wenzel et al. 2000), and low seawater $\delta^{18}O$ (Veizer et al. 1986). Our preferred interpretation, a combination of high tropical temperatures and low seawater $\delta^{18}O$ (Shields et al. 2003), would, according to geochemical models, necessitate greater low-temperature oceanic crust alteration in the pre-Mesozoic world (Wallmann 2001). A partly climatic interpretation for the Ordovician trend of increasing $\delta^{18}O$ is supported by the observation that, once the long-term trend to increasing calcite $\delta^{18}O$ through the Phanerozoic is removed, four icehouse-greenhouse climate cycles, which have been independently confirmed by paleoclimatic studies, can be recognized in the $\delta^{18}O$ record (Veizer et al. 2000).

$\delta^{13}C$ and $\delta^{18}O$ values tend to be higher by 1–2‰ from the Mid to Late Ordovician (figure 6.1), with reports of a positive $\delta^{13}C$ excursion of up to 3‰ during the early Late Ordovician (mid Caradoc) in the United States (e.g., Hatch et al. 1987). This excursion is associated with a sea level rise and faunal extinction and can be recognized in both carbonate and organic carbon (Pancost et al. 1999). Holmden et al. (1998) suggest that this excursion may correspond to a regional oceanographic event, restricted to units east of the transcontinental arch of the United States, but a similar $\delta^{13}C$ excursion in probably coeval rocks from Estonia (Ainsaar et al. 1999) suggests a more widespread, possibly global origin.

Many studies have identified a short interval of anomalously high $\delta^{13}C$ and $\delta^{18}O$ during the latest

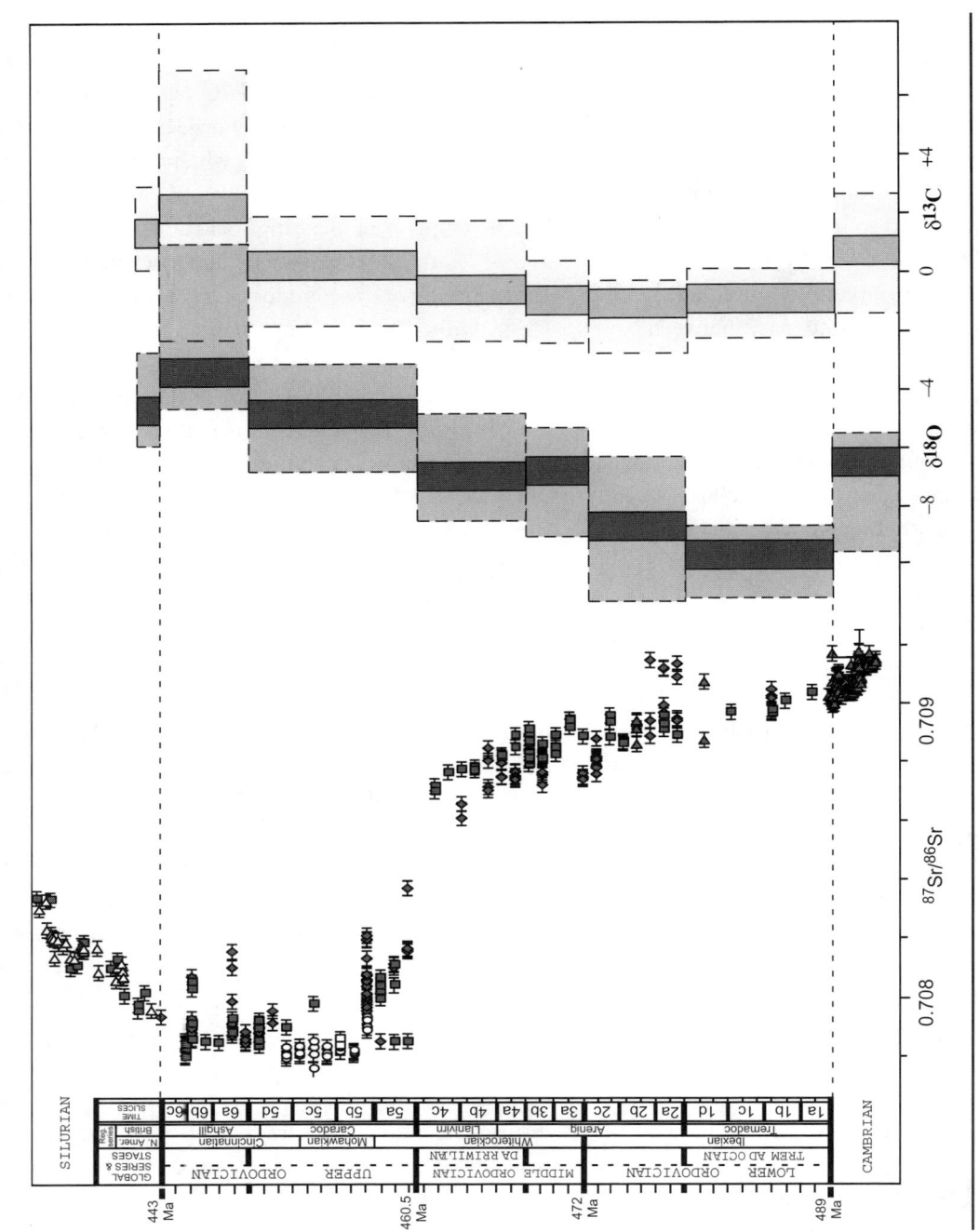

FIGURE 6.1. Strontium, carbon, and oxygen isotopic trends during the Ordovician incorporating all biostratigraphically constrainable brachiopod, conodont, and carbonate component data (compilation from Shields et al. 2003). Horizontal error bars correspond to a minimum analytical and geologic combined $^{87}Sr/^{86}Sr$ reproducibility of $\pm 25 \times 10^{-6}$. See Shields et al. 2003 for details of the associated symbols. Darker boxes for the carbon and oxygen data represent mean (± 1 s.d.), and the lighter boxes surrounded by dashed lines represent the absolute range of stable isotope values for each global stage.

Ordovician. The positive $\delta^{13}C$ excursion, ranging between 2‰ and 7‰, emerges from a background of generally rising mean $\delta^{13}C$ through the Ordovician and has been documented globally in both carbonate and organic carbon (e.g., Marshall and Middleton 1990; Finney et al. 1999; Kump et al. 1999). This transient excursion coincides with global faunal extinction, eustatic sea level fall, and glaciation (see chapter 9). Consistently positive $\delta^{13}C$ during mass extinction is not typical and may suggest that overall marine productivity was not significantly affected by this event or that the $\delta^{13}C$ record was reacting more to changes in source of ocean bicarbonate rather than bioproductivity (Kump et al. 1999). The associated, but short-lived, 2–3‰ positive $\delta^{18}O$ excursion is consistent with the rapid buildup of ice on the continent and generally lower tropical temperatures during the latest Ordovician glaciation.

Sulfur Isotopes

The sulfur isotopic composition of Ordovician seawater was until recently known only from sparse analyses of marine sulphate minerals (Thode and Monster 1965; Claypool et al. 1980; Fox and Videtich 1997). The last study derived a best estimate of Late Ordovician seawater of 25.5‰ ± 1.6‰ (2σ) by averaging their results with previously published data. Newly acquired $\delta^{34}S$ data from structurally substituted sulphate in biogenic calcite (Kampschulte 2001; Kampschulte and Strauss in press) indicate that seawater $\delta^{34}S$ decreased from >~30‰ to <~25‰ during the Ordovician Period. Such a decrease could indicate globally decreasing sulphate reduction rates or pyrite storage or both (Strauss 1997).

Conclusions

Strontium and neodymium isotope data reveal the Ordovician to cover an interval of generally waning continental influence and increasing dominance of juvenile volcanism over seawater composition. A major fall in seawater $^{87}Sr/^{86}Sr$ marks the Middle-Upper Ordovician boundary and can most plausibly be connected to enhanced volcanism at that time coupled with generally higher sea level. Carbon isotope data are far from straightforward to interpret, with the most enigmatic event being the Late Ordovician positive $\delta^{13}C$ excursion, which seems to be intimately connected with eustatic, climatic, and faunal change. A contemporaneous increase in $\delta^{18}O$ is confidently interpreted to reflect widespread glaciation and global cooling. However, anomalously low $\delta^{18}O$ values from the Lower and Middle Ordovician are harder to explain in terms of climate change and may instead reflect a ^{18}O-depleted ocean, which has implications for current models of ocean hydrothermal cycling. Sulfur isotope data are too scarce for any informative interpretation of their significance.

Future Directions

The Mid to Late Ordovician transition appears to be marked by significant but enigmatic changes in strontium, carbon, and oxygen isotope geochemistry, which may provide potential for global stratigraphic correlation and should help in understanding some of the controlling factors on seawater chemistry. More isotopic data on biostratigraphically constrained samples are required, especially for the Lower and Middle Ordovician.

7 Ordovician Oceans and Climate

Christopher R. Barnes

The Ordovician world was very different in many respects from that of today. Although many of the specific conditions are still speculative and need additional quantification, for the purpose of this chapter it is accepted that the prevailing conditions included the following (reference citations restricted by space limitations):

1. Significantly lower oxygen levels in the coupled ocean-atmosphere system, probably about 50 percent present atmospheric level (PAL) (Berner 2001)

2. A greenhouse climate state with high carbon dioxide levels of 8–18 × PAL (Berner 1994), with a reduction during the latest Caradoc–latest Llandovery glacial phases that peaked in the late Ashgill (Hirnantian stage) (Sheehan 2001b; Zhang and Barnes 2002)

3. A sluggish surface and deep ocean circulation system, due to the overall greenhouse state, in which circulation was driven more by midlatitude evaporation and subduction of higher salinity waters than the high-latitude thermohaline circulation pattern of today (Railsback et al. 1990)

4. Periods of extreme eustatic highstand included the largest (Caradoc) in the Phanerozoic (Hallam 1992). Broad epeiric seas were maintained for long periods; for those in low latitudes, high salinities were maintained locally producing evaporites.

5. Periods of increased volcanicity (Stillman 1984; Rogers and Van Staal in press), notably vast outpourings of volcanic ash, produced extensive bentonites in Laurentia, Baltica, and the Argentine Precordillera terrane. These included reputedly the largest eruptions of ash in the Phanerozoic (Huff et al. 1992).

6. A Mid Ordovician mantle superplume that probably developed in the late Middle Ordovician (chapter 8) would have increased temperature and modified deep ocean circulation and nutrient flux. This event may also explain the enhanced volcanism referred to earlier.

7. The Cambrian Evolutionary Fauna was replaced in the Early Ordovician by the Paleozoic Evolutionary Fauna (Sepkoski 1995), with a more than threefold increase in diversity (families, genera, species). There was an associated ecologic partitioning of the substrates and water column, an increase in tiering in and above the substrate, and new trophic structures.

■ Nature of Oceanographic and Climatic Events

The nature of Ordovician oceans and climates is poorly understood. In terms of the possible ocean circulation systems, some authors attempted to plot present-day circulation patterns on different paleogeographic reconstructions to unravel the pattern of paleobiogeography (e.g., McKerrow and Scotese 1990). A more sophisticated analysis was introduced by Wilde (1991), employing today's circulation system. Little attention was paid to the pattern of deep-sea circulation, except for progressive O_2 ventilation (Wilde and Berry 1982). Other workers have speculated on ocean circulation within a more regional context, such as for the Iapetus Ocean (e.g., Fortey and Cocks 1992).

Recently, some focus has been placed on the Ordovician coupled ocean-atmosphere system, especially the change from greenhouse to icehouse state in the Late Ordovician. Crowley and Baum (1995) and Gibbs et al. (1997) utilized paleoclimate modeling. A modified approach by Poussart et al. (1999) also generated models of both surface and deep ocean circulation for the Late Ordovician, as well as surface salinities. This last study showed that (a) polar sea ice would occur in the largely landless Northern Hemisphere; (b) there would be a distinct asymmetry in the deep-sea circulation, being much more extensive and vigorous in the Southern Hemisphere (42 percent increase over present day); and (c) an extensive Southern Hemisphere glaciation could be generated with carbon dioxide levels of 8–10 × PAL. The pattern of deep ocean circulation was driven by the subduction of dense saline waters produced in midlatitudes (Railsback et al. 1990).

The nature of the atmosphere is critical to life, being in a partial pressure relationship to the ocean. Berner (1994) argued for carbon dioxide levels in the 14–18 × range through the Ordovician, with O_2 levels in the range of 50 percent PAL (Berner 2001; Berner et al. 2003). Another more potent gas for influencing the greenhouse state is water vapor. During the Ordovician, sea level highstands were the greatest in the Phanerozoic (Frakes et al. 1992; Hallam 1992). Many of the cratons were submerged, most having been extensively peneplaned. Consequently, low-latitude cratons (Laurentia, Siberia, and the Australian and North Chinese parts of Gondwana) were subjected to high evaporation, producing evaporites locally. Extreme evaporation rates, far exceeding today's levels, would have produced an increase in cloudiness with a feedback to reduce incoming insolation. The net effect was a high greenhouse state, an absence of polar icecaps, and a warm climate with wide tropical and warm temperate marine belts.

Such warm climatic conditions led to a more sluggish shallow and deep ocean circulation than that of today, resulting in reduced upwelling and nutrient supply from deep ocean sources and in oceans that were superoligotrophic, or nutrient starved (Martin 1996; but see Wood 1995 for an alternative view). The lower O_2 levels would have tended to promote anoxic conditions in the deep ocean (Sarmiento 1992; Lasaga and Ohmoto 2002), with a profound effect on the marine biota and significantly affecting faunal distribution patterns and provincialism. It is postulated that the low threshold levels of nutrients and oxygen were reached repeatedly and that disruptions in ocean circulation and internal structure had significant effects on the biota.

The greenhouse state and the high carbon dioxide levels influenced the oceans in other ways, with the development of calcitic seas, in which early calcite cementation was enhanced, compared with aragonite (and aragonitic seas) of today. This promoted the expansion of rooted echinoderms, the periodic development of phenomenal pelmatozoan gardens, and the resultant regional pelmatozoan lithofacies (e.g., Middle Ordovician). These shallow, low-latitude epeiric seas had raised salinities through tens of millions of years. Their restricted nature was enhanced by the evolution of microbial/coral/stromatoporoid bioherms that rimmed much of the low-latitude margins (e.g., James et al. 1989; de Freitas and Mayr 1995; Webby 2002). These faunas became modified to survive under such conditions and contributed to the global partitioning of faunas into distinct faunal realms (e.g., McKerrow and Scotese 1990). Only during times of major regression, or cooling, did this partitioning weaken (e.g., in the Ashgill).

Another important factor in Early Paleozoic oceanography is the extent of sea level oscillations. This time interval has long been known for the significant variation in eustasy, including attaining the greatest extent of cratonic submergence in the entire Phanerozoic (Hallam 1992). Full effects of lower-order eustatic oscillations are discussed elsewhere (e.g., Zhang and Barnes 2002; and see chapter 10). Second- to fourth-order changes have been documented (e.g., Fortey 1984; Ross and Ross 1992). In the 100–200 million years (m.y.) after the breakup of the Rodinia and Pannotia supercontinents, most cratons suffered extensive peneplanation; only in parts of Gondwana and some European peri-Gondwanan terranes are there high rates of clastic sedimentation. The causes and variation in eustasy remain as intriguing puzzles, especially the rapidity of change in some intervals. Apart from those in the Late Ordovician–Early Silurian, there is little evidence of a glaciation to provide an explanation. The largest submergence, in the Caradoc, can be explained by a mantle superplume (chapter 8), and this could have also enhanced oceanic productivity.

■ Major Disruptions to Ocean Circulation and Climate Change: Their Effects on the Biotas

Several major oceanographic and climatic events during the Ordovician helped shape the pattern of overall biodiversity, summarized briefly as follows:

1. A global transgression (e.g., Fortey 1984; Ross and Ross 1992; Barnes et al. 1996) occurred near the base of each of five of the Ordovician stages (within the intervals of the time slices *TS*.1a, *TS*.2a, *TS*.4b, *TS*.5a–c, and *TS*.6a; chapter 2). Given the peneplaned nature of many cratons, this produced a series of pulses of invasion and adaption to the extensive shallow-water niches and major area effects on habitats.

2. The most extensive transgression occurred in the Caradoc, the largest of the entire Phanerozoic (Hallam 1992). This allowed maximum communication across the epeiric seas within which most of the earlier emergent domes and arches were submerged. High evaporation rates, internal brine accumulation within intracratonic basins (e.g., Hudson Bay Basin; Le Fèvre et al. 1976), and reduced land area would have affected the specialized biota as well as reducing the amount of chemical and mechanical erosion and hence inorganic nutrient supply.

3. The progressive rifting of the Gondwanan margin in the Late Cambrian–Early Ordovician shed a series of microcontinents that tracked across the Iapetus Ocean, along with Baltica, to amalgamate with volcanic arcs and to collide with or come close to the Laurentian margin at the end of the period (figure 7.1). These changes had the potential to strongly influence oceanic gyres and to create important gateways that could have had important effects on cli-

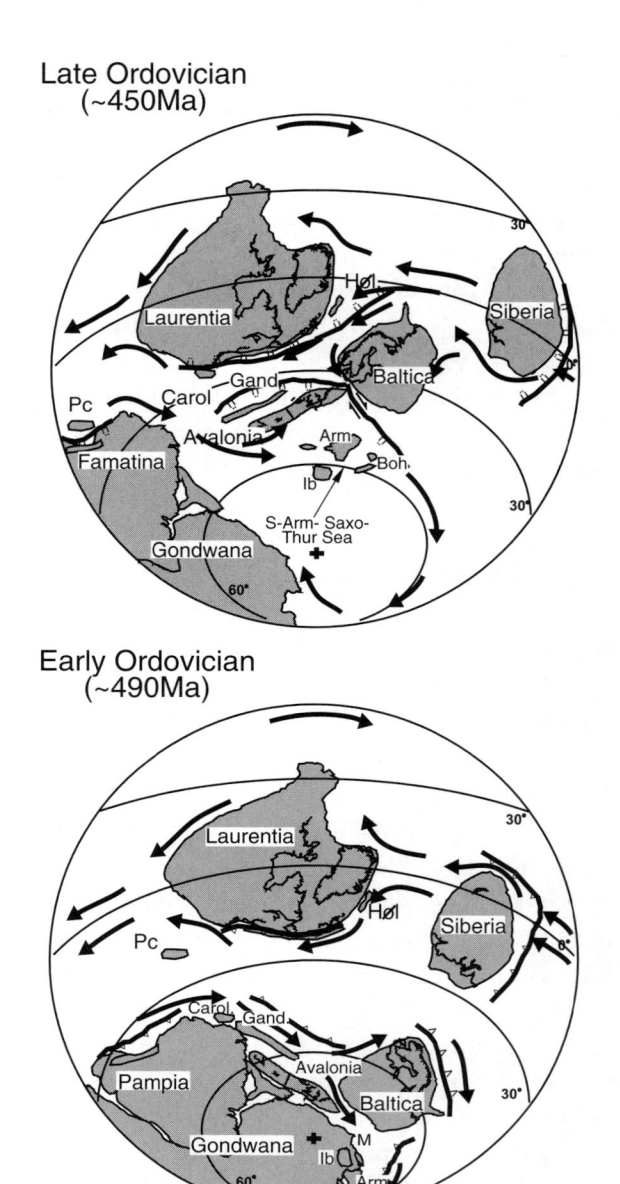

FIGURE 7.1. Paleogeographic reconstructions for the Early and Late Ordovician from Mac Niocaill (2001) showing the complex of terranes/microcontinents within the Iapetus Ocean and the interpreted surface ocean circulation patterns. Abbreviations: Arm = Armorica; Boh = Bohemia; Carol = Carolina Slate Belt; Gand = Gander; Høl = Hølonda; Ib = Iberia; Pc = Argentine Precordillera; S-Arm-Saxo-Thur = Southern Armorica-Saxony-Thuringia Sea.

mate (cf. Cenozoic Drake and Panama gateways). Without more stable and precise paleogeographic reconstructions (cf. Scotese and McKerrow 1991; Mac Niocaill et al. 1997; Scotese 1997; Van Staal et al. 1998; Mac Niocaill 2001), it will be some time before their effects can be established and quantified.

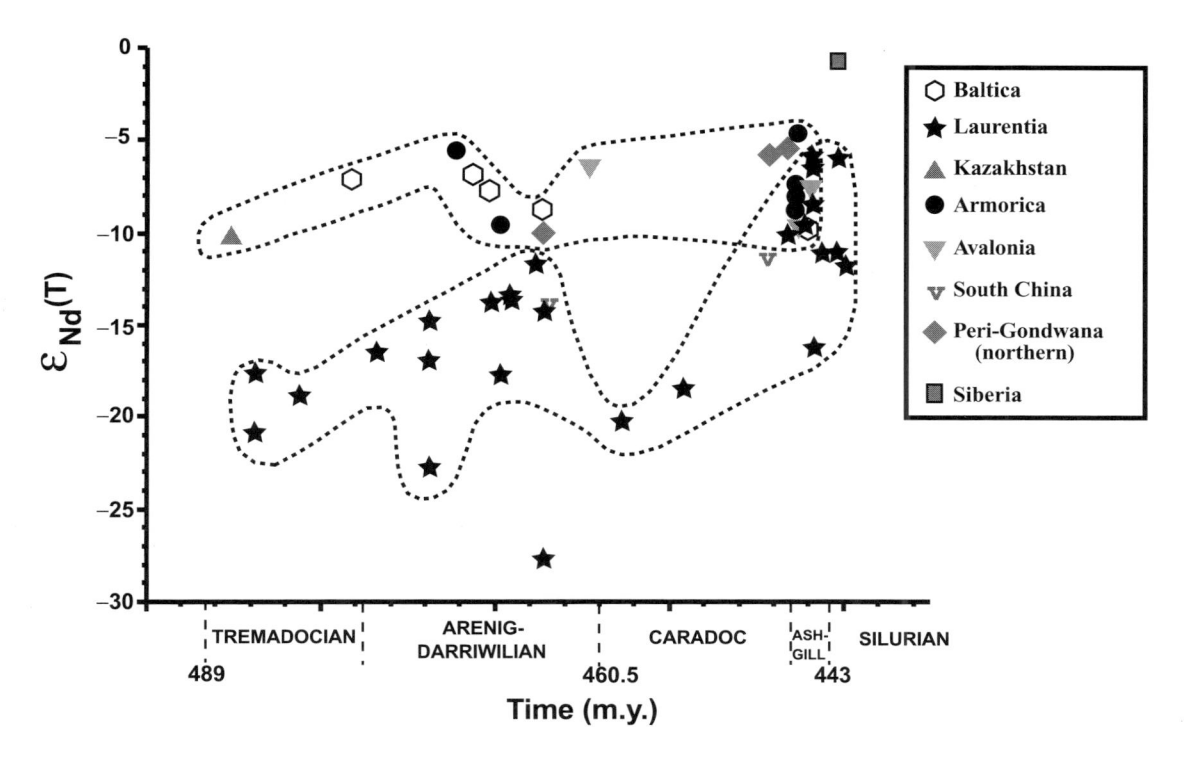

FIGURE 7.2. Summary plot of Nd isotope values from Ordovician conodonts showing, in particular, the isolation of Laurentian epeiric sea waters from those of the Iapetus region until the latest Ordovician (from Wright et al. 2002).

4. The Ordovician mantle superplume (chapter 8) would have had consequences for ocean temperatures, climates, and possibly regional deep oceanic circulation and nutrient supply (cf. Caldeira and Rampino 1991 and Larson 1991 for the Cretaceous).

5. The closure of the arcs and microcontinents against the Laurentian margin (figure 7.1) and the development of the subduction zones were responsible for the Taconic Orogeny. This was one of the largest tectonic events in the period, yet relatively small compared with later orogenic phases in the Paleozoic. Biotic effects remain uncertain, but Miller and Mao (1995) have argued for increased diversity as a consequence of new inshore niches associated with the clastic wedges of the foreland basins. These foreland basins received waters more similar to the marginal Iapetus Ocean than the interior epeiric seas, based on Nd isotope studies (Holmden et al. 1996; Wright et al. 2002) (figure 7.2).

6. The greatest oceanic and climatic disruption came at the end of the period with the glaciation in Gondwana. The major reorganization of oceanic circulation (Poussart et al. 1999; Wright et al. 2002) developed aggressive deep ocean circulation and ventilation, perhaps for the first time in the Ordovician. Carbon dioxide levels to 8–10 × PAL from 16–18 × PAL have been modeled (e.g., Crowley and Baum 1997; Gibbs et al. 1997; Kump et al. 1999; Poussart et al. 1999). The last authors also modeled the ocean surface and deep-water circulation. These changes combined to produce the terminal Ordovician mass extinction (Sheehan 2001b), the second most severe in the entire Phanerozoic.

■ Summary

Ocean and climate states were critical conditions of the Ordovician world that influenced the patterns of biogeography, evolution, and extinction of the marine biota. This world was characterized by a greenhouse climate state; sluggish ocean circulation; most of the landmass located in the Southern Hemisphere; a complex cluster of arcs, microcontinents, and cratons in the Iapetus Ocean; the Delamerian, Taconic, and Benambran orogenies; a probable mantle superplume event; widespread epeiric seas; and a terminal Ordovician Gondwanan glaciation that resulted in the second most severe mass extinction in the Phanerozoic. The pattern of changes in ocean and climate states partly explains the rapid diversification of the Paleozoic Evolutionary Fauna through the Early to Mid Ordovician epochs and its partial collapse at the end of the Ordovician Period.

ACKNOWLEDGMENTS

Continuing research grant support for Lower Paleozoic studies is acknowledged from the Natural Sciences and Engineering Research Council of Canada and the Canadian Institute for Advanced Research (Earth System Evolution Program).

8 Was There an Ordovician Superplume Event?

Christopher R. Barnes

A mantle superplume event occurs when hotter mantle material rises from near the core-mantle boundary as a continuous or discontinuous plume to intersect the base of the lithosphere (Condie 2001). At that point, it spreads out below the colder lithosphere as a mushroom-shaped head. Some penetration of material and associated gases may rise and leak through the lithosphere to produce volcanic activity. Upwelling superplumes, typically arising from the core-mantle boundary ("D" Zone) during periods of anomalous heat flux from the core, are associated with an interval of cessation of magnetic reversals. In recent years, the development of mantle tomography to image the temperature differentials within the mantle has shown two large antipodal superplumes that currently rise under the Pacific and under Africa (Gurnis 2001).

Superplumes are of a different scale than the smaller plumes, or hotspots, such as those in Iceland, the Azores, and Hawaii. These may produce a trail of volcanoes of progressively younger age as oceanic crust drifts over the relatively stable point source, such as the Emperor Chain over the past 40 million years (m.y.) for Hawaii and the trace of the Great Meteor Hotspot that leads through Montreal, the New England seamounts, to the Azores over the past 120 m.y. (Morgan 1981; Legall et al. 1982). Such ancient hotspot tracks imprint a thermal maturation record, and the subsequent thermal uplift leaves an erosional record that can be detected in the ancient stratigraphic record, such as in the Ordovician strata in eastern Canada (Legall et al. 1982; Nowlan and Barnes 1987). Plumes and superplumes may give rise to continental dike swarms and to the rifting and breakup of continents and supercontinents. In other areas, they give rise to massive outpourings of basic volcanics such as the Deccan Traps (latest Cretaceous) as India drifted over the Réunion Hotspot or the present-day Yellowstone Hotspot and associated Paleogene volcanic record in the Columbia Basalts. In extreme cases, such ancient volcanism appears to have coincided with, and possibly caused, mass extinctions such as that at the Permo-Triassic boundary, which has been correlated to the outpouring of the Siberian Traps that have a 251-million-year-old (Ma) assignment (Erwin 1993). For additional citations and a recent summary and detailed specific examples of mantle plumes, see Condie (2001) and Ernst and Buchan (2001, 2003).

The purpose of this chapter is to present evidence that favors the occurrence of a Mid Ordovician mantle superplume event and its possible effect on the global biota. Initially, it is worth noting the strong parallels with the widely accepted superplume event in the Mid Cretaceous. The latter was first proposed by Larson (1991), with additional details outlined by Caldeira and Rampino (1991). To summarize their findings for the purpose of comparison, the main evidence came from data from the Ocean Drilling Project (ODP) in areas such as the Ontong Java Plateau of the northwestern Pacific. Over an elliptical area some 3,000 km across, there is a record of abundant volcanic seamounts of Mid Cretaceous age. Their age span of 80–100 m.y. corresponds to an interval in the Mid Cretaceous when there were limited magnetic reversals (the Mid Cretaceous Superchron). These authors considered that the mushroom head of the superplume resulted in thermal doming of a vast area of the Pacific seafloor and that water was displaced from the ocean basins onto the adjacent marginal or interior seaways as the Albian-Cenomanian Transgression. This is regarded as the second largest transgressive event in the Phanerozoic (Hallam 1992). The thermal effects and the greenhouse gas contribution of emitted volcanic gases may have caused the widespread anoxic oceanic events expressed as black shales. These shales reflect high carbon burial resulting from periods of high organic productivity, which in turn reflect the added inorganic nutrient supply from both atmospheric and submarine volcanic emissions, such as by iron fertilization (Frogner et al. 2001). The Mid Cretaceous superplume event appears to have many parallels to the evidence, outlined in the next section, from the Ordovician.

■ Evidence for a Mid Ordovician Superplume

The concept of a Mid Ordovician mantle superplume was advanced provisionally by Barnes et al. (1996), Qing et al. (1998), and Condie (2001). The principal lines of evidence are detailed briefly in this chapter; none alone is confirmatory, but together they make a permissive case for a superplume event in the Mid Ordovician.

Lack of Magnetic Reversals

Johnson et al. (1995) interpreted all the known magnetic reversal data for the entire Phanerozoic. Although the data are not uniformly distributed, there were four intervals during which the earth's magnetic field showed limited reversal activity. One of these was in the Ordovician (figure 8.1) and was more pronounced than that for the Mid Cretaceous.

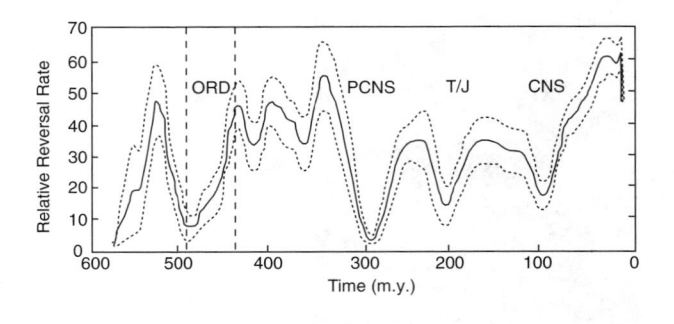

FIGURE 8.1. A plot of relative reversal rate (RRR; inner heavy line) calculated as a function of age for the Phanerozoic Global Paleomagnetic Database (from Johnson et al. 1995: figure 3), showing the lack of reversals within the Ordovician. Abbreviations: ORD = Ordovician; PCNS = Permo-Carboniferous Normal Superchron; T/J = Triassic-Jurassic; CNS = Cretaceous Normal Superchron.

Extensive Volcanism

Fischer (1984) showed that the Ordovician and Cretaceous were the two intervals with the most abundant record of volcanism in the Phanerozoic. A review of Ordovician volcanism was provided by Stillman (1984). It is clear that the Iapetus Ocean was a center of volcanic activity through the Ordovician. Rogers and Van Staal (in press) report gigantic mid Arenig (466 Ma) eruptions in the Popelogan Arc (now in New Brunswick). Volcanism was also expressed as widespread bentonites in the Mid through early Late Ordovician (time slice TS.3a–5c). Huff et al. (1992) estimated, based on a particular paleogeographic reconstruction, that the Deike-Millbrig bentonites (454 Ma) in eastern Laurentia and their equivalents in Baltica must have resulted from the ejection of about 1,000 km^3 of material—the largest eruptions known in the Phanerozoic. The record of bentonites is widespread in this interval in these areas as well as in the Argentine Precordillera, which was tracking from Laurentia across the Iapetus Ocean at this time (Astini et al. 1995). The progressive thickening of the bentonites to the southeast in eastern Laurentia is toward the present-day Carolinas. This may suggest that the primary source was the Carolina Slate Belt volcanic arc that was within the Iapetus (see chapter 7: figure 7.1). The complexity of volcanic arcs and peri-Laurentian terranes that existed is only recently beginning to be understood (Mac Niocaill et al. 1997; Van Staal et al. 1998; Mac Niocaill 2001).

However, it is clear from their detailed geochemistry that many of the volcanics are related to subduction and arc processes and tend to be dominantly felsic rather than basaltic, and it is not yet possible to assess if any of these remarkable volcanics may have been generated by a superplume. Other proxies of a superplume, such as giant dyke swarms, oceanic plateaus, komatiites, and flood basalts, have yet to be identified, but some may no longer be preserved.

Major Transgression

The Middle Ordovician in characterized by several major transgressions (McKerrow 1979; Fortey 1984; Ross and Ross 1992; Barnes et al. 1996), especially that of the basal Caradoc, regarded (Hallam 1992) as the most extensive in the entire Phanerozoic. As noted by Barnes (chapter 7), many of the inundated cratons were already extensively beveled by erosion, and so the actual amount of sea level increase may have been just 100–200 m, especially given that the terminal Ordovician eustatic sea level fall that caused all these epeiric seas to be drained was of this order of magnitude.

Strontium Isotope Excursion

Another indicator of enhanced mantle interaction with the hydrosphere is the major strontium isotope excursion to less radiogenic values at the Darriwilian-Caradoc boundary interval (c. 460 Ma). Qing et al. (1998) used conodont and brachiopod data to produce a refined strontium isotope curve for the Ordovician and Silurian. The major excursion, from values of 0.709 to 0.708, occurred over a short interval near the Darriwilian-Caradoc boundary. This result has been verified and refined by Shields and Veizer (chapter 6). This is equivalent to the total range of the strontium excursion through the entire Cenozoic. This shift could have been affected by the contribution of submarine mafic volcanism or a period of fast seafloor spreading and by the shutdown of the cratonic strontium contribution by the marine submergence. This event is unlikely to be influenced by the impact from the peak clastic contribution from the Taconic Orogeny, which occurs later in the Late Ordovician, whereas the timing, brevity, and scale of the excursion correlate closely with the basal Caradocian transgression.

Extensive Black Shales

The Darriwilian and Caradoc are well known as intervals of extensive black shales, indicative of perhaps both widespread marine anoxia and high organic productivity (Leggett 1980).

■ Implications for Marine Biota and the Rise of the Paleozoic Evolutionary Fauna

In considering other superplume events, Caldeira and Rampino (1991) and Condie (2001) have discussed their possible effects on the biota. The principal effect is related to the scale of environmental perturbation produced by the thermal increase and the gaseous, mineral, and trace element emanations. These emissions can be interpreted as beneficial to at least part of the food chain components by adding nutrients (e.g., iron, phosphorous), but other extreme loadings could be toxic to life and cause extinctions or mass extinctions (cf. end Permian; Erwin 1993). In the case of the Mid Ordovician superplume, the timing is more related to the rise of the Paleozoic Evolutionary Fauna and is too early to be tied to the end Ordovician mass extinction.

The effects of a superplume would likely span a few million years: from the initial rise, to the spreading mushroom head, to the leakage of volcanic and gaseous emissions. If the basal Caradoc transgression and the sharp excursion in the strontium isotope values are taken as a late-stage expression, the initial superplume-lithosphere interactions and early volcanism may be expected to have occurred within the mid Arenig to Darriwilian. It is in this very interval that the rise of the Paleozoic Evolutionary Fauna occurs, and its rise may have been supported, if not triggered, by the increased nutrients and warming generated by the superplume. The enhanced organic productivity may be expressed by the widespread black shale occurrences of Darriwilian and Caradoc age.

■ Summary

Evidence is assembled that together favors the presence of a mantle superplume event in the Mid Ordovician. There is a period of reduced magnetic reversal, a major transgression, strong greenhouse effect with high carbon dioxide levels, extensive volcanism but not felsic volcanism, widespread black shales, and a strontium isotope excursion reflecting increased mantle input. The strontium excursion is of short duration and occurs close to the Darriwilian-Caradoc boundary. The effects of the superplume could be expected to occur over several million years, starting perhaps from the early Darriwilian. This interval corresponds to that of the rise of the Paleozoic Evolutionary Fauna, whose success may have been assisted by a higher nutrient flux and increased niche area and diversity for the marine biota within the expanded epeiric seas.

ACKNOWLEDGMENTS

Continuing research grant support for Lower Paleozoic studies is acknowledged from the Natural Sciences and Engineering Research Council of Canada and the Canadian Institute for Advanced Research (Earth System Evolution Program).

9 End Ordovician Glaciation

Patrick J. Brenchley

The link between the extensive end Ordovician glaciation in North Africa and contemporaneous glacio-eustatic and faunal changes was first recognized by Berry and Boucot (1973). Subsequently it has been shown that in the two phases of extinction, which together constitute the mass extinction, an estimated 85 percent of species and 61 percent of genera disappeared (Jablonski 1991), drastically lowering the high levels of diversity that had been attained during the "Great Ordovician Biodiversification." During the first phase of extinction, at the start of the Hirnantian (uppermost stage of the Ashgill), a large fall in diversity occurred in most groups. However, a few groups, such as the conodonts, mainly survived the first phase of extinction but then suffered large losses in the second phase, in the mid to late Hirnantian. Other groups that had lost many species in the first phase suffered further losses in the second (Sheehan 2001b).

The very close correlation between the first phase of extinction and the onset of glaciation and between the second phase and waning glaciation strongly suggests that the associated environmental changes were major factors in causing the extinctions. The Late Ordovician stable isotope record shows a major positive excursion in both carbon and oxygen, which started in the earliest Hirnantian, rose to a peak followed by a plateau of values, and then fell to original values in the mid Hirnantian (figure 9.1). The positive shift of 3–4 percent in $\delta^{18}O$ values suggests a large growth in the volume of continental ice and a fall in seawater temperatures by possibly as much as 10°C (Brenchley et al. 1994). The synchronous positive excursion of $\delta^{13}C$ values by as much as 7‰ (per mil) indicates large changes in carbon cycling that are still poorly understood; they might be a result of changes in ocean structure and circulation, affecting levels of nutrients (Brenchley et al. 1995), or they might be related to changes in the flux of carbonates into the oceans (Kump et al. 1999). Synchronous with the start of the isotope excursions was the onset of a global fall of sea level of 50–100 m followed by a rise to previous levels in the mid Hirnantian (figure 9.1) (Brenchley et al. 1995).

Late Ordovician glacial deposits are now known to be widely spread across Gondwana (figure 9.2) (Hambrey 1985). They occur widely across North and West Africa and in Saudi Arabia, South Africa, and Argentina, where they are one of many glacial deposits that occur in a discontinuous belt of Lower Paleozoic rocks extending from Argentina to Ecuador (Astini 1999b). In peripheral areas of Gondwana largely glaciomarine deposits are recorded at scattered locations from Portugal in western Europe to the Czech Republic in the east (figure 9.2), (Robardet and Doré 1988; Hambrey 1985). The area of glacial activity in North and West Africa has been estimated to be $6–8 \times 10^6$ km (Hambrey 1985), while a later

FIGURE 9.1. Environmental changes associated with the two phases of the Late Ordovician mass extinction. Modified from Brenchley et al. (2001).

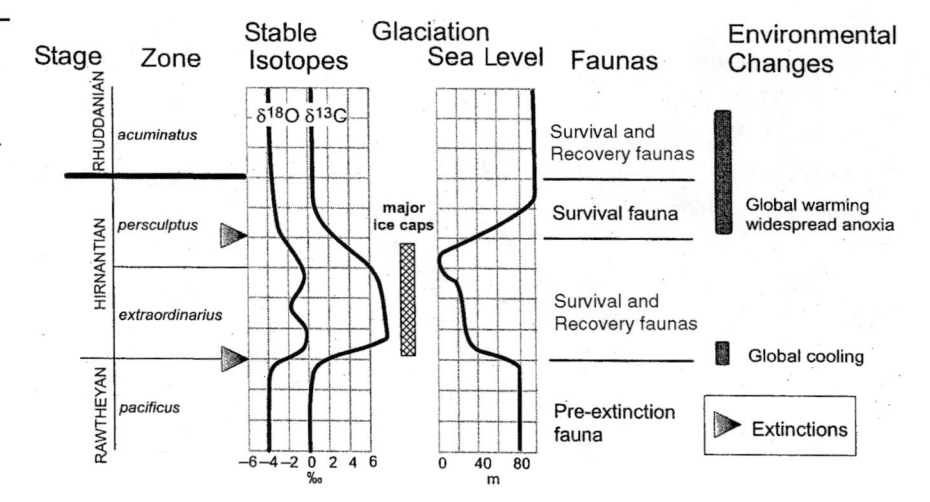

estimate of the area of the total ice cap was 11.8 × 10^6 km, which would be approximately the area of the Pleistocene Laurentide ice cap or the present East Antarctic ice sheet (Crowley and Baum 1991). Both suggestions probably considerably underestimate the size of the ice cap at its maximum extent (figure 9.2) by omitting the glacial deposits of Saudi Arabia, western South America, and South Africa. A larger area of continental ice would be consistent with the magnitude of the isotopic and bathymetric changes.

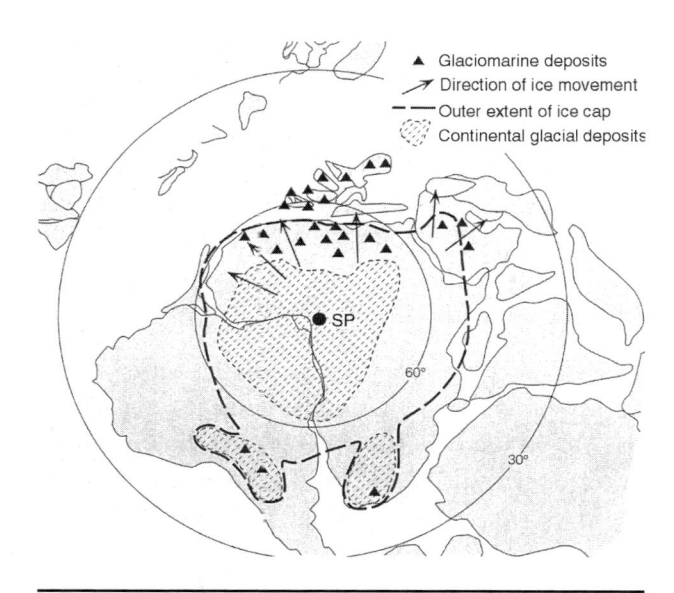

FIGURE 9.2. Paleogeographic reconstruction of Late Ordovician Gondwana, showing the interpreted position of the maximum extent of the South Polar ice cap (dashed line). Arrows show the direction of ice movement, after Beuf et al. (1971) and Vaslet (1990), and triangles show occurrences of glaciomarine diamictites. Modified from Paris et al. (1995).

The broader pattern of Late Ordovician climatic change is not closely constrained, though according to Frakes et al. (1992) a cool period started in the Caradoc and ended in the late Early Silurian, an interval of about 35 million years (m.y.). Oxygen isotope values that rose throughout the Ordovician to a peak in the latest Ordovician give some support to cooling in the Late Ordovician, though a substantial part of the overall rise is probably related to a change in seawater composition (Veizer et al. 1999). The time span of the glaciation itself is more controversial. In North Africa the early work (Beuf et al. 1971; Hambrey 1985) and more recently Hamoumi (1999) record the range of the glaciation as being from Caradoc to the early Silurian. However, those marginal Gondwanan marine diamictites that have been dated are demonstrably Hirnantian, and many of the widespread glacial deposits in North Africa are now also recognized as being Hirnantian on the evidence of the chitinozoan biostratigraphy (Paris et al. 1995) or the presence of the *Hirnantia* fauna (Legrand 1995). The controversy concerning the timing of the start of the glaciation centers most acutely on the sequence in Morocco that was interpreted by Hamoumi (1999) as being wholly glacially influenced; on the basis of a different sedimentologic interpretation of the critical sequences, Paris et al. (1995) and Sutcliffe et al. (2000) regard the glacial deposits as being wholly confined to the Hirnantian. The two disparate views are unresolved, but the wide distribution of Hirnantian glacial deposits, the contemporaneous large oxygen isotope excursion, and glacio-eustatic sea level changes all suggest that the main episode of glaciation was confined to a period in the early to mid Hirnantian, estimated to be between 0.5 and 1 m.y. (Brenchley et al. 1994).

The first phase of extinction occurred in a short interval at the onset of glaciation when, according to isotopic and sedimentologic evidence, there were substantial changes in the structure and circulation of the oceans and a rapid fall in both sea level and seawater temperatures associated with a contraction of climatic belts. These environmental changes would have affected a biota largely adapted to warmer, more stable conditions and appear to have caused higher levels of extinction among endemic faunas than cosmopolitan ones, resulting in a lower number of faunal provinces (Sheehan and Coorough 1990). The second phase of the extinction was associated with a different set of environmental changes. Marine temperatures were rising, sea level was falling, and regional anoxia was developing (Owen and Robertson 1995). Many of the taxa that survived the environmental changes in the first phase succumbed to the different changes in the second phase; the *Hirnantia* fauna that had flourished during the glacial interval also suffered substantial losses in this second phase (Owen and Robertson 1995).

All the kinds of environmental change that are coincident with the Late Ordovician mass extinction are a common part of geologic history (Brenchley and Harper 1998). Mostly they have not caused mass extinctions, but in other instances one or more of the environmental factors have been identified as the cause of a major extinction. It appears likely that temperature changes, allied with other coeval environmental effects, played an important part in the Late Ordovician mass extinction. It is tempting to see the particularly high rate of environmental change as being the one special aspect of the Late Ordovician that precipitated this mass extinction.

10 Ordovician Sea Level Changes: A Baltoscandian Perspective

Arne Thorshøj Nielsen

A detailed knowledge of sea level changes is likely to greatly assist the understanding of biodiversity patterns, adaptive radiations, and related topics because sea level exerts a strong first-order control on shifts in the marine environment. However, relatively little attention has been devoted to the reconstruction of Ordovician sea level changes. In order to improve this situation, an analysis of Baltoscandian successions was undertaken (full documentation, including interpretations of all age relationships and correlations is in preparation, Nielsen unpubl.). The scope of this chapter is to summarize this study and to make comparisons with the North American sea level curve documented by Ross and Ross (1992, 1995).

■ Depositional Model

The supply of clastics to the shallow sea covering most of Baltoscandia was extremely limited during the Ordovician, and carbonate production was slow owing to cold-water conditions. Average net accumulation rates were of the order of 1-9 mm/1,000 years (Sweden, East Baltic). Even in the distal foreland of the Caledonides (Oslo area), average accumulation rates were only 3–12 mm/1,000 years until the Ashgill, when they increased to 32 mm/1,000 years. Hence deposition did not, in general, affect the local sea level. Since the area was also largely tectonically inactive, the sea level changes recorded in the successions primarily reflect eustasy.

The earliest Ordovician was characterized by deposition of black, organic-rich shales. This facies was more widespread than in later periods as a result of low oxygen conditions. Sandstones were deposited nearshore. Deposition of condensed, "cool-water" limestones commenced in the late Tremadocian and characterized the middle shelf and then spread to the inner shelf by the early Mid Ordovician. The general offshore-nearshore lithofacies distribution across Sweden and Estonia is (1) dark graptolitic shales; (2) calcilutitic limestones; (3) calcarenitic limestones; (4) argillaceous limestones; (5) marls, mudstones, and siltstones; and (6) carbonate-rich quartz sands. The facies boundaries are gradational, and the seafloor is interpreted as extremely flat. "Warm-water" carbonates locally developed in facies 5–6 during the Late Ordovician, at which time Baltica probably had moved into the subtropics. However, depth of deposition likely also inhibited reef development. The three phases of carbonate mound growth in the mid Caradoc, late Caradoc, and Ashgill are each associated with lowstands.

The sedimentary pattern was almost identical across the Oslo-Scania-Bornholm area (figure 10.1) until the early Llanvirn, with uniform mudstone-dominated deep-shelf successions. From then on, sediment was

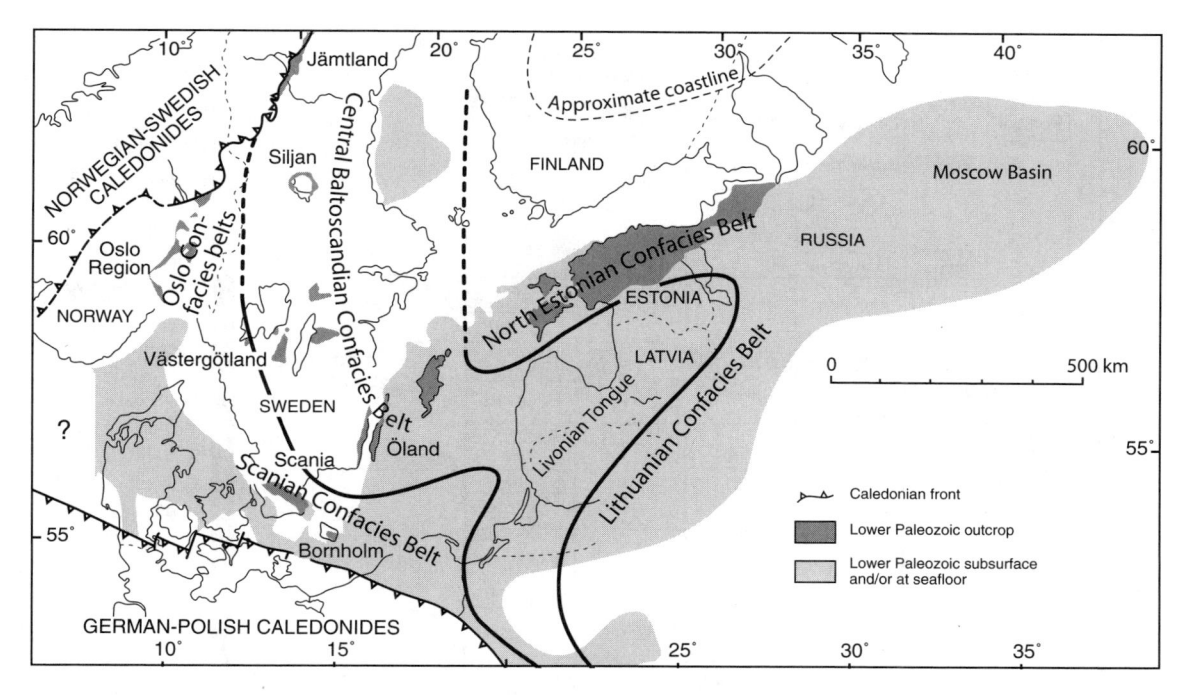

FIGURE 10.1. Generalized facies belts of Baltoscandia during the Mid and Late Ordovician; modified from Jaanusson (1995) and Nielsen (1995). Present-day distribution of Lower Palaeozoic rocks in Baltoscandia is also shown; originally these rocks covered most of the region. Names of districts and countries referred to are indicated.

transported from the collision zones within the Iapetus Ocean reaching western Baltica, so the Norwegian late Llanvirn to Ashgill successions were relatively expanded. The offshore to nearshore distribution of lithofacies in the Oslo region is (1) graptolitic dark to black mudstones, (2) dark to gray mudstones with thin, storm-generated sand beds, (3) nodular limestones intercalated with gray silty mudstone, followed shoreward by lithofacies similar to (5) and (6) mentioned earlier but thicker. Locally "warm-water" carbonates developed in shallow-water facies during the Late Ordovician. Depositional depth in the Ordovician was lower in the western districts of the Oslo region than in the Oslo-Asker district.

The present lithofacies interpretation differs from earlier analyses (Jaanusson 1973; Nestor and Einasto 1997; Dronov and Holmer 1999) by interpreting argillaceous limestones as relatively shallow-water lowstand deposits rather than deep-water facies transitional to mudstone/shale. The "dirty" carbonates reflect a marked increase in clastic supply during lowstands.

There has been substantial debate about depth of deposition of the limestones. This new study indicates that even the most nearshore part of the carbonate ramp now preserved (northwestern Estonia; figure 10.1) emerged only during the major lowstands, implying that much of the limestone deposition occurred at substantial depth, maybe even down to about 200 m.

Sea Level Changes

Lower Ordovician

The Alum Shale of the Upper Cambrian trilobite-based *Acerocare* Zone is preserved only in marginal areas of Scandinavia because of a low sea level ("Late Cambrian Lowstand Interval," culminating in the Acerocare Regressive Event; see figure 10.2). A relatively increased supply of sediment resulted in deposition of thin sand packages in nearshore areas (e.g., lower part of Kallavere Formation, Estonia).

Levels with high concentrations of graptolites (maximum drowning surfaces) indicate three or four flooding events in the lower part of the Tremadocian Alum Shale. The upper part of the sandy Kallavere Formation (Estonia) seems to have been deposited

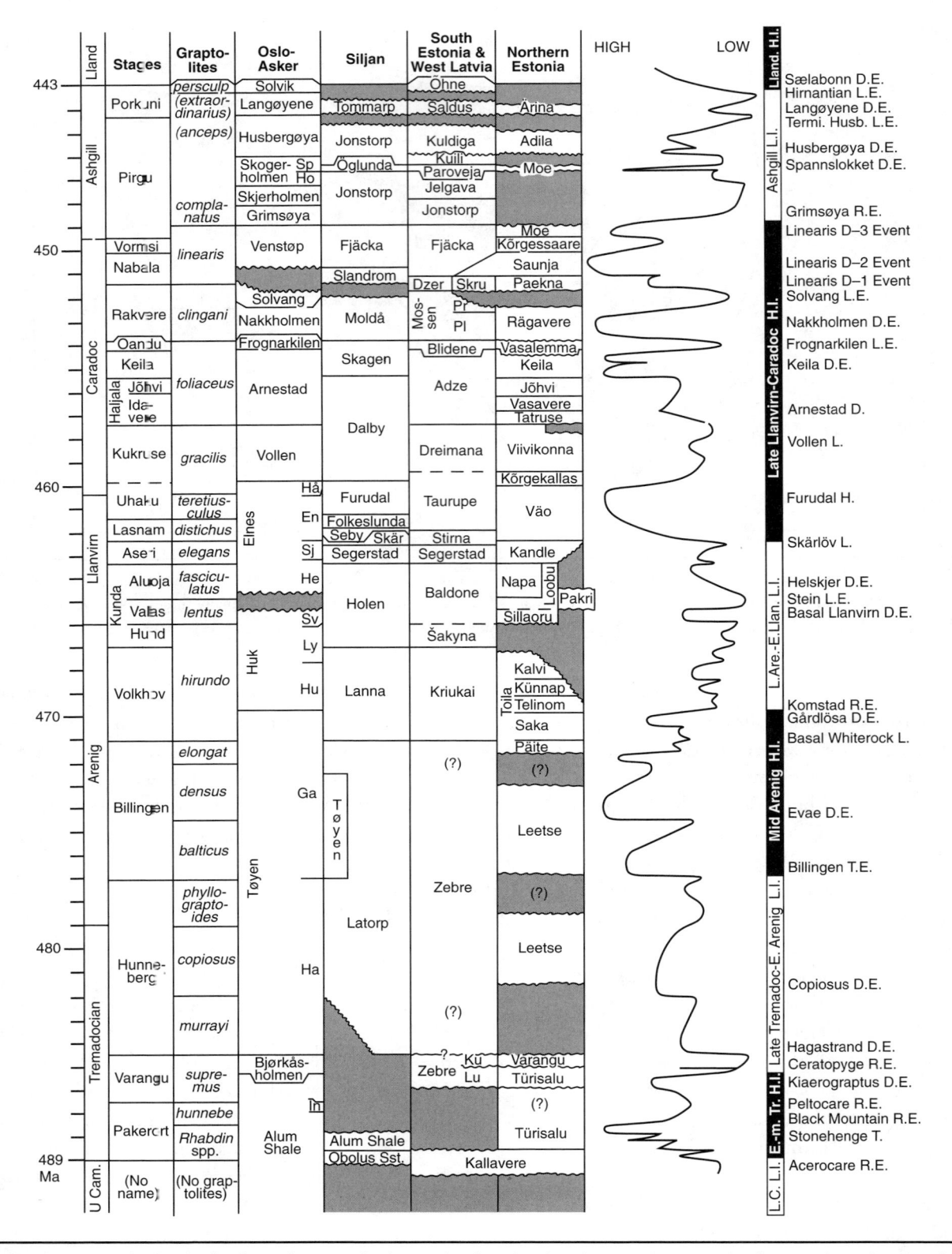

FIGURE 10.2. Sea level curve for the Ordovician of Baltoscandia plotted within the general stratigraphic framework employed in this volume, except that the Oandu stage is correlated with the *Diplograptus foliaceus* Zone (simplified from Nielsen unpubl.). The successions represent a depth transect across the shelf from the Oslo-Asker area (Norway) in the west to Estonia in the east (figure 10.1). The vertical columns show Baltoscandian stages, graptolite zones, and four generalized stratigraphic columns across Norway, Sweden, Estonia, and Latvia. The sea level changes were probably within the band of glacio-eustasy, that is, ranging through 250 m from high-

during the initial sea level rises, but the drowning eventually resulted in a spread of the Alum Shale facies into the East Baltic area (Türisalu Formation), notably during *Rhabdinopora socialis* time. Then followed a shallowing at the base of the graptolite-based *Adelograptus hunnebergensis* Zone, inferred mainly from paleontological evidence and equated with the Black Mountain Regressive Event (sensu Nicoll et al. 1992; Cooper 1999a). The sea level continued to fall through the *A. hunnebergensis* Zone to a minimum during the Peltocare Regressive Event (Erdtmann 1986; Nicoll et al. 1992).

A moderate sea level rise followed, accelerating within the uppermost part of the Alum Shale (Kiaerograptus Drowning Event), where graptolites again are concentrated in a thin interval of the Oslo area and elsewhere. However, the sandy lithology of the upper Türisalu Formation of Estonia points to shallower conditions than earlier in the Tremadocian.

The conspicuous Ceratopyge Regressive Event (CRE of Erdtmann 1986) terminated Alum Shale deposition, and limestones then spread across western Baltoscandia. The lowstand was locally associated with severe erosion of older strata. The CRE probably comprises two or maybe three smaller shallowing/drowning cycles. One of these is a remarkably short-lived and rapid drowning event (see figure 10.2), represented by a thin olenid trilobite marker horizon just below the main limestone of the Bjørkåsholmen Formation. The CRE concluded with the rapid Hagastrand Drowning Event.

In the Oslo area, the lower Tøyen Formation includes light-colored, bioturbated mudstone with storm sand beds (Hagastrand Member). This lithology suggests a moderately low sea level through the Hunneberg stage. Graptolitic shales (Galgeberg Member) then mark two prominent sea level rises (Billingen Transgressive Event and Evae Drowning Event; see figure 10.2). These events caused the spread of graptolitic mudstones into the central Swedish facies belt.

Sea level again abruptly lowered in the late part of the Billingen stage, at which time limestone deposition commenced in the East Baltic area. In addition, the base of the Volkhov is marked by a pronounced unconformity, suggesting another lowstand event. The composite late Billingen–early Volkhov shallowing (the Basal Whiterock Lowstand) is also recognizable as a light-colored interval of the Tøyen Shale in Oslo and Scandia.

Middle Ordovician

The sequence of events in the lower Middle Ordovician is detailed by Nielsen (1992b, 1995). The Basal Whiterock Lowstand is terminated by the Gårdlösa Drowning Event (figure 10.2). Shortly afterward the sea level dropped significantly again (the Komstad Regressive Event), with limestone facies spreading westward across Scandinavia. Concurrently, argillaceous limestone was deposited in Estonia and Russia. Overall, the sea level lowered through the Volkhov but was punctuated by moderate deepenings, revealed mainly by faunal data and corroborated by lithologic changes in the Oslo area and Öland. The base of the Kunda stage marks a major lowstand, with limestones becoming argillaceous over most of Sweden. An even more prominent lowstand marks the base of the trilobite-based *Asaphus raniceps* Zone (mid Kunda stage). The entire early Kunda is represented by a major unconformity in northwestern Estonia.

In the Oslo region (Svartodden Limestone) as well as parts of Sweden, a drowning in the lower part of

---

FIGURE 10.2. (*Continued*)
stand to lowstand extremes. Details of the abbreviated terminology for highstand (H.I.) and lowstand (L.I.) intervals, shown in the black and white column, respectively, to the right side of the figure, are as follows: L.C.L.I. = "Late Cambrian Lowstand Interval"; E.-m Tr.H.I. = "Early to Mid Tremadocian Highstand Interval"; Late Tremadoc–E. Arenig L.I. = "Late Tremadocian–Early Arenig Lowstand Interval"; L.Are.–E.Llan.L.I. = Late Arenig–Early Llanvirn Lowstand Interval"; Lland.H.I. = "Llandovery Highstand Interval." Other abbreviations: D. = Drowning; D.E. = Drowning Event; H. = Highstand; L. = Lowstand; L.E. = Lowstand Event; T. = Transgression; T.E. = Transgressive Event; R.E. = Regressive Event; Termi. Husb, L.E: = Terminal Husbergøya Lowstand Event. The key to various lithologic units identified only by their first two or more letters is as follows: Oslo-Asker: In = Incipiens Limestone; Ha = Hagastrand Member; Ga = Galgeberg Member; Hu = Hukodden Member; Ly = Lysaker Member; Sv = Svartodden Member; He = Helskjer Member; Sj = Sjøstrand Member; En = Engervik Member; Hå = Håkavik Member; Ho = Hovedøya Member; Sp = Spannslokket Member. Sweden: Skär = Skärlöv Limestone. Estonia: Lu = Lutrini Member; Ku = Kumbri Member; Pl = Plunge Member; Pr = Priekul Member; Dzer = Dzerbene Formation; Skru = Skrunda Formation; Telinom = Telinômme Member; Künnap = Künnapôhja Member.

the *A. raniceps* Zone (Basal Llanvirn Drowning Event) was associated with return to purer limestone deposition, while in western Scania, graptolitic shales succeed the Komstad Limestone. Shortly afterward the sea level dropped again, producing the Stein Lowstand Event, and limestone deposition spread far to the west into the Caledonides. The calcareous Pakri Sandstone of northwestern Estonia represents a lowstand wedge that formed near the coast, though the upper part of this formation seems to have been redeposited during the following transgression.

Sea level gradually rose in the late Kunda (Helskjer Drowning) with, eventually, black shale deposition resuming in the Oslo area; in Sweden this rise is reflected by a decreased grain size in the upper part of the Holen Limestone and change to a red color. At this stage a foreland basin developed in the Oslo area, amplifying the drowning. The Oslo succession exhibits a short-lived, gradual shallowing (Skärlöv Lowstand) with appearance of sandy storm beds and a change to lighter gray colors in the middle of the Sjøstrand Member (Elnes Formation); then higher in the Sjøstrand the color changes back to dark gray and black, suggesting deepening again. The lower part of the overlying Engervik Member shows a similar lithology. This thick black shale interval represents a highstand during the *distichus* and *teretiusculus* graptolite zones.

The same late Llanvirn sea level events are recognizable in the limestone successions of Sweden and Estonia. The upper part of the Segerstad Limestone becomes coarser grained and is succeeded by the impure Skärlöv Limestone, this latter marking the Skärlöv Lowstand. The following gradual deepening is recorded by the sequence of Seby (variegated calcarenites), Folkeslunda (mainly gray calcilutites), and Furudal (poorly fossiliferous calcilutites) limestones. Maximum drowning is referred to as the Furudal Highstand (figure 10.2).

In southern and central Estonia the Stirna Formation (argillaceous limestone) is considered to reflect the Skärlöv Lowstand. The overlying Taurupe Formation (mainly calcilutitic limestones) is interpreted as a highstand deposit. In North Estonia, an oolitic level in the upper part of the Kandle Formation (Hints 1997:68) is indicative of the Skärlöv Lowstand, while the overlying Väo Formation (calcarenites) records the rising sea level, leading to the Furudal Highstand (figure 10.2).

Upper Ordovician

The base of the graptolite-based *Nemagraptus gracilis* Zone correlates with the latter part of the Furudal Highstand, but most of the zone falls within the succeeding Vollen Lowstand. In the Oslo area the shaley Elnes Formation grades into the overlying lime-rich Vollen Formation. In Sweden the Furudal Limestone shows a gradual transition to more coarse-grained Dalby Limestone, and in Estonia the Kôrgekallas Formation (argillaceous calcarenitic limestones and marls) is transitional between the Väo Formation and Kukruse oil shales (Viivikonna Formation). The latter records significantly increased clastic influx (Nestor and Einasto 1997). These changes undoubtedly indicate a gradual major shallowing (the Vollen Lowstand; figure 10.2), and the prominent unconformity in northern Estonia above the Viivikonna Formation possibly reflects the lowstand climax.

Above the Vollen Formation (Oslo area) the less lime-rich Arnestad Formation reflects a deepening (Arnestad Drowning). In southern Estonia this drowning is recorded by the Adze Formation (calcarenitic limestones and marls). However, the impure limestones of the Jõhvi Formation in northern Estonia suggest that sea level rise was limited prior to the deposition of the Keila Formation (mainly calcilutites). This major Keila Drowning Event (figure 10.2) is also recorded by the fine-grained lower part of the Skagen Limestone (Sweden).

In the Oslo area the Arnestad Formation is followed by the limestone-rich Frognarkilen Formation, indicating a short-lived and conspicuous lowstand (Frognarkilen Lowstand Event) in the latest part of the graptolite-based *Diplograptus foliaceus* Zone. In the Siljan Region the lowstand event is signaled by the impure upper part of the Skagen Limestone and local biohermal Kullsberg Limestone. In the central and southern Estonian successions the Frognarkilen shallowing is also conspicuous, given the significant increases in clastics (Blidene and Variku formations). In northern Estonia the initial phase of the shallowing is recorded in the marly upper part of the Keila Formation. This is followed by the Vasalemma Formation with numerous bioherms. Locally, in other places, a prominent unconformity is present at this level.

A rapid and profound drowning event in the graptolite-based *Dicranograptus clingani* Zone is recorded by the Nakkholmen Formation of the Oslo area (pyritic graptolite shales). Coeval highstand deposits are the fine-grained Moldå Limestone of the Siljan area, graptolitic shale in the lower Mossen Formation of Västergötland, southern Estonia, and Latvia (= Plunge Member of Ainsaar and Meidla 2001), and the calcilutitic Rägavere Formation in central and northern Estonia (figure 10.2).

The gradual transition from the Nakkholmen Formation to the lime-rich Solvang Formation (Oslo area) indicates a relatively slow lowering of sea level that culminated in the significant Solvang Lowstand Event of intra-*clingani* zonal age. This major event was succeeded by a drowning that commenced in the latest *D. clingani* Zone and then accelerated rapidly into the graptolite-based *Pleurograptus linearis* Zone. A phosphorite marker level at the base of the overlying Venstøp Formation (Williams and Bruton 1983; Owen et al. 1990) is evidence of a pause in sedimentation. This maximum drowning surface broadly correlates with the early part of the *P. linearis* Zone but in most places also represents the latest part of the *D. clingani* Zone. The Venstøp Formation (black shales) is correlated with the middle to upper part of the *P. linearis* Zone (Williams and Bruton 1983).

There is little evidence for the Solvang Lowstand in the Siljan area, and an unconformity is inferred between the Moldå Formation and the Slandrom Limestone (figure 10.2). Clear indications of an unconformity exist between the Mossen Mudstone and the Bestorp Limestone of Västergötland. The *P. linearis* Zone is represented by the Slandrom Limestone (calcilutite) and overlying Fjäcka Shale (partly graptolitic) in the Siljan and Jämtland districts. This succession suggests that the drowning occurred in at least two steps (Linearis Drowning–1 and –2 events), with the latter being the more significant event. The dark, fine-grained Bestorp Limestone followed by Fjäcka Shale in Västergötland records the same succession of events.

In the central East Baltic area, the Priekul Member (greenish gray argillaceous marls with thin, intercalated calcarenitic limestones) of the upper Mossen Formation provides evidence of the shallowing culminating in the Solvang Lowstand Event. The overlying Dzerbene Formation (light-gray calcilutites) and following Fjäcka Shale of Latvia exhibit the same two phases of drowning (Linearis Drowning–1 and –2 events) as in Sweden.

A hiatus is inferred in northern and central Estonia as equivalent to the Solvang Lowstand (figure 10.2). The overlying, laterally equivalent Skrunda, Môntu, and Paekna formations are moderately shallow-water calcarenitic units that reflect the Linearis Drowning–1 Event. The Môntu and Paekna formations are succeeded by the Saunja Formation (calcilutites), and this latter is correlated with the Linearis Drowning–2 Event. The overlying Kôrgessaare Formation (calcarenitic limestones) and correlative Tudulinna Formation (argillaceous calcarenites and marls) exhibit evidence of the shallowing that terminated the Linearis Drowning–2 Event. The following Moe (the lower part) and laterally equivalent Tootsi formations (calcarenites and calcilutites) indicate the presence of a Linearis Drowning–3 Event. The Linearis Drowning–2 and –3 events are also discernible in the deep-water Fjäcka Shale and Venstøp Formation of Sweden and Norway.

The prominent Linearis highstands were followed by a significant shallowing (Grimsøya Regressive Event) that approximates the base of the graptolite-based *Dicellograptus complanatus* Zone (figure 10.2). The lower part of the Grimsøya Formation (Oslo area) is rich in nodular limestone, whereas the upper part consists of alternating bedded limestone/shale and siltstones indicating increased clastic supply. Shallowing continued into the Skjerholmen Formation with channeling and crossbedded sandstones reported from its upper half (Owen et al. 1990). The overlying Skogerholmen Formation has abundant limestones and thin, silty storm layers indicating slightly deeper depositional conditions.

Condensed, shale-dominated intercalations in the middle to late Ashgill of the Oslo area record three successive sea level rises, named the Spannslokket, Husbergøya, and Langøyene drowning events. The first two are the most conspicuous. Clastic supply increased significantly during the lowstands separating these highstand events. The upper part of the Husbergøya Formation and the Langøyene Formation are correlated with the British late Rawtheyan (Owen 1981) and Hirnantian (Brenchley and Cocks 1982), respectively. The sandstone capping the oolitic limestone of the upper part of the Langøyene Formation has been cut by large channels (Brenchley and Newall 1980) that indicate that the area was close to, probably partly above, sea level. A profound and rapid sea level rise took place during the graptolite-based *Glyptograptus persculptus* Zone. This drowning event (Sælabonn Drowning Event) extended into the Silurian and is recorded by the graptolitic mudstones of the Sælabonn and Solvik formations of Oslo.

Above the *Pleurograptus linearis* Zone, a major lowering of sea level is evident across Baltoscandia. In the Västergötland and Siljan districts, the graptolitic Fjäcka Shale is overlain by the Jonstorp Formation (argillaceous nodular limestone), recording a comparatively high clastic supply during lowstand conditions. Also locally the Boda Limestone mounds developed. The intercalated Öglunda Limestone (dense calcilutites), however, signals a short-lived, significant rise in sea level, most likely the Spannslokket Drowning. The Ulunda and Nittsjö formations (dark gray to black mudstones/gray limestones and marls) above the Jonstorp Formation in Västergötland may reflect the Husbergøya Drowning. The overlying Tommarp Mudstone presumably records the Langøyene Drowning, and a hiatus is inferred above this unit.

The Grimsøya Regressive Event is recorded in South Estonia by the shift to the Jonstorp Formation (mudstones, marls, and argillaceous limestones) and the following Jelgava Formation (marls). This extended lowstand is represented by a hiatus in northern Estonia. Then follow the laterally equivalent units—the Parovéja (mainly calcilutitic limestone), Oostriku (calcarenitic limestone), and upper part of the Moe (calcilutitic and calcarenitic limestone) formations—that probably represent the Spannslokket Drowning. The overlying Kuili Formation (marls with intercalated limestones) of southern Estonia records another drop in sea level, and an equivalent hiatus is inferred for this interval in central and northern Estonia. The succeeding Kuldiga Formation (argillaceous calcarenites, interbedded marls and limestones) and laterally equivalent Adila Formation (argillaceous calcarenitic limestone) may represent the Husbergøya Drowning. Then follows the Ärina (mainly calcarenitic and arenaceous limestones with small bioherms, partly dolomitic and with oolites) and laterally equivalent Saldus (silty and sandy limestones) formations that are taken to represent the Langøyene Drowning.

■ Correlation with North America

The only detailed Ordovician sea level curve available for comparison is the one published by Ross and Ross (1992, 1995), based on North American sections (figure 10.3).

Lower Ordovician

The late Skullrockian highstand indicated by Ross and Ross (1992, 1995) equates with the early Tremadocian sea level rise, and therefore their mid Skullrockian unconformity is ascribed to the terminal Cambrian Acerocare Regressive Event. The lowstand between the Skullrockian and Stairsian is correlated with the Baltoscandian *Adelograptus hunnebergensis* Zone lowstand (including the Black Mountain and Peltocare regressive events), hence establishing the Stairsian highstand as equivalent to the Kiaerograptus Drowning Event. The early Tremadocian highstand is interpreted to reach a higher sea level stand than the second (mid Tremadocian) highstand. This is also hinted at by Ross and Ross (1995) but not shown in figure 10.3. The break between the Stairsian and Tulean corresponds to the Ceratopyge Regressive Event. Therefore, a good general accord exists between the North American and Baltoscandian sea level curves through the "Early to Mid Tremadocian Highstand Interval."

Ross and Ross (1995) inferred a pronounced double-peaked highstand during the Tulean. It is uncertain whether the peaks match the moderate drownings at the base and the middle of the Hunneberg stage, within the Baltoscandian "Late Tremadocian–Early Arenig Lowstand Interval" (figures 10.2, 10.3). Overall, the agreement with the Scandinavian record is not good. The curve interpreted for the Canadian craton by Barnes (1984:58) is more closely comparable to the Baltoscandian curve.

The two Blackhillsian highstands (Ross and Ross 1995) are presumed to correlate with the Billingen Transgressive Event and Evae Drowning Event, respectively. Ross and Ross (1995) show the first event

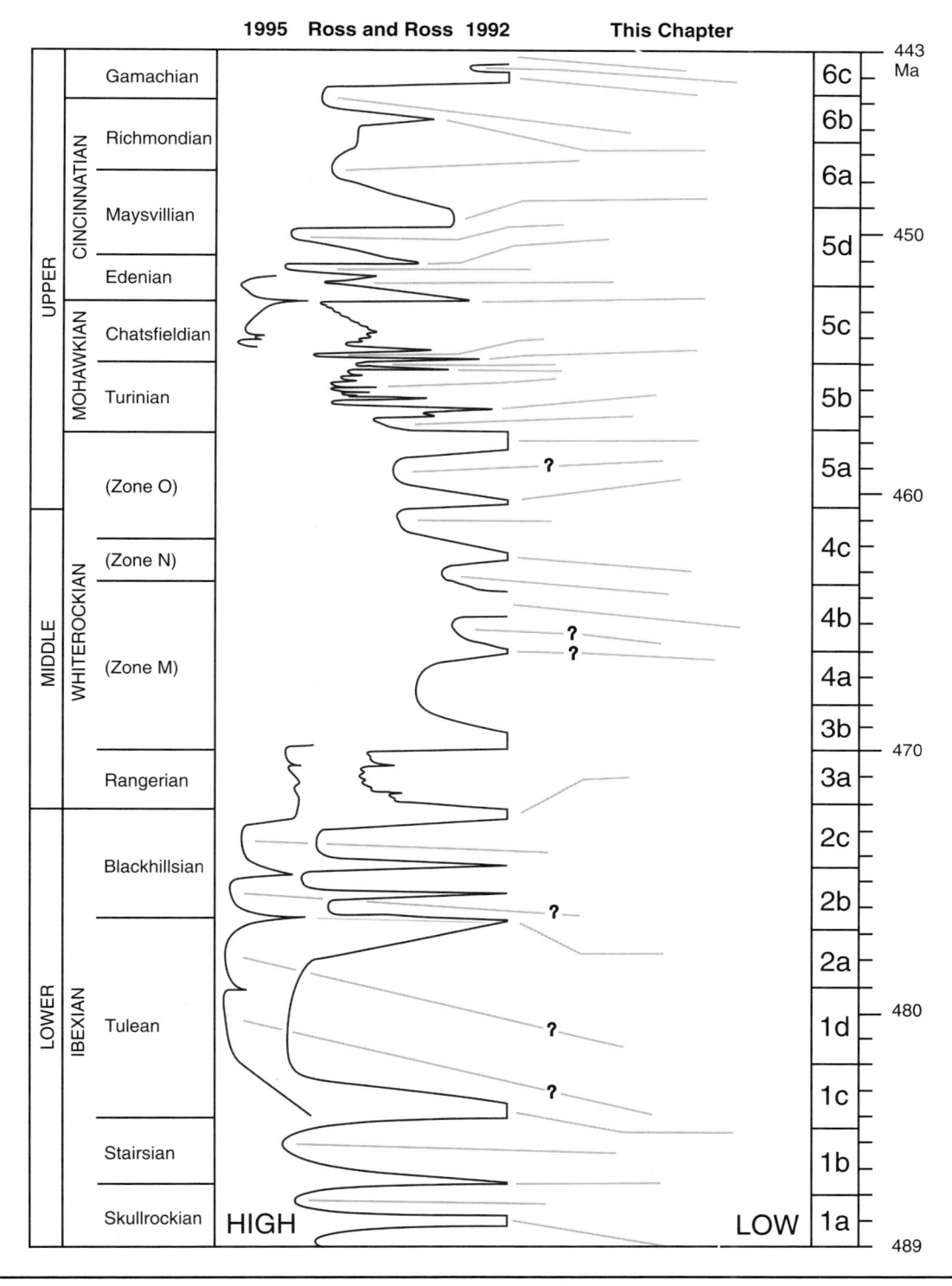

FIGURE 10.3. Comparison of North American and Baltoscandian sea level curves. The North American curve is from Ross and Ross (1992, 1995), adjusted approximately to the timescale used in figure 10.2. The North American stratigraphic framework is illustrated at the left side. Inferred correlations between the highstand and lowstand events are shown by the light-colored tie lines; less certain correlations also include question marks. Note that the time slices (chapter 2), as shown in the column to the right side, are tied only to the Baltoscandian curve.

as having the largest magnitude, but judging from Baltoscandia, the sea level was higher during the Evae Drowning.

Middle Ordovician

Ross and Ross (1992, 1995) interpreted the North American Whiterockian series as a lowstand interval. It correlates with the Baltoscandian "Late Arenig–Early Llanvirn Lowstand Interval," as well as the latest part of the "Mid Arenig Highstand Interval" and early part of the "Late Llanvirn–Caradoc Highstand Interval." The North American curve of Ross and Ross (1992) only broadly matches the Baltoscandian curve in the lower part of this interval, whereas a slightly better correlation exists through the upper part of the interval. The sea level was probably moderately high initially but then dropped to the lowest levels through the "Late Arenig–Early Llanvirn Lowstand Interval" to become higher at the end of the Whiterockian (during the lower part of the "Late Llanvirn–Caradoc Highstand Interval," prior to the Arnestad Drowning). However, sea level during the two mid to late Llanvirn highstands (in particular, during the Furudal Highstand) is shown to be significantly higher in Baltoscandia than in North America. On the other hand, the latest North American Whiterockian lowstand matches the early Caradoc Vollen Lowstand.

Correlation of the pre-Llanvirn events with the early Whiterockian is not particularly good. This may be a reflection of the incompleteness of successions on the American craton, which occasionally became emergent during this time (cf. Ross and Ross 1992). An attempt to correlate the relatively marginal Ibex section was published by Nielsen (1992b). However, it now seems more likely that the Juab/Kanosh shift with its sharp drop in sea level correlates with the middle of the upper Volkhov or, less likely, the base of the Kunda, when there was a marked increase in clastic input.

Upper Ordovician

The high sea level inferred for the Mohawkian and early Cincinnatian matches the latter part of the "Late Llanvirn–Caradoc Highstand Interval." The Arnestad Drowning correlates with the sea level rise at the base of the Mohawkian, but it seems that the amplitude is more exaggerated in the American curve (figure 10.3). It is agreed, however, that the sea level was high in the early part, then lowered (perhaps less so in Baltoscandia than in North America), and then again rose, reaching the highest level in the latter part of the Turinian (Keila Drowning). The Keila highstand was double-peaked, a pattern also recognizable in the Ross and Ross curve. A second shallowing is suggested by Ross and Ross (1995) in the later part of the highstand that is not recognized in Baltoscandia, presumably owing to a lack of resolution. The significant lowstand marking the Turinian-Chatsfieldian boundary equates with the Frognarkilen Lowstand.

Ross and Ross (1995) depict a profound drowning in the basal Chatsfieldian, which is equivalent to the Nakkholmen Drowning. No early *Dicranograptus clingani* events have been recognized in Baltoscandia, which may be due to relatively low resolution during the highstand interval. The following pronounced lowstand event at the Mohawkian/Cincinnatian boundary correlates with the Solvang Lowstand (figure 10.3), and the shallowing leading to this lowstand was also relatively gradual (Ross and Ross 1995). The sea level was probably slightly lower during the Solvang Lowstand than during the preceding Frognarkilen Lowstand (figure 10.3). Overall, there is good agreement between the North American and Baltoscandian curves for this main part of the Caradoc.

Ross and Ross (1992) also recognized the Linearis Drownings of the late Caradoc–early Ashgill and suggested that the composite highstand was triple peaked. Furthermore, the highstand peaks show the same relationships in North America and Baltoscandia—with the middle peak having the greatest amplitude and the amplitude of the upper peak greater than that of the lower peak. The "Ashgill Lowstand Interval" can also be identified in the Ross and Ross (1992) curve, and despite small differences in scaling, the relative curves are similar for this second-order lowstand. In Baltoscandia the Grimsøya Regressive Event was followed by three successively less intensive drowning events (figure 10.3). Ross and Ross (1992) also indicated that the first of these drownings was the largest, and it may be correlated with the Spannslokket Drowning. The amplitude of this drowning may have been even greater than that shown by Ross and Ross. If the correlations are correct, it may indicate that the unconformity on top of Fort Atkinson and Aleman formations is more significant than was indicated by Ross and Ross (1992:333).

■ Conclusions

The Ordovician was characterized by high sea levels, perhaps the highest in the Phanerozoic. A second-order pattern has been adopted for subdividing the Ordovician into three highstand and three lowstand "intervals" (figure 10.3). These, and the time-slice (*TS*) equivalents (chapter 2), are as follows: the "Early to Mid Tremadocian Highstand Interval" (*TS*.1a–lower1b), the "Late Tremadocian–Early Arenig Lowstand Interval" (upper *TS*.1b–2a), the "Mid Arenig Highstand Interval" (*TS*.2b–3a), the "Late Arenig–Early Llanvirn Lowstand Interval" (*TS*.3b–lower4b), the "Late Llanvirn–Caradoc Highstand Interval" (upper *TS*.4b–5d), and the "Ashgill Lowstand Interval" (*TS*.6a–6c).

Some of the major sea level changes seem to have been rapid and therefore have been referred to as events. They have potential for international correlation. It is, however, stressed that highstand events are easier to correlate to biostratigraphic indices (e.g., widely distributed deep-water faunas; see also Fortey 1984) than the lowstand events. In addition, the adaptive radiation of graptolites seems to have a direct relationship to drowning events. Lowstands leave conspicuous evidence in the shelfal sedimentary record and remain a primary tool for correlation of depositional sequences (Vail et al. 1977). However, shallow-water faunas tend to be more endemic and are therefore notoriously more difficult to correlate.

The Stonehenge (early Tremadocian), Evae (early Arenig), Nakkholmen (late Caradoc), and Linearis (latest Caradoc) drowning events reflect major sea level rises with excellent global correlation potential. The significant Furudal Highstand (late Llanvirn–early Caradoc) may also have correlation potential, but the drowning appears to have been gradual and is not designated an event. Further, the less conspicuous Kiaerograptus (late Tremadocian), Keila (upper part of the *Diplograptus foliaceus* Zone), and Spannslokket (intra-Ashgill) drownings could be useful for intercontinental correlation. In addition, the Ceratopyge, Frognarkilen, Solvang, Grimsøya, and terminal Ashgill lowstands may be widely recognizable on a global scale as shallowing events, or unconformities, whereas the progressive shallowings through the early Whiterockian seem difficult to correlate because of the composite nature of this lowstand interval.

At second-order level there is good agreement between the Baltoscandian curve and the Ross and Ross (1992, 1995) curve for the North American craton. It is also currently possible to match most of the third-order changes on either side of the Iapetus Ocean through the "Early to Mid Tremadocian Highstand Interval," as well as within the "Late Llanvirn–Caradoc Highstand Interval" and the "Ashgill Lowstand Interval." Correlation of the third-order changes during the "Late Tremadocian–Early Arenig Lowstand Interval," the "Mid Arenig Highstand Interval," and the "Late Arenig–Early Llanvirn Lowstand Interval" is less convincing. The late Tremadocian–Arenig curve inferred for the Canadian craton by Barnes (1984:58) shows a better overall match. The Baltoscandian curve seems to be more resolved for the lower half of the Ordovician, up to the Llanvirn, whereas the North American curve has a higher resolution during the Caradoc. Further work is needed to focus on the discrepancies. Currently it is possible to correlate about 30 sea level changes with a fair degree of confidence. The detailed match of late Caradoc and Ashgill sea level events attests to the potential of a high-resolution sequence stratigraphic correlation.

ACKNOWLEDGMENTS

D. A. T. Harper, Geological Museum, Copenhagen; David Bruton, Palaeontological Museum, Tøyen, Oslo; and Alan Owen, The University, Dundee, Scotland, are thanked for advice on Caradoc-Ashgill stratigraphy. Tonu Meidla, Tartu University, and Linda Hints, Tallinn Technical University, kindly shared their extensive knowledge on the Upper Ordovician Estonian sections and gave advice on the likelihood of alternative correlations. However, the correlations made in this chapter are entirely the responsibility of the author. Barry Webby, Macquarie University, and Ian Percival, Geological Survey of New South Wales, are thanked for advice on shortening the manuscript to meet the space limitations of the volume.

PART III Taxonomic Groups

11 Radiolarians

Paula J. Noble and Taniel Danelian

Ordovician radiolarian studies are in their infancy. Only a few dozen studies have been published worldwide, most of which focus on descriptive taxonomy. Little attention has been paid to discerning evolutionary patterns, most likely because much basic descriptive work remains to be done. This poses many challenges in estimating Ordovician radiolarian biodiversity patterns. Nonetheless, we provide a review of the existing data and make some qualitative assessments of which data are most appropriate for revealing these patterns. These data are compiled and used to estimate biodiversity throughout the Ordovician by applying some commonly used diversity measures.

■ Previous Work

Early taxonomic studies on Ordovician radiolarians were conducted on thin sections of radiolarian chert (e.g., Hinde 1890; Ruedemann and Wilson 1936). Unfortunately, many of the genus- and species-level characters cannot be seen in thin section. All but a few of the taxa described in these early works are rendered *nomen dubia*. In the next phase of work, the most significant contributions to Ordovician radiolarian studies were made by Boris Nazarov. Early in his career, Nazarov worked on thin-section material and devised ingenious statistical methods of estimating the number of spines and other three-dimensional aspects of the radiolarian test from multiple cross-sectional views (Nazarov 1975). At this time, many of the Ordovician sphaeroidal radiolarians were assigned to the family Entactiniidae. These studies still had their limitations, however, and it was not until the 1970s that Nazarov was able to use techniques of hydrofluoric etching to examine his chert faunas matrix-free (e.g., Nazarov and Ormiston 1984). Reexamination of the matrix-free chert faunas, coupled with studies on material etched from limestones (Nazarov and Popov 1980; Nazarov and Nõlvak 1983), lead to significant modifications to the existing taxonomic scheme, particularly the erection of the family Inaniguttidae and the transfer of many species previously assigned to the Entactiniidae (Nazarov and Ormiston 1984, 1993; Nazarov 1988). Throughout the 1980s and 1990s additional progress in Ordovician studies was made, largely in the continued description of matrix-free faunas from chert and carbonates (Goto et al. 1992; Wang 1993; H. Li 1995; Kozur et al. 1996; Aitchison 1998; Aitchison et al. 1998; Danelian 1999; Noble 2000; Danelian and Floyd 2001) and in their application to biostratigraphy (Noble and Aitchison 2000).

The majority of radiolarians are recovered from biogenic chert, although they may also be found in concretions or in limestone beds at sporadic intervals within open marine carbonate and shale sequences. Compared with chert faunas, concretion and limestone faunas are almost always better preserved and

TABLE 11.1. Number of Radiolarian Species and Genera per Assemblage

Time Slice (TS)	Locality	References	Lithology	ds$_{sample}$	dg$_{sample}$
1a or older	Little Port Complex, Newfoundland	Aitchison et al. 1998	Chert	2	2
1a	Antelope Range, Nevada	Kozur et al. 1996	Limestone	6	3
2a–2b	Akzal Mountains, Kazakhstan	Danelian and Popov 2003	Limestone	4	3
2c–3a	Ballantrae Complex, Scotland	Aitchison 1998	Chert	4	3
4a	Spitsbergen	Fortey and Holdsworth 1971	Limestone	5	4
4b	Kurchilik Formation (sample 19), Sarykumy (North Balkhash Region), Kazakhstan	Nazarov 1975; Nazarov and Popov 1980; Nazarov and Ormiston 1993	Limestone	12	6
4c	Bestamak Formation (sample 553a), Chagar River, Chingiz Range Kazakhstan	Nazarov 1975; Nazarov and Popov 1980; Nazarov and Ormiston 1993	Limestone	30	12
4c–5a	Crawford, Southern Uplands, Scotland	Danelian 1999; Danelian and Floyd 2001	Chert	9	4
5a	Pingliang, Gansu, China	Wang 1993	Chert	9	5
5d	Vinini Formation, Nevada	Noble 2000	Chert	5	5
5d–6a?	Lachlan NL15, New South Wales	Goto et al. 1992	Chert	13	6
5d–6a	Malongulli Formation, New South Wales	Webby and Blom 1986	Limestone	6	4
6b	Oisu, Chu-Ili Range, Kazakhstan	Nazarov 1975; Nazarov and Popov 1980; Nazarov and Ormiston 1993	Limestone	16	10
6b	Hanson Creek, Nevada	Dunham and Murphy 1986; Renz 1990	Limestone	11	8
5d–6a	Lachlan NL21, New South Wales	Goto et al. 1992	Chert	6	5

Note: Based on all currently available data. In most cases faunas recovered from limestones have higher diversity than chert faunas of the same age, owing to preservational biases. Exceptions to this are the Spitsbergen sample, which was described only provisionally, and the Malongulli Formation, which is very spiculitic and of shallower facies than optimal for radiolarian diversity.

provide better estimates of diversity (Blome and Reed 1993). Some valuable studies on limestone faunas are from the Tremadocian of Nevada, United States (Kozur et al. 1996), Arenig of Spitsbergen (Fortey and Holdsworth 1971), Arenig and Llanvirn of Kazakhstan (Nazarov and Popov 1980; Nazarov and Ormiston 1993; Danelian and Popov 2003), upper Caradoc and Ashgill of Estonia (Nazarov and Nõlvak 1983; Gorka 1994), eastern Australia (Webby and Blom 1986), and Nevada, United States (Renz 1990). The limestone studies usually exhibit greater species richness than that from coeval chert faunas (table 11.1). Exceptions to this rule are limestone samples from shallower facies that have lower species diversity owing to restricted marine conditions. Limestone faunas are also advantageous in that they commonly have precise age control from the co-occurrence of graptolites, trilobites, and conodonts. The age of chert faunas is typically less well constrained than that of limestones.

■ **Measures of Biodiversity**

Data were compiled from the literature, and to a lesser extent unpublished data were used from the Upper Ordovician of Nevada (Noble unpubl. data) and Lower to Middle Ordovician of Kazakhstan (Danelian unpubl. data). Data from poorly preserved assemblages and those assemblages lacking precise age control (within two time slices [*TS*]) were excluded so as not to overextend ranges. Species-level taxa described in open nomenclature were used when possible. For example, taxa that are adequately described but are not assignable to any formal species (e.g., *Xiphostylus* sp. A) were tallied into the total species diversity. Taxa with noted species affinity (sp. aff.) were counted as separate species unless their synonymy has been established. The biodiversity measures in Cooper (chapter 4) are followed. Note that we distinguish both species and generic diversity; thus, for example, dg_{sample} and ds_{sample} are number of genera and species per sample, respectively.

■ **Limitations of the Data**

Three factors affect the quality and usability of Ordovician radiolarian data. First, early papers limited to thin-section data are problematic. Most of the species recognized in thin section are *nomen dubia*, and we omitted them from this compilation. Second, significant number of taxa are published in open nomenclature, owing in part to the immature state of radiolarian taxonomic studies (i.e., Fortey and Holdsworth 1971; Webby and Blom 1986) and also to poor preservation (i.e., Goto et al. 1992; Aitchison 1998; Aitchison et al. 1998; Noble 2000). Treatment

TABLE 11.2. Radiolarian Taxon Diversity Measures

Time Slice (TS)	Number of Studies	Total Species Diversity ds_{tot}	Total Generic Diversity dg_{tot}	Normalized Species Diversity ds_{norm}
6c	0	-	-	-
6b	2	25	19	12.5
6a	4	20	18	12.5
5d	4	21	16	11.5
5c	0	4	11	3.5
5b	0	3	11	3.0
5a	3	26	11	15.0
4c	2	21	11	15.5
4b	0	35	12	17.5
4a	1	5	4	3.5
3b	0	2	2	2.0
3a	1	4	2	3.0
2c	4	7	3	4.0
2b	2	4	3	4.0
2a	0	4	3	2.0
1d	0	-	-	-
1c	0	-	-	-
1b	0	-	-	-
1a	1	6	3	3.0

of these taxa is discussed briefly in the previous section. Third, there are stratigraphic gaps representing intervals of time for which there are no data (see table 11.2, Number of Studies). Particularly problematic is the lack of data from the mid to late Tremadocian (*TS*.1b–d) and from the late Ashgill (*TS*.6c). Normalized species diversity measures are affected (made lower) by fewer range-through taxa occurring above and below these gaps. These factors must be kept in mind when evaluating biodiversity estimates.

■ Discussion

Diversity measures reveal a progressive increase in generic diversity throughout the Ordovician as new body plans are established (figure 11.1). Generic diversity reaches a maximum of 17 in the mid Ashgill. Species diversity peaks a little earlier, in the early Caradoc, reaching a maximum of 46. Three discrete diversification pulses are recognized and discussed in the following list. Both the timing and the abruptness of these pulses are subject to change as more data become available.

1. Lower Ordovician (*TS*.2a). Several new body plans make their first appearance, including the Inaniguttidae (figure 11.1). This group becomes increasingly important in Middle and Upper Ordovician assemblages. This diversification pulse is not noticed in the species and generic diversity numbers because there is a complete faunal turnover between the early Tremadocian and the early Arenig. The number of species and genera per sample in the Arenig is no higher than that recorded for the lower Tremadocian (table 11.1). A data gap in the middle and upper Tremadocian further obscures the timing and nature of this event.

2. Upper Middle Ordovician (*TS*.4b). Species per sample reach their highest (table 11.1), and the normalized species diversity is also at its highest (table 11.2). Most of the taxa are described from well-preserved limestone samples in Kazakhstan. Generic diversity also reflects this increase in biodiversity, with the total genera rising to 12.

3. Upper Ordovician (*TS*.5d). The final diversification pulse appears to follow a slight decline in biodiversity during the mid Caradoc and is marked by the appearance of seven new genera, including many groups that persist into the Silurian (figure 11.1; table 11.2). The decline, however, may be an artifact of a data gap in the middle Caradoc (*TS*.5b–c). Generic diversity reaches its highest point in the Ashgill (*TS*.6b) and appears to be a function of the lack of extinctions at the genus level going into the Late Ordovician.

It should be noted that data gaps affect the diversity numbers. Time slices following a data gap (i.e., *TS*.2a and 5d) have an artificially lowered normalized species diversity because the gap reduces the number of taxa that range through. Time slices for which there are no data show lowered diversity because taxa are limited to those that range through. A data gap within the Caradoc (*TS*.5b–c) produces an odd effect whereby the number of genera appears to exceed the number of species, an impossibility. Species tend to have shorter life spans than genera, creating a more marked drop in the species number during this data gap while the genera persist. In this instance, the generic diversity numbers appear to be the most reliable measure of diversity within the Caradoc.

■ Discussion of Clades

A less direct but useful means of gleaning information regarding biodiversity patterns is through examination of the appearance and duration of Ordovician radiolarian clades (figure 11.1). The currently known radiolarian assemblages from the Upper Cambrian

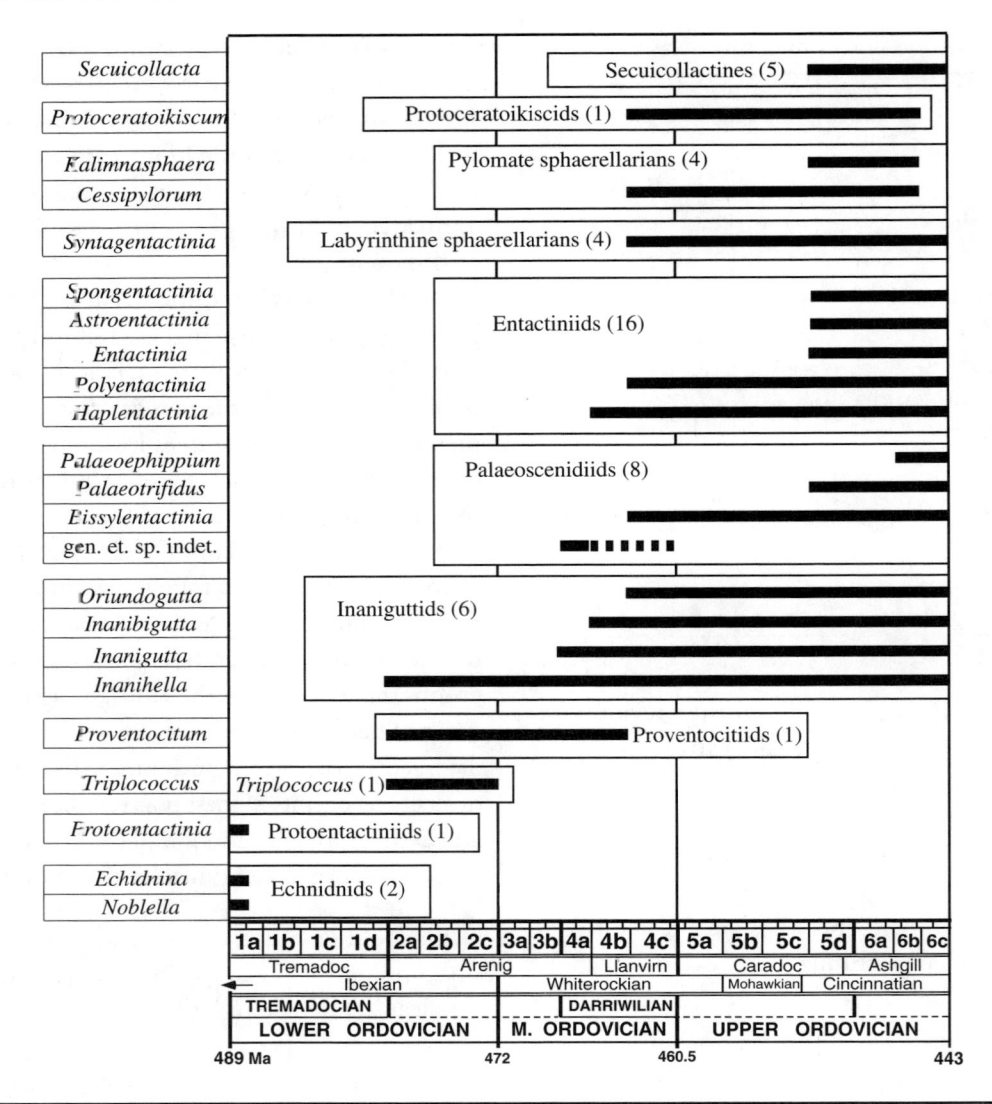

FIGURE 11.1. Range of Ordovician radiolarian clades and the contained genera. Clade names appear in larger boxes. Number in parentheses following the clade name is the total number of genera within that clade, including post-Ordovician genera. One exception is the Palaeoscenidiidae, for which the total number of genera includes only Paleozoic forms.

(Won and Iams 2002) and Tremadocian (Kozur et al. 1996) strata contain primitive radiolarian skeletons that are entirely different from typical Ordovician radiolarian body plans. A profound faunal turnover occurs between the lower Tremadocian and the lower Arenig. This pulse is not noticeable in the diversity numbers because there is no significant net increase in genera and species. None of the Tremadocian clades persist into the Arenig. This turnover is no doubt exaggerated by a large data gap in the middle and upper Tremadocian. Nonetheless, it is clear that several major Middle Ordovician clades became established by Arenig time. Clades such as the Inaniguttidae, Entactiniidae, and Palaeoscenididae continue to persist after the Arenig as some of the most dominant and abundant taxa from the Middle Ordovician through Lower Carboniferous.

The Middle Ordovician diversification coincides with the first appearance of two important genera—*Protoceratoikiscum* and *Cessipylorum* (both restricted to the upper Middle and Upper Ordovician)—and the first appearance of the labyrinthine sphaerellarian clade. Inaniguttid species diversity reaches its maximum at this time, and this clade becomes the dominant component of currently known Middle Ordovician assemblages.

The Upper Ordovician pulse is also apparent in figure 11.1, coinciding with the first appearance of the Secuicollactinae, radiations in the palaeoscenidiid, labyrinthine sphaerellarian, and pylomate sphaerellarian clades. No data are available for the uppermost Ashgill. However, it should be noted that of the 17 genera present at the end of the Ordovician, 14 range into the Lower Silurian.

■ Conclusions

Data for Ordovician radiolarians are spotty and more sporadic than those for other plankton groups, making a measure of biodiversity an estimation at best. Despite these shortcomings, data compiled from localities with precise age control reveal an overall increase in biodiversity throughout the Ordovician with three discernible pulses: one in the early Arenig (or possibly older), a second in the Darriwilian, and a third in the late Caradoc. Several Ordovician radiolarian clades become established by the late Arenig, including the Inaniguttidae and Palaeoscenidiidae, which persist through the Early Carboniferous. Additional clades make their first appearance in the late Mid Ordovician and the early Late Ordovician. Species diversity reaches its maximum in the Darriwilian to early Caradoc and drops slightly in the Ashgill. Generic diversity begins rising in the Arenig and continues to rise until it reaches its maximum in the Ashgill, showing a net increase throughout the Ordovician. Only one extinction is seen at the genus level from the Arenig through the middle Ashgill (*Proventocitum*). Likewise, the majority of the Ashgill genera persist into the Silurian, which implies that the end Ordovician (Hirnantian) extinction had no major impact. Although no Hirnantian data exist (*TS.6c*), Lower Silurian assemblages contain 14 of the 17 genera found prior to the Hirnantian (*TS.6b*). Overall, the Ordovician Period can be considered a time of substantial increase in biodiversity and establishes many important clades that persist into the Silurian and Devonian.

12 Sponges

Marcelo G. Carrera and J. Keith Rigby

Sponges include the most primitive multicellular organisms and have a record commencing in the Late Precambrian. All major sponge groups are represented as Cambrian fossils. Ordovician sponges are only moderately well known when compared with other groups of fossils, though they are known from most continents and 275 species have been described. This relatively limited known diversity may be due in part to the scarcity of taxonomic studies and in part to the phylum's conservative history.

Ordovician diversification is marked by the extensive radiation of some groups, whereas others remained limited and little diversified. Even though several sponge groups are moderately well known in the Ordovician, their origins from Cambrian elements are obscure, largely because of significant breaks between Middle Cambrian and Lower Ordovician sponge records.

Ordovician sponge distribution is known in varying degrees of detail for different areas. The limited stratigraphic information given in the older literature makes the assignment of some ranges very difficult. In addition, a particular problem is the correlation of northern European sponge associations, which are largely known only from Quaternary glacial erratic boulders. Microfossils found in the associated matrix indicate that most are of early Ashgill age (Hacht and Rhebergen 1997; Rhebergen et al. 2001). However, some limestone boulders in the same deposits are mid to late Caradoc in age. As a consequence, accurate stratigraphic positions cannot be determined for this probable carbonate platform association, limiting the detailed analysis.

We adopted two alternatives: a system in which the first appearances (FAD) of European sponges are allocated a range from time slice $TS.5c$ to the last appearances (LAD) during $TS.6a$. This system potentially overestimates the real diversity values. The other alternative is to exclude the European association from the database, which considerably underestimates the values. The resulting diversity curves, constructed using normalized diversity measures, are shown in figure 12.1. The curve showing relative distribution for the time intervals for the major sponge groups was calculated based on the first of the alternatives data (figure 12.2). The sponge database has been incorporated into the database managed by Arnold I. Miller (University of Cincinnati).

■ Demosponges

The origin of the class Demospongea is obscured by the poor record in the Lower Cambrian. Only relatively primitive-appearing sponges with simple, tubular, thin-walled skeletons made by monaxons have been reported (Rigby 1991). Similar forms and many additional monaxonid and probably keratose demosponges are well represented in the Middle Cambrian of the

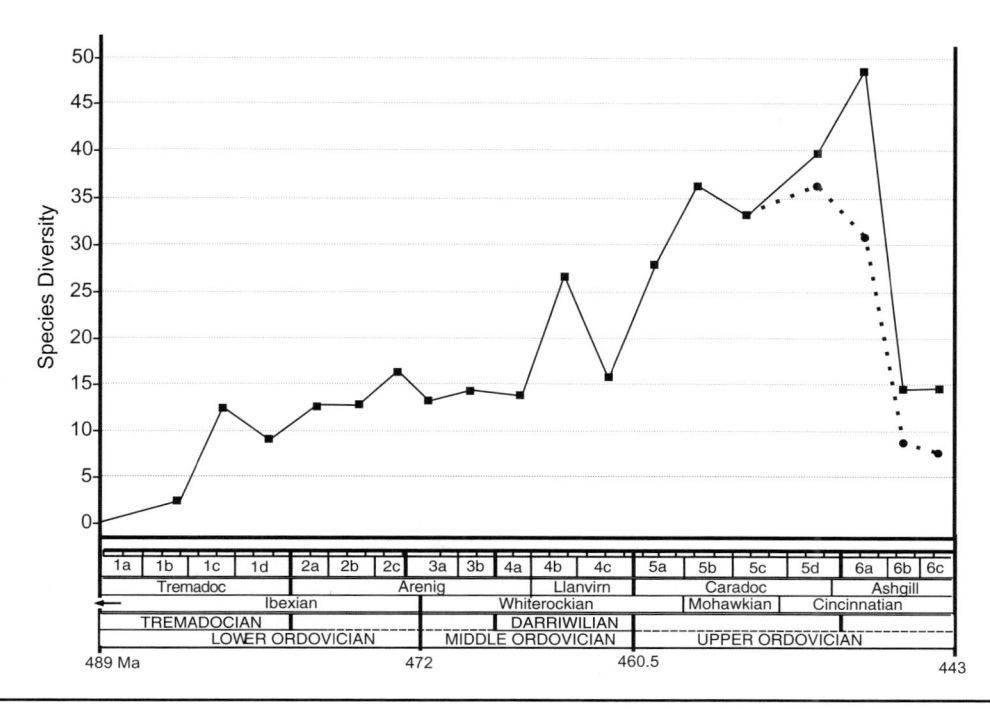

FIGURE 12.1. Normalized diversity curve for the Ordovician sponges at species level. Black line with rectangle points indicates diversity including mid Caradoc–early Ashgill northern European sponges. Dashed line with circle points shows diversity minus the northern European species.

Burgess Shale in western Canada and upper Middle Cambrian rocks in Utah (Rigby 1986). Monaxonid genera such as *Halichondrites* and *Choia* continued to exist and even proliferated during the Early Ordovician (Rigby 1991). In the Late Ordovician only two genera of the order Monaxonida (figure 12.2) have been reported from *TS*.5b and *TS*.6a (Rigby 1971). The post-Cambrian record of nonlithistid demosponges of the Paleozoic is poor, and so most of the fossils included within the demosponges belong to the order Lithistida.

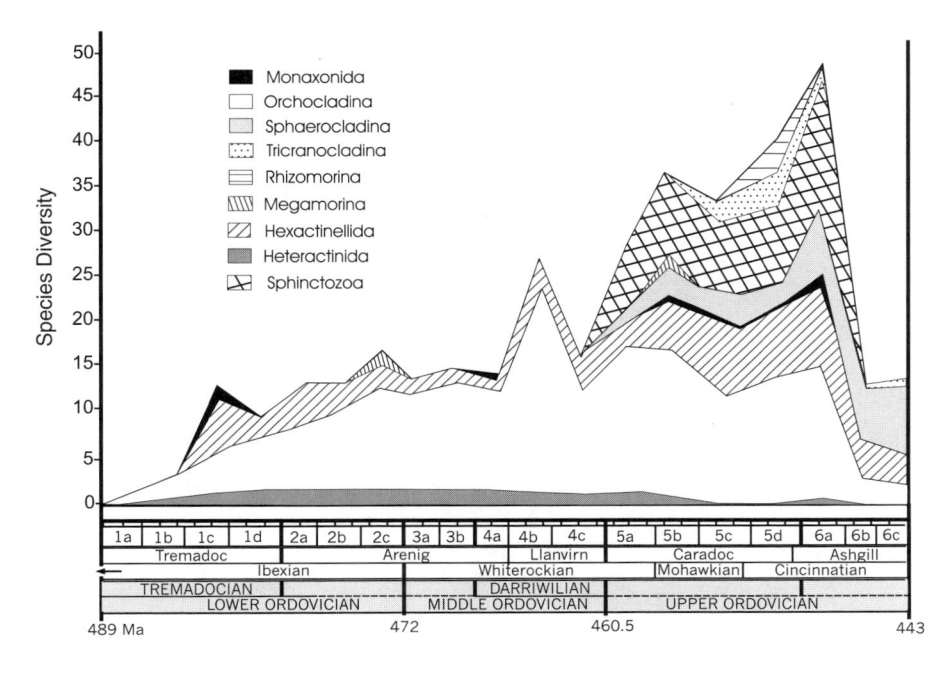

FIGURE 12.2. Normalized diversity curve for the Ordovician sponges at species level and contribution of major sponge groups to total diversity in the Ordovician. Most of the sponge groups are shown at suborder level. However, hexactinellids and sphinctozoans are different sponge classes, and monaxonids and heteractinids are sponge orders. The limited variability expressed in classes such as Sphinctozoa and Hexactinellida, compared with the lithistid suborders, enforced a differential comparison of groups.

Orchocladina

Only three genera of orchocladine lithisitids are known from the Cambrian, but the suborder expanded abruptly to more than 45 genera in the Ordovician. Approximately 10 genera are known from the Silurian and fewer in the Mid Paleozoic. Skeletons made principally of dendroclones or chiastoclones characterize the Orchocladina. Three families can be differentiated within the suborder based on skeletal arrangement. Those considered to be anthaspidellids have relatively regular, simple skeletal structure, typified by such genera as *Archaeoscyphia, Rhopalocoelia,* or *Calycocoelia*. The streptosolenids have a distinctly less well organized canal system and skeletal structure. *Lissocoelia* and *Streptosolen* characterize this group. The third group is the chiastoclonellids and is represented by only two Ordovician genera.

In the Early Ordovician record, a few genera of the Anthaspidellidae are associated with reef mounds, built by microbes, sponges, and calathiid receptaculitids. A sponge-microbial facies forming either bioherms or biostromes occurs in Lower Ordovician rocks deposited around the margins of Laurentia from Newfoundland to the Great Basin (Alberstadt and Repetski 1989 and references therein). Other reef mounds are known from the Argentine Precordillera (Cañas and Carrera 1993), Hubei Province, South China (Liu et al. 1997), and Siberia (Webby 1999).

Diversification of orchocladines increased by *TS*.2c and is recognized mainly from the Shingle Formation of the Egan Range in the Great Basin, western United States (Bassler 1941; Johns 1994), with the appearance of 11 species, and in the San Juan Formation, Argentine Precordillera, with 8 species (Carrera and Rigby 1999).

The peak of orchocladine diversification during *TS*.4b was more significant, given that 20 species occur in the San Juan Formation, Argentine Precordillera (Carrera and Rigby 1999), and 31 species in the Antelope Valley Limestone of Nevada, the Wah Wah Formation of Utah, and equivalents in the Mazourka Canyon, California (Bassler 1941; Greife and Langeheim 1963; Johns 1994).

Both diversification events were recorded in middle to distal carbonate platform environments and were coincident with sea level rises during the *Oepikodus evae* transgression (*TS*.2b–c) and *Lenodus variabilis–Eoplacognathus suecicus* transgression (*TS*.4a–b), respectively.

In the Caradoc (*TS*.5a–c) the rate of orchocladine expansion diminished, and only a few new genera appeared. The majority of Late Ordovician orchocladine genera had their origins in the Early and Mid Ordovician. No significant expansion in genus diversity occurs in the Late Ordovician; the slight recovery in diversity values evidenced in *TS*.5d–6a is mainly due to the appearance of new species of these older genera. Upper Ordovician orchocladine occurrences are localized in midcontinental, eastern, and western North America (Rigby et al. 1993; Carrera and Rigby 1999 and references therein), in Baltica (Hacht and Rhebergen 1997; Rhebergen et al. 2001), and in the Malongulli Formation of New South Wales, Australia (Rigby and Webby 1988).

Sphaerocladina

The suborder Sphaerocladina appeared in the Caradoc as typical spherical forms made of a tridimensional gridwork of sphaerocladine spicules. They are moderately rare forms in the Ordovician of North America but are present in great numbers in the Silurian. Abundant occurrences of these genera are reported from the Late Ordovician of northern Europe. The suborder probably became extinct toward the end of the Devonian.

Only two genera, *Astylospongia* and *Caryospongia*, appeared in the mid Caradoc of eastern North America, in the St. Lawrence Valley (Wilson 1948). Another three genera, *Phialaspongia, Camellaspongia,* and *Caliculospongia,* are recorded as endemic (Rigby and Bayer 1976) from the late Caradocian of the North American Midcontinent.

Several species of *Caryospongia*, two of *Astylospongia*, a single species of *Palaeomanon*, and the endemic *Syltrochos* have been reported from the middle Caradoc–lower Ashgill rocks of northern Europe (Hacht and Rhebergen 1997; Rhebergen et al. 2001). In New South Wales, Australia, only the endemic species *Astylostroma micra* is recorded from the late Caradoc–early Ashgill, *TS*.5d–6a (Rigby and Webby 1988).

Tricranocladina

The suborder is a conservative group that appeared in the Caradocian and existed until the end of the Permian. In general, outside Australia, the tricranocladines are an exceedingly conservative group. In the Late Ordovician Malongulli Formation of New South Wales, however, several species of the endemic genera *Palmatohindia, Arborohindia, Belubulaspongia, Fenestrospongia,* and *Mamelohindia* were recorded (Rigby and Webby 1988). The widespread species *Hindia sphaeroidalis* is also reported. In North America only two species of tricranocladines are recorded in the Ordovician of the North American Midcontinent, *H. sphaeroidalis* in the mid Caradoc to early Ashgill and *Cotylahindia panaca* in the late Caradoc–early Ashgill (*TS*.5d–6a). In northern Europe, only the widespread *H. sphaeroidalis* is recorded.

Rhizomorina

Until the recent discovery of major sponge faunas in the Ordovician of Australia (Rigby and Webby 1988) and in the Silurian of Arctic Canada (Rigby and Chatterton 1989), the Rhizomorina were an exclusively Upper Paleozoic group. Ordovician rhizomorines appear restricted in the late Caradoc–early Ashgill of the Malongulli Formation in New South Wales, Australia. The endemic genera *Lewinia, Boonderooia, Taplowia,* and *Warrigalia* and one species of the genus *Haplistion* have been recorded from there (Rigby and Webby 1988).

Megamorina

Discovery of the family Nexospongiidae allows differentiation of two lineages within the megamorinids (Carrera 1996). One line includes the family Saccospongiidae with skeletons typically composed of tracts of heloclones and monaxons. *Rugospongia* from the Argentine Precordillera, *Epiplastospongia* and *Saccospongia* from North America, and *Cliefdenospongia* from New South Wales, Australia, are included in this family. The other line has skeletons composed of irregularly distributed heloclones and monaxons that are not grouped in tracts. *Nexospongia* from *TS*.2b and 2c of the Argentine Precordillera is to date the only Paleozoic representative of this lineage. Megamorines recorded two small peaks of diversity—in *TS*.2c, related to the Argentinian genera, and *TS*.5b, related to North America and eastern Australia.

■ Hexactinellids

Members of the class Hexactinellida are represented by entire sponge skeletons in the Lower and Middle Cambrian (Rigby 1986; Rigby and Hou 1995 and references therein). The majority of these sponges resemble *Protospongia* and possess a single layer of parallel stauractines. *Multivasculatus,* a still more complex hexactinellid with two or more layers of parallel hexactines, appears in the Late Cambrian (Finks 1983).

Hexactinellid sponges are much less diverse than demosponges in the Ordovician, but they have more recognizable roots to Cambrian faunas, in that at least three families have a more continuous record from the Cambrian to the Ordovician. The Cambrian and Lower Ordovician families Protospongiidae and Hintzespongiidae are represented by relatively primitive hexactinellid assemblages, which apparently thrived in black shales at the margins of Ordovician continents. Thin-walled species with root tufts, resembling those of the Cambrian, are found in similar facies of the Ordovician, mainly shales or limestones of quiet waters. These are more advanced in the sense that they typically show at least a double layer of spicules (dermal and gastral).

Species of *Protospongia, Cyathophycus, Diagoniella, Acanthodictya,* and *Palaeosacus* have been described from the upper Tremadocian black shales of Little Métis, St. Lawrence Valley, eastern Canada (Dawson 1889). In addition, a more complex pelicaspongiid hexactinellid has been described recently from the upper Tremadocian of the Puna Region, northwestern Argentina (Carrera 1998). There are few known occurrences of Arenig hexactinellids.

Middle and Upper Ordovician records include a widespread development of the primitive Protospongiidae and Hintzespongiidae in North America (Utica Shale in New York, Trenton Group in Ohio, Tennessee, and Kentucky, as well as the Vinini Formation in Nevada).

Diversification was more significant in the family Brachiospongiidae recorded in North America and

Baltica (Wilson 1948; Hacht and Rhebergen 1997), in the family Pyruspongiidae from the Williston Basin, Manitoba (Rigby 1971), New South Wales, Australia (Rigby and Webby 1988), and in the family Pelicaspongiidae, which has one described species from Anticosti Island, Canada, and four species from New South Wales, Australia. The families Teganiidae and Pattersoniidae (e.g., Utica Shale in New York, and Bigsby and Bellevue limestones in Kentucky) also appeared in the Middle and Late Ordovician but overall show only a minor expansion.

Middle and Upper Ordovician hexactinellids have skeletons that are more complex and three-dimensional, and some such forms invaded the platform facies (Brachiospongioidea). One type is represented by *Pattersonia,* with a stout root tuft and no central cavity. The other, including *Brachiospongia* and *Pyruspongia,* has flat bottoms without root tufts and a large central cavity.

Discrete siliceous spicule form genera of mainly hexactinellid affinity have also been reported. For example, the form genera *Silicunculus, Kometia, Chelispongia,* and other problematic forms (e.g., *Pseudolancicula*) occur in rich and abundant assemblages of disarticulated spicules in the late Caradoc–early Ashgill Malongulli Formation of New South Wales, Australia (Webby and Trotter 1993).

Heteractinids

The heteractinids are a relatively minor group of exclusively Paleozoic calcareous sponges with records that can be traced back to the Early Cambrian, with *Eiffelia* from Asia and the same genus in the Middle Cambrian Burgess Shale of Canada (Rigby 1986). *Jawonya* and *Wagima* (Kruse 1987) from the Middle Cambrian of Australia were initially considered to be sphinctozoan genera but are now interpreted as heteractinids (Rigby 1991; Rigby et al. 1993) or two-walled sponges with no clear affinities (Kruse 1996; Debrenne and Reitner 2001).

Rigby (1991) visualized two different lineages for the evolutionary history of the Heteractinida. One leads from *Eiffelia* to the Carboniferous *Zangerlispongia,* including the Ordovician genera *Toquimiella* from North America and *Chilcaia* from the Argentine Precordillera (Carrera 1994). The sponges of this lineage have skeletons that are thin walled, composed of geometrically ranked sexiradiates and mainly diversified in the Early and Mid Ordovician.

The other lineage, with *Jawonya* as the stem genus, includes the Late Ordovician *Astraeoconus, Astraeospongium, Asteriospongia,* and *Constellatospongia* (Rigby 1991 and references therein). This group has densely packed irregularly oriented octactine-based skeletons and thin walls in *Jawonya* and *Astraeoconus* or moderately thick walls in *Astraeospongium, Asteriospongia,* and *Constellatospongia.*

Sphinctozoans

Sclerosponges of sphinctozoan grade have segmented irregularly proliferated chambers, arranged around a central cavity. They flourished mainly in the Late Paleozoic and Early Mesozoic, but the earliest known sphinctozoans have been reported from the Middle Cambrian of New South Wales (Pickett and Jell 1983).

Ordovician sphinctozoan sponges exhibit a distinctive biogeographic distribution restricted to Upper Ordovician fold-belt successions of the Paleo-Pacific Ocean rim. They occur in Alaska and northern California (Rigby et al. 1988), in New South Wales, Australia, and in successions in Asia at the margins of the North China Platform and the Altai fold-belt region of northern Xinjiang and the Altai-Sayan mountain belt of Salair, southwestern Siberia (Pickett and Webby 2000 and references therein).

Diversification Patterns

Ordovician sponge evolutionary development shows three major diversification peaks (figures 12.1 and 12.2). A first one, in the Middle Ordovician, occurred on carbonate platforms and includes mainly the diversification of one group, the lithistid suborder Orchocladina. However, this expansion in sponge diversification involves a more complex array of group evolutionary patterns. In the Lower and Middle Ordovician, a diversity peak of hexactinellids, monaxonids, and a few orchocladines occurred in *TS.*1c. Orchocladine diversity increased slightly in *TS.*1d–2a, and a more important rise occurred in *TS.*2c, with megamorinids also appearing as minor components. There was an important decline in *TS.*3b–4a before the great rise near the base of the Darriwilian (*TS.*4a–b).

The second and third diversification peaks (*TS*.5b and 6a respectively) in the Late Ordovician are greater in the sense that they involved not only greater diversity but also a wider range of taxa, the suborders Rhizomorina, Tricranocladina, Sphaerocladina, and Megamorina, as well as a main diversification of sphinctozoan and hexactinellid sponges. The orchocladines also recorded important peaks in species diversity in the mid Caradoc (*TS*.5a–b) and early Ashgill (*TS*.6a). However, diversity values of genera and species of sphinctozoans, tricranocladines, and sphaerocladines are more significant. The Late Ordovician sponge radiation was associated with mixed calcareous-siliciclastic platforms, foreland basins, and island arcs of tropical and subtropical areas.

Early Ordovician

Simple anthaspidellid forms of orchocladines first appeared in the Early Ordovician. A few widespread genera were contributors to the development of reef mounds. The rise of orchocladines coincided with a worldwide transgression documented in *TS*.1d–2a. Sea level rose, and large parts of the low-relief continental margins were inundated toward the end of the Early Ordovician (Ross and Ross 1992).

In the Lower Ordovician there was widespread addition of sponge-bearing reef mounds. Rather than being purely stromatolitic or thrombolitic, these bioherms had a sponge-microbial fabric. The hard substrate offered by the microbial structures supported a diverse metazoan benthic fauna, including lithistid sponges. After this initial success these lithistid-microbe reef mounds virtually disappeared at the end of the Early Ordovician. Only scattered examples continued to appear in the Mid Ordovician.

The appearance of microbial-sponge mounds had been favored by the abundance of suitable environments (carbonate shelf and ramps). The record of Late Cambrian and Early Ordovician rocks indicates that hard substrates, especially widespread hardgrounds, had become abundant in shallow-shelf environments by this time. This increase in hard substrates has been related to the change in seawater chemistry ("calcite" seas of Sandberg 1983). Appearance of lithistid sponges in these environments may also have been related to the late Tremadocian transgression and the incursions of nutrient-rich waters from deep basin and/or inner platform settings (Brunton and Dixon 1994).

Mid Ordovician

There was a widespread modification in morphology and body plans in sponges following the early Darriwilian (*TS*.4a–b). The dominant anthaspidellid forms of the Early Ordovician remained important, but a great variety of canal systems and complexity of skeletal nets appeared. These latter discoidal, globose, palmate, and digitate forms were all better adapted to a wide range of carbonate ramp and platform environments from nearshore to offshore settings.

This important diversification, which occurred mainly among the orchocladines, produced no significant geographic expansion. Until the Late Ordovician, sponges thrived across the carbonate platform belt of Laurentia from the Great Basin to the northern Appalachians and on the carbonate platforms in the Argentine Precordillera and China (figure 12.3). Faunal exchange between these areas was a common feature during the Early and Mid Ordovician (Carrera and Rigby 1999).

The Mid Ordovician brought about an increasing complexity in reefs. As a result, new Mid and Late Ordovician reef associations include a variety of baffling and encrusting organisms such as bryozoans, corals, and stromatoporoids, but lithistid sponges retain an accessory role in these younger bioherms.

Late Ordovician

In the Late Ordovician, sponges underwent maximum diversification, with the appearance of several groups of lithistids (suborders Sphaerocladina, Rhizomorina, and Tricranocladina) and of calcareous and hexactinellid sponges. A high proportion of the genera are endemic, indicating significant isolation and, as a result, marked provincialism in some areas (Carrera and Rigby 1999).

Diversification of lithistid suborders started in the mid Caradoc (*TS*.5b), with few representatives of sphaerocladines and tricranocladines in eastern North America and Baltica. A minor decrease is evidenced in *TS*.5c with a major peak in *TS*.6a. The suborder Rhizomorina is a characteristic, widely distributed group in the Late Paleozoic, but to date in the Ordovician, it is a suborder restricted to Australia. Its peak of diversification is restricted to *TS*.5d and 6a.

After their maximum diversity in *TS*.4b, orchocladines recorded two other maxima in the mid Caradoc *TS*.5b and early Ashgill *TS*.6a, following a somewhat similar pattern experienced by other lithistid suborders. In a different scenario, and restricted to western North America and eastern Australia, sphinctozoans registered a diversification peak in the mid Caradoc *TS*.5b with a decline in *TS*.5c–d and a major peak in *TS*.6a. Hexactinellids recorded a diversification peak with first genera inhabiting shallow platform facies in *TS*.4b to *TS*.5b, with a major peak in *TS*.5c and *TS*.6a.

Late Ordovician sponge radiation occurs predominantly in three main associations localized in Laurentia, Baltica, and island-arc terranes, now located in eastern Australia and the western North American Cordillera. Diversification of sponges in Baltica represents an important event restricted to the mid Caradoc and early Ashgill (*TS*.5c–6a); however, uncertainty of ages of the Baltic faunas prevents precise analysis. Radiation and migration patterns in Laurentia appear to be well established and provide some clues about patterns of sponge diversification and migration across the whole paleocontinent and the Iapetus Ocean.

The northern Appalachian sections record a long history of sponge development. Anthaspidellid orchocladines dominated the area from the Tremadocian to the Darriwilian, for example, in western Newfoundland, in the Romaine Formation of the Mingan Islands, and in Vermont and New York (Rigby and Desrochers 1995 and references therein). There was an addition of possible migrants from the Argentine Precordillera and the Great Basin in the early Caradoc, which raised to 20 the total number of anthaspidellid and streptosolenid orchocladines and megamorinids (figures 12.3 and 12.4). The first arrivals of sphaerocladines (*Astylospongia* and *Caryospongia*) to Laurentia, probably from Baltica, occurred in the mid Caradoc (*TS*.5b), in the Ottawa Group, St. Lawrence Valley (Wilson 1948).

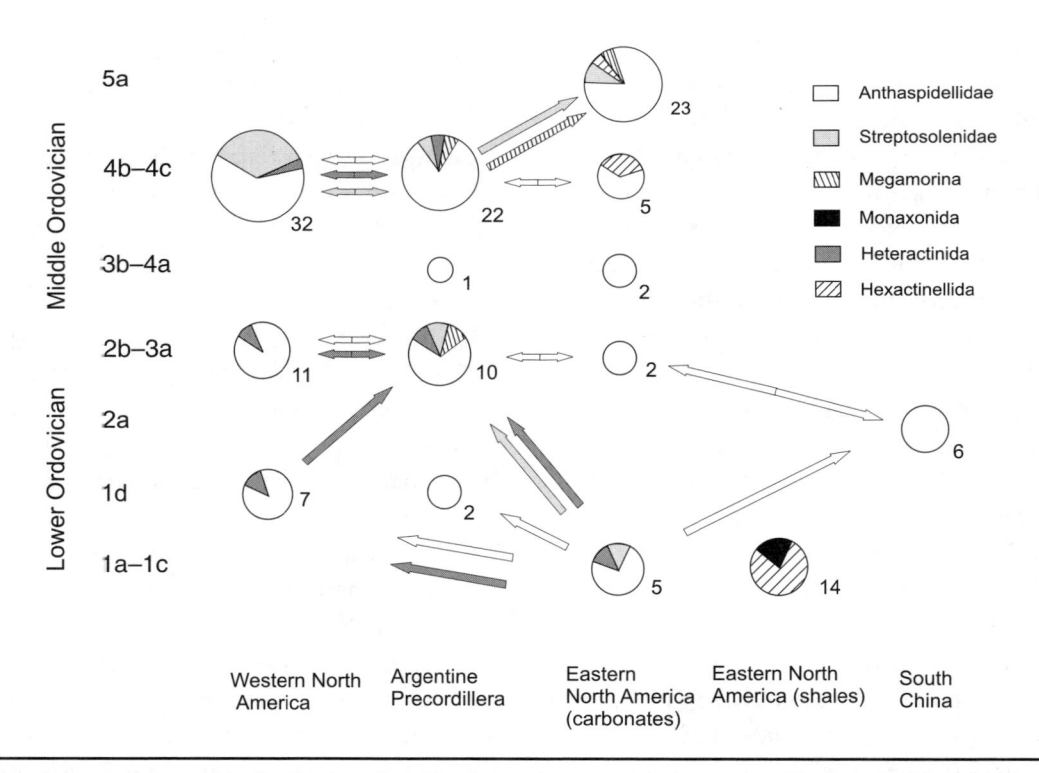

FIGURE 12.3. Patterns of geographic distribution, diversification, and major migration routes of Early and Mid Ordovician sponges. Sizes of circles and numbers indicate the numbers of species, and the relative sizes of the segments show the relative abundances of the sponge groups. Different arrows indicate possible migration routes of the various groups.

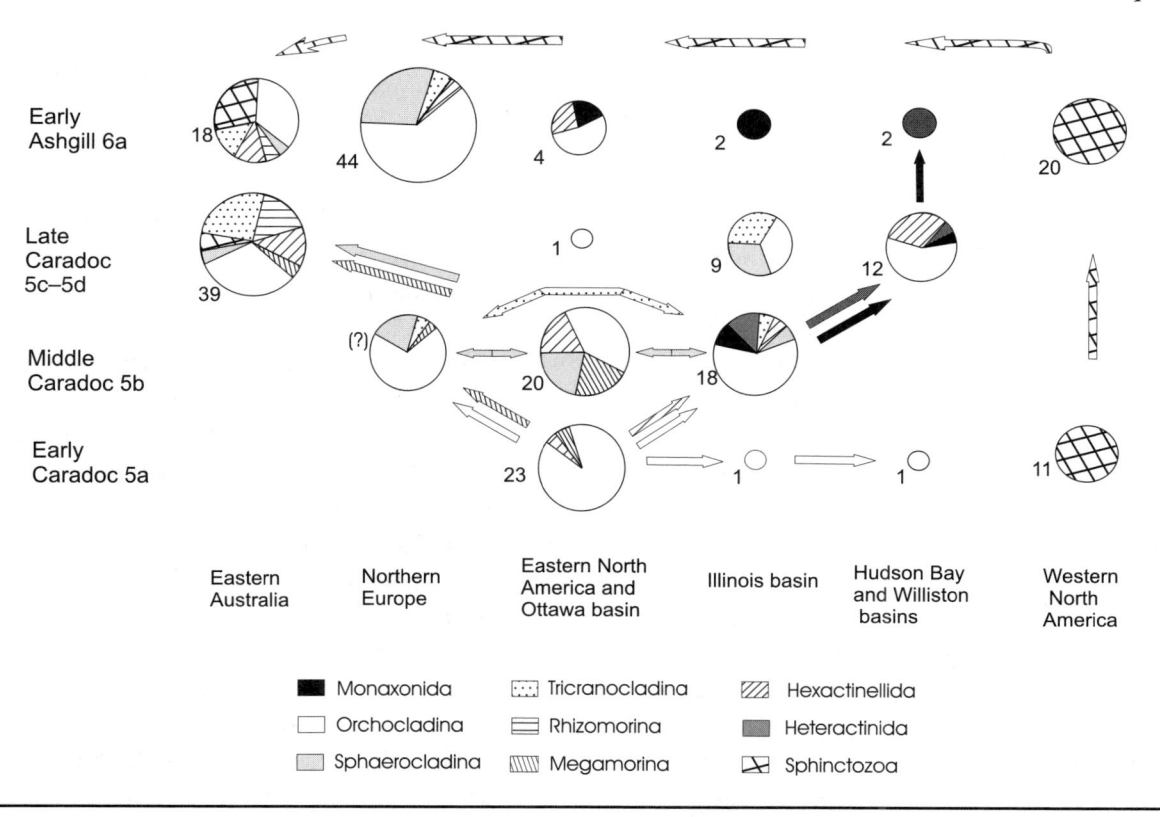

FIGURE 12.4. Patterns of geographic distribution, diversification, and major migration routes of Mid and Late Ordovician sponges. Sizes of circles and the numbers indicate the numbers of species, and the relative sizes of the segments represent the relative abundances of sponge groups. Different arrows indicate possible migration routes of the various groups.

The sponge fauna then migrated to the adjacent Williston and Illinois basins. This migration was favored by a transgression that also probably carried sponge larvae to nearby areas on the craton. Consequently, late Caradoc and early Ashgill faunas are widely distributed across the midcontinental interior (figure 12.4). The geographic dispersion of sphaerocladines and tricranocladines continued in the Silurian of Arctic Canada (Rigby and Chatterton 1989).

Representatives of tricranocladine and sphaerocladine suborders are, in general, spherical forms with centripetal skeletons and canal systems. The genus *Hindia* and its relatives are the most widely distributed Paleozoic sponges and commonly occur together, which suggests that the spherical forms were considerably more mobile than the associated asymmetrical genera (Rigby 1991; Carrera and Rigby 1999). These morphologies are best adapted to changing environments with shifting substrate, which allowed them to move into a wide range of environments, expanding geographic distribution of these sponges in the Late Ordovician.

Radiation and migratory patterns have also been recognized in the spread of Pacific faunas (Carrera and Rigby 1999) (figure 12.4). Ordovician sphinctozoans in eastern Australia (New South Wales; Pickett and Webby 2000) occur in an island-arc terrane, as do those in Alaska and in the Klamath Mountains, California (Rigby et al. 1988). They also occur in northwestern China (Chinese Altai Mountains) in island arcs, and recent finds have been made of *Cliefdenella*-like forms in Kazakhstan, also in an island-arc setting (Webby pers. comm.).

Sphinctozoans are very rare in platform associations of the world, many of which are still preserved, but the island-arc terranes have been largely subducted. Therefore, the positive record of sphinctozoans in the few preserved island arcs is overwhelming evidence that they must have diversified and dominated in such habitats, at least during the Ordovician.

Concluding Remarks

Globally, the Ordovician was marked by increased levels of tectonic activity. The Early Ordovician documents a final stage of extensive tectonism and the development of vast carbonate platforms. Volcanism and orogeny (e.g., Taconic Orogeny) increased substantially during Mid and Late Ordovician time, associated in part with the closing of the Iapetus Ocean.

Miller and Mao (1995) suggested that tectonic activity, mainly volcanism and orogeny, was responsible for an increase in marine biodiversity. They proposed that the increment of Mid and Late Ordovician biodiversity occurred mainly in tectonic active zones such as foreland basins and transition zones, whereas the generic richness of carbonate platforms remained fairly stable through the entire Ordovician. They based their assumptions in part on the increase in habitat partitioning and shifting substrates that developed during tectonic activity.

The peak of sponge diversity in the Mid Ordovician, although restricted to the orchocladines, occurred in carbonate platforms, whereas that of hexactinellids and monaxonids occurred mainly in deep-water environments. In the Late Ordovician the lithistid suborders Sphaerocladina, Rhizomorina, and Tricranocladina, the calcareous sphinctozoans, and the hexactinellid sponges are main representatives of the major diversity peak. Late Ordovician sponge radiation was localized in areas of active tectonism and orogeny (with the probable exception of Baltica, where sponges were apparently derived from a carbonate platform).

All these regions (including Baltica) record abundant sponge faunas, along with spicule-rich deposits or concentrations of opaline silica. They are unusual in the fossil record and document times and areas of high productivity. Such areas in modern seas correspond to areas of upwelling (Parrish 1982) where nutrients are brought up into the photic zone.

More or less contemporaneous sponge associations from midcontinental North America can be used as an example of differential diversification of sponges in two dissimilar geodynamic contexts. These sponge associations have partially comparable relationships with coral assemblages from the vast transcontinental Red River–Stony Mountain Province and, in its eastern part, the discrete Richmond Province (Elias 1995). The region, especially east of the Transcontinental Arch, was influenced by a combination of tectonic activity (Taconic Appalachians) and a related transgressive-regressive cycle. Within the main western part of the Red River–Stony Mountain Province, deposition on the continental shelf and in the epeiric sea was predominantly of carbonates (Williston and Hudson Bay basins). Orchocladines and some heteractinid sponges mainly occur in these regions, with only limited diversification. In provincial components east of the transcontinental arch (Illinois and Michigan basins), however, clastic deposition was significant (Elias 1995). It included the Maquoketa Group, which consists mainly of argillaceous material, probably principally derived from the Taconic upland. Diversification of lithistid tricranocladines, sphaerocladines, megamorines, and hexactinellids in this eastern region was important and occurred in a more tectonically active setting.

Diversification of sponges in the Ordovician shows a strong geographic and environmental overprint. Early and Mid Ordovician diversification follows a worldwide pattern. Curves of diversity peaks in the Great Basin and in the Appalachians are very similar to those of the Argentine Precordillera and China. In the Late Ordovician, however, paleogeography exerted a more direct influence, with peaks of diversity appearing to be slightly decoupled between these geographic regions (see figures 12.3 and 12.4). Ordovician radiation of sponges appears to be also influenced by the global sea level pattern (Ross and Ross 1992), with major peaks coinciding with high sea level intervals.

The sponge decline in Ashgill TS.6b predates the Hirnantian extinction. Recovery of sponge diversity in the Lower Silurian involved major groups that were dominant in the Ashgill diversification peak. Silurian assemblages are mainly composed of orchocladines, sphaerocladines, tricranocladines, and some hexactinellid groups.

Important changes in sponge classification and significant recent discoveries have occurred in the past 20 years. Radiation patterns and the compilation of sponges according to Sepkoski's evolutionary faunas (Sepkoski and Sheehan 1983), therefore, should be revised in light of these new taxonomic studies and the Ordovician data compiled in this work.

ACKNOWLEDGMENTS

We are grateful to John Pickett and Ronald Johns for their helpful, supportive comments in their reviews of the chapter. Appreciation is extended to the editors of the book, especially Barry D. Webby, for their useful comments on earlier versions of the manuscript. MGC acknowledges support from CONICET and ANPCyT.

13 Stromatoporoids

Barry D. Webby

Representatives of the class Stromatoporoidea Nicholson and Murie 1878 (*ex* order Nicholson and Murie 1878; emended Stearn 1980) are simple, large, aspiculate, Paleozoic calcified sponges that show an essential unity. The calcified basal skeletons have laminar, domical, bulbous, branching or columnar external form and internally a more or less continuous, monotonous skeletal meshwork of growth-normal and growth-parallel structural elements, as well as, in places dependent on preservation, canal-like aquiferous systems. The group as a whole comprises seven orders, 26 families, 120 genera, and approximately 1,190 described species (Stearn et al. 1999). The stromatoporoids were important frame-building contributors to Ordovician-Devonian reefs.

Only two stromatoporoid orders—the Labechiida and the Clathrodictyida—are confirmed as having an Ordovician biodiversity record (figure 13.1). The Labechiida is the most diversified and abundant Middle to Upper Ordovician group, with six families and 22 genera. Another family, the problematic Lower to Middle Ordovician Pulchrilaminidae, is doubtfully assigned to the Labechiida. The Clathrodictyida appeared (two families and four genera) in the Late Ordovician but became a far more diversified and widespread (cosmopolitan) group through the Silurian.

The Labechiida has a basal skeletal network of growth-parallel, upwardly convex cyst plates and interconnected growth-normal pillars (or denticles where confined to tops of cyst plates). A thin layer of soft tissue is presumed, by analogy with living coralline sponges, to have mantled the upper growing surface of the skeleton (Stearn 1975). Canal-like aquiferous systems are rarely preserved, but there are commonly upraised centers of growth (mamelons) in skeletons of the order. This suggests a differentiation of inhalent and exhalent water flows, with the radiating excurrent (astrorhizal) canals probably confined entirely to the centers, where the soft tissue is updomed over mamelons of the underlying skeleton.

The Clathrodictyida exhibits a markedly different skeletal meshwork. The prominent, laterally persistent laminae are planar in some genera and more zigzagged in others, with characteristically downwardly inflected pillars (Webby 1994; Nestor 1997). The skeletons also exhibit more common astrorhizae, suggesting that the canal systems and associated soft tissues penetrated the uppermost galleries. In consequence, the canals were more likely to become imprinted within the skeleton (Stearn and Pickett 1994).

Ordovician stromatoporoids occupied rather similar warm, shallow, well-circulated (non-turbid) reef to bank-type (and level-bottom) habitats in a variety of carbonate platform, shelf, and fringing island settings. They characteristically occur in reefs (Webby 2002) and in associations with bahamitic-type sediments. Jaanusson (1973) first established the association of Ordovician stromatoporoids, reefs, and

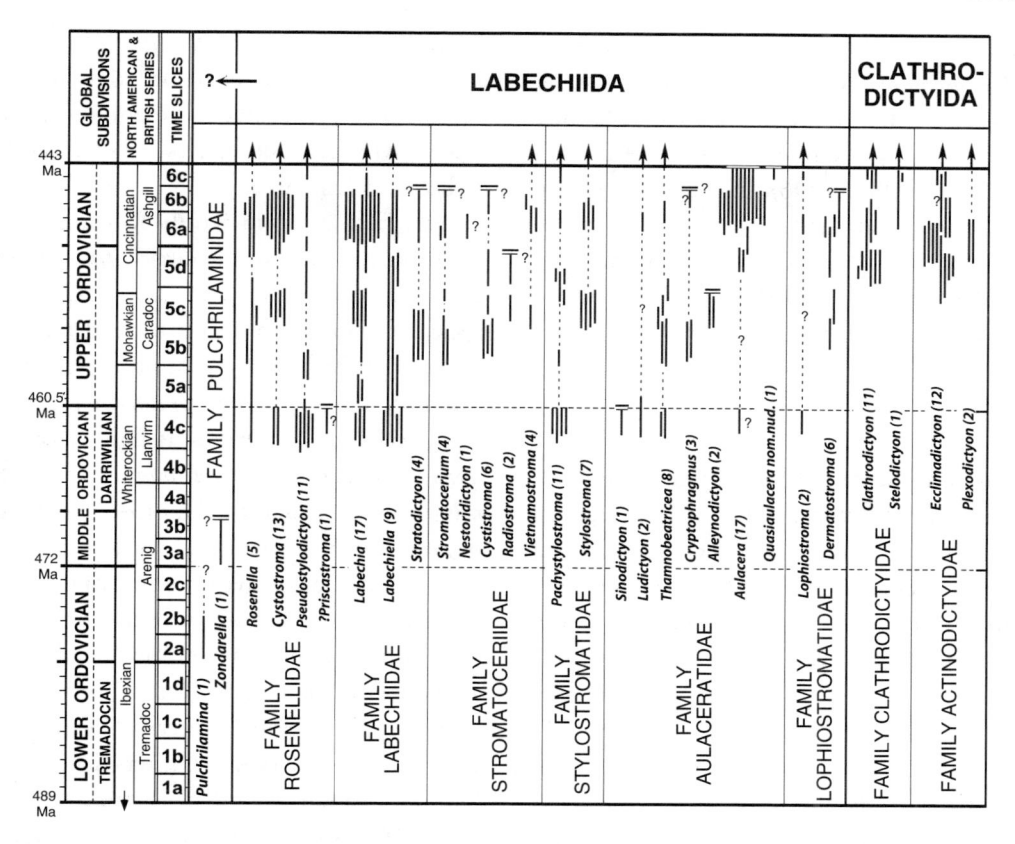

FIGURE 13.1. Range chart showing the temporal distribution of Ordovician stromatoporoids worldwide—named genera and species numbers (in parentheses). Close-spaced double horizontal lines represent the extinction levels of certain genera; the vertical arrows show the other genera that range upward into the Silurian; and each solid vertical line depicts the range of a species range within a genus—that is, representing the sampled or "recorded" diversity data, as distinct from the "range-through" data that include an extrapolation of the ranges of genera (shown by the fine, dotted vertical lines) across time slices where there are no sampled records.

bahamitic sediments in Baltoscandia as indicative of shallow, warm-water depositional conditions. Most of the stromatoporoids of these shallow, warm-water reef and bank-type associations have skeletons of laminar, domical, and bulbous types.

A second community (or biofacies) type is characterized by certain columnar growth forms. Included are a number of representatives of the family Aulaceratidae, especially the genus *Aulacera* (figure 13.1). Spectacular examples of these are preserved as "forests" of large broken aulaceratid specimens in the uppermost Ordovician successions of Anticosti Island, Canada (Cameron and Copper 1994). The huge tree-trunk-like skeletons are now incorporated in storm-generated deposits. They are interpreted by Cameron and Copper (1994) to have grown originally to heights of more than 2 m on soft muddy carbonate sediments of a carbonate ramp (or deeper platform) in water depths of at least 10 m (between fair-weather and storm-wave base). Their bases may have been stabilized by early cementation of the substrate. Nestor (1999) similarly differentiated the Late Ordovician stromatoporoid communities of Baltoscandia into two main community types—a high-energy shoal (and reef) assemblage (Benthic Assemblage [BA] 2 of Boucot 1975), and a moderate- to low-energy, open-shelf assemblage (BA 3).

Additionally there are other late Mid Ordovician to Late Ordovician aulaceratids, for example, branching, cylindrical genera (e.g., *Alleynodictyon*, *Thamnobeatricea*) that occur in bedded, muddy, carbonate successions that represent more restricted, onshore, "lagoonal-type" platform and island shelf settings. Successions of this type are developed in parts of North America, Russia, North China, New South Wales, and Tasmania. The aulaceratids most commonly occur in a low-diversity association with a few other stromatoporoids (*Cystostroma, Pseudostylodictyon*) and small dendroid tetradiid colonies (Semeniuk 1971; Webby 1992a). They probably occupied a BA 1–BA 2 position (Percival and Webby 1996). In a broader global biogeographic context, all these Ordovician stromatoporoid assemblages plot on map reconstructions in low paleolatitudes, between 30°N and 30°S of the paleoequator (Webby 1980, 1992b).

■ Distribution

The pulchrilaminids are the first large (submeter- to meter-sized) domical to irregularly shaped skeletal frame builders to appear in Ordovician reefs, but their distribution remained comparatively localized, to on-shelf reefal sites of North America and the Argentine Precordillera (Webby 2002). The genus *Pulchrilamina* (one species) occurs in North America and ranges through the late Early Ordovician (*TS*.2a–b; see time slices in chapter 2). It resembles labechiids in exhibiting mainly spar-filled latilaminae that alternate with sediment layers, and fine spinelike pillars that protrude above the tops of successive latilaminae into overlying sediment (Webby 1986). In the Argentine Precordillera, the other genus, *Zondarella* (one species), has a limited early Mid Ordovician (*TS*.3) range. These two reef formers disappeared at least 6 million years (m.y.) before the appearance of the first indubitable labechiids.

The first major diversification of the Labechiida occurred during the late Mid Ordovician (late Darriwilian; latter part of *TS*.4b to 4c), at the same time that other metazoans (e.g., corals, bryozoans) and the "solenoporacean" algae were rapidly diversifying, and many new and complex reefal community interrelationships were becoming established (Webby 2002). Five of the six labechiid families, and more than half of the labechiid genera (12 genera, differentiated into 27 species), made their first appearances within the late Darriwilian (figure 13.1). A number of these genera had a limited geographic spread and diversity (one or two species), but a few others (*Pseudostylodictyon, Pachystylostroma, Labechiella, Thamnobeatricea*) became more diversified, with up to five species being recorded. Most of these labechiid genera are long-ranging forms, extending at least into the Silurian, apart from a few columnar aulaceratids (*Aulacera, Thamnobeatricea, Sinodictyon*). This *TS*.4b–c labechiid radiation is recorded in the low paleolatitudes of carbonate platform and shelfal sites in North America, Siberia, North China–Korea, Malaysia, and Tasmania.

The last labechiid family (Stromatoceriidae) and another five labechiid genera (*Stratodictyon, Stromatocerium, Cystistroma, Cryptophragmus, Dermatostroma*) appeared during the early to mid Caradoc interval (*TS*.5a–b), followed by another four (*Stylostroma, Radiostroma, Vietnamostroma, Alleynodictyon*) in the next, mid Caradoc–earliest Ashgill interval (*TS*.5c–d). Two of these (*Vietnamostroma, Stylostroma*) are long-ranging forms, but all the other genera disappeared before the end of the Ordovician. This represents a second, more extensive diversification of the Labechiida that spanned virtually the entire Caradoc interval. This weaker radiation event is recorded more widely across low paleolatitudes, on carbonate platforms and shelf margins (North America, Baltoscandia, Urals, Kolyma, Siberia, North China–Korea, northwestern China, Tasmania) and in island arcs (central New South Wales). The genera *Labechia* (seven species), *Labechiella* (seven species), and *Stylostroma* (four species) are the most diversified Caradoc labechiids.

Initial radiation of the Clathrodictyida occurred during the mid to late Caradoc (*TS*.5c–d), with the genus *Ecclimadictyon* appearing first, then *Clathrodictyon*. Diversification of both genera was rapid through *TS*.5d with up to 10 species of *Ecclimadictyon* and 6 species of *Clathrodictyon* recorded. The two genera continued to be widely distributed through the latest Ordovician. Two other clathrodictyids also appear, in local occurrences—*Plexodictyon* (*TS*.5d) in New South Wales, and *Stelodictyon* (*TS*.6b) in Estonia.

The highest diversity of Ordovician stromatoporoid species occurs through the early to mid Ashgill interval (*TS*.6a–b), with both labechiids and clathrodictyids contributing significantly to this diversity increase. Of particular note is the great expansion of columnar aulaceratids, with up to 13 species of *Aulacera* recorded worldwide during *TS*.6a–b (figure 13.1). In the localized Vaureal–Ellis Bay succession on Anticosti Island (Canada) that ranges through *TS*.6b–c, five species of *Aulacera* are recorded (Bolton 1988).

A marked decline in labechiid diversity (with some associated extinctions) occurred through the late Ashgill, or Hirnantian (*TS*.6c), interval. The diversity

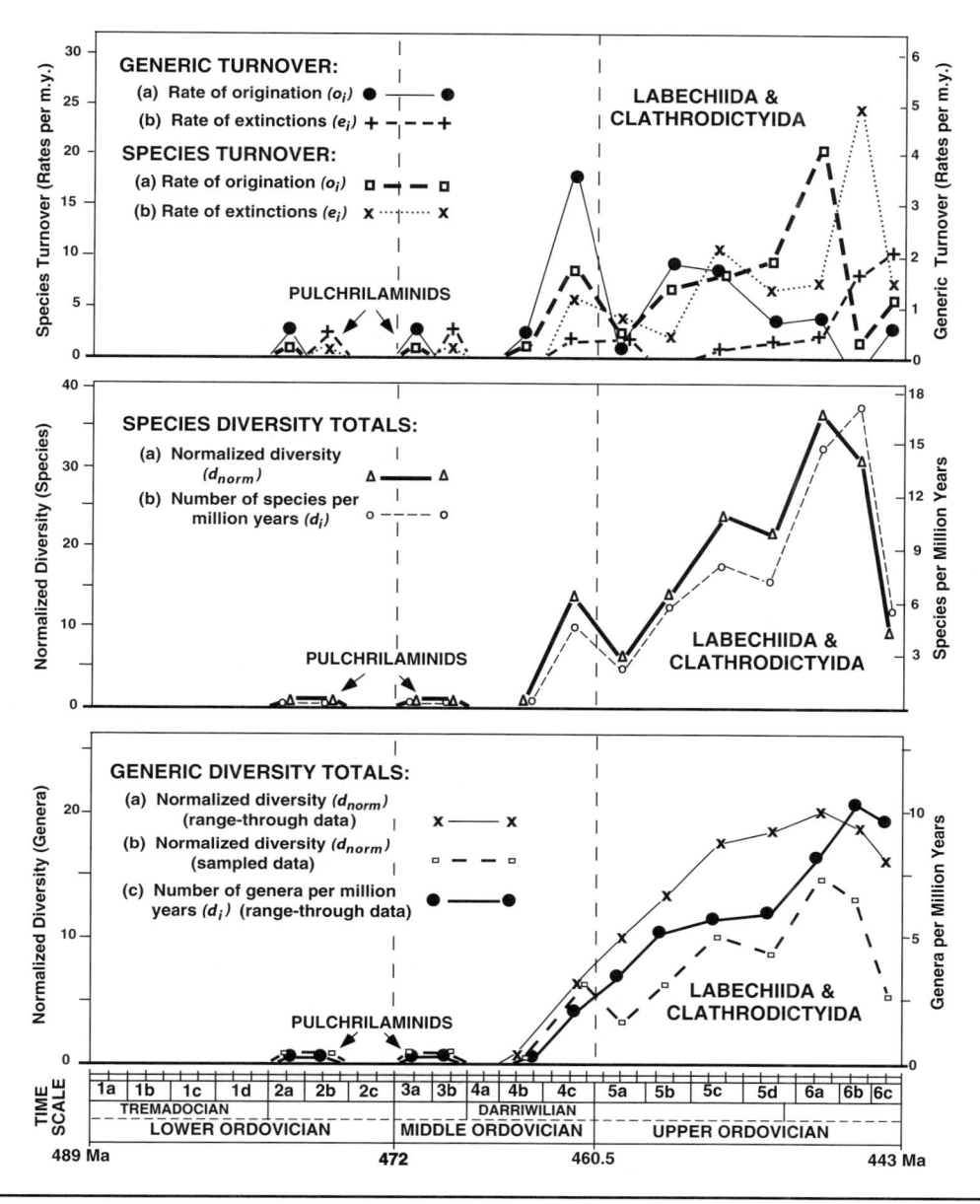

FIGURE 13.2. Diversity curves for the Stromatoporoidea, at both generic and species levels, through Ordovician time. The top diagram shows generic and species turnover—rates of appearances (o_i) and disappearances (e_i) of genera and species per million years. The middle and bottom diagrams depict the normalized species and genus diversity totals (d_{norm}) for each time slice and the number of species and genera per million years (d_i), respectively.

changes seem to coincide directly with the two distinct, short-term extinction pulses of Brenchley and Marshall (1999). The first (and more important) extinction event occurred at, or near, the base of the Hirnantian (across the transition from *TS*.6b–c) with dramatic decline, a loss of almost 70 percent of the species diversity (mainly labechiids and a few clathrodictyids), as well as the disappearance of five labechiid genera—*Stratodictyon, Stromatocerium, Cystistroma,* *Cryptophragmus,* and *Dermatostroma* (figures 13.1, 13.2). As the clathrodictyids became more important contributors to reef growth in the mid Hirnantian (e.g., Anticosti Island and Estonia), the labechiid component of these assemblages began to decline.

The second event, in the mid to late Hirnantian (late *TS*.6c), also includes loss of species, as well as the extinction of almost all the columnar aulaceratids (e.g., long-ranging genus *Aulacera* and short-lived

Quasiaulacera—this latter a *nomen nudum;* Copper 1999). *Ludictyon* was the only aulaceratid genus to survive into the Silurian. Representatives of all the labechiid families and half of the Ordovician genera persisted, but only as minor, mainly holdover components through the Silurian to the Late Devonian (Frasnian). After brief resurgence during the Famennian to a total of 13 genera (Webby 1993), the group finally disappeared at the Devonian-Carboniferous ("Hangenberg") extinction event. Clathrodictyids, after extensive radiation during the Early Silurian, became the dominant Silurian group and had an important, continuing role in the Devonian to the end Devonian event (Nestor 1997).

■ Diversity Patterns

Genera

The normalized generic biodiversity data have been assessed in two ways through the Ordovician time slices. The data showing the known (sampled) species ranges also record the known stratigraphic ranges of the various genera—the "recorded" generic data (figure 13.1). However, there are many gaps in continuity of these recorded genera across the time slices, and many of these temporary disappearances and reappearances reflect inadequate sampling, limited documentation, restricted number of paleoenvironments available, or other factors. Consequently, the generic data also need to be presented as a "range-through" diversity to give a more realistic picture of the diversity patterns.

The "recorded" normalized generic diversity (d_{norm}) exhibits an initial sharp rise during the late Darriwilian (*TS*.4c) to a value of 6.5 genera, or genera per million years (d_i) of 2.2, and then briefly declines in the earliest Caradoc (*TS*.5a), only to increase again to a modestly higher peak during the mid Caradoc (*TS*.5c), to 10 genera. Another limited decrease in diversity followed in the late Caradoc (*TS*.5d), then a rise to the maximum of diversity during the early to mid Ashgill interval (*TS*.6a–b), to a value of 14.5 genera, or d_i of 5.8. Finally, the diversity declined dramatically through the late Ashgill (5.5 genera, or d_i of 3.2). The generalized shape of the normalized generic diversity curve is similar to that shown by the species (figure 13.2, bottom).

The "range-through" diversity pattern differs in showing a gradual increase from the late Darriwilian (*TS*.4c) with a diversity of 6.5 genera and then a flattening of the curve through the mid to late Caradoc interval (*TS*.5c–d), when the diversity reaches 18 genera. A further gentle rise follows, to a high of 18.5 genera, or d_i of 10.2, in the mid Ashgill (*TS*.6b), and then decline during the late Ashgill (*TS*.6c).

The generic turnover rates show some similarities to the species data, but the Caradoc-Ashgill appearances do not track the disappearances like the species patterns (figure 13.2, top). Genus appearance rates (o_i) initially remained at background levels of 0.5 from the late Early Ordovician to the mid Darriwilian (through *TS*.2a–4b inclusive). But then in the late Darriwilian (*TS*.4c), an initially elevated rate of origination of labechiids occurred with appearance rates (o_i) of 3.7 (note the prominent spike in figure 13.2, top). A second, less intensive origination pulse followed in the early to mid Caradoc (*TS*.5b) with an o_i of 1.8. Appearance rates then decline steadily until near background levels were again attained during Ashgill time (*TS*.6a–c).

Genus disappearance rates (e_i) continued at background levels of 0.5 until the Ashgill (figure 13.2, top). Only during the mid to late Ashgill interval (*TS*.6b–c) was there a significant rise in the disappearance rates (e_i) to a peak of 2.1 during *TS*.6c. These pulses of generic extinction are associated with the two sharp cooling to warming events of the Hirnantian glaciation.

Species

The normalized species diversity (d_{norm}) and species numbers per million years (d_i) plots show a rather similar tracking pattern through late Mid to Late Ordovician time (figure 13.2, middle) except for the slightly divergent relationship at the culmination of the diversity increase in the Ashgill, between *TS*.6a and 6b. The diversity (d_{norm}) rises in three steplike stages: the first, a late Darriwilian peak of 14 species (only labechiids); the second, a mid Caradoc peak of 23.5 species (labechiids and the first clathrodictyids); and the third, the culmination of diversity increase in the early to mid Ashgill with some 37 species (both labechiids and clathrodictyids). The lowered levels of diversity across the intervening *TS*.5a and 5d (figure

13.2, middle) are more clearly related to slowing rates of origination than to localized peaks of extinction. The marked decline in diversity from TS.6b–c, on the other hand, is clearly related to species extinction events associated with the end Ordovician glaciation.

Unlike the generic turnover profiles, the species-level data show an approximately similar tracking pattern (figure 13.2, top). There are two distinct spikes of elevated species origination turnover (o_i), at 8.5 for the late Darriwilian (TS.4c) and 20.6 for the early Ashgill (TS.6a), with a more modest rise across the intervening Caradoc (TS.5b–d) interval (figure 15.2, top). The counterparts are the three steplike peaks of extinction turnover (e_i), of 5.2 for TS.4c, 10.7 for TS.5c, and 25 for TS.6b (figure 13.2, top). This latter set of peaks clearly depicts the species extinctions related to the first (early Hirnantian) glacial cooling event.

■ Discussion

Nestor and Stock (2001) regarded the Late Ordovician stromatoporoids (only labechiids and clathrodictyids) as temperature-sensitive organisms, based on their studies of North American and Baltoscandian Late Ordovician–Early Silurian genera. They noted the gradual extinction of genera through the late Caradoc to end Ordovician interval, relating the changes to global cooling. The extinctions of mainly labechiid genera were attributed to their failure to adapt to the cooler conditions. Stearn (1987), on the other hand, suggested that labechiids, at least those in the Late Devonian, were better adapted to cooler waters because of their short-lived Famennian resurgence after the end Frasnian ("Kellerwasser") faunal crisis. This event caused extinctions of other stromatoporoid groups and the dramatic decline of the Mid Paleozoic stromatoporoid–coral reef community. Sheehan (2001b) has noted that groups with a simple morphology such as graptolites, bryozoans, and acritarchs preferentially survived the end Ordovician extinctions, and stromatoporoids also exhibit this pattern. Some of the labechiid genera and all the clathrodictyid genera adapted to or otherwise survived extremes of cooling during the intensive, short-term glaciation at the end of the Ordovician.

The labechiids have a very patchy Silurian and Devonian record and do not seem to occur in associations separate from the nonlabechiid stromatoporoids. For example, in a particularly well exposed Lower Devonian stromatoporoid-bearing succession of North Queensland, Webby and Zhen (1997) found the labechiids as poorly preserved minor components in associations with the much better preserved, more abundant nonlabechiids at a number of horizons and localities. The labechiids were an important, though inconspicuous, component of the well-preserved Lower Devonian succession and therefore likely to be overlooked if the exposures are poor or the rocks metamorphosed or diagenetically altered. These factors also apply to the Ordovician record. Another factor in sequences of a particular environmental setting where gaps occur in the record may relate to the loss of the ability of the relatively simple labechiid organism to calcify its skeleton.

Nestor and Stock (2001) also commented on the differential preservation of labechiids and other stromatoporoids. They claimed that skeletal changes accompanied the gradual cooling in the Late Ordovician, with poorly calcified (aragonitic) labechiids through the TS.4c–5b interval and then the better-preserved, calcitic labechiids and clathrodictyids later, in TS.5c–6c interval. This interpretation, however, remains controversial, at least until more adequate documentation of the preservational and successional changes is provided. If Nestor and Stock's arguments are accepted, then the two main radiations of labechiid genera, the initial TS.4c event and, to a lesser extent, the TS.5c event (figure 13.2, top), may have occurred in the warmer conditions. However, labechiid species show the greatest diversification during the early to mid Ashgill (TS.6a–b), when Nestor and Stock claim the global climate to have been significantly cooler.

Unfortunately, there is as yet no precise record of climatic fluctuations through Late Ordovician time, though Veizer et al. (2000) have shown a rising linear trend of oxygen isotope values that suggests a progressive decline in tropical sea surface temperatures. An oceanic overturn probably accompanied the change from greenhouse to icehouse conditions toward the end Ordovician glaciation, but the timing of this turnover is uncertain (Berry and Wilde 1978). Berner's (1994) evidence that concentrations of atmospheric carbon dioxide remained high during the Ordovician (some 14 to 18 times present-day levels) remains difficult to reconcile with the view of progressive cooling leading into the end Ordovician glaciation.

The dramatic, successive sea level changes through the Late Ordovician (chapter 10) provide a means of interpreting in more detail the nature of the climatic change. Significant highs seem to have occurred in the *TS*.4c, 5b, 5c, 5d, and 6a–b intervals, moderate lows in the *TS*.5a, 5b–c, and 5c–d boundary intervals, and more extreme lows during *TS*.6a, the 6b–c boundary, and the 6c interval. These many short-term episodic drowning and shallowing events may represent alternating warming and cooling phases that were superimposed on an overall cooling through Late Ordovician time. The Baltoscandian curve (Nielsen, chapter 10) shows progressively lowered maximum lowstand peaks through Late Ordovician (*TS*.5a to 6c) time, and a complementary pattern of decline in highstand peaks from *TS*.5d to 6c also occurs. The North American sea level curve (Ross and Ross 1995) exhibits a broadly similar decline in highstand peaks through the late Caradoc–Ashgill (*TS*.5c to 6c) interval. Nielsen interpreted the Baltoscandian curve as dominantly eustatic. It seems unlikely, therefore, that the sea level changes were, to any significant extent, modified by the equatorially directed drift of Baltica through Mid to Late Ordovician time (Torsvik et al. 1992).

Nielsen depicted significant highstand episodes in the upper *TS*.4c, across the 5b–c boundary, in the mid 5c, mid 5d, and in 6a–b, and the lowstands, more moderate in *TS*.5a, lower 5c, and the 5c–d boundary intervals, to more extreme in *TS*.6a, across the 6b–c boundary, and in 6c. The initial steep rise in genus- and species-level stromatoporoid diversity coincides with the upper *TS*.4c highstand (Nielsen's Furudal Highstand), and the subsequent species-level decline seems to relate to the *TS*.5a lowstand (Vollen Lowstand). Moderate levels of genus- and species-level diversification occurred during the *TS*.5b and 5c highstand intervals (Keila and Nakkholmen drowning events), whereas the peaks of genus- and species-level diversity through the *TS*.6a–b interval (figure 13.2) may be best related to Nielsen's Spannslokket and Husbergøya drowning events (chapter 10: figure 10.2). In relation to declines (and extinctions) of stromatoporoid taxa (especially species) the most important lowstand intervals are across the *TS*.6b–c boundary interval and during 6c; these relate directly to the two main pulses associated with the end Ordovician glaciation.

14 Conulariids

Heyo Van Iten and Zdenka Vyhlasová

Conulariids are an extinct group of benthic marine cnidarians the extant nearest relatives of which are the stauromedusans or the coronatid scyphozoans (Van Iten 1992a, 1992b; Jerre 1994a, 1994b; Van Iten et al. 1996; Collins et al. 2000; Hughes et al. 2000). Conulariids (e.g., *Conularia* Sowerby 1821) having a pyramidal theca may form a monophyletic group that excludes morphologically similar taxa such as circoconulariids. Relatively small conulariids were sessile animals, attached to firm substrates at their apex, but conulariids (e.g., many specimens of *Metaconularia* Foerste 1928) whose thecae exceeded about 10 cm in length may have become recumbent at some point in their lives. Conulariid macrofossils generally are extremely rare, with many Ordovician species represented by fewer than 10 reposited specimens. Partly for this reason, the present survey focuses on the stratigraphic and paleogeographic occurrences of Ordovician genera. Data were obtained primarily from Sinclair (1940, 1941, 1942, 1944, 1946a, 1948, 1952), Hessland (1949), Wilson (1951), Sayar (1964), Serpagli (1970), Drygant (1971), Bender (1974), Babcock et al. (1987), Hergarten (1988), Jerre (1994b), Van Iten (1994), Van Iten et al. (1996), Brabcová (1999, 2000), Vizcaïno and Álvaro (2001), and Richardson and Babcock (2002). Ages of host strata in the Bohemian Massif (Czech Republic) and present-day cratonic North America were obtained mainly from Havlíček and Vaněk (1990), Bergström and Mitchell (1994), Witzke and Bunker (1996), Chlupáč (1998), and Leslie (2000). Our complete data set (Ordovician Conulariid Species and Genera [ORDOCON]) is contained in the database managed by Arnold I. Miller (University of Cincinnati).

■ Ordovician Genera

Ordovician strata have yielded 12 conulariid genera, the most of any system (Cambrian-Triassic) known to contain these fossils. All 12 genera first occur in the Ordovician. Six genera, *Archaeoconularia* Bouček 1939, *Conularia, Conulariella* Bouček 1928, *Eoconularia* Sinclair 1944, *Exoconularia* Sinclair 1952, and *Pseudoconularia* Bouček 1939, first occur in the Lower Ordovician. The Middle and Upper Ordovician together contain the first occurrences of *Anaconularia* Sinclair 1952 (monospecific), *Climacoconus* Sinclair 1942, *Conularina* Sinclair 1942, *Ctenoconularia* Sinclair 1952, *Glyptoconularia* Sinclair 1952 (monospecific), and *Metaconularia. Anaconularia, Conularina, Exoconularia,* and *Glyptoconularia* are known only from the Ordovician. (*Tasmanoconularia tuberosa* Parfrey 1982 from the Upper Ordovician of Tasmania probably is a species of *Conularia*.)

Archaeoconularia, Conularia, Eoconularia, Metaconularia, and *Pseudoconularia* also occur in the Silurian System (e.g., Jerre 1994b). In addition, *Climacoconus* has been documented from Lower Devonian strata in Germany (see Hergarten 1985: plate 4, figure 14). A single genus, *Baccaconularia* Hughes, Gunderson, and Weedon 2000, occurs in Upper Cambrian strata of Minnesota and Wisconsin (Hughes et al. 2000).

■ Stratigraphic Distribution: Bohemian Massif and Cratonic North America

Stratigraphic occurrence data for the Bohemian Massif and the present-day North American craton are reasonably precise (figure 14.1), although in the latter region a Middle Ordovician record is largely missing, and available samples of many taxa are very small. In Bohemia, *Archaeoconularia* first occurs in the middle Tremadocian (Mílina Formation), and *Conulariella* first occurs in the lower Arenig (Klabava Formation). *Metaconularia* first occurs in the middle Arenig (Klabava Formation). Seven of the 19 Ordovician species from Bohemia first occur in the uppermost Arenig (Šárka Formation). Three of these represent first appearances of *Conularia*, *Exoconularia*, and *Pseudoconularia*. Three species, including a species of *Anaconularia*, first occur in the lower Caradoc (Libeň Formation). The base of this interval also contains the last occurrences of six species, including all known (three) species of *Conulariella*. *Conularia* and *Exoconularia* last appear in the lower Ashgill (Bohdalec Formation), and *Anaconularia* is last represented in the middle Caradoc (Zahořany Formation). *Archaeoconularia* and *Metaconularia* are last recorded from the middle Ashgill (Kosov Formation), and *Pseudoconularia* last appears in the latest Ashgill.

Owing in part to very small sample sizes (20 of the 32 North American species shown in figure 14.1 are represented by only one or two reposited specimens [n]), the actual stratigraphic ranges of many North American conulariids are poorly constrained. *Climacoconus* and *Conularia*, by far the two most abundant genera (with at least 600 specimens reposited to date), first occur in the upper Whiterockian, in the Chazy Group of New York and Vermont. *Climacoconus* last appears in the upper Richmondian (Zone 4 of the Vauréal Formation, Anticosti Island), and *Conularia* last appears in the mid to late Richmondian. *Conularina* ($n < 10$ reposited specimens) ranges from the upper Whiterockian (Chazy Group, New York and Quebec) to the upper Turinian (Leray Formation, Quebec/Ontario). *Archaeoconularia* ($n < 10$ reposited specimens) first occurs in the uppermost Whiterockian–lowermost Turinian Holsten Formation of Tennessee and is last reported from the Edenian (upper Trenton Group, Quebec). *Ctenoconularia* ($n < 10$ reposited specimens) first occurs in the Turinian, in the Leray Formation of Quebec and the Platteville Formation of Minnesota, and last appears in the upper Maysvillian Corryville Formation of Ohio. *Metaconularia* also first occurs in the Turinian, in the Platteville Formation of Wisconsin and the Plattin Group of Missouri, and is last recorded from the mid to late Richmondian. *Eoconularia* ($n < 10$ reposited specimens) is first found in the Chatfieldian(?) (lower Verulam [?] Formation, Ontario), and may range upward into the Maysvillian/Richmondian (Georgian Bay Formation, Ontario). *Glyptoconularia* ($n < 10$ reposited specimens), currently known only from present-day cratonic North America, ranges from the Chatfieldian(?) (Dolgeville Facies[?] of the Trenton Group, New York) to the lower Richmondian (Elgin Member of the Maquoketa Formation, Iowa).

■ Paleogeographic Distribution

Laurentia

Present in uppermost Middle to Upper Ordovician strata are *Archaeoconularia*, *Climacoconus*, *Conularia*, *Conularina*, *Ctenoconularia*, *Eoconularia*, *Glyptoconularia*, *Metaconularia*, and *Pseudoconularia* (Sinclair 1940, 1941, 1942, 1944, 1946a, 1948, 1952; Wilson 1951; Babcock et al. 1987; Van Iten 1994; Van Iten et al. 1996; Richardson and Babcock 2002). With the exception of *Pseudoconularia*, all these genera occur in present-day cratonic North America, mainly in open-shelf and shelf-slope carbonates and basinal shales. In the Elgin Member of the Maquoketa Formation (lower Richmondian) of northeastern Iowa/southeastern Minnesota, the relative abundance ratio of reposited macrofossil specimens of *Conularia*, *Climacoconus*, *Glyptoconularia*, and *Metaconularia* is approximately 100:75:2:2 ($n = 333$ specimens; see Van Iten et al. 1996, table 1).

In the Girvan area of Scotland, siliciclastic mudstones of the Upper Ordovician Balclatchie and Drummock formations together yield *Archaeoconularia*, *Climacoconus*, *Conularia*, *Ctenoconularia*, and *Pseudoconularia* (Sinclair 1948).

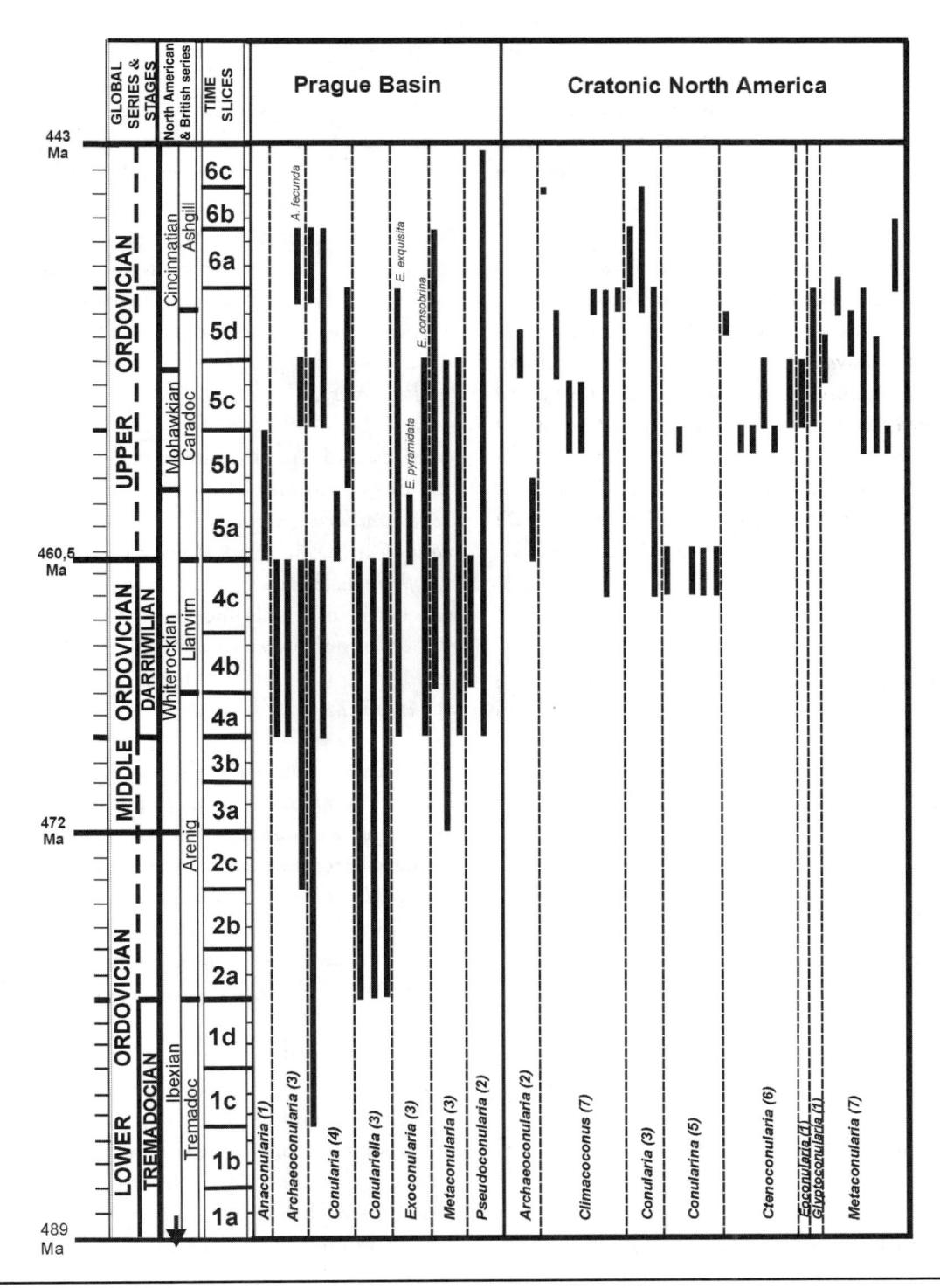

FIGURE 14.1. Approximate ranges of the 19 conulariid species in the Ordovician of the Bohemian Massif (Czech Republic) and all or most species of the nine genera known to occur in the Ordovician of present-day cratonic North America. Owing (in many cases) to very small sample sizes, the lengths of the North American "range" bars should not be interpreted as representing precise stratigraphic range estimates but rather as showing that a species existed during at least part of the time interval indicated for that species. Data for Bohemia are from Brabcová (1999, 2000). Data for present-day cratonic North America are from Sinclair (1940, 1941, 1942, 1944, 1948, 1952), Wilson (1951), Babcock et al. (1987), Van Iten (1994), and Van Iten et al. (1996), with ages of North American rock units mainly from Bergström and Mitchell (1994), Witzke and Bunker (1996), and Leslie (2000).

Baltica

Ordovician rocks of northern Europe collectively yield *Climacoconus, Conularia, Conularina, Exoconularia,* and *Pseudoconularia* (Sinclair 1948; Hessland 1949; Drygant 1971; Hergarten 1988; Jerre 1994b). Present in the Lower Ordovician are *Conularia* (Estonia and Sweden) and *Pseudoconularia* (Sweden) and possibly also *Archaeoconularia* (Estonia). Middle Ordovician limestones in southern Sweden yield *Conularia* and *Pseudoconularia*. Present in the Upper Ordovician, mainly in limestones, are *Conularia* (Estonia, Sweden, northwestern Ukraine), *Conularina* (Sweden), *Exoconularia* (Sweden and northwestern Ukraine), *Pseudoconularia* (Sweden), and *Climacoconus* (Sweden and Baltic Germany; see Hergarten 1988: figure 28a–b).

North Gondwana

Middle and Upper Ordovician siliciclastic strata of Jordan and Morocco collectively yield *Anaconularia, Archaeoconularia,* and *Metaconularia*. In addition, *Pseudoconularia* may be present in the Lower Ordovician of Morocco (Hessland 1949), and *Exoconularia* may occur in the Upper Ordovician *Conularia* Sandstone of Jordan (Bender 1974). *Anaconularia,* previously known only from the Bohemian Massif, also occurs in the Upper Ordovician (Upper Ktao Formation) of Morocco. The Middle Ordovician ("Llandeilian") sequence of Morocco contains *Archaeoconularia* (Termier and Termier 1947), and in Jordan *Metaconularia* occurs in the *Conularia* Sandstone.

■ Perunica

In addition to the Bohemian occurrences discussed earlier in this chapter, the Middle and Upper Ordovician sequences of Thuringia, possibly a part of Perunica, contain *Archaeoconularia* (Ashgill) and *Conularia* (Middle and Upper Ordovician) (Hergarten 1988). In the Bohemian Massif, *Archaeoconularia, Exoconularia,* and *Pseudoconularia* are the three most abundant genera, together accounting for about 70 percent of reposited Ordovician conulariid specimens ($n = 2,000$).

Armorica

The Upper Ordovician sequences of northwestern France and Sardinia collectively yield *Archaeoconularia* (Caradoc of Sardinia; Serpagli 1970) and *Exoconularia* (Upper Ordovician of northwestern France and Sardinia; Sinclair 1948; Serpagli 1970). In addition, Tremadocian mudstones in the southern Montagne Noire area (France) contain *Eoconularia* (*E. azaisi*; Vizcaïno and Álvaro 2001).

Pontides

Middle to Upper Ordovician oolitic ironstones in the Bosphorus area of northern Turkey contain *Archaeoconularia* (*A. fecunda* and four species of *Exoconularia* [Sayar 1964]). Four of these five species (*A. fecunda, E. consobrina, E. exquisita, E. pyramidata*) also occur in Middle and Upper Ordovician strata in Bohemia, northwestern France, and Thuringia.

Avalonia

Present in Ordovician strata of southern Britain are *Archaeoconularia, Conularia, Conularina, Eoconularia,* and *Exoconularia* (Sinclair 1948). Lower Ordovician (Tremadocian) slates of northern Wales yield *Exoconularia*. The Middle Ordovician (Llanvirn) contains *Archaeoconularia* and *Exoconularia,* and the Upper Ordovician (Caradoc) yields *Conularia, Conularina,* and *Eoconularia*.

■ Interpretative Summary

Understanding the diversification of conulariids is hindered by small sample sizes and by the paucity of recent work on the phylogeny and systematics of this group. Nevertheless, a number of intriguing and possibly robust temporal and spatial distribution patterns are evident in the data presented here. For example, the highest taxonomic and morphological diversities ever exhibited by conulariids may have been achieved during late Mid to early Late Ordovician times, when at least 11 highly disparate (morphologically) genera were present (figure 14.1). In addition, the spatio-temporal distribution of Ordovician genera suggests the existence of a fairly high degree of faunal provincialism throughout most of the period and may provide further corroboration of major aspects of current reconstructions of Ordovician paleogeography (chapter 5).

In the late Mid to early Late Ordovician, representatives of nine genera inhabited low-latitude epeiric seas on Laurentia. Six genera, including four (*Archaeoconularia, Conularia, Metaconularia,* and *Pseudoconularia*) present in Laurentia, inhabited Perunica, interpreted as a separate block situated at low middle latitudes between North Gondwana and Baltica. Interestingly, one of the Perunican genera, *Anaconularia,* currently is known only from Upper Ordovician strata in the Bohemian Massif and northwestern Africa, a part of North Gondwana. Ordovician species of *Climacoconus* currently are known only from uppermost Middle to Upper Ordovician carbonate rocks on Laurentia and Baltica, which by Late Ordovician times were situated in relatively close proximity to each other and to Avalonia, at low to low-middle latitudes. The genus *Conularina* is known exclusively from uppermost Middle to Upper Ordovician strata on these three blocks, which were separated from most other Late Ordovician terranes by the Rheic Ocean. To date, only three genera (*Anaconularia, Archaeoconularia,* and *Exoconularia*) have been documented from Upper Ordovician strata in the former high-latitude localities of northwestern Africa and Armorica, suggesting that during Late Ordovician times conulariids may have exhibited a latitudinal taxonomic diversity gradient.

ACKNOWLEDGMENTS

For helpful discussions and/or reviews of our work we thank David Bruton, Nigel Hughes, Axel Munnecke, David Rudkin, Enrico Serpagli, Barry Webby, and Brian Witzke.

15 Corals

*Barry D. Webby, Robert J. Elias, Graham A. Young,
Björn E. E. Neuman, and Dimitri Kaljo*

This contribution includes an overview of relationships between the various Paleozoic coral groups, a global diversity analysis of the exclusively Ordovician tetradiid corals, and regional analyses of the Laurentian corals, Baltoscandian rugose corals, and Australasian corals.

■ Overview (BDW)

The history of Paleozoic corals (phylum Cnidaria, subclass Zoantharia) has been much discussed, especially in relation to the origins and early evolutionary development of the two major groups, Tabulata and Rugosa. Sokolov (1955, 1962), Flower (1961), Weyer (1973), Flower and Duncan (1975), Scrutton (1979, 1984, 1988, 1997), Hill (1981), Neuman (1984), and Oliver (1996) have been at the forefront of these studies. The Ordovician faunas have been widely recognized as pivotal to any attempt to establish early diversification patterns in these groups, but unfortunately for many decades now, comparatively little attention has been given to fully documenting and revising these crucially important faunas. Moreover, until recently, the framework of an Ordovician timescale was not adequately resolved for global correlation. A much more reliable global timescale is now available, but the Ordovician coral faunas still need much critical reappraisal.

Scrutton (1979, 1984, 1997) has evaluated the problematic Early to Mid Cambrian record of coralomorphs, concluding that a number of Cambrian genera are skeletonized zoantharian corals but that none is in a direct line of descent to an Ordovician clade. Whether the nonskeletonized anemone ancestors that acquired skeletons in the Cambrian were related or unrelated to the stocks that gave rise to skeletonized Ordovician corals remains uncertain. Oliver (1996), however, has noted that the anemone group that represented the preskeletal "tabulates" would have been colonial. Skeletonization events probably occurred in anenome groups many times through the Early Paleozoic, at least twice during the Cambrian, and probably three or four times during the Ordovician. Through the Ordovician this resulted in successive radiations of the two major Paleozoic groups (Tabulata and Rugosa), the much smaller, more restricted Ordovician Tetradiida (see later in this chapter) and the very small, solitary, short-lived, and localized kilbuchophyllids of mid Caradoc age (Scrutton and Clarkson 1991).

The exclusively modular (colonial), probably mainly calcitic tabulates are the dominant Ordovician group. They exhibit comparatively smaller corallites, septa less well developed, usually undifferentiated, commonly spinose, and some exhibit mural pores, connecting tubules, or intercorallite coenenchymal skeletal structures. Increase is mainly lateral and non-parricidal (where the parent survives the budding event). On the other hand, the rugosans are solitary to modular (dendroid-cerioid), also with calcitic skeletons, and commonly with larger corallites; septa are inserted in four quadrants, typically in alternating orders (major and minor) and usually with axial tabulation (but rarely peripheral dissepiments in the Ordovician forms). The tetradiids (see later in this chapter) are a unified late Mid Ordovician to Late Ordovician group with a modular (massive to phaceloid, cateniform and ramose), possibly aragonitic skeleton and tetrameral symmetry. Increase is parricidal, axial, and quadripartite, with offsets produced by fusion of inner ends of the four laminar septa. The Kilbuchophyllida is a small scleractinamorph clade of solitary, button-shaped corals, with highly differentiated septa, of mid Caradoc age (Scrutton and Clarkson 1991).

Scrutton (1979, 1984, 1997) depicted the Ordovician generic-level coral diversification in range charts and plots showing overall, Paleozoic-wide diversification changes. These latter plots show the numbers of genera, first appearances, and extinctions of genera (as numbers and as percentages), as well as the changes in overall colony organization, through each epoch of Ordovician-Permian time. In these charts the subdivisions of Ordovician time used were based mainly on the *Treatise* (Hill 1981) and on American usage of "middle Ordovician," encompassing both Whiterockian and Mohawkian. With the base of the *Nemagraptus gracilis* Zone (Bergström et al. 2000) now ratified by the International Union of Geological Sciences (IUGS) and associated International Commission on Stratigraphy (ICS) at the base of the Upper Ordovician Series, the Mohawkian becomes part of the Upper Ordovician (see chapter 2). This means that the data incorporated in the upper half of Scrutton's "middle" Ordovician distributions (see Scrutton 1979: figure 3; 1984: text figure 1; 1988: figures 3.3, 3.4; 1997: figure 20A–C) now belong to the Upper Ordovician Series.

Early Tabulate Coral Diversification

The tabulate corals, as constituted here, mainly following Scrutton (1997), are the following suborders: Lichenariina, Auloporina, Sarcinulina, Heliolitina, Halysitina, and Favositina. In the Early Ordovician the only confirmed tabulates are lichenariinids. The earliest confirmed records are from the early to mid Tremadocian (*TS*.1a–b; see time slices in chapter 2: figures 2.1, 2.2), based on the occurrences of small, calcified modular skeletons of the genus *Lichenaria* in Newfoundland (Pratt and James 1982), Virginia (Bova and Read 1987), central Texas (*L. cloudi* Bassler), and Missouri (McLeod 1979) and possibly another species from a slightly higher Tremadocian horizon (*TS*.1c–d) in Maryland (Bassler 1950). *Lichenaria* is recognized as having had a short-lived, geographically restricted existence in the reef habitat in the early to mid Tremadocian of Newfoundland and Virginia (Webby 2002).

Sokolov (1962) wrote that auloporinids were probably widespread in the Early Ordovician but did not attract attention owing to their small size and poor state of preservation. No Early Ordovician records have yet been confirmed.

The Mid Ordovician is represented by a very sparse fauna, except for common occurrences in the latest part of the epoch, that is, through the late Darriwilian (*TS*.4c) interval. The only confirmed pre–late Darriwilian Mid Ordovician record is the first appearance of *Cryptolichenaria* on the Siberian Platform (Sokolov and Tesakov 1963), which correlates with the graptolite-based *D. hirundo* Zone, of early Mid Ordovician (*TS*.3a–4a) age. Another species of *Lichenaria* is recorded from the still younger Lehman Formation of western Utah by Rigby and Hintze (1977) and has a mid Whiterockian (Trilobite Zone N; i.e., mid Darriwilian = *TS*.4b) age.

Consequently, *Lichenaria* and, to a much lesser extent, *Cryptolichenaria* exhibit a long Ordovician history prior to the late Darriwilian (*TS*.4c). These limited records demonstrate that the tabulate coral clade remained an extremely poorly diversified group, with small colony sizes, tiny corallites, and low population densities, for at least 25 m.y., that is, for about half the length of the Ordovician Period, prior to its major diversification.

In Utah, the first *Eofletcheria* appeared just above the *Lichenaria*-bearing Lehman Formation, in beds of the Crystal Peak Dolomite, of late Darriwilian (*TS*.4c) age (Rigby and Hintze 1977). *Eofletcheria* is probably a descendant of *Lichenaria* (Scrutton 1984), and its appearance with other tabulate genera marks the start of the first really significant increase in tabulate coral diversity, in the latest Mid Ordovician. This late Darriwilian (*TS*.4c) expansion coincides with the great diversification of other groups of metazoans and algae and appearances of the first major Ordovician ("Chazy"-type) metazoan-dominated reef complexes (Webby 2002). Some of the tabulate corals became important frame-building and binding constitutents in these reef complexes. *Eofletcheria* is a constituent in the "Chazy"-type reefs, but forms such as the first sarcinulinid *Billingsaria* were more important reef binders and frame builders. *Billingsaria* occurs in the Chazy Group and equivalents of New York, Vermont, and Quebec, in equivalent successions of Virginia and Tennessee (Scrutton 1984), in the Krivolutskiy unit of the Siberian Platform (Sokolov 1955; Sokolov and Tesakov 1963), and in the Wahringa Limestone Member, New South Wales (Webby in Percival et al. 2001), all of confirmed late Darriwilian (*TS*.4c) age.

The genus *Saffordophyllum* is another important genus first recorded from the McLish Formation of Oklahoma (Flower 1961), in approximate correlatives of the Chazy Group (Ross et al. 1982) and therefore also of probable late Darriwilian (*TS*.4c) age. Scrutton (1979, 1984) viewed the genus as possibly ancestral to *Paleofavosites* and other favositinids, though Lee and Elias (2000) have recently cast doubt on this linkage. The favositinids became important only in the late Caradoc to Ashgill, from the *TS*.5d interval onward, and, relative to the heliolitinids, remained a smaller group in Ordovician time. The favositinids, however, diversified rapidly during the Silurian and Early to Mid Devonian, becoming the dominant Middle Paleozoic tabulate coral group (see their representation as "cerioid perforate" colonies in Scrutton 1988: figure 3.4B).

The early, aporous, cerioid, modular genus *Foerstephyllum* is yet another with links to *Lichenaria*, though it has been classified with the sarcinulinids (Hill 1981). It is recorded from the oldest Ordovician coral assemblage in the Billabong Creek Limestone of central New South Wales (Pickett and Percival 2001) of Gisbornian, that is, early Caradoc (*TS*.5a–b), age. The North American and North Chinese reefs of this interval include a number of frame-building tabulate corals such as *Lichenaria, Cryptolichenaria, Eofletcheria,* and *Foerstephyllum,* as well as first appearances of lichenariinid *Amsassia,* sarcinulinid *Lyopora,* and syringoporoid-like auloporinid *Labyrinthites* (Webby 2002).

In central New South Wales the major diversification of tabulate and tetradiid corals (see Australasian corals later in this chapter; also Pickett and Percival 2001: table 1) occurs in the mid Caradoc (late Mohawkian, *TS*.5c), with maximum numbers of tetradiid, sarcinulinid, and auloporinid (especially *Bajgolia*) species. In addition, the heliolitinids make their first appearances in this interval and include, significantly, rather abundant encrusting-type *Coccoseris* and *Acidolites* with septal trabeculae (Webby and Kruse 1984). This invites comparison with the slightly older-appearing species of *Billingsaria* (*B. spissa* Webby, in Percival et al. 2001) that also has an explanate growth form and much thickened septal trabeculae. The major part of the genus-level diversification of heliolitinids seems to have occurred during the mid to late Caradoc (*TS*.5c–d) interval. Scrutton's (1984) proposed derivation of *Coccoseris* from *Billingsaria* is supported by the New South Wales occurrences. The early heliolitinid diversification has been outlined previously by Webby and Kruse (1984: text figure 6), based on the excellent record of faunas through this *TS*.5c interval in the Australian succession (see later in this chapter).

It is also apparent from the New South Wales record that, as previously noted by Scrutton (1984), the sarcinulinid genus *Nyctopora* evolved directly from the earlier *Billingsaria* stock with loss of a columella and reduced dilation of vertical trabeculae through the *TS*.5b–c interval (see also Pickett in Pickett and Percival 2001).

The halysitinids are another important group that seem to have diversified mainly during the mid to late Caradoc—first appearing in the early Mohawkian (*TS*.5b) of North America (Scrutton 1984), based on the initial Canadian record of *Quepora,* followed by *Catenipora* in the late Mohawkian (*TS*.5c) and then the first "dimorphic" *Halysites,* possibly derived from a *Quepora* and recorded from a New South Wales horizon (Webby and Semeniuk 1969). This correlates with the lower part of the *TS*.5d interval (equivalent to late Caradoc or North American early Cincinnatian (Edenian-Maysvillian). The record of North American Cincinnatian (late Caradoc to

Ashgill) tabulate, tetradiid, and rugose corals is outlined by Elias and Young (later in this chapter).

Hill (1981) listed a total of 77 confirmed genera of Ordovician tabulate corals. Scrutton (1988: figure 3.4; 1997: figure 20A), in a generalized plot, presented a slightly lower overall genus-level diversity, with a progressive rise from the beginning of the Ordovician to a peak of near 60 genera during the mid Ashgill. A small decline (loss of only a few genera) followed toward the Ordovician-Silurian boundary. However, this plot does not truly reflect the nature of the diversity profile given that basically only two tabulate genera existed through the first half of the Ordovician (for nearly 25 m.y.). Scrutton's Ordovician data are subdivided into only three time units, representing his epoch-based subdivisions of Ordovician time, and this produces a rather crude result with a markedly smoothed-out diversity profile. Nevertheless, the plots do show important data, such as that, in terms of all the tabulate coral groups, the heliolitinids (those with characteristic "coenenchymal imperforate" colony form) have the greatest genus-level diversification during Ordovician time, expanding to a diversification peak in the mid Ashgill, though declining later. At their mid Ashgill diversification peak, the heliolitinids constituted approximately half of the total tabulate coral diversity (about 30 of the 60 genera represented).

Sepkoski (1995) provided a more precise diversity profile for the tabulate corals using 12 time divisions based on the British series. This profile also depicts a gradual diversity increase, but it does not commence until the late Arenig, initially rising gently but then progressively steepening to a sharp peak in the mid Ashgill with a total of 80 genera. A dramatic decline then followed to a low of about 25 genera at the Ordovician-Silurian boundary. Sepkoski (1995) noted that the coral radiation commenced slightly later in the Ordovician than for some other groups, such as the articulated brachiopods and echinoderms.

Early Rugose Coral Diversification

The earlier views that the rugose corals evolved from tabulates through lines of descent from early tabulates such as *Lichenaria,* auloporinids, and/or *Foerstephyllum* (Flower 1961; Flower and Duncan 1975; Sokolov 1962; Ivanovskii 1972; Webby 1971) have been challenged in more recent decades, for example, by Sytova (1977), Scrutton (1979, 1997), and Neuman (1984), among others. These latter authors rejected the view that an intimate relationship existed between early tabulates and early rugosans, preferring an independent origin for the early rugosans. Most recent comments are by Oliver (1996) and Scrutton (1997), who have referred to the Rugosa as probably being a monophyletic group and most likely derived from a solitary, soft-bodied anemone group that acquired the ability to secrete a calcified skeleton (i.e., presumably with an initial skeleton like a *Primitophyllum* or a *Lambeophyllum*). The view that the rugosans descended from the tabulate corals now appears less likely, but many late Darriwilian to mid Caradoc (*TS*4c–5c) rugose and tabulate coral genera have not yet been restudied to clarify their relationships.

From the revised correlations adopted here, the earliest records of the solitary rugosans, *Primitophyllum* and *Lambelasma* in Baltoscandia (see Neuman and Kaljo later in this chapter), occur in the *TS*.5b interval, while the solitary dendroid *Hillophyllum* from central New South Wales first appears in *TS*.5c (see Webby, Australasian corals, later in this chapter). Consequently, the Baltoscandian genera are not likely to be direct descendants of *Hillophyllum* as previously suggested (Webby 1971).

Scrutton (1997) stated that the earliest representatives of solitary and modular (dendroid to cerioid) rugose genera (*Primitophyllum, Hillophyllum, Lambeophyllum, Streptelasma,* and *Favistina*) all appeared about the same time, during the early Mohawkian (i.e., early, but not earliest, Caradoc = *TS*.5b). This view is substantially accepted, except for *Hillophyllum* (as noted earlier), which now seems to have appeared first, about one time slice later, during the *TS*.5c interval (that is, during the mid Caradoc/late Mohawkian).

The supposed "Chazyan" records of solitary rugose corals remain unsubstantiated and probably should now be discounted as in error. Welby (1961), in describing a new species of cerioid tabulate *Foerstephyllum* (*F. wissleri* Welby) from the Chazy Group limestones, listed two associated species as solitary rugosans, identifying them as *Streptelasma* cf. *S. expansum* Hall and *Lambeophyllum* cf. *L. profundum* (Hall). He also recorded the stromatoporoid *Stromatocerium* from this same Chazy *F. wissleri* type locality. The stromatoporoid was clearly misidentified, as this common "Blackriveran" (early Mohawkian) genus does not occur in the older, Chazy limestones (Kapp and Stearn 1975). No independent verification of Welby's report of rugose corals from the extensive outcrop belt of Chazy limestones across New York, Vermont, and Quebec has been made in the past four decades, so these supposed earlier, "Chazyan" records, close to the Middle to Upper Ordovician boundary (top *TS.*4c–5a interval), may be discounted in favor of early Mohawkian (*TS.*5b) "origins" of the group.

Also significant is that the major higher-level diversification into separate suborders occurred comparatively early in the history of the group (see Neuman and Kaljo later in this chapter; also Weyer 1973: figure 2; Scrutton 1997), mainly during the early to mid Caradoc (*TS.*5b–c). The main subdivision in suborders comprises (1) the Cystiphyllina with *Primitophyllum* and *Hiliophyllum,* (2) the Streptelasmatina with *Streptelasma* and *Leolasma,* (3) the Stauriina with *Favistina* and *Palaeophyllum,* and (4) the Calostylina with *Lambelasma* and probably *Lambeophyllum.*

Information about the overall generic-level diversity of the group is somewhat contradictory. Scrutton (1979:184) indicated that some 30 genera of rugosans had appeared by the end of the Ordovician and that these were dominantly streptelasmatinids. He also presented a generalized Paleozoic diversity curve for the Rugosa with a gradual rise to a total of about 60 genera at the end of the Ordovician (see Scrutton 1988: figure 3.3; 1997: figure 20B). Hill (1981), on the other hand, listed 49 valid Ordovician genera, a total close to Sepkoski's (1995) mid Ashgill diversity peak (about 47 genera). This latter was reached after a slow initial buildup through the Caradoc, then a steeper rise into the Ashgill, before a dramatic decline to about 20 genera across the Ordovician-Silurian boundary (Sepkoski 1995). Scrutton (1997) recorded about 70 percent generic extinction for both rugose and tabulate corals, across the Hirnantian interval (*TS.*6c). Presumably about half of this extinction relates to the onset of glaciation at the beginning of the Hirnantian, and the other half is associated with the second, late Hirnantian event. This latter apparently involved rapid warming, sea level rise, and spread of anoxic waters across shallow platform areas (chapter 9). Kaljo (1996) depicted tabulate corals in Estonia with similar extinction rates (between 60 and 67 percent) for the two events.

■ Tetradiid Corals (BDW)

The order Tetradiida Okulitch 1936 is the only exclusively Ordovician group of corals. Though the origins and relationships remain obscure, it seems likely that this small, unified, monophyletic group, with its modular skeleton, possibly mainly aragonitic, and distinctive quadripartite longitudinal mode of fission, represents an independent cnidarian clade that arose from an unknown anemone group about 465 million years (m.y.) ago (i.e., during the late Mid Ordovician). Perhaps a colonial group of anemones was ancestral to the tetradiids. The group flourished through the Late Ordovician (Caradoc and Ashgill)—about 20 m.y.—diversifying exclusively in tropical, shallow marine carbonate platform, shelf, and fringing island settings and establishing a wide range of growth morphologies in order to colonize successfully a variety of high- to low-energy settings in shallow, warm-water subtidal to intertidal habitats. In places, tetradiid-dominated small reefs and biostromes formed (Webby 2002), as well as microatolls (large, massive, flat-topped colonies), such as the *Porites* colonies of the modern reef flat (Webb 1997; Webby et al. 1997). Other colony forms, such as the erect, bundled columnar clusters of "*Phytopsis,*" were wave bafflers (Walker 1972). The severe global climatic fluctuations (overall cooling) of the late Ashgill appear, directly or indirectly, to have caused the decline of the group. However, the disappearance of the group coincides with the second (mid to late Hirnantian) mass extinction, interpreted as a "global warming" phase of the end Ordovician glaciation (chapter 9). The surviving low-diversity tetradiid fauna of shallow, equatorial waters failed to recover from one or more consequences of the global warming, probably the spread of anoxic waters as sea level rose.

Taxonomic Relationships

The Tetradiida as currently understood comprises two families, the Tetradiidae Nicholson 1879 and the Paleoalveolitidae Okultich 1935b (see Hill 1981). Approximately 90 percent of the 100 or so described species are representatives of the family Tetradiidae, and the other 10 percent belong to the family Paleoalveolitidae. The Tetradiidae developed branching (dendroid, phaceloid), cateniform, columnar (bundled), and massive (cerioid, phaceloid) colonies. The corallites of the phaceloid and dendroid branching forms tend to be more loosely aggregated with corallites linked in close associations only where medial growth offsets (quadripartite fission) occur. Columnar forms were characteristically formed of erect, compactly bundled corallites surrounded by a holotheca. Both massive-cerioid and massive-phaceloid colonies are superficially similar reef- and microatoll-forming types, but the cerioid forms have fused corallites (space filling), whereas the phaceloid forms exhibit non-space-filling corallites, the interspaces being infilled with sediment. The sediment was probably shed into the narrow intercorallite spaces of the colony during growth. Cateniform colonies are composed of chains or networks (also non-space-filling) that are partially or completely enclosed by sediment-filled spaces (lacunae). These sediment infills may have helped stabilize the colonies during growth (Scrutton 1998). All the growth forms appear to have developed from medial growth (i.e., offsetting from within the colony). Members of the family Paleoalveolitidae have massive to ramose colonies, also with medial growth. Commonly their outwardly flexed corallites exhibit alveolitoid shapes.

Two approaches have been employed to subdivide the Tetradiida at the generic level. The first proposed by Bassler (1950) involved retaining just two valid genera, *Tetradium* and *Paleoalveolites,* with the further subdivision of *Tetradium* into "species groups" based on the markedly different colony growth forms. Webby and Semeniuk (1971) employed this "species group" concept because the Australian tetradiids exhibit a wide range of growth forms that were demonstrably related to ecological factors, and they lacked a consistent pattern of differences in their internal morphology to justify a meaningful generic-level subdivision of the group.

Sokolov (1955), alternatively, argued that the main growth forms were of basic taxonomic importance and that they should have separate generic status. This alternative approach involved (1) emending the genus *Tetradium* Dana 1846 to include only species of the *fibratum* group, that is, those with massive cerioid colonies; (2) restoring the status and redesignating the genus *Phytopsis* Hall 1847 for the species of the *cellulosum* group, that is, for those composed of small, erect, bundled, columnar clusters; (3) proposing a new genus, *Rhabdotetradium,* for the various types of phaceloid coralla and long, meandering, free corallites, including the species of the *syringoporoides* group of Bassler, though, as noted by Sokolov (1955: 248), some of the species, particularly the small, simple dendroid species, are not typical of the new genus; and (4) introducing the genus *Paratetradium* to accommodate species of the *halysitoides* group, that is, the forms characterized by different sorts of chains or irregular networks, with corallites in two or more rows, and separated by sediment-filled spaces (lacunae). In addition, Sokolov synonymized the Paleoalveolitidae with the Tetradiidae, thus making *Paleoalveolites* also a member of the family Tetradiidae.

Hall (1847) originally based the genus *Phytopsis* on two species—he did not designate a type species. Hall's first described species was the trace fossil *P. tubulosum* (Hall 1847:38), and the second was the tetradiid coral *P. cellulosum* (Hall 1847:39). The trace fossil *P. tubulosum* was listed by Okulitch (1935a: 103) as a "worm," and so in 1935 the trace fossil name strictly had priority, and Häntzschel (1965:72) later fixed the trace fossil *P. tubulosum* as type species of *Phytopsis*. However, 11 years earlier Sokolov (1955) had made specific designation of *cellulosum* as the type species. A ruling from the International Commission on Zoological Nomenclature is now needed to establish whether Sokolov's designation of *Phytopsis* has priority over the trace fossil name.

The type species of Sokolov's (1955) genus *Rhabdotetradium*, *R. nobile* and allied species—*R. floriforme* Sokolov and Tesakov and *R. apertum* (Safford)—from the Siberian Platform, exhibit phaceloid coralla and have long, slender, subparallel to meandering and free, subquadrate corallites that in places join at the corners or may be weakly linked in more continuous, short chains. The Siberian material has been described and illustrated mainly from thin sections, with little other information about the specimens (e.g., growth form, size, or other relationships). Only one small fragmentary specimen of the type species, measuring 45 × 35 mm across and 35 mm in height, is illustrated (Sokolov 1955: plate 57, figures 3, 4), at least showing that it has a phaceloid-massive form.

The New South Wales colonies of *Tetradium cribriforme* (Etheridge 1909; also see Webby and Semeniuk 1971) have an internal morphology that is remarkably similar to the type species *R. nobile* but are represented by very large (near meter-sized) phaceloid-massive coralla (Webby et al. 1997: plate I). Apart from the slightly more wavy, cribriform appearance of the corallite walls, the New South Wales species is indistinguishable from the *Rhabdotetradium* type species, and it should therefore be assigned to the genus. Other closely related species have been described by Zhizhina (1956, 1966) from Eastern Taymyr and Vaigach Island (Arctic Russia).

Most of the small, bushy, open-branching dendroid colonies included by Bassler (1950) in his *syringoporoides* species group were also assigned to *Rhabdotetradium* by Sokolov (1955), though they do not exhibit the typical phaceloid-massive colonies with closely spaced, long, slender subparallel to meandering corallites. These include some very simple forms such as the earliest (?Darriwilian–earliest Caradoc) tetradiid species—for example, *R. syringoporoides* (Ulrich), *R. cylindricum* (Wilson)—and the last survivors of late Ashgill (Hirnantian) age (e.g., *R. frutex* Klaamann and *R.* sp. A of Young and Elias 1995). Sokolov's *Rhabdotetradium* has therefore been separated into two species groups for this biodiversity study (figure 15.1): the first group comprises the phaceloid-massive colony forms with closely spaced subparallel to meandering corallites such as the type species and *R. cribriforme* (here referred to as the *R. nobile/cribriforme* group); the second includes a wide range of small, simple, open dendroid forms (more or less the *R. syringoporoides* group of Bassler). A new genus is probably required to accommodate these latter forms.

The late Caradoc–Ashgill (Cincinnatian) tetradiids characteristically exhibit much better developed tabulae than their earlier (early to mid Caradoc) counterparts. The majority of these earlier tetradiid species (representatives of most genera, not just *Tetradium*) have rather poorly preserved tabulae. Copper and Morrison (1978) recognized the presence of well-preserved tabulae as diagnostic of the subgenus *Paenetetradium;* type species is *T. (P.) huronensis* Foord. The taxon is here tentatively considered a valid genus, derived from massive cerioid *Tetradium* (figure 15.1). The genus is well represented in the

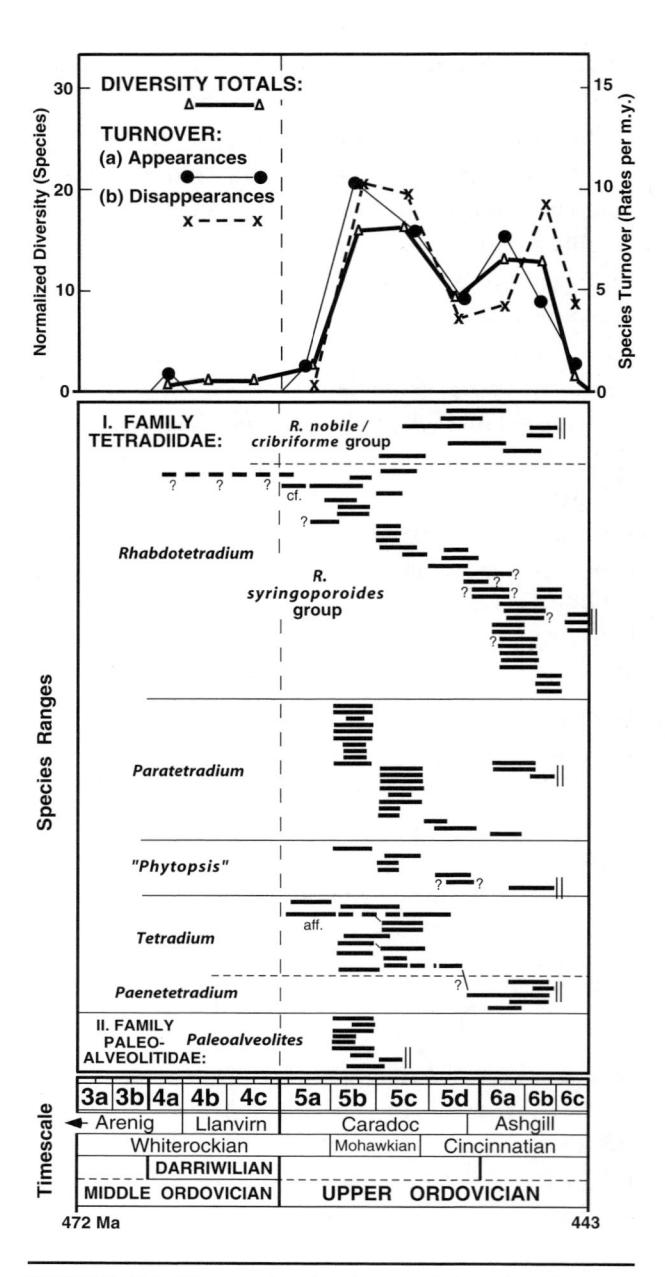

FIGURE 15.1. Diagram showing the recorded stratigraphic ranges of species of tetradiid coral families and genera (global data), represented by thick, solid lines, through Mid to Late Ordovician time (bottom part), and a plot (at the top) illustrating the normalized species diversity and turnover (appearance and disappearance) rates per m.y. for tetradiid coral species. The early (*TS.*4a–5a) records of *Rhabdotetradium* in North China require further confirmation, so they are depicted by a thick, dashed line and question marks. The closely spaced fine double lines at the right side of the range chart represent likely extinction levels for particular genera.

Richmondian successions of the eastern North American continental interior.

The genus *Paleoalveolites* comprises a number of mid Caradoc species from North America, including the type species, *P. carterensis* (Bassler), a species (*P. tasmaniense*) from Tasmania, and a doubtful species (?*P. explanatus* Pickett) from New South Wales.

Most recent workers (e.g., Copper and Morrison 1978; Hill 1981; G. A. Young 1995; Young and Elias 1995; Webb 1997) have adopted Sokolov's (1955, 1962) elevation of Bassler's species groups to full generic rank. I have tentatively adopted this approach for the present biodiversity survey, though with the qualifications enumerated earlier. A more exhaustive, globally based taxonomic, preservational, and ecologic survey of the tetradiid corals is required that will also, as a by-product, adequately determine the status of different growth forms relative to the internal morphological features for use in taxonomically subdividing the group.

Affinities of the tetradiid corals remain uncertain. Some specialists have advocated that they are representatives of the subclass Tabulata (Hill and Stumm 1956; Sokolov 1962; Webby and Semeniuk 1971; Hill 1981; Young and Elias 1995), whereas others (e.g., Okulitch 1936; Bassler 1950; Flower and Duncan 1975; Scrutton 1997; Elias and Young later in this chapter), with differing degrees of conviction, have suggested that they should not be classified with "true" tabulate corals. Tetradiids differ from the latter in exhibiting laminar septa with tetrameral symmetry and quadripartite longitudinal fission, and they probably secreted an aragonitic skeleton (Semeniuk 1971; Wendt 1989; Webby 1990; Scrutton 1997). G. A. Young (1995), however, suggested that a tetradiid species from Manitoba had an original composition of high-magnesium calcite because of its poor, non-dolomitized preservation compared with two other preservational types—well-preserved calcitic forms on the one hand and dolomitized on the other (the latter presumed by Young to be originally aragonitic).

Distribution

The earliest stratigraphically well-controlled occurrences of tetradiids are reported from successions in Maryland and adjacent states (United States) and in Ontario (Canada). Representatives of common "single tube" and clustered small branching colonies of *Rhabdotetradium syringoporoides*, *R.* cf. *syringoporoides*, massive cerioid *Tetradium marylandicus* Neuman, and *T.* aff. *fibratum* Safford are recorded by Neuman (1951) from the Newmarket Limestone of Maryland (more or less equivalent to the Chazyan Crown Point and Valcour formations, which equates to *TS*.5a). Another is single corallite *R. cylindricum* recorded from the Aylmer Limestone of Ontario, a localized unit that directly underlies the Lowville Limestone (probably correlating with a level close to the *TS*.5a–b boundary). The main "Blackriveran" (basal Mohawkian = base of *TS*.5b) diversification follows (figure 15.1), as recorded in the Lowville Limestone of New York, Quebec, and Ontario and Appalachian equivalents (Bassler 1950) with appearances of other distinctive generic elements, including *Paratetradium* (*P. clarkei*), "*Phytopsis*" ("*P.*" *cellulosum*), and *Paleoalveolites* (*P. carterensis*).

Rhabdotetradium apparently appeared earlier in China than elsewhere, judging from the records of the genus in the lower Majiagou Formation of Hebei, North China Platform, of possible early to middle Darriwilian (*TS*.4a–b) age, and the Sandaogou Formation of the southwestern platform margin of late Darriwilian to earliest Caradocian (*TS*.4c–5a) age (Lin 1983; Lin Baoyu pers. comm.). These occurrences in North China appear to be the earliest records of tetradiids (Lin and Webby 1989). All the other Chinese tetradiids are of much later (Ashgill) age.

The major Mohawkian (*TS*.5b–c) radiation event (figure 15.1) is recorded widely across low paleolatitudes on the carbonate platforms and shelf margins (North America, Siberian Platform, Scotland, Björnöya [south of Spitsbergen], North China, Southeast China, Tasmania) and in island arcs (central New South Wales) and involves the diversification of *Rhabdotetradium* (12 species), *Paratetradium* (18 species), "*Phytopsis*" (4 species), *Tetradium* (11 species), and *Paleoalveolites* (9 species). The North American tetradiids are the most diversified, and three of the genera (*Rhabdotetradium*, *Paratetradium*, and *Tetradium*) show evidence of two-phase diversification, a first pulse in the Turinian (*TS*.5b—"Blackriveran" equivalents) and then a second in the Chatfieldian (*TS*.5c—"Trentonian" correlatives). At the end of this radiation event there was a marked decline in species diversity from *TS*.5c to 5d, with disappearances of *Paleoalveolites* (*TS*.5c) and *Tetradium* (*TS*.5d) and a significantly lowered level of diversity of *Paratetradium* and *Rhabdotetradium* through the *TS*.5c–d interval.

The second main diversification of tetradiids occurs through the early to mid Ashgill (top *TS*.5d, 6a–b), with maximum diversification of *Rhabdotetradium* (especially the *R. syringoporoides* group—18 species). The prominently tabulated cerioid *Paenetetradium* (five species) also appears in North America. Again, these tetradiid faunas have a wide distribution in low paleolatitudes of continental platforms and margins, being recorded in North America, North Greenland, South Wales, Eastern Taymyr and Vaigach Island of Arctic Russia, ?Urals, Siberian Platform, Southwest Siberia, Northeast Russia, Western Mongolia, and North, Northwest, and Southeast China. *Paratetradium* and "*Phytopsis*" are in decline and along with *Paenetetradium* and representatives of the *R. nobile/cribriforme* group disappear prior to or across the time-slice transition from *TS*.6b to 6c, coincident with the initial major global cooling and establishment of the Late Ordovician glaciation (chapter 9). Only the elements of the small, single or branched (clustered) corallite tubes of the *R. syringoporoides* group survive in a few widely scattered localities of North America—in the continental interior (Young and Elias 1995) and shelf margins of Anticosti Island (Copper 1999)—and in Baltoscandia, from the Porkuni and equivalent limestones of Estonia (Klaamann 1966) and Norway (Hanken 1979). These are the last occurrences of tetradiids in the latest Ordovician (Hirnantian—*TS*.6c age). A second, major (mid to late Hirnantian) global event of the end Ordovician glaciation, the warming and melting of the ice caps (see chapter 9), immediately followed and may have had a major impact on the remaining widely dispersed localized relict stocks of tetradiid corals. As in at least one Holocene example (Levin 2002), the spread of waters of oxygen-minimum zones into shallower areas during a warming phase may have caused the final extinction of the group.

The diversification of the *R. nobile/cribriforme* group seems to have been somewhat different from that of other tetradiids. The earliest member of the group to appear was the New South Wales *R. cribriforme*, during the early Eastonian (lower *TS*.5c interval). On the Siberian Platform, an occurrence of *R. apertum* in the Mangazaysk horizon (mid to upper *TS*.5c) was followed by *R. nobile*, *R. floriforme,* and *R. apertum* in the succeeding Dolbor horizon (*TS*.5d). In the Chukotski Peninsula (Northwest Russia) and in Xinjiang (Northwest China), reports of *R. nobile* seem to be mainly from higher levels, within *TS*.6a (even extending into *TS*.6b). Other members of the species group in the Eastern Taymyr, Arctic Russia (e.g., Zhizhina's *R. subapertum*, *R. quadratum*, *R.? tessulatus,* and *R.? tessulatiformis*), are also recorded from higher horizons (*TS*.6a–b).

Diversity Patterns

Ordovician tetradiids seem initially to have maintained a restricted, background level of existence (one genus, one species) through most of the Darriwilian—*TS*.4a–c) but then commenced a rapid diversification in the early Caradoc (*TS*.5a to 5b) rising to a maximum of five "genera" with a peak of normalized diversity of 17 species by the mid Caradoc (*TS*.5b–c). Then, after a moderate decline to 9 species within the late Caradoc (*TS*.5d), the diversity rose again to an early to mid Ashgill (*TS*.6a–b) peak of 13 species. Finally, the species diversity declined dramatically toward the mid to late Hirnantian extinction level within *TS*.6c.

The species/genus ratios remain at a background level of 1 during the Darriwilian but rise to maxima of 6.2 species per genus in *TS*.5b–c and then to 6.25 and 6.0 species/genus, respectively, in *TS*.6a–b.

The species-level turnover profiles show appearance (origination) and disappearance (extinction) rates more or less tracking each other in two major, mid Caradoc and early-mid Ashgill, turnover spikes. These are in comparatively close conformity with the patterns of the two diversity maxima (figure 15.1, top). Species origination rates peak at 10.4 and 7.8 species per m.y. during *TS*.5b and 6a, respectively, and the complementary extinction rates exhibit maxima of 10.4–9.8 and 9.4 species per m.y. during the *TS*.5b–c and the *TS*.6b intervals, respectively. This latter extinction maximum seems to demonstrate that the first (early Hirnantian) glacial cooling event at the *TS*.6b–c boundary caused the most severe diversity decline of the tetradiid fauna, with extinction of the bulk of species (and almost all the genera). Only a few species of *Rhabdotetradium* were left by Hirnantian time (*TS*.6c), and these lived in isolated and reduced populations that were unable to survive the second (mid to late Hirnantian) global warming event.

■ Laurentian Corals (RJE, GAY)

Laurentia was an important center of coral origination and diversification in the Ordovician, but the early history of these faunas is poorly understood. The earliest coral-like fossils that may represent tabulates are Ibexian in age and have been identified as *Lichenaria* (Scrutton 1984). However, the earliest coral assignable with confidence to *Lichenaria* is late Whiterockian ("Chazyan") and perhaps slightly postdates several other tabulates (Laub 1984). The earliest reported rugosan is a solitary type from the late Whiterockian, identified as *Lambeophyllum;* this occurrence is unconfirmed. The earliest proven rugosan fauna, which includes solitary and colonial types, is early Mohawkian ("Blackriveran") (Neuman 1984).

Mohawkian corals are poorly known, especially in the present-day West, where sand was widely deposited (Duncan 1956). In the East, tabulates and rugosans were widespread in the carbonate facies (Bassler 1950); this was apparently an area of significant diversification. Beginning in the late Mohawkian, faunas were displaced westward owing to development of a foreland basin related to the Taconic Orogeny (Patzkowsky and Holland 1996). This set the stage for the better-known Cincinnatian faunas, described later in this chapter in a framework of biogeography and environmental setting.

Data

Four biogeographic divisions based on corals are recognized in the Cincinnatian of Laurentia (figure 15.2). At the species level, only the Edgewood provincial fauna has been studied in its entirety using modern taxonomic concepts and methods (McAuley and Elias 1990; Young and Elias 1995). Solitary rugosans of the Richmond Province are well known (Elias 1982; Elias et al. 1990), but coverage of colonial rugosans and tabulates is not comprehensive (e.g., Browne 1965; Jull 1976). The Red River–Stony

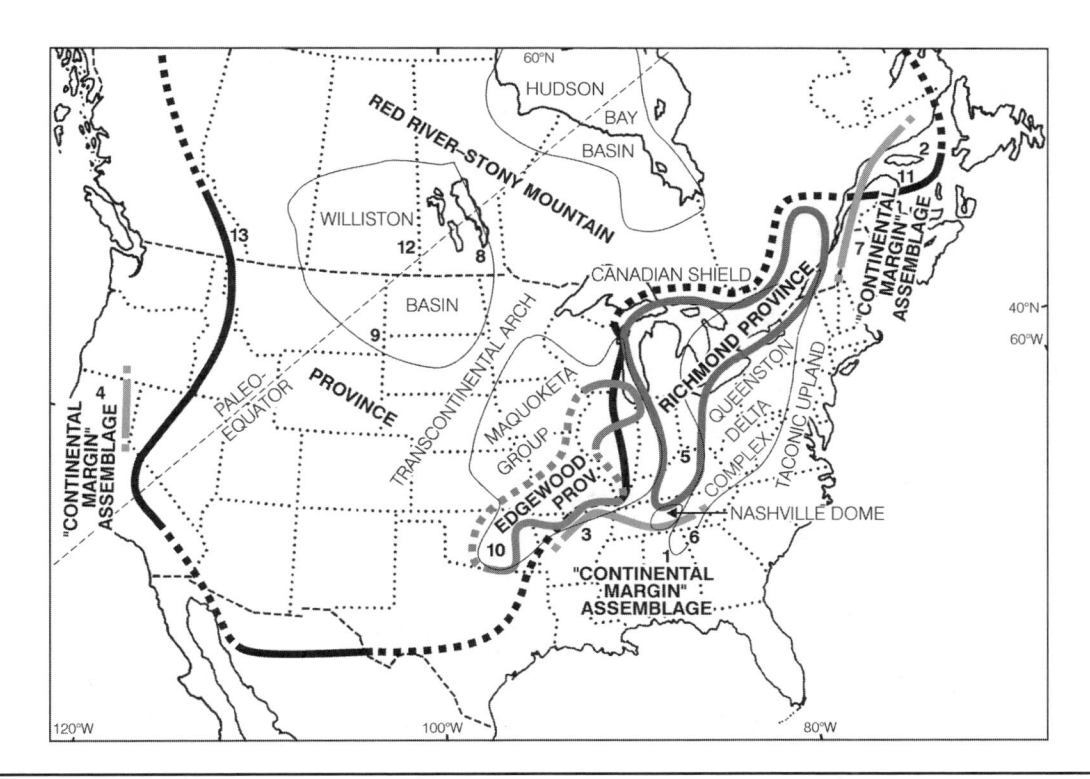

FIGURE 15.2. Cincinnatian coral biogeography (bold; dotted line = position uncertain) and geologic features in Laurentia. Locations referred to in text: 1 = Alabama; 2 = Anticosti Island; 3 = Arkansas; 4 = California; 5 = Cincinnati Arch; 6 = Georgia; 7 = Maine; 8 = Manitoba; 9 = Montana; 10 = Oklahoma; 11 = Percé; 12 = Saskatchewan; 13 = southern Canadian Rocky Mountains.

Mountain provincial fauna (e.g., Flower 1961; Nelson 1963, 1981; Elias 1985) and the "continental margin" assemblage (e.g., Dixon 1974; Bolton 1980; Elias et al. 1994) are incompletely known, particularly the colonial rugosans and tabulates and especially in remote areas. Clearly, biogeographic comparisons of diversity and determination of continental diversity at the species level are not feasible.

For this study, data are tabulated at the genus level (table 15.1). Identifications are based on our own work and our interpretation of descriptions and reports published by others. Problems regarding completeness of data, outlined earlier, are undoubtedly less significant when estimating generic diversity. A number of taxonomic problems affect determinations of diversity. For example, is *Favistina* distinct from *Cyathophylloides* (e.g., Flower 1961), or is it a junior synonym (e.g., Browne 1965; arbitrarily followed herein)? Our study includes tetradiids, but they are probably not tabulates (Scrutton 1997) and may not even be corals. Taxa representing the Red River–Stony Mountain Province and the "continental margin" assemblage are present in the Anticosti-Percé area of Quebec; these occurrences are credited to the Red River–Stony Mountain Province.

Determination of diversity during time slices (see chapter 2) is hindered by stratigraphic imprecision in the documentation of some coral occurrences and by problems concerning correlations and ages of strata. For example, in the Red River–Stony Mountain Province we follow the stratigraphic framework of Sweet (1979) and extrapolations therefrom (Elias 1991). Thus, strata are considered to range in age from mid Edenian to Gamachian (e.g., figure 15.3). Others, however, have suggested that the lower part of the succession is Richmondian (Bergström 1997) and that the Ordovician portion of the Stonewall Formation in Saskatchewan and Manitoba is Richmondian rather than Gamachian (Norford et al. 1998). We therefore simply assign definite pre-Gamachian coral occurrences to *TS*.5d through 6b because the mid to late Edenian and early Richmondian both fall within *TS*.5d (as does the intervening Maysvillian) and the Richmondian also includes *TS*.6a and 6b. Uncertainty about the Gamachian (*TS*.6c) age of corals in the Stonewall Formation is noted in table 15.1. The Ellis Bay Formation of Anticosti Island definitely represents *TS*.6c.

TABLE 15.1. Occurrences of Coral Genera in the Cincinnatian of Laurentia

Taxa	1	2	3	4
Rugosa				
Cystiphyllida				
Tryplasma	A,B?	—	—	—
Stauriida				
Bighornia	A,B?	—	—	—
Bodophyllum	A	—	B	A
Complexophyllum	A	—	—	—
Crenulites	A	A	—	—
Cyathophylloides	A,B	A	—	—
Deiracorallium	A	—	—	—
Grewingkia	A,B	A	B	A
Keelophyllum	—	—	B	—
Kenophyllum?	—	—	—	A
Lobocorallium	A	—	—	—
Neotryplasma	A	—	—	—
Palaeophyllum	A,B	A	B	A
Paliphyllum	A,B	—	—	—
Pycnostylus	—	—	B?	—
Rhegmaphyllum	—	—	—	A,B
Salvadorea	A,B	—	—	—
Streptelasma	A,B	A	B	A
Number of genera = 18	14	5	6	6
Tabulata				
Auloporida				
Aulopora	B	—	B	—
fletcheriellid gen. indet.	A,B?	—	—	—
Labyrinthites	A	—	—	—
Troedssonites	A	—	—	—
Favositida				
Agetolites?	A	—	—	—
Manipora	A	—	—	—
Paleofavosites	A,B	—	B	A
Saffordophyllum	A	—	—	—
Halysitida				
Catenipora	A,B	A	B	A
Halysites	—	—	B	—
Heliolitida				
Acidolites	B	—	B	A
Cyrtophyllum	A	—	—	—
Ellisites	A,B	A	—	—
Mcleodea	A	—	—	—
Plasmopora	—	—	B?	—
Plasmoporella	—	—	—	A
Pragnellia	A	—	—	—
Propora	A,B	—	B	A
Protaraea	A	A	B	—
Protrochiscolithus	A,B	—	—	—
Sibiriolites	A	—	—	—
Wormsipora	A	—	—	—
Sarcinulida				
Angopora?	B?	—	—	—
Calapoecia	A,B	—	—	—
Foerstephyllum	—	A	—	—
Nyctopora	A	A	—	—
Reuschia?	—	—	—	A
Sarcinula	A	—	—	—
Tollina	A	—	—	—
Trabeculites	A	—	—	—
Tetradiida				
Paratetradium	—	A	—	—
Rhabdotetradium	A,B	—	B	—
Tetradium	A	A	—	A
Number of genera = 33	27	8	9	7
Total number of genera = 51	41	13	15	13

Note: Age assignments: A = pre-Gamachian (within, but not necessarily throughout, *TS*.5d–6b); B = Gamachian (*TS*.6c). Biogeographic divisions: 1 = Red River–Stony Mountain Province including Anticosti-Percé area (B? indicates occurrences that may be pre-Gamachian rather than Gamachian); 2 = Richmond Province; 3 = Edgewood Province (B? indicates occurrences that may be early Rhuddanian rather than Gamachian); 4 = "continental margin" assemblage excluding Anticosti-Percé area.

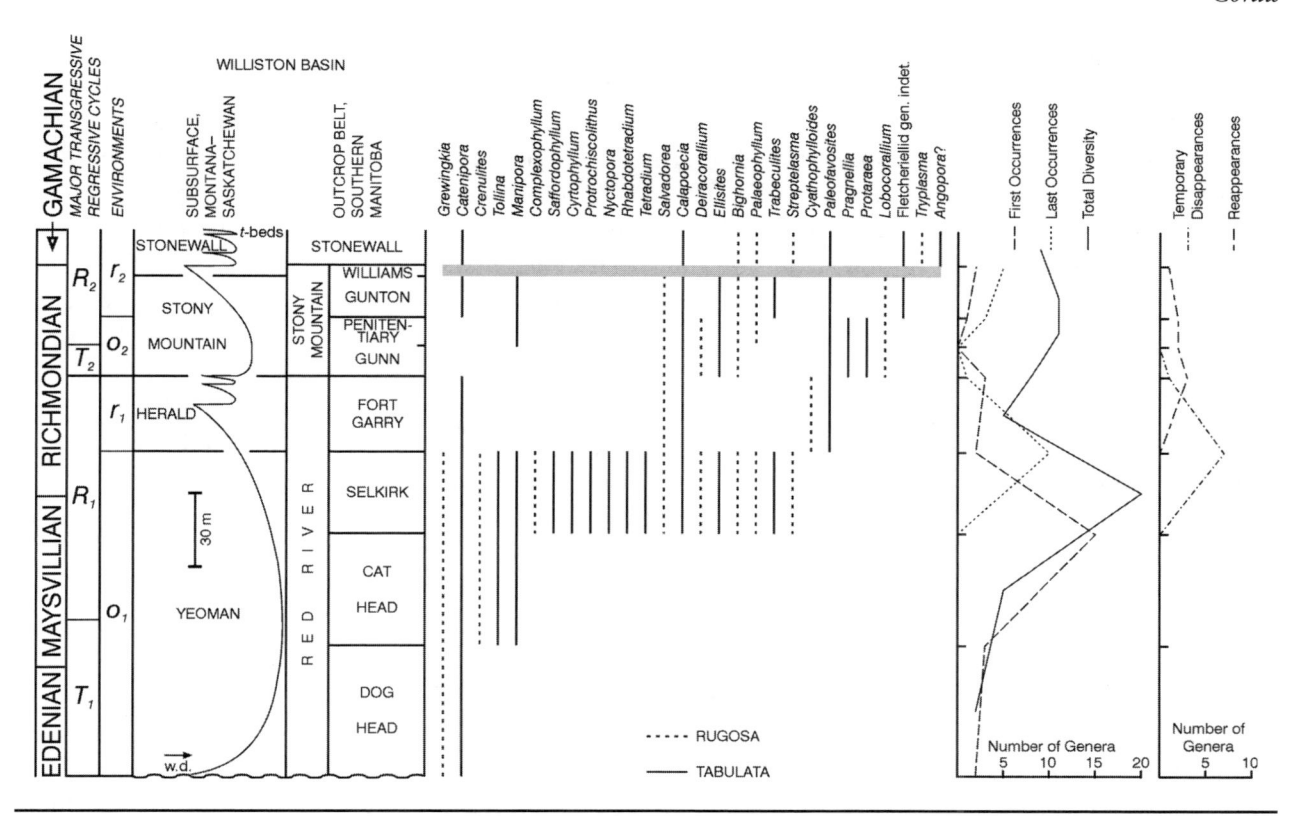

FIGURE 15.3. Cincinnatian coral diversity in the Williston Basin outcrop belt, southern Manitoba. The subsurface section (to scale shown) is based on a well in Montana, which Sweet (1979) integrated with his conodont-based graphic correlation of American western-midcontinent sections and which he related to Cincinnatian stages as shown on the left side of figure. The section apparently represents almost continuous deposition from mid Edenian through Gamachian (and into Early Silurian). The composite stratigraphic section in the outcrop belt of southern Manitoba (not to scale) shows the relationship of units to Sweet's stratigraphic framework. Major transgressive-regressive (T_1, R_1, T_2, R_2) cycles and open marine to restricted marine environments (o_1, r_1, o_2, r_2) follow Elias (1991); curve representing water depth (w.d., increasing in direction of arrow) based on sedimentologic and paleontologic data. Coral occurrences within a stratigraphic unit are shown as range lines from base to top of unit; the absence of data from Williams Member is due to insufficient exposure. For the purpose of the graphs on right side of figure, Gunton Member occurrences are extended to the top of Williams; first occurrences and reappearances are plotted at bases of units; last occurrences and temporary disappearances are plotted at tops of units; and total diversity is plotted at centers of units.

The Richmond Province is Richmondian in age; these corals are assigned to TS.5d through 6b. The Edgewood Province is considered to range in age from Gamachian to early Rhuddanian (earliest Silurian), but the position of the systemic boundary in some areas is uncertain. Most of the corals definitely occur in Gamachian strata and therefore represent TS.6c. Two, however, may not have appeared until the Rhuddanian; this uncertainty is noted in table 15.1. Corals of the "continental margin" assemblage that are considered to be Richmondian (or Ashgillian, pre-Hirnantian) are assigned to TS.5d through 6b. Those in the Gamachian represent TS.6c.

Red River–Stony Mountain Province

The Red River–Stony Mountain Province (Elias 1981), which existed for 7–9 m.y. during the Cincinnatian, occupied a vast region of predominantly carbonate deposition (figure 15.2). Temperature and salinity in the epicontinental sea were probably elevated in comparison with the ocean. Among the characteristic corals were solitary rugosans such as *Bighornia*, *Deiracorallium*, *Grewingkia* (endemic species including triangulate and trilobate forms), *Lobocorallium*, and *Salvadorea* and tabulates such as *Manipora*, *Saffordophyllum*, *Cyrtophyllum*, *Protrochiscolithus*, and *Trabeculites* (table 15.1). Unlike other provinces in the cratonic interior, the colonial rugosan *Palaeophyllum* and tabulate *Catenipora* were widespread.

Overall provincial generic diversity was high (41; 66 percent tabulates, 34 percent rugosans). The area of highest diversity was the Williston Basin (28); 20 genera were present during deposition of a single unit (Selkirk Member). On the stable continental shelf in the Anticosti-Percé area, taxa typical of this province occurred together with those representing the "continental margin" assemblage (e.g., tabulate *Propora*).

There were two major transgressive-regressive cycles in cratonic areas of the province (Elias 1991). In the Williston Basin, for example, relatively open marine environments prevailed during the transgressive phases and early in regressive phases (figure 15.3). Late in the regressive phases, conditions were somewhat restricted to evaporitic, with smaller scale cyclicity. The greatest water depths and most open marine conditions were attained during the first cycle, and seas were shallowest and most restricted toward the end of the second.

Changes in diversity were related to these major cycles. In the Williston Basin, overall generic diversity was greater in the first cycle (22) than in the second (17). In each cycle, diversity was low or moderate during the transgressive phase and reached a peak in the regressive phase before declining to low or moderate levels. During the first cycle, rising diversity involved introductions of genera, reaching a maximum in the Selkirk Member. This was followed by a high number of last occurrences as well as temporary disappearances and a low number of first occurrences. A rise in diversity during the second cycle involved reappearances as well as first occurrences. Last occurrences in the Gunton Member exceeded first occurrences and a reappearance in the Stonewall Formation, thereby reducing diversity in the latest Ordovician.

This province originated during the transgression of the first major cycle, which provided an opportunity for immigration from the craton east of the Transcontinental Arch and from the western continental margin (e.g., *Grewingkia*). In addition, some new genera apparently arose when the province originated or following the first transgressive maximum (e.g., *Deiracorallium*). *Lobocorallium* arose at the start of the second major cycle. During this transgression, the province expanded to the southern Canadian Rocky Mountains and Anticosti-Percé areas and spread eastward in the area of Maquoketa deposition. Late in the regressive phase of the second cycle, a few taxa that had previously been confined to areas near the continental margin appeared in the cratonic interior (e.g., *Tryplasma*), possibly reflecting cooling of the epicontinental sea. The sea withdrew from most areas during the major end Richmondian regression. The province was terminated in the Gamachian, when the last of its characteristic taxa disappeared. This was evidently due to loss of suitable habitat areas and decreasing temperature associated with the latest Ordovician glaciation in Gondwana.

Richmond Province

The Richmond Province (Elias 1982), of Richmondian age with a duration of about 5 m.y., coincided with a carbonate-siliciclastic platform (figure 15.2). To the west was deeper water in which Maquoketa Group shales accumulated; to the east was the Queenston Delta Complex. The Nashville Dome to the south, and probably the Canadian Shield to the north, rose above sea level. Runoff would have tended to reduce water salinity and raise nutrient levels on the platform.

The solitary rugosan *Grewingkia* (endemic species with circular cross sections) and tabulate *Foerstephyllum* were characteristic taxa (table 15.1). Compared with the situation in the other biogeographic divisions, tetradiid tabulates were far more common; favositid tabulates were absent. Overall provincial generic diversity was low (13; 62 percent tabulates, 38 percent rugosans). No area within the province had more than 10 genera; maximum diversity within a single unit was 10 (C5 sequence, Cincinnati Arch area).

This province originated and developed through significant immigrations to the region, although oceanographic and geographic barriers were limiting factors. Some corals were present in this vicinity earlier in the Cincinnatian (e.g., colonial rugosan *Cyathophylloides,* tetradiids). Others arrived from the Red River–Stony Mountain Province. The first important introduction coincided with an early Richmondian transgression. The most substantial immigration occurred during a major mid Richmondian transgression, when maximum diversity and distribution were attained as favorable environments became widespread. In the late Richmondian, diversity and distribution were limited. Westward progradation of the Queenston Delta reduced the size of the province, which disappeared during the major end Richmondian regression. Nutrient enrichment and related environmental destabilization in the shrinking epicontinental sea were probably significant factors in the associated regional extinctions.

Edgewood Province

The Edgewood Province (Elias 1982) existed for 2–3 m.y., from the Gamachian to early Rhuddanian (earliest Silurian). It coincided with a small epicontinental sea in which mainly carbonate sediments accumulated above the Maquoketa Group (figure 15.2). Gamachian units in the southern and central areas were likely deposited during minor northward transgressions from the ocean to the south, as sea level fluctuated during the glacial maximum. In the northern area, sedimentation occurred during the major transgression associated with deglaciation in latest Gamachian to Rhuddanian time. Fluctuating, elevated levels of nutrients and runoff from adjacent terrestrial areas likely contributed to environmental destabilization within the Edgewood sea (Elias and Young 1998).

This fauna was characterized by strong dominance of the solitary rugosan *Streptelasma,* followed by the tabulates *Paleofavosites* and *Propora* (table 15.1). Several genera did not occur in the other biogeographic divisions. In contrast with the situation in the other divisions, sarcinulid tabulates were absent. Provincial generic diversity was low (15; 60 percent tabulates, 40 percent rugosans). The geographic area and unit with the highest diversity (10 in both cases) were in Oklahoma (Keel Formation).

Corals in this province evidently immigrated from the continental margin or were derived from "continental margin" forms as suitable habitats shifted into the cratonic interior. Compared with faunas in the Richmondian epicontinental sea, Edgewood corals may have lived in somewhat cooler water. Environmental instability was likely instrumental in maintaining low overall diversity. This province was terminated with the appearance of a distinct Silurian assemblage in the late Rhuddanian.

"Continental Margin" Assemblage

A distinct "continental margin" assemblage (Elias 1989) of Richmondian-Gamachian (Ashgillian) age occurred around Laurentia (figure 15.2, table 15.1). The solitary rugosan *Grewingkia* (endemic species with indistinct cardinal septa and fossulae) was widespread. In contrast with the situation in the other biogeographic divisions, the tabulates *Catenipora* and *Propora* were both widespread. Excluding the Anticosti-Percé area, overall generic diversity was low (13; 54 percent tabulates, 46 percent rugosans).

"Continental margin" corals apparently favored open marine conditions, probably with somewhat lower salinity and temperature than were typical in epicontinental seas. In northern Alabama and northern Georgia, the solitary rugosans *Grewingkia* and *Rhegmaphyllum* occurred in shallow-water Richmondian environments. *Rhegmaphyllum* was the only coral in a shallow-water Gamachian setting in northern Arkansas. Faunas of low generic diversity (maximum eight per site) occurred in areas that were tectonically and volcanically active, such as northern California and northern Maine.

Continental Diversity

Considering Laurentia as a whole, 51 genera were present in the Cincinnatian (*TS*.5d–6c), assuming that *Pycnostylus* and *Plasmopora* appeared in the Gamachian (table 15.1). Factors contributing to this high overall diversity included the presence of epicontinental seas with environments suitable for corals and the development of provincialism related to environmentally distinct regions. A total of 45 genera appeared in the pre-Gamachian (*TS*.5d–6b), assuming that *Angopora?* did not appear until the Gamachian. Of these, 20 ranged into the Gamachian (*TS*.6c), assuming that *Tryplasma, Bighornia,* and fletcheriellid gen. indet. ranged into the Gamachian. In addition, 6 first appeared in the Gamachian. Thus, a total of 26 genera occurred in the Gamachian.

At the genus level, tabulates outnumbered rugosans in the Cincinnatian (65 percent tabulates, 35 percent rugosans). The ratios were similar for corals in the pre-Gamachian (64 percent tabulates, 36 percent rugosans) and for those that first appeared in the Gamachian (67 percent tabulates, 33 percent rugosans).

Rugosans, however, accounted for 50 percent of corals that ranged from the pre-Gamachian into the Gamachian. Therefore, the overall ratio of tabulates to rugosans dropped in the Gamachian (54 percent tabulates, 46 percent rugosans). Interestingly, this ratio was the same as that for pre-Gamachian corals of the "continental margin" assemblage, which apparently lived in relatively cool water. The continentwide shift to a lower tabulate-to-rugosan ratio in the Gamachian may be a reflection of cooler epicontinental seas during the latest Ordovician glacial age.

■ Baltoscandian Rugose Corals (BEEN, DK)

Ordovician rugose corals of Baltoscandia have attracted the attention of natural scientists since the early nineteenth century. The most important contributions are those of W. Dybowski and G. Lindström and, later, W. Scheffer, V. Reiman, and N. Spjeldnaes. The present knowledge about coral taxonomic diversity and distribution is based mainly on papers by Kaljo (1958, 1961), Neuman (1968, 1969, 1975, 1998), and Weyer (1973, 1982, 1983, 1984, 1997).

The Late Ordovician Baltoscandian basin was a pericratonic, partly gulflike sea on the western margin of Baltica, which at that time was drifting from midlatitudes to a subequatorial position. Suitable coral habitats (carbonate shelf in shallow sea, often with nearshore reefs) occurred along the shores of this basin. Nowadays these deposits form a continuous belt in North Estonia and westernmost Russia and more limited areas in Sweden (e.g., Dalarna, Östergötland and Västergötland, Jämtland) and in the Oslo Region (e.g., Asker, Ringerike, Hadeland) in Norway (for locations see chapter 10: figure 10.1).

Ordovician successions in Baltoscandia have been subdivided in considerable detail, and owing to the abundance of fossils in most of the rock units, correlation between sections of the different areas is especially well established (Nõlvak and Grahn 1993). The stratigraphic relationships of many of these Baltoscandian units are shown in Nielsen's figure 10.2 (chapter 10). In this summary the data are analyzed and presented according to the time slices (*TS*) used in this volume (see chapter 2, especially figures 2.1, 2.2). The local rugose-coral-bearing units that are considered as representing the corresponding time slices are specified in table 15.2.

Distribution and Diversity

The Baltoscandian rugose corals compiled in this study comprise 92 species belonging to 31 genera of the families Lambelasmatidae (11 genera), Streptelasmatidae (15 genera), and others (Calostylidae, Tryplas-

TABLE 15.2. Baltoscandian Rugose Coral-Bearing Stratigraphic Units and Localities Assigned to the Upper Ordovician Time Slices (*TS*.5b–6c) Inclusive

Time Slices	Estonia: Stages/Substages/ Formations	Sweden	Norway	Germany
TS.6c (1.7 m.y.)	Porkuni stage, and Ärina Formation	Loka Formation, Borenshult and Öjle Myr faunas (erratic boulders)	Langøyene and Langåra (upper) formations; Hirnantian stage in the Farsund drill core	
TS.6b (1.8 m.y.)	Pirgu stage (upper), and Adila Formation	Boda Limestone (upper)	Bønsnes, Herøya (upper), and Langåra (upper) formations; Rawtheyan stage in the Farsund drill core	
TS.6a (2.5 m.y.)	Pirgu stage (lower), and Moe Formation	Boda Limestone (lower)	Herøya (lower part) and Grimsøya formations, Tretaspis Limestone	
TS.5d (3 m.y.)	Nabala and Vormsi stages	Fjäcka Formation, Hulterstad and Laknäs faunas (erratic boulders)		Rügen erratic boulders
TS.5c (3 m.y.)	Keila Oandu (= Vasalemma), and Rakvere stages	Kullsberg, Macrourus, and Moldå limestones, Skagen Formation, and Gräsgård fauna (erratic boulders)	Solvang and Furuberget formations, Spaeronid Limestone	Macrourus erratic fauna
TS.5b (2.4 m.y.)	Haljala (= Idavere + Jõhvi substages)	Dalby Formation, Jämtland		

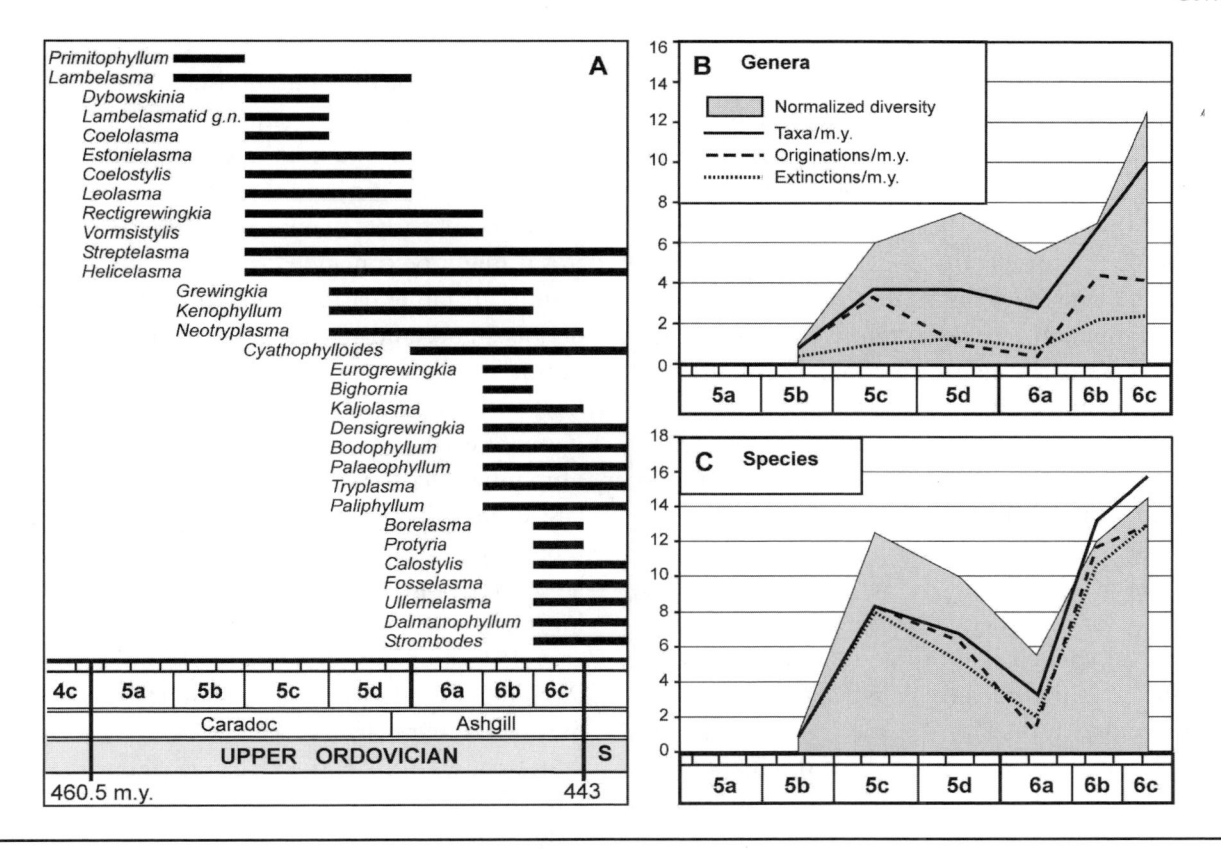

FIGURE 15.4. Distribution and diversity dynamics of the Baltoscandian Late Ordovician rugose corals. A, stratigraphical distribution of identified rugose coral genera. B, plot of diversity dynamics for genera. C, plot of diversity dynamics for species.

matidae, Paliphyllidae, Chonophyllidae, Stauridae). The latter five families are represented by only a few species at the end of the Ordovician Period (figure 15.4A). The following are features of importance:

1. The oldest rugose corals of the region (*Primitophyllum, Lambelasma*) appear within *TS*.5b in the lowermost middle Caradoc. These are rather simple lambelasmatid corals with monacanthine septa.

2. The Lambelasmatidae are dominant in *TS*.5b and 5c. The first Streptelasmatidae appear in *TS*.5c, but they are not common until the end of the Caradoc. Streptelasmatids prevail in the Ashgill (*TS*.6a–c), but the Hirnantian (*TS*.6c) is highlighted by the appearance of several new morphotypes (*Calostylis, Strombodes*, septofossulate streptelasmatid *Ullernelasma*; Neuman 1997) and a few new species of colonial genera (*Palaeophyllum, Protyria, Tryplasma*).

3. Through the entire Late Ordovician the taxonomic diversity of the Baltoscandian rugose corals steadily increased. Against the background of this general trend two intervals of greater changes were observed, separated by a declining interval in the early Ashgill (*TS*.6a). The first remarkable diversification took place in the mid Caradoc (*TS*.5c), when, mainly as a result of radiation in Lambelasmatidae, 10 new genera appeared. Yet the event was not a single change but a stepped process, which began with origination of 3–4 new genera in the early part, followed by a burst (5 new genera and 14 species appeared) in the middle of the time slice.

The second notable diversification event commenced in the Rawtheyan (*TS*.6b) and reached a peak with 17 genera in the early Hirnantian (*TS*.6c). This event was mainly caused by morphological innovations mentioned earlier but was also partly due to the appearance of coral families that were to become more widely distributed in the Silurian. All Hirnantian corals occurring in Estonia are definitely of early Hirnantian age; the same is likely but not so certain for the Scandinavian occurrences. A few genera (*Densigrewingkia, Helicelasma, Ullernelasma*), found in the topmost part of the corresponding interval in the Farsund drill core, may be of late Hirnantian age.

In terms of the mass extinction, which is usually connected with the Hirnantian glaciation and affected several groups of fossils including tabulates in the Baltic area and elsewhere, we might expect the rugosans to exhibit a similar pattern. Figure 15.4B, however, does not show this pattern clearly. In *TS*.5d, 6b, and 6c the turnover rates of the last occurrences are 1.3, 2.2, and 2.4. This means that during the Ashgill the extinction rate rises slightly, but not enough to call the process a mass extinction. The genus level innovations in *TS*.6b and 6c, which originated several new phylogenetic lines, were more important in the shaping of the rugose coral diversity than disappearance of several primitive stocks.

4. The species-level diversity (92 species identified; figure 15.4C) had the same general trend of changes as for the genera—an overall rise throughout the interval studied interrupted by a fall in the late Caradoc (*TS*.5d) and early Ashgill (*TS*.6a). The decline is more distinct at the species level than the genus level, even if the curves of the originations are rather similar. The species diversity rises most rapidly in *TS*.5c and 6b–c (figure 15.4C), but in the case of *TS*.6b–c the high rate is partly boosted by the shorter time slices. In the same time the number of species per time slice is almost identical for *TS*.5c, 6b, and 6c. Compared with equivalent generic values, the turnover rates of the species origination are much higher in the mid Caradoc, Rawtheyan, and Hirnantian, making the species normalized diversity changes more rapid.

The data for the disappearance of species are quite different. Major changes in rugose corals occurred in the latest Ordovician (Kaljo 1996). Many species disappeared in *TS*.5c, 6b, and 6c (with corresponding disappearances of 24, 19, and 22). In terms of the turnover rate of the disappearing species, the extinction is the most significant in the Rawtheyan and Hirnantian (*TS*.6b–c). A very high extinction rate is exhibited using the percentage of disappearing species to the number of species occurring in a time slice. For the time slices such percentages are as follows (*TS*.5b to 6c inclusive): 100, 96, 75, 63, 79, 81 percent.

To interpret these percentages correctly, we should keep in mind one more characteristic feature of Late Ordovician rugose corals, which in part also is influenced by the state and mode of the coral studies in the region. In the distribution list of species there are on average 3 species per genus, but 12 of the genera are monospecific and 6 others have only 2 species. Another characteristic is that only 5 species of 92 range through three consecutive time slices. From this it might be concluded that the species in Baltoscandia are rather narrowly determined and that their disappearance and replacement might be caused by normal evolutionary process determining the high level of the background disappearance. However, this does not exclude also "catastrophic" influence of the Hirnantian glaciation as a cause of the well-known late Ordovician mass extinction. This influence is more apparent at the species level than at the genus level. From the other side the early Ashgill diversity low may be explained also by facies constraints or incomplete sampling, or both. In Estonia the lower Pirgu rocks are clearly less coralliferous than the upper part of the stage. In Sweden the Boda Limestone is correlated with the entire Pirgu stage, but the corals have practically all been collected from the upper part only. This correlation may need to be revised if further work on the lower Boda corals reveals important differences.

Summary Remarks

The record of rugose corals commenced in Baltoscandia during the early to mid Caradoc (*TS*.5b) with appearances of two rather primitive genera. A couple of million years later the diversity rose considerably (11 genera identified in *TS*.5c, with 90 percent of them new). Then a second striking diversification event took place in the Rawtheyan and early Hirnantian with 21 genera appearing, more than 70 percent of them for the first time.

Radiation of the local phylogenetic lines played the main role in the diversification of the Baltoscandian rugose coral fauna. At some levels in the Caradoc, and later also, a few representatives of Laurentian genera (e.g., *Streptelasma, Helicelasma*) made an important addition to the diversity.

The disappearance rate of the Baltoscandian genera rose only slightly during the Late Ordovician. Therefore the late Caradoc (*TS*.5d) and late Ashgill (*TS*.6c) extinctions are weakly or not at all observed here among genera but are well expressed in diversity changes at the species level (figure 15.3B,C). The main part (more than 75 percent) of the Hirnantian genera continued into the lowermost Silurian.

■ Australasian Corals (BDW)

This section focuses on the Ordovician rugose and tabulate corals of eastern Australia and New Zealand that have been documented from contrasting tectonic settings, including part of East Gondwana (the Tasmanian Shelf), and a number of offshore, peri-Gondwanan arc-related terranes in central New South Wales (the Macquarie Arc, formerly the Macquarie Volcanic Belt of Webby 1976), northeastern New South Wales (Tamworth Terrane), central and northern Queensland (Anakie High, Broken River Embayment), and northwestern Nelson, New Zealand (Takaka Terrane).

The Australasian Ordovician corals have been studied principally by Etheridge (1909), Hill (1942, 1955, 1957), Webby and Semeniuk (1969, 1971), Webby (1971, 1972, 1977, 1988), Hall (1975), McLean and Webby (1976), Webby and Kruse (1984), Webby in Percival et al. (2001), and Pickett and Percival (2001). The bulk of these taxonomic descriptions remain valid with a few exceptions in the earlier literature. However, substantial components of the coral fauna remain incompletely described, and this is especially so in Tasmania, in northern and central Queensland, and in New Zealand, across the entire fauna, and important parts of the central New South Wales fauna (e.g., the heliolitinids and favositinids) also need further study.

Aspects of biogeography of these Australasian Ordovician coral faunas have been discussed previously (Webby 1987; Lin and Webby 1989; Webby in Webby et al. 2000), and the contrasting biodiversity profiles in Tasmanian and central New South Wales successions have been outlined (Percival and Webby 1999). A generalized similarity seems to exist in the Mid to Late Ordovician generic-level coral biodiversity across the entire region—between the East Gondwanan (Tasmanian) Shelf and offshore peri-Gondwanan arc-related terranes—with few genera so far confirmed as endemic, which is in striking contrast to the species-level coral data. At the present stage of knowledge of these coral faunas, only the generic-level diversity data can be adequately presented here. Much further work is needed to evaluate the species-level diversification fully.

These Australasian corals are composed entirely of shallow marine, warm-water assemblages dominated by compound (and a few solitary) rugosans (representatives of the suborders Cystiphyllina, Streptelasmatina, and Stauriina), the tabulates (mainly members of the suborders Sarcinulina, Heliolitina, Halysitina, Auloporina, and Favositina), and the order Tetradiida (see figure 15.5). They are characteristic of the equatorial "American-Siberian realm" (Kaljo and Klaamann 1973; Webby 1992b; Webby et al. 2000), being reported through mainly Upper Ordovician carbonate successions from the early Gisbornian (*TS*.5a), or a slightly older horizon (?latest Darriwilian—*TS*.4c), to the more numerous records through the late Gisbornian (*TS*.5b), the succeeding Eastonian (*TS*.5c to 5d), and the early Bolindian (*TS*.6a). Since the late 1960s an informal coral and stromatoporoid biostratigraphic scheme has been used for correlating the Ordovician limestones of eastern Australia, especially in central New South Wales (Webby 1969, 1975; Webby in Young and Laurie 1996). Faunas I–III (inclusive), which span the Eastonian stage, correlate with early to mid *TS*.5c, mid to late 5c, and 5d, respectively, and Fauna IV (of early Bolindian age) equates with *TS*.6a (figure 15.5).

Distribution

The oldest known Australasian corals are tabulates, recorded from a pre–Fauna I interval (mainly Gisbornian) of two separate central New South Wales localities. The first, at Wahringa (south of Wellington), is the earliest appearance of an Australian tabulate—the sarcinulid *Billingsaria spissa*—of latest Darriwilian or early Gisbornian (*TS*.5a) age, probably the latter (Webby in Percival et al. 2001). The second is at Gunningbland (west of Parkes), where further sarcinulinids (*Billingsaria domica, Foerstephyllum* spp.) and the first tetradiids (*Rhabdotetradium, Paleoalveolites*) appear. These come from Gisbornian, probably late Gisbornian (*TS*.5b) horizons (Pickett and Percival 2001). In Tasmania the first sarcinulinids and the tetradiids appear a little later during

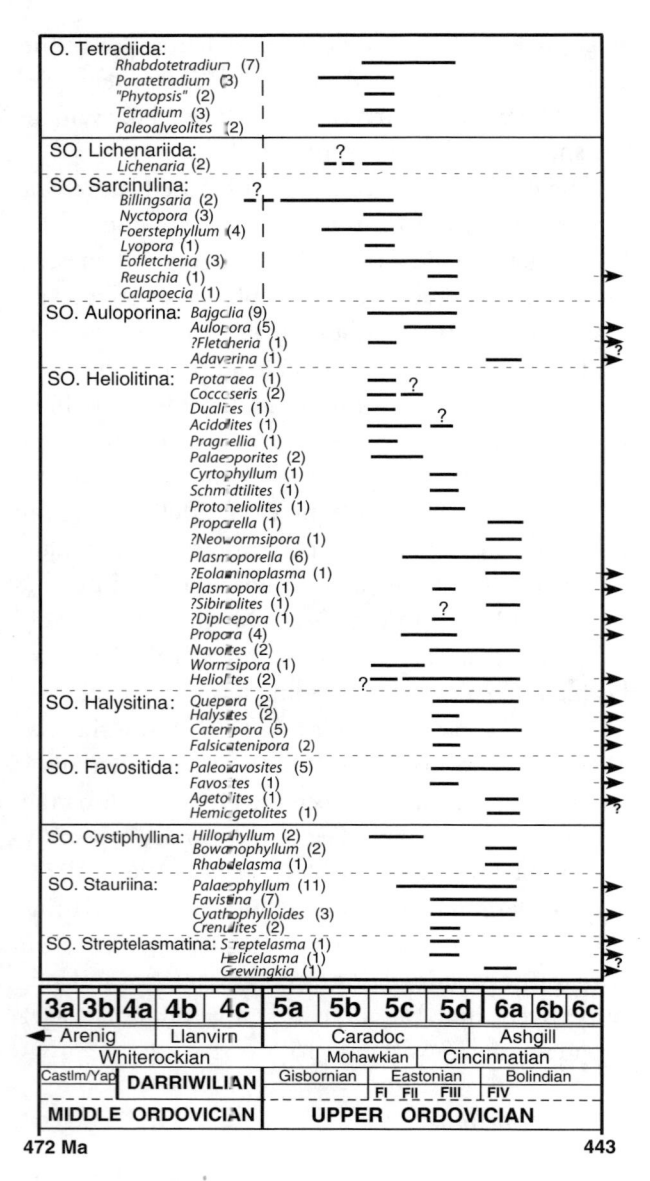

FIGURE 15.5. The temporal distribution of Ordovician coral genera (suborders of Tabulata and Rugosa, and the order Tetradiida) from Australasia. Species numbers are shown in parentheses. The timescale in the lower part of the diagram shows the stratigraphic relationships, the time slices, and the Australian regional coral-stromatoporoid biostratigraphic scheme (Faunas [F] I, II, III, and IV inclusive). Abbreviation: Castlm/Yap = Castlemainian to Yapeenian stage interval. Arrows depict the known genera that have extended ranges into the Silurian.

pearance during this early Eastonian (Fauna I) interval in both central New South Wales and Tasmania. Other major tabulate groups, such as the halysitinids and favositinids, appear somewhat later in all eastern Australian successions, that is, during the late Eastonian Fauna III (TS.5d). The spread of the auloporinids, heliolitinids, and favositinids continued into the early Bolindian (Fauna IV—TS.6a), but other main tabulate groups (lichenariinids, sarcinulinids) and the tetradiids were no longer present in the region (figure 15.5).

The rugose corals exhibit a rather different pattern, with the cystiphyllids (exhibiting acanthine septa) the earliest group to appear. The genus *Hillophyllum* first appeared in Fauna I (lower TS.5c) of central New South Wales, but not in Tasmania (Webby 1971), and then later, in Fauna IV (TS.6a), there were appearances of endemics (*Bowanophyllum, Rhabdelasma*), also restricted to central New South Wales (McLean and Webby 1976). The streptelasmatinids and stauriinids (with laminar septa), in contrast, are well represented in both New South Wales (central and northeastern regions) and Tasmania, from Fauna II (upper TS.5c) to IV (TS.6a). The colonial *Palaeophyllum* appeared first in Fauna II, then solitary (*Streptelasma, Helicelasma*) and colonial cerioid (*Favistina, Cyathophylloides, Crenulites*) in Fauna III, and finally solitary *Grewingkia* in Fauna IV (Webby 1972, 1988; McLean and Webby 1976).

In the Fossil Hill succession at Cliefden Caves (New South Wales; Macquarie Arc), occurrences within lower TS.5c may be differentiated broadly into three distinctive onshore-offshore community (or biofacies) types, from Benthic Assemblage (BA) 2–3 of Boucot (1975). From onshore there is a restricted marine, peritidal (lagoonal) inner-shelf succession (inner BA 2) exhibiting large, thick-shelled, inarticulated *Eodinobolus* shell banks with only two associated corals, both tetradiid species (Webby and Semeniuk 1971). These comprise small, slender *Rhabdotetradium* (*R. duplex*), as well as the thicketlike "*Phytopsis*" *variabile,* which possibly acted as a localized wave baffle (Webby and Percival 1983).

This is laterally equivalent to the higher-energy, midshelf, shoal, or reeflike environment of massive *Rhabdotetradium cribriforme* (Etheridge) and tabulate corals (*Nyctopora, Bajgolia, Dualites,* and *Hillophyllum*) and the stromatoporoid *Cystistroma,* which

the early Eastonian Fauna I (lower TS.5c; figure 15.5). The same taxonomic nomenclature adopted for the global survey of tetradiid corals by Webby earlier in this chapter is used here.

A number of other major groups (lichenariinids, auloporinids, and heliolitinids) make their first ap-

probably represents the outer part of BA 2. The large *R. cribriforme* colonies are mainly hemispherical, but locally they develop flattened tops like *Porites* microatolls of the modern reef flat (Kobluk and Noor 1990; Percival and Webby 1996; Webby et al. 1997) that suggests localized emergence of parts of the shoals. An alternative proposal that they formed in subtidal conditions by smothering of sediment does not explain how the upper surfaces of the hemispherical colonies initially became selectively flattened. The *R. cribriforme* biofacies has a diverse coralline fauna (up to 13 species), with large sizes, and commonly corals exhibit intergrowths with stromatoporoids, cyanobacteria, and/or bryozoans.

The third, more offshore biofacies has a similar high diversity but is developed in a quiet-water, level-bottom, BA 3 position. It occupied a subtidal, open- (or outer-) shelf site (Webby et al. 1997). The coral colonies have smaller sizes, and the tetradiids disappear, in contrast to the sarcinulids, which maintain their diversity levels, and the heliolitinids, which become more diverse.

In addition to the Ordovician coral localities in central New South Wales (see a recently compiled list of species in Pickett and Percival 2001) and in Tasmania (Hill 1955; Corbett and Banks 1974; Banks and Burrett 1980), there is an adequately documented, varied rugose and tabulate coral fauna (Fauna III age) in the separated fault blocks of the Tamworth Terrane, northeastern New South Wales (Hall 1975). The fauna (especially the occurrences of compound rugosan species of *Favistina, Cyathophylloides,* and *Crenulites*) suggests a North American "Red River" provincial affinity (Flower 1961; Webby et al. 2000:71). The coral faunas mentioned by Palmieri (1978) from the Emerald area of central Queensland (Anakie High) also closely resemble the Tamworth assemblages described by Hall (1975) and probably have a similar Fauna III age but are yet to be described.

In northern Queensland, an association of *Plasmoporella, Catenipora,* and *Agetolites* has been identified (Hill et al. 1969) and is of probable Fauna IV age (early Bolindian). A diverse heliolitinid component of this fauna has been studied by John S. Jell and Owen A. Dixon but remains unpublished. In New Zealand only two heliolitinid taxa and rugosan *Favistina* have been recorded from the Takaka Terrane in northwestern Nelson (Cooper 1968). No additional details are available, but the assemblage is likely to be late Eastonian in age.

Diversity Patterns

In assessing and measuring the diversity change in the Australasian corals it is important to emphasize again the preliminary and regional nature of the data set. Altogether some 45 genera (101 species) of tabulates and tetradids (on average 2.2 species per genus) have been recorded from Late Ordovician successions of Australasia, as well as 10 genera (31 species) of rugose corals (on average 3 species per genus) (figure 15.6). However, 22 tabulate and tetradiid genera are monotypic (i.e., 50 percent of the fauna), and among the rugose components, 4 of the 10 are monotypic (i.e., 40 percent of the fauna). The Heliolitina is the most diversified tabulate group at the generic level, with 20 genera (32 species). At the species level the most diversified genera (through parts of the *TS*.5c–6a interval—i.e., over a 5–7 m.y. duration) are the rugosans *Palaeophyllum* (11 species) and *Favistina* (7 species), the auloporinid *Bajgolia* (9 species), the heliolitinid *Plasmoporella* (6 species), and the tetradiid *Rhabdotetradium* (7 species). The genera have comparatively short temporal durations of up to 8 m.y. (figure 15.6).

Two main measures of genus-level diversity change have been employed: (1) to estimate the normalized diversity for the Australasian region by counting in

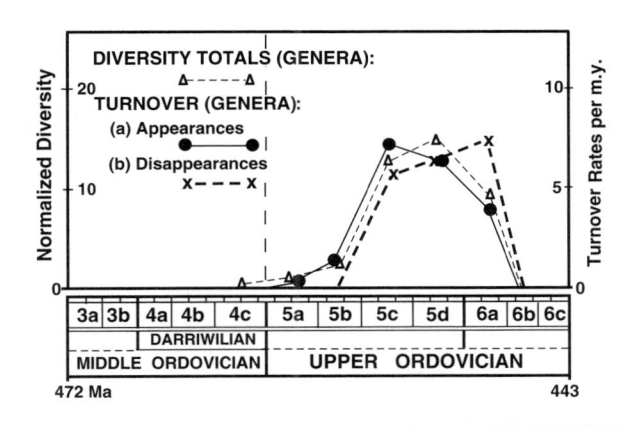

FIGURE 15.6. Ordovician diversity curve for Australasian generic data, showing the normalized diversity (range-through data across *TS*.4c to 6a only), and turnover rates per m.y. (appearances and disappearances).

each time slice the genera that range through the specific interval and adding half values for the other genera that appear, disappear, or are entirely confined to that interval; and (2) to measure the faunal turnover (inferred evolutionary change) within the region, based on rates of appearances and disappearances per m.y. However, in this type of regional treatment a distinction should be drawn between regional disappearances and worldwide disappearance events that may represent true extinctions. The former may, for example, be related to regional environmental changes as a result of major orogenic activity.

Only one maximum is shown by the normalized generic diversity data with the peak of diversity occurring during the late Eastonian (= late Caradoc), i.e., TS.5d (figure 15.6). Initially, in the latest Darriwilian to the Gisbornian (TS.5a–b), the rise in diversity is slow but then increases dramatically between TS.5b and 5c. The decline in diversity through the TS.5d–6b interval is equally rapid. This generic diversity is presented as a range-through record for the TS.4a–6a interval only—for the entire available Australasian Ordovician coral diversity record. No attempt was made in this biodiversity survey to compile as well the regional Silurian coral diversity profile in order to represent the range-through data across the TS.6b–c interval upward into the Silurian or to take account of reappearances of some genera during the Silurian. The record for Australasia with its single peak of genus-level diversity during the TS.5d interval contrasts with the patterns recorded in Laurentia and Baltica. Elias and Young (earlier in this chapter) depict comparable patterns for tabulate and rugosan corals with "total diversity" peaks in the late Maysvillian/early Richmondian (TS.5d) and latest Richmondian/Gamachian (TS.6b–c). In addition, Neuman and Kaljo (earlier in this chapter) show two maxima of normalized diversity for the Baltic rugosans, the first coinciding with the late Caradoc (TS.5d) interval and the second in the late Ashgill (TS.6c).

The loss of biodiversity in Australasia during TS.6b–c is attributed entirely to the intensity of environmental effects associated with latest Ordovician–Early Silurian Benambran Orogeny in southeastern Australia. Crustal-thickening events involving the closure, contraction, folding, metamorphism, and uplift of the former turbidite-infilled, western-Pacific-type Wagga marginal sea, together with magmatic events associated with the westward subduction of the adjacent (ocean-side) Macquarie volcanic arc (Fergusson and Fanning 2002), caused dramatic changes to configurations on land and in the adjacent seas of the entire region, with loss of suitable shallow carbonate shelf habitats for continued occupation and diversification of coral biotas. On the continental margin of East Gondwana (Tasmanian Shelf) the depositional facies switched from dominantly Eastonian carbonates with coral faunas (part of the Gordon Group) to siliciclastics (e.g., Westfield Sandstone) of late Bolindian age (Banks 1988), while across the cratonic interior of mainland Australia there was epeirogenic activity—the Rodingan Movement of uplift and erosion in the Amadeus, Ngalia, and Georgina Basins of central and northern Australia (Webby 1978)—as well as establishment of the great, mainly land-locked, Mediterranean-type evaporite basin in the Canning Basin of western Australia (Webby et al. 2000:103). The role of glacially induced end Ordovician mass extinctions is unknown because its effects were completely masked by the intensity of this orogenic activity in eastern Australia.

A general correspondence exists between the normalized diversity curve for the Australasian coral data set and the curves illustrating generic turnover, with the appearances and disappearances showing rather similar tracking patterns (figure 15.6). The rates of genus-level appearances per m.y. initially remain at background levels during TS.4c to 5a and then slowly increase to 1.4 genera per m.y. in TS.5b, followed by a rapid rise to a peak of 7.3 genera per m.y. in TS.5c, then a progressive decline through two successive time slices to 4 genera per m.y. during TS.6a. In contrast, the rate of disappearances increases from 5.7 genera per m.y. in TS.5c to a peak of 7.6 general per m.y. in TS.6a and then declines within that time slice, as the regional orogenic activity becomes more intense.

■ Concluding Remarks (BDW, RJE, BEEN, DK, GAY)

1. A number of different soft-bodied (solitary and colonial) anemone groups probably developed skeletons independently during the Ordovician Period, in order to explain the separate appearances of the different orders of tabulate, rugosan, tetradiid, and kilbuchophyllid corals. The tabulate and rugose corals ex-

hibit the greatest, worldwide, diversifications through the Paleozoic, in contrast to the tetradiid, corals that only participated in a much smaller, more restricted, exclusively Ordovician, global radiation event, and the solitary, button-shaped, kilbuchophyllid corals that had an even more limited, short-lived (mid Caradoc), and localized British record of occurrences. None of these groups shows close links to known Cambrian coralomorphs (Scrutton 1997; Debrenne and Reitner 2000).

2. The tabulate corals have a long, rather patchy, and low diversity record of occurrences through the Early and Mid Ordovician—only one or two genera of small, simple, and poorly differentiated colonial forms existing until near the end of the Mid Ordovician. Then, the group began to diversify rapidly during the late Darriwilian to mid Caradoc, including the appearances of a number of higher-level taxa (suborders) as a wide range of reef and nonreef habitats became available in the warm, shallow carbonate seas. The diversity of tabulates rose significantly through the Caradoc and continued to expand to a peak of more than 60 genera during the mid Ashgill. The rapid decline that followed was probably caused mainly by the global cooling and regressional events that accompanied the short-lived end Ordovician glaciation.

3. The rugose corals, on the other hand, have no confirmed Early to Mid Ordovician records—their major higher-level diversification did not occur until the Late Ordovician, and they then involved rapid expansion mainly during early to mid Caradoc time. From available global data, rugosan generic numbers appear to have risen steadily through the Caradoc, increasing to near 50 genera by the mid Ashgill, then declined dramatically to about half this number, probably owing to the global cooling and regression in association with the short-lived glaciation. However in some regions such as Baltoscandia there was a lower rate of extinction, for example, with about 75 percent of the 17 recorded latest Ordovician (Hirnantian) genera surviving into the Silurian.

4. The colonial tetradiids flourished in shallow, somewhat muddy carbonate habitats of equatorially disposed platforms, shelves, and island fringes of the Late Ordovician. The most notable diversification of genera occurred during the early to mid Caradoc, and the species-level diversification is recorded in two distinct peaks—in the mid Caradoc and early-mid Ashgill. The group failed to survive the end Ordovician extinction, probably mainly because it involved two pulses of short-lived environmental change—the group just survived the initial severe and rapid global cooling, but succumbed to the warming event with its associated sea level rise and spread of anoxic waters that followed.

5. Laurentia straddled the Ordovician equator and was one of the more significant coral diversification centers. The late Caradoc to Ashgill (Cincinnatian) coral faunas became differentiated zoogeographically into four discrete provinces, with the largest province developing across the greater part of the Laurentian platform, and exhibiting the highest diversity (about 40 genera; tabulate/rugosan ration of 2:1). Distinctive diversity patterns are depicted within the two major transgressive-regressive cycles of this province (for example, in the Williston Basin). The diversity levels were low to moderate during the transgressive pulses, then reached a maximum through the early to middle parts of regressive pulses, and then declined to low levels at the regressive peak as depositional conditions became more restrictive to evaporitic. The other provinces developed in more marginal parts of the Laurentian platform, and each exhibits a comparatively lower diversity (up to 10 genera). Two of these provinces exhibit faunas of latest Ordovician age, and these show a lower tabulate/rugose ratio of 3:2 than earlier.

6. The rugose coral faunas in Baltoscandia diversified continuously through the Late Ordovician, though at a more accelerated rate through the mid Caradoc and mid Ashgill intervals. These faunas occupied shallow shelf habitats of Baltica as the block drifted from the middle latitudes into the sub-tropics. The earlier (Caradoc) rugosans, which were all solitary forms, therefore lived in relatively slightly cooler waters of the middle latitudes. Colonial rugosans only started to appear in the later (Ashgill) associations when Baltica had moved into lower latitudes, that is, in all probability into somewhat warmer seas. A comparison of strictly correlative (late Caradoc to Ashgill) parts of the Baltoscandian rugose succession and the "equatorial" Laurentian rugose succession suggests that the Baltoscandian fauna of 27 genera is markedly more diverse than the equivalent Laurentian fauna with its 18 genera. This seems to suggest that the most diverse rugose faunas, especially those dominated by solitary

forms, are not necessarily restricted to the low latitudes. The late Ashgill rugosans of Baltoscandia are significant in including a number of new morphotypes that became influential in shaping later (Silurian) evolutionary development.

7. The mid Caradoc to early Ashgill coral faunas of eastern Australasia are partially older and partially correlative with the Laurentian Cincinnatian assemblages. These Australasian faunas are also characteristically a part of warm-water, "equatorial" assemblages, and have a higher overall diversity (about 55 genera). These partially older Australasian assemblages have a much higher tabulate/rugose ratio of 4:1, a relatively higher diversity of heliolitine corals than the Laurentian associations. Also the Australasian faunas include a much lesser number of rugosan taxa (10 genera, only a few being solitary forms) than in either Baltoscandian or Laurentian assemblages.

ACKNOWLEDGMENTS

Elias and Young both acknowledge financial support from the Natural Sciences and Engineering Research Council of Canada. Neuman and Kaljo thank O. Hints and A. Noor for technical assistance. Their study was partly supported by the Estonian Science Foundation (grant No. 5042).

16 Bryozoans

Paul D. Taylor and Andrej Ernst

Ordovician benthic assemblages commonly contain bryozoans, sometimes in rock-forming abundance, for example, the Cincinnatian of the U.S. Midwest. These colonial suspension feeders were often major components of Ordovician seabed communities and, like their living relatives, undoubtedly had important roles in sediment stabilization and binding, as well as providing habitats and sources of food for other benthic organisms. In the absence of unequivocal records of bryozoans in the Cambrian, the Ordovician holds considerable evolutionary significance for the phylum, not only in furnishing the oldest known bryozoans but also in recording the first of three major evolutionary radiations that can be recognized in the fossil record of the phylum (Taylor and Larwood 1990). According to Anstey and Pachut (1995:274), "Early Ordovician speciation apparently produced most of the clade branching and family-level derived character states in the entire Paleozoic. The morphological distances achieved by early speciation events, if the species record is not unusually biased, must have been much greater than at any other subsequent time."

Regrettably, however, research on Ordovician bryozoans has lagged well behind that on groups of comparable diversity and abundance. This can be partly explained by their lack of utility in Ordovician biostratigraphy. In addition, identification of bryozoans is not as straightforward as that of other groups—precisely oriented thin sections are required in order to identify most Ordovician bryozoans below subordinal level, and much of the scattered literature is old and fails to take into account the skeletal variability that typifies benthic colonial animals. Furthermore, most decalcified bryozoans do not preserve the internal skeletal characters needed for identification. Ordovician bryozoan faunas from some parts of the world remain virtually undescribed; Tuckey (1990), for example, remarked that relatively few faunas had been described from southern Europe, Africa, China, South America, and Australia. The paucity of work on Ordovician bryozoans has two major consequences with respect to this chapter. First, bryozoan phylogeny is poorly understood and has not yet been used to refine classification; therefore, monophyletic clades cannot be recognized simply by reference to the classification currently in use. Second, the occurrences of taxa in time and space are not adequately established. Consequently, taxic diversity patterns must be viewed as very provisional. With these caveats made explicit, our main aims in this chapter are to (1) summarize the early fossil record and phylogeny of bryozoans and (2) quantify and interpret taxic patterns of diversity change during the Ordovician.

■ Introduction

Good general accounts of bryozoans can be found in Ryland (1970) and McKinney and Jackson (1989);

Anstey and Pachut (1995) have analyzed the diversity and evolution of Paleozoic bryozoans, and Taylor (1999) has summarized the functional morphology of the hard skeleton. Bryozoans are almost universally accorded the status of a phylum. All modern representatives, of which there are perhaps 6,000 species, form colonies of modular individuals known as zooids. Individual zooids generally measure less than a millimeter in size, whereas colonies are typically centimeter-scale but can attain diameters of up to a meter by budding substantial numbers of clonal zooids as they grow. The asexual process of bryozoan colony growth is distinct from sexual reproduction, which in bryozoans entails the fertilization of an egg, typically followed by brooding of the embryo and release of a larva that swims away from the parent colony, settles on a hard or firm substrate (e.g., shell, rock, or seaweed), and metamorphoses into the founder zooid of a new colony. Clonal fragmentation is an alternative mode of new colony formation.

Feeding zooids in bryozoan colonies have a tentaculate lophophore with a mouth at the center. Beating of the cilia on the tentacles drives water containing suspended plankton toward the mouth. Retraction of the lophophore into the safety of the main body cavity takes place by extremely rapid contraction of a muscle fixed to the base of the lophophore. Whereas a few bryozoan species are monomorphic, most exhibit zooidal polymorphism in which zooids serving different functions (e.g., reproduction, defense, feeding, support) differ in morphology.

Using zooids of relatively uniform shape, bryozoans have evolved a multiplicity of different colony forms (see classification of Hageman et al. 1998), mostly by variations in budding pattern. A basic distinction can be made between encrusting colonies with zooids tightly adpressed to the substrate and erect colonies developing growth away from the substrate and into higher feeding tiers. Evolutionary convergence in colony form is a central theme of bryozoan evolution and can be related to the functional attributes of particular colony shapes (see McKinney and Jackson 1989). Thus, for example, netlike colonies of the Ordovician fenestrate *Chasmatopora*, the Cenozoic cyclostome *Retihornera*, and the Recent cheilostome *Sertella* may at first glance appear almost identical; closer inspection of the morphology of the zooids and the structure of the skeleton is needed to make a distinction. This convergent colony form permits the formation of a unidirectional colonial feeding current.

Mineralized skeletons characterize most Recent and fossil bryozoans. These epithelial secretions are always calcareous and in all known Paleozoic taxa are calcitic, usually low-Mg calcite but sometimes evidently neomorphosed from an original high-Mg calcite composition (Taylor and Wilson 1999). The skeletons of individual zooids (zooecia) vary from tubular to box-shaped and contain a terminal aperture or orifice from which the lophophore is protruded to feed. Tiny pores, providing soft tissue linkage between zooids, penetrate the skeletal walls of some bryozoan species. This linkage can be supplemented or replaced by soft tissues investing the outer surface of the colony. Pores are lacking in the great majority of Ordovician bryozoans, but skeletal evidence shows clearly that epithelia (and associated body cavity) were continuous from one zooid to the next over the outer ends of the skeletal walls in most Ordovician species. Among Recent bryozoans, two groups lack mineralized skeletons, the entirely freshwater class Phylactolaemata and the marine-freshwater order Ctenostomata.

■ Early Fossil Record

Published records of Cambrian bryozoans are either founded on misidentification or require substantial further reinvestigation before they can be accepted (Larwood and Taylor 1979). Cobbold (1931) described as a "polyzoon" (i.e., bryozoan) a decalcified fossil from the Lower Cambrian of Comley, Shropshire, England. Subsequently, Cobbold and Pocock (1934) described four specimens from the Lower Cambrian of Rushton, Shropshire, as "Polyzoa Gen. et Spp. indet." Cobbold's material is a heterogeneous assemblage of small pitted plates none of which warrant inclusion in the Bryozoa. Fritz (1947) erected the genus *Archaeotrypa* for two new species of putative bryozoans from the Dresbachian (late Mid to early Late Cambrian) of Alberta. The limited material available was redescribed by Kobluk (1984), who was noncommittal about the affinities of one of the species (*A. secunda*) but tentatively considered the type species (*A. prima*) to be either a cystoporate bryozoan or an echinoderm. The single crystal calcite structure of the skeletal units present in *A. prima* favors their interpretation as echinoderm ossicles rather than infilled bryozoan zooids.

The earliest unequivocal bryozoans are recorded by Hu and Spjeldnaes (1991) from the Fenshiang Formation of Hubei, China, which is of late Tremadocian age. Two genera ("*Hubeipora*" and "*Yichangopora*") and seven species are mentioned but all await full and formal description and have yet to be illustrated. These early species apparently share two features, a granular wall microstructure and large acanthostyles indenting the zooecial chambers. They are provisionally placed in the esthonioporine trepostome family Orbiporidae.

Although no additional bryozoans have yet been recorded from the Tremadocian, the succeeding Arenig furnishes numerous records (see Taylor and Curry 1985; Taylor and Cope 1987; Taylor and Rozhnov 1996; Pushkin and Popov 1999). These include examples of all five orders of stenolaemates, plus a ctenostome gymnolaemate (Taylor and Curry 1985; Anstey and Pachut 1995; Taylor and Rozhnov 1996; Todd 2000). Probably the earliest Arenig occurrence of a bryozoan is represented by a record of the trepostome *Orbipora* from the Bolahaul Member, Ogof Hên Formation of Camarthen, southern Wales (Taylor and Cope 1987). This horizon falls within the graptolite-based *T. phyllograptoides* Zone of the earliest Arenig (Fortey et al. 2000) and therefore within *TS*.2a (see figure 16.1 and times slices in chapter 2). The oldest diverse Arenig bryofauna was described by Pushkin and Popov (1999) from the Billingen stage (*TS*.2b–c) of the St. Petersburg area and comprises six species of trepostomes (esthonioporines and halloporines). Arenig bryozoans are also known in Utah: a rare example of the esthonioporine trepostome *Dianulites* has been described from the Fillmore Formation (Taylor and Wilson 1999), which may be approximately equivalent in age to the Billingen fauna. The earliest known fenestrate bryozoan—*Alwynopora*—comes from the Tourmakeady Formation of western Ireland (Taylor and Curry 1985), which belongs to the graptolite-based *D. simulans* Zone or *I. victoriae* Subzone (Fortey et al. 2000), *TS*.2c and 3a, respectively. Volkhov stage (*TS*.3a–4a) deposits of eastern Russia, Estonia, and Poland (as erratic blocks) contain more than 20 species of trepostomes, as well as the earliest examples of cyclostomes, cystoporates, and rhabdomesine and ptilodictyine cryptostomes (Bassler

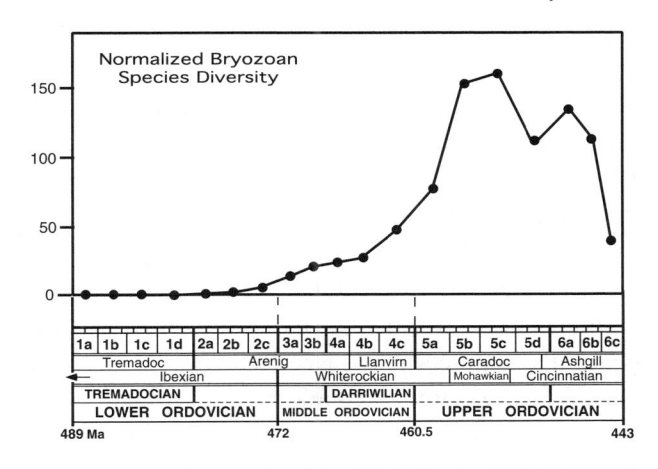

FIGURE 16.1. Bryozoan species diversity changes through 19 Ordovician time slices (*TS*) calculated as "normalized diversity."

1911; Modzalevskaya 1953; Männil 1959; Dzik 1981; Taylor and Rozhnov 1996).

With regard to the two bryozoan groups lacking mineralized skeletons, the statoblasts (dispersal and over-wintering bodies) of freshwater phylactolaemates have been recorded as far back as the Permian (Vinogradov 1996), but there are as yet no Ordovician or other Lower Paleozoic records for this group. Soft-bodied ctenostomes are represented in the fossil record by borings into calcareous substrates (see Pohowsky 1978) and by bioimmurations—that is, when natural molds are formed by organic overgrowth of other encrusters (see Taylor 1990a). Unequivocal ctenostome borings have been formally described from the "early Llanvirn" (*TS*.4b) of Spain (Mayoral et al. 1994), but an older occurrence from the Volkhov of the St. Petersburg area in Russia, noted by Taylor and Rozhnov (1996) and by Todd (2000), awaits description. The oldest recorded bioimmured ctenostomes date only from the Triassic (Todd and Hagdorn 1993), and the fossil record of bioimmurations may be strongly dependent on the occurrence of suitable overgrowing organisms such as oysters and serpulids. A body fossil from the Kunda (*TS*.4a–b) of the St. Petersburg area redescribed as a putative alcyonidiid ctenostome by Viskova and Ivantsov (1999) requires further study.

■ Phylogeny

Bryozoan phylogeny is not well understood, and molecular phylogenetics of the phylum has scarcely begun, with sequence data available for only a small number of species (Collins and Valentine 2001). Nevertheless, a few published cladistic analyses based on morphological data (Anstey 1990; Cuffey and Blake 1991; Anstey and Pachut 1995; P. D. Taylor 2000; Todd 2000) permit the tentative cladogram shown in figure 16.2 to be constructed by combining the topologies of these published cladograms. Two major clades of bryozoans are recognizable from anatomical evidence: Phylactolaemata and "marine bryozoans" (Gymnolaemata sensu Allman, *non* Borg). Whereas phylactolaemates are probably monophyletic, ctenostomes are interpreted as paraphyletic because all of the other groups of bryozoans with calcareous skeletons likely nest within them (Todd 2000). Ctenostome-grade bryozoans are thought to have been ancestral to the two marine bryozoan clades (class Stenolaemata and order Cheilostomata) possessing calcareous skeletons. The fossil record furnishes good evidence for the Jurassic origin of cheilostomes through the occurrence of bioimmured colonies of inferred stem group cheilostomes (Taylor 1990b). Less is known, however, about the origin of the stenolaemates, which must have appeared in the Ordovician or possibly earlier; a basal Ordovician origin is inferred later in this chapter.

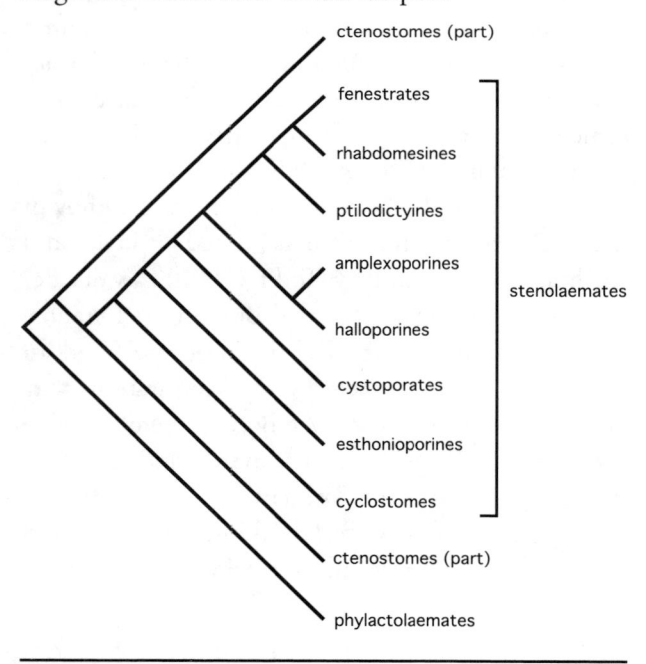

FIGURE 16.2. Cladogram summarizing inferred relationships between main bryozoan groups present in the Ordovician (compiled from heterogeneous sources).

As noted earlier, all five stenolaemate orders (Cryptostomata, Cyclostomata, Cystoporata, Fenestrata, and Trepostomata) commonly recognized by Western bryozoologists are present by the Arenig. The exact sequence of appearance of these orders in the fossil record does not provide a satisfactory means of understanding bryozoan phylogeny. Instead, cladistic analysis of morphology must be used. Cyclostomes are interpreted as the most primitive stenolaemates and are paraphyletic, the other four stenolaemate orders likely nesting within the Cyclostomata (P. D. Taylor 2000; for simplicity, figure 16.2 depicts cyclostomes as though monophyletic). Among cyclostomes, corynotrypids, in which only the exterior body walls of the uniserially arranged zooids are calcified, closely resemble certain ctenostomes in zooid shape and colony budding pattern (Taylor and Wilson 1994). The transition from a soft-bodied ctenostome to a corynotrypid cyclostome with calcified body walls necessitated a major change in the mechanism of tentacle protrusion (Taylor and Larwood 1979; Taylor 1981). In ctenostomes parietal muscles acting directly on the outer body wall compress the coelom, causing the tentacle sheath to evert and the tentacles to be protruded. In corynotrypids (and other stenolaemates) the outer body wall is rigidly calcified and not deformable by parietal muscles. Rather, the muscles involved in tentacle protrusion act on internal membranes, notably the membranous sac, to compress the coelom and bring about tentacle protrusion (Nielsen and Pedersen 1979). Acquiring this new mechanism of tentacle protrusion was a key evolutionary innovation for bryozoans. Not only did it permit calcification of the body walls, with all the protective and structural advantages that a hard skeleton normally brings, but the rigidity of the outer body walls also allowed a greater degree of zooid contiguity and the evolution of massive, multiserial colony forms (Taylor 1981). The spectacular evolutionary radiation of bryozoans in the Ordovician therefore reflects the combined and correlated taphonomic (preservability of calcareous skeletons) and ecologic changes (diversification of adaptive colony forms) that the phylum underwent at this crucial time in its evolution.

More advanced Paleozoic cyclostomes have calcified interior walls as well as calcified exterior frontal walls, and some possess pores in these walls (e.g., Buttler 1989). The remaining four stenolaemate orders form an unnamed "free-walled" clade distinguished mainly by the lack of calcified exterior frontal walls, a loss explicable in terms of zooidal paedomorphosis deleting a late stage of skeletal growth from the ontogeny of descendant species (Larwood and Taylor 1979; see also P. D. Taylor 2000). This dominant Paleozoic clade is also characterized by having skeletal rods (styles, etc.) within the zooidal walls (apparently secondarily lost in some species).

Relationships between free-walled Paleozoic stenolaemates are in need of study, and the scheme depicted in figure 16.2 is based on the cladogram of Paleozoic suborders published by Anstey and Pachut (1995: figure 8.1), omitting the Timanodictyina, which are not known until the Devonian. It should be noted that additional analyses of Anstey and Pachut (1995: figures 8.2 and 8.3) at the family level yielded trees with some topological differences, particularly affecting the status and position of the cystoporates. Trepostomes conventionally understood to include three suborders—Esthonioporina, Halloporina, and Amplexoporina—seem likely to be nonmonophyletic, with the esthonioporines occupying a basal position among the free-walled Paleozoic stenolaemates and the halloporines + amplexoporines positioned more crownward beyond the cystoporates (figure 16.2). The Cryptostomata are interpreted as an advanced, monophyletic clade comprising Ptilodictyina ("bifoliate cryptostomes"), Rhabdomesina ("stick-like cryptostomes"), and Fenestrata (i.e., Phylloporina + Fenestellina; see McKinney 2000), with the Ptilodictyina forming the sister group of the two latter taxa.

■ Taxic Diversity Patterns

Data and Methods

We have assembled a database of bryozoan occurrences in the Ordovician by trawling the primary literature; this database is available on request as a series of Excel files, one for each of the major taxonomic groups shown in figure 16.2. Wherever possible, we have assigned species occurrences to the 19 time slices (*TS*) recognized by Webby et al. (chapter 2) with durations of 1.5–3.0 million years (m.y.) (mean

TABLE 16.1. Summary of the Ordovician Bryozoan Taxic Database

	Genera	Genera Surviving into the Silurian	Species	Species Allocated to a Time Slice
Cyclostomata	16	9	35	35
Esthonioporina	6	0[a]	28	25
Cystoporata	23	6	65	57
Halloporina	33	15	550	443
Amplexoporina	15	10	97	75
Ptilodictyina	43	12	234	167
Rhabdomesina	17	10	57	33
Fenestrata	16	5	54	45
Totals	169	67	1,120	880

[a]Although no Ordovician esthonioporine genera survive into the Silurian, one Devonian genus (*Chondraulus* Duncan 1939) was assigned to this suborder by Astrova (1978).

2.4 m.y.). Table 16.1 summarizes the taxonomic composition of our database and the extent to which it has been possible to place species into time slices. In total, 78 percent of the 1,110 species in the database have been precisely dated. The database includes only Ordovician bryozoans with calcified skeletons (i.e., stenolaemates); the relatively few soft-bodied ctenostomes are not recorded.

For each time slice, normalized diversity (d_{norm}), following Cooper (chapter 4), has been calculated for both species and genera. The measure minimizes the distorting effects of variations in the durations of the time slices. The calculation involves assigning a full unit value to a taxon that ranges through the time slice from the preceding to the following time slice, a half value to a taxon that has one end of its range extending beyond the time-slice boundary and the other confined within the time slice, and another half value to a taxon confirmed completely to the time slice. For genera we have calculated d_{norm} twice, first for data that extrapolate ranges between time slices (i.e., including Lazarus taxa that are recorded both before and after but not within the time slice in question), and second for data without range extrapolations (i.e., excluding Lazarus taxa).

Per capita origination and extinction rates have been calculated using the d_{norm} values. As d_{norm} values effectively normalize for time, these rates of turnover are directly comparable between time slices regardless of differences in their durations.

Occurrences of taxa have also been placed within the four Ordovician bryozoan provinces—North

American, Baltic, Siberian, and Mediterranean—recognized by Tuckey (1990) on the basis of a multivariate statistical analysis (detrended correspondence analysis) of bryozoan paleogeographic distributions. This has enabled us to assess changing levels of provinciality through the Ordovician, as well as variations in the higher taxonomic composition between provinces.

Species and Genus Patterns

Figure 16.1 depicts the species diversification pattern for stenolaemate bryozoans during the Ordovician. Species diversity was very low throughout the Early Ordovician before increasing at an accelerating rate through the Middle and early Late Ordovician. Peak species diversity was achieved in *TS*.5c of the Caradoc, after which there was a large decline into *TS*.5d followed by a slight rise and then a large decline into *TS*.6c, the final time slice of the Ashgill. Logarithmic transformation of the species diversity values (figure 16.3) demonstrates the close match of the main phase of diversification (*TS*.2c–5c) to an exponential pattern—the data points fit almost exactly ($r = 0.951$), the straight line signifying exponential diversification.

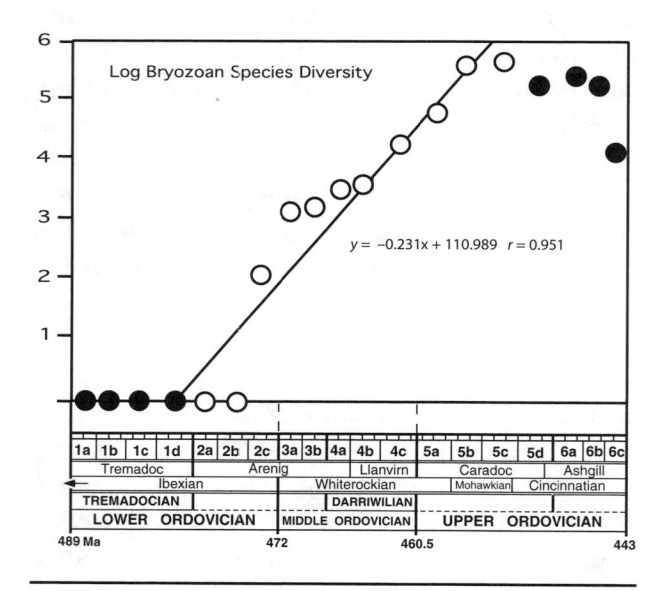

FIGURE 16.3. Logarithmic plot ($\ln_e$) showing bryozoan species diversification through 19 Ordovician time slices. The linear regression line is fitted only to the points shown as unfilled circles; r is the regression coefficient.

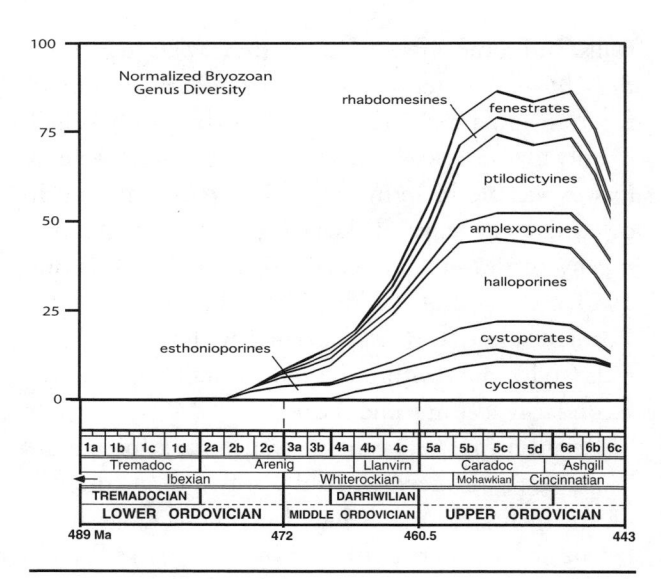

FIGURE 16.4. Bryozoan generic diversification pattern among major taxonomic groups through 19 Ordovician time slices. Calculated as normalized diversity and inclusive of Lazarus genera (i.e., ranges extrapolated).

The generic pattern of diversity change (figure 16.4) is essentially the same as the species pattern, although the declines in diversity at the end of the Ordovician and, more particularly, between *TS*.5c and 5d of the Caradoc are less pronounced. Subdivision of the generic data into the clades shown in figure 16.2 shows some variations in the diversification patterns of the eight major groups of stenolaemates. Esthonioporines were proportionally more numerous during the early radiation of the phylum, although absolute numbers are low. Halloporine trepostomes were the dominant clade overall, particularly from *TS*.4a to 5b (Darriwilian–mid Caradoc). Ptilodictyine cryptostomes and cystoporates attained maximum relative diversity in *TS*.5c and 5d respectively; cyclostomes, amplexoporine trepostomes, and fenestrates all peaked in relative diversity during the early to mid Ashgill (*TS*.6a–b). The genus *Chondraulus* from the Devonian is the only esthonioporine bryozoan known in rocks of post-Ordovician age (Astrova 1978).

Turnover

Figure 16.5 summarizes per capita generic turnover changes. Origination rate shows an overall decline from the highest values in *TS*.2a–c (early to

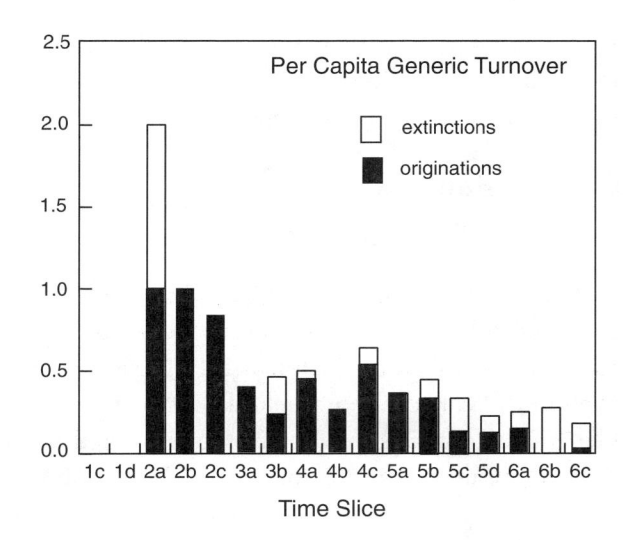

FIGURE 16.5. Changes in per capita generic turnover (origination and extinction) rates through Ordovician time slices (see figures 16.1, 16.3, 16.4).

mate genera survived into the Silurian (table 16.1), with amplexoporine trepostomes, rhabdomesine cryptostomes, and cyclostomes exhibiting the highest relative levels of per capita survivorship.

Paleogeographic Patterns

Partitioning the eight major stenolaemate clades into the four Ordovician bryozoan provinces recognized by Tuckey (1990) reveals some striking differences in clade representation according to province (figure 16.6). Halloporine trepostomes are the most species-rich group in all provinces apart from Siberia, where ptilodictyine cryptostomes dominate. By contrast, ptilodictyines are conspicuously depauperate in the Mediterranean Province. Esthonioporines are uncommon in all provinces apart from the Baltic, and cyclostomes have not been recorded in the Siberian Province. Investigation is required of the extent to which these patterns, especially the dominance of ptilodictyines in Siberia, are due to monographic biases. Additionally, the patterns undoubtedly reflect the ages of bryozoan faunas present in each province; for example, the high proportion of esthonioporines in the Baltic Province correlates with the existence of relatively old bryozoan faunas containing abundant examples of this primitive group of stenolaemates.

mid Arenig), punctuated by a subsidiary peak in *TS*.4c (late Darriwilian). Extinction rate fluctuates through time, with the highest value in *TS*.2a, probably an artifact of small sample size at this early stage of diversification, and other highs in *TS*.3b, 5c, 6b, and 6c. The extinction highs in *TS*.6b and 6c contribute to the major drop in bryozoan diversity at the end of the Ordovician. In total, some 67 stenolae-

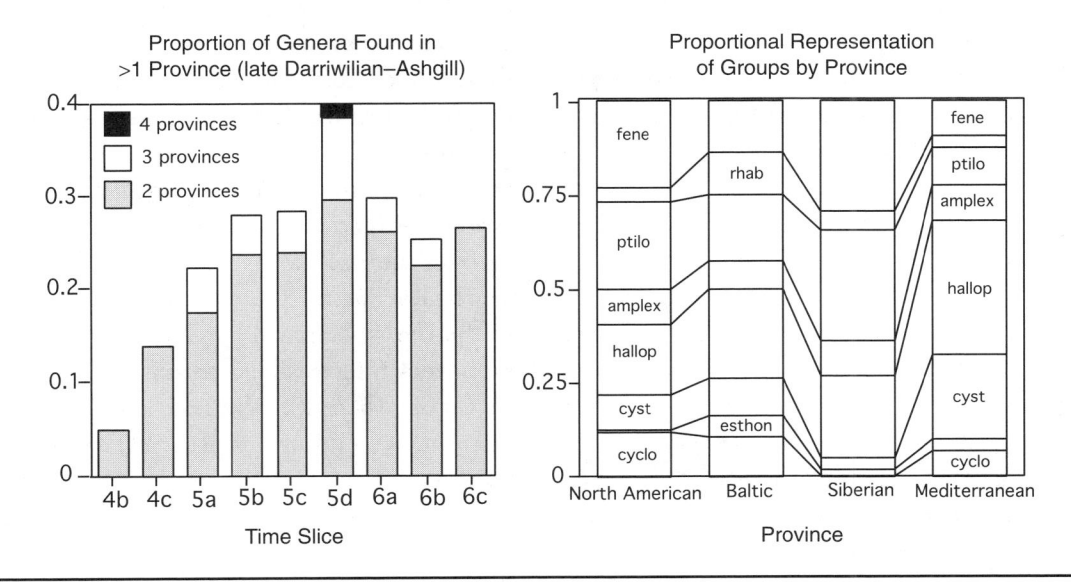

FIGURE 16.6. Geographic aspects of Ordovician bryozoan radiation. Frequency histogram on the left shows varying levels of provinciality in terms of the proportion of genera recorded in two or more of bryozoan provinces through Ordovician *TS*.4b to 6c (see figures 16.1, 16.3, 16.4). Differences in the relative proportions of genera belonging to eight major stenolaemate clades in these four provinces are depicted in the diagram on the right.

The number of endemic genera in each province is 19 for North America, 41 for the Baltic, 7 for Siberia, and 1 for the Mediterranean. Changing patterns of provinciality are evident when the numbers of genera present in two, three, or all four provinces are quantified through time (figure 16.6). The largest spread of genera between provinces occurred in *TS*.5d (latest Caradoc–earliest Ashgill), the only interval for which a genus (*Hallopora*) has been recorded from all four provinces. Pending a more sophisticated analysis of these distributional data, it appears that bryozoan provinciality was greatest in Arenig times and declined progressively into the earliest Ashgill before increasing again toward the end of Ordovician time (cf. Tuckey 1990).

Completeness of the Fossil Record

One of several different methods (Paul 1998) of evaluating the completeness of the fossil record is gap analysis in which the proportion of Lazarus taxa per time interval is calculated. Figure 16.7 depicts the results for stenolaemate genera through the time slices of the Ordovician. An increasingly high proportion of Lazarus taxa characterizes the later part of the Ordovician, indicating a poor fossil record, culminating in *TS*.6c, where more than half of the genera known to be present by range extrapolation are missing from the record. This finding implies that the decline in bryozoan diversity during the late Caradoc to Ashgill (figures 16.1 and 16.3) is partly due to deficiencies in the fossil record. A mediocre record is also apparent for *TS*.3a–4a of the early to mid Arenig, although the lower diversity here means that less significance can be attached to this finding.

■ Discussion

Patterns of Radiation

Three earlier papers have depicted the pattern of bryozoan generic radiation during the Ordovician (Taylor and Larwood 1990: figure 10.2; Anstey and Pachut 1995: figure 8.5; Sepkoski 1995: figure 2). Generic range charts are also given by Ross and Ross (1996), but diversities are not tallied. These analyses, employing overlapping global databases and varying in their stratigraphic resolution, reveal patterns almost identical to the pattern shown here in figure 16.4. All show a prolonged diversification for most of the Ordovician before a diversity decline at the end of the period, although the Sepkoski (1995) curve shows an unexpected second peak in diversity at the very end of the Ordovician.

Species-level assessment of Ordovician diversity changes has not previously been attempted at this stratigraphic precision (cf. Horowitz and Pachut 2000). Our results indicate that the increase in diversity from mid Arenig to mid Caradoc followed an almost perfectly exponential trajectory (figure 16.3), suggesting unconstrained diversification. If diversification was truly exponential, it can be predicted that stenolaemate bryozoans were present as early as the first or second Ordovician time slices (*TS*.1a or 1b) in the Tremadocian, which is where the line of regression of figure 16.3 intersects the *y*-axis. This offers hope for the future discovery of skeletally preserved bryozoans following a more thorough searching of rocks of this age.

The leveling-off and decline in diversity at the end of the Ordovician can be viewed in the contexts of (1) a worsening fossil record as revealed by gap analysis (figure 16.7); (2) the Late Ordovician mass extinction (see Tuckey and Anstey 1992); and (3) the beginnings of a diversity plateau that characterized the remainder of the Paleozoic (Taylor and Larwood 1990: figure 10.1), arguably indicating attainment of the equilibrium diversity level identified by Sepkoski (1979) for marine families as a whole.

Several authors have remarked on the emergence of most higher taxa (orders and suborders) of bry-

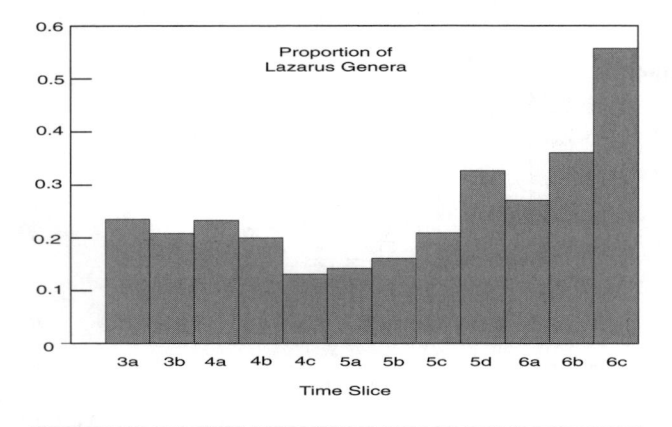

FIGURE 16.7. Proportion of Lazarus genera of bryozoans for *TS*.3a to 6c (see figures 16.1, 16.3, 16.4).

Erratum
This page replaces page 91.

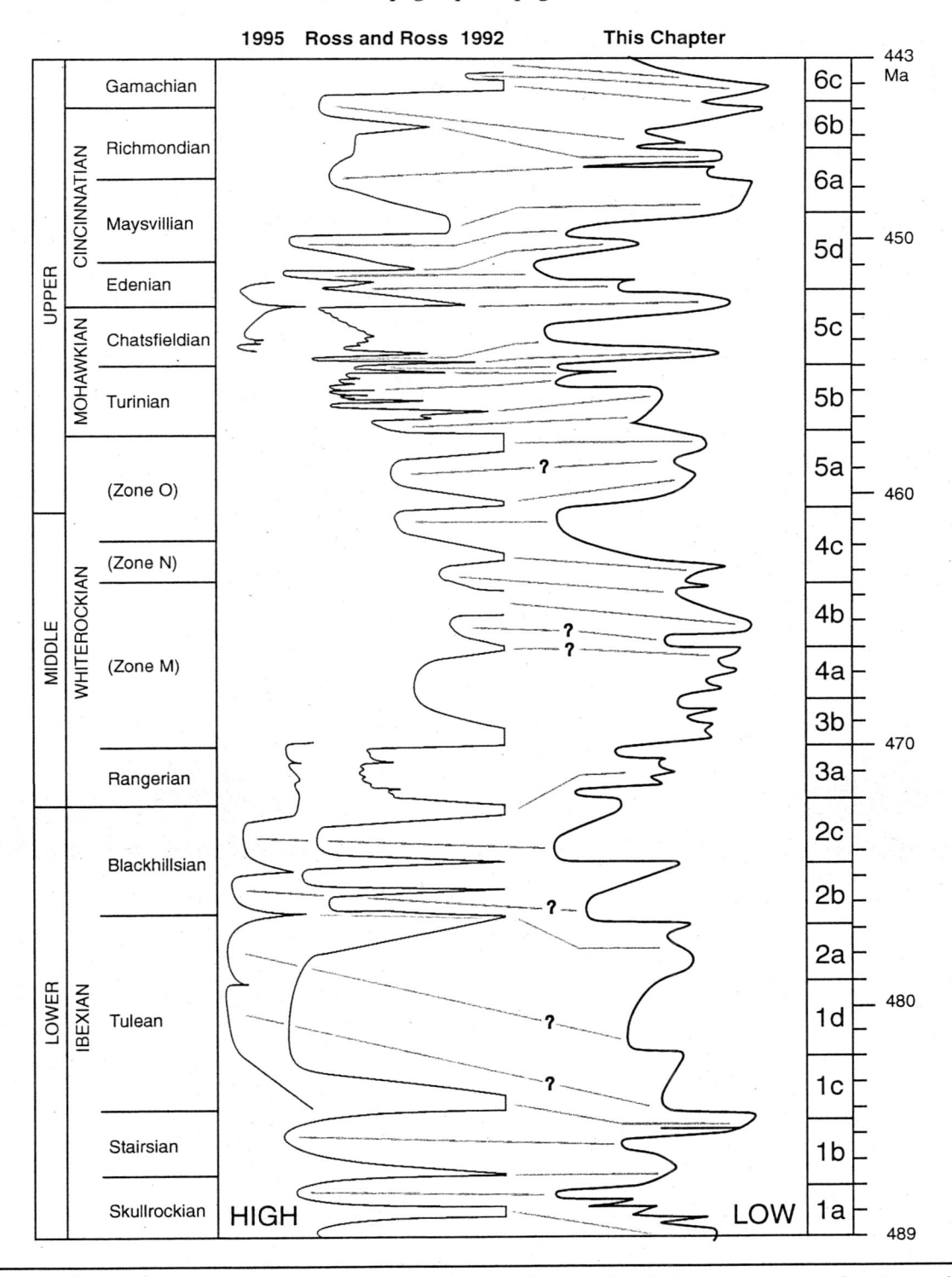

FIGURE 10.3. Comparison of North American and Baltoscandian sea level curves. The North American curve is from Ross and Ross (1992, 1995), adjusted approximately to the timescale used in figure 10.2. The North American stratigraphic framework is illustrated at the left side. Inferred correlations between the highstand and lowstand events are shown by the light-colored tie lines; less certain correlations also include question marks. Note that the time slices (chapter 2), as shown in the column to the right side, are tied only to the Baltoscandian curve.

ozoans during the Ordovician (e.g., Taylor and Curry 1985; Anstey and Pachut 1995; Markov et al. 1998). Hierarchical taxonomic structure alone is capable of explaining why diversification at high taxonomic levels occurred so early in the evolution of the phylum. The sudden appearance of higher taxa is paralleled by numerous evolutionary innovations, including taxon-defining apomorphies. The subsequent evolution of Paleozoic stenolaemate bryozoans is to a large extent based on recombination (Markov et al. 1998) and/or parallel origin of characters first seen in the Ordovician, hence the high levels of convergence exhibited by Paleozoic bryozoans (e.g., Anstey and Pachut 1995). Most of the major colony growth forms found among Paleozoic bryozoans appeared during the Ordovician radiation of the phylum (Larwood and Taylor 1979), and their main ecologic roles were also established at this time (Ross 1984).

There are clear similarities between the bryozoan radiation pattern and those published for other groups, for example, "articulate" brachiopods, echinoderms, corals, gastropods, bivalves, and conodonts (see Sepkoski 1995 and chapters in this volume). Although some of these groups (e.g., brachiopods, echinoderms, conodonts) commenced radiation before the Bryozoa, they all show a sustained pattern of diversity increase until the Caradoc or Ashgill, followed by a diversity drop close to or at the end of the Ordovician.

Processes Driving the Radiation

According to Sepkoski and Sheehan (1983), there is no obvious physical trigger, such as major sea level change, for the Ordovician Radiation of marine animals. The advent of "Calcite Seas" in the Early Ordovician facilitated the rapid formation of carbonate hardgrounds and would have increased ecologic habitats and evolutionary opportunities for bryozoans, echinoderms, and other benthic organisms utilizing hard substrates for attachment (Wilson et al. 1992; Rozhnov 2001). Whether this was sufficient to trigger explosive evolutionary radiation in bryozoans is contestable, however. Ross (1985) suggested that climatic changes may have been involved in the Ordovician radiation of bryozoans, with diversity increasing when the climate became warmer and declining as it became cooler, particularly during the Ashgill. Anstey (1986) hypothesized that the loss, related to global cooling and eustatic lowering of sea level, of a highly diverse region—the Cincinnati Province—was the chief factor in the extinction of bryozoans during the Late Ordovician, at least within North America. Eustatic sea level changes have also been implicated more widely in Ordovician (and Paleozoic) diversity patterns of bryozoans by Ross and Ross (1996). Transgressions led to diversity increases, as in the early Caradoc, and regressions to decreases, as at the end of the Ashgill. The post-Tremadocian match between the diversity pattern found here (figures 16.1 and 16.4) and the sea level curve published by Ross and Ross (1996: figure 1) is not exact. Although the Caradoc-Ashgill patterns are well correlated, the low sea levels that pertained during most of the Arenig and throughout the Darriwilian correspond with a marked phase of bryozoan diversification.

The control of bryozoan diversity by an extrinsic factor or factors during the Ordovician is attractive in view of the parallel patterns exhibited by other phyla, but it is difficult to ignore the role of intrinsic changes in driving the radiation of the Bryozoa. The acquisition of a mineralized skeleton was of profound importance to the Bryozoa in paving the way for morphological and ecologic expansion not possible in primitive, soft-bodied bryozoans. This key evolutionary innovation may have been sufficient to trigger the seemingly unconstrained exponential diversification of bryozoans that persisted until mid Caradoc times (figure 16.3). Analogies with cheilostome bryozoans, which radiated explosively in the Cretaceous, point to another possible key evolutionary innovation (Taylor and Larwood 1990). Cheilostome radiation commenced immediately after the first skeletal evidence for larval brooding, suggesting a macroevolutionary model whereby the advent of short-lived brooded larvae led to decreased gene flow between populations and a dramatic increase in the rate of allopatric speciation (Taylor 1988). Unfortunately, this model is not testable in stenolaemate bryozoans until skeletal criteria for distinguishing brooding species have been established.

Future Research

Many questions remain to be addressed regarding the Ordovician radiation of bryozoans. These demand further analysis of already published data as well as new empirical studies of fossil bryozoans.

A fuller appreciation of macroevolutionary patterns can be obtained by looking at changes in assemblage diversity and biomass, as has been done by McKinney et al. (2001) for post-Paleozoic bryozoans. Research is needed to see whether these changes correlate with or are decoupled (cf. Droser et al. 2000) from the global radiation pattern found here.

Further cladistic analyses are required of Ordovician bryozoans to define clades with more confidence and precision. It will then be possible, for example, to study the dynamics of individual clades during the Ordovician radiation and to study temporal, biogeographic, and environmental patterns of parallel character acquisition in different clades.

Quantification of disparity changes among bryozoans during the Ordovician has yet to be undertaken. Such research on morphospace occupancy would provide an additional perspective on the Ordovician radiation of bryozoans, as well as a potentially instructive comparison with published findings on the radiation of a noncolonial phylum (Arthropoda) during an earlier (Cambrian) major radiation (Fortey et al. 1996).

Concerning possibilities of new fossil discoveries, extrapolation of the exponential species diversification curve (figure 16.3) predicts that stenolaemate bryozoans should occur back to the base of the Ordovician. Carbonate platforms of Tremadocian age, not unlike those known to be inhabited by bryozoans later in the Ordovician, have so far failed to produce any definite records of bryozoans. It may be that the earliest stenolaemates lived in more nearshore clastic environments and migrated offshore with time, in accordance with a general pattern described for the Ordovician by Sepkoski and Sheehan (1983). Clastic facies are rarely studied for bryozoans because of the typically decalcified preservation of their shelly faunas. Importantly, the oldest known occurrence of bryozoans outside China is a decalcified esthonioporine from nearshore clastics of early Arenig age in Wales (Taylor and Cope 1987). The most primitive stenolaemates, corynotrypid cyclostomes, are less likely to be noticed because of the small size of their branched, encrusting colonies.

■ Conclusions

The Bryozoa are a diverse phylum of benthic, colonial, suspension-feeding metazoans. They have no unequivocal Cambrian fossil record—the oldest fossil bryozoans are recorded from the Tremadocian. A major radiation of bryozoans occurred during the Ordovician, with the rapid appearance of all but one of the orders of marine bryozoans generally recognized in the Phanerozoic, the attainment of a family diversity level that was maintained as a plateau for the remainder of the Paleozoic, and a fossil record comprising more than 1,000 described species. Acquisition of a calcareous skeleton was very important, not merely for providing fossilizable hard parts but in allowing morphological and ecologic expansion of the phylum. Analysis of taxic patterns of diversity change and turnover through the Ordovician based on 19 time slices (each between 1.5 and 3 m.y. in duration) shows an exponential increase in species diversity from mid Arenig to mid Caradoc, followed by a species (and genus) diversity decline that is particularly pronounced at the very end of the Ordovician, where deficiencies in the fossil record are implied by the high proportion of Lazarus genera. Some variations in the pattern of radiation are evident according to major taxonomic group and biogeographic province.

ACKNOWLEDGMENTS

We thank Barry Webby for inviting us to participate in IGCP 410 and the Deutsche Forschungsgemeinschaft (DFG) for funding one of us (AE) through a fellowship.

17 Brachiopods

*David A. T. Harper, L. Robin M. Cocks, Leonid E. Popov,
Peter M. Sheehan, Michael G. Bassett, Paul Copper,
Lars E. Holmer, Jisuo Jin, and Rong Jia-yu*

The Ordovician brachiopod radiation was the most marked interval of diversification in the entire history of the phylum (Harper and Rong 2001). Apart from the initial acquisition of hard shells in the earliest Cambrian, the Ordovician was the most important period for brachiopod evolution and diversification in the whole Phanerozoic with the continued exponential increase in numbers of brachiopod genera seeded during the Cambrian (Patzkowsky 1995a). A series of stepwise radiations across most of the major orders helped set the agenda for much of life on the Paleozoic seafloor. In particular, the articulated brachiopods, now termed the subphylum Rhynchonelliformea, diversified from the 4 superfamilies Billingselloidea, Orthoidea, Plectorthoidea, and Porambonitoidea present at the end of the Cambrian into the 19 superfamilies (Eichwaldioidea, Strophomenoidea, Plectambonitoidea, Chonetoidea, Chilidiopsoidea, Triplesioidea, Skenidioidea, Orthoidea, Plectorthoidea, Dalmanelloidea, Enteletoidea, Porambonitoidea, Pentameroidea, Stricklandioidea, Ancistrorhynchoidea, Camarotoechioidea, Atrypoidea, Athyridoidea, and Cyrtinoidea) all present by the end of the period; only the Billingselloidea, Polytoechioidea (Tremadocian to Caradoc), and Clitambonitoidea (Arenig-Ashgill) became extinct within the Ordovician.

By the end of the Ordovician, with the exception of the spinose productoids, superrecumbent strophalosioids, the coral-like richthofenioids, and the bizarre lyttonioids, the majority of morphological innovations had already appeared (Harper and Wright 1996). Key morphological adaptations promoted both taxonomic variety and a spectrum of lifestyles despite the relatively constrained skeletal architecture of the brachiopod animal. The radiation progressed with the expansion of brachiopod clades into outer-shelf and slope settings together with carbonate platform and reef environments. Such colonizations involved particular specializations (Copper 2001a). Nevertheless, across the phylum macroevolutionary changes were apparently decoupled from fluctuations in taxon abundance and taxonomic diversity (Droser et al. 1997); moreover, ecologic changes were not necessarily linked to crashes or hikes in taxonomic diversity (Droser et al. 2000).

The origins for the radiation are unclear, in part owing to the time lag between the appearance of macroevolutionary innovations and subsequent apparent diversifications. The role of marginal and intra-oceanic sites, many located within Early Paleozoic mountain belts or destroyed by subduction-related processes, may have been important during the early stages of the radiation (Harper and Mac Niocaill 2002). During the radiation, a shift in ecologic bias occurred during the early Darriwilian, particularly on cratons such as Laurentia, from communities with abundant trilobites to those dominated by brachiopods

(Droser and Sheehan 1997b). Bambachian megaguilds were filled, and new paleocommunities were generated as the radiation progressed (Bottjer et al. 2001).

■ Taxonomic Diversification

The taxonomic component of the diversification involved an expansion of taxa at all levels, but the escalation is most marked at the generic level. For example, nearly 30 articulated genera are known from the lower Tremadocian, whereas a total of 185 genera have been recorded from lower to middle Ashgill horizons. Nevertheless, the radiation was not uniform, with peaks and troughs varying considerably across the different superfamilies. The maximum diversity of individual assemblages increased significantly from fewer than 10 taxa in the Upper Cambrian to more than 30 genera in many Late Ordovician paleocommunities. The diversification of the nonarticulates is less obvious; nonetheless, more than 150 Ordovician genera are known.

Phylogeny

A relatively stable, cladistically based classification (Williams et al. 1996) has been developed within the framework of the revised *Treatise* volumes for the phylum (Kaesler 2000). The broad-frame phylogeny recognizes three subclasses: the linguliformeans, the craniiformeans, and the rhynchonelliformeans. Generally the first two divisions include the "inarticulates" of previous classifications, whereas the last division embodies the concept of the "articulate" brachiopod. The Ordovician Radiation involved "nonarticulated" Lingulida, Acrotretida, and Siphonotretida together with the Craniopsida, Craniida, and Trimerellida; the paterinates were present during the event but did not contribute significantly to the Ordovician biodiversifications. The "articulated" groups included the Strophomenida, Triplesiidina, Billingsellidina, Clitambonitidina, Protorthida, Orthida, Pentamerida, Rhynchonellida, Atrypida, and Athyridida; the Productida (Chonetoidea), Orthotetida, and Spiriferida were present at low diversities prior to radiations during the Silurian, whereas the enigmatic Dictyonellida were a very minor part of the later Ordovician brachiopod fauna.

Approach and Methodology

The database has been developed from a series of range charts. For virtually all the groups discussed, up-to-date information is available through research associated with the revised volumes of part H (Brachiopoda) of the *Treatise on Invertebrate Paleontology*. In this respect both brachiopod phylogeny and taxonomy are state of the art. A stable stratigraphic framework is provided by the continuing researches of the IUGS Subcommission on Ordovician Stratigraphy. Nevertheless, there are, of course, error limits in both range and geographic data.

The known stratigraphic distributions of all genera have been plotted within a framework of 19 chronostratigraphic and geochronologic intervals developed for the Ordovician System (e.g., figure 17.1). Standing diversity has been plotted for most of the major groups together with *first* and *last* appearance information. Moreover, for some groups diversity counts for each stratigraphic interval have been corrected. The main trends, however, are clearer when the raw data are displayed graphically. The "endpoint correction" inserts the value of 0.5 for a "first appearance datum" (FAD) or a "last appearance datum" (LAD) in a given sample, whereas a value of 0.33 is inserted for the occurrence of both a FAD and a LAD. Corrections were calculated with PAST software (Hammer et al. 2001) (http://folk.uio.no/ohammer/past/). The genus origination/extinction rates, plotted as per lineage million years (Lma), were calculated as the number of generic originations/extinctions within a given stratigraphic interval, divided by the total generic diversity within the interval, divided by the chronologic duration of the corresponding time interval (see Patzkowsky and Holland 1997 for methodology and discussion).

Linguliformean and Craniiformean Brachiopods (LEP, LEH, MGB)

The linguliformean and craniiformean brachiopods together represent a relatively numerically insignificant component of the Ordovician fauna as a whole. Nevertheless, these two groups formerly regarded as "inarticulates" are each rather different in their environmental requirements and dispersal potential, defined mainly by differences in their larval

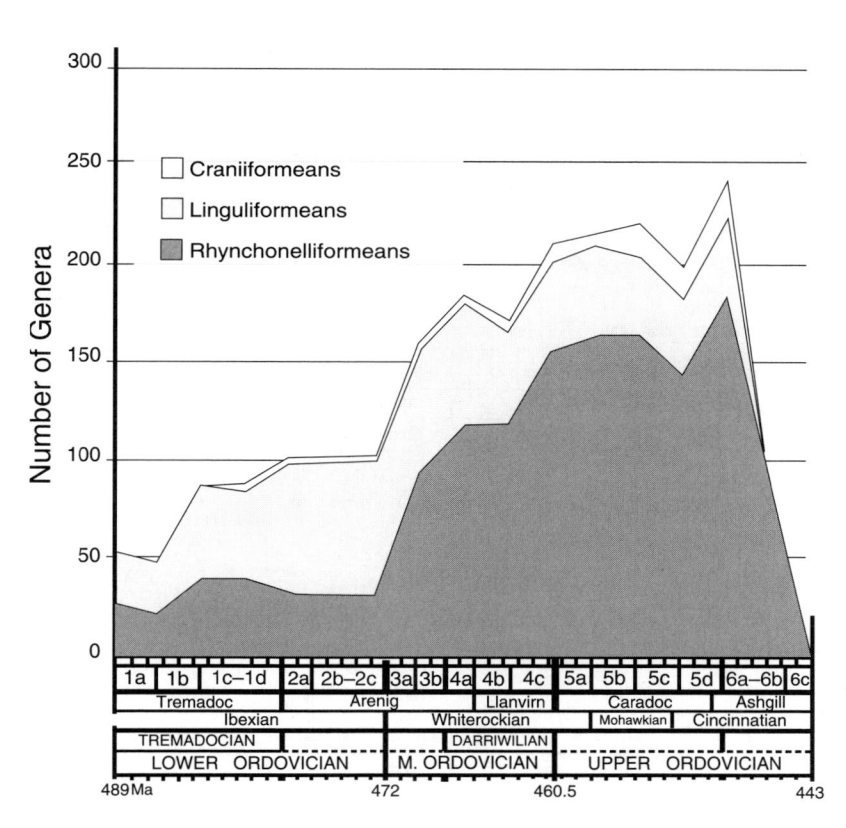

FIGURE 17.1. Ordovician standing diversity patterns of the linguliformeans, craniiformeans, and rhynchonelliformeans in terms of numbers of genera.

development. Linguliformean genera with their planktotrophic larval phases usually show a more widespread and even biogeographic distribution during the Ordovician mostly controlled by climatic factors and similar to the distributional patterns of pelagic animals such as conodonts, whereas craniiformeans with their lecithotrophic larvae that fed on yolk were probably unable to cross even narrow oceanic tracts, and their patterns of biogeographic distribution and diversification in Ordovician depended strongly on the relative position of the ancient continents. The linguliformeans represent, however, together with trilobites, the two most distinctive benthic groups of the Cambrian Evolutionary Fauna (EF) (Sepkoski 1981a).

The Cambrian record of the craniiformeans is very sparse (Popov et al. 1999b). Only two craniopside genera were reported from the Early and Mid Cambrian, followed by a gap in known stratigraphic distributions until their reappearance in the Late Cambrian and Tremadocian. All three orders of craniiformeans (Craniida, Craniopsida, and Trimerellida) are characterized by the lack of a pedicle at all growth stages and instead pursued a free-lying, cemented, or encrusting life habit. Larvae of the Recent genus *Novocrania* have a free-swimming stage of very short duration, and this condition had probably evolved very early in the history of group, as is evident from the biogeographic patterns and life strategies of the group already established in the Ordovician. Major biodiversification and biogeographic expansion of the group have occurred since the Mid Ordovician (Popov et al. 1999b); thus they, in contrast to the linguliformeans, have a diversity history similar to that of the Paleozoic Evolutionary Fauna (EF).

Linguliformea

The Lingulida and Acrotretida had already achieved considerable evolutionary success during the initial Cambrian radiation of the metazoans and were common in all types of marine environments by the end of the Cambrian. Their Ordovician record includes 150 genera (Lingulida—85 genera; Acrotretida—47 genera; Siphonotretida—15 genera; and Paterinida—3 genera). Geologic time-frequency distributions showing plots of generic diversity within linguliformean orders, together with the numbers of first and last appearances of Ordovician genera, are depicted in figure 17.2. Origination and extinction rates

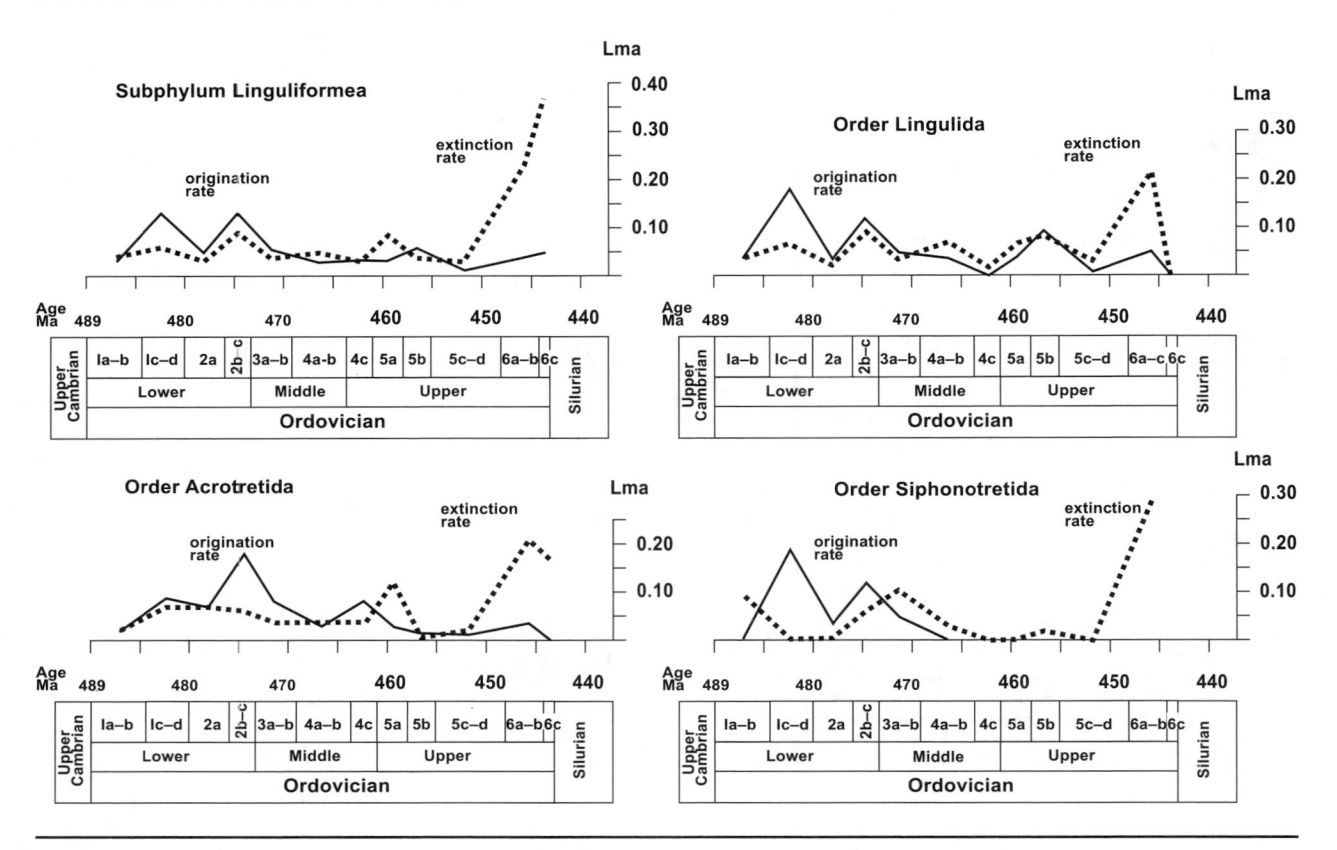

FIGURE 17.2. Extinction and origination rates (clockwise from top left) of the linguliformeans, lingulides, siphonotretides, and acrotretides (after Bassett et al. 1999b); see text for calculation of the rates (Lma).

for genera of the Lingulida, Siphonotretida, and Acrotretida are also plotted in figure 17.2. There is a considerable decline in generic diversity in the second half of the Late Cambrian from the *Proconodontus* Zone (Bassett et al. 1999b; Holmer et al. 2001). Linguliformean faunas on the outer shelf were most affected, and more than 90 percent of the Late Cambrian genera became extinct by the beginning of the Ordovician. In the early Tremadocian (*TS*.1a–b; see time slices in chapter 2), acrotretide diversity of the microbrachiopod assemblages was reduced to one to three genera, and the *Broggeria* fauna, adapted to dysaerobic environments, spread across high and temperate latitudes following an expansion of the areas of black shale deposition (Popov and Holmer 1994, 1995).

Recovery of linguliformean faunas on all the major Early Paleozoic platforms occurred during the late Tremadocian–early Arenig (*TS*.1b–2c, or equivalent to the *Paltodus deltifer* to *Oepikodus evae* conodont zones), and by the mid Arenig (*TS*.3a) their generic diversity had been restored to the maximum observed in the Cambrian. The recovery characteristics were different for the faunas that inhabited the nearshore and outer-shelf environments.

The distinctive feature of the newly emerging microbrachiopod assemblages, which spread widely across the outer shelf and basin, during the later Ordovician is abundance of the acrotretide families Ephippelasmatidae, Eoconulidae, Scaphelasmatidae, Torynelasmatidae, and Biernatidae together with the lingulides *Rowellella, Elliptoglossa,* and *Paterula* and siphonotretides (Popov and Holmer 1994, 1995). Environments inhabited by these linguliformean faunas also developed as refugia for the last paterinides, where they survived until the terminal Ordovician extinction. Presently available data from Kazakhstanian and the Central Asian terranes (Holmer et al. 2000; Popov and Holmer 1995) suggest that the Ephippelasmatidae, Eoconulidae, Scaphelasmatidae, and possibly Torynelasmatidae, which constitute the core of the Ordovician micromorphic brachiopod faunas, evolved first in an island-arc setting located in the temperate and low latitudes between Baltica and East Gondwana.

The first pulse of diversification of the micromorphic linguliformeans in the outer shelf is recorded in the late Tremadocian (*TS*.1c), when this fauna suddenly appeared on the opposite sides of Baltica (Baltoscandian basin and South Urals). They sustain a second pulse of generic biodiversification during the mid Arenig (*TS*.2c), when they are recorded for the first time in Laurentia (Krause and Rowell 1975) and South China (Zhang 1995); but in general their taxonomic structure at the family level and generic diversity remained relatively stable until the mid Ashgill (*TS*.6a–b) (Wright and McClean 1991; Bassett et al. 1999b; Popov 2000b). The diversity of the lingulide family Elkaniidae, which was most characteristic of outer-shelf environments during the Early and Mid Ordovician, had decreased by the Late Ordovician to the single genus *Tilasia,* which survived until the mid Ashgill. The terminal Ordovician extinction strongly affected the composition of the micromorphic brachiopod assemblages at the generic level, but it is not so clear at the family level. Among acrotretide families only Eoconulidae and Ephippelasmatidae became extinct, but a number of Silurian genera belonging to other families (Acrotretidae, Biernatidae, Scaphelasmatidae, and Torynelasmatidae) were reduced dramatically, and each family is represented by only a single genus in the Silurian. Although the siphonotretides survived the terminal Ordovician extinction, they became very rare in Silurian and Devonian rocks (Mergl 2001).

During the Mid and Late Cambrian, linguliformeans, especially the obolids and zhanatellids, became widespread across the shallow clastic shelves. In particular, they developed adaptations that permitted colonization of mobile sandy substrates, where they retained their dominant position through the Early Paleozoic. These nearshore lingulide faunas were partly affected by regional extinctions in Baltica and Laurentia near the Cambrian-Ordovician boundary. It is also likely that the low- to temperate-latitude lingulide faunas of West Gondwana (including North Africa, the Middle East, Avalonia, and the Perunica microcontinent [= Bohemia]) played an important role in the origin of the Ordovician shallow-shelf assemblages. A good example is the *Thysanotos-Leptembolon* fauna, which probably evolved in some peri-Gondwanan settings and soon spread across Baltica (Popov and Holmer 1994; Mergl 1997). High origination rates of the lingulides and siphonotretides during the late Tremadocian (*TS*.1d) were caused mainly by the proliferation of the shallow-shelf assemblages (figure 17.2). The Darriwilian (*TS*.4a–c) was the interval of significant reorganization of shallow-shelf assemblages with increased turnover rates (figure 17.2), when epibenthic lingulides were completely replaced by taxa with a burrowing habit. It occurs well before the first appearance of bivalves on the clastic shelves of the Kazakhstanian terranes and the shallow carbonate shelves of Baltica; thus, it is rather unlikely that increased bioturbation promoted by the colonization of bivalves is responsible for the extinction of the epibenthic lingulide and siphonotretide taxa. During that time some shallow-shelf lingulide assemblages (e.g., the *Hyperobolus-Talasotreta* association) indicated a considerable shift toward basinal environments (Holmer et al. 1996), but in general this change is not characteristic of the Ordovician linguliformean faunas.

Despite a persistence of the family stocks, somewhere between 50 and 60 percent of the total number of genera disappeared from linguliformean assemblages during the Llanvirn and early Caradoc (*TS*.4a–c, 5a) (figures 17.2, 17.3). The number of siphonotretides dropped to six genera at the beginning of the Llanvirn (*TS*.4a–b) (figure 17.2). In nearshore and shallow-shelf environments, faunas of epibenthic lingulides of the families Obolidae and Zhanatellidae were replaced completely by lingulide-mollusk assemblages, in which lingulides, adapted to life in burrows, became the predominant brachiopods (Bassett et al. 1999b).

Most of the evolutionary novelties that defined morphological and anatomic characters of the orders and superfamilies, as well as major adaptations, evolved in linguliformeans well before the beginning of the Ordovician. The only exception is the first appearance of the superfamily Discinoidea in the early Arenig, which represents the last significant evolutionary event in lingulide history. In contrast with the majority of other lingulides, they acquired a combination of features such as an elongate pedicle track posterior to the ventral umbo covered anteriorly by a listrium, holoperipheral growth of the ventral valve (family Trematidae) or both valves (family Discinidae), and complete reduction of pseudointerareas, apart from a narrow strip retained occasionally in some early genera.

In the acrotretides the most significant morphological novelty was the elaboration of the dorsal median septum, involving the development of spines and widely variable morphologies of the surmounting plate that evolved in several genera of the acrotretide families Acrotretidae (e.g., *Fascicoma*), Ephippelasmatidae (e.g., *Numericoma*), and Torynelasmatidae (e.g., *Cristicoma* and *Polylasma*) during the mid to late Arenig and early Llanvirn (*TS*.3c–4b) (Holmer and Popov 2000; Nazarov and Popov 1980).

Adaptation to a burrowing life mode possibly evolved sometime in the Late Cambrian from an escape strategy within the epibenthic lingulides that inhabited mobile sandy substrates (Popov et al. 1989). This shell morphology, as well as the burrowing habit, probably evolved convergently and slightly diachronously in a number of lingulide lineages, since taxa with both finely pitted and smooth larval and postlarval shells are known. However, most of the characteristic features of the infaunal lingulides—for example, convergent *vascula lateralia* in both valves and the complete reduction of the dorsal *vascula media*—evolved later, sometime in the Llanvirn.

During the Ordovician radiation of linguliformeans, most of the first-order morphological innovations occurred during the Early Ordovician. From the Arenig onward, linguliformeans were insignificant components of benthic assemblages, which became dominated by filtrators such as the rhynchonelliformean brachiopods, bryozoans, and echinoderms. However, some linguliformeans continued to form distinctive parts of the benthos in marginal environments such as eutrophic basins, mobile sands on shallow shelves, and abyssal basins. But in general, post-Arenig linguliformeans had much lower, near-background genus origination rates (0.04–0.07 Lma) compared with the other major groups of Sepkoski's (1981a) Paleozoic EF, including, for example, the rhynchonelliformean brachiopods (Patzkowsky and Holland 1997).

Craniiformea

The oldest known Ordovician craniiformeans, not yet formally described, are those from the lower Arenig of South Tien Shan in Central Asia, where they are represented already by two craniid genera including an encrusting *Petrocrania* associated with the rhynchonelliformean brachiopods *Clarkella* and *Tritoechia* and a linguliformean assemblage very similar to the early Arenig assemblages (*TS*.2a–b) of Baltica (Holmer et al. 2000). Another possible island-arc setting where craniides already occur during the Arenig is the Hinggan Ling Mountains of northeastern China, where *Celidocrania* occurs together with other late Arenig brachiopods of mixed Avalonian and Baltoscandian affinity (Liu et al. 1985; Neuman and Harper 1992).

Craniides had already colonized Baltica in the mid Arenig (*TS*.3a–b or Volkhov regional stage), where they are represented by *Pseudocrania* and *Petrocrania*. There is also a report on the occurrence of a craniopside from the lower Llanvirn of Öland, Sweden. The environments inhabited by craniides at that time were located within shallow pericontinental seas (BA 3 and 4), associated with storm-generated sedimentation and widespread hardgrounds.

During the "Llandeilo" (*TS*.4c–5a, *Husteograptus teretiusculus* and *Nemagraptus gracilis* graptolite zones), craniides diversified in Baltica to four genera. About the same time the earliest-proven craniopside *Pseudopholidops* became common in the Kukruse stage, and also the earliest trimerellides emerged in an island-arc setting between Baltica and East Gondwana, now incorporated in the Kazakhstanian orogen. The occurrence of three trimerellide genera in Chingiz-Tarbagatay is of particular significance. Here, for the first time, they formed a prominent nucleus of low-diversity/high-density shallow marine (BA 3) benthic assemblages inhabiting carbonate sands and muds in close proximity to carbonate mud mounds. At present there is no pre-Caradoc Ordovician record of craniiformeans from mainland Gondwana, Laurentia, Avalonia, and Siberia.

The early Caradoc (*TS*.5a) was the time of craniiformean diversification; taxa spread widely across the temperate and low latitudes (figure 17.3). There was a marked radiation of trimerellides in East Gondwana (Australia), and *Eodinobolus* reached Laurentia. In both regions they became dominant taxa in low-diversity, shallow-water (BA 2–3) assemblages (Norford and Steele 1969; Percival 1995). Such faunas remained notably absent in high-latitude West Gondwana.

craniides also reached Bohemia (*Petrocrania*), and craniopsides (*Paracraniops*) and craniides (*Orthisocrania*) with strong Baltic affinities appeared in Avalonia.

The late Caradoc–early Ashgill (figure 17.3; *TS*.5c–d, 6a–b) was an interval of maximum Ordovician radiation and expansion of the craniiformeans. Trimerellides spread from Kazakhstanian terranes toward Baltica and Siberia (Popov et al. 1997). Diverse trimerellide faunas appeared suddenly on the Yangtze Platform in the early Ashgill. Their ecology and origin are poorly known, but high endemicity at the generic level may suggest that they evolved over a considerable period of time in relative isolation (Popov et al. 1997). High-latitude Gondwana again remained unaffected by this late Caradoc–early Ashgill event.

A significant decline in the diversity of craniiformean faunas took place in the mid Ashgill (upper *TS*.6a–b; graptolite-based *Dicellograptus anceps* Zone), affecting faunas across Kazakhstanian terranes, the Yangtze Platform, and East Gondwana. They disappeared completely from Siberia. Only the enigmatic *Gasconsia* suggests a pattern of geographic expansion in Kazakhstan and Baltoscandia, following the general distribution of the *Holorhynchus* brachiopod fauna. Extinction of Ordovician trimerellide faunas across Baltica, Laurentia, and the Kazakhstanian terranes took place during the Hirnantian. Among the craniopsides and craniides only *Petrocrania* and *Pseudopholidops* are relatively widespread as rare components of the *Hirnantia* brachiopod assemblages.

Among the craniiformeans only the Trimerellida achieved considerable evolutionary success during the Ordovician, whereas the Craniopsida and Craniida remained insignificant parts of benthic assemblages dominated by filtrators. Separation of the Craniida and Trimerellida from the craniiformean stem group, which possibly consisted of craniopsides, was linked initially to adaptations to high-energy, shallow-water environments. The origin of an encrusting habit in the craniides made it possible for them to inhabit hardgrounds and rocky bottoms in such environments of the shallow shelf, which were otherwise hardly accessible to most Early Paleozoic benthic animals. The great increase in shell size in trimerellides also gave them the ability to colonize mainly shallow-water habitats affected by seasonal storms (Webby and Percival 1983). The ability to undertake

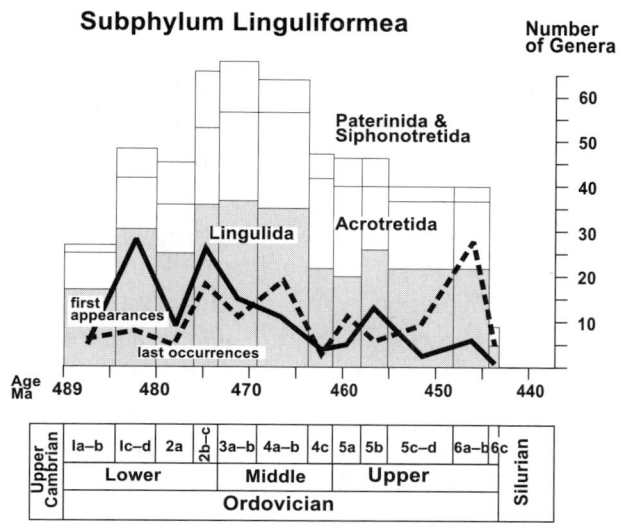

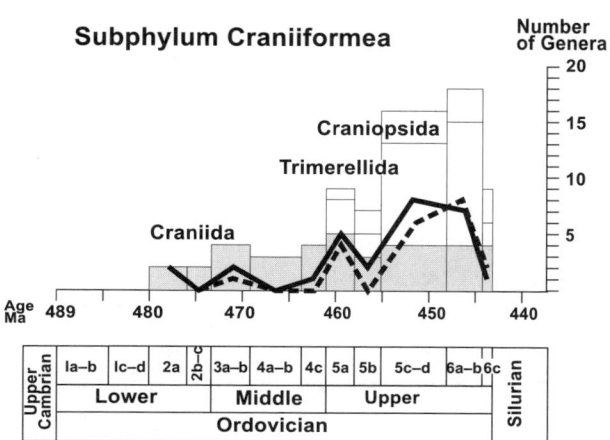

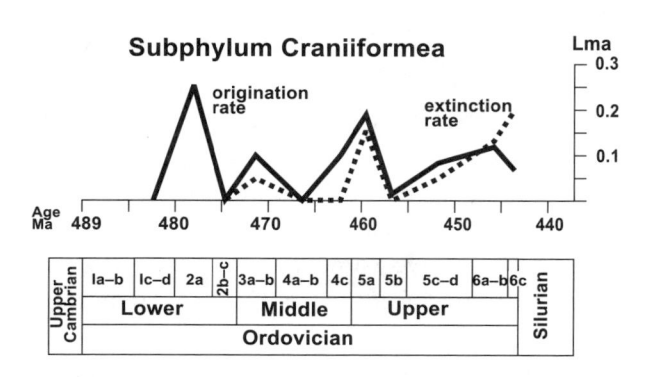

FIGURE 17.3. From top, absolute abundance of linguliformean orders (after Bassett et al. 1999b), absolute abundance of craniiformean orders, and extinction and origination rates across the craniiformeans (after Popov et al. 1999b).

From the early to mid Caradoc (*TS*.5b), craniides and craniopsides appeared in Laurentia, where they became relatively common in shallow marine biofacies (BA 2–3) on carbonate platforms. By that time

migration was severely limited in Ordovician craniiformeans, and their appearance across various plates and on ancient island arcs was markedly diachronous (Popov et al. 1999b: figure 1). Craniopsides and craniides remain unknown from Siberia if we exclude the Taimyr Peninsula (Cocks and Fortey 1998). Their appearance in the Caradoc of Avalonia was close to the time of its collision with Baltica. The Ordovician occurrences of trimerellides suggest a distinct pattern of island-arc settlement and biogeographic isolation, as well as indicating the possible position of Laurentia in proximity to East Gondwana sometime during the Darriwilian to early Caradoc (Popov et al. 1997).

Rhynchonelliformea

Strophomenata

Strophomenida (LRMC). The two superfamilies within the order Strophomenida, the Strophomenoidea and the Plectambonitoidea, both have their first representatives in Ordovician rocks (Cocks and Rong 2000). The plectambonitoids were the earlier, perhaps originating from the Billingselloidea during the Tremadocian and differing from the latter principally in having a fibrous pseudopunctate shell. The Strophomenoidea have a pseudopunctate shell, but laminar with a bilobed cardinal process, and were probably derived from the Plectambonitoidea in the Arenig, with the earliest known strophomenoid from the South China plate. So far as is now known, 95 plectambonitoid and 80 strophomenoid genera and subgenera were present in the Ordovician (figures 17.4, 17.5). Both superfamilies survived the end Ordovician turnovers, although with initially reduced diversity, with the Plectambonitoidea becoming extinct in the Late Devonian at the Frasnian-Famennian boundary and the Strophomenoidea continuing on into the Late Carboniferous.

Plectambonitoids appear to have originated in midshelf (BA 3) communities on the low-latitude terrane of the Altai-Sayan, with *Akelina* as the first known genus, and migrated to more temperate paleocontinents such as Baltica by the Arenig. In the latter half to two-thirds of the Ordovician, plectambonitoid genera expanded rapidly (figures 17.4, 17.5), reaching their all-time highest number of first appearances (FADs) in the early Llanvirn (*TS*.4b), and in particular became well established and often even abundant in shallower-water (BA 2) communities as well as those farther offshore. They also spread across the paleolatitudes, with *Aegiromena* forming a notable constituent of the West Gondwanan and peri-Gondwanan communities at high paleolatitudes.

Strophomenoids radiated more slowly than plectambonitoids, reaching their first peak in the early Caradoc (*TS*.5a), and mostly occurred in midshelf (BA 3 and 4) communities throughout their history. A second peak of FADs in both superfamilies (and the highest for Ordovician strophomenoids) occurred in the early Ashgill (*TS*.6a). Last appearances (LADs) (figure 17.4) for the plectambonitoids had an early sharp peak in the early Llanvirn (*TS*.4b), probably reflecting the relatively large number of endemic genera known from that period, but the highest number of LADs was in the early Ashgill (*TS*.6a). The strophomenoids had a more even pattern of LADs, but with some peaking in the early part of the late Caradoc (*TS*.5c) and during the early Ashgill (*TS*.6a). Only four strophomenoid genera are actually recorded from the latest Ordovician Hirnantian stage (*TS*.6c), but it is instructive to note that five Ordovician strophomenoid taxa are known as Lazarus taxa from the Silurian, even though none has yet been recovered from rocks of Hirnantian age. Similarly, although only seven plectambonitoids are known from the Hirnantian, further four Ordovician Lazarus taxa are recorded from the Silurian. It is noteworthy that, in the uppermost Ordovician rocks and continuing on into the Silurian, most plectambonitoids, for example, the abundant *Eoplectodonta*, thrived best in deeper-shelf (BA 4 and 5) communities and were comparatively rare onshore.

Although relatively uncommon during the first half of the Ordovician (figures 17.4, 17.5), both the Plectambonitoidea and the Strophomenoidea radiated quickly during the second half and often dominated some of the assemblages and communities in which they are found. A maximum of 43 genera or subgenera are recorded from a single time slice for each of the superfamilies, during the mid Caradoc (*TS*.5b) for the Plectambonitoidea and the early Ashgill (*TS*.6a) for the Strophomenoidea. In the Silurian, plectambonitoids dwindled in both numbers and diversity, in contrast to the strophomenoids, whose denticulated forms underwent further substantial radiations.

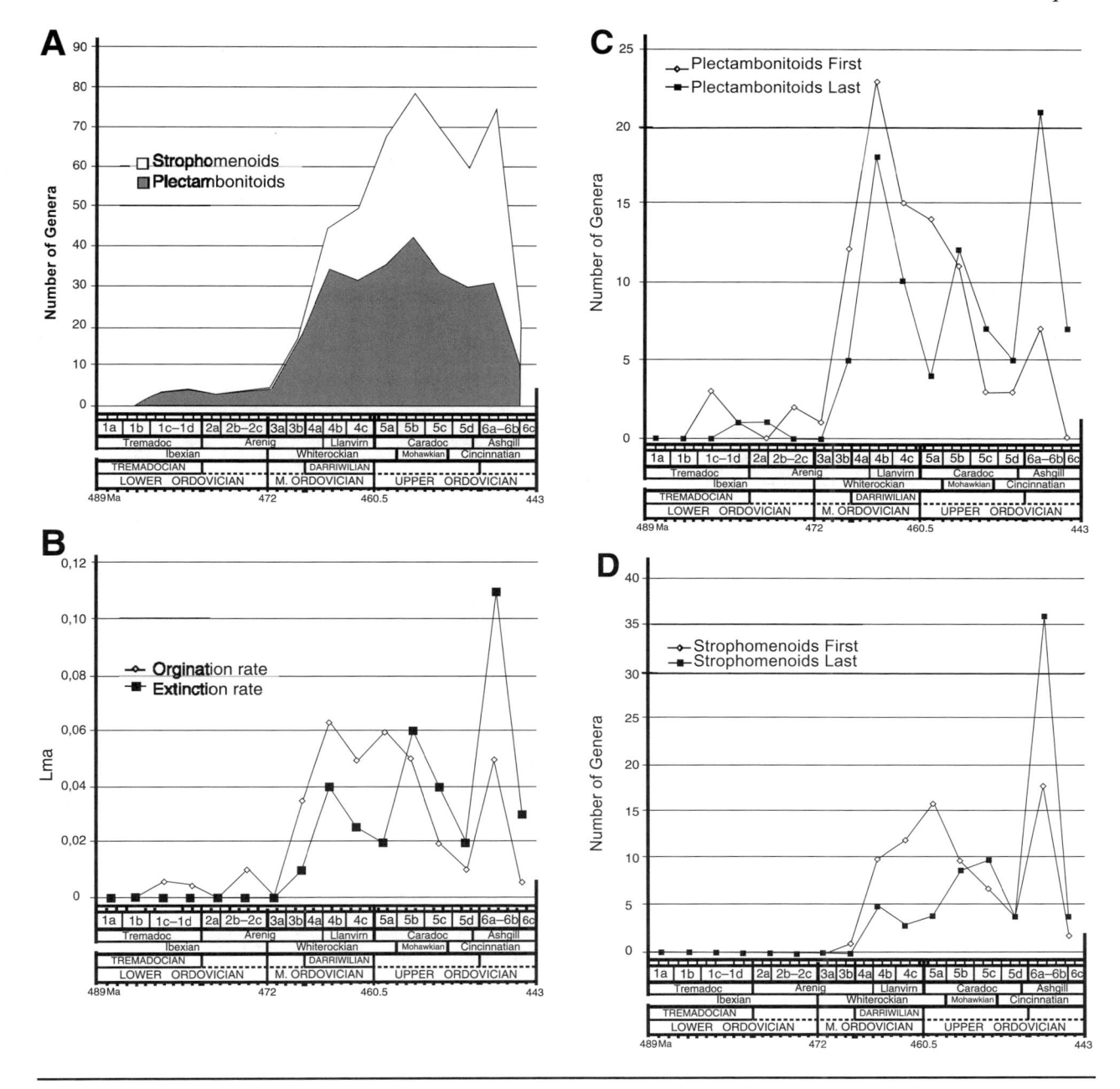

FIGURE 17.4. Strophomenide diversity. A, absolute abundances of the two superfamilies of strophomenide brachiopod. B, extinction and origination rates across the strophomenides. C, first and last appearances of the plectambonitoideans. D, first and last appearances of the strophomenoideans.

Chonetida and Orthotetida (LRMC). Although themselves relatively minor components of the Ordovician faunas, the era witnessed the origins of two orders of brachiopods that became numerous and diverse in later rocks. The Chonetida is represented by the single genus *Archaeochonetes* from the Ordovician and has been found in the Ordovician only from the middle Ashgill of Anticosti Island, Canada, although it is also recorded from the Lower Silurian. The origins of the Chonetida are obscure in detail, although they almost certainly lie within the order Strophomenida: the ancestor was probably a plectambonitoidean within the family Sowerbyellidae, a family that has a cardinal process and internal brachial

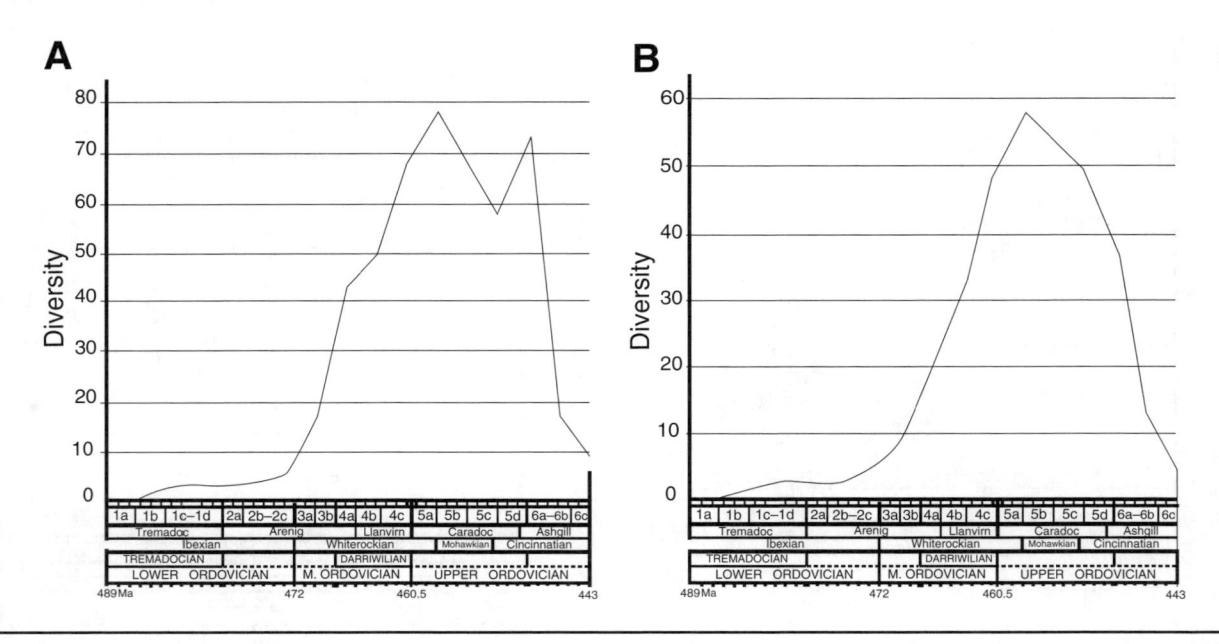

FIGURE 17.5. Comparison of standing diversity (A) and corrected diversity (B) of strophomenide brachiopods measured in numbers of genera.

valve septa and other structures comparable to those of early chonetides but that lacked the characteristic chonetide spines on the hinge line.

Ordovician Orthotetida all possess impunctate shells (in contrast to Silurian and later genera within the order), and it is uncertain whether they evolved from impunctate orthoids or pseudopunctate strophomenoids. There are four genera within the orthotetid Chilidiopsoidea in the Ordovician, and these are first represented by *Gacella*, first known from the middle Caradoc Chu-Ili terrane in Kazakhstan and from the upper Caradoc of Laurentia and elsewhere. In addition, the enigmatic Eocramatiidae, which is here also included within the Orthotetida (in contrast to other workers, who have included it within the Plectambonitoidea), includes only two rather rare genera, *Eocramatia*, with first and last appearances in TS.4b, and the closely related *Neocramatia*, recorded only from TS.5d and 6a (Harper 1989). Thus the order did not diversify much in generic terms during the Ordovician, but it became much increased in abundance, with *Coolinia* being a common constituent of the late Ashgill *Hirnantia* Fauna, which had a very wide distribution. Like the strophomenoids, the orthotetids were chiefly to be found in the middle shelf, principally in BA 3 and BA 4. *Coolinia* and *Fardenia* both survived into the Silurian, during which the order underwent considerable radiation (Williams and Brunton 2000).

Triplesiidina (DATH). This small but distinctive group, characterized by an unusual cardinalia, has 13 Ordovician representatives (Wright 1993, 2000). On the basis of shell morphology and especially the possession of a laminar secondary shell fabric, the triplesiidines have been closely related to the orthotetidines within the strophomenide clade. The group shows an extraordinary morphological variability in shell outline, profiles, and external ornament. The earliest taxon, *Onychoplecia*, occurs in the lower Llanvirn (TS.4b) rocks of western Newfoundland within the Toquima–Table Head Province. Nevertheless, both *Oxoplecia* and *Triplesia* have been described from the upper Llanvirn (TS.4c) rocks of Wales (Avalonia). The group underwent moderate radiations during the Caradoc and Ashgill, becoming widespread but never abundant.

Billingsellida (DATH). The order Billingsellida is probably the least stable of the main taxonomic groupings. The order currently contains the Billingselloidea, Clitambonitoidea, and the Polytoechioidea (Williams and Harper 2000a). Opinion is divided on whether the billingsellides are monophyletic (Williams et al. 1996) or polyphyletic (Popov et al. 2001).

Within the billingsellidines, only *Billingsella, Cymbithyris, Eosotrematorthis, Kozhuchinella,* and *Xenorthis* have Lower Ordovician records, with only *Eosotrematorthis* reported from the Arenig (Williams and Harper 2000a). The group was widespread but with most occurrences at low latitudes.

Two superfamilies, the Clitambonitoidea and the Polytoechioidea (Wright and Rubel 1996; Rubel and Wright 2000), are recognized within the clitambonitidines. The morphological features of the clitambonitoids suggest derivation from a protorthoid ancestor such as *Arctohedra,* rather than from within the billingsellide stem group, whereas most members of the polytoechioids may have evolved directly from the billingselloids and developed a quite different biogeographic distribution.

The Clitambonitidina, apart from the polytoechioids, were a dominantly Baltoscandian group with all genera confined to the Ordovician. A massive diversification of the clitambonitoidines at the base of the Arenig (*TS*.2a) was mainly restricted to the Baltic Province and involved the clitambonitoids, preferring the cool-water carbonate facies of the Baltoscandian basin. By the end of the Caradoc (*TS*.5d) only four genera remained, with two, *Kullervo* and *Vellamo,* having relatively widespread distributions. The majority of the polytoechioids appeared during the Arenig-Llanvirn interval and with the notable exceptions of *Protambonites* and *Tritoechia* were most characteristic of low latitudes. The last representatives, *Admixtella* and *Peritritoechia,* continued into the lower Caradoc in Central Asia and South China, respectively.

Rhynchonellata

Protorthida (DATH). The superfamily Skenidioidea represents this small but important group during the Ordovician (Williams and Harper 2000b). Genera are characterized by the apomorphies of a free spondylium and brachiophore plates that converge to form a plate underlying the notothyrial margin; both features were prominent in Early Cambrian taxa but also are typical of the post-Cambrian skenidioideans (Williams and Harper 2000b). The group includes five Ordovician taxa, first appearing during the Arenig in the circum-Iapetus region and China. The skenidiids were never common and conspicuous elements of the Ordovician radiations, but during the later Ordovician the group anticipated the deeper-water habitats preferred by the family during the Silurian.

Orthida (DATH). The pedunculate Orthida together with the recumbent Strophomenida dominated the brachiopod faunas of the Ordovician. Both groups underwent significant and sustained radiations during the period (figures 17.6, 17.7), most typically associated with siliciclastic environments. The Orthida probably represents the "modern articulated" brachiopod (Williams et al. 1996), and during the period the group was associated with a spectrum of habitats from nearshore to deep-sea environments together with carbonate mudmound and reef facies.

The order Orthida (Williams and Harper 2000c) has been divided into two suborders, the Orthidina (impunctate orthides) and the Dalmanellidina (punctate orthides). The first has a significant Cambrian history. The second appeared during the Arenig; its origin within and relationship to specific orthidines remain uncertain. The greater orthide group contains more than 300 genera grouped into 46 families; orthide history encompassed the entire Paleozoic, although few taxa occur during the later parts of the era. Radiations occurred as a series of stepwise diversifications simulating ecologic displacements by successive superfamilies during the Early Paleozoic (Harper and Gallagher 2001); peaks in diversification are matched by expansions in morphological disparity (Harper and Gallagher 2001). The orthides, however, attained a maximum diversity during the Ordovician, accelerated by the Ordovician radiation event (figures 17.6, 17.7); nearly 90 orthide genera, dominated by the orthoids and dalmanelloids, characterize the Caradoc interval (Harper and Gallagher 2001).

The Orthidina, comprising the Orthoidea and Plectorthoidea, appeared first, probably during the Early Cambrian (Williams and Harper 2000c). By the late Tremadocian, nearly 20 genera of orthidines, dominated by the orthoids, were present, many occupying shallow-water siliciclastic environments at high to temperate latitudes. The plectorthoids, however, were more common at low latitudes, commonly associated with carbonate environments. The first major radiation of the group occurred during the early Darriwilian (*TS*.4a) with an escalation in the numbers of orthoids and plectorthoids and the

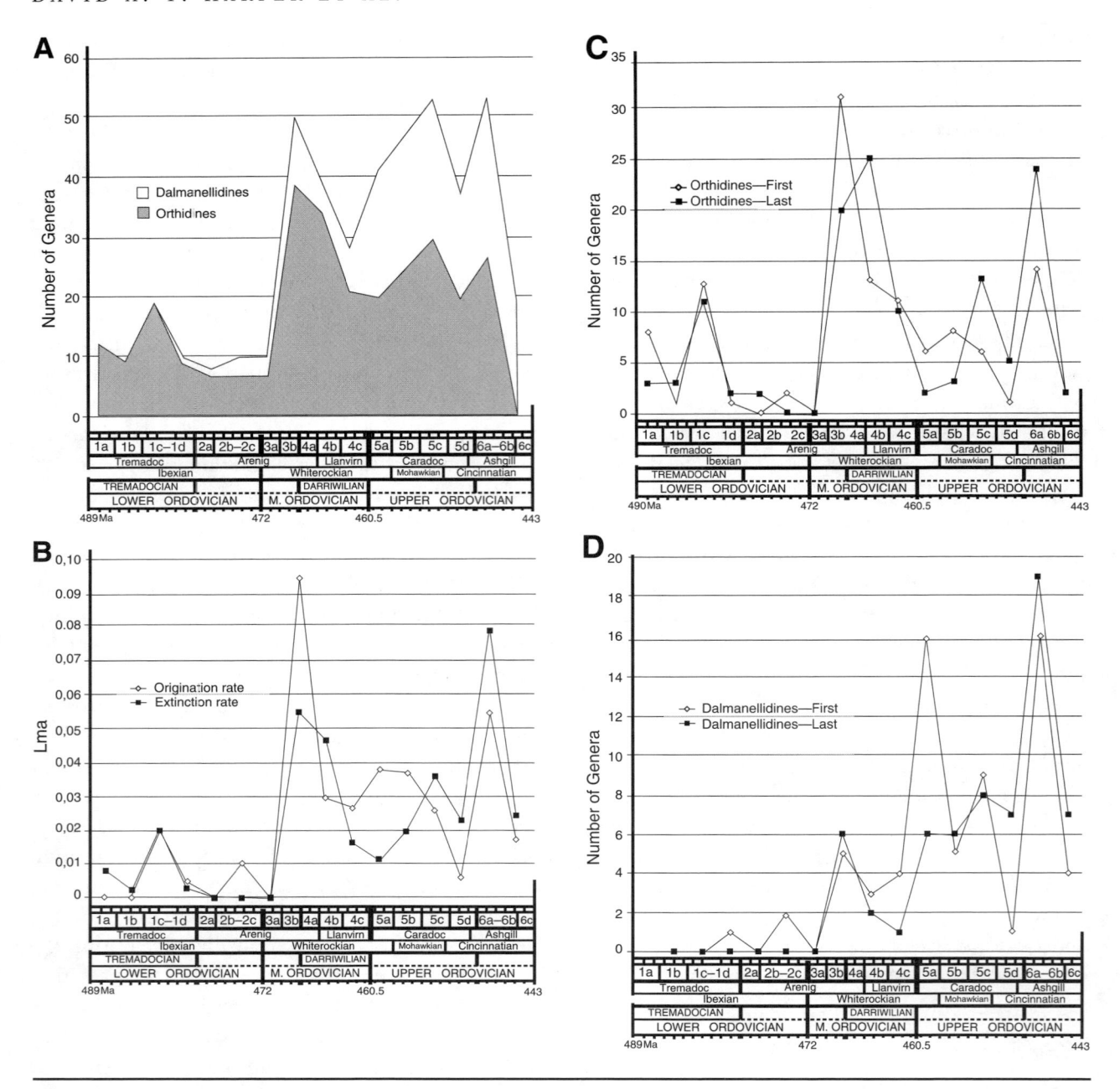

FIGURE 17.6. Orthide diversity. A, absolute abundances of the two suborders of orthide brachiopod. B, extinction and origination rates across the orthides. C, first and last appearances of the orthidines. D, first and last appearances of the dalmanellidines.

development of the dalmanellidine superfamilies, the dalamanelloids and enteletoids. Data suggest that this was the largest and most profound diversification with a standing count of 50 genera. Analysis of first and last appearances within the group nevertheless suggests that many were relatively short-lived stocks (figure 17.6C, D).

Diversity again peaked during the mid Caradoc, with more than 50 genera. The dominant components of this event were the dalmanellidines, particularly the dalmanelloids. Although a significant number of genera (16) appeared during the early Caradoc (*TS*.5a), the dominance of the group during the mid Caradoc is accentuated by a further diversification of both dalmanelloids and enteletoids and an increased decline among the impunctate groups.

The dalmanellidines (Harper 2000) maintained their position during the later Ordovician, reaching

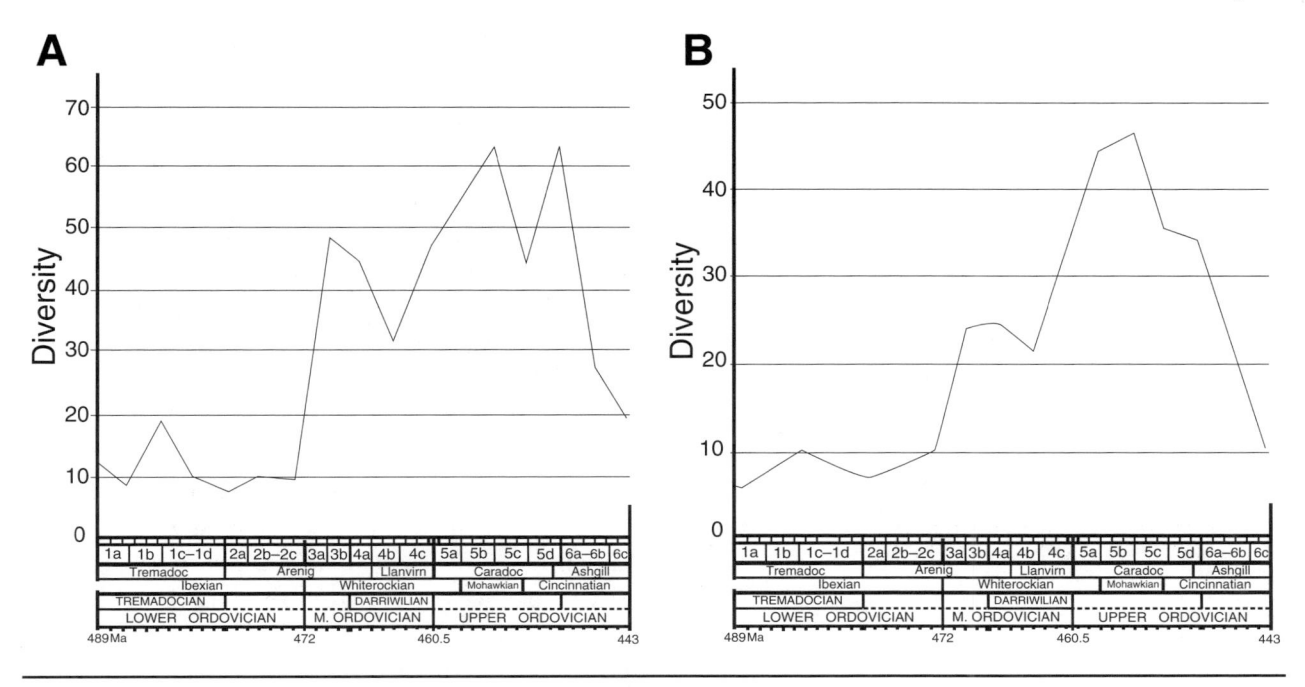

FIGURE 17.7. Comparison of standing diversity (A) and corrected diversity (B) of orthide brachiopods measured in numbers of genera.

near parity with the orthidines during the early to mid Ashgill (*TS.*6a–b), when a number of short-lived taxa, signaled by increased first and last appearances, dominated the punctate orthides (figure 17.6C, D). During the Hirnantian (*TS.*6c) diversity dropped to fewer than 20 genera, matching diversities during the late Tremadocian. Assemblages were characterized by punctate orthides, the impunctate forms never recovering their dominant role after the end Ordovician extinction event.

Pentamerida (JJ). The Ordovician was marked by the presence of two quite different pentameride groups (Carlson 1996). During the Early and Mid Ordovician only the Syntrophiidina were present. The first Pentameridina did not evolve until the Ashgill. The largest radiation occurred in the order (figure 17.8) during the Tremadocian (*TS.*1a–c); only four genera, *Bobinella, Mesonomia, Plectostrophia,* and *Palaeostrophia,* continue through from the underlying Cambrian. There are more modest radiations during the early Arenig, Llanvirn, and mid Ashgill. Tremadocian faunas show high rates of origination and relatively low extinction rates. High rates of origination continue through the early Arenig, followed by low-diversity, holdover faunas during the mid Arenig. The most significant turnover, however, occurs at the Arenig-Llanvirn boundary (*TS.*4a–b transition), when only four taxa continue over the boundary. The Caradoc (*TS.*5a–d) was characterized by relatively low diversities and a steady state of origins and extinctions. In contrast, the low-diversity Ashgill (*TS.*6a–c) pentameride fauna suffered high rates of extinction and was marked by a relatively large number of short-lived, endemic genera. Pentameride genera were generally cosmopolitan during the Tremadocian to the early Caradoc. In contrast, Ashgill taxa exhibit much greater degrees of provinciality.

Rhynchonellida (JJ). The rhynchonellide brachiopods first appeared during the Darriwilian (*TS.*4a–c) as low diversity faunas characterized by small shells occurring in the cool-water seas of the continental margins (Bassett et al. 1999c). Nevertheless, by the latest Darriwilian (*TS.*4c) the group had appeared in a variety of disjunct paleoplates and arc systems. The most important turnover occurred during the late Caradoc (figure 17.8B), near the base of the graptolite-based *linearis* Zone (*TS.*5c–d boundary). The taxa that originated during the early to mid Caradoc (*TS.*5a–b) were mainly cosmopolitan or widespread in the tropics and adjacent areas. In contrast, those originating during the late Caradoc (*TS.*5d) are endemic or show relatively high degrees

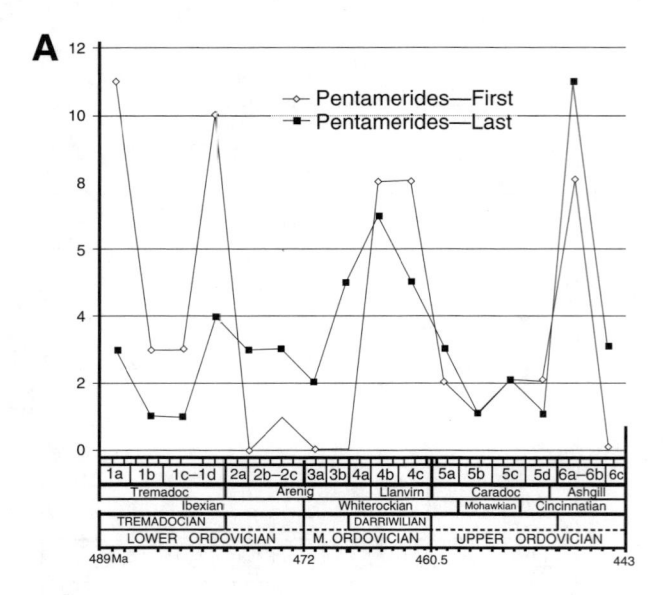

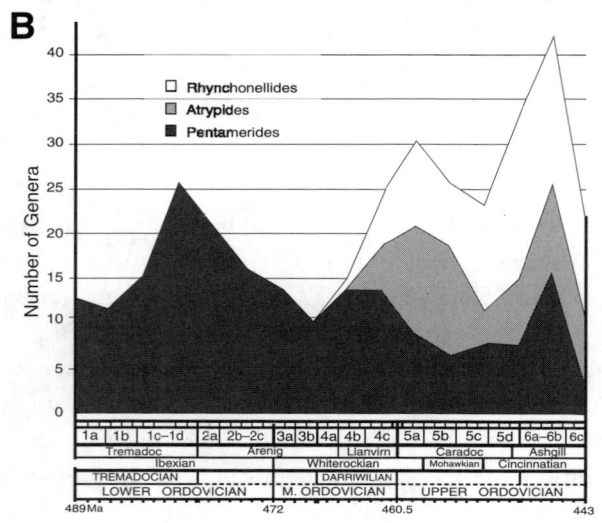

FIGURE 17.8. Pentameride, atrypide, and rhynchonellide biodiversity. A, first and last appearances of the pentameride taxa measured in numbers of genera. B, standing diversities of the cyrtomatodont atrypides and rhynchonellides together with the advanced deltidiodont pentamerides.

of provinciality. During the late Caradoc and early Ashgill, rhynchonellides invaded the epicontinental seas of both Laurentia and Siberia, becoming the dominant components of the shallower-water brachiopod fauna. During much of the Ashgill (*TS.*6a–b) the rhynchonellides exhibited remarkably low rates of origin and high rates of extinction. New genera mainly appeared during the Hirnantian, when short-lived, possibly opportunist forms such as *Plectothyrella* and *Thebesia* developed surprisingly widespread distributions within the *Hirnantia* and Edgewood faunas, respectively.

Atrypida (PC). The Atrypida originated during a late Llanvirn (Llandeilian) radiation (figure 17.8B). Throughout the history of the order its taxa were confined to the tropics, usually occupying subtidal carbonate shelf and ramp environments, commonly in shallow water (Copper 2001a). The first major radiation of the group occurred during the Caradoc and involved 13 genera. The Caradoc interval provided the opportunity for considerable experimentation within the order with the development of a variety of lophophore types and their supports. The origin of the group is still unclear, although smooth and slightly costate forms, *Manespira* and *Anazyga* respectively, may have occurred in the Llanvirn (Llandeilian). Nevertheless, by the later Caradoc to the mid Ashgill, anazygids such as *Catazyga* and *Zygospira* were very abundant on and around Laurentia, commonly forming shell pavements (Copper 2001b). *Zygospira* was most common in shallow-water facies, whereas *Catazyga* generally occupied deeper-water habitats. During the mid to late Ashgill, atrypid diversity declined; both *Catazyga* and *Zygospira* have yet to be reported from definitive Hirnantian rocks, although both subfamilies to which they belong reappeared during the Llandovery. In contrast, the spirigerinids diversified during the late Ashgill, replacing the early to mid Caradoc *Pectenospira* and *Sulcatospira*, both associated with carbonate mound facies in Kazakhstan. The spirigerinids increased in abundance and diversity in North America and Europe during the Hirnantian, displacing the incumbent anazygids (Copper and Gourvennec 1996). Septatrypinids such as *Webbyspira* and *Idiospira,* together with cyclospirids such as *Rozmanospira* and *Cyclospira*, evolved simultaneously during the Caradoc and Ashgill to complete the diversity of the group. The latter family probably originated in China or Kazakhstan and migrated westward during the late Caradoc (Popov et al. 1999a).

Spiriferida (JR). Traditionally spiriferide origins were tracked back to the earliest Silurian. However, recent investigations suggest that the order originated during the Ashgill. *Eospirifer* from the middle Ashgill of East China is the earliest known representative of the order Spiriferida (Rong et al. 1994).

The major apomorphies of the spiriferides were established by the mid Ashgill, including a strophic shell with well-developed interarea and a brachidium of laterally directed spirialial cones (Carter et al. 1994). This is thus the last major taxonomic group of Brachiopoda to appear in the Ordovician. The taxon *Eospirifer praecursor* Rong et al. 1994 occupied a wide range of environments from BA 2 to BA 5. It occurred very abundantly in BA 2 and also occurred rarely as part of the *Foliomena* Fauna (BA 5).

Athyridida (JR). To date, information is sparse on the origin and early evolution of the athyridides. Nevertheless, the earliest athyridides occur in Upper Ordovician rocks, represented by a few genera whose mutual relationships are not well known. The earliest known athyridide is *Kellerella* Nikitin et al. 1996, which appeared in Kazakhstan during the *multidens* Zone (*TS*.5b, early to mid Caradoc). At that time, the Kazakhstanian terranes were located as small microplates and island arcs adjacent to East Gondwana. Although the early Caradoc was a time of rapid morphological diversification in atrypids, a brachidium with spiral jugal processes had just appeared in the earliest known athyridids. They are impunctate and smooth (Alvarez et al. 1998) and include taxa lacking the septalium (*Kellerella*) and those possessing a well-developed septalium supported by a median septum (*Nikolaispira*). The fauna is associated with carbonate buildups and is characterized by a predominance of both *Kellerella ditissima* and *Nikolaispira rasilis* (Nikitin et al. 1996).

The third genus of this major group, *Hindella* (= *Cryptothyrella*) is common in Hirnantian and Lower to Middle Llandovery rocks in the Northern Hemisphere. The earliest occurrence of *Hindella* may be in middle Ashgill rocks in Europe. However, a monospecific assemblage of *Cryptothyrella?* sp. occurs in strata, suggestive of BA 3–4, in the lower Ashgill of the Kerman and Tabas regions, east-central Iran (Bassett et al. 1999a). This early occurrence of the athyridides is significant; the group has not been reported in the coeval high-latitude faunas of West Gondwana.

Fu (1982) established two genera, *Apheathyris* and *Weibeia*, from the Caradoc rocks of northwestern China. *Apheathyris*, despite reservations (Nikitin and Popov 1996), is retained within the athyridides. The systematic position of *Weibeia* is, however, less secure because the characters of the brachidium are currently unknown and, moreover, *Weibeia* possesses a pair of long, parallel brachial plates, a feature atypical of the athyridides. *Hyattidina* and *Whitfieldella* occur in the higher Ordovician. The earliest species of *Hyattidina* is not well known, with *Hyattidina? sulcata* possibly the earliest record of this genus, from the Kiln Mudstones, Girvan, Scotland (Williams 1962). The earliest representatives of *Whitfieldella* are reported from the *Hirnantia* Fauna (*TS*.6c) in southwestern China (Rong 1979). Alvarez et al. (1998) summarized these data in their revised classification of the order.

Based on these data, the athyridides probably originated in the regions of Kazakhstan, North China, and Qaidam, possibly in the early Caradoc (*TS*.5b). They may have been derived from an atrypide group by a substantial change of the direction of spiralia.

■ Ecologic Diversification (PMS, DATH)

The Ordovician Radiation marked the transition from the Cambrian EF to the Paleozoic EF (Sepkoski and Sheehan 1983; Sepkoski 1991a). Increasing diversity was accompanied by the rise to ecologic dominance of many new groups of organisms with the complexity of the ecosystem increasing markedly, especially in local communities. New adaptive strategies allowed some organisms to move into previously unoccupied habitats, and many more groups invaded habitats that had been occupied by fewer clades in the Cambrian EF.

Compared with the many taxonomic studies of the Ordovician Radiation, relatively few studies have addressed ecologic changes among animals during the radiation, and this remains a fertile area of research (Droser and Sheehan 1997b). Ecologic changes can be grouped into autecologic changes, as new morphologies evolved that better adapted animals to the physical environment, and synecological changes, as animals adapted to increasing competition and predation.

Competition for food may have been a driving force in the Ordovician Radiation; the success of brachiopods may have been due in great measure to their low metabolic rates (Rhodes and Thompson 1993), which allowed them to be successful in oceans of the Early Paleozoic, when food was in short supply

(Bambach 1993; Martin 1998). Low food requirements were also characteristic of many other dominant early Palaeozoic groups, such as echinoderms, bryozoans, and sponges. Brachiopods have among the lowest metabolic rates of any marine organisms; the summary of Peck (2001:102) is worth repeating: "these (metabolic) attributes make them less able to compete for resources when they are in plentiful supply, but make them very strong competitors in a world where resources are limited or temporally restricted. They can survive where others starve from lack of resource supply, making them very resilient in difficult times." Furthermore, brachiopods may have been well suited to exploit the Ordovician seas because their calcitic shells preadapted them to life in the calcite seas of the period (Stanley and Hardie 1999).

The addition of new kinds of carnivores also increased community complexity; many species faced increased competition for food and at the same time were prey for increasingly efficient and specialized carnivores. These pressures must have driven adaptations in many directions, but the general trend was toward increased specializations and decreasing niche sizes. Communities from across marine environmental transects increased in diversity (Sepkoski 1988; Patzkowsky 1995c), with the exception of high-stress settings, such as nearshore environments, where organisms tended to be broadly niched and diversity changed little throughout the Phanerozoic.

Bambach (1983) recognized 20 distinct megaguilds in the modern marine ecosystem; suspension feeders, herbivores, carnivores, and detritus feeders occurred across pelagic, infaunal, and epifaunal habitats that were further subdivided into mobile or sedentary lifestyles that were in turn subdivided into low or high tiers. Of the 20 megaguilds in the modern oceans, only 10 were occupied in the Cambrian EF, and this increased to 14 in the Paleozoic EF. Many new clades entered the various megaguilds during the Ordovician (figures 17.9, 17.10).

The implications of differing community complexity and biodiversity are currently being examined in conservation biology. Increased complexity enhances the amount of resources captured and available to the community as a whole (Fridley 2001; Cardinale et al. 2002). This increase is driven especially by more efficient utilization of food and resources as organisms become more specialized for particular food-gathering strategies. Thus, there may have been a positive feedback in the Ordovician radiation that enhanced the amount of resources available to a community as diversity and complexity increased.

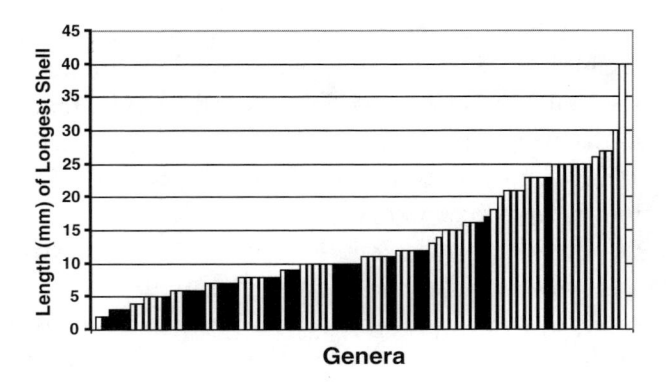

FIGURE 17.9. Increase in size of Plectorthoidea and Orthoidea brachiopods through the Ordovician radiation. The graph compares the largest individuals of genera illustrated in the *Treatise* (Williams and Harper 2000c) prior to the radiation (Upper Cambrian and Tremadocian, 25 genera, black columns) with those after the radiation (Caradoc and Ashgill, 53 genera, white columns). An increase in maximum size of brachiopods is indicated, although the method provides only a coarse estimate of the nature of this increase.

Ecologic aspects of the Ordovician radiation of brachiopods may have been decoupled from their phylogenetic radiation (Droser et al. 1997; Bottjer et al. 2001). Clades may become major players in local communities by diversifying either before or after they increase in abundance in local communities. In addition, since there is a strong geographic component to the radiation (Miller and Mao 1998; Harper and Sandy 2001), it may not be possible to understand global patterns until many regions are examined (Droser and Sheehan 1997b). For example, bivalve mollusks diversified and became common members of communities in high latitudes during the Arenig before they appeared in North America (Babin 1995; chapter 20).

Food Resources and Brachiopod Size Ranges

Phytoplankton-based food resources in the oceans increased significantly in the Mid Ordovician (Tappan and Loeblich 1973; Tappan 1986; Vermeij 1987; Bambach 1993), but the level of food resources was still far below modern levels. Many of the ecologic

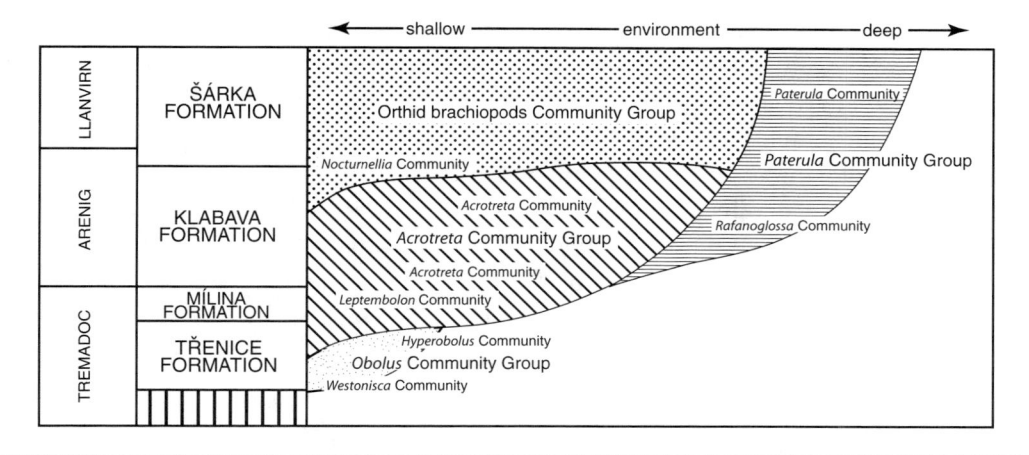

FIGURE 17.10. Time environment diagram of the onshore-offshore pattern of the Ordovician Radiation in the Prague Basin (after Mergl 1999).

changes of the radiation may have been associated with increased food supply, and this may have been crucial for the expansion to proceed. Brachiopods increased in size during the radiation (figure 17.9), possibly facilitated by increasing food supplies, a trend documented by Bambach (1993) for other organisms. Similarly, the increasing abundance of brachiopods in shell beds after the radiation commenced (Li and Droser 1999) signaled increased numbers of brachiopods, perhaps again facilitated by increasing food.

Orthide brachiopod clades that were present both before and after the radiation increased substantially in maximum size. Figure 17.9 compares the largest individuals of each genus of the Plectorthoidea and Orthoidea illustrated in the *Treatise* (Williams and Harper 2000c) from the Upper Cambrian and lowest Ordovician (Tremadocian) with those from the Caradoc and Ashgill. Although the method is imprecise, the trend is so clear that it is almost certainly robust. *Plaesiomys* from the Upper Ordovician of the Mackenzie Mountains of northwestern Canada reached a diameter of 60 mm. Other clades of brachiopods that originated during the radiation attained similar or larger sizes. In Laurentia, for example, the strophomenides *Nasutimena, Tetraphalerella,* and *Oepikina* from the epicontinental seas of Laurentia attained sizes of 60–80 mm (Jin 2001; Jin and Zhan 2001). Moreover, several members of the rhynchonelloid genera *Hiscobeccus* and *Lepidocyclus* attained lengths of 30–40 mm during the Ashgill. Larger brachiopods were not simply exceptions but were common, especially in open marine communities. In addition, Jin (2001) has shown that epicontinental-sea brachiopods were typically much larger than their congeneric counterparts on the shelf margins of Laurentia. The alternative hypothesis that this is simply a classic example of Cope's Rule is difficult to test. Nevertheless, the gradual increase in size during clade history up to a maximum, predicted by the rule, is not typical of the Early Paleozoic history of the Orthida; Cambrian and Early Ordovician orthides had a rather constant size prior to a short interval of marked size increase during the radiation.

Although food supply increased, it was still far below that available to the Modern EF (Bambach 1993); brachiopods, bryozoans, and echinoderms would still have maintained the advantages afforded by their low metabolic rates. Large size can have many selective advantages, such as the ability to occupy space and resist predation, but these advantages could be realized only if there were sufficient energy sources for the animals to succeed at these sizes.

Autecology

As each of the new brachiopod clades evolved, a range of morphological innovations appeared. These innovations improved the adaptation of each species to the local environment. Ushatinskaya (2000) has provided a clear understanding of the autecology of Cambrian brachiopods. The linguliformean brachiopods, which were part of the Cambrian EF, had expanded to fill most of the lifestyles of later nonarticulates.

The Cambrian articulated brachiopods, however, had not diversified morphologically and were primarily biconvex, epifaunal, and attached by a pedicle.

Two of the most significant morphological changes during the Ordovician Radiation were the development of planar shells in the dominant lineages of post-Cambrian strophomenides and the evolution of cyrtomatodont teeth. Strophomenides adopted planar profiles in which opposite valves were closely opposed. When strophomenides abandoned the globose, biconvex shell, they abandoned a feeding strategy that enclosed the feeding currents and allowed an animal with very little biomass to occupy a relatively large space (Peck 2001). The planar shells had very little space for filter-feeding currents, and the strophomenides underwent a complete overhaul of the feeding method preferred by most brachiopods. The planar shells also permitted strophomenides to adopt a recumbent lifestyle (Leighton and Savarese 1996). This free-lying, reclining megaguild (Bambach 1983) is lacking in modern brachiopods, and as a result there is considerable uncertainty as to how these organisms functioned. Most Ordovician strophomenides probably lived concave upward on fine-grained substrates, with most utilizing "snowshoe" adaptations or, among geniculate forms, iceberg adaptations (Thayer 1975). Many other strophomenides lived on firm substrates, and they were often common in reefs.

The cyrtomatodont teeth provided more substantial hinging of the shells, possibly partly as a means to inhibit predators and protect the animal from desiccation and the effects of turbulence. The mechanical hinge also lessened the expenditure of muscular energy needed to keep the shells aligned. Growth of the cyrtomatodont ball-and-socket articulation required new styles of shell formation including an emphasis on resorption of shell material to allow the socket to increase in size during growth. The cyrtomatodont teeth appeared first in rhynchonellides during the Darriwilian. Atrypides followed, then athyridides in the early Caradoc and spiriferides in the Ashgill.

Many morphological changes were adaptations that had less extreme implications for brachiopod lifestyles. Some species developed adaptations that served multiple functions; for example, the evolution of plications served a dual purpose of preventing large particles from entering the chamber while at the same time increasing the strength of the shell. Evolution of "snowshoe morphologies" (Thayer 1975) allowed brachiopods to inhabit soft substrates. These are two examples of the morphological innovations that permitted brachiopods to compete with other benthos for resources, defend against predation, and interact with the physical environment. Each of the radiating clades had many unique morphological innovations, many of which have been widely discussed (Rudwick 1970; Alexander 2001; Leighton 2001). However, the sequence of appearance of innovations during the radiation has not been addressed. For example, do morphological innovations appear before, after, or during increases in community complexity, as evidenced by increasing numbers of species and variety of guilds in a given habitat?

Synecology

The environment, which was driving evolution, included both physical and biotic factors. There were probably modest changes in the physical environment during the radiation, and some brachiopods moved into niches that previously had not been occupied by brachiopods. However, the biotic environment was in the throes of a massive reorganization and expansion. As organisms diversified and autecologic changes accumulated, each species in the local communities was faced with a new synecological setting with increased competition for resources and increased predation.

The relative simplicity of Cambrian EF communities contrasts with the ecologic diversity of those following the Ordovician Radiation (see Cocks and McKerrow, figures 2 and 11, in McKerrow 1978). The Ordovician Radiation set the style of communities for the remainder of the Paleozoic. The similarity of ecologic style was recognized by Walker and Laporte (1970) and later formalized by Sepkoski (1981a, 1991b) as the Paleozoic EF. Boucot (1975, 1983) divided the Paleozoic EF into a series of Ecologic Evolutionary Units during which communities were stable in the sense that diversity and ecologic structure were maintained. Only occasionally did groups move into new Benthic Assemblages (BAs). Species-level diversity and the various morphological types present in the community remained constant for tens of millions of years (e.g., Watkins et al. 2000).

By the end of the Ordovician Radiation brachiopods were established as important, often dominant members of communities in open marine settings. The communities were arrayed along environmental gradients (Springer and Bambach 1985), conveniently grouped into a series of six progressively deeper BAs (Boucot 1975). Although arranged from shallow (BA 1) to deep (BA 6), the assemblages did not have a consistent depth from region to region or sharp boundaries; rather, the BAs were controlled by parameters of the physical environment, including decreasing wave energy with depth, depth of penetration of sunlight, temperature gradients, nutrient supply, and oxygen levels. This pattern, which developed during the radiation, was maintained through the remainder of the Paleozoic (Boucot 1983).

Changes in benthic communities from the Cambrian EF to the Paleozoic EF are readily apparent. However, limited knowledge about the relative timing of changes makes it impossible to choose between two hypotheses: (1) the radiation involved displacement of the Cambrian EF by members of a more competitive Paleozoic EF, or (2) a physical disruption of the Cambrian EF allowed radiation of the Paleozoic EF despite the advantages provided by incumbency to the Cambrian EF (Sepkoski 1991b). In fact, there is still uncertainty as to whether the Cambrian EF actually declined or if it continued in preradiation abundance and diversity but was simply less conspicuous because of a vastly more common and more diverse Paleozoic EF (Westrop et al. 1995).

Synecology of the Rhynchonelliformean Brachiopods

Increasing taxonomic diversity by itself is a major component of the radiation. But taxonomic diversity, for example, the number of families within orders, also provides a proxy for ecologic diversity because species within families commonly share many morphological features that determine their lifestyles. The enormous increase in numbers of orders, together with the numbers of families and genera (figure 17.1), indicates an enormous increase in the different lifestyles of rhynchonelliformean brachiopods, which were incorporated into the marine ecosystem during the Ordovician Radiation. The linguliformean fauna developed as a holdover from the Cambrian EF, while the craniiformeans played a relatively obscure role in the radiation as a whole (see Popov et al. earlier in this chapter). The rhynchonelliformeans, however, dominated the event.

Distribution of modern articulated brachiopods is entirely by nonplanktotrophic reproduction, producing relatively brief larval stages. Groups with nonplanktotrophic larvae have smaller numbers (lower fecundity) than those of planktotrophic groups. Nonplanktotrophic larvae and low fecundity probably also were typical of Ordovician orthide brachiopods, based on pallial markings (Law and Thayer 1990). However, Brunton and Cocks (1996) noted that some strophomenides had developed a small shell (protegular node) when their larvae settled, implying a prolonged larval feeding stage such as that in modern nonarticulated brachiopods. In terms of a radiation, the brief larval stages of nonplanktotrophic groups should have resulted in more narrow distributions than planktotrophic groups. Nonplanktotrophs might speciate more readily than planktotrophs because of their more restricted geographic ranges; this could have promoted rapid radiation (Jablonski 1986). A comparison of the rates of radiation of planktotrophic and nonplanktotrophic groups (both brachiopods and other phyla) in a single region during the Ordovician could provide a test for these hypotheses.

Although nonplanktotrophic larval stages may have aided brachiopods during the radiation, it may also have been part of their ultimate decline. Valentine and Jablonski (1983) suggested that the decline of post-Paleozoic brachiopods might have been related to developmental inflexibility, which resulted in their replacement by bivalves, with a greater developmental flexibility.

Open Marine Environments

More than 50 morphologically distinct ecologic guilds have been identified in open marine communities of the Paleozoic EF (e.g., Bambach 1983; Watkins 1991, 1993). Many more guilds have yet to be recognized. After the radiation most brachiopods belonged to one of seven guilds (Bambach 1983; Watkins 1991, 1993): (1) cap-shaped "inarticulates," (2) infaunal "inarticulates," (3) strongly biconvex pedunculate, (4) strongly biconvex alate pedunculate, (5) planoconvex to weakly biconvex pedunculate, (6) relatively flat free-lying, and (7) inflated free-lying articulates.

There were also many minor guilds, such as cemented forms, brachiopods that attached to echinoderms in upper tiers and inhabitants of cryptic spaces. Additional brachiopod guilds, such as spinose, semi-infaunal productids, appeared later in the Paleozoic. The guilds presumably parceled ecospace so that their members were in greater competition among themselves than with members of other guilds. Not all guilds are present in any given local community, but many guilds have several species.

Prior to the radiation, orthides and pentamerides were the most common articulated brachiopods, with billingsellids less common. The primary rhynchonelliformean brachiopod orders that diversified during the Ordovician Radiation (the Orthida and Strophomenida) had contrasting lifestyles. Among the orthides only the orthidines were prominent prior to the radiation, and they maintained and even increased in diversity during the radiation. The dalmanellidines radiated substantially, becoming as diverse as the orthidines. Other radiating groups included the Atrypida and Rhynchonellida (figure 17.8), Orthotetida, and Athyridida. The Billingsellida (sensu stricto) expanded during the early phases of the radiation but then declined through the Ordovician. The Pentamerida declined from preradiation levels but staged a comeback in the Silurian. The Spiriferida appeared later in the Ordovician and radiated in the Silurian. Each of these newly radiating groups had evolved distinct morphological characteristics that allowed them to exploit the environment in different ways. A result of this radiation was that the number of brachiopod guilds in communities increased substantially.

Marginal Environments

Extreme or marginal environments were occupied by low-diversity communities. Shallow, restricted environments with soft substrates often were colonized by the *Lingula* Community type characterizing BA 1 of Boucot (1983). Shallow-water environments subjected to heavy wave action typically have low diversity faunas of broadly niched species capable of withstanding unpredictable environmental changes such as desiccation, temperature variations, variations in normal waves, and occasional storms. Diversity has remained low in these environments throughout the Phanerozoic (Bambach 1983), probably because of the need to maintain relatively broad niches and tolerances to live successfully in unpredictable settings.

At the other end of the marine spectrum, BA 6 communities lived in dysaerobic, relatively deep offshore environments that were transitional between open marine assemblages and the graptolite shale environments that commonly had insufficient oxygen for benthic animals. An example of this is the *Foliomena* Community type (Sheehan 1973; Cocks and Rong 1988; Harper and Rong 1995; Rong et al. 1999), with small, thin-shelled brachiopods commonly with soft-substrate morphologies that were associated with vagile trilobites.

Predation

Predation is an important aspect of synecological interaction. The effects can be highly varied, in some cases excluding groups from a region and in other cases culling dominants and thus providing opportunities for less competitive groups to survive. Crushed brachiopod shells are not uncommon following the Ordovician Radiation, and many examples of repair of damaged shells have been described, indicating that predation was dominated by durophagous predators (arthropods and possibly nautiloids and asteroids), together with fungal borers (Alexander 1986). Drilling, if present, was not common until the Devonian, when fishes and possibly gastropods became important predators (Leighton 2001).

A variety of morphological changes may have been adaptations against predators. Examples include increasing shell strength with a fold and sulcus, plications, cyrtomatodont tooth structures, protective structures such as spines, coverings of the pedicle opening, punctae, and chemical defenses that made the animals inedible.

Pace of Synecological Change

The transition from communities dominated by the Cambrian EF to communities dominated by the Paleozoic EF moved progressively from shallow to deep water across the shelf and upper slope (Sepkoski and Sheehan 1983; Sepkoski and Miller 1985). The progressive dominance of brachiopods in more offshore environments has been independently confirmed by studies of the Ordovician brachiopods of the Anglo-

Welsh region (Lockley 1993) and marginal Laurentia (Patzkowsky 1995b). The pace of onshore to offshore synecological change seems to have been moderately rapid. Time-environment diagrams generally show that the change from Cambrian EF communities to Paleozoic EF communities across shelf environments began in the Arenig and was completed during the Caradoc in North America (Sepkoski and Sheehan 1983: figure 7). It may have been more rapid in Bohemia, where it was completed in the Llanvirn (Mergl 1999; and see figure 17.10). In Argentina the change took place in the early Arenig, possibly during a short interval at the end of the early Arenig (Waisfeld et al. 1999).

Few synecological field studies have actually examined the Ordovician Radiation. Li and Droser (1999) showed that brachiopod abundance increased rapidly near the base of the Whiterockian stage in North America. Finnegan and Droser (2001) found that in western North America, although brachiopod abundance increased rapidly, the diversity of level-bottom communities lagged during the early Whiterockian. Brachiopod-dominated communities were characterized by relatively low diversity despite the presence of very abundant brachiopods. Furthermore, few of the new clades that appeared during the radiation were present in the initial communities. By the end of the Whiterockian, community diversity of brachiopods had increased substantially, and the new clades were present.

Paleoecologic Levels

Comparing ecologic changes at different times in the geologic record can be accomplished only if there is a standardized method of assessing changes. A series of four standardized paleoecologic levels of change has been developed (Droser et al. 1997, 2000; Bottjer et al. 2001). These levels are ordered from those with the farthest-reaching ecologic change to those that primarily involve local communities.

First-level changes represent the advent of a new ecosystem. The colonization of land was proceeding during the Ordovician, but this first-level change did not involve brachiopods. Second-level changes involve appearance or changes in the ecologic dominants in the ecosystem. The change from trilobite-dominated level-bottom communities of the Cambrian EF to domination by brachiopods, echinoderms, bryozoans, and other groups of the Paleozoic EF (figures 17.10, 17.11) was a second-level change. The change in shell morphology and feeding methods of the strophomenides was a second-level change that allowed them to move into ecospace previously unoccupied by brachiopods. Increasing levels of predation and the way brachiopods responded to predation are also second-level changes that significantly restructured communities and interactions at the local level.

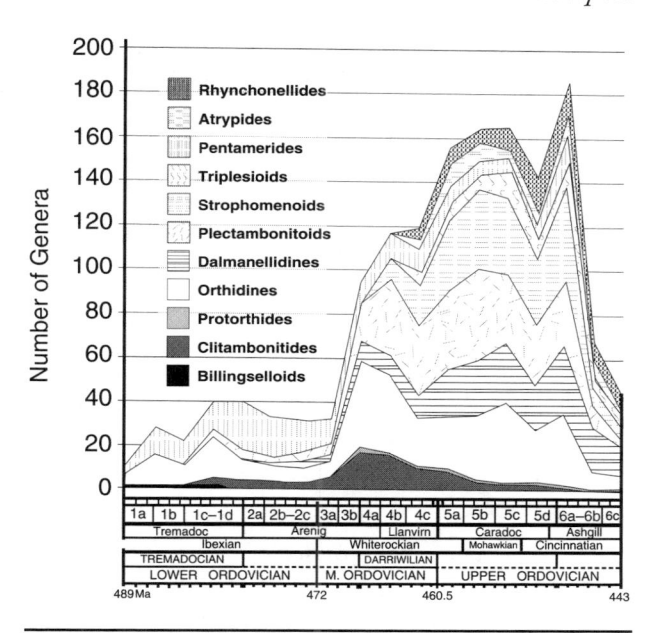

FIGURE 17.11. Standing diversities of all major rhynchonelliformean groups compared; measured as numbers of genera.

Much of the story of the radiation of brachiopods took place at the third level, when megaguilds were filled to Paleozoic EF levels (Bottjer et al. 2001). A third-level increase in epifaunal tiering, largely produced by echinoderms (Ausich and Bottjer 1982), may have affected brachiopods, since food was removed from the water column prior to reaching base-level brachiopods. Fourth-level changes involved formation of new paleocommunities, diversification of the newly evolved clades, and increase in maximum size of brachiopods.

■ Conclusions

The Ordovician radiation of the brachiopods was dominated by the expansion of rhynchonelliformean clades (figure 17.11), the linguliformeans and craniiformeans playing more minor but nevertheless significant roles in the event. The linguliformeans were already an important part of the Cambrian fauna but generally migrated into deep-water habitats during the Ordovician, when the acrotretides became the dominant group.

The craniiformeans had a limited migrational capacity and apparently moved diachronously across the various Ordovician paleoplates and arc systems. Following a pre-Caradoc history associated with Baltica and various terranes located between Baltica and East Gondwana, the group diversified across a range of latitudes.

The two main groups of deltidiodont rhynchonelliformeans, the biconvex and pedunculate orthides together with the plano- to concavo-convex recumbent strophomenides, participated in a series of stepwise radiations. Mid Ordovician diversification involved both the displacement and expansion of many clades into deep-water habitats (Harper et al. 1999a, 1999b).

Among the cyrtomatodonts, the earliest occurrences of the orders Athyridida, Atrypida, and Spiriferida were concentrated in low-latitudinal areas between the Baltica and East Gondwana, that is, somewhere between equatorial Gondwana and the Baltica, during the Caradoc (see, e.g., Jaanusson 1979; Popov et al. 1997). This region was occupied by a number of peri-Gondwanan plates and volcanic island arcs (such as North and South China together with the Kazakhstanian terranes), providing a locus for the origin and initial radiation of the earliest spire-bearing brachipods. The rhynchonellides, however, may have originated on Laurentia, radiating during the Mid and Late Ordovician across a spectrum of shallow-water carbonate environments.

The complex taxonomic aspects of the radiation are clear signals of an equally complicated ecologic event involving the transition from the Cambrian to Paleozoic evolutionary faunas, the availability of nutrients, the diversification of predators, and life in a calcitic sea. The hypothesis that the expansion of clades prompted the occupation of ecospace by both *alpha* (within community) and *beta* (between community) diversity (Sepkoski 1988) requires rigorous investigation. Nevertheless, during the event the number of megaguilds increased while a number of new clades entered the megaguild structure. This development of a range of new adaptive strategies among the Brachiopoda markedly changed the Paleozoic seascape.

ACKNOWLEDGMENTS

Harper thanks the Danish Natural Science Research Council (SNF) for support and Anne Haastrup Hansen (Geological Museum, Copenhagen) for redrafting most of the figures. Research was also supported by grants (EAR-9706736, EAR-9910198) to Sheehan from the U.S. National Science Foundation. Popov and Bassett acknowledge support from the Royal Society and the National Museum of Wales. The chapter was improved by constructive and thoughtful reviews from Mark Patzkowsky (Pennsylvania) and Tony (A. D.) Wright (Leicester) together with editorial comments from Barry Webby.

18 Polyplacophoran and Symmetrical Univalve Mollusks

Lesley Cherns, David M. Rohr, and Jiří Frýda

Generalized taxonomic relationships and biodiversity patterns of the Ordovician representatives of two small molluscan groups—the sclerite-bearing polyplacophorans (or chitons), and the symmetrical univalves (Tryblidiida and Bellerophontida)—are discussed here.

■ Polyplacophorans (LC)

The polyplacophorans (or chitons) have a long fossil record (Cambrian–Recent) but are rare fossils and are mostly known from isolated intermediate sclerites. Head and tail plates, which commonly differ markedly in morphology, size, and ornament from the series of intermediate plates, are notably few in the fossil record. A similar complement of eight sclerites to Recent polyplacophorans is apparent from rare articulated specimens (e.g., Rolfe 1981) and also a comparable paleoecology for most in shallow marine facies (e.g., Smith and Toomey 1964; Runnegar et al. 1979; Cherns 1999). The generally poor preservation potential of rocky shore paleoenvironments probably contributes to the sparse fossil record.

The pattern of Ordovician chiton diversification is distorted by detailed records limited to only a few faunas. Those major faunas come from the Lower Ordovician of the United States and Australia (e.g., Bergenhayn 1960; Smith and Toomey 1964; Runnegar et al. 1979; Stinchcomb and Darrough 1995). Other records are fairly sparse and isolated but include significant Upper Ordovician genera from Scotland (e.g., Rolfe 1981). The phosphatic, debatably polyplacophoran *Cobcrephora* from Australia (Bischoff 1981), from an allochthonous block of probable Late Ordovician age within the Lower Devonian Cuga Burga Volcanics (I. G. Percival pers. comm.), remains a problematic taxon.

The higher taxonomy of polyplacophorans is far from settled. In the current scheme, almost all Early Paleozoic chitons are included in the subclass Paleoloricata (Cambrian–Upper Cretaceous), whose members lack an articulamentum and hence sutural laminae and insertion plates (Smith and Hoare 1987). In the subclass Neoloricata (Carboniferous–Recent), which includes all Recent chitons, only some Carboniferous lepidopleurine families lack insertion plates. Carboniferous lepidopleurid species previously assigned to *Helminthochiton* were reassigned to *Gryphochiton* because the Silurian type species *H. griffithi* and also Middle Ordovician *H.? aequivoca* apparently lack both sutural laminae and insertion plates (Smith and Hoare 1987). Hoare (2000: figure 5) proposed the paleoloricate family Helminthochitonidae, including Early to Mid Paleozoic *Helminthochiton* species, as a possible stem group for later neoloricates.

Discussion of the early history and origin of polyplacophorans centers around genera of Upper

Cambrian–Lower Ordovician multiplated organisms that are questionably chitons, such as *Matthevia*, *Hemithecella*, and *Preacanthochiton*. Lower Cambrian microgenera from China (Yu 1987) were rejected as polyplacophorans by Qian and Bengtson (1989; also Sirenko 1997; Hoare 2000). Separate higher-level molluscan taxa were proposed for *Matthevia* (Yochelson 1966) and *Hemithecella* (Stinchcomb and Darrough 1995). Stinchcomb and Darrough (1995) established several new genera for elongate sclerites associated with *Hemithecella* but wished to exclude all these and also *Preacanthochiton* (Bergenhayn 1960) from the Polyplacophora because of what they recognized as "monoplacophoran-like multiple muscle scars" (Stinchcomb and Darrough 1995:52) and some valve asymmetry. However, Runnegar et al. (1979) argued that Upper Cambrian *Matthevia* represented the earliest known polyplacophoran and that its tall conical valves became replaced by flatter, elongate, and overlapping valves through younger *Hemithecella* to *Chelodes*. *Matthevia*, *Hemithecella*, and *Preacanthochiton* are accepted here as paleoloricates (Runnegar et al. 1979; Smith and Hoare 1987; Hoare 2000), although Sirenko (1997) proposed their exclusion from his classification into two orders and eight families. Hoare's (2000) modified classification (from Sirenko 1997), including all those genera, has only four families present in the Ordovician.

Paleozoic chiton phylogeny still has much to resolve, and the recent description of a plated aplacophoran from the Silurian (Sutton et al. 2001) has further widened the debate. The data in figure 18.1 include only published species and follow Hoare's (2000) scheme except that the Septemchitonidae is restored. The elongate, robust sclerites of Tremadocian *Hemithecella*, *Conodia*, *Robustum*, and *Calceochiton*, together with *Chelodes* and *Eochelodes*, are in the broad grouping of the Mattheviidae (figure 18.1A–F). *Chelodes* is long-ranging and occurs widely in Ordovician and later Silurian rocks. Tremadocian *Preacanthochiton* (figure 18.1G), which has small conical plates with a pustular ornament, in the Preacanthochitonidae, was placed by Runnegar et al. (1979) outside its *Matthevia-Chelodes* lineage. In the Gotlandochitonidae, sclerites are wider than long, with weak to clearly defined shell areas. *Gotlandochiton*, *Paleochiton*, and *Ivoechiton* (figure 18.1H–J) occur in an important chiton assemblage from the early Arenig (Smith and Toomey 1964). The Helminthochitonidae (Sirenko 1997; figure 18.1K, L) includes *Kindbladochiton* from that same Arenig assemblage (Smith and Toomey 1964) and *Helminthochiton?*. The Septemchitonidae includes Upper Ordovician *Septemchiton* and *Solenocaris* (figure 18.1M, N), which have very elongate, narrow sclerites. *Priscochiton* (figure 18.1O), based on very limited material, lacks family assignment.

The pattern of Ordovician diversification in figure 18.1 shows that the Early Ordovician was an important period of polyplacophoran radiation, particularly on the low-latitude Laurentian margin. Upper Cambrian–lower Tremadocian assemblages of *Matthevia*, *Hemithecella*, *Preacanthochiton*, *Conodia*, and *Robustum* occur in stromatolitic, shallow marine carbonates and dolomites (Yochelson 1966; Runnegar et al. 1979; Stinchcomb and Darrough 1995). Similar paleoenvironments were associated with Smith and Toomey's (1964) diverse lower Arenig Laurentian assemblage and on the Gondwanan margin for *Chelodes whitehousei* from Australia (Runnegar et al. 1979). The Lower Ordovician shows the highest diversity of chitons with 10 genera and 20 species. This peak of diversity compares with the initial radiation of bivalves, except that chitons were diversifying on the low-latitude Laurentian carbonate margin, whereas bivalves were confined to higher latitudes and siliciclastic-dominated settings on Gondwana and Avalonia (e.g., chapter 20).

The data from the Middle Ordovician are currently restricted to very few, isolated records. *Chelodes? mirabilis* from Alabama represents Laurentia, and *Helminthochiton? aequivoca* from Bohemia represents a peri-Gondwanan terrane, Perunica (chapter 5).

The Upper Ordovician shows more diversity, with five genera and nine species. *Solenocaris* includes the first record from Baltica. Laurentian chitons are represented by *Solenocaris*, *Chelodes*, and *Septemchiton* from the Midland Valley of Scotland, the last two also from the United States. *Eochelodes* from Bohemia represents Perunica. Isolated and localized occurrences preclude useful paleoenvironmental interpretation. There are no records of polyplacophorans from late Ashgill rocks.

The Silurian record of polyplacophorans shows diverse assemblages particularly from shallow carbonates on Gotland (Bergenhayn 1955; Cherns 1998a,

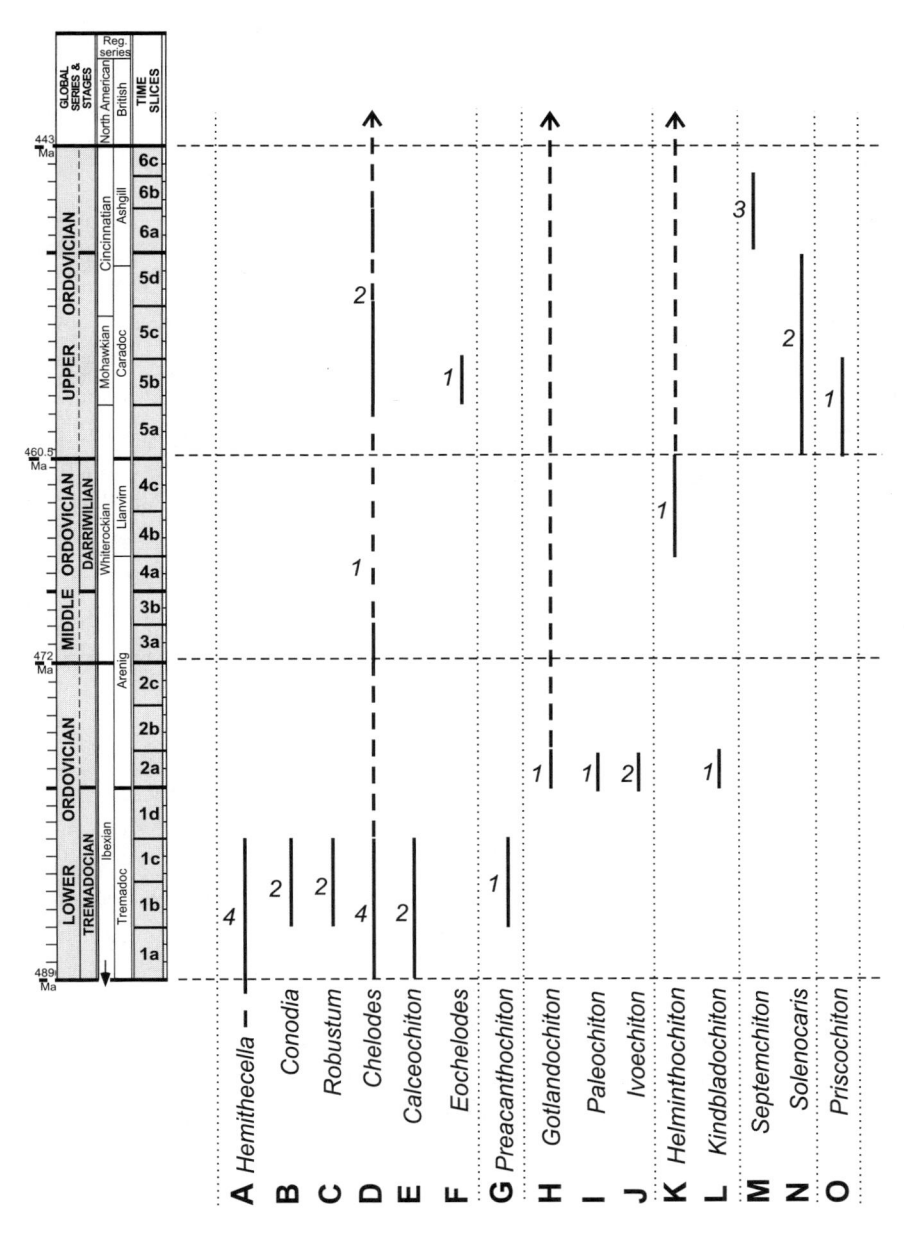

FIGURE 18.1. Range chart for genera of Ordovician polyplacophorans. Information on ranges is given as accurately as possible, for most genera to one time slice (*TS*), and ranges are shown to time-slice boundaries. Numbers against each line indicate species recognized for Lower, Middle, and Upper Ordovician parts of ranges, as appropriate. Dotted lines signify gaps within ranges through which a genus can be assumed to continue. Family assignments: A–F, Mattheviidae; G, Preacanthochitonidae; H–J, Gotlandochitonidae, K–L, Helminthochitonidae; M–N, Septemchitonidae; O, indeterminate family.

Tryblidiids and Bellerophontids (DMR, JF)

Paleozoic symmetrical univalved mollusks (Tryblidiida and Bellerophontida) are one of the most discussed groups of the phylum Mollusca. Correct evaluation of phylogenetic relations of these ancient mollusks to other molluscan groups is crucial for our understanding of the evolution of all molluscan classes of the subphylum Cyrtosoma. Opinions on their phylogenetic relations were recently analyzed in several papers (e.g., Starobogatov 1970; Peel 1991a, 1999b; Wahlman 1992; Geyer 1994). However, these analyses resulted in quite different phylogenetic models and caused the erection of many new order- and class-level names. In addition, their subsequent usage and generic content have often been very different, and they have disregarded their original diagnoses (e.g., Monoplacophora and Tergomya). Tryblidiida with cap-shaped shells and isostrophically coiled Bellerophontida belong to the largest groups among the Ordovician symmetrical univalved mollusks. In

1998b). Though still rare, these chiton-rich assemblages emphasize the bias thrown on analyses of diversity where the fossil record otherwise is very sparse.

addition, several minor molluscan groups such as Helcionelloida (= Eomonoplacophora), Cyrtolitida, and Pelagiellida are known also from Ordovician strata and probably played a significant role in early molluscan phylogeny (see Starobogatov 1970; Geyer 1994; and Gubanov and Peel 2000 for discussion).

Because of the absence of a generally accepted classification of the Paleozoic symmetrical univalved mollusks, we analyzed only the two largest Ordovician groups, Tryblidiida and Bellerophontida, which can be easily distinguished by their shell morphologies. We (DMR, JF) feel confident that both groups are probably polyphyletic as suggested by many authors (see later in this chapter and chapter 19). Nevertheless, their more detailed analysis is impossible without future reevaluation of all available data as well as reconsideration of the significance of some characters (e.g., muscle scars). The main goal here is to illustrate the Ordovician biodiversity patterns of Tryblidiida and Bellerophontida and to point out some problems in the evaluation of their phylogenetic relationships. Refer to the next chapter in this volume on the Gastropoda for the definitions of diversity parameters.

Tryblidiida (= "Monoplacophora")

The Ordovician groups placed in Tryblidiida by Geyer (1994) as well as Achinacelloidea are united here in one group. The systematic position of the latter group is still uncertain, and it has been considered to belong to the class Gastropoda (e.g., Knight and Yochelson 1958; Peel and Horný 1999) or to the "Monoplacophora" (Horný 1963; Runnegar and Jell 1976). At present there is no reliable evidence about whether members of both these groups, Tryblidiida and Achinacelloidea, are torted or untorted. As correctly noted by Harper and Rollins (1982, 2000) and Wanninger et al. (2000), the usage of muscle scar patterns as a key for determination of torsion in fossil mollusks has no basis in fact. In addition, the belief that symmetrically arranged multiple muscle scars are evidence for symmetrical, segmented anatomic arrangement of the shell body is false. These two speculations were evoked by discoveries of living "monoplacophorans" (Neopilinidae), which seem to confirm a theoretical concept of a stem group for the conchiferan mollusks as expressed by Odhner (in Wenz 1940). These facts also opened a controversial debate over which characters of neopilinid anatomy are derived and which are primary (see Haszprunar and Schaefer 1997 and references therein). Anatomy of the modern Neopilinidae seems to be a result of ancient molluscan organization as well as specialization. Interpretation of the body plan in the Paleozoic Tryblidioidae as being closely similar to that of the living Neopilinidae is thus problematic. Moreover, the integration of both groups into one natural group is uncertain because there is no reliable evidence that they are really closely related. As shown earlier in this chapter, we are very far from being able to determine natural groups within the Ordovician mollusks with cap-shaped shells.

The Ordovician Tryblidiida are known from equatorial as well as high-latitude regions. The majority of genera were described from the paleotropical realm (Laurentia, Baltica, and Siberia), and they had very limited geographic ranges.

The quantitative diversity pattern of the Ordovician Tryblidiida cannot be described in detail because of a lack of detailed stratigraphic data for the majority (mainly Siberia and Baltica). Their genera are often based on limited material (sometimes even on a single shell with well-preserved muscle scars). The Ordovician Tryblidiida reached their highest diversity during the Early Ordovician (18 genera), and most genera were restricted to this time interval. During the Mid and Late Ordovician their diversity is slightly lower (14 and 13 genera, respectively). Only about one-quarter of the genera crossed the Ordovician/Silurian boundary.

Bellerophontida (Amphigastropoda)

The class-level assignment of bellerophontoidean mollusks has often been discussed during the past 50 years. Considerations about the morphology of soft parts of these coiled, bilaterally symmetrical mollusks divided paleontologists into several groups. The question essentially resolves to whether all or part or none of the bellerophontiform mollusks were untorted, exogastrically oriented monoplacophorans or torted, endogastrically oriented gastropods (see Yochelson 1967; Harper and Rollins 1982, 2000; Peel 1991a, 1991b; Wahlman 1992; Frýda 1999b, for review). Recently Frýda (1999b) described a multiwhorled

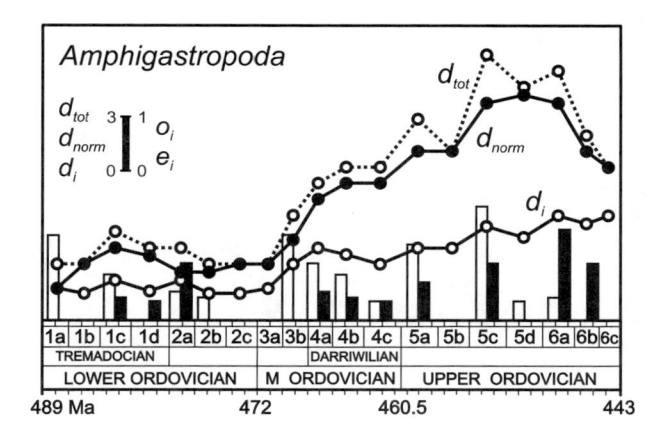

FIGURE 18.2. Plot illustrating generic biodiversity and turnover rates of Amphigastropoda (bellerophontiform mollusks). Lines show the biodiversity (d_{tot} = total diversity; d_{norm} = normalized biodiversity, and d_i = number of genera per m.y.). Bars show turnover rates (white = rate of originations per m.y. [o_i], black = rate of extinctions per m.y. [e_i]). For definitions of the parameters used, see chapter 19.

protoconch in the type genus *Bellerophon* and interpreted it as a true larval shell. This fact suggests that this genus (and thus, also the Bellerophontoidea, Amphigastropoda) does not belong to the Archaeogastropoda (Frýda 1999c, 2001). On the other hand, some new data on the protoconch morphology indicate that the Paleozoic bellerophontiform mollusks represent a polyphyletic group (Frýda 1999c). Recently Harper and Rollins (2000) concluded that the bellerophontoideans and the coiled and high-domed "monoplacophorans" were gastropods.

The Early Paleozoic bellerophontoidean mollusks were a very successful group and occupied a wide range of depositional environments (Peel 1977, 1978; Wahlman 1992). In his excellent monograph, Wahlman (1992) summarized knowledge of the paleoecology of Ordovician bellerophontoidean mollusks and clearly showed that they led a variety of modes of life. This fact probably explains their cosmopolitan distribution.

The oldest bellerophontoidean mollusks (family Sinuitidae) are known from Upper Cambrian strata. Their generic diversity was relatively low until the early Mid Ordovician. From approximately the beginning of the Darriwilian (*TS*.3b and 4a; see time slices in chapter 2), the number of genera per million years (m.y.) (d_i) continually increased to the end of the Ordovician (figure 18.2). The mid Late Ordovician (*TS*.5c–6a) was the time of the highest diversity of the Ordovician bellerophontoidean mollusks (d_{norm} close to 14). The rate of originations per m.y. (o_i) was variable through the Ordovician with distinct peaks at the early Tremadocian (*TS*.1a), early Mid Ordovician (*TS*.3b), and early (*TS*.5a) and mid (*TS*.5c) Late Ordovician (figure 18.2). The rate of extinctions per m.y. (e_i) reached the highest values at the beginning of Arenig (*TS*.2a) and during the Ashgill (*TS*.6a and 6b). At present it is difficult to make a more detailed analysis of the turnover rates because the bellerophontoidean mollusks probably represent a polyphyletic group (see earlier in this chapter).

It is concluded that the Ordovician symmetrical univalved mollusks form a relatively large group (about 75 genera), even though the majority of them are imperfectly known. They include several morphological groups (see Wahlman 1992 and Geyer 1994 for review) for which neither the phylogenetic relationships nor their relationships to extant molluscan groups are known. Biodiversity of the Tryblidiida seems to have decreased slightly during the Ordovician, and it was strongly affected by the extinction event close to the Ordovician/Silurian boundary. On the other hand, the number of the genera belonging to the Bellerophontida increased through the Ordovician. Their diversity pattern shows similarities with some gastropod groups (see chapter 19).

19 Gastropods

Jiří Frýda and David M. Rohr

Gastropods are one of the most diverse groups of animals and include more than 100,000 living species. The number of fossil species must have been much higher, and estimates of gastropod diversity suggest about 13,000 extant and fossil genera (Bieler 1992). Gastropods' rich and long-ranging fossil record (more than 500 million years [m.y.]), coupled with their occurrence in almost all marine, freshwater, and terrestrial environments, makes them a unique animal group for evolutionary, ecologic, and biogeographic investigations. Even the early evolution of the class Gastropoda seems to be well documented by rich fossil material from Ordovician strata. However, attempting a more detailed analysis of Paleozoic gastropods has revealed quite a serious problem, the core of which lies in the limited compatibility between neontological and paleontological taxonomic systems of the class Gastropoda. In contrast to the classification of living gastropods, that of fossil gastropods is based on only a limited number of characters, observable in their fossilized hard body parts. Correct determination of phylogenetic relationships in fossil gastropods decreases considerably with increase in their geologic age because of their uncertain relationships to modern gastropod groups and the frequent homoplastic similarity of their shells. Thus, it is quite evident that paleontologists, having only the possibility to evaluate a limited number of shell features, have very often worked with artificial groups, which in reality do not illustrate the natural phylogenetic relationships among fossil gastropods.

Higher classification of Ordovician gastropods has traditionally been based only on evaluation of their teleoconch characters (Wenz 1938–1944; Knight et al. 1960; Pchelintsev and Korobkov 1960). Certainly the teleoconch characters bear some phylogenetic signals, but because of frequent homoplasy, these features cannot produce a reliable evaluation of the higher taxonomic position of the fossil gastropods. Modern evaluation of the gastropod fossil record is also beginning to use high-level taxonomically significant characters such as the pattern of the gastropods' early ontogeny, which has been reflected in their protoconch morphology. However, some recent discoveries such as the presence of openly coiled protoconchs in some Paleozoic gastropod groups (Hynda 1986; Dzik 1994c; Frýda and Manda 1997; Bandel and Frýda 1998, 1999; Frýda 1999a, 1999b; and Nützel et al. 2000 and references therein) seem to complicate understanding of gastropods' early evolution. Another very serious problem is the lack of modern data about their stratigraphic ranges. Many large Ordovician gastropod faunas have not been studied, or they rely on descriptions that were published about 100 years ago. These facts considerably limit analysis of the biodiversity of Ordovician gastropods. In this chapter, we analyze their generic diversity and discuss definition, phylogenetic relationships, and paleobiogeography of the major groups of Ordovician gastropods in order to illustrate the diversity patterns at higher taxonomic levels.

■ Diversity Measures

As shown in detail later in this chapter, there are many unsolved problems in the higher taxonomy of Ordovician gastropods. Classifications inferred only from teleoconch characters do not fit well together. For example, compare the commonly used alpha taxonomic approach of Knight et al. (1960) and Pchelintsev and Korobkov (1960) with results of the phylogenetic analyses of Wagner (1995a, 1995a). In addition, there is considerable inconsistency between the classifications inferred from teleoconch and protoconch characters (Frýda 1999b). Because of the lack of a generally accepted classification of Ordovician gastropods, a relatively simple method is chosen for the description of the diversity patterns. The analysis is based on generic taxa, and only generally accepted higher taxa are used. Modifications of generic content of some higher taxa used for our analysis are mentioned in the text.

There are many possible ways to measure biodiversity of fossil animals. Selection of appropriate parameters for quantification of their diversity depends on the longevity of the taxa relative to the duration of the time units used (e.g., Archibald 1993; Cooper 1999c). The present analysis of Ordovician gastropod generic diversity is based on more than 2,500 records of Ordovician gastropods belonging to about 140 genera. Nineteen time slices (*TS*) through the Ordovician are used in order to compare our results directly with those of other clades, even though this seems to be too detailed for generic analysis. In general, most Ordovician gastropod genera have ranges much longer (average = 14 m.y.; median = 12 m.y.) than the 1.5–3 m.y. durations of the time slices used. Only about 6 percent of the Ordovician gastropod genera had ranges equal to or shorter than those of the time slices. For these reasons the following parameters were used for the analysis of biodiversity: (a) *total diversity* (d_{tot}), defined as the total number of genera that are recorded from the time slice; (b) *normalized diversity* (d_{norm}), defined as the number of gastropod genera ranging through the time slice plus half the number of genera confined to the slice or ranging beyond the time slice but originating or ending within it; and (c) *genera per m.y.* (d_i), defined as the number of gastropod genera present within the time slice divided by its duration. Because of longevity of Ordovician gastropod genera, the d_{norm} values are very close to the d_{tot} values. Turnover rates were characterized by two parameters: (a) *rate of originations* (o_i) *per m.y.* or *rate of extinction* (e_i) *per m.y.*, defined as the total number of genera originating or going extinct within the time slice divided by its duration, and (b) *mean per capita origination* (o_{di}) or *extinction* (e_{di}) *rates,* defined as the number of originations or extinctions divided by the total number of genera present and by the duration of the time slice in m.y.

■ Classification

Phylogenetic relationships among fossil and living gastropod higher taxa are a widely discussed problem in recent paleontological and neontological literature (Golikov and Starobogatov 1975, 1988; Salvini-Plawen 1980, 1990; McLean 1981, 1984; Yochelson 1984; Haszprunar 1988, 1993; Ponder and Warén 1988; Peel 1991a, 1991b; Wagner 1995a; Biggelaar and Haszprunar 1996; Bandel 1997; Nützel 1997, 2002; Ponder and Lindberg 1997; Bandel and Frýda 1998, 1999; Frýda and Blodgett 2001, references therein). During the past 20 years great progress in the evaluation of phylogenetic relationships of modern gastropod groups has been made (see Ponder and Lindberg 1997 for review). This progress, based on evaluation of a large number of morphological and anatomic data, has brought relative taxonomic stability to the class Gastropoda. The class may be subdivided (figure 19.1) into Patellogastropoda (= Docoglossa), Archaeogastropoda sensu stricto, Neritimorpha, Caenogastropoda, and Heterobranchia, which are diagnosed by many synapomorphies (see Ponder and Lindberg 1997 for references). The validity of some of these groups is also strongly supported by recent studies based on molecular data (Colgan et al. 2000). Unfortunately, diagnostic features of these higher taxa involve mainly soft-body characters, which could not be used in the fossil record. Thus, there is a relatively stable taxonomic system for living gastropods, which has very limited application to the more numerous fossil gastropods. Fortunately, recent neontological studies have shown that some features

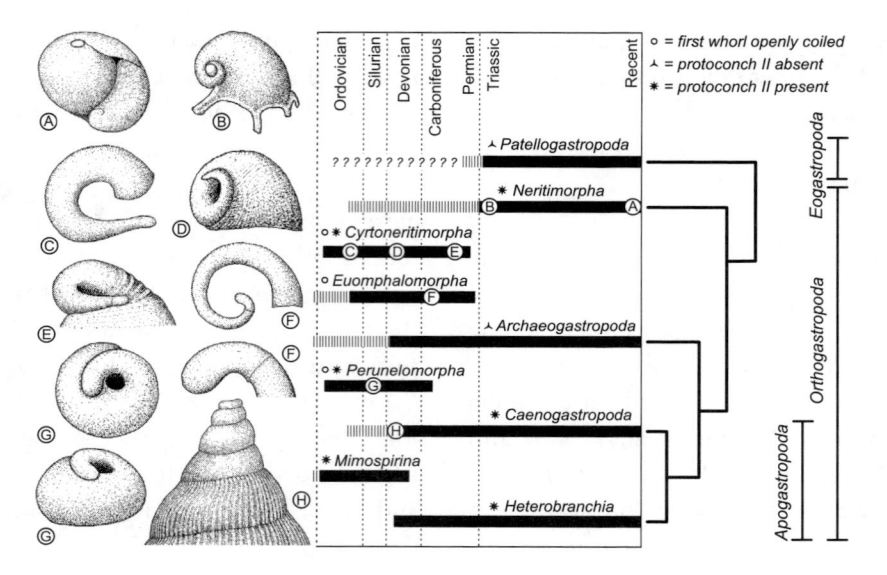

FIGURE 19.1. Stratigraphic ranges of main gastropod groups based on protoconch morphology (black bars). Shaded bars show stratigraphic ranges inferred from teleoconch features. Presumed phylogenetic relationships of the higher taxa of gastropods are shown on the right side (see text). Characteristic protoconchs are drawn on the left side. A, larval shell of modern members of the Neritidae. B, larva shell of Late Triassic *Pseudorthonychia* Bandel and Frýda 1999. C, Late Ordovician cyrtoneritimorph larval shell. D, Early Devonian cyrtoneritimorph *Vltaviela* Frýda and Manda 1997. E, larval shell of the Carboniferous *Orthonychia* Hall 1843. F, euomphalomorph protoconch of the Early Carboniferous *Serpulospira* Cossmann 1916. G, Late Silurian perunelomorph larval shell. H, Early Devonian subulitid larval shell. (Sketches made according to Dzik 1994c; Yoo 1994; Frýda and Manda 1997; Bandel and Frýda 1999; Frýda 1999b, 2001.)

such as the nature of early shell development may be applied even in fossil gastropods. Some of modern higher taxa have been identified by their typical protoconch morphology even in Paleozoic strata (see Frýda 1999b, 2001; Nützel et al. 2000; and Frýda and Blodgett 2001 for reviews). However, during Early and Mid Paleozoic time there were also several gastropod groups such as the Bellerophontoidea, Macluritoidea, Euomphaloidea, Peruneloidea, Platyceratoidea, Subulitoidea, Helcionelloida, Cyrtoneritimorpha, and Mimospirina, which are not easily connected with modern gastropod higher taxa.

■ Major Groups

The last complete revision of high-level classification of Paleozoic gastropods (Knight et al. 1960) in many cases followed the older classification of Wenz (1938–1944) and placed the great majority of the Ordovician gastropods in the Archaeogastropoda Thiele 1925. Only two superfamilies, Loxonematoidea Koken 1889 and Subulitoidae Lindström 1884, were placed in the Caenogastropoda Cox 1959. The present state of knowledge regarding the definition, phylogenetic relationships, and paleobiogeography of the major groups of Ordovician gastropods is discussed in the following sections. Such an approach allows us to make a more stable analysis of their diversity patterns.

Patellogastropoda (Eogastropoda)

Ponder and Lindberg (1995) divided the class Gastropoda into two groups, Eogastropoda and Orthogastropoda. Eogastropoda, including Patellogastropoda (= Docoglossa), have been considered to represent the first gastropod offshoot (see Ponder and Lindberg 1995, 1997, and references therein). However, there is no undoubted evidence for Paleozoic Patellogastropoda. Until now, the oldest known limpet belonging without doubt to the Patellogastropoda comes from Triassic strata (Hedegaard et al. 1997). Considerations about the nature of shell in the Paleozoic Patellogastropoda divide neontologists and paleontologists into two groups, who dispute whether the ancestors of the Patellogastropoda had anisostrophically coiled shells or bilaterally symmetrical limpet shells like their post-Paleozoic descendants (see Ponder and Lindberg 1997). Limpetlike shell morphology was evolved independently among numerous living and fossil groups of Gastropoda, as well as in the Tryblidiida (= Monoplacophora). Thus, it is almost impossible to determine the higher taxonomic position of limpetlike fossil gastropods only on the basis of their teleoconch characters. Some Paleozoic limpets belong to the extinct group Cyrtoneritimorpha (Neritimorpha), with a typical fish-hooklike protoconch (Bandel and Frýda 1999; Frýda 1999b), and to the Pragoscutulidae Frýda 1998 (?Caenogastropoda; Frýda 2001), but the higher taxonomic position of the remaining majority of Paleozoic limpets is still uncertain. Many may belong to the class Tryblidiida (= Monoplacophora). Even if the Paleozoic ancestors of Patellogastropoda had a coiled shell, they still could not be recognized among Paleozoic gastropods.

Archaeogastropoda

The concept of the Archaeogastropoda has been changed many times, and different usages may be found even in the most recent studies (e.g., Golikov and Starobogatov 1975; McLean 1981; Bandel 1982; Salvini-Plawen and Haszprunar 1987; Hickman 1988; Haszprunar 1993; Biggelaar and Haszprunar 1996; and Ponder and Lindberg 1997 and references therein). Among the Paleozoic gastropods, the Archaeogastropoda have been considered to be the most commonly occurring gastropod group (figure 19.2C–H). The oldest presumed Archaeogastropoda based on teleoconch morphology are known from the Late Cambrian (Knight et al. 1960; Tracey et al. 1993). The oldest evidence for their occurrence based on the typical protoconch type comes from the Early Devonian (figure 19.1; Frýda and Bandel 1997; Frýda and Manda 1997). As mentioned by several authors (see Tracey et al. 1993 and Bandel and Frýda 1996 for discussion), family-level classification of the Archaeogastropoda requires complete revision and therefore is not used here.

As a dominant group within the Early Paleozoic Gastropoda, the Ordovician Archaeogastropoda had a cosmopolitan distribution. In contrast to modern slit-bearing Archaeogastropoda, their Paleozoic representatives occupied a wide range of depositional environments (Peel 1984).

To facilitate analysis of diversity patterns and turnover rates within Ordovician Archaeogastropoda, its component genera are divided into two groups: (1) slit- or sinus-bearing Archaeogastropoda (or "Selenimorpha") (figure 19.2C–F) and (2) those without this apertural feature (or "Trochomorpha") (figure 19.2G, H). Ordovician gastropod genera placed by Knight et al. (1960) in any of 22 families of the superfamily Pleurotomarioidea Swainson 1840 are included in the first group together with genera assigned to the superfamily Murchisonioidea Koken 1896. The latter superfamily, which is very common in the Paleozoic, has variously been classified as belonging to the Archaeogastropoda (Knight et al. 1960) or Caenogastropoda (Ponder and Warén 1988) or as a group with uncertain higher position (Tracey et al. 1993). Yochelson (1984) suggested placing the Murchisonioidea close to the Pleurotomarioidea. Recent discoveries of well-preserved protoconchs in Early and Mid Devonian murchisonioideans, including the type genus *Murchisonia,* provided undoubted evidence that these gastropods belong to the Archaeogastropoda (Frýda and Manda 1997). The second analyzed group (Archaeogastropoda without apertural sinus or slit) unites Ordovician genera placed by Knight et al. (1960) in Trochonematoidea Zittel 1895, Euomphalopteridae Koken 1896, Holopeidae Wenz 1938, and Microdomatoidea Wenz 1938. Placement of the last two groups within the Archaeogastropoda was supported by protoconch morphology in some of their Mid and Late Paleozoic members (Frýda and Bandel 1997; Bandel and Nützel pers. comm.).

Slit-Bearing Archaeogastropoda ("Selenimorpha")

The oldest slit- or sinus-bearing Archaeogastropoda are known from the latest Cambrian (Tracey et al. 1993). Their generic diversity (figure 19.3B) increased moderately from the beginning of the Ordovician until the Mid to Late Ordovician (*TS*.4c–5a), when the generic diversity increased at a much greater rate. The normalized diversity (d_{norm}) reaches its highest values (figure 19.3B) during the late Late Ordovician (*TS*.6a; d_{norm} close to 20). The number of genera per m.y. (d_i) increased during the Ordovician and reached its maximum value (about 12) at the end of the Ordovician. Rate of originations (o_i)

FIGURE 19.2. Examples of main groups of Ordovician gastropods and bellerophontiform mollusks. A, B, *Tropidodiscus*, side views, ×2, late Early Ordovician, Missouri—bellerophontiform mollusks (Amphigastropoda). C, D, *Trochonemella*, ×1, views of sinus and selenizone, Late Ordovician, Alaska, USNM 422342 and 422347; E, *Ectomaria*, ×2.6, Late Ordovician, Alaska, USNM 422360; F, *Ectomaria*, ×2, Late Ordovician, Alaska, USNM422363. C, D, E, and F are examples of the Archaeogastropoda ("Selenimorpha"). G, H, *Slehoferia*, top and apertural views, ×4, Caradoc, Czech Republic, YA 2602—Archaeogastropoda ("Trochomorpha"). I, *Angulospira*, ×2, Whiterockian, Nevada, USNM 485257—Mimospirina. J, K, *Lytospira*, upper and lower surfaces, ×1.3, Whiterockian, USNM 473944; L–N, *Malayaspira*, ×1.3, Arenig, Malaysia, USNM topotype 473953. J, K, and L–N are examples of the Euomphaloidea (Euomphalomorpha). O, P, *Monitorella*, apertural and top views ×1, and Q, interior of operculum, ×1.3, Whiterockian, Newfoundland, NFM 321-322—Macluritoidea. R, S, *Cyclonema*, apertural and side views, ×0.7, Late Ordovician, Tennessee—Neritimorpha (Platyceratidae). T, *Subulites (Fusispira)*, apertural view, ×1.6, Late Ordovician, Alaska, USNM 422371—Caenogastropoda (Subulitoidea). USNM = U.S. National Museum, Washington, D.C.; NFM = Newfoundland Museum, Canada; YA = Czech Geological Survey, Prague.

reached its maximum values close to the Tremadocian-Arenig boundary (*TS*.2a), early Mid Ordovician (*TS*.3b), and during the mid Caradoc (*TS*.5b–c). Rate of extinctions (e_i) was relatively high between the mid Early Ordovician (*TS*.1d) and early Mid Ordovician (*TS*.3b), as well as during the Ashgill (*TS*.6a–c). The Late Ordovician may be characterized as a period of fast radiation of the slit-bearing Archaeogastropoda. The number of genera doubled from the early Late Ordovician (*TS*.5b) to the late Late Ordovician (*TS*.6a). This observation fits well with the fact that the slit-bearing Archaeogastropoda are a dominant gastropod group in Silurian strata.

Archaeogastropoda Without Slit ("Trochomorpha")

In contrast to the "Selenimorpha," there is no reliable evidence for trochomorph Archaeogastropoda in Lower Ordovician strata. During the Mid Ordovician the "Trochomorpha" had low diversity until the beginning of the Late Ordovician, when the number of genera started to increase rapidly, reaching their high-

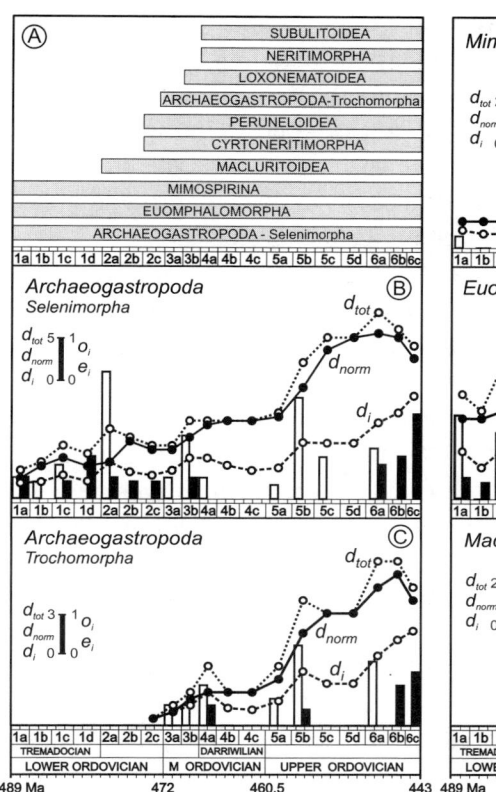

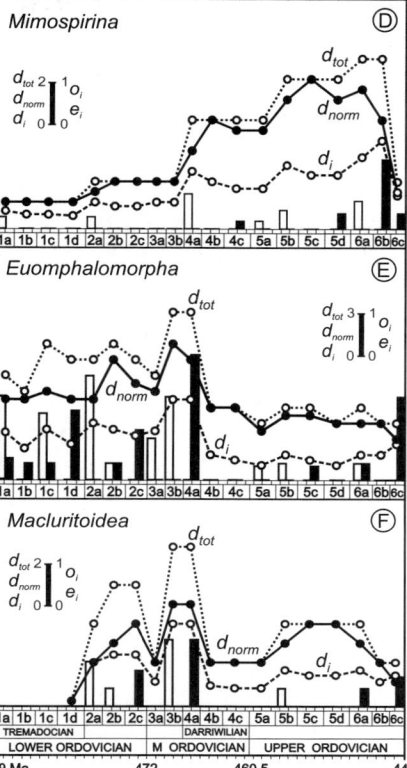

FIGURE 19.3. A, stratigraphic ranges of the main Ordovician gastropod groups. B–F, generic biodiversity and origination and extinction rates of Archaeogastropoda, Mimospirina, Euomphalomorpha, and Macluritoidea. Lines show the diversity (d_{tot} = total diversity, d_{norm} = normalized diversity; and d_i = number of genera per m.y.). Bars show origination and extinction rates (white = rate of originations [o_i], black = rate of extinctions [e_i]). For definition of these parameters, see text.

est value in the late Late Ordovician (figure 19.3C). Arenig (TS.3a–4a), early Late Ordovician (TS.5a–b), and early Ashgill (TS.6a) are time intervals with the highest origination rate. The pattern of diversification in Mid and Late Ordovician "Trochomorpha" fits well with that of the "Selenimorpha."

Mimospirina

Mimospirina was established by Dzik (1983) as a new suborder of the Archaeogastropoda uniting members of the extinct families Clisospiridae and Onychochilidae (figure 19.2I). Opinions on the higher taxonomic position of both families have often changed (see Frýda 1992 and Frýda and Rohr 1999 for review). McLean (1981) suggested that the members of the superfamilies Macluritoidea and Clisospiroidea do not belong to his suborder Euomphalina but represent lineages apart from this group. On the other hand, Linsley and Kier (1984), on the basis of a functional analysis, proposed uniting the Clisospiroidea, Macluritoidea, and possibly the Euomphaloidea into their new class Paragastropoda. The class Paragastropoda was considered to represent untorted mollusks.

Discoveries during the past decade revealed new data on the early shell morphology of several gastropod groups, which have enabled a reevaluation of the phylogenetic relationships of the Mimospirina. Quite differing protoconch morphology of members of the Clisospiroidea (Dzik 1983; Frýda and Rohr 1999) and Euomphaloidea (Yoo 1994; Bandel and Frýda 1998; Nützel 2002) suggests that the class Paragastropoda is an artificial group with no zoological validity (Frýda 1999b). These facts also support Dzik's (1983) concept of the Mimospirina. The multiwhorled protoconch in *Mimospira* represents a true larval shell (protoconch II), and its large size and shape suggest the presence of a nonplanktotrophic larval stage during which the larval shell was formed. This interpretation argues against the inclusion of Mimospirina within the Archaeogastropoda (Frýda 1999b, 2001).

The Mimospirina are relatively rare in Cambrian-Devonian marine communities. The highest diversity of the Ordovician clisospirids and onychochilids is found in the Baltic fauna. Paleogeographic distribution of the Ordovician Mimospirina was summarized by Frýda and Rohr (1999).

The generic diversity of the Mimospirina was relatively low up to the early Mid Ordovician (*TS*.3b). From approximately the beginning of the Darriwilian (*TS*.4a), the normalized diversity (d_{norm}) started to increase strongly, attaining its highest value in the middle of the Late Ordovician (figure 19.3D). The number of genera per m.y. (d_i) increased during the Ordovician and reached its maximum value in the later Late Ordovician (*TS*.6b). The latest Ordovician (*TS*.6c) represents a time of sharp decrease in both normalized diversity and number of genera per m.y. The rate of originations (o_i) is highest at the beginning of the Darriwilian (*TS*.4a) and in the late Late Ordovician (*TS*.6a), followed by a distinct extinction event in the latest Ordovician (*TS*.6b–c).

Euomphaloidea (Euomphalomorpha)

Yochelson (1956) interpreted the Euomphaloidea (figure 19.2J–M) as derived from the Macluritoidea in Early Ordovician time and assigned three families (Euomphalidae Helicotomidae, and Omphalotrochidae) to this superfamily. This concept was followed by Knight et al. (1960) and many paleontologists subsequently. McLean (1981) considered the deep-sea hot vent limpet *Neomphalus* McLean 1981 to be related to extinct Euomphaloidea. For this reason, McLean (1981) united members of the modern Neomphaloidea with the Paleozoic Euomphaloidea and placed them in the Archaeogastropoda. Yoo (1994) and Bandel and Frýda (1998) found an unusual protoconch morphology in some Devonian and Carboniferous euomphaloidean genera, which distinguishes them from members of extant gastropod groups (Patellogastropoda, Archaeogastropoda, Neritimorpha, Caenogastropoda, and Heterobranchia). For this reason, Bandel and Frýda (1998) established a new taxon, Euomphalomorpha, which is considered to represent an independent gastropod group, known only from the Paleozoic (Cambrian–Permian). Recently, Nützel (2002) confirmed earlier observations (Yoo 1994; Bandel and Frýda 1998) on the nature of the protoconch and shape of the boundary between protoconch and teleoconch. He also noted some affinities among Euomphaloidea, Docoglossa, Cocculiniformia, and Neomphalidae. In the Ordovician Euomphalomorpha we include the following families: Euomphalidae de Koninck 1881, Helicotomidae Wenz 1938, and Ophiletidae Knight 1956.

The Euomphalomorpha are one of the dominant groups of Paleozoic gastropods. During Ordovician time they are known mainly from paleotropical regions. Their highest biodiversity is found in carbonate-dominated facies of Laurentia and Baltica.

Euomphaloidea (figures 19.1, 19.3E) are known from the Late Cambrian to the Late Permian. During the Early and early Mid Ordovician the number of genera per m.y. (d_i) slowly increased and reached its maximum value in the early Mid Ordovician (*TS*.3b and 4a). Subsequently, the number of general per m.y. dropped (*TS*.4b) and remained approximately constant until the end of the Ordovician (figure 19.3). The curve of the normalized diversity (d_{norm}) shows two peaks (early Arenig, *TS*.2b, and early Mid Ordovician, *TS*.3b–4a). Both the rates of originations (o_i) and extinctions (e_i) are generally higher during the Early and early Mid Ordovician than in the later Ordovician (figure 19.3E). The beginning of Arenig (*TS*.2a) and its mid to later part (*TS*.3a–b) are times with the highest origination rates. On the other hand, the rate of extinctions (e_i) is highest during the early Darriwilian (*TS*.4a) and latest Ordovician.

Macluritoidea

Although Macluritoidea (figure 19.2N, O) belong to the best-known group of Ordovician gastropods (Rohr 1979, 1994; Rohr and Gubanov 1997), little is known about their higher taxonomic position. Knight et al. (1960) placed Macluritoidea Fisher 1885, together with Euomphaloidea de Koninck 1881, in the suborder Macluritina Cox and Knight 1960 of the Archaeogastropoda. The concept of Macluritina in uniting Macluritoidea and Euomphaloidea was later criticized by several authors (Morris and Cleevely 1981; Dzik 1983; Linsley and Kier 1984; Yochelson 1984).

As shown by numerous studies (Rohr 1979; Blodgett et al. 1987; Rohr et al. 1992; Gubanov and Rohr 1995; Gubanov and Tait 1998), Macluritoidea lived in warm, shallow marine waters in both carbonate and siliciclastic facies and are known from all Ordovician paleocontinents situated in tropical regions. More detailed data on the paleogeographic distribution of the Ordovician Macluritoidea may be found in Gubanov and Rohr (1995).

The oldest Macluritoidea (in the sense of Gubanov and Rohr 1995) are known from the early Arenig (*TS*.2a). They are restricted to the Ordovician and reached their highest diversity in the early Mid Ordovician (*TS*.3b–4a). In the mid Darriwilian (figure 21.3F) their generic diversity rapidly dropped and remained low until the Ashgill, when the Macluritoidea became extinct. The pattern of diversification of the Macluritoidea resembles that of the Ordovician Euomphaloidea.

Neritimorpha

Bandel (1982) pointed out that the Neritimorpha Golikov and Starobogatov 1975 represent an independent gastropod group characterized by unique protoconch morphology. This opinion seems to be supported by the study of gastropod cleavage patterns (Biggelaar and Haszprunar 1996) and by cladistic analysis of a large number of morphological and anatomic data of modern gastropods (Ponder and Lindberg 1997). According to the latter authors, the Neritimorpha represent a sister group to Archaeogastropoda and Apogastropoda (Caenogastropoda and Heterobranchia). Besides several synapomorphies including mainly soft body characters (see Ponder and Lindberg 1997), the modern and fossil Neritimorpha may be characterized by a typical, strongly convolute protoconch (Bandel 1982). The oldest undoubted evidence for this protoconch type is known from Triassic strata (Bandel and Frýda 1999). Presumed Paleozoic members of Neritimorpha (see Knight et al. 1960) have two different types of protoconch (figure 19.1): (1) closely coiled (but not convolute) protoconch (e.g., Platyceratidae, Plagiothyridae, Naticopsidae, and Oriostomatoidea), and (2) openly coiled, fishhooklike protoconch (Vltaviellidae and Orthonychiidae). The first group may represent ancestors of modern Neritimorpha (Cycloneritimorpha). Presumed Paleozoic members of Neritimorpha developing the fishhooklike larval shell (figure 19.1) were united into the taxon Cyrtoneritimorpha (Frýda 1998, 1999b; Bandel and Frýda 1999; Frýda and Heidelberger 2003).

The Ordovician Neritimorpha (figure 19.2P, Q) are known from tropical as well as high-latitude regions. Ordovician Cyrtoneritimorpha are currently known only from Baltica (Bockelie and Yochelson 1979; Dzik 1994c).

On the basis of their teleoconch shape, the first presumed Neritimorpha (Platyceratidae and Oriostomatidae) appeared during the Mid Ordovician. On the other hand, isolated protoconchs of the Cyrtoneritimorpha are known in upper Lower Ordovician strata (Bockelie and Yochelson 1979). Dzik (1994c: figure 22a) illustrated cyrtoneritimorph protoconchs also in Middle Ordovician strata. Thus, the existence of Cyrtoneritimorpha in the Ordovician is beyond doubt, but there is no certain record of Neritimorpha with a strongly convolute protoconch (see discussion in Bandel and Frýda 1999). Because of the lack of data on Ordovician neritimorph gastropods, a detailed quantitative analysis of their diversity pattern and turnover rates would be meaningless.

Caenogastropoda (Subulitoidea and Peruneloidea)

Salvini-Plawen and Haszprunar (1987) introduced the term Apogastropoda for the Caenogastropoda and basal Heterobranchia. Haszprunar (1988) suggested that the first caenogastropods and heterobranchs had a common ancestor. Although there is a rich fossil record for the Late Paleozoic Caenogastropoda (see Nützel et al. 2000 for references), evidence for pre-Carboniferous Caenogastropoda and Heterobranchia is very poor (Frýda and Blodgett 2001). The discovery of the first presumed pre-Carboniferous caenogastropod was reported from the Early Devonian (*Pragoscutula wareni* Frýda 1998; see Frýda 2001: figure 1). The true larval shell (protoconch II) of the Caenogastropoda was also recently documented in some members of the Devonian (Frýda 2001: figures 3, 4) and Carboniferous (Nützel et al. 2000) Subulitoidea. The latter superfamily includes true Caenogastropoda with characteristically closely coiled larval shells. On the other hand, some Devonian gastropod genera closely resembling subulitoidean gastropods (Chuchlinidae of Peruneloidea Frýda and Bandel 1997) developed true larval shells with an openly coiled first whorl (Frýda and Manda 1997; Frýda 1999a). According to Frýda (1999b, 2001), the Ordovician-Carboniferous Peruneloidea may be considered to be an independent gastropod group at the ordinal level (Perunelomorpha), which represents the ancestral group (or basal group) of the Caenogastropoda or even the whole Apogastropoda (figure 19.1).

Ordovician Subulitoidea (figure 19.2R) are known mainly from tropical regions (Laurentia and Baltica). Distribution of the Ordovician Peruneloidea is imperfectly known, with the only available data from Baltica (Bockelie and Yochelson 1979; Hynda 1986; Dzik 1994c).

The presumed members of the Subulitoidea Lindström 1884 developed two types of protoconchs: (1) closely coiled larval shells like modern Caenogastropoda (Nützel et al. 2000; Frýda 2001), or (2) larval shells with an openly coiled first whorl (Peruneloidea; see Frýda 2001 for references). The oldest members of the first group are known from the Early Devonian. However, members of the second group are known to be present from the Early Ordovician to the Carboniferous (figures 19.1, 19.3A). Unfortunately, no information exists concerning the morphology of the protoconch of the type genus *Subulites* Emmons 1842, and so the relationships of the Subulitoidea and Peruneloidea cannot be resolved. Regardless of this taxonomic problem, both of the latter groups are well known from the Ordovician (the first on the basis of teleoconchs and the second by the discovery of typical protoconchs). Both gastropod groups developed true larval shells, and their morphology supports their assignment to the Caenogastropoda. Like the Neritimorpha, there is good qualitative evidence for the existence of the Caenogastropoda (or their ancestors) in the Ordovician (figures 19.3A, 19.4D), but the lack of data precludes a detailed quantitative analysis of their diversity pattern.

■ Discussion

To date, only a few papers (Sepkoski 1995; Wagner 1995b) have investigated diversity patterns among the Ordovician gastropods. As in these studies, we found that their diversity (figure 19.4A) was increasing from the earliest Ordovician until the Late Ordovician, when it dropped drastically close to the Ordovician/Silurian boundary. In contrast to previous studies, our analysis has been based on much more detailed stratigraphic data, which has revealed some new and more detailed observations. The mid–Late Ordovician (*TS*.5c–6b) is the time of highest diversity. Another peak of high diversity is found in the lowermost Darriwilian (*TS*.4a). Both the rates of originations (o_i and o_{di}) show four distinct, regularly spaced peaks (figure 19.4B, C). More detailed analysis of this phenomenon shows that each of these peaks has contributions from different higher taxa coming from different regions. A relatively high rate of origination (see o_{di} in figure 19.4C)

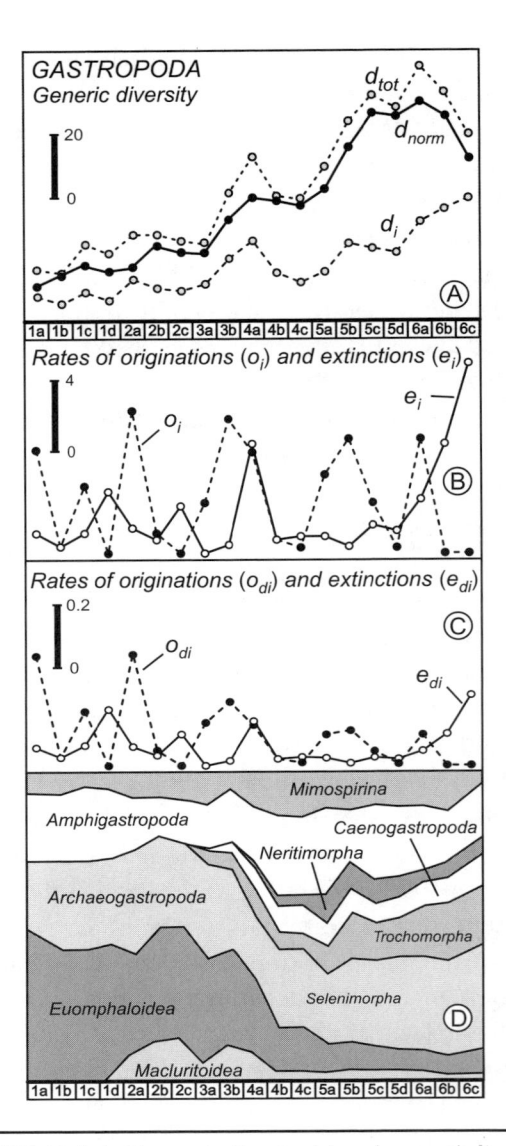

FIGURE 19.4. A–C, generic diversity (d_{tot}, d_{norm}, and d_i) of the Ordovician gastropods and their rates of originations (o_i and o_{di}) and extinctions (e_i and e_{di}). (For definition of these parameters, see text.). D, relative richness of the major gastropod taxa.

also occurs at the base of the Tremadocian. Unfortunately, both very low total diversity and the lack of exact stratigraphic data for Tremadocian gastropod distribution preclude a more detailed analysis. The extinction rates through the Ordovician show a much simpler pattern. The highest value was reached close to the Ordovician/Silurian boundary. This result is in full concordance with previous observations (Sepkoski 1995; Wagner 1995b). Our analysis has also revealed a very strong peak of the extinction rates at the beginning of the Darriwilian (*TS*.4a). The bioevents mentioned earlier are discussed in detail in the next sections.

Early Arenig Origination Event

The first major peak of origination rates recognized in the Ordovician coincides with the base of the Arenig (*TS*.2a—graptolite-based *Tetragraptus approximatus* Zone). Both rates of originations (o_i and o_{di}) reached maximum values, 8.5 and 0.5, respectively. The increase in generic diversity as well as changes in generic composition of both the slit-bearing Archaeogastropoda and Euomphaloidea made the most significant contribution to this peak (figure 19.4B, C). This faunal change is coupled with the latest Tremadocian (*TS*.1d) extinction event, during which 11 genera became extinct. Barnes et al. (1995) noted a first-order regression/transgression couplet close to the Tremadocian-Arenig boundary. A major regression could be responsible for the extinction event, which was followed by a period (early Arenig, *TS*.2a) of adaptive radiation during rapid transgression. Barnes et al. (1996) suggested that during transgression the shallow platform waters attained a higher level of oxygenation, which allowed the rapid radiation of a new Ordovician fauna, and this fauna became differentiated from the Cambrian fauna, which was adapted to relatively low levels of oxygenation. Currently, it is impossible to test this hypothesis because of a lack of data on the environmental distribution of Late Cambrian and Early Ordovician gastropods. Nevertheless, besides the trilobites, graptolites, and conodonts (Barnes et al. 1996), gastropods are an additional animal group showing significant faunal change at the base of the Arenig (*TS*.2a—*T. approximatus* Zone).

Earliest Darriwilian Origination/Extinction Events

The interval close to the lower boundary of the Darriwilian exhibits high turnover rates (figure 19.4). Origination rates (o_i and o_{di}) reached maximum values just before the Darriwilian (*TS*.3b) and remained high during the earliest Darriwilian (*TS*.4a). At the same time, the extinction rates increased strongly to a local maximum (early Darriwilian, *TS*.4a). This couplet of earliest Darriwilian origination/extinction events corresponds to the most distinct faunal change within the Ordovician (28 new gastropod genera originated, and 14 genera became extinct). The earliest Darriwilian origination event influenced the diversity of all Ordovician gastropod groups (figures 19.3, 19.4). The first representatives of more advanced gastropod groups (Loxonematoidea, Subulitoidea, figure 19.3A) also appeared during this origination event. The earliest Darriwilian origination/extinction events probably had a crucial influence on the subsequent evolution of the Euomphaloidea and Macluritoidea. The latter superfamilies reached their highest diversity close to the lower boundary of the Darriwilian (*TS*.3b–4a). Subsequently their diversities decreased strongly (i.e., 6 of 11 euomphaloidean genera became extinct). During the rest of the Ordovician the proportion of both Euomphaloidea and Macluritoidea to the total gastropod diversity decreased (figure 19.4). The former superfamily underwent the next of its several radiation events during the Mid and Late Paleozoic. On the other hand, the Macluritoidea definitely became extinct during the mass extinction event close to the Ordovician/Silurian boundary. The similarity in the diversity and turnover patterns of the Macluritoidea and Euomphaloidea (figure 19.3E, F) may suggest similarities in some of their life strategies or perhaps some of their phylogenetic links (see discussion earlier in this chapter).

A similar though slightly younger bioevent (the so-called Basal Llanvirn Bio-Event) was described by Barnes et al. (1996), which distinctly influenced the diversity of trilobites, graptolites, and conodonts. On the other hand, Sepkoski (1995) pointed out that the Ordovician diversification of typical Paleozoic groups really started in the late Arenig (i.e., in the earliest

Darriwilian). The data here provide evidence that the Gastropoda also belong to these groups. Barnes et al. (1996) noted that at the level of the so-called Basal Llanvirn Bio-Event a major excursion in the $^{87}Sr/^{86}Sr$ isotope curve occurred. These authors suggested a possible relation between the bioevent and the increased seafloor spreading and ridge activity inferred from the $^{87}Sr/^{86}Sr$ isotope curve. However, new data on marine $^{87}Sr/^{86}Sr$ isotope evolution (e.g., Veizer et al. 1999) show that the decrease in the $^{87}Sr/^{86}Sr$ values started much earlier (latest Tremadocian) and continued up to the Late Ordovician. In contrast to that, our analysis reveals a short period of very strong increase in extinction rates at the beginning of the Darriwilian (*TS*.4a). An increase of extinction rates at this time was previously observed in some other faunal groups (e.g., Sepkoski 1995). It is difficult at this time to speculate on the reasons for the early Darriwilian extinction event. However, it is interesting to note that the age of the relatively large Ames impact structure (about 470 Ma) in Laurentia (Koeberl et al. 2001) agrees well with the age of the extinction event. Future detailed paleontological, climatic, and isotopic studies may shed new light on the reasons of this bioevent.

Early Caradoc Origination Event

The early Caradoc (*TS*.5a–b) also represented an interval with a distinct rise in origination rates (figures 19.3, 19.4). In contrast to the previous bioevents, it is not coupled with a significant extinction event. The typical feature of this bioevent is the considerable acceleration of the radiation (31 gastropod genera had their first appearances, and only 5 genera became extinct) among the majority of gastropod groups ("selenimorph" and "trochomorph" Archaeogastropoda, Mimospirina, Neritimorpha, Amphigastropoda, and Caenogastropoda) after the late Darriwilian stasis. Several authors (see Barnes et al. 1995; Sepkoski 1995; and Bassett et al. 1999c for reviews) have earlier mentioned an origination bioevent that occurred during the early Caradoc. Barnes et al. (1996) suggested that the latter bioevent corresponds to a significant regressive phase followed by mid Caradoc transgression inundating most cratons.

Ashgill Origination/Extinction Events

After a short stasis in the middle of the Late Ordovician (figure 19.4), the origination rates increased again from the beginning of the Ashgill (*TS*.6a). During this origination event, 17 new genera appeared, belonging mainly to the Archaeogastropoda and Mimospirina. After this event, no new generic taxa have been documented as originating in the remainder of the Ordovician. On the other hand, extinction rates (e_i and e_{di}) started to increase dramatically until they peaked at the end of the Ordovician (figure 19.4B, C). During the Ashgill 38 gastropod genera disappeared, and this crisis affected all gastropod groups. A detailed analysis of extinctions in different environments and paleogeographic regions should be the subject of a future study. At present it should be noted that the exact time of the highest extinction rate for each gastropod group seems to differ slightly (figure 19.3), even though the extinction curve for all gastropods (figure 19.4B, C) shows a simple ascending pattern through the Ashgill. This may reflect multiple crises during the uppermost Ordovician as suggested by Copper (2001c).

■ Conclusions

During the past several years views on the dynamics of evolution have changed markedly (see Brett and Baird 1995 and Sheehan 2001a for references). It seems now that there were geologically long intervals of faunal stability (stasis), which lasted several m.y. and display little taxonomic and ecologic change. These periods of stability were bounded by shorter periods of evolutionary and ecologic turnover, characterized by high levels of speciation and extinction. These evolutionary patterns have been documented for many marine animal groups (e.g., Boucot 1975, 1983; Patzkowsky and Holland 1997). Although our analysis of diversity and turnover of the Ordovician gastropods also revealed a similar pattern (figure 19.4), it shows that the periods of stability were of lengths similar to the intervals with high turnover rates, in contrast to the studies of many other marine groups. This pattern may be because gastropods belong to the groups with lower turnover rates (Sepkoski 1995). It could also be that this pattern is an artifact of the poor quality of stratigraphic data available for some Ordovician gastropod taxa. Nevertheless, disregarding relative lengths of stases, it is beyond doubt that the evolutionary pattern of the Ordovician gastropods shows periods of relative stability separated by periods with high levels of turnover rates.

The Ordovician was a time of substantial growth in gastropod taxonomic richness. Even though our knowledge of the Tremadocian gastropod fauna is still poor, this time interval may be characterized by relatively high origination rates. The first significant extinction event was observed at the end of Tremadocian (*TS.*1c). The following early Arenig origination bioevent (*TS.*2a—*Tetragraptus approximatus* Zone) represents the most important Ordovician origination affecting the gastropods. The subsequent earliest Darriwilian origination/extinction couplet corresponds to the most distinct generic change within the Ordovician, causing significant acceleration of development and radiation in some gastropod groups (Archaeogastropoda, Mimospirina, Neritimorpha, Amphigastropoda, and Caenogastropoda?) as well as a crisis for the Euomphaloidea and Macluritoidea. The earliest Darriwilian origination/extinction bioevents also considerably changed the relative abundance of the higher gastropod taxa (figure 19.4D). After the mid to late Darriwilian stasis, the early Caradoc origination bioevent (*TS.*5a–b) again distinctly accelerated radiation of the majority of gastropod groups. In contrast to the previous bioevent, no significant changes in the relative abundance of higher gastropod taxa were observed during the early Caradoc and the early Ashgill origination bioevents. The latest Ordovician was a time of the highest extinction rates, affecting all the gastropod groups.

For a better understanding of the biodiversity patterns of Ordovician gastropods, future efforts should be focused on the study of their species-level diversity, as well as on a comparison of their diversities and turnover rates in different environments and paleogeographic regions. This however, cannot be attained without future intensive, specimen-based studies.

ACKNOWLEDGMENTS

This study was supported by the Grant Agency of the Czech Republic (grant 205/01/0143), Alexander von Humboldt-Stiftung, and IGCP project no. 410. Rohr's work was supported in part by grants from the National Geographic Society, Committee for Research and Exploration, and from Sul Ross State University, Faculty Research Enhancement Program. We also acknowledge Ian Percival (Sydney), Arthur J. Boucot, and Robert B. Blodgett (Oregon) for their helpful, critical reviews of this chapter.

20 Bivalve and Rostroconch Mollusks

John C. W. Cope

Bivalves and rostroconchs are often similar in shape, and some rostroconchs were identified originally as bivalves. The apparently bivalved condition of the rostroconchs is, however, illusory, as they develop from a univalved protoconch as opposed to the bivalved protoconch of bivalves. Rostroconchs are therefore pseudobivalved; they appear to have developed from laterally compressed monoplacaphorans in the Early Cambrian and are believed in turn to have given rise to the bivalves, also in the Early Cambrian. The two groups share a laterally compressed body organization that clearly had a relatively straight gut with mouth and anus well separated at the anterior and posterior of the shell, respectively. The molluscan head was very poorly developed or absent, and sensory functions were concentrated in the mantle margins. Although there is no radula in bivalves, it is possible that some rostroconchs had this rasping tonguelike organ. The similarities between these two classes led Runnegar and Pojeta (1974) to combine them in the subclass Diasoma, together with the class Scaphopoda. The latter class has been recorded in the past from the Ordovician, but more recent work has disproved these Ordovician records, and it is now clear that the scaphopods were derived from the conocardioid rostroconchs in the Devonian (Engeser and Riedel 1996).

■ Bivalvia

Bivalves are one of the least well-known groups of Ordovician fossils, primarily because they are often extremely rare, especially in the Early and Mid Ordovician; however, in particular facies and at individual times they become locally the dominant invertebrate group, even from the Early Ordovician (Cope 1996b). Although a handful of taxa are known from Lower and Middle Cambrian rocks, there are no undoubted Upper Cambrian bivalves, and thus all Lower Ordovician forms are cryptogenic. Pojeta (1980) figured a possible Upper Cambrian bivalve from Maryland, but this specimen shows no features unique to bivalves, and Hinz-Schallreuter (2000) has pointed out that Berg-Madsen's (1987) claimed Upper Cambrian specimen of *Tuarangia* could equally be from the Middle Cambrian, from which the genus was already known.

Pojeta (1971) recorded some 1,400 species of Ordovician bivalves, but in compiling the data for this chapter it has not proved possible to use species, since many of the primary descriptions (particularly of Laurentian species) are now in need of major revision. The specific names include many synonyms, and many species were inadequately founded. Some genera, too, are ill founded, and without reexamination of the type material, doubts exist concerning their interpretation.

This problem has been compounded by assigning bivalves from other paleocontinents to these doubtful taxa. The same criticism can be leveled at many of the forms figured in the nineteenth century from classical areas of Europe including Bohemia and Brittany, and only since the latter part of the twentieth century have monographic descriptions been more rigorous and included details of the number of specimens, their repository and registered number, and the precise age. From such data useful analysis can result, and it can now be shown that latitude-dependent differences exist in Ordovician bivalve faunas (Babin 1993, 2000; Cope and Babin 1999; Cope 2002).

Some generic names have been so widely applied that they have become meaningless. A prime example is the nuculoid *Ctenodonta*. The Upper Ordovician type species, *C. nasuta* (Hall), is very distinctive, but the name has been applied to nuculoids of totally different shape and to those lacking musculature or dentition; Pojeta (1971) recorded that 183 species had been assigned to the genus. Herein all records of the genus prior to those from the Middle Ordovician, and those that show no kinship to the type species, have been discounted. Genera that remain uninterpretable through inadequate original description, such as *Allodesma* Ulrich 1894, are also omitted.

Specific names, too, cause problems because of the differing levels of treatment by various authors. The description of 20 species belonging to one genus from one locality would not be accepted today, but this is the level of splitting adopted by more than one author in the nineteenth century and first half of the twentieth. Pojeta (1971) noted that of the 1,400 specific names he had recorded for Ordovician bivalves, more than 700 were accounted for by only 16 genera; thus true estimates of the number of species are hard to make, and a "best guess" approach has been used to try to determine the specific diversity as it would be interpreted today.

Collection failure remains a significant factor in bivalve diversity calculations and is evident in several groups. For instance, the earliest arcoidean, *Catamarcaia*, is known from the "Middle Arenig" of Argentina (*TS*.3a); the next youngest arcoidean is from the Lower Llandovery, demonstrating a total lacuna in early arcoidean history during the Mid and Late Ordovician.

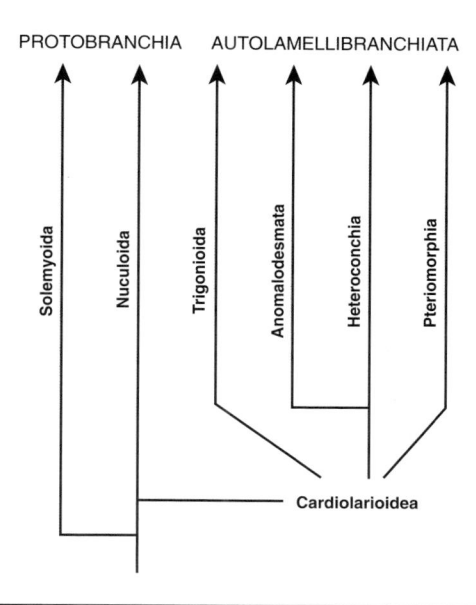

FIGURE 20.1. Classification of the Bivalvia adopted herein, showing the probable phylogenetic links between the major bivalve groups. The two major subdivisions (horizontal) are subclasses, and the vertical groups are superorders. The superfamily Cardiolarioidea is at present not assigned to any higher-level taxon. After Cope (2002). Reproduced by permission of the Geological Society of London.

The classification of bivalves used herein is that of Cope (2000, 2002), depicted in figure 20.1.

Lower Ordovician faunas are very rare, Tremadocian faunas (*TS*.1a–d) being certainly known only from Argentina (Harrington 1938), the Montagne Noire (Babin 1982), Central Australia (Pojeta and Gilbert-Tomlinson 1977), and Morocco (pers. observ. 2001). By the early Arenig (*TS*.2a) bivalve faunas are known from six localities worldwide; this early Arenig increase in abundance was mirrored by increases in diversity and size. Cope (1995) correlated this with the evolution of the feeding gill and identified a group of bivalves, the Cardiolarioidea, that links the protobranch bivalves with the autolamellibranch forms in which the gills are used for feeding as well as respiration (Cope 1995, 1997b). Early Ordovician bivalves are confined entirely to the Gondwanan continent, including Avalonia (Cope and Babin 1999), although they range from very high latitudes (e.g., Morocco) to equatorial latitudes (e.g., Central Australia).

There are 37 recorded genera and an estimated 60 Early Ordovician species. All the genera are new, and the explosive nature of the Early Ordovician bivalve diversification can be gauged from the fact that these genera belong to no fewer than 18 families. The major diversification period was in the early Arenig (*TS*.2a), the most important faunas coming from South Wales, with 20 species belonging to 18 genera (Cope 1996b), and the Montagne Noire, with 9 species belonging to 9 genera (Babin 1982); this represents a major advance over the total of probably 6 known species worldwide in the late Tremadocian (*TS*.1d).

By the Mid Ordovician bivalves had become much more abundant locally; Babin and Gutiérrez-Marco (1991) recorded them from 87 Spanish localities, but elsewhere around the Gondwanan margins, as in Avalonia, bivalves remained generally rare, and Cope (1999) described Anglo-Welsh Middle Ordovician bivalves essentially from two localities. Worldwide, Middle Ordovician bivalves are known from more than 100 localities. Bivalves were still largely restricted to Gondwana and Avalonia in the Middle Ordovician (Cope and Babin 1999); unequivocal Middle Ordovician bivalves are otherwise known only from Baltica (two species) and the margin of Laurentia on Svalbard (one species). The many bivalve records from the Middle Ordovician of the remainder of Laurentia, Siberia, and Bohemia have all proved to belong to the Upper Ordovician (*TS*.5a or later). Kazakhstan, however, has three latest *TS*.4c species (L. E. Popov pers. comm.). In the Middle Ordovician the number of genera increased to 53, of which 35 are new and 18 are holdover genera; these include an estimated 120 species. The 18 recorded families include 4 new ones, but 3 families known from both the Lower and the Upper Ordovician are as yet unknown from the Middle Ordovician.

In the Late Ordovician, bivalves finally reached the Laurentian and Baltoscandian carbonate platforms, providing ideal epifaunal habitats where a second major radiation took place (Cope and Babin 1999). Greatest diversification occurred in pteriomorph groups, but the exclusively infaunal nuculoids also diversified more here, perhaps taking advantage of the warmer low-latitude waters (Cope 2002). There was a dramatic increase to 93 genera (63 new) that include an estimated 350–500 species, though exact numbers are hard to establish. The 20 families include only 2 new ones; thus familial diversity remained little different throughout the Ordovician Period after the early Arenig (*TS*.2a) explosive diversification. Forty-nine genera (53 percent) are pteriomorphians, and 24 genera (26 percent) are nuculoids; by far the largest number of species is also among the pteriomorphians and may amount to three-quarters of Upper Ordovician species. Upper Ordovician bivalve faunas are dominated by low-latitude faunas. Few species are known from high latitudes in the Late Ordovician, as the faunas are poorly known. In contrast, Laurentian Upper Ordovician faunas are in need of major taxonomic revision to reinterpret the material and arrive at a realistic number of species.

Nuculoids

Nuculoids are represented by nine genera described hitherto from the Lower Ordovician, although some, originally assigned to *Ctenodonta*, require new generic names. The earliest are present in the basal Ordovician in Argentina (*TS*.1a—Harrington 1938) and are possibly indirect descendants of the Cambrian taxodont genera *Pojetaia* and *Tuarangia*, both of which occur in the Middle Cambrian. Most of the Lower Ordovician species are from high latitudes. In the Middle Ordovician the number of genera so far described is 17; of these, 13 are new and 4 holdover genera. Four genera are known only from low-latitude Gondwana. In the Upper Ordovician there are 24 nuculoid genera, of which 14 are new and 10 holdover (figure 20.2); the new genera are principally low-latitude forms from Laurentia, Siberia, and Gondwana, reflecting the greater diversity of nuculoids at low latitudes (Cope 2002).

Solemyoids

Solemyoids are anteriorly elongate protobranchs that are among the rarest Ordovician bivalves, and there is controversy about their origins. Cope (1996b) described the earliest known form, *Ovatoconcha*, from the early Arenig (*TS*.2a) of South Wales, but previously Pojeta (1988) had postulated an origin from ctenodontid nuculoids in the Late Ordovician and assigned the genera *Dystactella* and *Psiloconcha* to the group. As pointed out by Cope (2000), however,

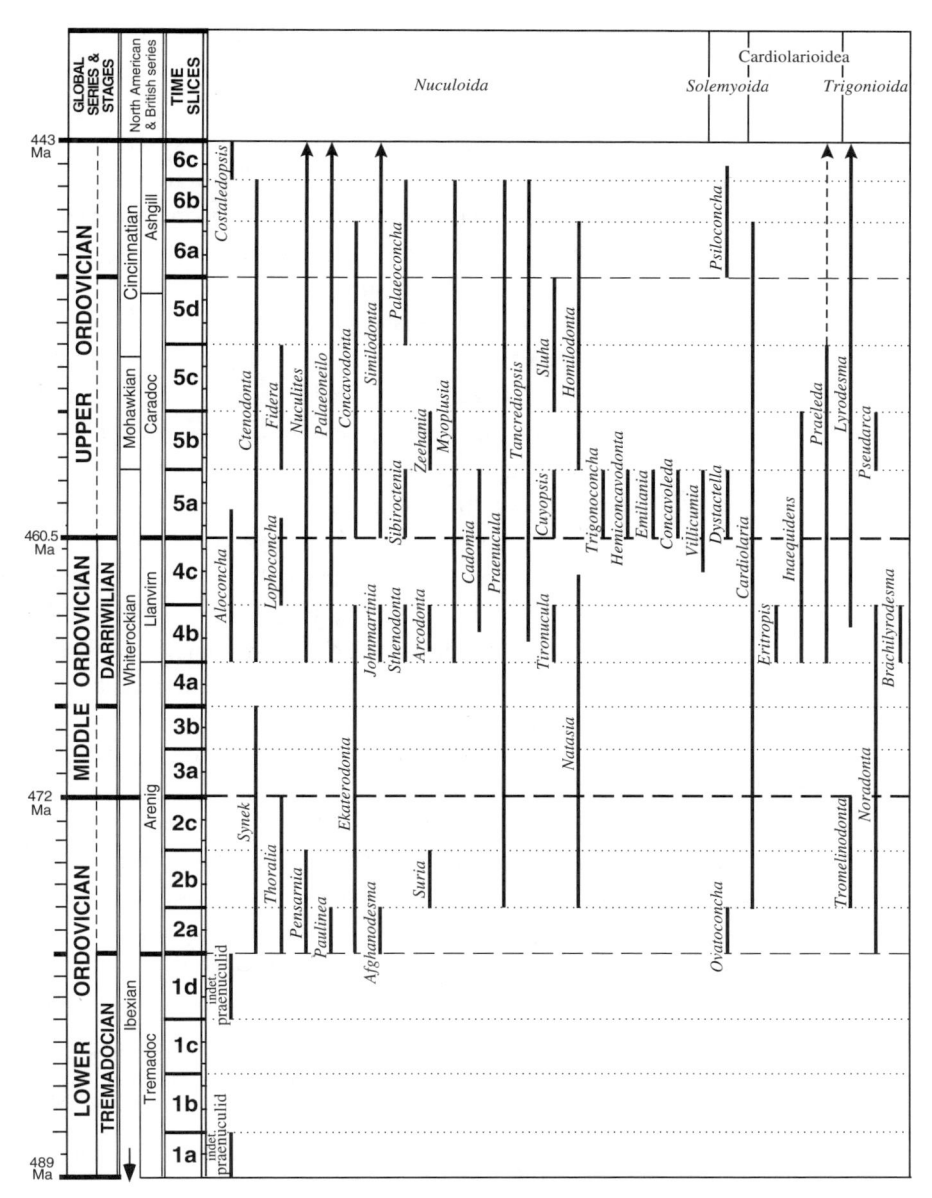

FIGURE 20.2. Range chart of the genera of Ordovician Bivalvia belonging to the Nuculoida, Solemyoida, Cardiolarioidea, and Trigonioida. In most cases the actual range *within* a time slice is not known, and the range is taken to the nearest upper and lower time-slice boundaries. Dotted lines signify ranges for which a genus must have survived but for which no record yet exists. Higher level taxa shown in subheadings are orders (in italics) and superfamilies (in roman).

the earliest true *Ctenodonta* species, from the Middle Ordovician, all have umbones that are well to the anterior, and all lack ligamental nymphs that both Pojeta (1988) and Waller (1990, 1998) believed to be synapomorphies linking the ctenodontids and the solemyoids; thus ligamental nymphs originated independently in the two groups, and the claimed intermediate genus *Dystactella* is more likely to be a persistent intermediate stock, with the true origins of the solemyoids lying unknown in the earliest Ordovician (or latest Cambrian). There are so far no solemyoids known from the Middle Ordovician.

Cardiolarioids

The Cardiolarioids are a small group of taxodont bivalves with hinges designed for wider valve opening while maintaining closer articulation of the valves than is possible with nuculoid hinges. Cope (1995) suggested that this hinge design was directly linked to the development of the filibranch gill and that the wider valve opening possibly would facilitate disposal of pseudofeces. These forms, which represent the earliest autolamellibranch bivalves, are now included in the Lower Ordovician to Lower Devonian superfamily

Cardiolarioidea (Cope 2000). *Cardiolaria* is the only form yet described from the Lower Ordovician and persists into the Middle Ordovician, where three new genera appear; of these four genera, three survive into the Upper Ordovician. Pojeta (1971, 1978), Pojeta and Gilbert-Tomlinson (1977), Tunnicliff (1982), and Cope (1996a) assigned (or assigned with a query) various bivalves to the genus *Deceptrix* Fuchs. Cope (1997b) showed that *Deceptrix* was a Silurian-Devonian cardiolarioid and that the Ordovician forms assigned to it were praenuculids (nuculoids).

Trigonioids

Trigonioids are distinctive bivalves with a prismato-nacreous shell that became most important in the Mesozoic. Throughout their geologic history they appear to have lived successfully in high-energy, coarse-grained sands. The earliest, *Noradonta,* has dentition linking it to *Cardiolaria,* and from this the other Lower Ordovician genus, *Tromelinodonta,* can be derived. The latter is restricted to the Lower Ordovician but gave rise to the Middle Ordovician to Silurian genus *Lyrodesma. Noradonta* persisted into the Middle Ordovician in Australia, where an endemic genus *Brachilyrodesma* also occurs. In the Upper Ordovician, in addition to the almost cosmopolitan *Lyrodesma,* the elongate *Pseudarca* occurs; the latter is known from both Brittany and Wales (figure 20.2).

Heteroconchs

Heteroconchs are bivalves that have their hinge teeth divided into cardinal, radiating from beneath the umbones, and lateral teeth, parallel to the dorsal margin of the shell. They may have originated from ancestors such as the cardiolarioids by fusion of the posterior teeth of the latter, already presaged in the cardiolarioid genus *Eritropis,* and lengthening of the anterior teeth. They include forms with a variety of shell structures including prismato-nacreous in some of the earlier forms and crossed-lamellar in many of the later ones; heteroconchs are the dominant group in Recent seas. A single genus, *Babinka,* is known from the late Tremadocian (*TS.*1d), but by the succeeding early Arenig (*TS.*2a) there is a wide variety of styles of dentition in the additional seven genera. In *TS.*2b is an unnamed genus, listed as "*Actinodonta,*" that refers to French forms (Barrois 1891; Babin 1966) identified with the Silurian genus *Actinodonta.* The majority of these heteroconchs have ventrally radiating cardinal teeth and can be assigned to the family Cycloconchidae. The Glyptarcoidea have teeth radiating in the opposite direction and are close to the origin of pteriomorphians; this was either from the heteroconchs or directly from the ancestral cardiolarioid stock. *Glyptarca* itself has a dentition remarkably similar to that of *Cardiolaria,* as noted by Cope (1995). Cope (2000) retained the glyptarcoids in the heteroconchs, but the cladistic analysis of Carter et al. (2000) suggested to them that they were pteriomorphs. It is noteworthy that all the Lower Ordovician heteroconchs are from high latitudes.

In the Middle Ordovician there are 12 genera of heteroconchs. Of these, 6 genera are new and 6 others are holdover genera. Although some species are remarkably common in certain faunas, others are much rarer, and the Spanish Middle Ordovician *Ananterodonta* is so far known from only a single specimen. There was now a low-latitude heteroconch, the Australian *Copidens,* which appears identical to the form figured by Guo (1988) as *Zadimerodia* from East Yunnan, China. Heteroconchs seem predominantly characteristic of high to intermediate latitudes and, as shown by Cope (2002), completely dominate the faunas at these latitudes.

Heteroconchs are reduced to six genera in the Upper Ordovician, of which three are new (figure 20.3). Upper Ordovician faunas are well known from low-latitude areas where heteroconchs are not to be expected in great numbers. Of the six recorded genera, three are from low-latitude areas; the three high-latitude genus are all holdover genera. Because there has been little work on high-latitude Upper Ordovician bivalve faunas, the depletion in heteroconchs in the Upper Ordovician may be more apparent than real.

Anomalodesmatans

Anomalodesmatans are burrowing bivalves with prismato-nacreous shells; throughout their geologic history they have constituted only a minor part of bivalve faunas. Dentition is frequently absent, and accurate valve articulation is provided by a well-located ligament; some forms have internal calcareous struc-

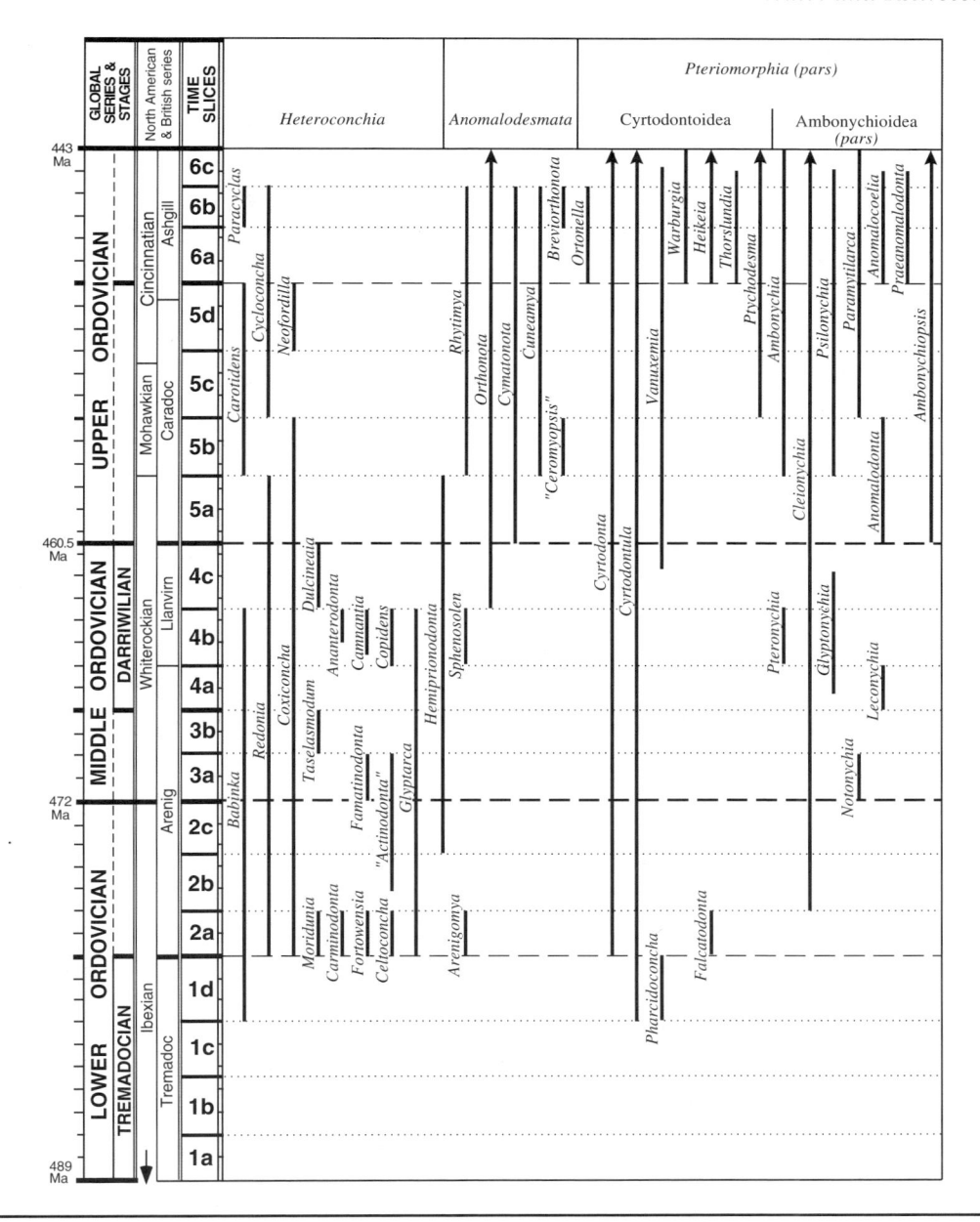

FIGURE 20.3. Range chart of the genera of Ordovician Bivalvia belonging to the Heteroconchia, Anomalodesmata, and Pteriomorphia (part). In most cases the actual range *within* a time slice is not known, and the range is taken to the nearest upper and lower time-slice boundaries. Note that *Carotidens* is a pterioid and should appear in figure 20.4. For identification of higher level taxa in subheadings, see figure 20.2.

tures that may have helped valve articulation. Fine-scale tuberculation of the shell is another characteristic of the anomalodesmatans. The earliest genus and sole Lower Ordovician form is *Arenigomya,* from South Wales (Avalonia). There are two new Middle Ordovician genera, while in the Upper Ordovician the number of genera has risen to seven, six of which are new. All the Middle and Upper Ordovician genera are from low-latitude areas. The origins of the anomalodesmatans are obscure, but they may have been derived from early heteroconchs by loss of teeth.

Pteriomorphians

The pteriomorphians are the most diverse groups of Ordovician bivalves; most have a three-layered shell with a calcitic outer layer and have multiple ligament insertions. They are frequently byssate. The earliest is a *Goniophora* (or possibly the closely related *Goniphorina;* see Sánchez 1997). *Goniophora* and *Goniphorina* are modiolopsids, a group of predominantly infaunal bivalves that have been frequently classified in the past with the heteroconchs but shown recently (Carter and Seed 1998) to include forms with multiple insertions of nonparivincular ligaments, the latter being a characteristic of many pteriomorphians. Confusion also resulted from the linking of modiolopsids with the similarly shaped but later modiomorphids; the latter are now regarded as anomalodesmatans (Fang and Morris 1997). Modiolopsids became well diversified in the Lower Ordovician with two genera known from the upper Tremadocian of Australia and six genera known from the Arenig, including *Goniophora;* this form (Cope 1996b) may, like the Argentinian form discussed earlier, be a *Goniophorina.* Both are recorded as *Goniophora* on figure 20.4. It thus seems that this group was dispersed around the Gondwanan continent.

Middle Ordovician modiolopsids were at a similar level of diversity with five recorded genera, three of which were holdover genera and two new genera, one from Australia and one from Wales. The dramatic change came within the Upper Ordovician whence 21 genera of modiolopsids have been recorded; this explosive increase occurred at low latitudes, not only on carbonate platforms such as on Laurentia and Baltica but also in Siberia and Kazakhstania. All these paleocontinents produced endemic modiolopsids. Of these 21 genera 15 are new, four are holdover genera, and two genera, *Parallelodus* and *Eurymya,* are known from the Lower Ordovician but not the Middle. Two genera (*Corallidomus* and *Semicorallidomus*) developed the habit of excavating living space (crypts) within stromatoporoids or bryozoans. With the Hirnantian regression and exposure of the carbonate platforms, both genera became extinct, and with them died the habit of boring among bivalves, which was not to be reevolved until the Late Jurassic.

Cyrtodontids are pteriomorphians that have anterior teeth very like those of the glyptarcoids, but unlike the latter cyrtodontids have an edentulous subumbonal lacuna on the hinge plate; they also have multiple ligamental insertions, but the latter have not been recorded (?preserved) in some of the earlier examples. The earliest forms are from the late Tremadocian (*TS.*1d) of Australia, whence *Cyrtodontula* and *Pharcidoconcha* were described by Pojeta and Gilbert-Tomlinson (1977); the latter genus also occurs in similar rocks in South China (Hsü and Ma 1948). *Cytrtodontula* also occurs in the early Arenig (*TS.*2a) of South Wales, where it is accompanied by *Cyrtodonta* and the strongly ribbed *Falcatodonta* (Cope 1996b). Cyrtodontids have a poor Middle Ordovician record; apart from four species of *Cyrtodonta* from Australia and two questionable records of *Cyrtodontula* from high latitudes in Spain and Morocco, the only other occurrence seems to be a late *TS.*4c occurrence of *Vanuxemia* from Kazakhstan (L. E. Popov pers. comm.). In the Upper Ordovician low-latitude carbonate shelves clearly provided a favorable environment for cyrtodonts, as in addition to *Cyrtodonta, Cyrtodontula,* and *Vanuxemia* there are five new genera and a large number of species. Pojeta (1971) recorded 116 species of *Cyrtodonta*—clearly an improbable total of real species but indicative of the group's high diversity in the Upper Ordovician.

Reduction of the anterior of the shell distinguishes the ambonychiids. The earliest are *Cleionychia* from the middle Arenig (*TS.*2c) of South Wales (Cope 1996b) and *Notonychia,* from the "Middle Arenig" (*TS.*3a) of Argentina (Sánchez 2001), possibly suggesting that this group originated at high latitudes, although it was clearly much more suited to low latitudes. The Middle Ordovician yields five genera, of which one is a holdover genus; three new genera are endemic to Australia and one to Argentina. In the Upper Ordovician there are 12 genera of ambonychiids, together with a large number of species, reflecting the low-latitude carbonate platforms as an ideal habitat for epifaunal bivalves. The end Ordovician regressions caused major extinctions on the carbonate platforms, and the ambonychiids were reduced to some five genera in the Silurian.

Pterioideans are normally inequivalve bivalves with a flatter right valve than left, though, as they are descended from equivalve ancestors, the time of origin of this is not clear. The earliest is a single convex right valve of a *Palaeopteria* from the lower Arenig of South Wales (Cope 1996b). The only Middle Ordovician genus yet recorded is the Australian *Denticelox*. In the Upper Ordovician *Palaeopteria* reappears, and there are four new genera. Pterioideans became more important forms in the Upper Paleozoic.

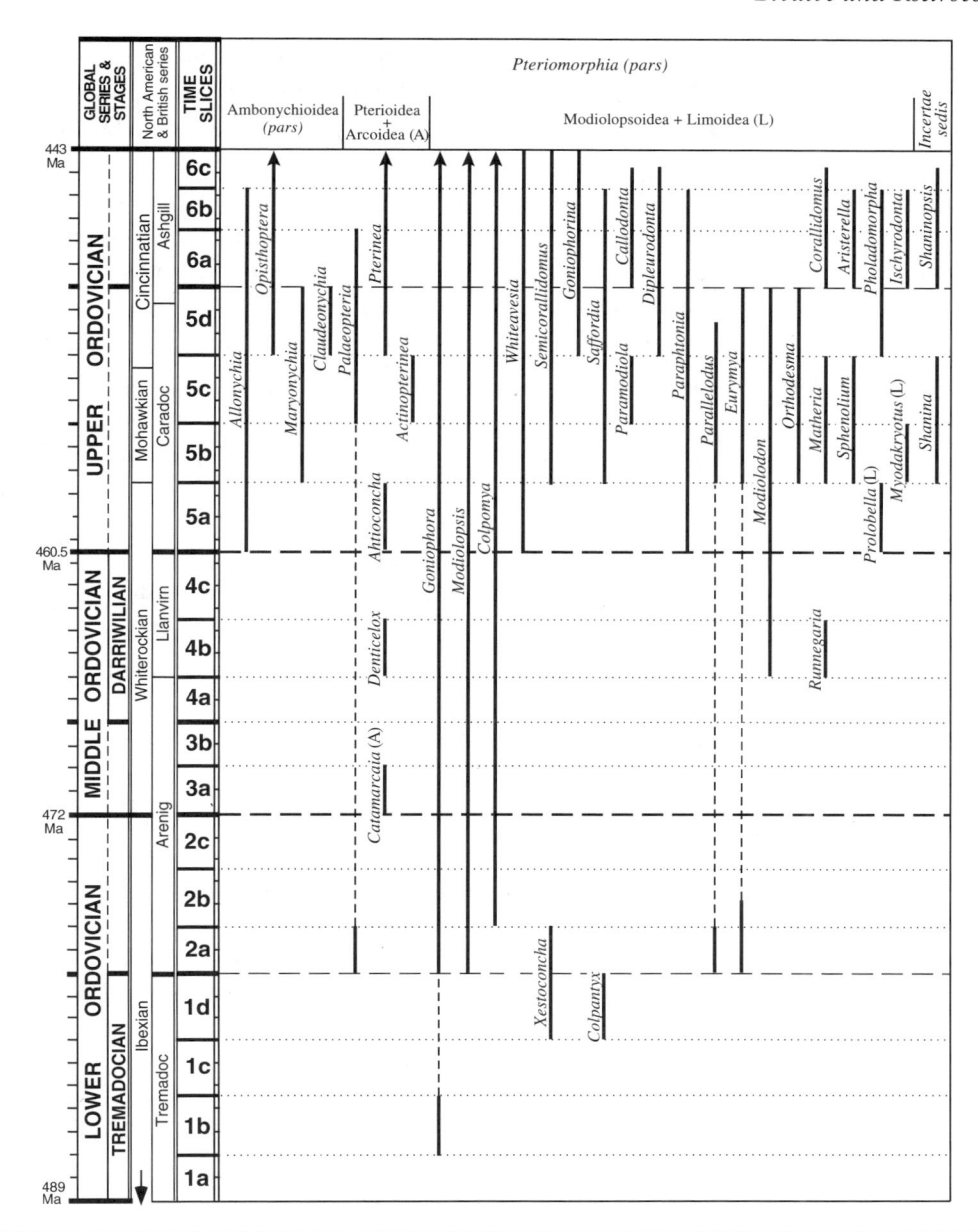

FIGURE 20.4. Range chart of the genera of Ordovician Bivalvia belonging to the Pteriomorphia (continued). In most cases the actual range *within* a time slice is not known, and the range is taken to the nearest upper and lower time-slice boundaries. Systematic position of the genera *Shanina* and *Shaninopsis* is uncertain, but both are probably pteriomorphians. Dotted lines signify ranges for which a genus must have survived but for which no record yet exists. Note that the pteroid *Carotidens* is incorrectly shown under the heteroconchs in figure 20.3. For identification of higher level taxa in subheadings, see figure 20.2. Abbreviations: (A) = Arcoidea; (B) = Limoidea.

The rarest Ordovician pteriomorphians are the arcoideans; they became much more important in the Upper Paleozoic and Mesozoic. They differ from other pteriomorphians in their two-layered shell structure. Their only Ordovician representative known hitherto is the genus *Catamarcaia* from the "Middle Arenig" (*TS*.3a) of Argentina (Sánchez and Babin 1993). As suggested by Cope (1997a, 1997b) and Ratter and Cope (1998), *Catamarcaia* could be derived from a glyptarcoid heteroconch. Collection failure is clearly responsible for the disjunct record of the early arcoids; the next example, *Alytodonta*, is known from a single Lower Llandovery specimen. *Catamarcaia* cannot be accommodated into the family Frejidae, which includes all Silurian arcoids, so Cope (2000) proposed the family Catamarcaiadae.

Ratter and Cope (1998) suggested that the exceptional rarity of arcoids was due to their preference for close inshore environments, which are not often preserved.

Two Upper Ordovician genera, *Myodakryotus* Tunnicliff 1987 and *Prolobella* Ulrich 1894 (*TS*.5b–c), are probably the earliest representatives of another group of pteriomorphians that became much more important later, the limoids. They were probably derived from the cyrtodonts and like them, but unlike later limoids, are dimyarian.

■ Rostroconchia

The Rostroconchia were the last class of the phylum Mollusca to be recognized. Although rostroconchs had been described since the early years of the nineteenth century, there was no recognition that they belonged to a single molluscan class. Much of the earlier literature had recognized them as members of three distinct groups of fossils: the ribeirioids, the eopteriids, and the conocardioids, based on the genera *Ribeiria* Sharpe 1853, *Eopteria* Billings 1865, and *Conocardium* Bronn 1835, respectively. The members of the last group were regarded as bivalves, usually allied to the cardioids, although by the time the bivalve volume of the *Treatise* (Branson et al. 1969) was published, they were treated as of uncertain bivalve subclass. The *Treatise* assigned *Eopteria* to the subclass Cryptodonta of the Bivalvia (Newell and LaRocque 1969), but both Kobayashi (1933, 1954) and Morris (1967) placed it with the ribeirioids; Pojeta (1971) however, preferred to group it with the conocardioids. Following Schubert and Waagen (1904), the ribeirioids were treated as arthropods, and this was followed by Kobayashi (1933) and most other earlier-twentieth-century workers. However, each of the three groups mentioned earlier has a calcareous shell bearing growth increments, developed from a protoconch, together with muscle scars, which also show growth increments. Such features are molluscan rather than arthropodan, and their recognition led to the concept of the class Rostroconchia by Pojeta et al. (1972) as a new class of bivalved mollusks, developed from a univalved protoconch. A major description and review of the class was published by Pojeta and Runnegar (1976) and a review of its geographic distribution by Pojeta (1979).

There are, however, differences between the numbers of species recorded herein and those produced by Pojeta (1979). There are several reasons for this. First, Pojeta's (1979) numbers of species are inflated by inclusion of *nomina nuda,* the counting of subspecies as species, and the inclusion—as separate species—of species that were treated as synonymous by Pojeta and Runnegar (1976). In addition, Pojeta recorded as Middle Ordovician many species that would now be regarded as Upper Ordovician. Thus, from the high number of species recorded in the Lower Ordovician, there is a dramatic drop in the Middle Ordovician, followed by something of a recovery in the Upper Ordovician.

Analysis of the distribution of species of rostroconchs shows that in the Early Ordovician there were 54 valid species distributed among 17 genera. This number has certainly been inflated by the records from two localities, particularly Manchuria, whence Kobayashi (1933) described many largely endemic genera and species. The other locality is in central Australia, whence Pojeta et al. (1977) described several endemic genera. The greatest generic diversification is evident in the Tremadocian, with no fewer than 16 genera, of which 9 are confined to that stage. There are some 21 Tremadocian species (figure 20.5).

In the Middle Ordovician 10 of the 15 species recorded belong to genera that have survived from the Lower Ordovician. Three species belong to genera making their first appearance and that range into the Upper Ordovician; the remaining two species belong to genera restricted to the Middle Ordovician (figure 20.5).

The Upper Ordovician shows a considerable increase in rostroconch diversity compared with the Middle Ordovician; although the number of genera has been reduced to seven, these contain 38 species. No new genera appear in the Late Ordovician, and of these, four survive into the latest Ordovician (*TS*.6c). Of the 38 species, some 26 are found on the Laurentian carbonate platform, and a further 7 species are known from the Baltoscandian carbonate platform. The remaining 5 species include 1 from low-latitude Gondwanan carbonates (Tasmania), 1 from midlatitude carbonates (Kazakhstan), 2 from Bohemian clastic rocks, and 1 from Avalonian clastics. There are no substantiated records of rostroconchs from uppermost Ordovician (Hirnantian) rocks; this

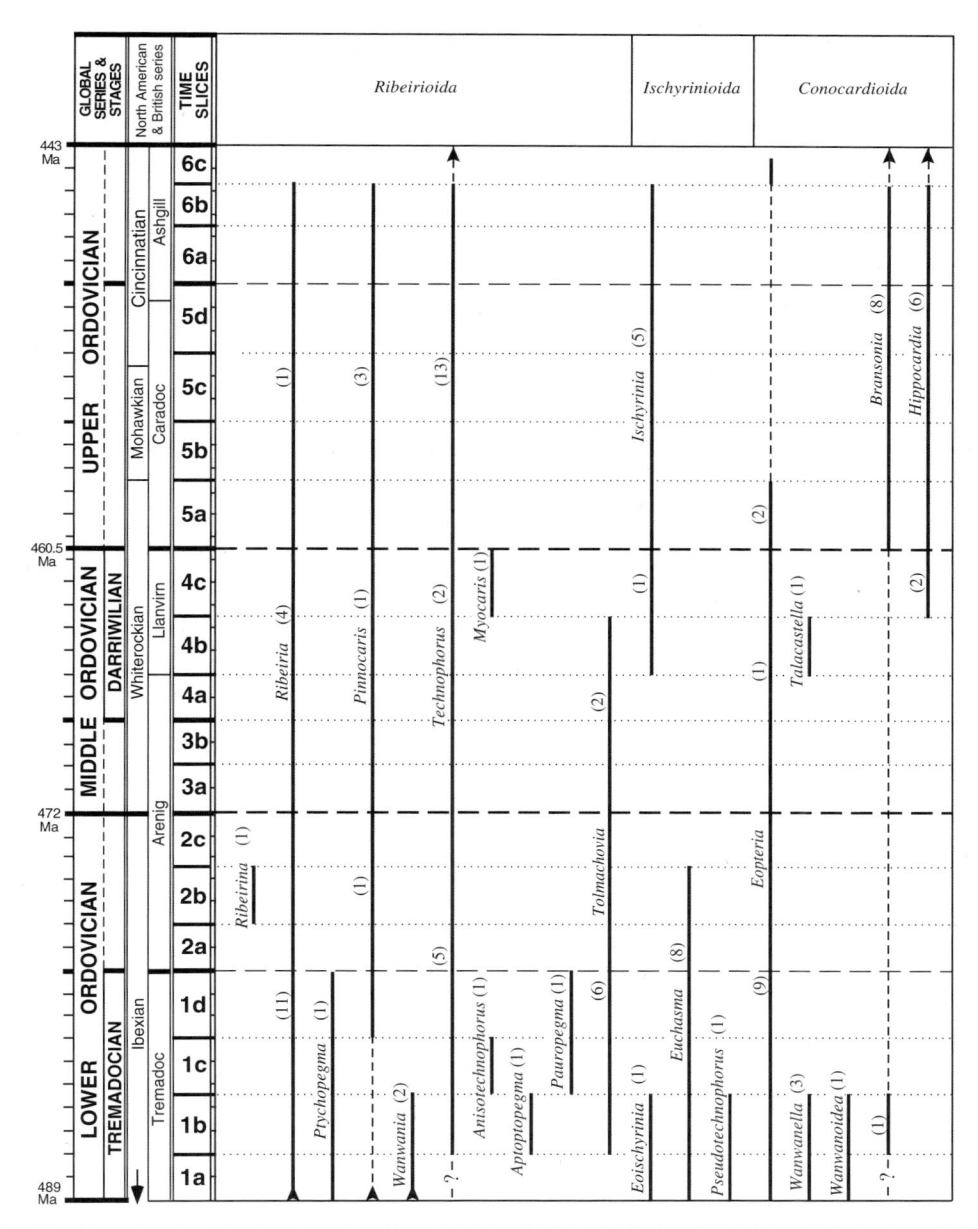

FIGURE 20.5. Range chart of the genera of Ordovician rostroconchs. Information on ranges is for the most part accurate to one time slice, and ranges are taken to the nearest time-slice boundary unless clear evidence exists to the contrary. Numbers on each line indicate the number of valid species recognized for Lower, Middle, or Upper Ordovician (as appropriate). Dotted lines signify ranges for which a genus must have survived but for which no record yet exists. For identification of higher level taxa in subheadings, see figure 20.2.

must be due to collection failure, as rostroconchs continue through to the end of the Permian Period.

Several conclusions may be drawn from examining these records of distribution of the rostroconchs. Generic diversity decreased throughout the Ordovician, and the number of monotypic genera decreased from six in the Early Ordovician to two in the Mid Ordovician; there are none in the Late Ordovician. Specific diversity decreased sharply from Early to Mid Ordovician, but then increased again in the Late Ordovician but failed to regain the level of Early Ordovician diversity (figure 20.5).

Ribeirioids

Of the order Ribeirioida, the Ribeiriidae, with a long Cambrian history, are more important in the Early Ordovician than later and are unknown from the Hirnantian (latest Ordovician, *TS*.6c). *Ribeiria* itself showed a decline through the Ordovician and is probably represented by a single species in the Late Ordovician, but *Pinnocaris,* which appeared in the latest Cambrian and has so far not been recorded from the earliest Ordovician, reappeared in the mid part of the Early Ordovician and reached its maximum diversity (three species) in the Late Ordovician. The Technophoridae, with 13 species in the later part of the Early Ordovician, declined to 5 species in the Mid Ordovician and then increased to 13 species (all species of *Technophorus*) in the Late Ordovician. They, too, are unknown from the Hirnantian stage (latest *TS*.6c) but, unlike other ribeirioids, survive into the Early Silurian (Zhang 1984).

Ischyrinioids

The monotypic order Ischyrinioida is restricted to the Ordovician. The family Ischyriniidae appears in the earliest Ordovician (*TS*.1a) with two species belonging to two genera. There is then a major break in the record of the family, as there are no records of it from the end of 1b to 4b in the later part of the Mid Ordovician, whence there is one record of an unidentified species of *Ischyrinia*. The genus is represented by five species in the Late Ordovician, though no ischyrinioids are recorded from the latest Ordovician (*TS*.6c).

Conocardioids

The order Conocardioida is well represented from the earliest Ordovician. The monotypic superfamily Eopterioidea includes the family Eopteriidae, which includes 21 Lower Ordovician species belonging to four genera. One of these, *Eopteria,* is represented by one species in the Middle Ordovician, but there is only one other Middle Ordovician eopteriid, the South American monotypic genus *Talacastella.* Eopteriids are known from a single early Late Ordovician (*TS*.5a) species of *Eopteria* from North America and a single specimen from the mid Ashgill (mid-*TS*.6c) of Kazkhstan (Popov et al. 2003). The superfamily Conocardioidea has two families with representatives in the Ordovician. The Bransoniidae is represented by the genus *Bransonia*. This has a very disjunct range; a single species appeared in the early Tremadocian (*TS*.1a) of Australia, and the genus is then unknown until the Late Ordovician (*TS*.5a–6b), from which eight further species have been recorded. The Hippocardiidae appeared in the latest Mid Ordovician (*TS*.4c) with two species and in the Late Ordovician (*TS*.5a–6b) are represented by six further species. Neither *Bransonia* nor *Hippocardia* has been recorded from the latest Late Ordovician (latest *TS*.6c), but both are known from Silurian and later rocks.

■ Concluding Remarks

In summary, it appears that there were two phases to the Ordovician diversification of the Bivalvia. Knowledge of the first in the Lower Ordovician is hampered by the scarcity of faunas of that age, particularly from the Tremadocian. This initial explosive radiation, which may have followed on the evolution of the feeding gill, produced no fewer than 18 families representing most bivalve groups. The confinement of bivalves to Gondwana and closely bordering areas produced a steady increase in diversity through the Middle Ordovician. The irregular shape of the diversity curves reflects major publications (which in turn depend on individual faunas being found). This explains the peaks in *TS*.2a (Babin 1982; Cope 1996b) and in *TS*.4a (Pojeta and Gilbert-Tomlinson 1977; Babin and Gutiérrez-Marco 1991; Cope 1999). Other faunas may in time fill these gaps; for instance, a new fauna from West Yunnan, China (Fang and Cope in press), would raise the total generic diversity for *TS*.3a from 21 to 31, providing close to a linear increase from *TS*.2a to 4a. When bivalves managed to migrate to low-latitude carbonate platform areas, in the earliest Late Ordovician, there was major further diversification, especially within the pteriomorphians and the anomalodesmatans (see figure 20.6A), but heteroconchs were apparently markedly reduced; this latter effect may be the result of poor knowledge of high-latitude Late Ordovician faunas. The last diversification followed on from adoption of semi-infaunal and epifaunal habits and resulted in a dramatic increase in both generic and specific diversity, especially within pteriomorphians. The latest Ordovician eustatic regression, resulting from the growth of continental ice sheets, produced a major extinction among Ordovician bivalves, partic-

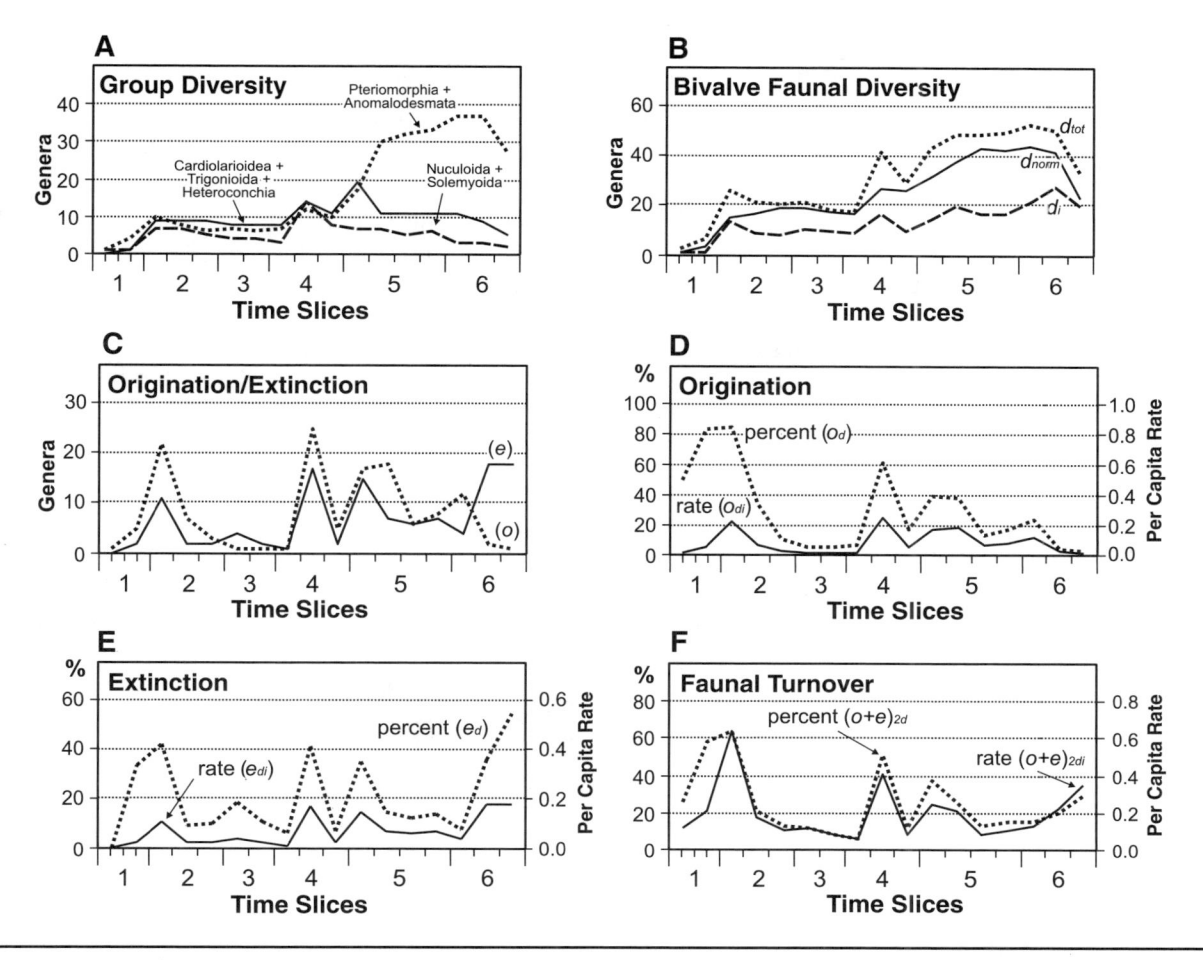

FIGURE 20.6. Bivalve diversity measures. Time slices are accorded equal duration, apart from *TS*.1a–d, for which data are so sparse that those for *TS*.1a and 1b have been combined, as have those for *TS*.1c and 1d. Figures 20.6A and 20.6B should be compared. Note that the *TS*.4b diversity peak (figure 20.6B) is accounted for by equal proportions of the three bivalve group combinations (figure 20.6A), while the *TS*.5b–6b peak (figure 20.6B) is accounted for solely by the dramatic increase in Pteriomorphia and Anomalodesmata (figure 20.6A).

ularly among the pteriomorphian groups that had exploited the newly available ecologic niches on the low-latitude carbonate platforms.

It is clear that the record of rostroconchs is deficient in several parts of their range; this is exemplified by the disjunct ranges of several taxa and the lack of any material from the Hirnantian (latest *TS*.6c). Clearly rostroconchs were profoundly affected by the Hirnantian regression; one species of ribeirioid is known from the Early Silurian and, surprisingly, appears to be the only record of a rostroconch from the Llandovery Series. The Conocardioidea, however, survived later into the Silurian and are recorded from the Wenlock Series upward to the end of the Paleozoic.

Pojeta (1979) suggested that an Ordovician decline in rostroconchs was correlated with bivalve diversification; he maintained that in competition between two infaunal groups, the superior burrowing powers of the Bivalvia gave them a decided advantage. Examination of the records of the rostroconchs (figure 20.7) and bivalves (figure 20.6) shows, however, that this idea can no longer be countenanced. First, the initial Early Ordovician bivalve diversification took place around the Gondwanan margins, where there were few rostroconchs. Second, the early Mid Ordovician rostroconch decline was largely extra-Gondwanan, whereas bivalves were still almost exclusively Gondwanan. Third, the Late Ordovician increase in rostroconch diversity, particularly on the

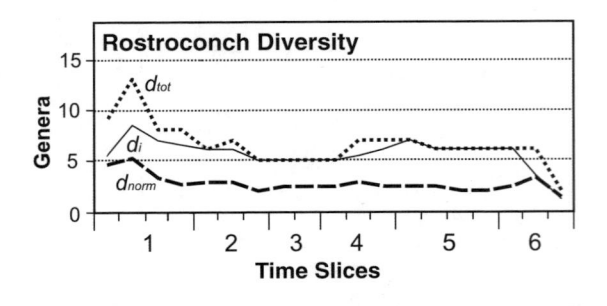

FIGURE 20.7. Rostroconch diversity. Time slices have been accorded equal duration. Since the decline is rostroconch genera in *TS*.1, generic diversity remained remarkably consistent (see figure 20.5).

Laurentian shelves, coincided with the arrival of bivalves on these shelves where the latter rapidly diversified. Rostroconchs were infaunal, but the Late Ordovician bivalve diversity increase was predominantly (though not exclusively) among epifaunal groups. Competition between bivalves and rostroconchs was clearly not a significant factor in rostroconch distribution through most of the Ordovician; the first time there could have been such competition was in the Late Ordovician, when significant numbers of bivalves and rostroconchs coexisted. Both classes were profoundly affected by the latest Ordovician glacially induced regression; bivalves largely recovered from this, but rostroconchs were reduced to one genus of ribeirioids, which survived through to the Llandovery Series, and one superfamily of conocardioids, which persisted through to the end of the Permian Period.

ACKNOWLEDGMENTS

I am grateful for the assistance of L. C. Norton (National Museum of Wales) in drafting figures 20.1–20.5 and A. P. Rogers (Cardiff University) for help with figures 20.6–20.7.

21 Nautiloid Cephalopods

Robert C. Frey, Matilde S. Beresi, David H. Evans, Alan H. King, and Ian G. Percival

The Ordovician Period marks the time of the great radiation of nautiloids, which proliferated from a single order at the beginning of the Ordovician (489 Ma) to a maximum of at least nine orders by the early Late Ordovician (455 Ma). It is also in the Ordovician that the greatest diversity of shell form and structures occurs in these externally shelled cephalopods (Flower 1976; Holland 1987; Teichert 1988). Most of these various shell architectures were new solutions to the problem of buoyancy control in these mobile mollusks (Crick 1988). Ordovician nautiloids are represented in strata exposed from the Arctic Circle to Tasmania and have been identified from lithofacies indicative of diverse marine environments. In the Ordovician, however, nautiloids are most abundant, display maximum diversity, and reach their greatest size in shallow marine carbonate platform facies deposited under tropical or subtropical climatic conditions (Flower 1976).

■ Status of Nautiloid Systematics

Although much has been achieved over the past century as far as the study of systematics of Ordovician nautiloids is concerned (Flower 1976; Teichert 1988), a number of issues remain to be dealt with, especially the validity of many published species. The tendency for some workers in the past to split nautiloids excessively at the species level was due to a lack of established criteria to define and differentiate taxa, and description of new species based on incomplete material. This often resulted in the application of several species names to different portions of the shell of the same organism. Another complication is the possibility of sexual dimorphism in fossil nautiloids, leading to the potential for different sexes of the same species to be identified as separate species. As a result of these unresolved difficulties at the species level, this review of nautiloid diversity in the Ordovician utilizes genus-level taxa, which are much better established. The higher-level taxonomic classification used here is modified from the phylogenetic diagrams of Flower (1988) and Wade (1988). This includes the recognition of the Pseudorthocerida as an order distinct from the Orthocerida; the merging of the Barrandeocerida with the Tarphycerida; and the recognition of an additional new order, the Dissidocerida, proposed by Zhuravleva (1994) since publication of the papers just cited. Family-level taxa include those listed in the compilation by King (1993).

■ Regional Distribution

Knowledge of the ranges and distribution of Ordovician nautiloid genera in specific geographic regions is highly variable. Ordovician nautiloid faunas have been extensively studied and well documented for North America (Flower 1976); Baltic Europe

(Flower 1976; King 1999), Bohemia (Marek 1999), Russia and Siberia (Balashov 1962), and Australia (Percival in Webby et al. 2000). Chinese Ordovician nautiloid faunas, both for North China and Korea and for South China, are relatively well established (Chen 1995), although in much of the older literature (Kobayashi 1927; Yu 1930; Endo 1932) many Chinese forms were ascribed to established North American and Baltic taxa when at least some of these genera represent new taxa.

Nautiloid faunas from classic Ordovician localities in the United Kingdom and other portions of the Avalonian microcontinent are inadequately known because of poor preservation of the nondescript longiconic nautiloids that dominate faunas in these shale and mudstone strata. Current work in South America, especially from Ordovician sections in Argentina and Bolivia, suggest that diverse, well-preserved faunas are present that are yet to be documented properly. Study of these South American faunas is also complicated by the disparate geologic histories of the Argentine Precordillera and Gondwana portions of northern Argentina and Bolivia (Astini 1998; Benedetto 1998). Also relatively poorly understood are Ordovician nautiloid faunas from southern Europe, the Middle East, Central Asia, and Southeast Asia, which are known only from a limited number of systematic studies.

■ Measured Parameters

Nautiloid diversity patterns were tabulated using normalized diversity after Cooper (chapter 4). Also calculated were turnover rates per million years, which record the number of genera originating or ending in each time unit divided by its duration. Figures 21.1 and 21.2 graphically display normalized diversity and turnover rate data (both originations and extinctions) for nautiloid genera for each of the identified time slices (*TS*) recognized here as subdividing the Ordovician Period.

■ Summary Trends in Nautiloid Diversity

The interpretation of general trends in nautiloid diversity in the Ordovician based on the faunas studied here is limited by a number of factors. The authors have compiled available data on the geologic

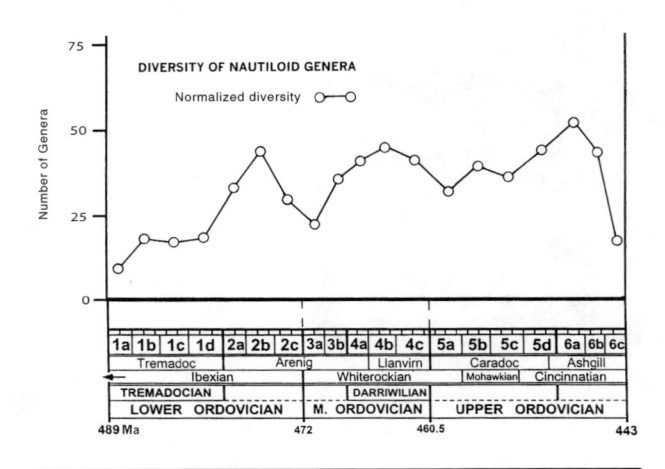

FIGURE 21.1. Normalized diversity of nautiloid genera through Ordovician time.

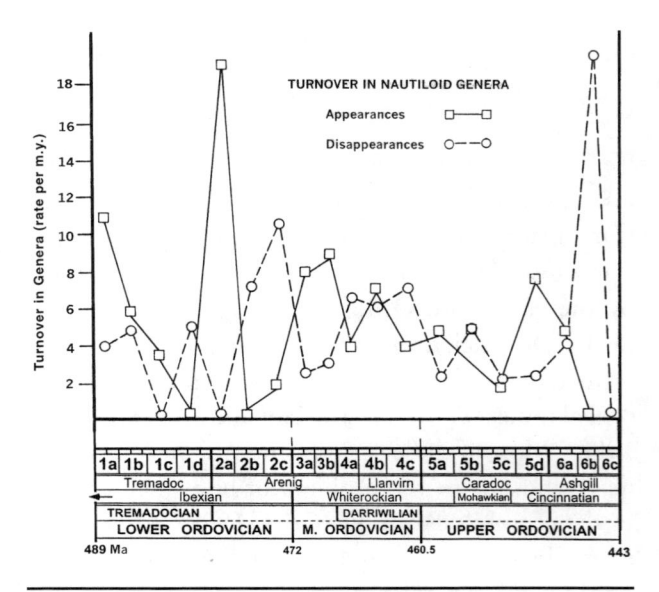

FIGURE 21.2. Originations and extinctions of nautiloid genera through Ordovician time.

and geographic occurrence of Ordovician nautiloid order-, family-, and genus-rank taxa from North America (Frey), the Baltic region (King), Avalonia (Evans), South America (Beresi), and Australia (Percival). The absence of data for significant geographic regions, especially the Siberian Platform and North and South China, calls into question interpretations of global trends in nautiloid diversity. Diversity trends, especially for mobile continent-bearing plates such as Avalonia, the Baltic region, and the Argentine Precordillera, appear to have been strongly influenced by regional geologic and biogeographic events. These factors and the effects of incomplete data for some regions and time periods (South America: entire Ordovician; North America: *TS*.4b–c) limit our confidence in the broad-scale diversity trends identified here. These uncertainties and limitations being stated, some general trends in nautiloid diversity in the Ordovician can be identified from these data. In support of the trends identified here, there is general agreement between these tabulations and standing generic diversities presented by Crick (1990), with the possible exception of the Late Ordovician (*TS*.6a–b).

Early Ordovician (TS.1a–2c)

Information on nautiloid diversity in the Early Ordovician is based primarily on faunas described from North America and from Western Australia. Both of these areas were sites of tropical-subtropical platform carbonate deposition in the Early Ordovician. Studied faunas in *TS*.1 (Tremadocian) mark the first great expansion of the cephalopods with the evolution and diversification of a number of small longicones and cyrtocones belonging to the order Ellesmerocerida. These earliest Ordovician faunas are dominated by members of the family Ellesmoceratidae. Early representatives of the Baltoceratidae, Bassleroceratidae, and Protocycloceratidae and the orders Endocerida and Tarphycerida evolve later during *TS*.1b–d (Flower 1964, 1976).

Faunas representative of *TS*.2a–c (early Arenig equivalents) are enriched by the diversification of the ellesmerocerid families Baltoceratidae, Cyclostomiceratidae, and Protocycloceratidae. The orders Endocerida (breviconic Piloceratidae, longiconic Proterocameroceratidae) and the coiled Tarphycerida (Tarphyceratidae) underwent rapid expansion, and the first members of the orders Orthocerida, Dissidocerida, and Actinocerida occur. Twenty-two new genera are added during *TS*.2a (figure 21.2), leading to an Early Ordovician maximum of diversity in *TS*.2b. Australasian faunas include a diverse group of endemic ellesmerocerids, endocerids, and the oldest members of the order Actinocerida, plus more cosmopolitan tarphycerid genera (Teichert and Glenister 1954). The end of the Early Ordovician (*TS*.2c) exhibits a slight decrease in diversity resulting from the extinction of a number of ellesmerocerid genera, primarily members of the Ellesmoceratidae that had been persistent for much of this time interval (figures 21.1, 21.2).

Mid Ordovician (TS.3a–4c)

According to Flower (1976:528), the Mid Ordovician marks "a definitive point in the evolution of the Nautiloidea" with the expansion of the orders Endocerida, Actinocerida, and Orthocerida and the first occurrences of the orders Pseudorthocerida, Oncocerida, and Discosorida. The Tarphyceratidae decline dramatically, except for relic forms that persist in the Baltic region. The Trocholitidae and Lituitidae diversify as the Tarphyceratidae decline. Documented faunas fall into two distinctive biogeographic groups (Flower 1976):

1. An equatorial carbonate platform fauna represented by faunas from North America, North China, Korea, and Australia, which is distinguished by a collection of robust endocerids with complex endosiphuncular structures (Allotrioceratidae, Endoceratidae, Padunoceratidae), large actinocerids belonging to the Wutinoceratidae, a variety of rather simple orthocerids, robust coiled tarphycerids of the Trocholitidae (*Litoceras, Plectolites*), and primitive oncocerids.

2. A higher-latitude, Gondwanan fauna represented by genera from both clastic and carbonate strata in the Baltic region, South China, southern Europe, Avalonia, and portions of South America. This fauna is distinguished by specialized ellesmerocerids with greatly thickened connecting rings (Bathmoceratidae), a number of slender longiconic endocerids (*Dideroceras* and *Suecoceras*), the cyrtoconic endocerid *Cyrtendoceras,* true *Orthoceras* and its close relatives (Orthoceratidae), the long-ranging, widespread coiled trocholitid genus *Discoceras,* and numerous members of the variably uncoiled Lituitidae. An undescribed Gondwanan fauna from southern Bolivia also includes large endocerids (Padunoceratidae) and actinocerids that appear to be assignable to the Georginidae (Flower 1988).

Nautiloid diversity reaches a second Ordovician maximum in *TS*.4b, primarily owing to the presence of a well-documented diverse Baltic fauna (Sweet 1958; King 1999). Many of the Mid Ordovician genera from both paleogeographic groups appear to have suffered extinction at the end of *TS*.4c (figure 21.1).

These extinctions may in part be the result of a global sea level drop at this time (Chen 1991) that is often overprinted by a post-Darriwilian unconformity, as is the case in North America (Flower 1970).

Late Ordovician (TS.5a–6c)

The early Late Ordovician (*TS.*5—Caradoc equivalent) is marked by continued diversification of the Orthocerida and the Pseudorthocerida and the rapid expansion of the Oncocerida. New families added include large actinocerids belonging to the Actinoceratidae, the Gonioceratidae, and the Armenoceratidae, plus large tarphycerids with cyrtochoanitic siphuncles (Barrandeoceratidae, Apsidoceratidae). Two new orders, the Ascocerida, with their curious truncated mature phragmocones, and the cyrtoconic and breviconic Discosorida, diversify during this time interval. However, diversity within the Endocerida, Trocholitidae, and Lituitidae declines significantly. The Ellesmerocerida, with the exception of a few baltoceratid genera and the peculiar curved brevicone *Cyrtocerina*, have disappeared.

Some of the geographic distinctions between nautiloid faunas disappear as Baltica and Avalonia approached Laurentia (Mac Niocall et al. 1997). A common fauna developed in association with carbonate platform facies distributed across a vast "tropical realm" extending from Greenland to Mexico (Flower 1976) and spreading eastward into now adjacent portions of the Baltic region (Strand 1933) and approaching portions of the Avalonia microcontinent (Evans 1993). This diverse "Arctic Ordovician Fauna" included often gigantic individuals belonging to the Endocerida, Dissidocerida (Narthecoceratidae), Actinocerida, Tarphycerida (Apsidoceratidae), Ascocerida, Oncocerida (Diestoceratidae), and Discosorida (Cyrtogomphoceratidae, Westonoceratidae) (Foerste 1929).

Elsewhere, South American faunas appear to decline dramatically as a result of drift into higher latitudes coupled with the onset of continental glaciation in Gondwana at the end of the Ordovician. Rare longiconic orthocerids occur associated with a "*Hirnantia*" shelly fauna in marine mudstone facies (Sánchez et al. 1991). Australian faunas persisted in carbonate platform facies centered in Tasmania and consisted of curious mixtures of cosmopolitan genera (*Beloitoceras, Gorbyoceras, Discoceras*) associated with distinctly Tasmanian taxa belonging to the endemic Gouldoceratidae (order Discosorida) (Stait 1988).

The later Late Ordovician (*TS.*6—Ashgill equivalent) demonstrates a continuation of diversity trends set in earlier in the Late Ordovician and marks the maximum in terms of generic diversity for nautiloids in the Ordovician. There is a major extinction of nautiloid genera and families at the end of the Ordovician (figure 21.2), associated with Late Ordovician glaciation events (chapter 9). A marked diversity decline occurs between *TS.*6b and 6c (figure 21.1). However, relic Ordovician faunas of endocerids, orthocerids, dissidocerids, actinocerids, and discosorids continued on into the Early Silurian (Rhuddanian-Aeronian), evidently finding refugia in carbonate platform habitats in Laurentia (North America, Greenland, the Arctic archipelago) (Frey and Holland 1995).

■ Comparisons with Published Biodiversity Analyses

Earlier we noted general (though not exact) agreement between our data and standing generic diversities presented by Crick (1990), with the possible exception of the Late Ordovician (*TS.*6a–b). A subsequent analysis by Sepkoski (1995) of a number of the major Ordovician clade groups includes a graphic representation of nautiloid radiation and extinction within the period, which in places also differs in detail from the generic diversity trends depicted here (figure 21.1).

For example, for the Arenig interval, Sepkoski counted 125 genera, and Crick listed 178 genera. Our compilation for the Arenig (*TS.*2a–b, 3a–b, 4a) totaled 105 genera, an estimate that seems to be generally in line with Sepkoski's data (and to a lesser extent Crick's tabulation), especially given the absence of generic information in our analysis from the Siberian Platform, former Soviet Central Asia, and North and South China. We recognize an Arenig biodiversity maximum near the end of the Lower Ordovician (*TS.*2b), resulting from a burst of evolutionary activity leading to the development of many new taxa that started in *TS.*2a. Data compiled for the Arenig interval in North America (by Frey, herein) included a block of new genera and species from unpublished manuscripts by the late Rousseau Flower, describing faunas from the El Paso region of Texas and New

Mexico and from Newfoundland. These data push the origins of many taxa, principally members of several endocerid families (especially the Piloceratidae), back to a point earlier in the Arenig rather than appearing toward the end of the interval as has been traditionally believed (reflected by Sepkoski's maxima at the Arenig-Llanvirn boundary).

For the Llanvirn (*TS*.4b–c), Crick (1990) tabulated 131 genera, whereas our compilation listed only 75 genera. This interval marked the occurrence of our second maximum of 43 genera. Llanvirn nautiloids are best documented from two regions—the Baltic, which is covered in our compilation, and two Chinese blocks, neither of which are included here. Absence of these Chinese taxa could very well account for the evident discrepancy between our numbers for total diversity and those of Crick.

Sepkoski (1995) recognized another maximum at the base of the Caradoc. We have a minor peak in diversity (37 genera) in roughly the same place. Crick (1990) recorded 135 genera from the entire Caradoc interval, not significantly different from our total of 125 genera. However, our compilation includes the Edenian (*TS*.5d) in the Caradoc, with its exceptionally diverse Laurentian Arctic Ordovician nautiloid fauna. Our compilation with regard to total Ashgill nautiloid genera was 65 (*TS*.6a–c). This is approximately half the total number of genera tabulated by Crick (1990) for his Ashgill interval (129 genera). According to Crick's data, the most diverse Ashgill faunas were from North America (95 genera) and the Baltic region (56 genera). Both areas were included in our compilation but with significantly different results: North America (65 genera) and the Baltic region (37 genera). If *TS*.5d is added to the Ashgill interval (as probably was the case with Crick's data), generic diversity climbs to 81 for North America and 39 for the Baltic, closing the gap somewhat but still leaving significant differences. As Ashgill faunas are well documented for both regions, these discrepancies are somewhat more difficult to explain.

A further maximum was identified by Sepkoski in the late (but not latest) Ashgill. Although we recognize a maximum of diversity (47 genera) somewhat earlier in the Ashgill (*TS*.6a), this discrepancy probably relates to slight differences in the boundaries of the time intervals used in the respective compilations.

With regard to these apparent differences between the compilations of Ordovician nautiloid diversity by Sepkoski (1995) and trends based on the diversity calculations used here, there are a number of possible explanations. The most likely are as follows. First, Sepkoski's compilation was based on worldwide data, whereas the present analysis was limited to North America, South America, the Baltic region, Avalonia, and Australia; significant faunas from the Siberian Platform, Central Asia, and both Chinese blocks are missing from our compilation. Second, differences in correlations used in these respective studies could have a significant effect on diversity per time interval tabulations. Finally, both Sepkoski's and Crick's tabulations are of simple generic diversity, whereas our compilation employed a "normalized" measure of generic diversity. Comparison of these two biodiversity measures for the sample time intervals indicates that our normalized generic diversity represents only two-thirds of the total number of genera present.

However, it is important to note here that in our compilation of Ordovician nautiloid genera, some subjective discretion was used in tabulating generic diversities. Not included in our counts were a number of questionable occurrences; also eliminated were erroneous assignments of Ordovician forms to Silurian and Devonian genera. These include orthocerid genera such as *Cycloceras, Dawsonoceras, Geisonoceras, Geisonocerina, Leurocycloceras, Michelinoceras, Protokionoceras, Spyroceras,* and *Virgoceras,* as well as oncocerid genera such as *Cyrtoceras, Gomphoceras,* and *Mixosiphonoceras*. Another possible reason for apparently lower generic diversity involved the removal from our lists of a number of forms now generally agreed to be synonyms of other recognized genera. For example, the endocerid genera *Cyclendoceras* Grabau and Shimer, *Foerstellites* Kobayashi, and *Nanno* Clarke are all regarded as synonyms of *Endoceras* Hall.

In conclusion, all the factors discussed here have played a role in causing the apparent differences between our compilation and those of previous authors including Crick (1990) and Sepkoski (1995). Each of these tabulations (including that presented here) represents a snapshot of Ordovician nautiloid diversity as it was known at the time of the compilation, each with its own set of limitations and factors that affect its accuracy and completeness.

22 Tube-Shaped Incertae Sedis

*John M. Malinky, Mark A. Wilson,
Lars E. Holmer, and Hubert Lardeux*

The hyoliths, cornulitids, coleoloids, sphenothallids, bryoniids, and tentaculitids included in this chapter represent unrelated, exclusively Paleozoic benthic and pelagic groups of organisms, with radial to bilateral symmetry, solitary (or rarely clustered), tube- or cone-shaped shells, and calcitic (possibly in some, originally aragonitic), phosphatic, and organic-walled shell preservation. The conical-shelled, operculate hyoliths are a moderately diverse Ordovician group, and though they did not reach the peak of abundance and diversity attained previously, in the Cambrian, they did become well diversified in higher-latitude cooler waters of the Mediterranean Province. The smaller groups comprising the mainly calcitic cone-shaped solitary coleoloids, the branched (compound) sphenothallids, and the phosphatic bryoniids with tubes and disklike attachments also have an earlier (Cambrian) history. This contrasts with the solitary conoidal, annulated shells of cornulitids (with characteristic longitudinal striae and vesicular wall) and tentaculitids (with a multilayered wall and internal septa), which do not appear until Late Ordovician, specifically mid Caradoc (= North American Mohawkian), time.

■ Hyoliths (JMM)

The class Hyolitha Marek 1963 (phylum Mollusca) encompasses calcareous, operculate, conical-shelled organisms, ranging from Early Cambrian to at least Mid Permian (Fisher 1962). The class includes two main groups, the order Hyolithida Syssoiev 1957 (Early Cambrian to Mid Permian), and the order Orthothecida Marek 1966 (Early Cambrian to Mid Devonian). The former is distinguished by the presence of a projection, or ligula, along the ventral rim (but see Kruse 1997) of the aperture and curvilinear structures known as helens, which protruded from the aperture even when the operculum was closed. These structures either may have functioned to stabilize the animal or may have permitted limited movement on the seafloor. The interior of the operculum possesses radially arranged structures known as clavicles, which occur as either a single pair or up to seven pairs in some taxa. The Orthothecida lacks the helens and ligula along the apertural rim, having instead a planar aperture or in some taxa an apertural rim with indentations, and the interior of the operculum lacks clavicles and other structures seen in the Hyolithida. The affinity of the Hyolitha remains a matter of controversy. Many early workers supported molluscan affinity, based largely on similarity in shell structure between hyoliths and other mollusks (Marek and Yochelson 1976). In contrast, other studies suggest separate phylum status (Runnegar et al. 1975). The issue of hyolith affinity remains unresolved, although herein hyoliths are regarded as mollusks.

Various aspects of hyolith ecology have long been controversial. A planktic mode of existence has been favored by many early workers owing to the supposed affinity of that group to modern pteropods (see Fisher 1962 for summary), whereas more recent investigations strongly suggest that these organisms likely were epifaunal (Marek and Yochelson 1976) or perhaps semi-infaunal. Hyoliths have long been interpreted as deposit-feeding organisms, an observation supported for the Orthothecida by the nature of the intestine (Marek and Yochelson 1976). More recently the Hyolithida have been reinterpreted as suspension-feeding organisms with the aperture oriented toward oncoming currents (Marek et al. 1997).

Hyoliths are found in all normal marine facies. Abundance and diversity attain a maximum in the Cambrian, followed by a major decline, perhaps due to increased competition from gastropods and other benthos. Only widely scattered, sporadic occurrences usually consisting of one or several individuals are known in post-Cambrian Paleozoic rocks. Kammer et al. (1986) considered the hyoliths to be a relict of the Cambrian Evolutionary Fauna (Sepkoski 1981a), with post-Cambrian occurrences largely confined to facies from stressed marine environments, where competition from other normal marine benthos would presumably have been less. This observation is supported by the abnormally high numbers of 76 hyoliths from an ooid shoal facies in the Mississippian of Iowa (Malinky and Sixt 1990) and several hundred from oxygen-deficient gray shale facies in the Midcontinent Pennsylvanian (Malinky and Mapes 1983). The Ordovician hyoliths discussed herein are an apparent exception, but high diversity and abundance in Sweden are due at least partly to a concentration of these fossils at a diastem (Dronov and Holmer 1999) and in Bohemia to unusually good circumstances of preservation (Marek 1967, 1989). Data on diversity of Ordovician hyoliths are given later in this chapter.

One aspect of hyolith paleontology that has long been neglected is the recognition of geographic and stratigraphic biodiversity patterns within the group. Marek (1976) was the first worker to provide a modern synthesis that emphasized but was not restricted to Ordovician hyoliths from Bohemia (Czech Republic) and later added an updated version (Marek 1989). Major impediments to formulating accurate statements on hyolith biodiversity include uncertainty about the morphological boundaries of the group, although this applies more to the Cambrian tubular fossils from Siberia and China than to any Ordovician specimens. Furthermore, with the exception of Marek (1976, 1989), whose works largely utilized recently collected material, workers were compelled to rely on museum collections assembled since the nineteenth century because in general hyoliths are such rare fossils. As a result, in many cases the stratigraphic level, and in some instances even the locality, cannot be identified with certainty (Malinky 2002). Usually the occurrences can be recorded only in terms of broader regional stratigraphic groupings. As a result, the number of taxa within a time slice cannot be determined with certainty; nor can they be related to particular environmental events.

Hyolith species are rarely found in Ordovician sequences, and they characteristically constitute a low-diversity component of a more diverse assemblage. Possibly this is a consequence of their low species abundances. Hyolith biodiversity is far more difficult to assess because of the overall paucity of specimens compared with other marine Ordovician invertebrates. A further complication arises from the quality of preservation of these fossils, which greatly limits how adequately some species can be identified. The skeletal mineralogy of the group has long been recognized as consisting of some unstable material (Holm 1893), possibly aragonite (Marek and Yochelson 1976), which would account for the overall poor preservation.

The present summary focuses mainly on material from areas of central Europe (Bohemia) and Morocco and from Scandinavia, within limits of the Mediterranean and Baltic faunal provinces, respectively (see Marek 1976). The species from these two provincially distinct regions have been revised recently using modern taxonomic principles and the stratigraphic control is far better than that of material from elsewhere. The present contribution is a revised version of Marek's (1976) study expanded to include recently restudied hyolith taxa (e.g., Malinky 1990, 2002). Marek's (1976) recognition of separate Mediterranean and Baltic hyolith provinces is further reenforced by data presented herein (figure 22.1). Significantly, the geographic distribution of some taxa is now known with much greater precision than their temporal distribution, which is the main subject of this contribution. Many Ordovician hyolith species listed by

FIGURE 22.1. Stratigraphic distribution of hyolith genera, with number of species known for each genus given in each column, including species in open nomenclature. Distributions are shown only in those intervals in which genera can be associated with a specific province. For example, *Hyolithes* ranges into the Lower Devonian, but because the fossil record of that genus is incompletely known, it is uncertain whether it was still present in Baltica in the Late Ordovician or whether by that time it had migrated to the Mediterranean Province. Orthothecids are indicated by an asterisk next to the name of the genus. Note that for the Baltic Province the Tremadoc has been interpreted as having a more abbreviated scope than in traditional British usage, now equivalent to the Pakerort and Varangu stages combined, while the Latorp is now represented by the Hunneberg and Billingen stages combined (see figure 2.1).

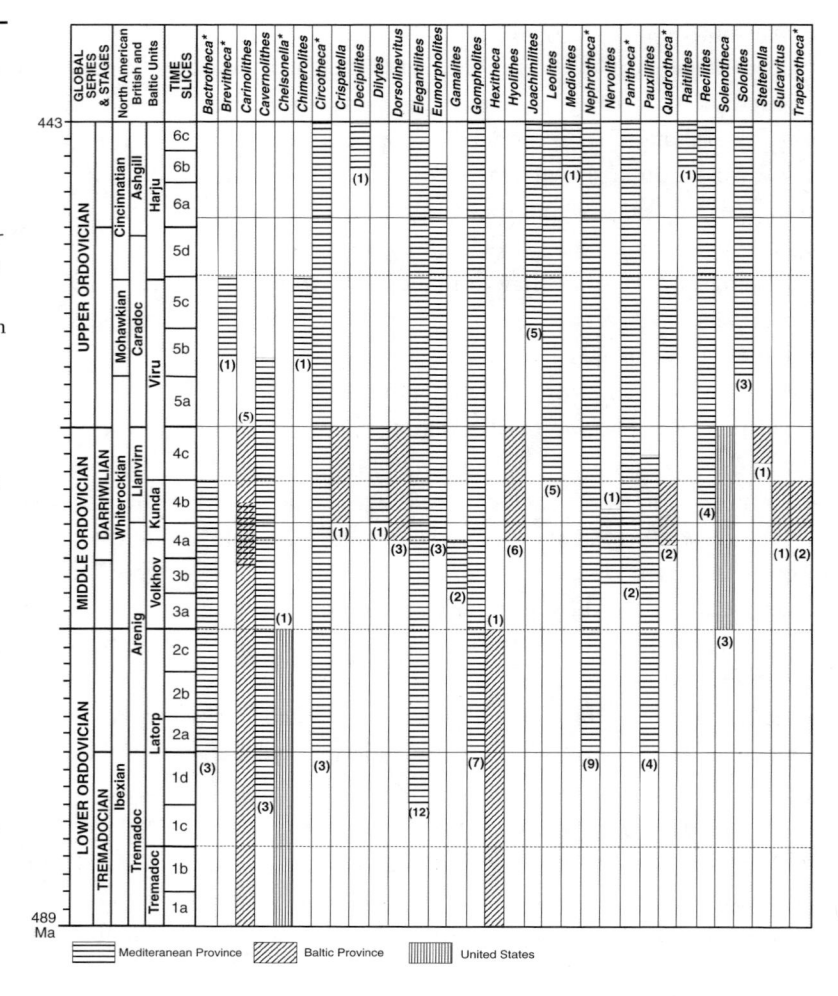

Sinclair (1946b) are in need of revision and therefore cannot be incorporated into the present biogeographic and biodiversity schemes.

Distribution and Diversity Patterns

Ordovician hyoliths have a worldwide distribution, with occurrences reported from all continents except Antarctica, but except for Europe and North Africa, the number of species described is small. Sinclair's (1946b) compilation consisted of about 25 species worldwide. That number has substantially increased since then primarily owing to the works of Marek cited herein.

Generic Patterns

In terms of the two distinct biogeographic provinces Marek (1976) recognized that 23 hyolith genera were present in the Mediterranean Province and 8 genera in the Baltic Province, with only 2 genera, *Carinolithes* and *Quadrotheca*, common to both in the Ordovician (figure 22.1). *Circotheca* occurs in both provinces, but it is exclusively a Mediterranean form in the Ordovician and solely a Baltic element in the Cambrian (Berg-Madsen and Malinky 1999). Marek (1976) stated that *Circotheca* also occurs in Germany and Bolivia, but these occurrences await confirmation. *Carinolithes* also occurs in Britain as *Hyolithus pennatuloides* (Malinky unpubl.), along with *Leolites* and *Recilites* (Malinky 2003). Revision of the few specimens reported from North America indicates that only two genera are recognizable—*Chelsonella* (Lower Ordovician) and *Solenotheca* (Middle Ordovician)—and these have a Laurentian distribution.

A marked endemicity is exhibited by the hyolith faunas of the Mediterranean and Baltic provinces.

This becomes even more evident with the incorporation of both newer hyolith taxa and taxa from other regions into the present paleogeographic/biodiversity summation. The few exceptions are *Carinolithes*, which occurs in both provinces but is represented by different species in each and has a much longer stratigraphic distribution in the Baltic Province (figure 22.1). *Hyolithes* is exclusively a Baltic form in the Ordovician but is found in Bohemia during the Silurian and Devonian. *Quadrotheca* occurs in both provinces, but at different times in the Ordovician.

Ten hyolith genera are known in the Early Ordovician, with a significant overall increase to 23 in the Mid Ordovician. Nearly two-thirds of the Mid Ordovician fauna is composed of Mediterranean generic components, and most of the remainder of the fauna consists of Baltic genera. All of the 17 reported from the Late Ordovician belong to the Mediterranean Province. The significant loss of hyolith diversity from the Baltic Province may perhaps be linked to the progressive cross-latitude drift of Baltica though Ordovician time, with the lithospheric plate moving from higher South Polar paleolatitudes to lower latitudes by Late Ordovician time (Torsvik et al. 1990).

The Mediterranean and Baltic provinces were only very weakly differentiated in the Early Ordovician (especially Tremadocian; *TS*.1a–d; see figures 22.1 and 22.2 and time slices in chapter 2), only becoming really well developed during the Mid Ordovician. Initially in the Tremadocian the hyolith generic diversity was low, comprising only two Mediterranean genera, *Cavernolithes* and *Elegantilites* from Morocco (Marek 1983), and the Baltic genera *Carinolithes* and *Hexitheca* from Sweden (Berg-Madsen and Malinky 1999; Malinky and Berg-Madsen 1999). One Lower Ordovician genus, *Chelsonella*, is recorded from Laurentia.

In the Arenig (*TS*.2a–4a), a more diverse Mediterranean fauna comprising *Bactrotheca, Cavernolithes, Circotheca, Elegantilites, Eumorpholites, Gamalites, Gompholites, Nephrotheca, Nervolites, Panitheca,* and *Pauxillites* is recorded from Bohemia, with similar occurrences in Morocco in the same interval (Marek 1976, 1983, 1989). Many of the Mediterranean Arenig genera have extended ranges in Bohemia, occurring along with the new Llanvirn (*TS*.4b–c) components *Dilytes, Leolites,* and *Recilites*, resulting in a slightly higher generic diversity.

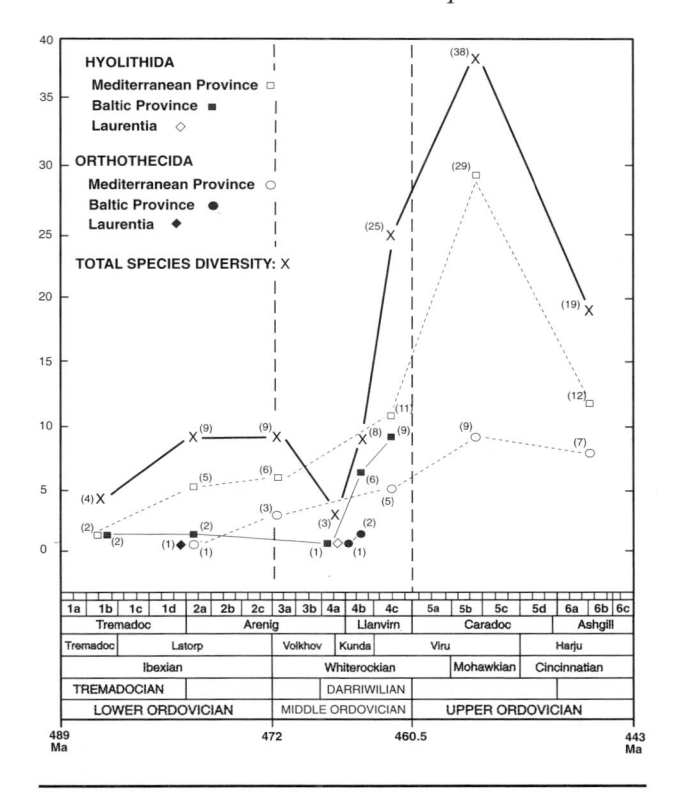

FIGURE 22.2. Number of species through time. Species occurrences are given in terms of broad regional stratigraphic units rather than time slices because of imprecisely known stratigraphic data for most species. For the Mediterranean Province, Tremadoc refers to *TS*.1a–d; Arenig: *TS*.2a–4a; Llanvirn: *TS*.4b–c; Caradoc: *TS*.5a–c; and Ashgill: *TS*.5d–6c. For the Baltic Province, Tremadoc is *TS*.1a–b; Latorp: *TS*.1c–2c; Volkhov: *TS*.3a–4a; Kunda: *TS*.4a–b; Viru: *TS*.4c–5c; and Harju: *TS*.5c–6c.

The provincially distinct Baltic hyoliths become well differentiated only during the Mid Ordovician. *Hexitheca* disappears at the Lower Ordovician–Middle Ordovician boundary, whereas *Carinolithes* persists throughout the entire Mid Ordovician (with migration to the Mediterranean Province during the Mid Ordovician). *Quadrotheca* appears near the top of the Volkhov (*TS*.4a), whereas *Dorsolinevitus, Hyolithes, Sulcavitus,* and *Trapezotheca* appear slightly later though in the same time slice near the Volkhov-Kunda boundary. *Crispatella* first appears in the lower Kunda (*TS*.4b–c), and *Stelterella* is confined to the Viru (*TS*.4c). *Solenotheca* is the sole recognizable Mid Ordovician hyolith genus from Laurentia.

In the Caradoc (*TS*.5a–d), all recognizable hyolith genera are confined to the Mediterranean Province. A total of 14 genera, including *Brevitheca* and *Chimerolites*, which are restricted to the Caradoc, are found in this interval. *Quadrotheca* occurs in the Mediterranean Province in the Late Ordovician but is known from the Mid Ordovician in the Baltic Province. Hyoliths were reported from the Late Ordovician of Sweden (Holm 1893), but none of those specimens may be confidently assigned to a genus (Malinky unpubl.). *Decipilites, Mediolites,* and *Railites* are the youngest Mediterranean hyolith genera to appear and are confined to the Ashgill (*TS*.6a–c). No Ordovician genera are known from the Late Ordovician in Laurentia.

Species Diversity

The endemic nature of hyolith species is even more pronounced than that of the genera, with no species common to both Mediterranean and Baltic provinces or to any other region. As with the hyolith genera, the apparent temporal restriction of species is partly an artifact of taxonomic treatment, as many taxa from outside the Baltic and Mediterranean provinces await reexamination. The possibility definitely exists that even species may have wide geographic distributions, as suggested by the Mid Cambrian *Contitheca cor* from Sweden, which seems to be present in Korea as *Hyolithes kotoi*. At present, all known Ordovician species appear to be confined to their respective provinces.

The pattern of hyolith species biodiversity (figure 22.2) follows that of the genera in that diversity is lowest in the Tremadocian (six species), followed by an increase to eight at the Tremadoc-Arenig boundary. Diversity increases slightly at the Lower Ordovician–Middle Ordovician boundary to nine species. The Tremadocian species consist of four Baltic and two Mediterranean elements, whereas among the later Early Ordovician species only two come from the Baltic, with all others from the Mediterranean Province. One Laurentian species extends through the entire Early Ordovician.

In the Mid Ordovician, two Baltic genera are known from near the Volkhov-Kunda (*TS*.4a) boundary, with the total increasing to eight in the Kunda. Hyolith biodiversity substantially increases in late Mid Ordovician, with 16 species from the Mediterranean Province and nine from the Baltic. The highest level of diversity is attained in the Upper Ordovician (Caradoc/Viru) with a total of 38 species, followed by a decrease to half that number in the Ashgill. All Late Ordovician species are from the Mediterranean Province.

■ Cornulitids, Coleoloids, and Sphenothallids (MAW)

Cornulitids, coleoloids, and sphenothallids are extinct tube-dwelling groups sometimes abundant in Ordovician faunas. They are not related to one another, at least not closely, and are treated here together for convenience. The systematic positions of these three groups are poorly understood, and their preserved skeletons are so simple that distinguishing species is difficult, but recent work has shown that they can be important for paleoecologic and paleoenvironmental interpretations, especially in the Ordovician.

Cornulitids

The family Cornulitidae Fisher 1962 currently consists of four genera (three in the Ordovician) and about 45 species of small to medium-sized (2–80 mm) calcareous tubes variously ornamented with concentric rings and longitudinal striae. Although they resemble "worm tubes" such as those of serpulids and spirorbids, the vesicular microstructure of cornulitid walls is very different from the laminar microstructure of annelid tubes, leaving their placement in a higher classification uncertain (Fisher 1962). Cornulitids are most commonly found cemented to hard substrates such as shells, carbonate hardgrounds, and rock grounds, but a few occur as unattached single tubes or clusters (Morris and Felton 1993).

The earliest cornulitids occur in the Upper Ordovician (Mohawkian), and the latest members of the family are found in the Tennessean of the Lower Carboniferous (Richards 1974). Ordovician cornulitids were almost certainly cosmopolitan in their distribution but thus far have been reported only from North America and Europe.

The three Ordovician cornulitid genera are

Cornulites Schlotheim 1820. Solitary conoidal tubes with prominent concentric rings and longitudinal striae in adult forms; vesicular walls thick; tubes reach 80 mm in length and 20 mm in diameter at aperture; Upper Ordovician (Mohawkian) to

Middle Devonian and possibly Lower Carboniferous (Fisher 1962).

Conchicolites Nicholson 1872. Conoidal tubes found solitary or in clusters; tubes attached at the narrow tip, which is slightly curved; short, imbricated rings but no longitudinal striae; vesicular walls thin; tubes up to 13 mm long, with diameters up to 3 mm; Upper Ordovician (Mohawkian) to Lower Devonian (Fisher 1962; Richards 1974).

Cornulitella Howell 1952. Solitary conoidal tubes attached along one side; rings prominent on all sides except the attached one; no longitudinal striae; thick vesicular walls; tubes up to 13 mm long and 3 mm in diameter; Upper Ordovician to Lower Carboniferous (Fisher 1962).

The paleoecology of Ordovician cornulitids has been extensively treated by Morris and Rollins (1971), Richards (1974), and Morris and Felton (1993). Most cornulitids attached to hard substrates and were undoubtedly filter feeders, but a few lived unattached in soft sediments, leading Richards (1974:515) to consider them as possible "experiments in deposit feeding." Morris and Felton (1993) showed a convincing symbiosis between the crinoid *Glyptocrinus,* platyceratid gastropods, and *Cornulites* in the Cincinnatian of the upper midwestern United States.

Coleoloids

The family Coleolidae Fisher 1962 was placed "provisionally" in the phylum Mollusca by Fisher (1962). Coleoloids are elongate cones made of calcium carbonate (calcite, but some may have been aragonitic) ranging from 0.5 to 75 mm long. They usually have a slight curve toward the closed apex. The tubes are unornamented, or have longitudinal striae, or bear oblique ridges. The shell walls are relatively thick and laminated, with some interiors in most. About 20 species have been described in seven coleoloid genera (Fisher 1962), but several of these may belong to other groups of conoidal organisms.

The earliest coleoloids appear in the Lower Cambrian, and the last of them are found in Carboniferous rocks. Coleoloids are usually very rare, but they are occasionally found in oriented masses (Fisher 1962; Yochelson 1968).

Three coleoloid genera have been reported from Ordovician rocks. One of them, *Polylopia,* was later removed from the coleoloids by Yochelson (1968). It is included here for historical reasons and to give it a temporary home for the discussion of Ordovician biodiversity.

Paoshanella Yin 1937. Compressed cone, lenticular in cross section; longitudinal striations; up to 60 mm long; Lower Ordovician of China (Fisher 1962).

Polylopia Clark 1925. Straight cone with longitudinal ribbing and circular cross section; up to 3 cm long (Fisher 1962); ?Middle Ordovician of North America. *Polylopia* was originally described as "multilayered"; the name itself is derived from the Greek for "many layers of tree bark" (Clark 1925). Yochelson (1968) demonstrated, though, that this was a misinterpretation of depositionally nested, single-layer cones. Yochelson (1968) made a strong case that *Polylopia* is a mollusk and, based on more tenuous morphological and paleoecologic evidence, suggested that it might be a hyolith.

Salopiella Cobbold 1921. Small (up to 3 mm long) with oblique steplike ridges around the tube on the interior as well as the exterior; Lower Cambrian of England (Cobbold 1921) and "lower part of Middle Ordovician" of Volynia, Ukraine (Hynda 1973:250). Fisher (1962:W134) considered the coleoloid placement of *Salopiella* "uncertain" because the interior of the shell is not smooth.

Polylopia is the only one of these genera for which we have paleoecologic information. Yochelson (1968) interpreted *Polylopia* as a benthic organism living in very shallow nearshore marine waters. It may have carried its shell erect, aided by gas inside the closed apex.

Sphenothallids

Sphenothallids consist of a single genus (*Sphenothallus* Hall 1847) with a complicated taxonomic history reviewed most recently by Zhu et al. (2000). Their amended diagnosis describes *Sphenothallus* as an elongate, slender theca, single or branched, with a subconical holdfast and two longitudinal thickenings extending up the theca; the theca is composed of apatite or organic material, and it is lamellar with the lamellae parallel to the theca surface; the thecae range in length from less than 2 mm to tens of mm.

Sphenothallus occurs from the Lower Cambrian to the Permian in North and South America, Europe, and southern China (Zhu et al. 2000). It is found throughout the Ordovician, beginning in the Tremadocian (Choi 1990), with the most common occurrences in the Cincinnatian of North America (Bodenbender et al. 1989; Bolton 1994; Neal and Hannibal 2000). Most commonly sphenothallids are seen in the Ordovician as clusters of holdfasts (Bodenbender et al. 1989; Neal and Hannibal 2000).

The systematic placement of *Sphenothallus* is uncertain. Two hypotheses are current: that they were "worms" of some sort (i.e., Mason and Yochelson 1985), or that they were cnidarians (i.e., Van Iten et al. 1992). The branching, clonal nature of *Sphenothallus*, along with the similarity of its walls to those of conulariids, favors the cnidarian hypothesis.

Sphenothallus was apparently a gregarious, euryoxic, and eurytopic organism comfortable in a variety of marine conditions, from soft muds to shells and hardgrounds (Neal and Hannibal 2000).

■ Byroniids (LEH)

Byroniids (order Byroniida Bischoff 1989) are phosphatic and/or organic tube-shaped fossils that were attached to the substrate by a disk (generally between 0.1 and 0.8 mm in diameter); they have been described from most continents and range in age from the Cambrian to the Permian (Bischoff 1989). Fragmented remains of the isolated phosphatic attachment disks (figure 22.3A) have generally been referred to *Phosphannulus* Müller, Nogami, and Lenz 1974, but this genus is now considered to be a junior synonym of *Byronia* Matthew (see Bischoff 1989: 477). In the Early Paleozoic, these attachment disks have generally been referred to a single species, *Byronia universalis* (Müller, Nogami, and Lenz 1974), which ranges from the Late Cambrian to the Late Devonian. The attachment disks of byroniids are closely similar to those of *Sphenothallus* Hall, but the internal structure of the tube appears to be different (see Wilson earlier in this chapter).

Ordovician Byroniids

Byroniids appear to represent a moderately common constituent in Ordovician shallow-water carbonate deposits from across the world, but to date there are only a very limited number of published

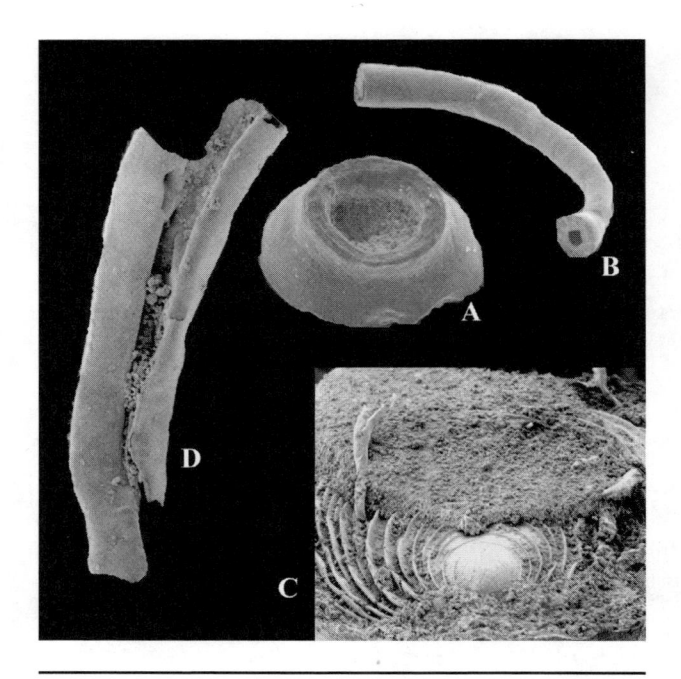

FIGURE 22.3. *Byronia universalis* Müller et al. (1974). A, B, Folkeslunda Limestone (sample DLK-Fo-5; Holmer 1989: figure 9A), Lasnamägi regional stage, Dalarna, Sweden. C, D, Viivikonna Formation, Kukruse regional stage, Kohtla, Estonia. A, isolated attachment disk, ×35. B, curved tube and attachment disk, ×10.5. C, tubes and attachment disks on dorsal valve of *Schizotreta* sp., ×2.9. D, isolated tube, ×10.5.

accounts, and there is no basis for analyzing the patterns of their distribution. The first account of an Ordovician byroniid seems to have been by Öpik (1930), who described and illustrated the occurrence of numerous attachment disks (Öpik 1930:31, figure 11, plate 5:2) of what are evidently "*Phosphannulus*-type" byroniids (probably representing *Byronia universalis* in a wide sense) from the Kukruse beds (lower Upper Ordovician, *TS*.5a) in Estonia (figure 22.3B, C). Koslowski (1967) described well-preserved organic-walled byroniids from Baltic erratic boulders of Caradoc age occurring in Poland (*TS*.5; see also Bischoff 1989). Warn (1974) extended the record to North America and described material now referred to byroniids from the Late Ordovician of Ohio. Müller et al. (1974) completed a comprehensive study of the Early Paleozoic record of "*Phosphannulus*-type" byroniids; they recorded *B. universalis* from the Late Cambrian–Early Ordovician of Wyoming, Iran, and Sweden. In Sweden there is an almost continuous record of forms referable to *B. universalis,* which ranges at least from the Tremadocian (*TS*.1b; Müller et al. 1974) through the Volkhov-Kunda (*TS*.3a–4b; Müller et al. 1974; Eisenack 1978; and Holmer unpubl.) and the Aseri-Kukruse regional stages (*TS*.4c–5a; Müller et al. 1974; Holmer unpubl.; figure 22.3A, B herein) to the late Caradoc (*TS*.5b–c; Holmer 1986, 1987). Much of the Baltic material of byroniids is well preserved and includes both the attachment disk and the long tube (e.g., Holmer 1987: figure 1K, L; figure 22.3A, B herein). Bischoff (1989) described an extensive material of Early Paleozoic Australian byroniids including a single species from the Late Ordovician of New South Wales; he also reviewed the entire record of Early Paleozoic byroniids. In Canada, byroniids have been noted from the Cambrian-Ordovician (Landing et al. 1980) and the Ordovician-Silurian boundary beds (Nowlan et al. 1988) in the Northwest Territories.

Mode of Life and Zoological Affinity

The remains of byroniids are generally found as a by-product of etching for conodonts or other acid-resistant microfossils (e.g. Bischoff 1989; Müller et al. 1974), and thus any evidence of the host or substrate to which the attachment disk was cemented is lost. The best-known cases of still-attached byroniids are from crinoid stems, where "*Phosphannulus*-type" byroniids with phosphatic disks and tubes are surrounded and embedded within the stereom of the stem, forming so-called stem galls; the record of this association ranges from the Late Ordovician (Warn 1974) to the Permian (Welch 1976; see also Werle et al. 1984). Müller et al. (1974) described two possible attached disks of *B. universalis* sitting on an indeterminate phosphatic fragment and a poorly preserved lingulate brachiopod. The association between lingulate brachiopods and attached byroniids is also known from the Kukruse beds (lower Upper Ordovician, *TS*.5a) in Estonia, where dorsal valves of the discinid brachiopod *Schizotreta* are covered by numerous disks and tubes of *B. universalis* (figure 22.3C). However, the byroniids do not appear to have been host-specific, since identical attached phosphatic disks have also been reported occurring on trilobites and other types of brachiopods from the same level in Estonia by Öpik (1930: figure 11, plate 5:2).

The byroniids have been considered to represent a variety of different animal groups but most commonly referred to some kind of tube-forming worm (e.g., Müller et al. 1974). However, Bischoff (1989) summarized and reviewed the contrasting views on the zoological affinities of byroniids and concluded that the tube and attachment disks of byroniids are most similar to the attached thecae of the polypoid stage of modern coronate scyphozoans (see also Glaessner 1971, 1984; Bischoff 1978). The Byroniida can be considered as an extinct order of the thecate scyphopolyps (Bischoff 1989).

■ Tentaculitids (HL)

If one accepts the most generally adopted classification of tentaculitids (e.g., Lardeux 1969; Larsson 1979), namely, as the class Tentaculitoidea Lyashenko 1957 consisting of three orders—Tentaculitida Lyashenko 1955, Homoctenida Boucek 1964, and Dacryocorarida Fisher 1962—it may be suggested that no occurrences of indisputable Ordovician tentaculitids have been published yet.

These claims rest on two bases: some authors class as tentaculitids certain shells that undoubtedly resemble the genus *Tentaculites* but whose state of preservation is insufficient to allow this attribution with any certainty; and others, more frequently, assign shells of the cornulitid group (e.g., *Cornulites, Conchicolites*) to the tentaculitids.

Since the insightful contribution of Yochelson (1961, 2000), followed by Fisher (1962), it has been generally agreed that the hyolithids must be distinguished from the tentaculitids. Fisher (1962), for his part, distinguishes between tentaculitids and cornulitids, for which he created a new family, placed in an "uncertain" order and class. Boucek (1964), however, took a different view, proposing a new class, Tentaculita, which combined the orders Tentaculitida, Homoctenida, Dacryoconarida, Coleolida, and Cornulitida. It is true that some cornulitids are strongly reminiscent of tentaculitids, a fact that was first noted a long time ago (e.g., Hall 1847; Barrande 1867). Nevertheless, Boucek's suggestion has not gained general acceptance (see Lardeux 1969; Larsson 1979).

The assumption of the presence of tentaculitids at the beginning of the Early Ordovician rests on the discovery of small conical shells with ringed exteriors, but lacking their apical and oral regions, from the Chepultepec Limestone (early Tremadocian age) of the Shenandoah Valley (Virginia, United States). These shells were called *Tentaculites lowndoni* by Fisher and Young 1955 (but named "*Tentaculites*" *lowndoni* by Fisher in Downie et al. 1967). I have had the opportunity to examine a specimen of this species in the Smithsonian Institution (ref. U.S.N.M. 33403), and my judgment is that its state of preservation is too poor for it to be identified as a tentaculitid.

It would appear that it is on the basis of the work of Fisher (1962) that several authors cite the presence of tentaculitids in the Ordovician. However, a closer reading of Fisher's text reveals that he very clearly allows that the appearance of *Tentaculites* and *Uniconus* in the Lower Ordovician is subject to doubt, although this is not represented in the distribution table he provided (Fisher 1962: figure 53). We see this doubt though, on pages 110 and 113, where the Lower Ordovician distributions of the two genera are marked by question marks.

The only species he represents as appearing uncontroversially in the Ordovician (p. 111) is "*Tentaculites anglicus* Salter" from the Caradoc of England, but this may be a cornulitid, as already suggested by Boucek (1964:50).

It is true that the "*Tentaculites*" *anglicus* shell is nearly as straight as that of true tentaculitids, but its marked longitudinal costulation, a feature of cornulitids, should not be confused with the discreet microcostulation of tentaculitids (Lardeux 1969: plate IX, figure 1).

From reports in the literature it may be suggested that the first uncontroversial tentaculitids appeared toward the middle of the Silurian (see Larsson 1979). However, Bergström (1996) has confirmed the presence of tentaculitids in the Upper Ordovician rocks of Ohio. The two illustrated species are relatively common in strata of Richmond age (i.e., ranging through from Waynesville to Whitewater formations). One of the species, *Tentaculites richmondensis* Miller, exhibits straight to slightly curved, tapering conchs, 25 to 30 mm long and 2 to 3 mm wide, with distinctive transverse rings that gradually increase in size from proximal to distal ends (no bulbs), and the proximal portions of conchs show internal septa. Only the well-preserved specimens show an ornamentation of fine longitudinal striae. The other widespread species,

T. sterlingensis Meek and Worthen, is much smaller and has more densely spaced transverse rings. Tentaculitids also occur in slightly older Cincinnatian rocks, for example, in the Kope Formation of Ohio (S. M. Bergström pers.comm.). These are of Edenian (late *TS*.5c–early 5d), or late Caradoc, age.

In addition, in recent discoveries, as yet unpublished, Kent Larsson (pers. comm.) has identified Caradoc tentaculitids from the Onny Valley of southern Shropshire, England, including specimens of "*Tentaculites*" *anglicus*, with all the typical features of the tentaculitid group. They exhibit multilayered, unattached, conical conchs with transverse rings, internal septa, and closed apical ends that are tapered without bulbs. Furthermore, these Ordovician tentaculitids possess a moderately distinct longitudinal costulation, and so this feature is not typical only of cornulitids, and they may also show a slight curvature apically, like many Silurian tentaculitids. Larsson has concluded, based on this Onny fauna, that true tentaculitids were present by mid Caradoc time.

Additionally, Kent Larsson found Ashgill tentaculitid assemblages in Sweden and Norway, and some of these specimens have a large size (40 to 60 mm in length). Another occurrence of Ashgill tentaculitids (also currently undocumented) is in thin-bedded limestones (unit 6) of the Tsagaandel Formation at Tsagaandel Hill, west of Bayankhongor, central Mongolia (B. D. Webby pers. comm.; for stratigraphic context, see Minjin 2001).

Consequently, the tentaculitids first appeared, like many other Paleozoic groups, during the Caradoc and have a continuing presence in the Ashgill. The earlier Ordovician records of tentaculitids, however, remain unsubstantiated. Affinities of tentaculitids remain uncertain; possibly they represent an independent group long separated from major phyla such as the Annelida and Mollusca.

ACKNOWLEDGMENTS

Kent Larsson (Lund, Sweden) kindly allowed his new and significant confirmation of occurrences of Late Ordovician tentaculitids in the British Isles and Baltoscandia to be mentioned here.

23 Worms, Wormlike and Sclerite-Bearing Taxa

Olle Hints, Mats Eriksson, Anette E. S. Högström, Petr Kraft, and Oliver Lehnert

The biodiversity of four wormlike groups is outlined in this chapter. The first is a larger group, the jaw-bearing polychaetes, which are represented by scolecodonts, the organic-walled elements of their jaw apparatuses. The others are smaller, more problematic groups. The machaeridians are most commonly preserved as isolated calcitic sclerite, derived from the dorsal exoskeletal armor of an unknown, bilaterally symmetrical wormlike organism. The palaeoscolecideans and chaetognaths (arrow worms) are phosphatic, identified mainly by their microfossil remains and a few by more complete body fossils. Characteristic microfossil elements of palaeoscolecidans are platelike tubercles, and the chaetognaths exhibit spinelike grasping elements that suggest the chaetognaths may have close links to protoconodonts.

■ Jawed Polychaetes (Scolecodonts) (OH, ME)

Scolecodonts, the jaws of polychaete worms (Annelida, Polychaeta), are common microfossils in Ordovician sedimentary rocks. They are composed of yellow to black organic material allowing acid preparation techniques to be used for their extraction.

The first report on Ordovician scolecodonts is by Hinde (1879), who described specimens from Canada. Since then, Ordovician scolecodonts have been recovered from many regions of the world, for example, Australia (Furey-Greig 1999), Kazakhstan (Klenina 1989), China (Gao 1980), South America (Ottone and Holfeltz 1992), North America (e.g., Stauffer 1933; Eller 1945 1969; Bergman 1998; Eriksson and Bergman 1998, 2001), and Europe (e.g., Kielan-Jaworowska 1966; Szaniawski 1970; Hints 1998). However, except for the last two regions, the existing data are far too limited to allow detailed diversity analyses. Hence, this chapter is focused on the Baltic region and North America.

Preliminary data on interregional distribution of jaw-bearing polychaetes indicate that there were close links between the faunas of different continents during the Ordovician (Hints et al. 2000; Eriksson and Bergman 2001). Most genera were common to Laurentia and Baltica, but the majority of the few species recovered from other regions apparently also belong to these widespread genera. There are, however, a number of Baltic genera that have not yet been identified in Laurentian deposits and vice versa. At the species level, on the other hand, relatively few forms are present in more than one continent. Thus, currently available species-level diversity data characterize a particular region, or a continent, but cannot yet be used for the entire world. Consequently, the global diversity of jawed polychaetes can be assessed in a meaningful way only at the genus level, and even then, the preliminary nature of the data must be stressed.

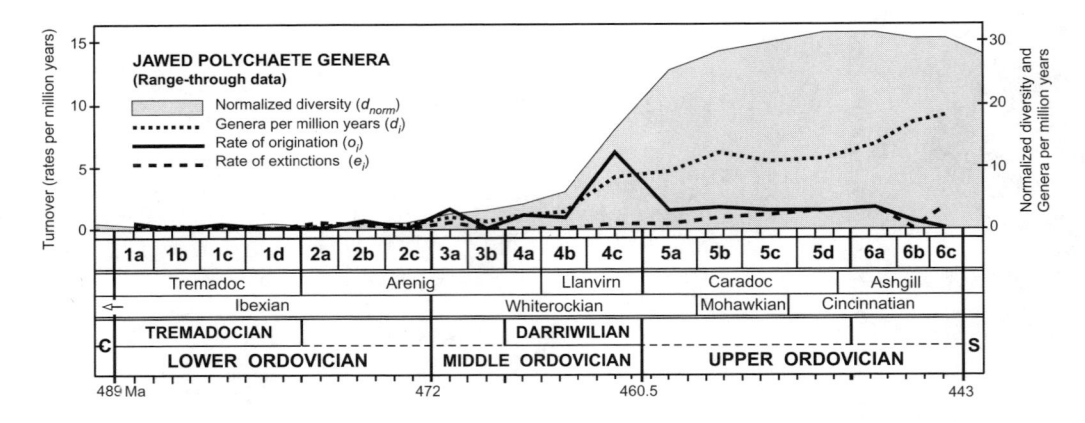

FIGURE 23.1. Genus-level diversity pattern of Ordovician jawed polychaetes. Stratigraphic scale is given in accordance with Webby et al. (chapter 2).

A multielement taxonomy has been applied here for analyzing the distribution and diversity patterns of jawed polychaetes. Taxa recognized using the outdated single-element taxonomy are of little use without careful revisions of old literature and the corresponding type collections (Eriksson and Bergman 1998).

The taxonomic richness was calculated using normalized diversity and taxa per million years. The normalized diversity (number of taxa ranging through the time unit, plus half the number of taxa appearing or disappearing in it, plus half the number of those confined to the unit) differs from total generic diversity only by producing slightly lower values and a smoother curve (figure 23.1). In addition, origination and extinction rates per million years were calculated. In all cases, range-through data rather than actually recorded data were used.

Approximately 50 Ordovician genera, including those undergoing taxonomic revision and those that lack formally proposed generic names, are currently known. The number of families is debatable, but some 15–20 families were probably present in the Ordovician. The total number of species is even more difficult to estimate. In the Baltic region, 150 multielement species are known. The number of species seems to be slightly lower in North America, and some 70–100 species have thus far been recovered, based mainly on material from the midcontinent region. These numbers will most likely increase with ongoing sampling and investigations combined with revisions of the pioneer scolecodont literature.

The oldest known scolecodonts have been recovered from the uppermost Cambrian (S. H. Williams pers. comm.) in Newfoundland. The same area has yielded some of the oldest known Ordovician scolecodonts (Underhay and Williams 1995; Williams et al. 1999), but Early Ordovician specimens have also been found in Estonia by O. Hints (unpubl. data) and in China (R. Brocke pers. comm.). The collections from *TS*.1–2 (see time slices in chapter 2) have not yet been thoroughly studied, but it seems that the diversity of jawed polychaetes is very low through this interval. The limited number of jaw apparatuses recovered or reconstructed allows at least three distinct genera to be distinguished. It is currently premature to evaluate the higher-level taxonomy of early polychaetes, although it seems that some of these forms are related to xanioprionids and conjungaspids and others are very similar to *Lunoprionella,* a common genus in younger Ordovician strata. More advanced forms, such as those with labidognath and prionognath type apparatuses (e.g., Kielan-Jaworowska 1966), have not been recorded in this interval.

In *TS*.3a, members of *Oenonites* and *Mochtyella* are first recorded (Underhay and Williams 1995). These genera, particularly the former, commonly dominate the assemblages in younger Ordovician strata as well as in the Silurian (Eriksson 1997; Hints 2000). At least by this time (*TS*.3a) the evolution of jaw-bearing polychaetes had passed some important milestones, and the main polychaete lineages had become differentiated.

The diversity remains rather low until *TS*.4, when a major increase in species diversity, as well as in abundance, is recorded and the number of genera increased very rapidly. Moreover, this interval marks the first appearance of most of the genera that became common in younger strata. A major component of the rapid increase in normalized diversity occurs during the *TS*.4c–5a interval (figure 23.1). Since the genera appearing in *TS*.4c and 5a commonly range through the remaining part of the Ordovician, the diversity curve changes only very little in later time slices, displaying a slight increase until *TS*.5d–6a and a subsequent decrease in *TS*.6b–c. It is difficult to ascertain whether minor fluctuations are meaningful. Since most genera occurring in the uppermost Ordovician range into the Silurian, there is also no distinct drop in diversity at the Ordovician-Silurian boundary, which is marked by a significant diversity drop in many other fossil groups.

Genera per million years gives a pattern slightly different from the normalized diversity curve (figure 23.1). It shows much smoother increasing trend with some slightly steeper parts in *TS*.4c and 5b. However, as most genera occur in several time slices, the taxonomic richness in shorter time slices, such as *TS*.6b and 6c, becomes somewhat overestimated using this measure.

The turnover estimates are in good accordance with the normalized diversity curve. There is a small peak in origination rate in *TS*.3a, but the major peak is confined to *TS*.4c. After that the origination rate remains very stable until a decrease in *TS*.6b and 6c (see figure 23.1). The extinction rate, in contrast, has no clearly defined peaks. That is, jawed polychaetes appear to have been relatively unaffected by some of the events causing severe extinctions in many other fossil groups.

Composite species-level data, such as those provided by Hints (2000) for the Baltic region, may display higher complexity, especially in turnover rates. However, as is the case in global genus-level data, the species-level taxonomic richness in the Baltic region remained relatively stable after the main diversity rise (Hints 2000: figure 2).

There is some question as to how well the revealed pattern corresponds to the actual diversification of the group and whether conclusions can be drawn based on such a patchy data set. It is indeed very likely that the number of genera will increase and that the ranges of many taxa will be extended with results from ongoing investigations. Therefore, the present diversity estimates should be viewed as preliminary rather than comprehensive (i.e., the diversity was at least not lower than here indicated). The record is especially incomplete for the Lower and lower Middle Ordovician (*TS*.1–3). Further data from this interval will be invaluable, and they will likely change the current picture to some extent.

The present knowledge about the diversity patterns of Ordovician jaw-bearing polychaetes can be summarized as follows:

1. Jawed polychaetes originated in pre-Ordovician time.

2. Several advanced taxa appeared in *TS*.3, but the most significant increase in diversity and abundance seems to have occurred in the Middle Ordovician (*TS*.4), after which genus-level diversity remained relatively stable without distinct peaks in origination or extinction rates.

3. Most of the Ordovician genera are apparently long-ranging, and many of them extend into the Silurian.

4. An intercontinental faunal exchange is indicated, and most families and many genera were common to Laurentia and the Baltic region. However, there are taxonomic differences in the faunas between these paleocontinents, especially at the species level.

5. The abundance and great diversity of scolecodonts in the Ordovician warrant further investigation of this group.

■ Sclerite-Bearing Machaeridians (AESH)

Sclerite-bearing taxa of varying types abound in Lower Paleozoic rocks from many parts of the world. Several have a very restricted stratigraphic range, and a number of better-known groups, such as the tommotiids and halkieriids, occur only in the Cambrian. The problematic machaeridians, however, have a substantial stratigraphic distribution (Lower Ordovician to Middle Permian). Additionally, their disarticulated remains are being recognized and reported in increasing numbers (Ordovician examples include Dzik 1994b; Ebbestad and Högström 1999; Högström and Droser 2001). All scleritome-forming taxa suffer essentially similar preservational problems: the sclerites become widely dispersed after the animal dies, and isolated sclerites are thus by far the most common type of preservation. Hence, the distribution of isolated sclerites is the main indicator of their spatial and temporal

FIGURE 23.2. A selection of different Ordovician machaeridian types. A, Middle Ordovician *Lepidocoleus ulrichi* Withers 1926, anterior end of a partial scleritome, from the Trenton Group, Prosser Limestone, Cannon Falls, Minnesota, United States, ×5.6. B, *Plumulites* sp. from the upper Ashgill of the Taimyr Peninsula, Arctic Russia, ×8.0. C, *Plumulites* sp. from the Ashgill Boda Limestone of central Sweden, ×2.1. D, E, *Lepidocoleus suecicus* Moberg 1914, from the Ashgill of Bohemia, latex casts, both ×6.3. F, G, minute plumulitid-type sclerites with extreme marginal spines from the Ashgill, Fjäcka Shale, Koängen boring, Scania, Sweden; F, ×16.7; G, ×10.5. H, I, plumulitid-type machaeridians from the upper Ibexian (Arenig) Al Rose Formation of the Basin and Range, Inyo Mountains, California, United States, both ×4.5.

distribution as well as of their faunal importance and environmental preferences. However, complete scleritomes are necessary for accurate reconstructions and for understanding the morphology and function of the scleritome. In general a machaeridian scleritome consists of two or four longitudinal series of posteriorly overlapping and imbricating calcitic sclerites (see figure 23.2A). These are arranged in transverse segments of varying numbers, from 14 to well over 60 (Högström and Taylor 2001). The three major families (Lepidocoleidae, Plumulitidae, and Turrilepadidae), grouped in the two orders (Lepidocoleomorpha and Turrilepadomorpha), are primarily distinguished on the basis of scleritome arrangement (Jell 1979).

Machaeridians have a wide environmental distribution that ranges from deep offshore shelves to shallow epicontinental seas and includes apparent epibenthic as well as semi-infaunal modes of life (Högström 2000). An example of the former may be the extremely spiny and minute sclerites of an undescribed plumulitidlike machaeridian from the Upper Ordovician Fjäcka Shale of Sweden (figure 23.2F, G).

When the organism is reconstructed, the resulting "snowshoe effect" conferred by these spines can be interpreted as a possible adaptation to life on unstable soft substrates. Terrace-shaped rugae reminiscent of burrowing sculpture found in some machaeridians suggest a possible semi-infaunal lifestyle (for example, see figure 23.2A, D–E). In addition, a large collection (~100 specimens) of well-articulated lepidocoleid machaeridians from the Lower Devonian (Lochkovian) of Oklahoma are preserved in a manner indicating rapid burial and possible in situ preservation within fine-grained muds. More unusual is the occurrence, from carbonate mud mounds of the Upper Ordovician Boda Limestone of Sweden, of rare sclerites of a very large plumulitid machaeridian (figure 23.3C), in an environment otherwise seemingly devoid of machaeridians.

Ordovician Taxa

Despite the increasing number of reported machaeridians, their record is still "spotty" and with taxonomic uncertainties largely caused by preservation and relatively few specimens. It is, however, valuable to produce a general compilation of machaeridian distribution through the Ordovician, based on work published so far as well as on personal observations and communications.

Tremadocian

The oldest confirmed machaeridian sclerites are of plumulitid type from the Tremadocian Dumugol Formation (Dongjeom area) of South Korea (Kobayashi and Hamada 1976; Choi and Kim 1989), where *Plumulites gumunsoensis* Choi and Kim 1989 was quoted as a fairly common but overlooked component of the fauna in the Dumugol Formation. From the same formation *Plumulites primus* Kobayashi 1934 is known (Kobayashi and Hamada 1976). The Bjørkåsholmen Formation of Norway may contain the oldest Baltic machaeridians (B. Funke pers. comm.).

Arenig

The early to mid Arenig shows signs of the first larger radiation of machaeridians with several new occurrences of plumulitid-type machaeridians, for example, in the Al Rose Formation of the Inyo Mountains of the Basin and Range, California (figure 23.2H, I) (pers. observ.). Barrande (1872) described the earliest machaeridians from Bohemia (four species of *Plumulites*), but there is a slight degree of uncertainty in their stratigraphic placement, and some may be of late Tremadocian age. Toward the upper Arenig, machaeridians are known also from northeastern North America (Clark 1924), China (Kobayashi and Hamada 1976), and the Holy Cross Mountains of Poland (Dzik 1994b). In addition, as yet undescribed material from the St. Petersburg district consists of what may be the earliest lepidocoleid-type machaeridians (pers. observ.), belonging to the Baltoscandian middle and upper Volkhov stage (Egerquist 1999).

Llanvirn

The trend from the Arenig continues into the Llanvirn (mid to late Darriwilian) with an increase in the number of taxa and their distribution. A species of *Plumulites* occurs in abundance in the Kanosh Formation of the Ibex area in the Great Basin of Utah and Nevada. Machaeridians also show a continuous distribution through the Holy Cross Mountains sequence (Dzik 1994b). The first machaeridians from Morocco (Chauvel 1967) appear in the Llanvirn and show a continued Gondwanan distribution, and the interval also marks the continued appearance of South Korean and Chinese material (Kobayashi and Hamada 1976). Eastern North America continues to show the same increasing trend with additional plumulitid machaeridians as well as the first lepidocoleids (Hall and Whitfield 1875; Withers 1926). One of the first reports with certain Baltic affinity is from the Kukruse stage in Estonia (Withers 1921), and they also continue to occur in Bohemia.

Caradoc

The Late Ordovician exhibits taxa persisting from the Llanvirn, for example, from the Holy Cross Mountains (Dzik 1994b), but also from South Korea (Kobayashi and Hamada 1976), North America (Withers 1926), and Morocco (Chauvel 1967). New taxa are recorded from among other areas—Sweden, Bohemia, North America, and Scotland (Barrande 1872; Moberg 1914; Withers 1926). Overall the Late Ordovician is a time when species diversity increases and machaeridians appear to thrive in many settings.

Ashgill

The uppermost Ordovician contains faunas with diverse and numerous machaeridian remains (Dzik 1994b; Ebbestad and Högström 1999). In Europe several taxa from the Holy Cross Mountains continue into the Ashgill (Dzik 1994b), as well as Bohemian and Baltic taxa (Barrande 1872; Withers 1926). The Baltic fauna, especially the Fjäcka Shale, shows a diverse and conspicuous machaeridian component, where they may be the numerically dominant taxon in some beds. In addition to plumulitids and lepidocoleid machaeridians there are examples of new minute forms with extreme marginal spines mentioned earlier (figure 23.2F, G). The North American fauna continues to exhibit *Lepidocoleus jamesi* (Hall and Whitfield 1875) and *Lepidocoleus strictus* Withers 1926, as well as components that at this point appear relatively similar to Baltic types. The occurrence in China of two subspecies of a Baltic species indicates a possible faunal connection between these areas during the Ashgill (Wu 1990). Additional Malaysian and South Korean taxa occur in the Upper Ordovician as well, giving this region a continuous record from the Tremadocian (Kobayashi and Hamada 1976). Moroccan machaeridians are seemingly diverse and abundant (Chauvel 1967) but in urgent need of further studies. In Russia *Plumulites* sp. occurs just below the Ordovician-Silurian boundary layers of the Taimyr Peninsula (figure 23.2B) (pers. observ.).

Discussion

The general impression of machaeridian diversification and distribution through the Ordovician and onward is one of continuing success. Although the group is not one of the most common, its substantial stratigraphic distribution and environmental and geographic spread suggest very successful organisms. Uncertainties and problems do remain with the machaeridian record, owing to both preservation and lack of study. However, an increased focus on certain areas has shown that machaeridians very likely were a more conspicuous component of Paleozoic faunas than previously recognized. This is well exemplified, for example, by the recently discovered machaeridians in the Basin and Range of California and the Great Basin of Utah and Nevada. Work in progress there has revealed large numbers of sclerites from the previously mentioned Al Rose (late Ibexian/Arenig) and Kanosh (mid to late Whiterockian/mid to late Darriwilian) formations and additional (but rarer) isolated machaeridian sclerites from the Nine Mile (Whiterock Canyon, Nevada: late Ibexian/Arenig) and Wahwah (Ibex area, Utah: latest Ibexian to early Whiterockian/Arenig) formations (pers. observs.). Yet another example where machaeridian remains are turning up in large numbers and occasionally constitute one of the most numerous of the taxa is the Ashgill Fjäcka Shale of central Sweden (Ebbestad and Högström 1999), even though machaeridians have been known from this level since the beginning of the 1900s (Moberg 1914). *Lepidocoleus suecicus* Moberg 1914, common in the Fjäcka Shale, is also found at similar stratigraphic levels in Bohemia (figure 23.2D, E), indicating a possible closer relationship to Baltica than was previously thought. Compared with the diversity in both the Al Rose and Kanosh formations, that of the Fjäcka Shale is much higher, with a few numerous forms and a number of rarer ones. This also seems to be the general trend, with a substantially higher diversity toward the Late Ordovician. In addition, faunas in North America and Europe grow increasingly similar toward the end of the Ordovician as a result of the closure of the Iapetus Ocean.

The apparent conservative morphology of machaeridian scleritomes through time is quite striking. This is especially evident in plumulitid machaeridians, in which articulated Ordovician specimens differ little from later forms, for example, from the Devonian (Jell 1979; Rudkin 2001). This is at least partly reflected by a low higher-level diversity, in which most differences between forms lie in the sclerites themselves and not in the overall organization of the scleritome.

The number of new finds from various parts of the world gives hope that the poor resolution of the machaeridian record will improve significantly in the near future. In addition, more studies of continuous sections such as the one by Dzik (1994b) from the Holy Cross Mountains will help to clarify the picture of the distribution and turnover rates of machaeridians as well as their role in different communities.

■ Palaeoscolecidans and Chaetognaths (PK, OL)

Palaeoscolecidans and chaetognaths (arrow worms) belong to separate phyla. Soft-body preservation is extremely rare in both groups. This fact strongly biases our picture of the diversity. However, isolated phosphatic microelements—present in each of these groups—provide a basis for studies of their biostratigraphic and paleogeographic distribution.

Palaeoscolecidans

Palaeoscolecidans have been assigned to a number of different phyla. Most authors consider them as annelids (e.g., Robison 1969; Glaessner 1979). Recently, new opinions have emerged and led to a proposal that their systematic position is as an extinct group of nematomorphs (Hou and Bergström 1994) or more probably priapulids (e.g., Conway Morris 1997). Their body is composed of tuberculate annuli with plates of different shape and distribution.

Preservation of body fossils is due to the phosphatized cuticle (probably secondary). The phosphatic tubercles (plates) are the heaviest mineralized parts, and as a consequence of postmortem decay they are often found isolated (see, e.g., Hinz et al. 1990). These microfossils have been described as *Hadimopanella* Gedik (= *Lenargyrion* Bengtson), *Kaimenella* Märss, and *Milaculum* Müller, while some sets of integrated microelements are referred to *Utahphospha* Müller and Miller. The microelements have an independent paratomic classification. Naturally, attempts to correlate both modes of preservation have been published. For example, Müller and Hinz-Schallreuter (1993) assigned several fragments of body surface to *Hadimopanella* and *Milaculum* because of the typical morphology of tubercles (plates). On the other hand, van den Boogaard (1989) described isolated tubercles (plates) as ?*Palaeoscolex*, based on similarities to material from Bohemia (figure 23.3, small shaded area of "body fossilized" column of Palaeoscolecida).

Palaeoscolecidans range from Lower Cambrian to Silurian (Ludlow). In the Ordovician, body fossils have been described from the lower Tremadocian of Wales (*Palaeoscolex* Whittard 1953; see also Conway Morris 1997), lower Arenig to the lowermost Caradoc of Bohemia (?*Palaeoscolex*, *Plasmuscolex*, and *Gamascolex*; Kraft and Mergl 1989), and the lowermost Cincinnatian of Kentucky and Ohio ("*Protoscolex*"; for overview and comment, see Conway Morris 1977; Conway Morris et al. 1982).

Isolated phosphatic tubercles (plates) have been recorded from Ordovician rocks in different parts of Laurentia and Baltica. There is also considerable potential for information in many conodont collections (e.g., at different universities and in the Canadian and U.S. Geological Surveys), but this material has not yet been studied in detail.

Palaeoscolecidans had a wide paleogeographic distribution from tropical warm-water environments (Laurentia) to temperate-water areas (peri-Gondwana). Although the group ranges throughout the whole Ordovician, occurrences reach a maximum in the Lower and Middle Ordovician (see figure 23.3). The diversity of palaeoscolecidans seems to decrease from Cambrian through the Ordovician until they become extinct in the Silurian. This can be illustrated by the number of described genera: their record includes 15 genera of bodily preserved worms and 4 genera of tubercles (plates) from the Cambrian, 4 (or maybe 5) and 4 genera of worms and plates, respectively, from the Ordovician, and only 1 and 2 genera, respectively, from the Silurian.

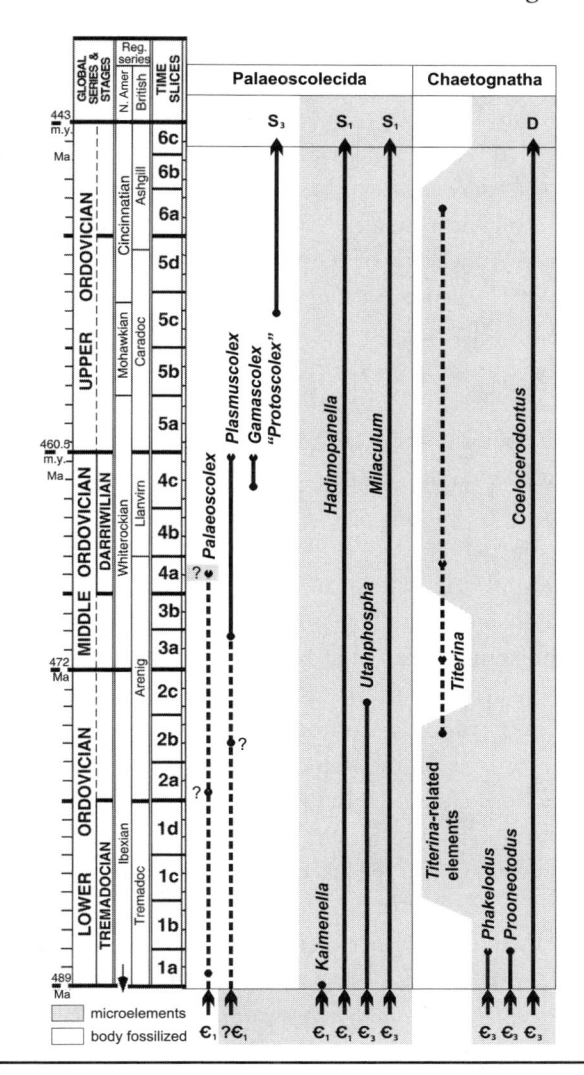

FIGURE 23.3. Stratigraphic ranges of palaeoscolecidans and chaetognaths. Bodily preserved taxa are in the white columns, microelements are shaded. Points mark only important appearance data. Abbreviations: Lower Cambrian ($\mathcal{C}_1$), Upper Cambrian ($\mathcal{C}_3$), Lower Silurian (S_1), Upper Siluarian (S_3), Devonian (D).

Chaetognaths

This phylum probably appeared in the Cambrian (e.g., Szaniawski 1982). Today there are some 250 species in this phylum, but soft-body preservation is extraordinarily rare throughout the fossil record. There are reports from the Lower Cambrian (*Protosagitta* Hu, in Chen et al. 2002), the Lower to Middle Ordovician, and the upper Pennsylvanian (Schram 1973). The Ordovician material includes three specimens referred to *Titerina rokycanensis* Kraft and Mergl 1989 (see also Kraft et al. 1999). Globular microfossils, interpreted as eggs of *Titerina* or a related genus, are known from the Arenig and lower Llanvirn of Argentina (Heuse et al. 1996). This species is unusual in the possession of a prominent pair of grasping spines. Almost identical isolated apparatuses have been documented from the Upper Ordovician of the Canadian Arctic (McCracken and Nowlan 1989). They may belong to *Titerina* or a closely related chaetognath genus.

Despite the paucity of this record, it has been repeatedly advocated that protoconodonts represent the grasping spines of chaetognaths, and we consider they may represent an extinct order of this phylum. Grasping spines in Cambrian and Ordovician protoconodonts are comparable in shape and size to those in modern chaetognath apparatuses and have a similar outline in cross section. In addition, sets of protoconodonts preserved as complete apparatuses have been recorded from the Upper Cambrian, and in particular the affinity between the apparatus of *Phakelodus tenuis* (Müller) and that of the recent pelagic chaetognath *Pseudosagitta maxima* (Conant) has been discussed by Szaniawski (1987).

In comparison, chaetognath teeth are very small, and in this sense only Szaniawski's (1996) report of small teeth attached to an Upper Cambrian *Phakelodus elongatus* (Zhang) cluster shows evidence for the systematic position of protoconodonts within this group. In addition, clusters of the thin-walled elements of *Coelocerodontus* Ethington, with deep basal cavities, may be attributed to protoconodonts. Their clusters have an apparatus architecture similar to that of *Phakelodus* Miller (cf. Szaniawski 1998), and the elements are also similar to grasping spines of *Titerina*. *Coelocerodontus* is well known from the Upper Cambrian through Upper Ordovician; the youngest reports are from the Devonian (e.g., Telford 1975).

Protoconodonts are widespread, diverse, and most common in the Upper Cambrian, but their number decreases at the end of the Cambrian and during the earliest Ordovician. Body fossils and protoconodonts testify to a cosmopolitan distribution of chaetognaths during the Cambro-Ordovician. Their overall diversity during the Ordovician is difficult to estimate. Additionally, the puzzle of isolated protoconodont elements and clusters and their long stratigraphic ranges indicate that chaetognath diversity was probably low in the Ordovician.

ACKNOWLEDGMENTS

Anette Högström acknowledges support from the Swedish Research Council (Vetenskapsrådet), the National Science Foundation through Mary Droser and the European Community Marie Curie Foundation. Simon Conway Morris and Barry Webby are thanked for their helpful and valuable comments.

24 Trilobites

Jonathan M. Adrain, Gregory D. Edgecombe, Richard A. Fortey, Øyvind Hammer, John R. Laurie, Timothy McCormick, Alan W. Owen, Beatriz G. Waisfeld, Barry D. Webby, Stephen R. Westrop, and Zhou Zhi-yi

The taxonomic diversity history of Ordovician trilobites has been explored on a broad global scale by Adrain et al. (1998), who provided an estimate based on a fourfold division of Ordovician time. Adrain and Westrop (2000) subsequently published a trilobite diversity curve based on nine Ordovician and five Early Silurian sampling intervals, and the pattern of trilobite alpha (within-habitat) diversity during the Ordovician has been documented by Westrop and Adrain (1998) and Adrain et al. (2000). The nature of the Ordovician radiation of trilobites is further characterized herein, through a new global analysis at finer resolution and by documenting geographic and environmental patterns in the data. Regional diversity curves are presented and discussed for Australasia, South America, Avalonia, Baltica, and South China, reflecting a spectrum of tectonic and paleogeographic settings through the Ordovician. Finally, the development of trilobite biofacies through the period is assessed in the context of the global and regional patterns of biodiversity change.

■ Global Patterns (JMA, SRW, RAF)

The Data Set

Data on the taxonomy and temporal and geographic distribution of Ordovician and Silurian trilobites have been compiled by J. M. Adrain beginning in 1997. In their assessment of post-Cambrian trilobite diversity and evolutionary faunas, Adrain et al. (1998) used a database of 1,241 recorded genera, of which they accepted 945 as meaningful taxa—842 are Ordovician genera. At that stage, it was possible to present data only at a relatively coarse series-level resolution.

In order to achieve further insight, the resolution of the data set needed to be increased. Effectively, this required that (1) a workable set of intervals be developed that could be applied with a minimum of uncertainty to data from all parts of the world and (2) the data set be extended to species level in order to document genus ranges and geographic occurrence accurately through time. The sampling intervals chosen (see Adrain and Westrop 2000: note 30) were a necessary compromise between the high precision available from some paleocontinents (e.g., Laurentia, Baltica, Australasia, Avalonia, and parts of Gondwana) and the sometimes very coarse resolution in others (e.g., Siberia, the central Asian terranes, and much of South America). One alternative was to adopt a highly resolved scheme and then deal with less-resolved data according to an error distribution formula (the procedure used by Sepkoski 1986; see also Sepkoski and Koch 1996 as a result of dependence on secondary data sources). Although such a system is reasonable for a combined global diversity estimate, it does not allow effective comparison of geographic regions by time interval. We therefore adopted a system in which data had a direct empirical assignment to a sampling bin, despite the limitations imposed by poorly sampled regions. The result is a ninefold division of Ordovician time, based on major correlative biohorizons identified by Webby (1995, 1998), corresponding approximately to the level of stage or subseries. In terms of the 19 time slices (*TS*) adopted for this book, our intervals are less well resolved but at least directly match particular time-slice boundaries. Relationship of the intervals is as follows: O1 = (*TS*.1a, 1b); O2 = (*TS*.1c, 1d); O3 = (*TS*.2a, 2b, 2c); O4 = (*TS*.3a, 3b); O5 = (*TS*.4a, 4b, 4c); O6 = (*TS*.5a); O7 = (*TS*.5b, 5c); O8 = (*TS*.5d); O9 = (*TS*.6a, 6b, 6c).

Compilation to the species level is now well advanced. Earliest and latest occurrences have been documented, and there is now considerable confidence in the genus ranges. Overall, the species compilation is sufficiently complete for preliminary analysis, and species data are used to document the latitudinal distribution of families later in this chapter. By 2000, the number of genera recorded had risen to 1383, with 994 accepted. Because about two-thirds of the added names were synonyms or *dubia* taken from more obscure primary literature, the effective database had increased in size by only 5.7 percent. This is the database analyzed in the present work. A small number of accepted new genera of Ordovician trilobites that were published in 2000 and 2001 are excluded from the present work in favor of concordance with the database used by Adrain and Westrop (2000).

Changes in the data set since 1998 include some taxonomic reassignment. Bathyurids and bathyurellids were united in a family Bathyuridae in 1998 but are considered separate families for purposes of this analysis. A family Panderiidae was recognized in 1998, but the group is considered a subfamily of Bathyurellidae in the present data set.

The current global trilobite diversity curve based on these data is shown in figure 24.1. A listing of taxa, synonyms, and stratigraphic ranges of the genus data set is available on request from J. M. Adrain.

A New Global Analysis

The main conclusion of Adrain et al. (1998) was that a significant portion of post-Cambrian trilobites experienced rapid diversification during the Ordovician Radiation in a fashion nearly identical to that of Sepkoski's (1981a) Paleozoic Evolutionary Fauna. The terms "Ibex Fauna" and "Whiterock Fauna" were introduced. The former comprised a cohort of families that peaked early, declined, and were eradicated before or during the end Ordovician mass extinction; the latter had low Early Ordovician diversity, diversified during the Ordovician Radiation, and contained all families that survived the end Ordovician mass extinction (a group termed the "Silurian Fauna").

Even though the fourfold division used by Adrain et al. (1998) clearly documented major unrecognized features of the trilobite record—Mid Ordovician diversification and its link to end Ordovician survival—a more stratigraphically resolved analysis of the current data set is desirable to test the cohesiveness of the new evolutionary faunas (essentially to see if they break down into discrete components) and better refine the timing of diversification events.

The analysis was carried out using the same protocols. Families were grouped according to similarity in their genus diversity through the Ordovician sampling intervals. Silurian data were not used in the analysis. At this increased level of resolution, the data

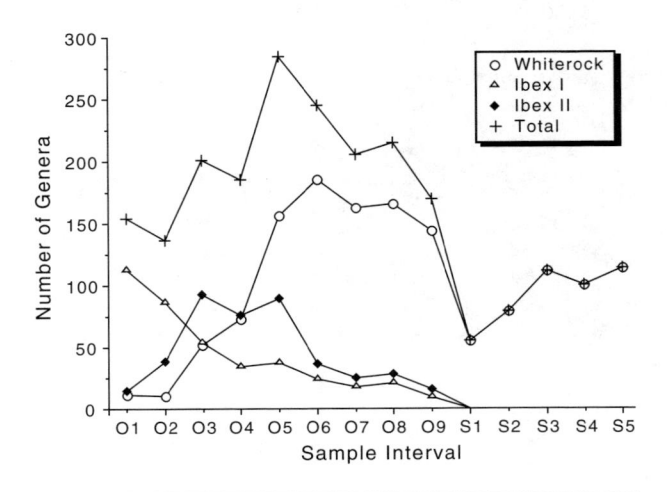

FIGURE 24.1. Ordovician and Early Silurian trilobite genus diversity. Whiterock, Ibex I, and Ibex II are evolutionary faunas defined by hierarchical cluster analysis (see figure 24.2). Sample intervals are as defined by Adrain and Westrop (2000:112, note 30). It should be noted that sample intervals O1–O3, O4–O5, and O6–O9 are equivalent to the tripartite Ordovician division, respectively, Lower (or Early), Middle (or Mid), and Upper (or Late) Ordovician, used in this volume. Intervals O1 and O5 also correlate with global Stages Tremadocian and Darriwilian, respectively.

became subject to edge effects. In particular, there were eight families with single occurrences (in some cases of a single specimen) in the earliest part of O1. These taxa essentially have no Ordovician history but, when admitted to the analysis, cluster together with 100 percent similarity and exert undue influence on the pattern of similarity of the remaining (interesting) taxa that have a genuine Ordovician history. These "singleton" taxa ("Dokimokephalidae," Idahoiidae, Lichakephalidae, Nepeidae, Norwoodiidae, Papyriaspididae, Plethopeltidae, and Solenopleuridae), restricted in occurrence to O1, were therefore excluded from the analysis.

The results of the cluster analysis are shown in figure 24.2. The main features are as follows:

1. The Ibex Fauna includes two distinct clusters with different diversity trajectories (see figure 24.1) during the O1–O3 intervals (lumped together as the Ibexian by Adrain et al. 1998). One group, termed "Ibex Fauna I," had high O1 diversity but steadily declined afterward. The Olenidae and Ceratopygidae are typical of this fauna. A second group, termed "Ibex Fauna II," had very low O1 diversity but radiated rapidly during O2, peaked during O3, and declined after O5. The Asaphidae and Bathyuridae are typical families of this fauna.

2. The Whiterock Fauna composition remains almost exactly as described in 1998, though there is a small amount of changed membership. The low-diversity families Dionididae and Bohemillidae, which had clustered with the Whiterock Fauna in the four-interval analysis, now move to Ibex Fauna II. Isocolidae, which had clustered with the Ibex Fauna, and Harpetidae, the sole unclustered family, both now move to the Whiterock Fauna. Twenty-two of 24 families are common to the 1998 and present versions of the Whiterock Fauna.

3. With increased stratigraphic resolution, it is evident that some Whiterock Fauna families began to radiate during the O3 interval (e.g., Trinucleidae, Raphiophoridae, Cyclopygidae), whereas others began to radiate during the early Whiterockian O4 (e.g., Illaenidae, Encrinuridae, Odontopleuridae). There is a strong geographic component to this distinction, which is discussed later in this chapter.

4. The two most surprising results in the analysis of Adrain et al. (1998) are confirmed. First, the Whiterock Fauna—the majority of post-Cambrian trilobites—experienced an Ordovician radiation (figure 24.1) much like that of the Paleozoic Evolutionary Fauna. Second, Ordovician diversity history remains an extremely accurate predictor of end Ordovician fate. All families of both Ibex Fauna I and Ibex Fauna II became extinct either before or during the end Ordovician mass extinction, whereas 18 of 26 Whiterock Fauna families survived into the Silurian. There is a strong geographic component to this Whiterock Fauna survivorship, also discussed later herein.

Trilobite Radiation by Realm

Modern work on global Ordovician trilobite biogeography (using cladistic methods to search for common historical signals as opposed to phenetic methods to compare taxonomic lists) has not been attempted and is sorely needed. The best estimate of global pattern around the time of the radiation remains Whittington and Hughes's (1972) classic quantitative study of taxic distribution, which used multidimensional scaling in the first comprehensive attempt to define major biogeographic areas in the Early Ordovician. Their analysis has largely been supported by subsequent work (Ross 1975; Fortey and Morris 1982; Cocks and Fortey 1982, 1988; Neuman 1984; Fortey and Cocks 1986, 1992; Fortey et al. 1989; Fortey and Mellish 1992), with the main additions being attention to the concept of biofacies and accounting for the effects of levels of endemicity varying with environment. Current concepts of broad area relationships during the Early Ordovician have been summarized by Cocks and Fortey (1990). The level of precision in well-studied areas such as eastern Avalonia and western Baltica is now high, though data remain sparse in many other parts of the world. Nevertheless, it is clear that during the time of the radiation, trilobites occupied at least four distinct biogeographic realms (figure 24.3).

An equatorial Bathyurid Realm (Bathyurid Province of Whittington and Hughes 1972) includes Laurentia and Siberia/Kolyma and possibly parts of Kazakhstan (as far as is known; see Apollonov 1975 for summary) and North China (Zhou and Fortey 1986). Data adequate for quantitative analysis have been compiled only for Laurentia. The most characteristic pre-radiation faunal elements are the endemic bathyurids

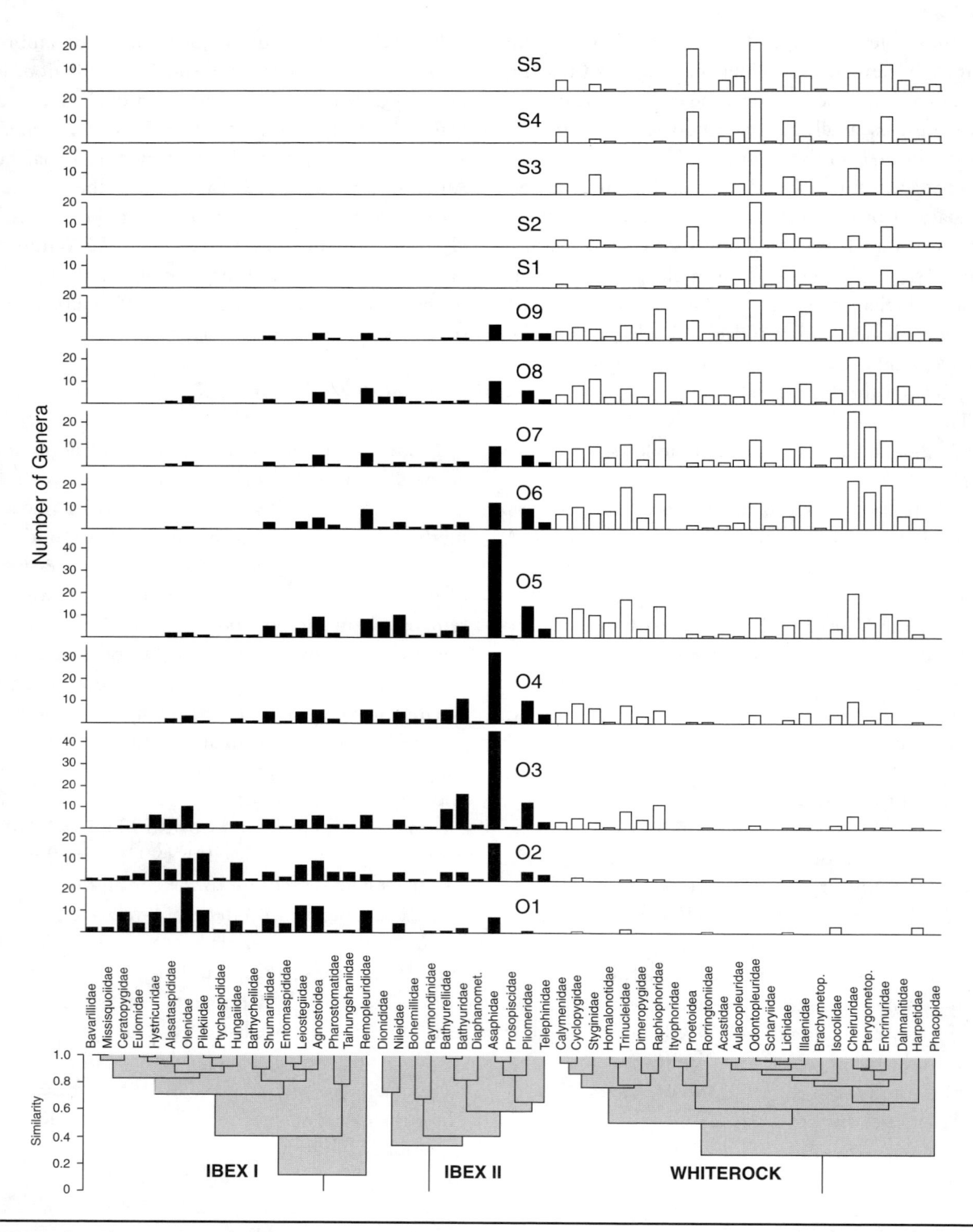

FIGURE 24.2. Cluster analysis of Ordovician trilobite families, with plots of their diversity through time. Clustering was based on Ordovician diversity only (intervals O1–O9). Taxa were clustered using as variables the number of genera in each of the nine Ordovician biostratigraphic intervals. The Pearson product-moment correlation coefficient was used as the index of similarity, and the clusters were formed using the average linkage method.

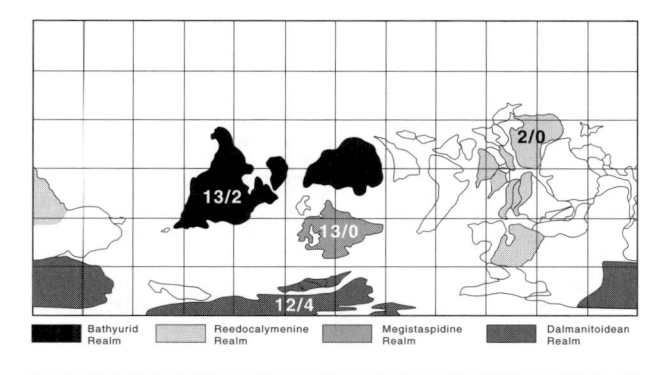

FIGURE 24.3. Four biogeographic realms during the time of the Ordovician Radiation, as defined by Whittington and Hughes (1972). Kazakhstan and North China are left unshaded because their affinities are not definite, but they may belong to the Bathyurid Realm. Numbers indicate the number of Whiterock Fauna clades occurring in the realm at the time of the radiation and the number endemic to that realm (see table 24.1).

and cybelopsine pliomerids, though other taxa, such as dimeropygids and hystricurids, have their distributions concentrated in this realm (e.g., Ross 1951; Hintze 1953; Whittington 1963, 1965; Fortey 1979, 1980; Adrain and Fortey 1997). Thirteen major Whiterock Fauna clades were present in the Bathyurid Realm (table 24.1) at the time of onset of the Ordovician Radiation, of which two were endemic.

A southern midlatitude Megistapidine Realm occupies Baltica (Asaphid Province of Whittington and Hughes 1972) and is marked by a striking endemic radiation of megistaspidine asaphids (Jaanusson 1953a, 1953b, 1956; Tjernvik 1956; Tjernvik and Johansson 1980; Nielsen 1995). Other elements of Baltic Arenig/Llanvirn faunas are shared with Laurentia (*Celmus, Nileus, Raymondaspis, Illaenus*), with which there are clearly the strongest faunal ties. Thirteen major Whiterock Fauna clades were present in the Megistaspidine Realm at the onset of radiation (table 24.1), but none were endemic to the realm.

A temperate southern Dalmanitoidean Realm (*Selenopeltis* Province of Whittington and Hughes 1972) includes Avalonia and parts of Gondwana (e.g., Armorica, Perunica, present-day North Africa). In keeping with its high-latitude position, this area has the fewest links with the other major realms and is dominated by cyclopygids, ogygiocarinine asaphids, trinucleids, reedocalymenine calymenids, and early dalmanitids (e.g., Hammann 1974, 1983; Fortey and Owens 1978, 1987; Henry 1980; Rabano 1990). Twelve major Whiterock Fauna clades were present in the Dalmanitoidean Realm at the onset of radiation (table 24.1), of which four were endemic.

A Reedocalymenine Realm also has tropical/equatorial distribution and comprises part of Gondwana, including South China, Australia, and much of South America (Asaphopsis Province of Whittington and Hughes 1972). The Hungaiidae (= Dikelokephalinidae; see Ludvigsen and Westrop in Ludvigsen et al. 1989 for discussion) is restricted to this

TABLE 24.1. Distribution by Faunal Realm of Whiterock Fauna Families and Subfamilies During the Onset of the Ordovician Radiation (intervals O3 and O4)

Taxon	Bathyurid	Megistaspidine	Dalmanitoidean	Reedocalymenine
Calymenidae:				
Calymeninae	√	√	√	
Colpocorpyinae			√	
Reedocalymeninae			√	
Cheiruridae	√	√	√	
Cyclopygidae		√	√	
Dalmanitidae			√	
Dimeropygidae	√	√		
Encrinuridae	√	√		
Homalonotidae			√	
Illaenidae	√	√	√	√
Isocolidae	√	√	√	
Lichidae	√	√		
Odontopleuridae:				
Ceratocephalinae	√			
Selenopeltinae	√	√	√	
Proetoidea	√			
Pterygometopidae		√		
Raphiophoridae	√	√	√	√
Styginidae	√	√		
Trinucleidae		√	√	

realm during the Arenig. Benthic platform faunas are highly endemic (e.g., Harrington and Leanza 1957; Fortey and Shergold 1984; Jell and Stait 1985b; Laurie and Shergold 1996a, 1996b). Pelagic telephinids, however (Cocks and Fortey 1990: figure 3), are shared with equatorial Laurentia and may be difficult to distinguish between continents even at the species level (Fortey 1975b; McCormick and Fortey 1999). Strikingly, only two Whiterock Fauna families, Raphiophoridae and Illaenidae, were present in the Reedocalymenine Realm during the onset of radiation. These are also the only two Whiterock Fauna groups with a global, fully cosmopolitan distribution during this time.

There is therefore, for reasons thus far unknown, no evidence for a significant Ordovician radiation of trilobites in the Reedocalymenine Realm, and it was not until considerably after the radiation elsewhere (Edgecombe, Webby and Laurie, later in this chapter) that many Whiterock Fauna groups appeared in Australia. We therefore exclude this realm from consideration and concentrate on latitudinal patterns between the remaining three, which were positioned at low, intermediate, and high latitudes at the time of the radiation.

High- versus Low-Latitude Radiation and End Ordovician Extinction

Trilobites are an exemplar taxon for groups hard hit by the end Ordovician mass extinction. By all estimates, including our own (figure 24.1), trilobites lost around half of their global taxic diversity during the event. The question of selectivity at a major mass extinction is always of interest. Are extinguished groups related, and different from survivors, in some particular trait or pattern? Chatterton and Speyer (1989), for example, claimed that the larvae of some trilobites were benthic, whereas others were pelagic, and that groups with the latter suffered greater extinction. Adrain et al. (1998) related extinction propensity to clade size, showing that the end Ordovician event preferentially removed clades whose latest Ordovician genus diversity was low. Is there any geographic component to end Ordovician extinction patterns?

For Ordovician trilobites, geographic patterns are masked by a burst of cosmopolitanism during the

TABLE 24.2. Latitudinal Distribution of Whiterock Fauna Families and Subfamilies During the Time of the Radiation Contrasted with End Ordovician Fate

Taxon	Low	Middle	High	End Ordovician
Calymenidae:				
Calymeninae	50	10	40	Survival
Colpocoryphinae	0	11	89	EXTINCT
Reedocalymeninae	6	29	65	EXTINCT
Cyclopygidae	4	13	83	EXTINCT
Styginidae	36	60	4	Survival
Homalonotidae	0	15	85	Survival
Trinucleidae	14	26	60	EXTINCT
Dimeropygidae	81	19	0	EXTINCT
Raphiophoridae	68	19	13	Survival
Proetoidea	60	0	40	Survival
Odontopleuridae	31	26	43	Survival
Lichidae	20	55	25	Survival
Illaenidae	49	32	19	Survival
Isocolidae	45	45	10	EXTINCT
Cheiruridae:				
Cheirurinae	63	26	11	Survival
Acanthoparyphinae	0	100	0	Survival
Cyrtometopinae	0	100	0	EXTINCT
Deiphoninae	75	25	0	Survival
Eccoptochilinae	0	26	74	EXTINCT
Sphaerexochinae	94	6	0	Survival
Pterygometopidae	23	73	4	Survival
Encrinuridae	62	19	19	Survival
Dalmanitidae	0	0	100	Survival

[a]This early distribution is calculated as a percentage of total species present during intervals 03, 04, and 05, at low, middle, and high latitudes.

Ashgill in response to climatic cooling. During this time, many groups that had shown strong high-latitude endemicity achieved wider, low-latitude distributions. For example, dalmanitids, chasmopine pterygometopids, reedocalymenine calymenids, homalonotids, and cyclopygids, among others, became widespread just prior to the extinction.

Based on the summary presented here, Ordovician radiations of trilobites clearly occurred with a substantial latitudinal component (table 24.2). Some radiating groups were initially endemic or nearly endemic to high-latitude Gondwana and others to low-latitude Laurentia (and possibly Siberia/Kolyma, though data are very sparse).

This tabulation demonstrates that a strong majority (12 of 15) of clades that had their origin and early diversification centered in the Bathyurid Realm or Megistaspidine Realm (i.e., had a majority or plurality of their species diversity occurring there) went on to survive the end Ordovician mass extinction and contribute to the Silurian Fauna. Further, of the 15 clades centered in either the Bathyurid or Megistaspidine realm, 14 are next most common in the other of these realms—that is, they have an extremely strong or exclusive intermediate and low-latitude distribution. The only exception is Encrinuridae, reflecting the high-latitude diversification of the dindymenines. In contrast, of eight clades that had their origin and early diversification centered in the Dalmanitoidean Realm, only three survived the end Ordovician mass extinction. The difference in survival between clades whose diversification occurred in low to intermediate latitudes and those with origins in high latitudes is statistically significant. A G-test with Williams's correction for small sample sizes (Sokal and Rohlf 1981) rejected the null hypothesis of independence of survival from latitudinal distribution at the .05 level.

Hence, for whatever reason, a significant majority of clades that would survive the extinction and form the Silurian Fauna had their first occurrence and their radiation heavily concentrated in intermediate and low latitudes, and most were centered in the Bathyurid Realm.

Development of the Whiterock Fauna in Laurentia

It is important to test for any environmental signal to the emergence of the Whiterock Fauna in all three realms in which a significant radiation of trilobites occurred. However, adequate data are currently limited to the low-latitude Bathyurid Realm, which contributed more than half of the clades destined to form the Silurian Fauna. Here, the focus is on the environmental pattern of the radiation in Laurentia because this continent contributes the overwhelming amount of data for the Bathyurid Realm and also has the best available environmental information.

Concern for Ordovician environmental distributions and their potential overprint on geographic patterns was pioneered by Fortey (1975a) in his qualitative analysis of Spitsbergen faunas. Ludvigsen (1978) applied quantitative techniques to biofacies analysis in the Middle Ordovician of northwestern Canada, and Q- and R-mode hierarchical cluster analysis of trilobite biofacies is now becoming routine (e.g., Ludvigsen and Westrop 1983; Westrop 1986; Ludvigsen et al. 1989; Westrop and Cuggy 1999). Fortey's biofacies scheme is intuitive and has been widely supported, but it has not been put on a formal analytical footing, a task that we attempt here.

Ideally, biofacies analysis should be undertaken with quantitative sampling data, so that relative abundance of taxa can participate in clustering. Such data are only occasionally available (summarized on a global scale by Westrop and Adrain [1998] and Adrain et al. [2000]) and are lacking for most well-described faunas from Laurentia at the time of onset of the Ordovician Radiation. As a result, simple presence/absence data were used, with the acknowledgment that information on relative abundance may in the future greatly enhance the analysis.

The results from analysis of 37 Laurentian collections from the O4 interval are shown in figure 24.4. Most taxa are genera, although occasionally subfamilies or even families are used where data were sparse. "Singleton" taxa restricted to a single collection were not scored. Analyses were performed with SPSS v.10.0 for the Macintosh computer (SPSS 2000). The analysis confirms the general applicability of Fortey's (1975a) scheme, yielding four progressively deeper-water biofacies labeled (following Fortey) Bathyurid, Illaenid-Cheirurid, Nileid, and Olenid. The Bathyurid biofacies includes collections from shallow subtidal environments (i.e., above-average storm wavebase) common on the craton interior. The Illaenid-Cheirurid biofacies occurs in deep subtidal and buildup collections, mostly derived from the craton margin. The Nileid and Olenid biofacies occur in progressively deeper-water, mostly marginal environments.

Does the Whiterock Fauna contribute more substantially to one or more of these biofacies? As demonstrated by figure 24.5, the majority of species occurring in the Illaenid-Cheirurid biofacies belong to Whiterock Fauna groups, whereas the Ibex faunas dominate in all other biofacies. The pattern is more striking still when only the Silurian Fauna (excluding the Raphiophoridae, as in Adrain et al. 1998) is plotted—groups that survived the end Ordovician were overwhelmingly concentrated in the Illaenid-Cheirurid biofacies. Hence, the trilobite radiation in Laurentia was initiated in marginal deep subtidal and buildup environments fringing the craton.

The pattern of origin and spread of the Whiterock Fauna in Laurentia is shown in figure 24.6. Early Ibexian faunas were completely dominated by elements of the Ibex faunas across the environmental spectrum. By the late Ibexian, the Whiterock Fauna had appeared in all environments but dominated

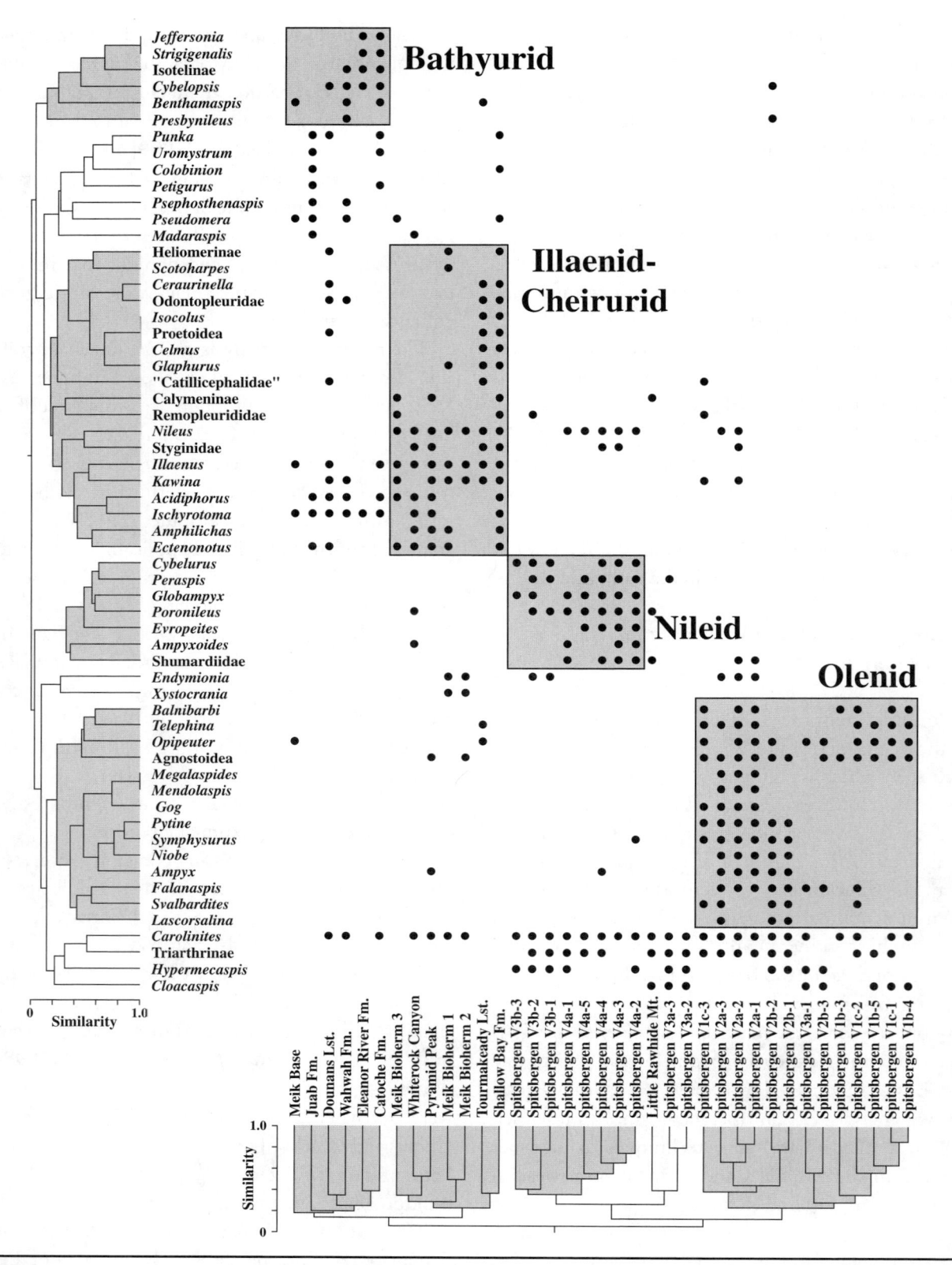

FIGURE 24.4. Definition of Laurentian trilobite biofacies during the onset of the Ordovician Radiation. Data matrix with horizons in Q-mode clustering order and taxa in R-mode clustering order. Data are presence/absence. Jaccard's coefficient was used as the index of similarity, and the clusters were formed using the average linkage method. Biofacies are defined using the intersection of Q- and R-mode clusters. Catoche Formation, western Newfoundland (late Ibexian); Dounans Limestone, Highland Border Complex, Scotland (late Ibexian); Eleanor River Formation, Ellesmere Island, Canadian Arctic (late Ibexian); Juab Formation, Ibex area, western Utah (early

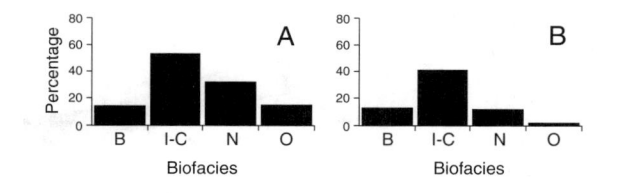

FIGURE 24.5. A, percentage of total species occurrence in each early Whiterockian Laurentian biofacies that is contributed by the Whiterock Fauna (B = Bathyurid Biofacies; I-C = Illaenid-Cheirurid Biofacies; N = Nileid Biofacies; O = Olenid Biofacies). B, percentage of total species occurrence contributed by the Silurian Fauna, excluding Raphiophoridae (see Adrain et al. 1998).

none. Transition to dominance by the Whiterock Fauna first occurred in deep subtidal and buildup environments during the early Whiterockian—at a time when the Ordovician Radiation of the Paleozoic Evolutionary Fauna was taking place. The Whiterock Fauna subsequently spread both on- and offshore, dominating all environments by the Late Ordovician and of course completely dominating Silurian environments with the extinction of all Ibex faunal elements by the end Ordovician.

Conclusions and Prospects from the Global Analysis

1. Ordovician trilobite families display one of three major diversification trajectories: earliest Ordovician success followed by steady and rapid decline and extinction prior to or at the end Ordovician (Ibex Fauna I); Early Ordovician radiation followed by Mid Ordovician decline and extinction prior to or at the end Ordovician (Ibex Fauna II); or Mid Ordovician radiation and Late Ordovician success, with substantial survivorship at the end Ordovician (Whiterock Fauna).

2. There was an Ordovician radiation of trilobites in all parts of the world except the tropical Gondwanan Reedocalymenine Realm. The radiation occurred at the same time as, and with dynamics similar to, that of the Paleozoic Evolutionary Fauna.

3. Whiterock Fauna clades that diversified at high latitudes were less likely to survive the end Ordovician mass extinction, whereas almost all of those that diversified at low latitudes survived.

4. In Laurentia, the Whiterock Fauna rose to dominance first in marginal deep subtidal and buildup environments and spread later to progressively dominate more on- and offshore environments.

Future research could be focused on improving all the analyses with more and better data and completely stepping the level of global resolution down to species level. Greater global stratigraphic resolution using trilobites seems unlikely in the foreseeable future owing to the limited study currently being undertaken in major areas. However, further progress could certainly be made in continents with good records. Particular questions requiring further investigation include the following:

1. What is the biofacies pattern of the transition in the Megistaspidine and Dalmanitoidean realms? Relative abundance data are even more sparse than for Laurentia (see Adrain et al. 2000), but it should be possible to collect presence/absence data.

2. What does the marginal Laurentian emergence of the Whiterock Fauna signify? Is there a major extrinsic cause for the trilobite radiation?

3. Are the radiations in the three realms one, two, or three events? The high-latitude radiation in the Dalmanitoidean Realm seems to precede that in the Bathyurid Realm (Owen and McCormick 2003). Did it have a different cause?

■ Regional Patterns in Australia and New Zealand (GDE, BDW, JRL)

Ordovician trilobite faunas of Australasia were reviewed by Webby and Edgecombe in Webby et al. (2000), with particular reference to their biogeographic affinities. Most of the data in that synthesis,

FIGURE 24.4. (*Continued*) Whiterockian); Little Rawhide Mountain, Antelope Valley Limestone, central Nevada (late Ibexian); Meik Base = Meiklejohn Bioherm basal beds (early Whiterockian), southwestern Nevada; Meik Bioherm 1–3 = Meiklejohn Bioherm, separate collections from biohermal beds (early Whiterockian); Pyramid Peak, Death Valley, California (early Whiterockian); Shallow Bay Formation, western Newfoundland (early Whiterockian); Spitsbergen V1–V4, various stratigraphic levels in the Valhallfonna Formation, drawn from Fortey (1980: figure 1), ranging from late Ibexian to early Whiterockian; Tourmakeady Limestone, western Ireland (early Whiterockian); Wahwah Formation, Ibex area, western Utah (late Ibexian); Whiterock Canyon, Antelope Valley Limestone, east-central Nevada (early Whiterockian).

FIGURE 24.6. Representative Laurentian faunas through time and along an environmental gradient, showing the pattern of changeover from the Ibex faunas (black) to the Whiterock Fauna (white), with number of species recorded from each fauna.

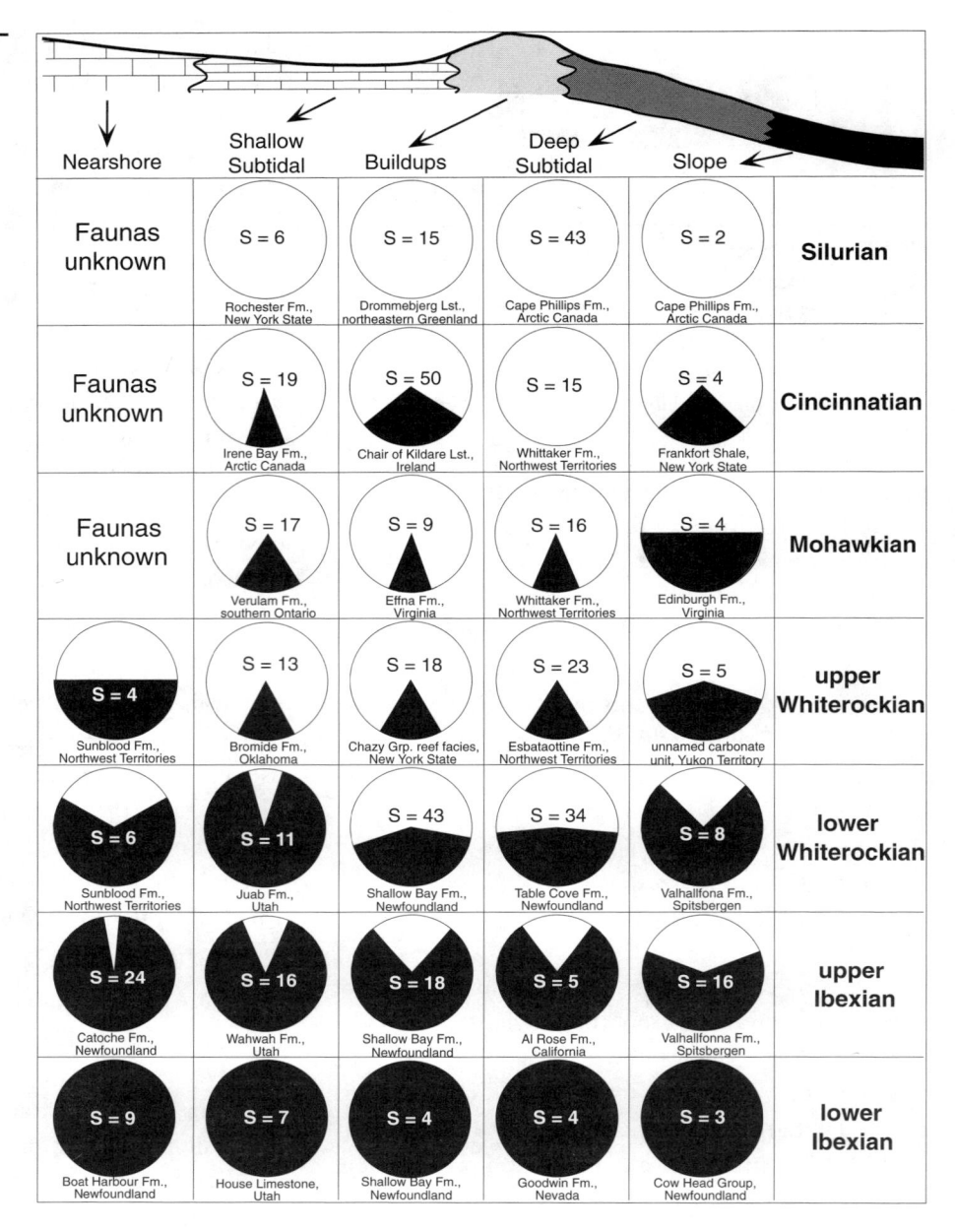

as well as the present analysis, are from Australia (see Webby and Nicoll 1989: appendix 2 for a species list). New Zealand is represented only by Lower Ordovician faunas of the Patriarch Formation and Summit Limestone in the Takaka Terrane, South Island (Wright et al. 1994) and a few species from the Upper Ordovician Golden Bay Group of the adjacent Buller Terrane.

The trilobite record in Australia samples a broad range of biofacies. For example, Lancefieldian/Tremadocian assemblages include platform sediments representing the inner detrital belt (Shergold 1991) and peritidal carbonates (Shergold 1975), as well as an outer detrital biofacies sampled in western New South Wales (Webby et al. 1988) and New Zealand (Wright et al. 1994). A succession of Ordovician faunas outlined by Webby et al. (2000) drew heavily on the biogeographic sensitivity of trilobites. The Lower Ordovician in Australasia was divided (in ascending order) into *Hysterolenus, Australoharpes, Koraipsis, Chosenia,* and *Encrinurella* faunas, all having affinities to North China and most having affinities to the Sibumasu Terrane. Relationships to China and Sibumasu are maintained through the Middle Ordovician *Railtonella* and *Prosopiscus* faunas, the latter also having links to Laurentia and the Precordillera of Argentina. A Gisbornian *Incaia* fauna in New Zealand

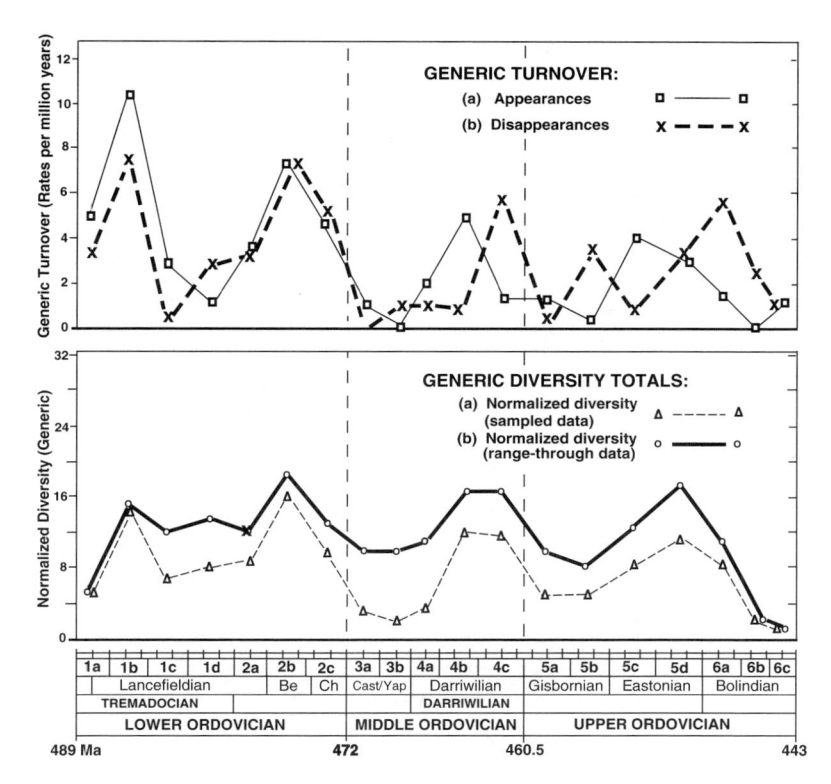

FIGURE 24.7. Generic turnover (appearance and disappearance rate per million years) and normalized generic diversity curves for Ordovician trilobites of Australia and New Zealand. Normalized diversity is plotted for sampled and range-through data. Be = Bendigonian; Ch = Chewtonian; Cast = Castlemanian; Yap = Yapeenian.

has South American affinities, whereas Eastonian-Bolindian trilobites of Australia (*Eokosovopeltis-Pliomerina* fauna) display a clear association with Kazakhstan, North China, and terranes near the margin of Gondwana. Elements of the cosmopolitan *Hirnantia* Fauna (brachiopods and trilobites) have been recorded in Victoria and Tasmania.

Diversity data for Australia are affected by biases in geographic representation through the Ordovician. The Canning Basin in Western Australia (Legg 1976; Laurie and Shergold 1996a, 1996b) has a full record for the Early and Mid Ordovician but was a site of nonmarine and evaporite sedimentation in the Late Ordovician. Late Ordovician data are derived largely from the island-arc complexes of central-west New South Wales and platform carbonate sequence of Tasmania. Most known Ordovician faunas in Australasia have been described, and we have included unpublished data for the Amadeus Basin, Tasmania, and central-west New South Wales to eliminate bias.

The diversity curves in figure 24.7 have been compiled at the generic level, but because a majority of genera are represented by single species, discrepancies between generic and specific diversity are not a major issue. The relatively short temporal duration of many genera is reflected in a general correspondence between appearances and disappearances through most of the Ordovician, and longer-ranging genera appear to be randomly distributed through time. Intervals in which disappearances substantially exceed appearances are *TS*.4c (late Darriwilian) and *TS*.6a (early Bolindian). Qing et al. (1998), Shields and Veizer (chapter 6), and Barnes (chapter 8) have demonstrated a dramatic decline in strontium isotope ratios across the late Darriwilian to early Caradoc (*TS*.4c–5a) interval. Qing et al. (1998) and Barnes link this change to a possible global mantle plume event with associated orogenic quiescence and/or major flooding (the early Caradoc transgression of Cocks and Fortey 1988). Shields and Veizer attribute the cause to increased ocean-ridge spreading rates. In contrast, the *TS*.6a decline seems to have been related more to regional than to global factors, with the Benambran Orogeny causing major tectonic changes and volcanic activity, especially to eastern Australia (Webby and Percival in Webby et al. 2000:109).

Range-through and sampled diversity both show four maxima (listed in descending order using the range-through data) in the Bendigonian (*TS*.2b), Eastonian (*TS*.5d), Darriwilian (*TS*.4b-4c), and Lancefieldian (*TS*.1b). Range-through diversity particularly exceeds sampled diversity through two intervals of the Ordovician, *TS*.1c–2a and *TS*.3a–4a. The latter (Castlemainian–early Darriwilian) is one of three marked minima in diversity by either measure, the others being in the Gisbornian (*TS*.5a–b) and in the globally depauperate late Bolindian (*TS*.6b–c). The diversity low through the *TS*.3a–4a interval may be at least in part an artifact of incomplete sampling, limited exposure, and/or restriction of favorable trilobite-bearing facies (Georgina Basin and Tasmania), and/or breaks (disconformities) that represent localized emergence through a part of this *TS*.3a–4a interval (Canning Basin).

Adrain et al. (earlier in this chapter) observed geographic variation in the expansion of their Whiterock Fauna (Adrain et al. 1998). Owen and McCormick (later in this chapter) noted an early (early Arenig) diversification of elements of this fauna in Avalonia, whereas it diversified relatively late in South America (Waisfeld, later in this chapter). Lower Ordovician faunas representing the inner detrital biofacies throughout Australia (e.g., Jell 1985; Jell and Stait 1985a, 1985b; Shergold 1991) are composed almost exclusively of members of the Ibex Fauna (e.g., asaphids, pilekiids, pliomerids, and hystricurids), like that of the diverse, highly endemic Bendigonian succession of the Canning Basin, where asaphids, pliomerids, and telephinids predominate (Laurie and Shergold 1996a, 1996b). In the Canning Basin, elements of the Whiterock Fauna such as illaenids and raphiophorids first rival the diversity of the Ibex Fauna in the Darriwilian (*TS*.4b–c; Legg 1976). Whiterock taxa, notably encrinurids, cheirurids, lichids, styginids, illaenids, and trinucleids, become dominant for the first time in the Eastonian faunas of New South Wales and Tasmania. Interval *TS*.5c (early Eastonian) is marked by generic appearances greatly exceeding disappearances, which appears to correspond to the expansion of the Whiterock Fauna.

Ibex elements (shumardiids, agnostids, telephinids) remain the sole or dominant component only in some Late Ordovician outer detrital biofacies (e.g., Shoemaker Beds in Tasmania: Burrett et al. 1983; Oakdale Formation in New South Wales: Webby 1974). This is, interestingly, opposite to the pattern for South America (Waisfeld, later in this chapter), where Ibex taxa preferentially survive in shallow-water settings. The near absence in the Australian record of particular shallow-water Ibex groups such as bathyurids may figure in the difference between the South American and Australian patterns. The domination of the Whiterock Fauna in Australia is even later than in South America. Although the paucity of early Mid Ordovician (Castlemainian–early Darriwilian) trilobite faunas is a problem for understanding this transition, even later Darriwilian asemblages (e.g., Nora Formation in the Georgina Basin: Fortey and Shergold 1984) are composed almost exclusively of Ibex taxa such as asaphids, leiostegiids, dikelokephalinids, telephinids, and prosopiscids.

■ Regional Patterns in South America (BGW)

Trilobite records in the Ordovician of South America are derived from three different geodynamic and environmental settings: pericratonic platforms developed along the southwestern edge of Gondwana (Andean belt), volcanic island-arc settings parautochthonous to the Gondwanan margin (Famatina Range and western Puna), and an allochthonous Laurentian-derived terrane (Argentine Precordillera).

The Andean belt records deposition in shallow marine siliciclastic shelves. In northern South America, Ordovician deposits are poorly known, and trilobite records are sparse. In contrast, better-known and trilobite-rich successions, mostly early Tremadocian to mid Arenig in age, crop out in the Central Andean basin (Cordillera Oriental of Peru, Bolivia, and northwestern Argentina). Trilobite species diversity is high in the early Tremadocian to mid Arenig (figure 24.8). Tremadocian successions are composed of transgressive deposits that frequently represent dysaerobic environments, punctuated by regressive episodes (cf. Moya 1988) represented by intertidal and shallow subtidal settings. The greatest number of species occurs in the early Tremadocian, associated with the global rise in sea level and the development of mud-dominated platforms, dominated by relatively widespread olenids and several families of agnostoids.

High diversity was maintained in the late Tremadocian to the early and mid Arenig, with a remarkable radiation of endemic asaphids in the early Arenig (*T. approximatus* to *B. deflexus* graptolite zones) associated with dysaerobic facies in distal parts of the inner shelf (Waisfeld et al. 1999). In the mid Arenig (uppermost *B. deflexus* and *D. bifidus* graptolite zones),

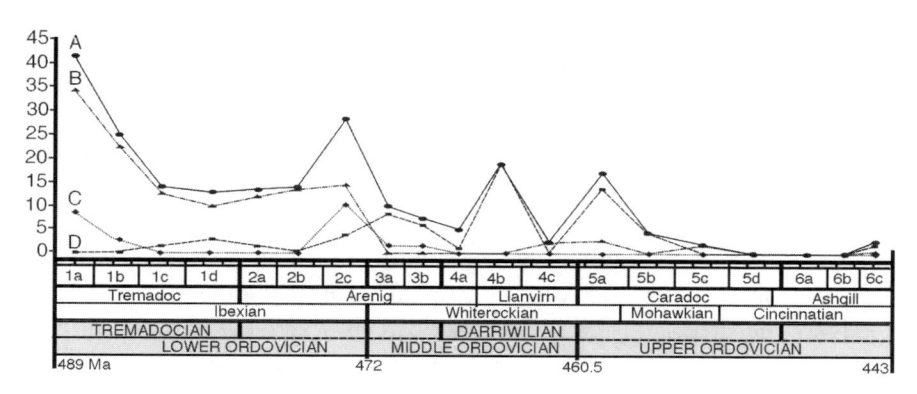

FIGURE 24.8. Normalized diversity of trilobite species in South America. A, aggregate diversity curve for all the basins. B, diversity curve for the Andean belt. C, diversity curve for Famatina Ranges and Western Puna. D, diversity curve for Argentine Precordillera.

in shallower and more aerobic waters of the inner shelf, there was a radiation of endemic trinucleids, raphiophorids, pliomerids, nileids, and asaphids. Diversity increase is also related to the occurrence of a few immigrants from northwestern Gondwana (Waisfeld 1995). Post-Arenig trilobite records in the Andean belt are sparse and restricted to isolated outcrops and, except in Argentina, mostly lack accurate chronological constraints. A few asaphids, along with calymenids, homalonotids, and trinucleids, are recorded in Llanvirn to Caradoc strata.

In Argentina there is evidence for active vulcanism during the mid Arenig in the Famatina Range and during the early Tremadocian and the mid Arenig in the western Puna region (Astini and Benedetto 1996; Kouhkarsky et al. 1996; Mángano and Buatois 1996). Trilobite information is still preliminary, particularly in the latter region. In the Famatina Range, trilobites exhibit the highest diversity in siliciclastic and volcaniclastic deposits of the proximal- to middle-shelf environment, associated with an increase in volcanic activity (conodont-based *O. evae* Zone). No particular radiation of endemic forms is recognized in this basin, but immigrants from West Gondwana coexist with East Gondwanan and Baltic forms (Vaccari 1995; Waisfeld and Vaccari 1996).

The Argentine Precordillera terrane rifted from Laurentia in the Early Cambrian and drifted from low to high latitudes during the Ordovician (Astini et al. 1995). From the latest Tremadocian, open marine carbonates prevailed, yielding a relatively complete record of the benthic fauna. Late Tremadocian to early Llanvirn carbonate sedimentation is represented by inner- and middle-ramp settings (Cañas 1999). Trilobite diversity is relatively low and composed largely of representatives of the Ibex Fauna.

The base of the Whiterockian is within this carbonate succession, but it is not marked by particular expansion of families belonging to the Whiterock Fauna. Endemic genera are restricted to a single pliomerid in the mid Arenig and a single scutelluinid in the early Llanvirn (Vaccari 2003) both from the middle ramp. Both are associated with immigrants from the low- and intermediate-latitude Bathyurid Province of East Gondwana and Baltica (cf. Vaccari 1995).

A change from the middle to distal ramp and a shift in facies from exclusively carbonate to mixed carbonate-siliciclastic sedimentation took place diachronously from late in the mid Arenig to the early Llanvirn. This facies change is linked to a tectonic shift associated with the generation of subsiding depocenters and with global sea level rise in the early Llanvirn (mid Darriwilian) (Astini 1999a; Cañas 1999). This shift was critical in the expansion of trilobites. The earliest records (conodont-based *B. navis* to *M. parva* zones) of trilobites in the distal ramp environment comprise relatively widespread elements (nileids, olenids, raphiophorids, etc.), but in the early Llanvirn (conodont-based *E. suecicus* Zone) these are associated with a radiation of new genera, particularly Raphiophoridae, and also Trinucleidae and Toernquistiidae. Families belonging to the Whiterock Fauna account for 40 percent of the fauna.

Caradoc and younger deposits show a strong facies differentiation in the Precordillera. Trilobites are particularly restricted to carbonate remnants that persisted locally after carbonate sedimentation was drowned in most parts of the basin. Their occurrence and diversification appear to be strongly controlled by environmental constraints imposed by the regional topography and local tectonics. An early Caradoc (graptolite-based *N. gracilis* and *C. bicornis* zones) peak in trilobite diversity is coincident with one of these carbonate remnants. Trilobites occur in slope-apron deposits in either autochthonous deep-water limestones (hemipelagites) or slightly shallower water resedimented carbonates (cf. Astini 1995). New species of earlier endemic raphiophorids and toernquistiids occur, together with endemic encrinurids and a remarkable radiation of trinucleids. The latitudinal position of the Precordillera during the Caradoc is still debated, with mixed biogeographic affinities of the fauna indicating proximity to the Gondwanan margin and a possible location at intermediate latitude (Benedetto 1998; Benedetto et al. 1999). Trilobite records in the late Caradoc to Ashgill of the Precordillera are scarce, with only a few homalonotids and dalmanitids associated with the *Hirnantia* Fauna.

Biodiversification of trilobites in South America shows a strong biogeographic and environmental overprint. In the Andean belt, local radiation of forms is remarkable in the early and mid Arenig among the representatives of the Ibex Fauna. The Whiterock Fauna is well developed in the Precordillera, in contrast to the Andean belt, where Middle and Upper Ordovician records are limited. The Whiterock Fauna diversified diachronously from late in the mid Arenig to the early Llanvirn, tracking the progressive development of distal ramp settings in the basin, while trilobite families of coeval shallower settings are still dominated by representatives of the Ibex Fauna. This pattern is similar to that reported by Adrain et al. (1998) in Laurentia. However, the initial development of the Whiterock Fauna appeared to take place slightly later. This could be a result of the appropriate distal facies being developed slightly later in the basin.

■ Regional Patterns in the Anglo-Welsh Sector of Avalonia (AWO, TMcC)

The Anglo-Welsh area contains the most complete Ordovician successions from Avalonia. They record the history of that microcontinental terrane from its Early Ordovician location on the intermediate to high-latitude Gondwanan margin, probably close to West Africa (McNamara et al. 2001), through its early Mid Ordovician rifting and northward drift leading to its collision with Baltica in the Late Ordovician and the Laurentian margin in the Early Silurian (Cocks et al. 1997; van Staal et al. 1998; Cocks 2000). The Anglo-Welsh Ordovician trilobite faunas are well documented, and their temporal and spatial diversity patterns have been investigated using a literature-based relational database in which the occurrences of species at localities are linked to an array of taxonomic, geographic, and stratigraphic data (Owen and McCormick 1999; McCormick and Owen 2001).

We have described elsewhere the patterns of trilobite diversity change in the Ordovician of the Welsh basin at genus and species level (McCormick and Owen 2001) and the whole Anglo-Welsh area at genus level (Owen and McCormick 2003). In doing so, we have demonstrated that members of the Whiterock Fauna became the dominant component of the Avalonian trilobite fauna earlier (early Arenig) than its rise to dominance in global terms and that they were fairly evenly distributed through the whole spectrum of shelf to upper-slope environments. These analyses were undertaken using the stages recognized in what is the historical type area for the Ordovician (Fortey et al. 1995, 2000) as the "time slices" and utilizing a simple measure of diversity that counted as unity the occurrence of a taxon within a given stage or inferred (in the case of "range-through" analyses) to have been present in that stage because of its presence in the preceding and succeeding intervals.

We present here species- and genus-level curves (figure 24.9) using the normalized diversity measure recommended for the IGCP 410 clade analyses and equating the Anglo-Welsh stratigraphy as closely as possible to the time slices recommended for the international project (chapter 2). Few of the boundaries between the latter divisions match exactly the chronostratigraphic or biostratigraphic boundaries defining the units within which the data were compiled, and the equivalencies used are shown in the caption to figure 24.9. The normalized diversity measure per time slice counts the number of taxa occurring within the slice and also found above and below it, plus half the number of those taxa that either originated or became extinct during that interval plus half the number of taxa restricted to the slice.

Two sets of curves have been computed, one based on recorded occurrences within each slice and one that infers the existence of taxa within an interval on the basis of the "range-through" principle (see earlier in this chapter). The stratigraphic age of some of

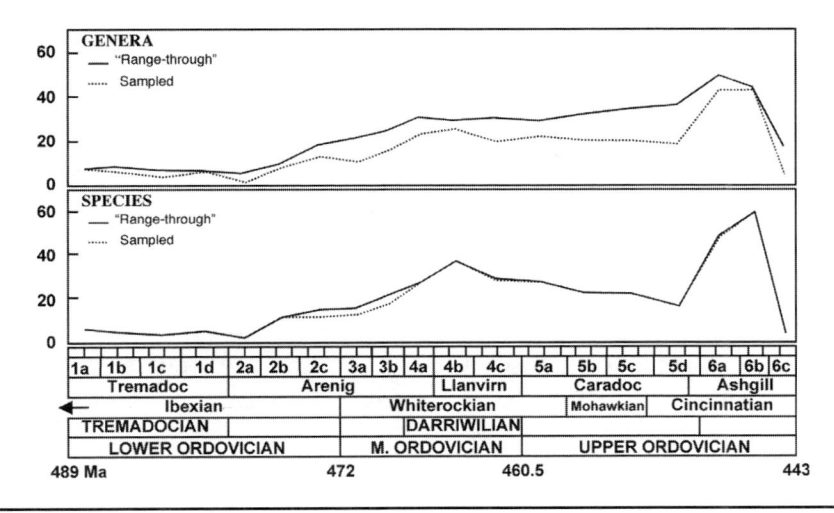

FIGURE 24.9. Sampled and range-through trilobite normalized biodiversity curves for the Anglo-Welsh sector of Avalonia. Note that the boundaries between many of the time slices recommended by Webby et al. (chapter 2) cannot be recognized in the Anglo-Welsh area or that the shelly faunal data could not be compiled relative to them. The time slices as used here have the following equivalencies in the Anglo-Welsh chronostratigraphy: TS.1a = Cressagian; TS.1b = early Migneintian; TS.1c = mid Migneintian; TS.1d = late Migneintian; TS.2a = early Moridunian; TS.2b = late Moridunian; TS.2c = Whitlandian; TS.3a = early Fennian; TS.3b = mid Fennian; TS.4a = late Fennian; TS.4b = early Abereiddian (= graptolite-based *D. artus* Zone); TS.4c = late Abereiddian + Llandeilian; TS.5a = Aurelucian; TS.5b = Burrellian; TS.5c = Cheneyan + Streffordian; TS.5d = Pusgillian; TS.6a = Cautleyan; TS.6b = Rawtheyan; TS.6c = Hirnantian.

the records in the database is known only to two or (rarely) more stages, in which case the taxa are counted as 0.25 in any resultant time slice above or below the unequivocal range of the taxa concerned and, in the case of the "sampled" data, 0.5 where the uncertain record fills in part of the known range of the taxon. The difference between the sampled and range-through curves is particularly marked in intervals where the rock succession, and therefore the sample coverage, includes only a limited set of biofacies. Moreover, because many species are confined to a single stage (e.g., Thomas et al. 1984) whereas most genera have a much longer range, the lower weighting placed on taxa restricted to a time slice in the diversity index produces the apparently anomalous situation of there seeming to be more genera than species in many time slices.

Compared with our earlier, genus-level analysis (Owen and McCormick 2003), the twofold division of the Moridunian stage produces a curve showing a more even rise in diversity through the Arenig, and the combining of the graptolite-based *murchisoni* Zone and Llandeilian stage and the Cheneyan and Streffordian stages to comprise TS.4c and TS.5c, respectively, produces smoother "sampled" diversity curves. Otherwise, patterns emerge from the analysis that are similar to those obtained earlier, testifying to the robustness of the signals. Bootstrap tests (multiple random resampling of the data in each time interval—see Gilinsky and Bambach 1986 for other examples) of the sample data in our earlier analyses show that (1) the number of samples or the range of biofacies preserved within a time unit can have a strong influence on the sampled diversity and (2) the range-through curves are probably a closer reflection of the true picture (McCormick and Owen 2001; Owen and McCormick 2003).

The range-through diversity curves presented herein (figure 24.9) show a rise through the Arenig to a late Arenig–early Caradoc (TS.4a–5a) plateau at genus level but a slight early Llanvirn (TS.4b) peak at species level. McCormick and Owen (2001) suggested that elevated levels of species richness of genera in the late Arenig–early Llanvirn in the Welsh basin may be linked to the rifting of Avalonia from Gondwana at that time (Cocks et al. 1997). Species-level diversity apparently fell during the Caradoc to earliest Ashgill (TS.5a–d), but this is to some extent an artifact of a combination of the restriction in range of preserved shelf biofacies as the basin deepened and the absence of the very deepest water trilobite biofacies from the Anglo-Welsh area. The latter, the cyclopygid-atheoptic association, was composed largely of long-ranging genera that reappeared in the Ashgill, and hence the range-through genus-level data show a trend contrary to the species data. The mid Ashgill (Cautleyan-Rawtheyan) peak in diversity in the Anglo-Welsh area contrasts markedly with the global curves (Adrain et al., earlier in this chapter), which show a maximum in the late Arenig and Llanvirn and a considerable decline thereafter. The Cautleyan-Rawtheyan peak in both genus and species diversity and the high species-to-genus ratio reflect the extreme heterogeneity of the environment and hence the wide spectrum of trilobite biofacies preserved in the Anglo-Welsh area prior to the extinctions that led to the Hirnantian diversity crash (Owen and McCormick 2003).

■ Regional Patterns in Baltica (ØH)

Baltoscandia provides the most complete record for the biodiversity history of Baltica during the Ordovician. It is believed that the region as a whole preserves a fairly complete stratigraphic sequence, although formations can be condensed, in particular in the Baltic countries.

The genus- and species-level diversity curves presented here (figure 24.10) have been derived from a larger database (openly available on the Internet at http://asaphus.uio.no), covering all major fossil groups from the Ordovician of Baltoscandia (Hammer in press). Owing to the nature of the literature, especially the older publications, it has not been possible to collect sufficient data about individual samples. The basic unit of the database is therefore first and last occurrences of a species at one locality, according to one publication. As far as possible, the "first-appearance datums" (FADs) and "last-appearance datums" (LADs) are then converted to apparent, calibrated ages according to the timescale used in this volume and correlations with local zones according to recent literature. In cases in which only the formation is known, the approximate ages of the lower and upper formation boundaries are used as the FAD and LAD, potentially overestimating real ranges and also by necessity disregarding diachronous lithological boundaries. The diversity estimates within each time slice (chapter 2) are then made using the range-through assumption and the normalized diversity count, whereby taxa having their FAD and/or LAD within a time slice count as only one-half of a unit within that slice. An alternative estimate involves counting taxa restricted to a time slice as one-third of a unit. There is no correction for the different durations of the time slices. Even though sample data have not been collected, it is to some degree possible to estimate sampling coverage by constructing "artificial samples" each consisting of taxa registered within a fixed time duration (e.g., one million years) at one locality. These quasi samples can then be subjected to bootstrap tests and other randomization methods (Hammer in press).

At the time of writing, the database consists of 10,340 stratigraphic ranges at localities, taken from 141 publications. Some 962 species and 259 genera of trilobites are included, with a total of 2,691 range entries. The trilobite taxonomy has been revised ac-

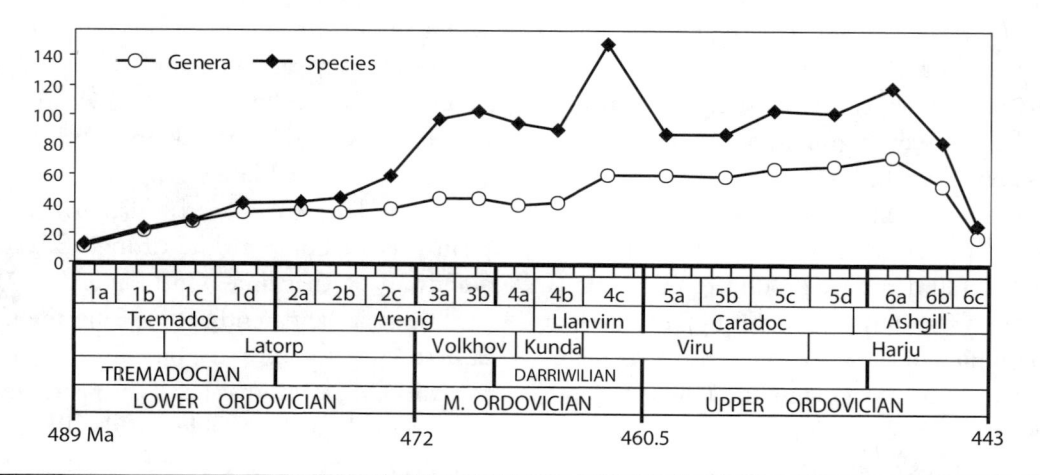

FIGURE 24.10. Range-through, normalized trilobite diversity curves for Baltoscandia, representing Baltica.

cording to Bruton et al. (1997), but the taxonomic uncertainties in the material are still rather extensive. It can only be hoped that the taxonomic problems are relatively unsystematic and will therefore deteriorate the signals, rather than producing false ones (Benton 1999).

The main feature of the curves is a more or less even, substantial increase in trilobite biodiversity throughout the period. In addition to global evolutionary trends, such a pattern may also have been influenced by local factors. Baltica drifted from a mid- to high-latitude position in the Southern Hemisphere in the Early Ordovician to low latitudes by the end of the Ordovician (Torsvik et al. 1992). In the context of the present-day latitudinal diversity gradient (Rosenzweig 1995), this movement may have contributed to the increase in diversity that is observed in all Baltoscandian fossil groups through the Ordovician. However, for trilobites, the diversity increase is less clear when the data are subjected to randomization tests (Hammer in press).

At the specific level, a steep increase in trilobite diversity is observed in *TS*.3a (mid Arenig) and is strongly supported by the randomization tests. This diversification event close to the Ibexian-Whiterockian boundary is contemporaneous with the beginning of a probably quite protracted regression in the area (Nielsen 1995:61).

A diversity peak in the species curve in the upper Llanvirn (*TS*.4c) is mainly due to Estonian data (in particular, those of Rõõmusoks 1970) and is not observed in the curves for Norway or Sweden. Trilobite diversity reached its all-time high in the early Ashgill (*TS*.6a) before dropping significantly during the late Ashgill. Even taking into account the artifacts produced by not counting Lazarus taxa, this end Ordovician decrease in diversity seems as dramatic as elsewhere in the world.

■ Regional Patterns in South China (ZYZ)

The South China Block exhibits extensive exposures of Ordovician deposits, with the most complete sequences and occurrences of fossil groups in China. The strata are well documented and the stratigraphic units highly resolved, with many selected as stratotypes for classifying and correlating the Ordovician of China. As reviewed by Cocks (2001), the South China Block was situated in low-latitude zones along the western margin of Gondwana during the Ordovician. Cocks and Torsvik (chapter 5) further considered the South China Block as a peri-Gondwanan terrane that, like the Sibumasu Terrane, drifted from intermediate to low latitudes outboard of various Himalayan fragments close to the Indian part of Gondwana.

Ordovician trilobites are well recorded (e.g., Lu 1975), and faunas display a progressive on-shelf to off-shelf transition in composition and diversity from present-day west either to the southeast or to the north and northeast of the block, with benthic forms most diversified in the shallower outer shelf and mesopelagic cyclopygids mainly distributed in the deeper outer-shelf and off-shelf slope (see Zhou et al. 1999, 2001, 2003; Yuan et al. 2000). Trilobite faunas exhibit close relationships to those of Australasia, Kazakhstan, the North China and Tarim blocks, and the Sibumasu Terrane on the one hand and to the Middle East, southern, central and western Europe, and the Indochina Terrane on the other (Zhou and Dean 1989; Zhou et al. 1998a, 1998b), providing evidence of links between the different Ordovician Gondwanan and peri-Gondwanan faunas of higher to lower latitudinal zones.

In the absence of an existing database of the geologic and geographic distributions of the Ordovician trilobites in the South China Block, the genus diversity curves (figure 24.11) presented here are derived from a genus-range chart that will be described more fully in a future paper. The chart was compiled on the basis of data from the literature and collections made recently by Zhiyi Zhou, Zhiqiang Zhou, and Wenwei Yuan from 36 measured Ordovician sections along a bathymetric gradient in South China. The basic data comprise 220 taxonomically valid trilobite genera. Of these the Asaphida are the predominate group, represented by up to 41.6 percent of the entire South Chinese Ordovician trilobite generic component.

The diversity data are presented using the unified Ordovician timescale with subdivisions into 19 time slices and correlations to equivalent Chinese chronostratigraphic intervals as outlined by Webby et al. (chapter 2). The data were calculated using normalized diversity measures, as recommended by Cooper (chapter 4).

Range-through diversity maintains a relationship roughly similar to the sampled diversity through the

FIGURE 24.11. Sampled and range-through normalized generic biodiversity curves for Ordovician trilobites of the South China Block.

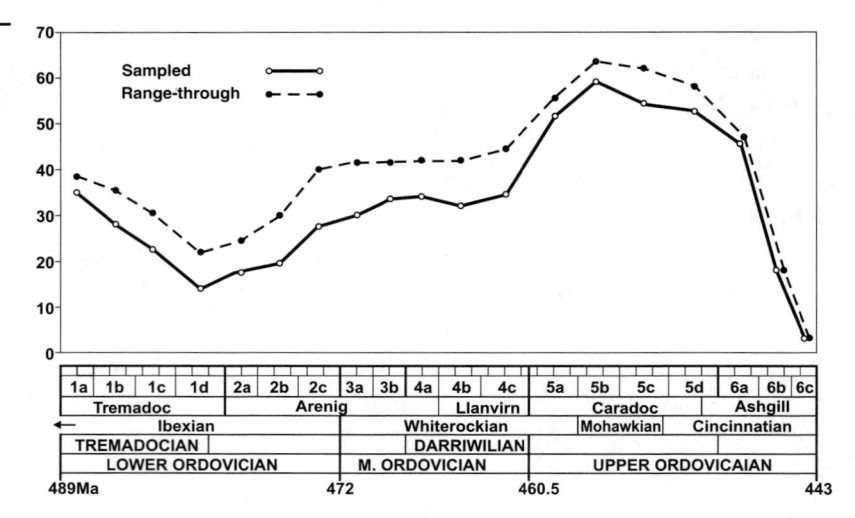

Early to Mid Ordovician (figure 24.11), but then the differences between the two sets of values become less marked during the Caradoc and eventually more or less coincide through the Ashgill (*TS*.6a–c interval). The overall diversity trend shows an initial sharp rise in *TS*.1a and then a decline to a low in *TS*.1d. This is followed by a rise to a broadly flattened to slightly elevated peak during the Mid Ordovician centered on *TS*.4a. Then a more intense radiation occurred during the Caradoc–early Ashgill (TS.5a–6a interval) with peak diversification in *TS*.5b, followed by a dramatic decline through the middle Ashgill to *TS*.6c. Although the limited *TS*.1a diversity increase in the early Tremadocian seems comparable to that of the South American, particularly the Andean Belt, plot (figure 26.8), the sharp diversity decrease in the Hirnantian is synchronous worldwide. The more or less steady rise in diversity shown here from the Arenig to early Ashgill is also revealed in Avalonia (figure 24.9) and Baltica (figure 24.10), especially the latter, where a similar Caradoc–early Ashgill diversity plateau also occurs. However, as a whole, the Ordovician trilobite biodiversity curves displayed in the different regions do not match one another very closely.

In summary, the diversity plot of South China shows peaks in the earliest Tremadocian, early Darriwilian, and Caradoc and two minima at the end of the Tremadocian and in the Hirnantian. The mid Ashgill diversity maximum illustrated by Owen and McCormick using Avalonian data (figure 24.9) is not depicted in South China, probably because there is a paucity of trilobite-rich rocks through this interval. Mechanisms that triggered the Ordovician biodiversity alternation may have involved a range of geologic, geographic, climatic, and oceanographic factors (Webby 2000). With the exception of the end Ordovician glaciation that caused the extinction of Ibex Fauna I and II and the decrease in diversity of the Whiterock Fauna (Adrain, Westrop, and Fortey, earlier in this chapter), the trilobite diversity changes exhibited in South China seem related generally to the sea level fluctuations delineated by Fortey (1984) and Ross and Ross (1992). High diversities seem to be associated with transgressive phases, and low diversities coincide with regressive intervals. The Caradoc diversity maximum may, for example, be connected with the Ordovician climax of transgression that took place in China (Zhou et al. 1989, 1992), as it did elsewhere in the world (Fortey 1984). It remains to be determined whether the correspondence between sea level and diversity change represents cause and effect or whether the diversity curve includes sample biases linked to sea level change (e.g., see Smith 2001).

In the South China Block, taxa of Ibex Fauna I had high diversity in *TS*.1a, and most of them belong to the *Hysterolenus* fauna of latest Cambrian–earliest Tremadocian, including mainly agnostids, dikelokephalinids, ceratopygids, kainellids, and olenids. The diversity of Ibex Fauna I declined steeply to *TS*.1d, and from that point on the related forms never became diversified again (figure 24.12B). The Ibex Fauna II (chiefly Asaphidae and Nileidae) peaked in *TS*.2b, but otherwise its diversity was uniformly low through the Ordovician (figure 24.12B); members of it were only proportionally higher in *TS*.1d–3a

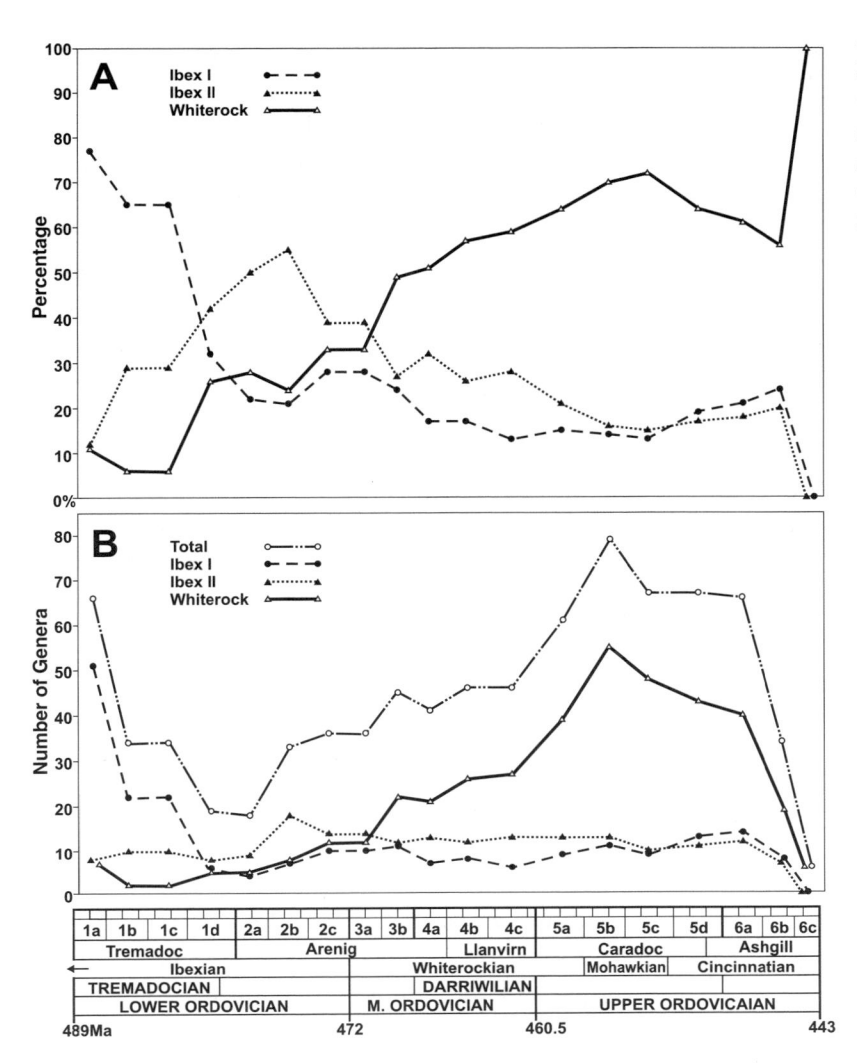

FIGURE 24.12. Proportion (A) and diversity (B) of Ordovician trilobite genera belonging respectively to the Ibex I, Ibex II, and Whiterock faunas (Adrain, Fortey, and Westrop, this chapter) in each of 19 sampled intervals of the South China Block.

(figure 24.12A), suggesting that this fauna had once radiated here during the latest Tremadocian to early late Arenig. Both Ibex I and II faunas became extinct just prior to the end Ordovician mass extinction.

The main radiation of the Whiterock Fauna is recorded from the Mid Ordovician onward, with the peak of diversification attained during *TS*.5b (figure 24.12B). A slight decline in diversity followed through to *TS*.6a, and then there was a rapid decrease to the Hirnantian (figure 24.12B). The fauna was proportionally dominant over that of either the Ibex Fauna I or II from *TS*.3b (49 percent of total fauna) to *TS*.6c (100 percent) (figure 24.12A), consisting of mainly cyclopygids, cheirurids, raphiophorids, trinucleids, illaenids, calymenids, and isocolids. Most of the members were outer-shelf dwellers. The Whiterock Fauna survived the two-phase end Ordovician mass extinction and, after a short period of recovery, reappeared in South China as the Silurian Fauna with representatives of up to 11 of its families during the mid Llandovery (mid to late Aeronian). In South China the main diversification of the Whiterock Fauna seems to have commenced during *TS*.3b, that is, early in the Mid Ordovician (just prior to the Darriwilian), when elements of the fauna occupied most environmental niches for the first time. This included the diversification of the mesopelagic cyclopygids as the Cyclopygid biofacies (see Fortey, later in this chapter) became established in South China during *TS*.3b.

■ **Adaptive Deployment (RAF)**

Biofacies profiles with their accompanying suites of trilobite faunas have now been recognized in the Ordovician for all the major paleocontinents. Although these follow a broadly shallow- to deep-water trajectory, the factors controlling their distribution may be only secondarily related to depth per se—for example, the level of oxygenation present at the sediment surface may be the prime influence on which organisms are present in any given location, and characteristic trilobites adapted to deoxygenated habitats may have a depth spread. The range of niches occupied by the trilobites is an important part of the Ordovician Radiation. The taxonomic constitution and guild distribution of trilobite "communities" during the Ordovician are considered here along with the ways in which they differed from those in the Cambrian and Silurian. In so doing, assumptions must be made about life habits, which are by no means universally agreed. Fortey and Owens (1990a) identified a series of typical morphologies ("morphotypes") repeatedly adopted by trilobites that subsequently (Fortey and Owens 1999) was extended to recognize feeding habits and habitat type. According to their criteria, the Ordovician was a time when more disparate trilobite taxa adopted a wider variety of morphotypes than at any other time, before or after (see also Foote 1991). This model is adopted here as the basis for discussion, although it would be surprising indeed if there were no modifications to this scenario in the future; nonetheless, it provides an explicit basis for this summary.

Pelagic Biofacies

Ordovician pelagic trilobites, mostly bearing enlarged ("hypertrophied") eyes, were polyphyletically derived. The pelagic biotope was more richly populated in the Ordovician than in the Late Cambrian and was never reestablished in the Silurian or later, and thus it is a characteristic component of the Ordovician Radiation. The taxa involved belong to the Ibex Fauna and Whiterock Fauna of Adrain et al. (1998). Pelagic trilobites included both epipelagic and mesopelagic species (McCormick and Fortey 1998). The former included some of the most widespread of all trilobites across biofacies; the latter were particularly characteristic of sites marginal to Ordovician paleocontinents, to which they were confined until late in the Ordovician.

The mesopelagic community was preserved in the Cyclopygid biofacies and was remarkable for its stability. The earliest record is in the Tremadocian of Argentina. The early history of the biofacies is entirely peri-Gondwanan. From the Arenig (Fortey and Owens 1987) to the Ashgill (Apollonov 1974) the Cyclopygid biofacies is taxonomically conservative, with the eponymous family dominating the trilobites. Fewer than 10 genera are present, and 6 (*Cyclopyge, Microparia, Degamella, Sagavia, Ellipsotaphrus,* and a pricyclopygine) are to be found in virtually all these faunas. The fact that rarer, but distinctive, genera such as *Gastropolus* Whittard also have Arenig-Ashgill ranges indicates that the whole biotope may eventually be known virtually throughout the Ordovician. These genera are so conservative in morphology that it can be difficult to discriminate an Ashgill from an Arenig species. Cyclopygids are accompanied by the bizarre pelagic bohemillids—now considered aberrant remopleuridioids—and by the enigmatic *Cremastoglottos;* the latter having as long a stratigraphic range as any cyclopygid. The mesopelagic habitat therefore persisted without a temporal break virtually throughout the Ordovician. The Cyclopygid biofacies did not survive into the Silurian, another line of evidence proving an oceanic crisis at the end of the Ashgill. It had no successor, and so it is absolutely diagnostic of the Ordovician Radiation. Pre-Ashgill Ordovician occurrences were in Avalonia, South America, central Europe, China, and Kazakhstan; in the Ashgill, typical Cylopygid biofacies are known from Girvan, Scotland, and from Quebec, proving that it had crossed the Iapetus remnant into tropical paleolatitudes by then. *Symphysops* is known from "mound" faunas at that time (Dean 1974), and it is conceivable that the cyclopygids had extended their bathymetric range toward the end of their history. Although very occasional cyclopyids can be found in shallower biofacies in the early Ordovician, they are mostly "stragglers."

The epipelagic biotope is typified by Telephinidae (*Carolinites, Oopsites, Opipeuter, Telephina,* and *Phorocephala*). The first three named did not survive the Mid Ordovician (Whiterockian) and were pan-tropical. One species (*Carolinites genacinaca*) has been described from North America, Siberia, China, and Australia (McCormick and Fortey 1999). *Telephina* and *Phorocephala* have early Laurentian records but apparently extended their ranges into high paleolatitudes by the Late Ordovician. No epipelagic type survived into the Silurian.

Olenid Biofacies

The olenid biofacies is typified by sulfide-rich, laminated black shale/dark limestone lithologies that accumulated under critically low oxygen conditions and included a restricted fauna adapted to these (Henningsmoen 1957), possibly including specialists capable of living symbiotically with sulfur bacteria (Fortey 2000). The biofacies continues uninterrupted through the Cambrian-Ordovician boundary and is thus composed of Cambrian-style taxa. Olenidae, often with many segments, low convexity, and wide thoraces, constitute the eponymous family. It is controversial whether the agnostoids that may co-occur with them were co-benthic. Other taxa were recruited into the Olenid biofacies through the Ordovician and at the same time assumed olenimorph morphological features, including alsataspidids (*Seleneceme*), Dionididae (*Aethedionide*), and remopleuridids (*Robergia*). Multiplication of segments in some of these forms is spectacular. All are of Ibex Fauna type. It has been recognized that some members of this biofacies (e.g., *Bienvillia, Parabolinella, Hypermecaspis*) were more or less independent of paleogeography. The Olenid biofacies typically is low gamma diversity/high individuals of species.

The terminal Ordovician crisis that extirpated the Cyclopygidae also eliminated the Olenid biofacies. Deep-sea oxygenated water entrained in oceanic overturn may have been the crucial factor in eliminating the appropriate environment. However, before the end of the Llandovery, olenimorphs (e.g., *Aulacopleura*) had reappeared, and so this trilobite habitat and biofacies is not uniquely Cambro-Ordovician.

Filter-Feeding Trilobites

Generally small trilobites having a vaulted cephalic chamber flanked by genal prolongations, thorax suspended above the sediment surface, weak axial musculature, "elevated" hypostome, and (usually) reduced eyes have been interpreted as having lived by filtering edible particles from suspension in a feeding chamber. Fortey and Owens (1990a) termed this the "trinucleimorph" design, and indeed this distinctive morphology is exemplified by the Trinucleidae, a diagnostically Ordovician family. However, it did not first appear in trinucleids, and more than one Cambrian family likely to be only remotely related to Trinucleidae also included genera that showed trinucleimorph design. In the Ordovician, Raphiophoridae, Harpetidae, and Dionididae also typically adopted this morphology. Of these, the first two named are known to be present in the Tremadocian and have plausible Cambrian relatives; trinucleids and dionidids are known from Arenig and younger strata. Rarely, similar morphologies were adopted from other families; for example, the bathyurid *Madaraspis* was a Laurentian endemic found in strata otherwise lacking species with trinucleimorph design.

The early Trinucleidae were overwhelmingly Gondwanan in distribution, achieving more global distribution by the later Llanvirn (late Whiterockian). Raphiophoridae, Harpetidae, and Dionididae seem to be pandemic from the first, and certain genera belonging to these families (e.g., *Ampyx, Dionide*) are as widely distributed. Among harpetids some genera (*Eoharpes*) are confined broadly to high paleolatitudes, others (*Hibbertia*) to low ones. By contrast, trinucleids tend to endemicity. Avalonia, for example, has at least 12 endemic genera (admittedly they may be "oversplit"); other endemic trinucleids are present on the Precordillera terrane of Argentina, which likewise enjoyed an independent history as a microplate. Some raphiophorids are as local: peculiar few-segmented forms such as *Taklamakania* and the distinctive *Bulbaspis* are confined to eastern Gondwana and common only in Tarim and Kazakh terranes. Elevated total diversity curves for these trilobites at times of terrane separation will be influenced by the addition of such local taxa. However, high endemicity sits ill with the premise that the asaphoid larvae of trinucleimorphs were planktonic in habit (Chatterton and Speyer 1989), unless this planktonic phase was exceptionally short lived.

Suspension-feeding trilobites were not confined to one biofacies but are commonest in outer, or at least quieter, shelf environments, which is to be expected given their substrate preferences. They are typical of the Raphiophorid biofacies around Gondwana (Wales: Fortey and Owens 1978; Argentina: Waisfeld 1995) and Nileid biofacies or equivalent in Laurentia, South China, and Baltica. Although they have reduced eyes or are blind, they may be found associated with other trilobites bearing normal eyes. They may be abundant:

Ampyxina is commonly dominant on shaley limestones of Mid Ordovician age in Virginia, and the present author has observed black limestones of slightly younger age largely composed of *Taklamakania* in northeastern Kazakhstan and Tarim (see also Zhou et al. 1994).

The great majority of Ordovician suspension feeders were members of the Ibex Fauna (Adrain et al. 1998; Adrain, Westrop, and Fortey, earlier in this chapter). Only one trinucleoid genus, *Raphiophorus*, survived the end Ordovician extinction, persisting until the Ludlow. Apart from the long-ranging harpetids, no other candidate for this life habit is known from the Llandovery, but later proetide-derived genera such as *Cordania* probably adopted it, and so it is not peculiarly Ordovician. However, it is true to say that these kinds of trilobites never again achieved the numerical abundance that they did during the Ordovician biodiversification event.

Particle Feeders

Small to middle-sized trilobites with natant hypostomes have been identified as sediment ingesters or particle feeders or both. They are abundant in the Cambrian among the paraphyletic "ptychoparioids." During that time they may be found in any water depth. In the earlier Ordovician (Ibexian) the habit is represented mostly among Cambrian-style "survivors," and in Laurentia, Siberia, North China, and Australia it is exemplified by trilobites traditionally assigned to the Hystricurinae. These are mostly found in rather shallow water deposits and may be associated with carbonate muds and silts. Some of the related Dimeropygidae may have had similar habits. In the Tremadocian of Gondwana, in fine-grained clastic deposits for the most part, some olenids that extended beyond the Olenid biofacies, and the eulomatids, were probably the ecologic equivalents of the Laurentian particle feeders. However, after the early Arenig there is a dearth of such forms in Gondwana, the reason for which is unclear. An important change occurs at the base of the Middle Ordovician, when Proetoidea and Aulacopleuroidea (Whiterock Fauna) appear with this morphology and remain almost its sole exemplars until the extinction of the Trilobita. It is clear that these trilobites must have had Cambrian ancestors, but the appearance, radiation, and spread of these small trilobites through the Mid and Late Ordovician faunas remains a striking phenomenon. They were little affected by the end Ordovician extinction event and are familiar components of Silurian faunas. They also appear to have been tolerant of a variety of biofacies but are most varied and numerous in carbonate "mound" biofacies such as the Boda Limestone (Ashgill) in Sweden.

Predators/Scavengers

Trilobites with large, rigidly attached, often buttressed hypostomes with modified posterior borders (forks, burrs, and the like) are attributed to this life mode. There is clearly a variety of specializations within this general category of which we have as yet only speculative ideas. Phacopoids, with specialized visual systems, are considered together with highly distinctive odontopleurids, and it is likely that there were subdivisions with regard to nocturnal or diurnal habits, prey type, and so on, about which we know nothing. However, it is clear that the largest trilobites are of this type. Asaphids, in particular, include very large and robust trilobites, consistent with a position near the top of the food chain, and these are joined by Lichida in the later Ordovician, both groups having pronounced hypostomal forks. The largest trilobites in the Ordovician trilobite faunas of Bohemia (*Nobiliasaphus*), Avalonia (*Basilicus*), Iberia (*Uralichas*), North China (*Eoisotelus*), and Laurentia (*Isotelus*) conform to this type.

Trilobites of this giant kind are apparently found in inshore habitats, but predator/scavenger morphology can be found in any biofacies except the typical Olenid biofacies. The Early to Mid Ordovician inshore *Neseuretus* biofacies of Gondwana (and its temporal successors) is dominated by trilobites with conterminant hypostomes having calymenoid, dalmanitoid, and asaphoid affinities. Hammann (1985) suggested that some trilobites with elevated eyes such as homalonotids may have been capable of burrowing. In deeper-shelf biofacies these groups are accompanied by trilobites with other feeding modes, continuing downslope to the atheloptic biofacies, in which the exemplars are often blind or have much reduced eyes. In Early Ordovician Laurentia, cheiruroids (Pliomeridae, Cheiruridae) accompany asaphids in the shallower environments and are joined by calymenoids, Phacopina, Encrinuridae, Odontopleurida, and Lichida in younger strata. Asaphids are particularly varied in the shelf limestones of the Baltic paleocontinent.

Asaphida are part of the Ibex Fauna, which diminished progressively in importance through the Ordovician. However, asaphids (*Isotelus*) remain conspicuous in shelf limestones in the Cincinnatian of North America. None survived the end Ordovician extinction event, and the proliferation of this group across Ordovician shelf seas can be regarded as a characteristic part of the "great biodiversification." However, it is striking that the components of the Whiterock Fauna of Adrain et al. (1998) that survived the Ordovician-Silurian event included an array of suborders/families with predator/scavenger morphology that were among the most important components of post-Ordovician faunas: dalmanitids, phacopids, cheirurids, encrinurids, styginids, Lichida, and Odontopleurida among them (figure 24.2). Hence trilobites with attached hypostomes—with the addition of the natant Proetida—survived the Ordovician to provide the basis for subsequent trilobite evolution.

Atheloptic Trilobites

Fortey and Owens (1987) coined this term for a "community" of deep-water trilobites that were blind or with eyes much reduced, the majority of which also had close relatives with normal eyes inhabiting more shoreward paleoenvironments. In the Ordovician they were often accompanied by such trilobites as shumardiids and raphiophorids, which lacked eyes in the whole clade. Shumardiids were the last of the miniaturized (at about a millimeter in length) benthic trilobites, which were more diverse in the Cambrian and underwent a modest Ordovician radiation; they may have been particle feeders. They did not survive the Ordovician. The earliest atheloptic assemblage is Tremadocian from northern England (Rushton 1988). Examples are known from all the "series": Arenig of South Wales, Llanvirn of Bohemia, Caradoc of Kazakhstan, and Ashgill of North Wales. A blind dalmanitoid, *Songxites,* is known from the Hirnantian of Dob's Linn, Scotland, and Ireland (Siveter et al. 1980).

The atheloptic assemblage of genera typically includes a mixture of Ibex Fauna clades (e.g., nileids such as *Illaenopsis,* shumardiids) and Whiterock ones (dindymenines, cheirurines, dalmanitoids). This biofacies, however, was apparently erased at the end of the Ordovician for a period lasting at least through the earlier half of the Silurian. Atheloptic biofacies are well developed again in the Devonian, but none of the proetides or phacopides occupying it then is closely related to the Ordovician examples.

■ Summary (JMA, GDE, RAF, ØH, JRL, TMcC, AWO, BGW, BDW, SRW, ZYZ)

Cluster analysis of all Ordovician trilobite families using nine biostratigraphic intervals shows that groups followed one of three diversity trajectories. Ibex Fauna I was successful during the Tremadocian and then rapidly declined; Ibex Fauna II radiated during the Arenig and then declined rapidly; the Whiterock Fauna radiated during the Ordovician diversification and was successful through the Ordovician. All the families that survived the end Ordovician mass extinction were members of the Whiterock Fauna. An Ordovician radiation of trilobites occurred globally, except for the tropics of Gondwana, and there was a strong geographic pattern to the radiation. Groups centered in high-latitude Gondwana began to radiate early in the Arenig, and the majority did not survive the end Ordovician. Tropical Laurentian (and Siberian) groups radiated during the late Arenig/early Whiterockian, and almost all survived the extinction. The Whiterock Fauna first dominated craton-margin environments in Laurentia and then spread onshore and offshore during the rest of the Ordovician. Regional diversity curves for Australasia, South America, Avalonia, Baltica, and South China reflect some of the paleogeographic subtleties of the biodiversity change of Ordovician trilobites across a spectrum of tectonic and latitudinal settings.

Ordovician trilobites occupied a variety of benthic niches in shallow to deep-water habitats and colonized the open seas. The pelagic habitat was remarkably persistent but did not survive the end Ordovician extinction event. The Olenid biofacies and atheloptic habitat were similarly affected, but their ecologic equivalents are known (with unrelated taxa) from younger strata. Among shelf faunas, the Ordovician was characterized by a proliferation of suspension-feeding genera; proetides were the principal deposit feeders from the Whiterockian onward. Members of the Ibex Fauna dominated the pelagic and Olenid biofacies and included also some of the largest predatory/scavenger taxa. The majority of trilobites of the Whiterock Fauna that survived the end Ordovician extinction had attached hypostomes.

ACKNOWLEDGMENTS

We are grateful to Bob Owens and John Shergold for their helpful and supportive comments on an earlier version of this chapter. J. M. Adrain and S. R. Westrop's research is supported by NSF grant EAR 9973065. A. W. Owen and T. McCormick's work on the Anglo-Welsh trilobites was funded by NERC Grant GR3/11834, which is gratefully acknowledged. B. G. Waisfeld acknowledges support from CONICET, ANPCyT, and Fundación Antorchas. Z.-Y. Zhou's research on Chinese trilobites was financially supported by the Major State Basic Research Development Program (No. G2000077700).

25 Eurypterids, Phyllocarids, and Ostracodes

Simon J. Braddy, Victor P. Tollerton Jr., Patrick R. Racheboeuf, and Roger Schallreuter

Overviews of three arthropod groups, the eurypterids, phyllocarids, and ostracodes, are included here. Ordovician eurypterids are rare. Reliable records, mostly from North America, include megalograptids and rarer carcinosomatids, stylonurids, and erieopterids. Of the 31 species described from the Ordovician of New York State, only one is a eurypterid, one is a phyllocarid, and the remainder comprise indistinguishable remains (pseudofossils). Eurypterid taxonomy and phylogeny are in a state of flux; calculations of diversity measures are thus inappropriate.

Ordovician phyllocarid crustaceans are widely distributed in graptolitic black shale facies, but documentation of the group remains inadequate. Biodiversity studies are limited to a preliminary survey of the phyllocarids from the Ordovician of South America presented herein.

Ostracodes are small crustaceans whose body is enclosed within a carapace consisting of two valves normally of calcareous composition, except for planktonic species with mainly organic shells. The body is strongly compressed laterally and also reduced in length to consist of head and thorax only; the abdomen is missing or only rudimentary, and the number of appendages is reduced to seven, the smallest number among all crustaceans. Extant ostracodes inhabit nearly all aquatic environments. Ordovician representatives appear to be confined to marine settings (although this may be limited by lack of recognition of lacustrine habitats of the time); to date, true pelagic forms have not been recognized.

Eurypterids (SJB, VPT)

The order Eurypterida forms a small (<250 species) but morphologically diverse clade (see Tollerton 1989) of Paleozoic chelicerates, based on two autapomorphies: a metastoma and median abdominal appendage, although both of these features also occur in chasmataspids, a group distinguished from eurypterids by their pre/postabdomen opisthosomal division of three/nine visible segments (most eurypterids except megalograptids have seven/five segments). Stylonurid eurypterids have traditionally been interpreted as the most primitive (Størmer 1955), based on the possession of a pediform sixth prosomal appendage. Caster and Kjellesvig-Waering (1964), however, suggested that the megalograptids may be more primitive, based on presumed plesiomorphic features (e.g., morphology of the carapace, doublure, appendages, and opisthosoma), but conceded that it was difficult to infer the primitive eurypterid condition.

Eurypterids are very scarce in Ordovician strata (table 25.1). They appear to have radiated rapidly, but many families reported to have originated in the Ordovician (e.g., Selden 1993: figure 17.1) are based on inadequate material from New York State (see later

TABLE 25.1. Reliable and Unverified Occurrences of Ordovician Eurypterida

Taxon, Reference (and details of material if uncertain)	Stratigraphy, Locality	Series/Stage (time slice)
Onychopterella augusti Braddy et al. 1995	Soom Shale Member, South Africa	Ashgill (*TS*.6b)
Megalograptus ohioensis Caster and Kjellesvig-Waering 1955	Elkhorn Member, Richmond Formation, Ohio	Ashgill (*TS*.6b)
Eocarcinosoma batrachophthalmus Caster and Kjellesvig-Waering 1964	Elkhorn Member, Richmond Formation, Ohio	Ashgill (*TS*.6b)
Megalograptus sp. *in* Caster and Kjellesvig-Waering 1964	Whitewater Formation, Ohio	Ashgill (*TS*.6b)
Megalograptus shideleri Caster and Kjellesvig-Waering 1964	Saluda Formation, Ohio	Ashgill (*TS*.6b)
Megalograptus welchi Miller 1874	Liberty Formation, Ohio	Ashgill (*TS*.6a–b)
Megalograptus williamsae Caster and Kjellesvig-Waering 1964	Waynesville Formation, Ohio	Ashgill (*TS*.6a)
?Eurypterida indet. *in* Elias 1980 Doubtful eurypterid as calcareous	Stony Mountain Formation, Manitoba, Canada	Ashgill (*TS*.6b)
Eurypterida indet. ?New genus (S. Donato pers. comm.)	Georgian Bay Formation, Manitoulin Island, Canada	Upper Ordovician (?)
?Eurypterida indet. (cf. *Onychopterella*) (S. Yanbin pers. comm.) Three incomplete specimens, one an orthocone nautiloid	?, Shanxi Province, northwestern China	Upper Ordovician (?)
?*Megalograptus* sp. Waines 1997	Martinsburg Formation, New York	Caradoc (*TS*.5c–d)
?*Megalograptus* sp. *in* Lehmann and Pope 1987	Martinsburg Formation, Swatara Gap, Pennsylvania	Caradoc (*TS*.5c–d)
Carcinosoma sp. *in* Foerste 1916	Richmond Formation, near Chambly, Quebec	Upper Ordovician (?)
?Eurypterida indet. *in* Little 1936 Prosoma and appendage	Collingwood Formation, Labrador, Canada	Caradoc (*TS*.5c–d)
Echinognathus clevelandi (Walcott) 1882	Utica Shale, New York	Caradoc (*TS*.5c–d)
Megalograptus alveolatus (Shuler) 1915 Three podomeres of a spiniferous appendage	Bays Formation, Virginia	Caradoc (*TS*.5b)
Brachyopterus stubblefieldi (Størmer) 1951	Shales below Breiden Dolerite, Criggen, Wales	Caradoc (*TS*.5b)
?Eurypterida indet. *in* Chlupáč 1999 Cuticle fragments	Letná Formation, Bohemia	Caradoc (*TS*.5b)
?Eurypterida indet. *in* Ramos and Blasco 1975 Prosoma and isolated appendage podomeres	Yerba Loca Formation, near Iachal, Argentina	Caradoc (?)
?Eurypterida indet. *in* Toro and Perez 1978 Tergites and a telson (inorganic or phyllocarid *Caryocaris*)	?, Bolivia	Darriwilian (?)

in this section on eurypterids). Given such limited data and poor resolution of eurypterid phylogeny, it is inappropriate to calculate diversity measures.

Depitout's (1962) suggestion that eurypterids originated in Laurentia during the Ordovician is dubious, given that it is based on the inadequate New York State material. *Onychopterella* from South Africa, *Brachyopterus* from Wales, and *Megalograptus* from Virginia provide the earliest reliable records of eurypterids from Gondwana, Baltica, and Laurentia, respectively (table 25.1). Together with indeterminate remains from South America and China (table 25.1) and scattered occurrences in North America, eurypterids are seemingly far more widespread than previously appreciated. Plotnick's (1983) suggestion that eurypterids were a highly diverse and spatially segregated fauna should be viewed with caution because of their poor early fossil record and widespread taxonomic oversplitting (particularly at species level).

Eurypterid paleoecology has been variously interpreted (see Braddy 2001 for review). Ordovician eurypterids have been interpreted as the most marine, associated with typical marine faunas (Plotnick 1983); for example, trilobites, nautiloids, and conodonts occur with *Megalograptus ohioensis* (table 25.1). Although this is consistent with the "transition" hypothesis (i.e., throughout the Paleozoic eurypterids moved from marine into brackish and freshwater settings), there are so few reliable Ordovician eurypterids that this scheme is based on limited data for this time period. Eurypterids were also able to undertake short amphibious excursions onto land, perhaps as part of their life cycle.

Body Fossils

The inorganic nature of most of the Ordovician eurypterid species from New York State, described by Walcott (1882), Clarke and Ruedemann (1912), Ruedemann (1942 and references therein), and Flower (1945), was recognized by Plotnick (1983) and Tollerton and Landing (1994). Continued study by one of us (VPT unpubl.) has revealed that of the 17 genera and 31 species described between 1882 and 1962, only 1—*Echinognathus clevelandi* (Walcott) 1882—is eurypterid. Material described by Ruedemann (1916) as this species consists of inorganic sedimentary structures. *Buffalopterus verrucosus* Kjellesvig-Waering and Heubusch 1962, described as a broad eurypterid telson, is actually a phyllocarid carapace of the genus *Mytocaris* (VPT unpubl.). The remaining material comprises regular to irregular outlines of siltstone with uneven parting and thin cross-laminations, siltstone intraclasts, or small patches of slickensides on sheared siltstone. Recently, more megalograptid material has been found in New York State (table 25.1).

The Ordovician eurypterids of Ohio are well known. Most are species of *Megalograptus*, although some taxonomic questions remain regarding this material that may lead to synonymy. *Eocarcinosoma batrachophthalmus* Caster and Kjellesvig-Waering 1964 requires restudy (it may be a juvenile *Megalograptus*) but is provisionally accepted here as reliable. The remaining North American Ordovician eurypterids have a fragmentary and questionable record (table 25.1).

Ordovician eurypterids from outside North America are exceedingly rare. The oldest unequivocal eurypterid is *Brachyopterus stubblefieldi* Størmer 1951; the youngest is *Onychopterella augusti* Braddy et al. 1995. The remainder, from South America and China (table 25.1), require restudy. Large fragments of arthropod cuticle, from the lower Caradoc Letná Formation of Bohemia, were questionably assigned to Eurypterida by Chlupáč (1999).

Trace Fossils

Ordovician eurypterids may also be inferred from their fossilized trackways. One trackway, from the Caradoc Hudson River Shales (upper Chazyan-Mohawkian) of New York State, preserved on a mud-cracked surface (a subaerially exposed mudflat), was originally attributed to a pterygotid, the outer tracks produced by their enlarged chelicerae (Sharpe 1932). As it is unlikely that the chelicerae had a locomotory function and pterygotids are unknown from the Ordovician, the actual producer remains uncertain. Trackways from the Graafwater (Lower Ordovician) and Peninsula (Upper Ordovician) formations of the Western Cape, South Africa (interpreted as a distal fluvial floodplain setting with an intermittent marine influence), were attributed to an onychopterellan by Braddy and Almond (1999). The diverse ichnofauna of the Tumblagooda Sandstone (?Upper Ordovician, based on conodont dating of the overlying Ajana Formation; Mory et al. 1998) of Western Australia also contains some very large eurypterid trackways (Trewin and McNamara 1995).

■ Phyllocarid Crustaceans (PRR)

Ordovician representatives of the subclass Phyllocarida (superclass Crustacea, class Malacostraca) are most commonly preserved in black shales, usually associated with graptolites, that is, in offshore pelagic environments. However, occurrences are usually considered only as minor faunal components, though a few papers have documented abundant as well as exceptionally well-preserved specimens. The generally small-sized exoskeleton and very thin cuticle often result in their remains becoming diagenetically or tectonically distorted (or both). The scarcity of complete, articulated, specimens adds to taxonomical difficulties. Some genera and most species have been described using carapaces only or isolated tailpieces or even a single, isolated furcal ramus. The limited number of reliable taxonomic characters hinders determination and explains why such a large number of occurrences are placed in open nomenclature, mainly as *Caryocaris* sp. These are not considered herein.

Recent papers (Chlupáč 1970; Jell 1980; Shen 1986; Aceñolaza and Esteban 1996; Racheboeuf et al. 2000; Vannier et al. 2003), as well as investigations in progress in South America, show that Ordovician phyllocarids are much more abundant and diversified than previously thought and suggest they may be of real biostratigraphic and biogeographic interest.

In 1853 the first Ordovician phyllocarids were described (see table 25.2). Rolfe (1969) recognized three genera belonging to the families Ceratiocarididae (namely, ?*Ceratiocaris* M'Coy 1849 and *Caryocaris* Salter 1863) and Aristozoidae (?*Aristozoe* Barrande 1872). *Ceratiocaris* was questioned because it was based mainly on the uncertain affinities of the morphology and nature of isolated spines in *C. primula* Barrande 1872 (see Chlupáč 1970), and Rolfe (1969:320) also doubted the interpretation of *Aristozoe*.

Rolfe (1969) listed six other genera that were considered to be of uncertain order and family: *Galenocaris* Wells 1944, *Lebesconteia* Jones and Woodward 1899, *Mytocaris* Chlupáč 1970, *Nothozoe* Barrande 1872, *Saccocaris* Salter 1873, and *Trigonocarys* Barrois

TABLE 25.2. Annotated List of Described Ordovician Phyllocarid Taxa

Date, Name and Author	Distribution	Comments
1853 *Hymenocaris vermicauda* Salter	Lower Ordovician, Wales, United Kingdom	[?Phyllocarid]
1853 *Dithyrocaris? longicauda* Sharpe	Ashgill, Serra de Buçaco, Portugal	[?=*Ceratiocaris*]
1863 *Caryocaris wrightii* Salter	Arenig-Llanvirn, United Kingdom, Europe, Australia	Type species
1872 *Ceratiocaris primula* Barrande	Ashgill, Czech Republic	[?Phyllocarid]
1872 *Nothozoe pollens* Barrande	Middle Ordovician, Czech Republic	[?Phyllocarid]
1873 *Saccocaris major* Salter	Lower Ordovician, Wales, United Kingdom	[?Phyllocarid]
1876 *Caryocaris marrii* Hicks	Arenig, Wales, United Kingdom	[=*C. wrighti*][a]
1891 *Saccocaris minor* Jones and Woodward	Arenig, Wales, United Kingdom	[?Phyllocarid]
1892 *Lingulocaris maccoyi* Etheridge	Lower Ordovician, New Zealand, Australia	[=*Rhinopterocaris*]
1892 *Protocimex siluricus* Moberg	Lower Ordovician, Sweden	[?=*Caryocaris*]
1896 *Caryocaris curvilatus* Gurley	Lower Ordovician Nevada, USA	[=*C. maccoyi*]
1896 *Caryocaris oblongus* Gurley	Point Levis, Canada	
1896 *Dawsonia monodon* Gurley	Point Levis, Canada	[=Furcal rami of *Caryocaris*]
1896 *Dawsonia tridens* Gurley	Point Levis, Canada	[=Furcal rami of *Caryocaris*]
1903 *Caryocaris augusta* Chapman	Ordovician, Victoria, Australia	[=*C. wrighti*][a]
1903 *Saccocaris tetragona* Chapman	Lower Ordovician, Victoria, Australia	[?Phyllocarid]
1906 *Ceratiocaris scanicus* Moberg and Segerberg	Lower Ordovician, Sweden	[?=*Caryocaris*][a]
1907 *Lamprocaris micans* Novak nomen nudum	Czech Republic	[=*C. wrighti*][a]
1912 *Caryocaris kilbridensis* Woodward	Arenig, Ireland	[=*C. wrighti*][a]
1919 *Caryocaris silicula* Bassler	Lower Ordovician, Maryland, USA	
1919 *Lamprocaris novaki* Zeliszko	Darriwilian, Czech Republic	[=*C. wrighti*][a]
1919 *Caryocaris barrandei* Zeliszko	Darriwilian, Czech Republic	
1931 *Caryocaris acuta* Bulman	?Caradoc, Peru	[?=*Pumilocaris*]
1934 *Caryocaris minima* Chapman	?Lower Ordovician, New Zealand	
1934 *Rhinopterocaris bulmani* Chapman	Lower Ordovician, New Zealand	[dubious taxon]
1934 *Caryocaris raymondi* Ruedemann	Athens Shale, Tennessee, Alabama, USA	
1934 *Caryocaris silurica* Ruedemann	Princes of Wales Island, Alaska, USA	[?Phyllocarid]
1935 *Caryocaris magnus* Ruedemann	Middle Ordovician, Henry House Shale, Oklahoma USA	[?]
1935 *Caryocaris oklahomensis* Ruedemann	Middle Ordovician, Henry House Shale, Oklahoma USA	[?]
1944 *Galenocaris campbelli* Wells	Upper Ordovician, Maquoketa Shale, USA	
1970 *C. (Rhinopterocaris) subula* Chlupáč	Darriwilian, Czech Republic	
1970 *Mytocaris kloucheki* Chlupáč	Arenig, Czech Republic	[?Phyllocarid]
1970 "*Nothozoe*" *barrandei* Chlupáč	Lower Caradoc, Czech Republic	[?Phyllocarid]
1980 *Caryocaris stewarti* Jell	Middle Tremadocian, Victoria, Australia	
1986 *Caryocaris zhejiangensis* Shen	Arenig, southeastern China	
1996 *C. bodenbenderi* Aceñolaza and Esteban	Tremadocian, Famatina, Argentina	[=Furcal rami of *Caryocaris*]
2000 *C. delicatus* Racheboeuff and Vannier	Darriwilian, Precordillera, Argentina	
2000 *Pumilocaris granulosus* Racheboeuff and Vannier	Early Caradoc, Precordillera, Argentina	

Note: This list of Ordovician taxa is not exhaustive. It includes all established and questionable phyllocarid taxa in the *Treatise* (Rolfe 1969), as well as taxa described since 1969, but does not include taxa removed previously from the Phyllocarida, as well as the numerous forms left in open nomenclature, mainly those assigned to *Caryocaris* sp.
[a] See Chlupáč 1970.

1891. Most of these generic names were poorly defined and illustrated. Another genus, *Rhinopterocaris* Chapman 1903, was either regarded as a subgenus of *Caryocaris* (Chlupáč 1970) or synonymized with *Caryocaris* (Racheboeuf et al. 2000). This latter also included the new ceratiocaridid genus *Pumilocaris* Racheboeuf, Vannier, and Ortega 2000.

All Ordovician taxa described as phyllocarids are listed in table 25.2, comprising 13 genera and 38 species. The forms left in open nomenclature are not included. The genus *Trigonocarys* Barrois 1891 is excluded from phyllocarids (Racheboeuf 1994). Four other genera (*Hymenocaris, Nothozoe, Saccocaris,* and *Mytocaris*) are doubtfully assigned to phyllocarids, and 1 genus (*Dawsonia*) has been described from isolated furcal rami of *Caryocaris*. At the species level, 2 are nonphyllocarid taxa, 9 are doubtful phyllocarid species, and another 7 are placed in synonymy. This leaves 9 valid genera (plus 4 doubtful) and 22 species (plus 9 doubtful). Most of these taxa are in need of revision, and several genera and species have to be reconsidered, redefined, and/or synonymized. Moreover, those belonging to *Caryocaris,* within the family Ceratiocarididae, now also appear to be uncertain and in need of further discussion.

All these factors make it currently difficult to establish the precise diversity levels of Ordovician phyllocarids. The group remains poorly known, and diversity patterns cannot be properly evaluated without a complete, exhaustive revision of all previously described taxa. Moreover, the ages of most taxa based on available graptolite zonations remain imprecise. However, it has been possible here, based on recent work, to present a tentative assessment of the biodiversity of phyllocarid crustaceans from the Ordovician of South America (Bolivia, Peru, and Argentina).

Distribution

Ordovician phyllocarids first occur in the Tremadocian (Australia, Argentina), and they are known from the Arenig to Ashgill, though they show a variable diversity through time. This variation appears to be correlated directly with the development of the graptolitic black shale facies and is represented mainly by the genus *Caryocaris*. Representatives of the *Caryocaris* group may be locally abundant, and bedding planes are often crowded with variably disarticulated carapaces and tailpieces associated with graptolites. The abundance and morphology of their exoskeletal pieces (thin carapace; leaflike, expanded furcal rami), coupled with their worldwide distribution, strongly suggest that they were capable of swimming. High concentrations of specimens on bedding planes suggest that phyllocarids congregated in schools like modern-day fish species.

In Europe *Caryocaris* has been described (or found recently) in Great Britain (from the lower Arenig graptolite-based *Tetragraptus phyllograptoides* Zone, *TS*.2a, to the Darriwilian *Didymograptus artus* Zone, *TS*.4b; see time slices in chapter 2), Ireland (Arenig, *D. ?varicosus* Zone, *TS*.2b–c), Norway (lower Arenig *T. phyllograptoides* Zone, *TS*.2a), Belgium (Darriwilian *D. artus* Zone, *TS*.4b, of Condroz), and southern Sweden and Germany (Darriwilian *D. artus* to *D. murchisoni* zones, *TS*.4b–c), as well as in Portugal (probable Ashgill of the Serra de Buçaco, *TS*.6a–b). In Bohemia, phyllocarids range from the lower to middle Arenig, *TS*.2a–3b (Klabava Formation) to the Ashgill, *TS*.6a–b (Králův Dvůr Formation); representatives of the genus *Caryocaris* (*Rhinopterocaris*) range from the Arenig, *TS*.2a, to the Darriwilian, *TS*.4c (Chlupáč 1970).

In North America, phyllocarids are known from both Canada and the United States. In Alaska and the western United States, *Caryocaris*, including *C. curvilata* Gurley 1896, ranges from the graptolite-based *Expansograptus hirundo* (*TS*.4a) to the *Nemagraptus gracilis* (*TS*.5a) zones, that is, Darriwilian–lower Caradoc in age (Churkin 1966). In South America (Argentina, Bolivia, Peru), *Caryocaris* ranges from the Tremadocian (*Adelograptus ? tenellus* or *A. victoriae* zones, *TS*.1b) to the Caradoc *Climacograptus bicornis* Zone, *TS*.5b (Racheboeuf et al. 2000). In Argentina it occurs abundantly in the Precordillera, as well as in the Famatina Range and in the La Puna area. Several new taxa, distinct from *Caryocaris*, have been collected recently from this latter area (see later in this chapter).

In China, *C. zhejiangensis* ranges from the graptolite-based *Azygograptus suecicus/Didymograptus abnormis* Zone (*TS*.2c) to the lower part of the *Cardiograptus/D. nexus* Zone (*TS*.3b), that is, middle to upper Arenig (Shen 1986). In Australia, *C. stewarti* is restricted to the Lancefieldian La1 Zone, probably lower to middle Tremadocian (*TS*.1a–b) in age (Jell 1979). In New Zealand the phyllocarids appear to be mostly Early Ordovician in age.

Hence, representatives of the genus *Caryocaris* range from the Tremadocian to the lower Caradoc (top of the *N. gracilis* Zone, *TS*.5a) and then disappear before the end Ordovician faunal crisis.

The latest Ordovician phyllocarid representatives include *Dithyrocaris longicauda* Sharpe 1853 from the Ashgill of Portugal and *Ceratiocaris primula* Barrande 1872 from the Ashgill of Bohemia, though they are still of relatively uncertain taxonomic affinity. They possibly belong to the family Ceratiocarididae and in this way are possible ancestors of the larger nektobenthic Silurian forms. Much older, possibly ancestral ceratiocaridids from the Arenig have been found recently in Argentina.

South American Diversity Record

Because of the need for revision of the ages of most previously recorded taxa, a global evaluation of phyllocarid diversity cannot be made, although it appears that the most abundant faunas and the highest diversities occur in black, pelagic, graptolite-bearing shales and siltstones. An attempt is made to outline the diversity record of South American taxa, based on recently collected materials (figure 25.1).

In South America, Ordovician phyllocarids have been reported from Peru, Bolivia, and Argentina. The Argentinian sequence provides the most complete Tremadocian to Caradoc record. The relative thickness of graptolite-bearing beds in northern Argentina (e.g., up to 1,500 meters for the Arenig succession in the La Puna area) as well as in southern Bolivia provides exceptional collecting conditions and a precise

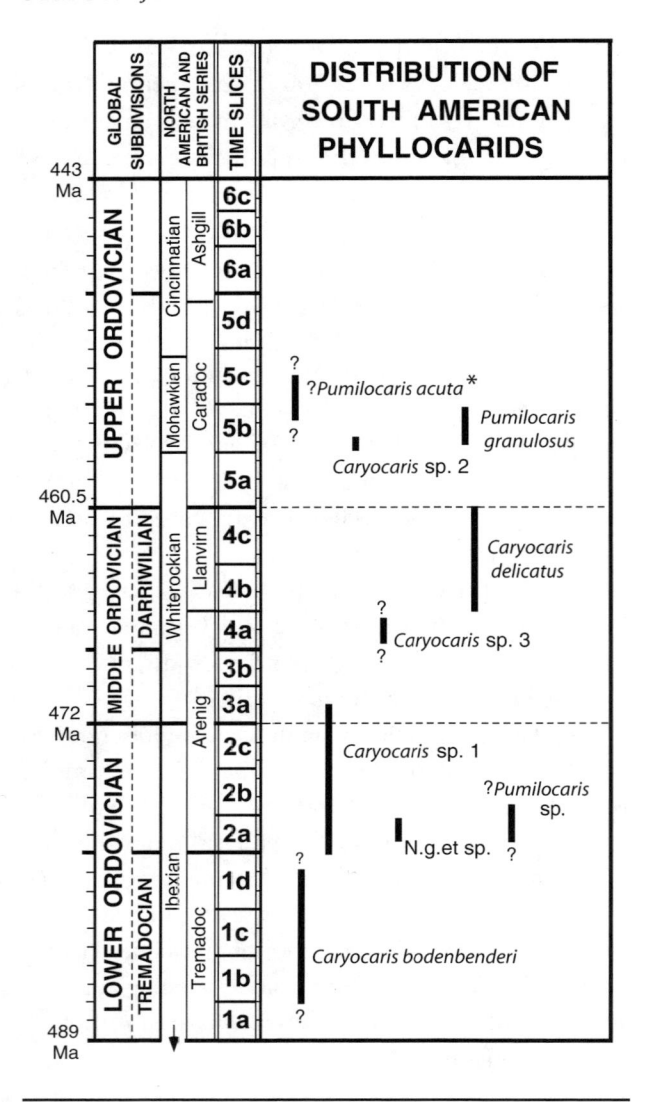

FIGURE 25.1. Vertical range of South American (Argentina, Bolivia, Peru) Ordovician phyllocarids (Racheboeuf et al. 2000). The asterisk indicates a species that is restricted to the Upper Ordovician of Peru.

record of occurrences associated with graptolites. Although phyllocarids are mainly represented by species of the genus *Caryocaris,* other genera occur. Their respective vertical distribution and overall diversity are shown in figure 25.1.

The first maximum of normalized diversity is in the Lower Ordovician, corresponding to *TS.*2a and 2b (i.e., lower Arenig); the second maximum is in the Upper Ordovician, associated with *TS.*5b of the lower to middle Caradoc. The actual disappearance of phyllocarids in the middle to upper Caradoc (*TS.*5c) is probably more related to environmental changes (a lack of deep-water graptolite-bearing beds) than to a true biologic event.

The detailed study of abundant, newly collected specimens reveals that morphological changes of the carapace are always matched by morphological changes of the tailpiece. Moreover, these changes in matching characters are remarkably uniform within the studied samples (more than 50 specimens each). All these correlated morphological characters strongly suggest that if specimens are carefully collected, particularly with regard to graptolite zones, they can be of real stratigraphic value by correlating with graptolites when the latter are missing, contrary to the views of Rushton and Williams (1996:110).

■ Ostracodes (RS)

Early Ostracode Evolutionary History

The origin of ostracodes has yet to be resolved. In the Cambrian, two groups of bivalved arthropods with phosphatic shells, the Bradoriida and the Phosphatocopida, superficially resemble ostracodes and have long been considered as their remote ancestors. However, the soft anatomy (e.g., appendage structure) of these two groups shows major differences from that of extant ostracodes. This led several authors (e.g., Hou et al. 1996; Walossek 1999; Siveter et al. 2001b) to place both the Bradoriida and the Phosphatocopida outside the ostracode lineage. Phosphatocopids and bradoriids were highly diversified during the Cambrian and had a virtually cosmopolitan distribution. Only very few representatives (including *Septadella, Eremos, Ludvigsenites, Chegetella,* and a few Chinese taxa) of these two groups survived into the Lower Ordovician.

The first unequivocally accepted ostracodes appear in the Ordovician, during which a great radiation of the group took place, so that by the end of this time all extant ostracode orders were present. There are only a few reports of "real" Tremadocian ostracodes, and the entire Lower Ordovician record of the group remains poor. Diversification started by the Mid Ordovician and reached its maximum in the mid to Late Ordovician. Toward the end of the Ordovician, ostracode diversity decreased markedly, possibly as a result of the glaciation. Most important during the Ordovician is the order Beyrichiocopa and its suborder Palaeocopa. The latter are characterized by a special kind of sexual dimorphism (velar dimorphism) that is unknown in Recent ostracodes. In Ordovician palaeocopes, velar dimorphism affects the development of a so-called antrum that is typically developed as a brood pouch outside the carapace proper. The original antrum of the morphs forms a more or less sausagelike sculpture (botulus), which in more advanced taxa is differentiated into several loculi. Loculi-bearing taxa are relatively widespread in the Upper Ordovician. However, they developed polyphyletically, with first appearances around the Middle to Upper Ordovician boundary.

Constraints on Biodiversification Analysis

Analysis of biodiversification trends among Ordovician ostracodes is hindered by inhomogeneous knowledge of their worldwide distribution. Ordovician ostracodes are known from all continents except Antarctica. Whereas more than 2,500 species have been described from the Northern Hemisphere (mainly from Europe and North America), the number of Ordovician ostracodes known from the Southern Hemisphere is extremely low (fewer than 100 species). From Asia only the ostracodes of Siberia are relatively well documented.

Different means of preservation and resulting variation in sample preparation methods also influence ostracode studies. The richest faunas come from samples dissolved by hydrochloric or hydrofluoric acid. One sample of the so-called Öjlemyrflint, for example, may contain at least 70 species (Schallreuter 1987), and the whole bed has more than 110 species (Schallreuter 1986). Acid preparation also ensures collection of the smaller faunal elements. Such material is generally much rarer or even missing in samples of rock material from which the ostracodes were extracted by physical methods. In such samples the amount of carapaces is generally higher, and the less sculptured forms are relatively more common (see, e.g., Meidla 1996).

Biodiversity of Selected Regions

Baltica

Ordovician ostracode faunas are best known and attain their maximum worldwide diversity in the Baltoscandian region. Ostracodes were investigated in or reported from Norway (Henningsmoen 1953, 1954a, 1954b; Qvale1980), Sweden (Jaanusson 1957), Denmark (Tinn and Meidla 2001), Finland (from glacial erratic boulders known as "geschiebes": Martinsson 1956; Nõlvak et al. 1995), Ingria (Melnikova 1999), Estonia (Sarv 1959; Meidla 1996), Latvia (Gailite, in Ul'st et al. 1982), Lithuania (Sidaravichiene 1992), Belorussia (Ropot and Pushkin 1987), East Prussia (Sztejn 1985), East Poland (Sztejn 1985), and "geschiebes" of northern central Europe (Schallreuter 1986, 1987, 1993a, 1994).

Biodiversity in the region is strongly influenced by facies, with highest diversity in the limestone facies. In the Scanian Confacies Belt, which is characterized by clastic sediments (mostly shales), ostracode diversity is rather low. The silicified limestones known as Backsteinkalk boulders, below the Big Bentonite Bed (BBB) of northern Germany (top *TS*.5b), have yielded more than 60 ostracode species, whereas the contemporaneous Sularp Shale in Scania displays only a quarter of that diversity with attendant low abundance. In Sweden the BBB is now referred to as the Kinnekulle K-bentonite bed (Bergström et al. 1995). Depth (associated with salinity and temperature variations) seems to be another factor influencing the diversity of ostracodes: the two different limestone facies of Baltoscandia (Central Baltoscandian and North Estonian confacies belts) have about equal numbers of species, but the species themselves are different, with only about 25 percent common to both regions.

During the Lower Ordovician, the diversity is rather poor. The oldest known species, *Nanopsis nanella* (Moberg and Segerberg 1906), comes from the *Ceratopyge* Limestone/Shale (Norwegian 3aα–γ unit, or in modern stratigraphic usages including the Bjørkåsholmen Formation of the mid to late Tremadocian, equivalent to *TS*.1b–c) of the Oslo region, Isle of Öland and Scania. A relatively diverse fauna from the Billingen (*TS*.2b–c) of Ingria (St. Petersburg region) was recently described by Melnikova (1999). The oldest horizon (Mäekula Member; conodont-based *Prioniodus elegans* Zone, *TS*.2b) yielded five species, the overlying Vasil'kovo and Päite members six further species. They were assigned to five families of at least four orders/suborders (Palaeocopa, Binodicopa, ?Eridostraca, and Platycopa sensu lato), which indicates that the ostracode radiation had already commenced in the Tremadocian-early Arenig.

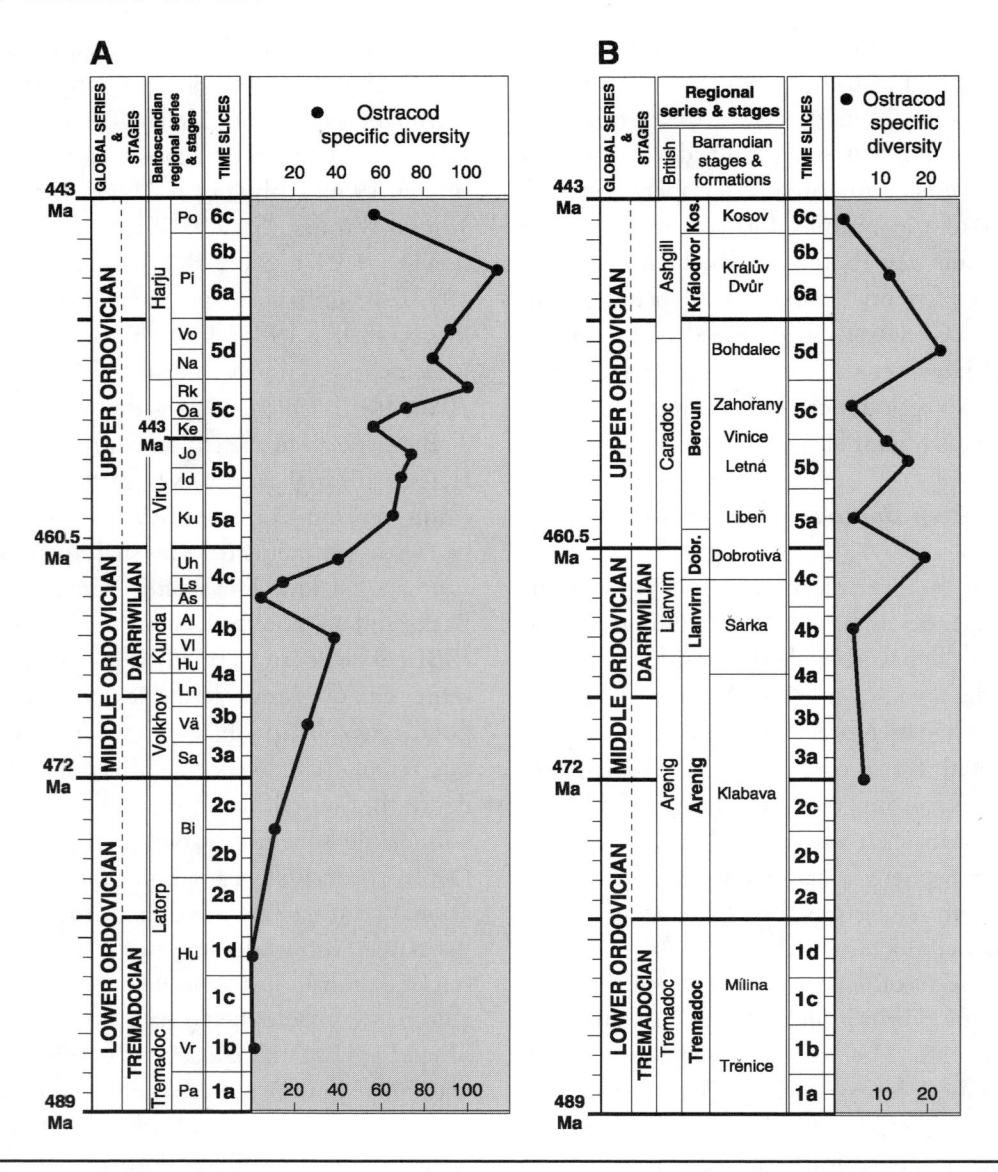

FIGURE 25.2. Diversity of Ordovician ostracodes in (A) Baltica and (B) the Barrandian area of the Bohemian Massif (Perunica). For key to abbreviations of the Baltoscandian names, see figure 2.1. Details of the names used in the Barrandean area (Prague Basin) are in Havlíček (1998, fig. 16).

In the Mid Ordovician (figure 25.2A) diversity increased considerably to 60 species in *TS*.4c), but did not reach its maximum until the Late Ordovician (229 species in *TS*.5c, and 176 species in *TS*.5d). Although diversity fluctuated considerably within a relatively short interval, such as in the Middle Ordovician of Central Sweden, overall diversity was generally not very high in any part of the region (Tinn and Meidla 2001). In the Kukruse stage (*TS*.5a), a first maximum was reached with 66 recorded species, and a similarly high level was attained in the Haljala stage (*TS*.5b). The effect on the fauna of the "catastrophic" Kinnekulle K-bentonite ash fall across the Johvi-Keila boundary (*TS*.5b–c) was relatively limited. In the Central Baltoscandian Confacies Belt, more than half (and potentially two-thirds) of the species survived the event. The highest diversity in Estonia is known from the Pirgu stage reaching 114 species in *TS*.6a–b, and in Lithuania from the Nabala stage (*TS*.5d), coincident with a relative increase of the non-palaeocopes and the first appearance of a cruminate palaeocope (*Fallaticella*). Glaciation at the end of the

Ordovician is thought to be responsible for a collapse in species diversity to 57 species in *TS*.6c, with no species surviving into the Silurian (although the ranges of some genera, e.g., *Semibolbina,* span the boundary).

Elsewhere in Baltica, more than 50 Ordovician ostracode species have been recorded recently by Melnikova (in Melnikova and Dmitrovskaja 1997) from the Moscow Basin. Approximately two-thirds of these occur in the upper Middle and lower Upper Ordovician. From Podolia, Krandievsky (1969) mentioned ostracodes from three Upper Ordovician horizons. The fauna, with 40 species distributed among 31 genera, was described by Abushik and Sarv (1983), who correlated the beds with the Nabala stage of Estonia (i.e., lower part of *TS*.5d) on the basis of the ostracodes. Ordovician ostracodes of the Holy Cross Mountains were described by Olempska (1994). The investigated section of the Mójcza Limestone comprises the entire stratigraphic sequence from the Kunda to the Jõhvi, through Nabala interval. From Novaya Zemlya or Vajgach (which are attributed to Baltica in some paleogeographic reconstructions; see Lehnert et al. 1999; chapter 5), 31 new species of Ordovician ostracodes have so far been described by Schallreuter et al. (2001); this represents about half of the known diversity of the collections. Glebovskaja (1949) published three late Llanvirn (*TS*.4c) species from the nearby island of Vajgach.

Perunica

Ordovician ostracodes of central Europe are preserved in mostly clastic facies, often as steinkerns and external molds only. In a few cases the ostracodes can be prepared by Wetzel's method with hydrofluoric acid; or, if the rock material is soft, they can be simply washed out. In Bohemia ostracodes are known from nearly all stages (except the Tremadocian), but the species diversity differs markedly (figure 25.2B). In the Lower and Middle Ordovician the diversity is low (Klabava Formation [*TS*.2a–4a pro parte] with ~6 species, and Šárka Formation [top *TS*.4a to early *TS*.4c] with ~4 species); but it increases in the upper Middle and Upper Ordovician to relatively higher levels (Dobrotivá Formation [late *TS*.4c–5a] with ~19 species, and Bohdalec Formation [*TS*.5c–d] with ~23 species) (Schallreuter and Krúta 2001a, 2001b). However, the diversity is never as high as in Baltica. A significant drop in diversity is registered in the uppermost Ordovician Kosov Formation, to two species in *TS*.6c.

Avalonia

Of the limited number of ostracode faunas known from the Ordovician of Britain, most species were described during the second half of the nineteeth century, and so their revision is long overdue. The most important paper published in more recent time is the monograph of Jones (1986, 1987) on "Llandeilo" and Caradoc beyrichiocope ostracodes from England and Wales. Members of the three suborders of the Beyrichiocopa (Palaeocopa, Binodicopa, and Eridostraca) account for some 90 percent of the fauna. The remainder belong mainly to the order Podocopa. A characteristic group of the region is the gunnaropsines, represented by only two species (*Harperopsis ohensis* Schallreuter et al. 2000 from the Oxhe Formation in Belgium and Bohemia) outside the British Isles. Ostracodes are virtually absent from the Arenig and most of the Llanvirn in England and Wales. However, the highest diversity (more than 40 species) occurs in the upper Llanvirn (*TS*.4c equivalent). More than half of these range into the Costonian substage (upper Aurelucian stage, i.e., upper part of *TS*.5a) but are unknown above the base of the Harnagian substage (lower Burrellian stage, i.e., base of *TS*.5b), the limiting factor here being presumably facies related (graptolitic shales). British Ashgill ostracode faunas are mainly undescribed as yet. Floyd et al. (1999) published a low-diversity marine ostracode fauna from the Girvan district (southwestern Scotland); a well-preserved silicified assemblage of more than 25 ostracode species is known from an Ashgill sample at Dryslwyn Castell, Dyfed (Jones 1987); and another Ashgill ostracode fauna from northern England (Cautley district) with more than 30 species has been described recently by Williams et al. (2001).

An epiplanktic ostracode fauna living presumably in floating algae of early Llanvirn age from Westphalia (Ebbe anticline, Rhenish Massif) has recently been described by Schallreuter and Koch (1999). The fauna is, like the accompanying trilobites, relatively rich in species but poor in individuals. A special feature is the preservation of most carapaces juxtaposed in a butterfly position.

Laurentia

Ordovician ostracodes are widespread in North America as well as Greenland (Schallreuter and Siveter 1985: text figure 3). Many of the early published species require revision, and recent descriptions of complete faunas exist from a few regions only, mainly from Canada. Copeland (1974, 1982, 1989) documented ostracodes of Whiterockian to Cincinnatian ages from the District of Mackenzie, where the oldest fauna (Sunblood Formation) yielded 3 leperditiocopes, 6 beyrichiocopes, and 2 podocopes. The succeeding Esbataottine Formation contained 1 archaeocope, 4 leperditiocopes, 21 beyrichiocopes, and 5 podocopes. The fauna from the lower Whittaker Formation (Cincinnatian) consists of 16 beyrichiocope and 5 podocope species. From Ontario, faunas of lower Upper Ordovician age were described by Copeland (1965 in Steele and Sinclair 1971) from Lake Timiskaming, Bucke (Liskeard) Formation (more than 70 species), and Braeside (Ottawa Valley) (16 species). Late Cincinnatian ostracodes from Anticosti Island (Quebec) include 29 species from the Vauréal Formation (Copeland 1970) and 39 species from the overlying Ellis Bay Formation (Copeland 1973). From the District of Franklin (Melville Peninsula, Baffin Island), Copeland (1977) identified 43 late Mohawkian ostracode species (TS.5c).

Ostracodes of Whiterockian age of the Ibex area, Utah, described by Berdan (1976, 1988) consist of 8 leperditiocopes, 30 beyrichiocopes, and only 1 podocope. Apart from the great number of leperditiocopes, it is remarkable that there is only 1 palaeocope among the beyrichiocopes. Leperditiocopes are also rather common in Kentucky and nearby states, from where Berdan (1984) described 17 species, whereas the number of beyrichiocopes and podocopes reaches only 53 (Warshauer and Berdan 1982). The frequency of the leperditiocopes is a special feature of Laurentia. With more than 130 species, the ostracode faunas of the Simpson Group from Oklahoma are the most diverse recorded from the lower to middle Upper Ordovician of North America, but only a few taxa have been revised (Williams and Vannier 1995).

Kazakhstania

Ordovician ostracodes from several localities in Kazakhstan were documented by Melnikova (1980, 1986). The Upper Ordovician fauna from the Akkerme Peninsula, Lake Balkhash, comprises 7 species. Faunas from several other localities came from the Llanvirn, Caradoc, and Ashgill. The highest diversity was discovered in the upper Caradoc or lower Ashgill (TS.5d or 6a) Dulankara Formation with 40 species; in other horizons the number of species varies between 4 and 16. Palaeocopes are rare in all horizons.

Siberia and Mongolia

More than 250 ostracode species have been described from localities in the Siberian Platform and northeastern Siberia but not from complete stratigraphic sections. Kolosnitsyna (in Ogienko et al. 1974) described 42 species from the Siberian Platform; the oldest species came from the Ust'kut Formation (Tremadocian: TS.1d), 36 species from the Krivolukian (Llanvirn: TS.4c), 10 species from the Chertovskian (lower Upper Ordovician: TS.5a). From the Ajchal' region, Kolosnitsyna (1984) published 15 species, 9 from the middle Upper Ordovician. In Sette-Daban (Kanygin 1971), the highest diversity among ostracodes occurs in the upper Llanvirn (TS.4c) and decreases in the Caradoc, whereas in the Selennyakh Range the highest diversity is in Caradoc faunas. On the Chukotka Peninsula (Kanygin 1977) only 37 ostracode species are known from the Isseten Formation (Llanvirn-Caradoc, TS.4c–5a).

Nine endemic species of Ashgill ostracodes from Mongolia (Mount Melden-Teg) were described by Melnikova (1978).

China

Few palaeocopes but relatively many bradorines have been documented from different parts of China (Hubei, Zhejiang, Shaanxi, Hunan, Liaoning), with 90 species described as new and more than 30 species left in open nomenclature (e.g., Shu 1990 documented the bradoriids *Zhexiella, Praeaechmina,* and *Polycostalis* from the Lower Ordovician of Zhejiang). A relatively high number of Tremadocian ostracodes are known (Hou 1953), but only relatively few species from other Lower Ordovician intervals. The richest fauna described (Sun 1988) is of Late Ordovician age.

Peri-Gondwana

In the microcontinent of Thuringia, ostracode faunas are extremely rare. Best preserved is the silicified fauna from the so-called Kalkbank, the only limestone member of the Thuringian Ordovician succession and described by Knüpfer (1968). Another silicified fauna was obtained from a limestone boulder of the Lederschiefer Member (*TS*.6c) (Blumenstengel 1965). Both faunas are of Late Ordovician age and comparably diverse (~25 and ~22 species, respectively).

A fauna of latest Ordovician age from Cellon in the Carnic Alps remains incompletely known (Schallreuter 1990). Most striking is the occurrence of the same Baltic species as in Bohemia (see earlier in this chapter).

Vannier (1986a, 1986b) described several beyrichiocope ostracodes from the Arenig-Caradoc of Ibero-Armorica. Characteristic for the region is the dominance of binodicopes, with 23 species versus 20 palaeocope species, as in Gondwana. Podocopes are present but apparently rare (Gutiérrez-Marco et al. 1996).

From Turkey (Istanbul area) only three species of Late Ordovician age are known (Sayar and Schallreuter 1989).

Gondwana

The ostracode diversity of Gondwana is generally very low throughout the whole Ordovician. A characteristic feature of beyrichiopcope faunas is that the palaeocopes are generally outnumbered by binodicopes and their closely related eridostracans.

Knowledge of Ordovician ostracodes from Africa is extremely poor. Only recently, the first Ordovician ostracode from South Africa, which is also the first known Ordovician myodocope, has been reported (Siveter et al. 2001a). In North Africa, a few ostracode species are reported from Libya (Vannier 1986b; Schallreuter and Hinz-Schallreuter 1998) and Morocco (Termier and Termier 1950; Gutiérrez-Marco et al. 1997).

From South America, Ordovician ostracodes are known from Peru (Hughes et al. 1980), Bolivia (Pribyl 1984), and Argentina (Pribyl 1996; Schallreuter 1999). Endemic species (55, among 24 genera) are recorded with binodicopes dominant, as in Australia. Palaeocope ostracodes do occur, but their diversity is comparatively low.

In Australia, ostracodes occur in all series of the Ordovician from all states except for Victoria and South Australia, but knowledge of the faunas remains rather limited. Currently, only 13 species are known, among them one of the oldest palaeocope ostracodes (*Eopilla ingelorae* Schallreuter 1993b from the lower Emanuel Formation of the Canning Basin), which occurs in the upper Tremadocian (*TS*.1d). Most of the other species are Late Ordovician (*TS*.5a–d) in age. Most important are the Pillinae (represented by four described genera), which outside Australia occur only in South America and maybe in China and Siberia.

Synthesis

Although Ordovician ostracodes occur worldwide (except in Antarctica), faunas are well studied from only a few regions of the Northern Hemisphere (Baltica, western Europe, Siberia, North America). Despite the poor knowledge of ostracode faunas from other parts of the world, some general conclusions can be drawn. Radiation of the ostracodes must have started in the Tremadocian, but only a few ostracodes from this time span are currently known (from Scandinavia, China, and Australia). Representatives of several orders are present by the Arenig, and certain Arenig palaeocopes had already attained a fairly high level of evolutionary development, as shown by dimorphism and reduced quadri-lobation (see Melnikova 1999). During the Llanvirn the diversity increased considerably but reached its maximum in the Late Ordovician. Glaciation at the end of the Ordovician led to a remarkable reduction in the number of species. The region with the greatest diversity is Baltica, followed by North America. There, members of all Ordovician ostracode orders (except the myodocopes) are present, with the dimorphic palaeocopes being the most important. This group is still not known with certainty from the Southern Hemisphere. Pelagic ostracodes have not been documented, except for an epiplanktic fauna from Westphalia.

ACKNOWLEDGMENTS

Roger Schallreuter thanks the following for their help in completion of the ostracode section of this chapter: Florentin Paris (Rennes) and Ian Percival (Sydney) for assistance during its preparation and Jean Vannier (Lyon) for his constructive review.

26 Crinozoan, Blastozoan, Echinozoan, Asterozoan, and Homalozoan Echinoderms

James Sprinkle and Thomas E. Guensburg

Echinoderms rapidly evolved and underwent a major diversification during the Ordovician Period. After they radiated into eight or nine classes in the Early and Mid Cambrian to produce the members of the Cambrian Evolutionary Fauna (CEF), echinoderm diversity declined during the Late Cambrian to only four to five classes (Sumrall et al. 1997). However, a much larger echinoderm radiation began in the Early Ordovician, when many new echinoderm groups appeared and older groups expanded as part of the Paleozoic Evolutionary Fauna (PEF). The number of echinoderm classes approximately doubled, and by the early Late Ordovician, all 21 echinoderm classes had appeared, echinoderms had reached their highest standing class diversity for the entire Phanerozoic, and they had attained the first of three Paleozoic peaks in generic and specific diversity (Sprinkle 1980). Echinoderms that appeared during this Ordovician Radiation then dominated Paleozoic benthic marine environments for the next 200 million years (m.y.).

Some PEF echinoderm classes continued to diversify after the Ordovician, dominating the Paleozoic record of echinoderms (such as crinoids and blastoids), but most dwindled during the Mid and Late Paleozoic to eventual extinction (e.g., rhombiferans, diploporans, and stylophorans). Extant echinoderm classes that appeared in the Ordovician Radiation (starfish, brittle stars, echinoids, and holothurians) diversified only moderately during the remainder of the Paleozoic but survived the severe Permo-Triassic extinction and then diversified much more during the Mesozoic and Cenozoic. They are considered members of the Modern Evolutionary Fauna (MEF), which dominates the present-day echinoderm diversity, along with the surviving PEF crinoids.

The record for the Ordovician Radiation of echinoderms is best preserved and has been most extensively studied on the paleocontinents of Laurentia, Baltica, and the northern parts of Gondwana. The echinoderm record for all these areas has long been adequately known for the Mid and redefined Late Ordovician (e.g., see Eckert 1988), but recent collecting in western Laurentia and some other areas has now added considerable new diversity to Early Ordovician echinoderms (Guensburg and Sprinkle 1992, 1994, 2000, 2001; Sprinkle and Wahlman 1994; Sprinkle and Guensburg 1995, 1997; Sumrall et al. 2001). Small collections of Ordovician echinoderms have come from other areas such as northeastern Gondwana (India, Burma), eastern Gondwana (Australia), western Gondwana (southern Mexico, Bolivia, western Argentina), and a few other areas (e.g., China, Iran, Korea) that were parts of other continents or were island arcs.

Echinoderms have a high-Mg calcite skeleton (a relatively stable depositional mineral) made up of sutured (tesselate) or imbricate single-crystal plates covering the body or forming segmented or solid appendages (stem, arms, spines) used for attachment, feeding, or protection. The individual plates, although usually small, are easily recognized both in the field, because they break along reflective cleavage planes, and in thin section, because of their distinctive microporous internal structure composed of mineralized stereom and originally tissue-filled stroma, now commonly filled with matrix or secondary calcite. Echinoderms buried in shales, siltstones, or sandstones often have the calcite plates dissolved away either during diagenesis or more commonly during present-day weathering. This leaves a mold of the exterior or interior thecal surface, which can be cast in latex or other media to give an exact replica of the original plating. Many Ordovician occurrences are known with this latter type of preservation (see examples in Chauvel 1966; Paul 1973, 1984). Echinoderms are sometimes found silicified in medium- to thick-bedded limestones, and complete specimens or abundant separate plates and stems can then be extracted by acid etching (for two Ordovician examples, see Sprinkle 1973a, 1973b).

Echinoderm debris is fairly common in many marine Ordovician units, but the overall echinoderm contribution to shell beds appears to have slowly declined through Ordovician time as other PEF groups (especially brachiopods and bryozoans) diversified (see Droser et al. 1996, for Early to Mid Ordovician patterns) and as burrowers churning through the sediment became more common (Droser and Bottjer 1989). Echinoderms tended to be gregarious and are preserved intact only under optimal conditions (Sprinkle 1982). Because the plated skeleton rapidly disarticulated if the animals were not buried quickly, they typically show a patchy distribution in the rock record, with small clusters of complete specimens in favorable settings and only scattered debris elsewhere. Some echinoderm species and even whole classes are rarely found in Ordovician shelf faunas, known only from one or a few occurrences. Echinoderms are most common in interbedded thin limestone and shale beds that were deposited in shallow cratonic seas; they are also common in thick-bedded limestones but are less easily collected as complete specimens. Complete echinoderms are locally concentrated immediately above hardgrounds where their attachment disks may be abundant (Wilson et al. 1992). They are also associated with thin bryozoan or sponge biostromes (Sprinkle 1974), accumulations of large brachiopod or bivalve shells (Waddington 1980), or small to large metazoan or stromatolitic mounds, where they are sometimes found draped over the mound tops or as abundant debris in the steeply dipping grainstone flank beds or in channels between the frame-building organisms (Sprinkle and Guensburg 1995). Echinoderms are less common in Ordovician deltaic, slope, or deep-water basinal deposits, although some occurrences in each facies are known. Macrocystellid rhombiferans (Paul 1984), diploporans (Chauvel 1978), and cornute and mitrate stylophorans (Ubaghs 1968b) are notably more common in these facies.

■ Diversity Plots for Echinoderms

The diversity data are plotted in four diagrams (figures 26.1, 26.2, 26.3, 26.4) showing species-level alpha, beta, and gamma diversity (Sepkoski 1988) for the commonly used echinoderm subphyla (figure 26.1) and for class or order-sized groups, most of which are thought to be monophyletic clades, within the subphyla (figures 26.2, 26.3, 26.4). The only change from a standard Linnaean classification (see Ubaghs 1968a; Sprinkle 1980) is that homoiosteleans (solutans) and Cambrian homosteleans (cinctans) have recently been shown to be blastozoans closely related to eocrinoids (David et al. 2000). Homoiosteleans have therefore been removed from the Homalozoa in figure 26.4 and plotted with the blastozoan eocrinoids in figure 26.3, leaving only cornute and mitrate stylophorans of uncertain origin assigned to the Ordovician Homalozoa.

Sprinkle (with the assistance of Guensburg and several other workers listed in the acknowledgments at the end of this chapter) plotted ranges for the figures using three large Excel spreadsheets, plus data from selected faunal lists or regional echinoderm monographs that were tabulated separately. A total of 820 species occurrences were tabulated on the spreadsheets, and 396 were added from the lists and monographs, for a total of 1,216 species of Ordovician echinoderms. One of the few previous attempts to analyze species-level diversity for a major group of fossil echinoderms was Moore (1952) using crinoids, and we have now nearly tripled the number of Ordovician crinoids to 516 species compared with

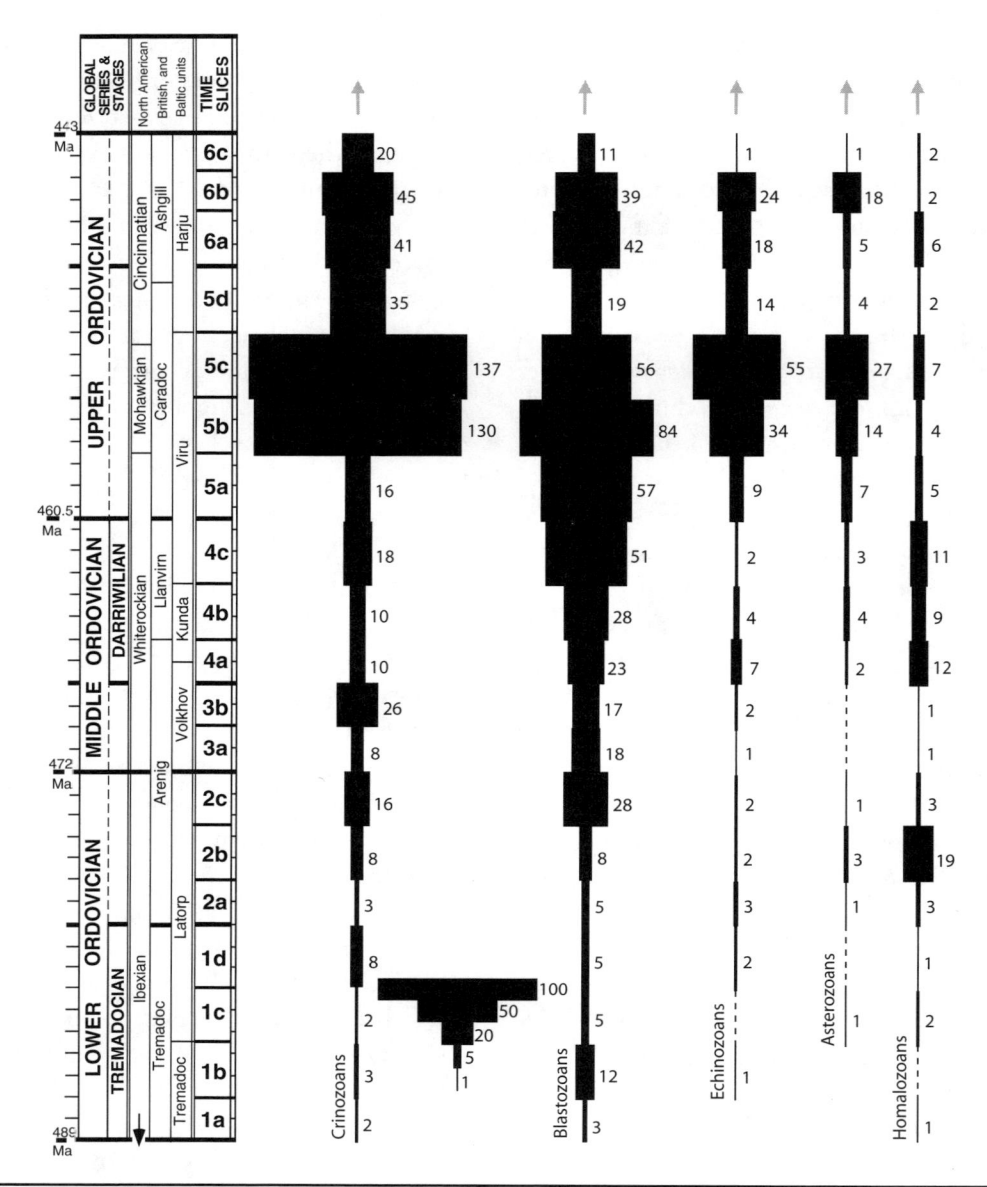

FIGURE 26.1. Species-level total diversity throughout the Ordovician of the five echinoderm subphyla: the crinozoans (7 clades), blastozoans (11 clades), echinozoans (7 clades), asterozoans (3 clades), and homalozoans (2 clades). Data are plotted on a standardized Ordovician timescale divided into 19 time slices. Arrows at the top indicate that the subphylum continued into the Silurian; dashed lines indicate gaps in the record; and width of the bar scale at bottom indicates species diversity.

Moore's about 170 species. In most cases, ranges were tabulated at the species level, including all named species that appeared to be valid on cursory examination. Only two named subspecies per species were included in the tabulations, as a compromise between listing them all or not listing any subspecies, but this affected only about 10 species that have numerous subspecies named by two to three authors. Unnamed species ("*Aristocystites* sp.") were included if the author, or the reviser, or our own check of the illustrations indicated that this taxon was likely a distinct species. About 60 unnamed or informally named new taxa were included, many listed in Sprinkle (1971) or in Sprinkle and Guensburg (1997: plate 1C), from unpublished work of our students (e.g., Lewis 1982), from faunal lists including a few new taxa from Baltic Russia and Estonia (supplied by Sergei Rozhnov), or from a recent abstract by Ausich and Copper (2002). This increased the number of echinoderms known from the Early and earliest Mid

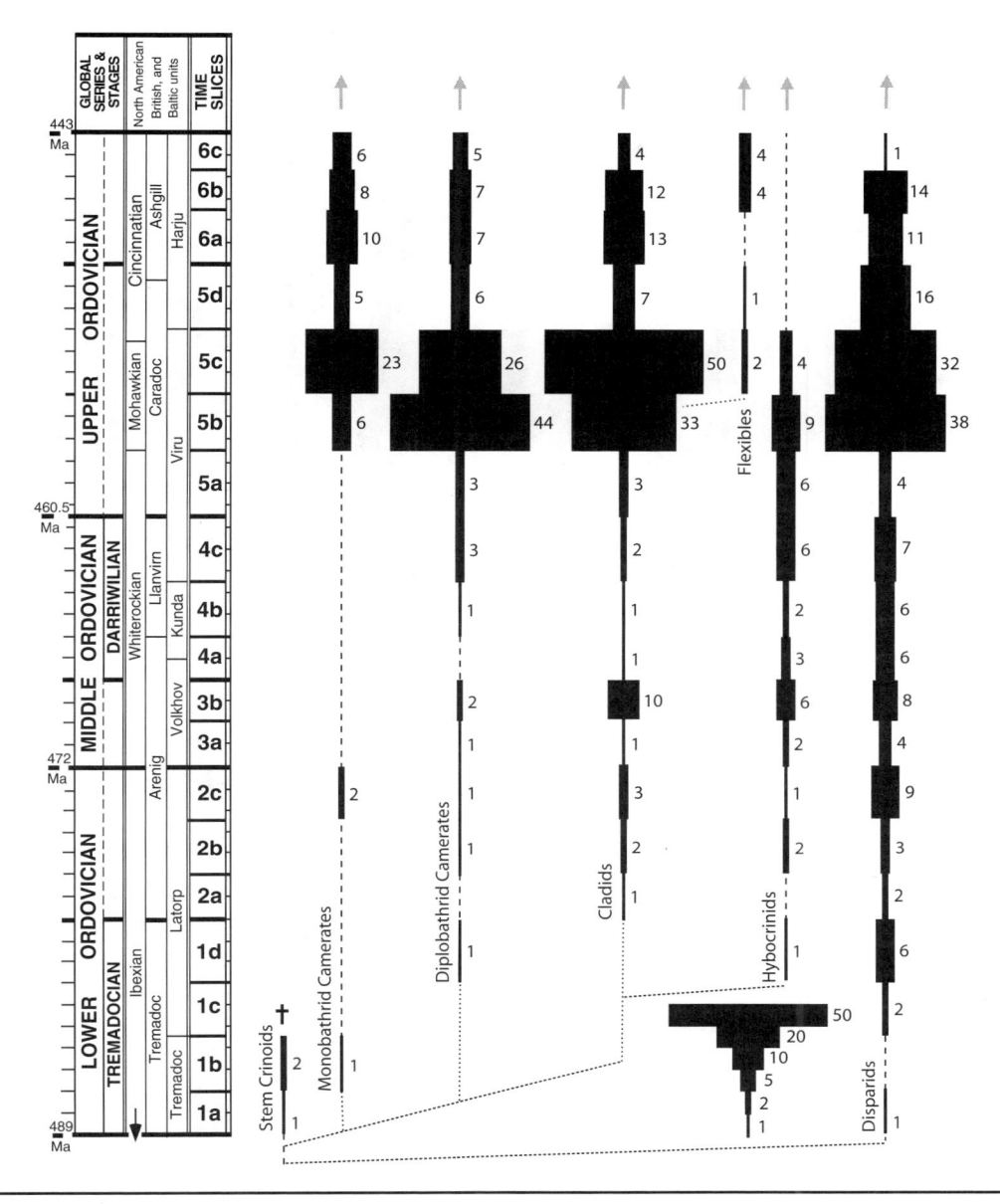

FIGURE 26.2. Crinozoan (crinoid) total diversity in the Ordovician divided into seven clades (stem crinoids, monobathrid camerates, diplobathrid camerates, cladids, flexibles, hybocrinids, and disparids). Dotted lines indicate preferred phylogeny and estimated branching times. Crosses indicate that the clade became extinct at this time. Other symbols and bar scale same as in figure 26.1.

Ordovician and from the latest Ordovician, when diversity levels were relatively low.

The most comprehensive and up-to-date group, faunal, or regional publications on Ordovician echinoderms were tabulated, including recent larger works such as the three *Treatise* volumes on echinoderms edited by Moore (1966, 1968) and Moore and Teichert (1978), plus major works by Chauvel (1966, 1978), Parsley (1970, 1991, 1998), Ubaghs (1970), Paul (1972, 1973, 1984, 1988, 1997), Brower (1973, 1995b), Sprinkle (1973a, 1982, 1995), Brower and Veinus (1974, 1978), Kolata (1975, 1982), Parsley and Mintz (1975), Bell (1976), Warn and Strimple (1977), Smith and Paul (1982), Guensburg (1984), Kolata et al. (1987), Eckert (1988), Smith (1988a), Donovan (1989), Rozhnov (1989), Simms et al. (1993), Guensburg and Sprinkle (1994, 2003), Donovan et al. (1996), Sprinkle and Guensburg (1997), Ausich (1998), Frest et al. (1999), Prokop and Petr (1999), and Vizcaïno and Lefebvre (1999). Several

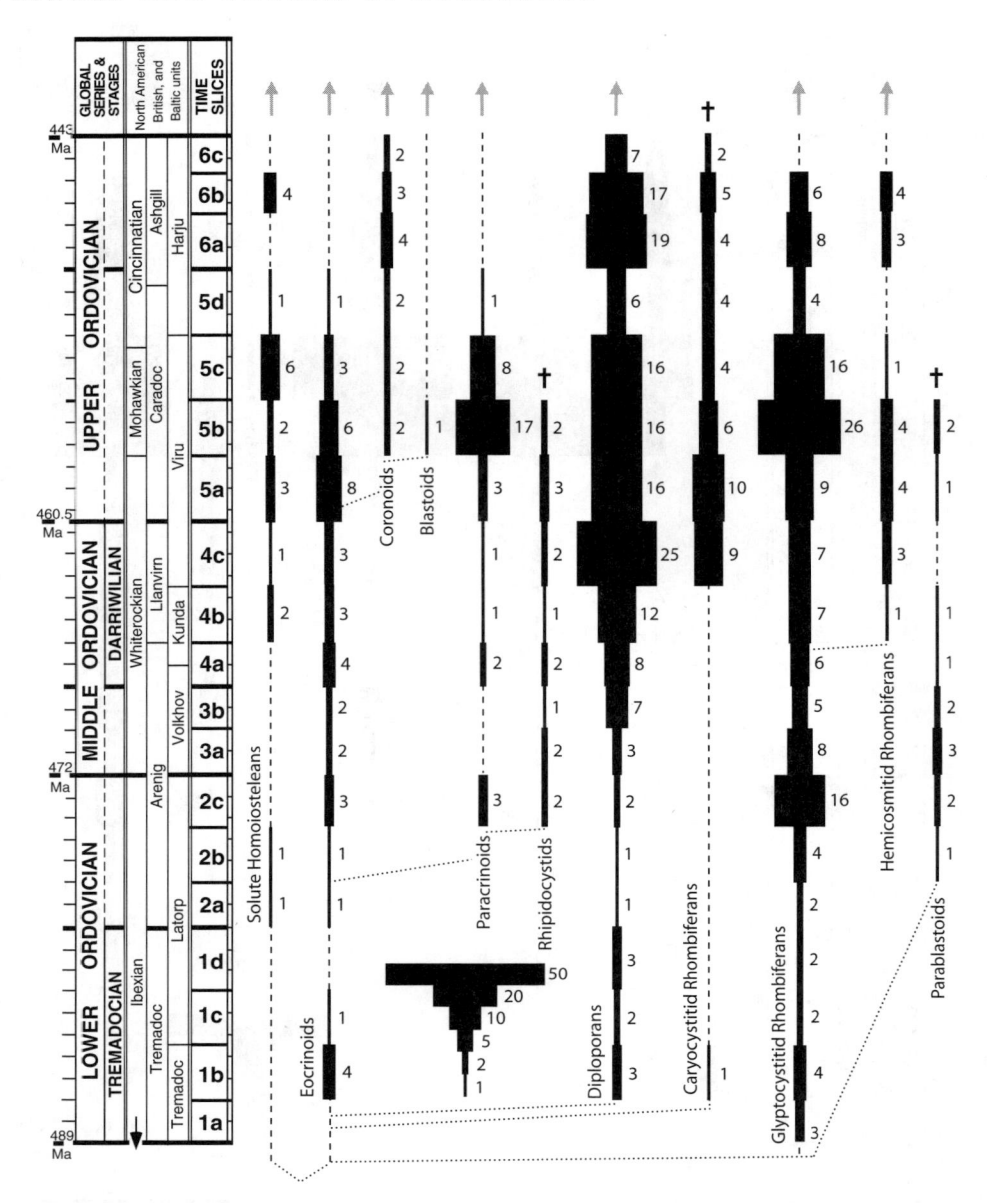

FIGURE 26.3. Blastozoan total diversity in the Ordovician divided into 11 clades (solute homoiosteleans, eocrinoids, coronoids, blastoids, paracrinoids, rhipidocystids, diploporans, caryocystitid rhombiferans, glyptocystitid rhombiferans, hemicosmitid rhombiferans, and parablastoids). Symbols and bar scale same as in figures 26.1 and 26.2.

hundred shorter papers describing one or a few echinoderm species were also consulted, including series of papers by authors such as Ubaghs (1960, 1963, 1969, 1979, 1991, 1994); Chauvel (1969, 1977, 1981); Chauvel and Le Menn (1973, 1979); Kolata (1973, 1976, 1983, 1986); Bockelie (1974, 1981a, 1981b, 1981c, 1982a, 1982b, 1984); Chauvel et al. (1975); Chauvel and Melendez (1978); Frest et al. (1979, 1980); Kolata and Guensburg (1979); Frest and Strimple (1982); Kolata and Jollie (1982); Brower and Strimple (1983); and Brower (1992a, 1992b, 1994, 1995a, 1996, 1997, 2001). A few older summaries or monographs, such as Bassler and Moodey (1943), Moore and Laudon (1943), Regnéll (1945), Wilson (1946), and Ramsbottom (1961), provided useful information about faunas that were poorly known or had not been revised recently. We did not personally evaluate in detail the validity of echinoderm species

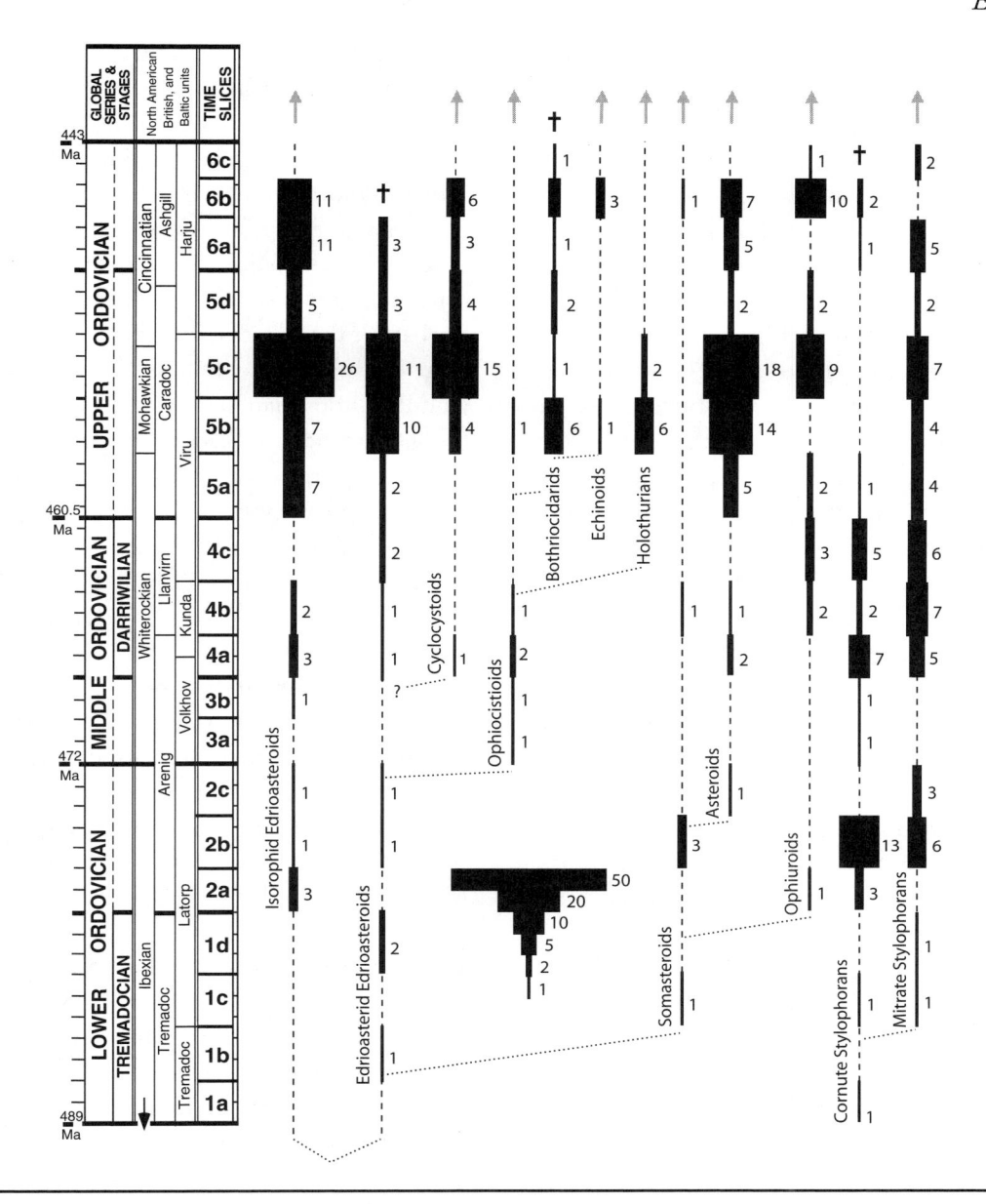

FIGURE 26.4. Echinozoan, asterozoan, and homalozoan total diversity in the Ordovician divided into 12 clades (echinozoans: isorophid edrioasteroids, edrioasterid edrioasteroids, cyclocystoids, ophiocistioids, bothriocidarids, echinoids, and holothurians; asterozoans: somasteroids, asteroids, and ophiuroids; homalozoans: cornute stylophorans and mitrate stylophorans). Symbols and bar scale same as in figures 26.1 and 26.2.

that we found in the literature, visit museums to examine type or reference specimens, or run computerized literature searches, and so there were probably synonyms, misassigned species, and also some publications containing taxa that we missed entirely. These three problems were probably minor in small and well-studied groups but were more severe in very large and/or poorly understood groups, such as crinoids and diploporans, in which we estimate as many as 20 percent of the taxa either were missed or were duplicates (synonyms). These two unfortunate problems did not compensate for each other, either, because missing taxa would most likely be randomly distributed whereas synonyms would be concentrated at times of highest diversity. The diversity plots in figures 26.1, 26.2, 26.3, and 26.4 should be viewed as only a first attempt to analyze the diversity of echinoderms at the species level during a critical early time in their evolutionary history; we hope to continue adding to and improving the accuracy of the database.

Ranges were plotted on a standardized Ordovician timescale divided into 19 time slices (seven Early, five Mid, and seven Late Ordovician), each ranging in length from 1.7 to 3 m.y. (see chapters 2 and 3). If the age of an echinoderm-bearing formation was well known based on trilobite, graptolite, conodont, or chitinozoan zones, then it was usually easy to assign the echinoderm species to one particular time slice. About 82 percent of the echinoderm species listed on the spreadsheets occurred in only one time slice, indicating that the known record for these species was mainly from a single occurrence or several closely spaced occurrences. The other 18 percent of species typically occurred in two to three adjacent intervals, although one species ranged through six time slices spanning nearly 14 m.y., and a few species had disjunct ranges with first and last occurrences separated by (usually small) gaps. In some cases we had to make arbitrary decisions to assign the rather generalized listed range of a species (such as "Caradoc" or even "Late Ordovician") to the smaller time slices that we were using. However, if an author indicated that numerous specimens of a species had been collected from different parts of a thick section representing the entire "Caradoc," we assigned that species to the entire range (in this example, the four time slices, *TS*.5a–d).

In terms of the percentages of completeness of the echinoderm fossil record in the Ordovician, the 30 clades plotted over the 19 time slices in figures 26.2, 26.3, and 26.4 had on average 32 percent gaps (implying 68 percent occurrences), or a range from 0 percent gaps (4 clades) to 83 percent gaps (1 clade). We counted time slices occupied from the clade's first appearance in the Ordovician to either its extinction or the end of the period. Many clades were missing from Hirnantian *TS*.6c at the end of the Ordovician, when there was a major sea level drop and a mass extinction (Sheehan 2001b); this was counted as a gap if the clade continued into the Silurian (23 out of 30 echinoderm clades). Several closely related clades had very divergent occurrence patterns, even over the same time span. For example, coronoids (figure 28.2) had a complete occurrence pattern with 0 percent gaps from Late Ordovician *TS*.5b to 6c, in contrast to closely related blastoids, which had only a single occurrence and 83 percent gaps during this same interval. Both clades reappeared in the Silurian, but surprisingly blastoids became much more successful and long-lasting than coronoids (Brett et al. 1983).

The global and regional time-stratigraphic units and their boundaries in this timescale should be familiar to most Ordovician echinoderm workers, with the possible exception of two fairly recent changes. One is that the Middle Ordovician–Upper Ordovician boundary has now been officially designated at the base of the Caradoc (figure 26.1, left), much lower than in the past, so that familiar North American Middle Ordovician stages, such as the Shermanian (now Chatfieldian), Blackriveran (now Turinian), and even Chazyan, are now in the lower part of the Upper Ordovician (Webby et al. 1991; Bergström 1995; Leslie and Bergström 1995). The other exception involves the Arenig; previously all of it was considered Lower Ordovician by most British and Baltoscandian workers, but now the Lower Ordovician–Middle Ordovician boundary has been placed almost in the middle of the Arenig (figure 26.1, left). Because of this change and other correlation problems, we had difficulties early in this project trying to correlate echinoderm-bearing units of Arenig age near this boundary between Baltic Russia and the United States (Sprinkle et al. 1999).

■ Diversity Levels and Patterns

Late Cambrian CEF echinoderms are scarce and low diversity, typically represented by only four classes: eocrinoids, cornute stylophorans, solute homoiosteleans, and edrioasteroids (Sumrall et al. 1997). Echinoderm faunas from this interval have been termed the eocrinoid-stylophoran fauna after the two most abundant echinoderm groups (Guensburg and Sprinkle 2000). However, Ubaghs (1999) described two transitional eocrinoids or stem-group glyptocystitid rhombiferans from the latest Cambrian of southern France, which may mark the initial radiation of that new PEF clade.

Crinozoans, blastozoans, and homalozoans all appeared or reappeared at the base of the Ordovician in *TS*.1a (figure 26.1). The earliest echinozoan (an edrioasteroid) reappeared soon after in *TS*.1b, and the first asterozoan (a somasteroid) appeared slightly later in *TS*.1c, still within the Early Ordovician. Most newly appearing PEF and continuing CEF classes and orders within the subphyla (21 of 30) also diversified during the Early Ordovician (figures 26.2, 26.3, 26.4; Smith 1988a), but a few clades appeared considerably later, in the Mid (two clades) or Late Ordovician (seven clades). These include hemicosmitid rhombiferans and ophiocistioids (both members of the PEF), which appeared in the Mid Ordovician (figures 26.3, 26.4), and flexible crinoids, coronoids plus blastoids, cyclocystoids (all representatives of the PEF), bothriocidarids plus echinoids (then PEF, now considered members of the MEF), and most likely holothurians based on sclerites (then either CEF? or PEF, now MEF), which appeared (or reappeared) in the early Late Ordovician (figures 26.2, 26.3, 26.4). These new groups, along with CEF holdovers, more than doubled the previous standing diversity of class- and order-sized echinoderm clades. This new and enlarged Ordovician echinoderm fauna has been called the crinoid-rhombiferan fauna (Guensburg and Sprinkle 2000), based on two of the largest groups. All echinoderm classes (including holothurians; see Reich 2001) were present in the fossil record (or had already become extinct) by the early Late Ordovician. This time interval (Mohawkian in Laurentia, late Viruan in Baltica, and mid Caradoc in North Gondwana) is represented by the largest number of co-occurring echinoderm classes for the entire Phanerozoic (Sprinkle 1980), as well as the diversity peak for Ordovician echinoderm genera and species (figure 26.1).

Most echinoderm clades show very little change in diversity at the Early Ordovician–Mid Ordovician boundary (Ibexian-Whiterockian boundary or mid Arenig) and relatively little change in diversity at the Mid Ordovician–Late Ordovician boundary (late Whiterockian or Darriwilian-Caradoc boundary). Droser et al. (1996:130) argued that the Ibexian-Whiterockian boundary represents the "pivotal point in the Ordovician radiation" and that "for many groups, but not all, this [pivotal] point is at or near the base of the Middle Ordovician." However, of the 30 clades of echinoderms plotted on figures 26.2, 26.3, and 26.4, only diploporans (figure 26.3) show a buildup in diversity from the base of the Whiterockian (TS.3a) to the end of the Mid Ordovician (TS.4c), when the group has its highest diversity, before decreasing somewhat through the Late Ordovician. One other clade, ophiocistioids (figure 26.4), first appeared at the base of the Whiterockian but remained at a low level of diversity in the following four time slices and then has a gap in the record until the Late Silurian. A few other clades, such as closely related paracrinoids and rhipidocystids (figure 26.3) and asteroids (figure 26.4), originated in TS.2c just before this boundary but maintained low diversity until sharply increasing much later in the Mohawkian (TS.5b). No other new echinoderm clades appeared at this boundary, and 16 of the 17 echinoderm clades already present near the Ibexian-Whiterockian boundary show low diversity. The seventeenth clade (glyptocystitid rhombiferans; figure 26.3) shows an intermediate peak in diversity immediately before the boundary and then decreases in diversity for several time slices before rising again to an even higher peak in the Mohawkian.

The Ibexian-Whiterockian boundary was a major sequence boundary on the Laurentian craton (Sloss 1963), showing a rapid drop in sea level in TS.2c followed by a slow return of epeiric seas in TS.3–4 in the early and mid Whiterockian. Other shelly fossil groups, such as brachiopods, bryozoans, corals, nautiloids, and ostracods, responded to this gradual return of more widespread epeiric seas by radiating to much higher diversity levels in the Whiterockian (Droser et al. 1996), but echinoderms did not show this same response. One factor contributing to the low echinoderm diversity during TS.3a–b is that echinoderms were almost completely absent from deeper-water clastic sections, such as those of Bohemia and southern France (Prokop and Petr 1999; Vizcaïno and Lefebvre 1999), in contrast to other parts of these long-duration sections where echinoderms were much more common.

Echinoderms also show little change at the Mid Ordovician–Late Ordovician boundary, when most clades exhibit low to moderate diversity levels (figure 26.1). A few clades became somewhat more diverse while others declined, but the overall change in diversity was slight. However, there was a dramatic change only one to two time slices later (TS.5b–c; i.e., Mohawkian or mid Caradoc), when six new echinoderm clades appeared and many other clades (especially in the crinoids, see figure 26.2) underwent a huge expansion to the highest standing diversities reached in the Ordovician. Before this interval, crinoids had been less common than several clades of stemmed or cemented blastozoans (compare columns in figure 26.1), but crinoids had a much larger expansion in the Mohawkian (figure 26.2) than the various clades of blastozoans (figure 26.3). At this time, three blastozoan clades expanded moderately, two clades contracted, and two small clades (rhipidocystids and parablastoids) became extinct. The echinozoans and asterozoans (figure 26.4) also underwent a major increase in diversity similar to that of the crinoids during the Mohawkian, in contrast to the homalozoans, with the cornutes almost disappearing, whereas the mitrates showed little change across the interval (figure 26.4).

After their rise to dominance in the Mohawkian, crinoids declined in diversity, as did the other echinoderm subphyla (figure 26.1), but remained nearly equal in diversity to blastozoans through the end of the Ordovician. After the end Ordovician Hirnantian extinction, crinoids rediversified more quickly during the Early Silurian than blastozoan or other echinoderm clades. Crinoids then became more and more diverse and dominant throughout the rest of the Paleozoic as other competing suspension-feeding blastozoans and echinozoans gradually declined or became extinct and no other echinoderm classes rose to take the place of these declining groups (see Sprinkle 1980: figure 1). This dominance by crinoids that began in the Mohawkian did not end until their near extinction at the Permo-Triassic boundary some 200 m.y. later.

The Mohawkian (mid Caradoc) echinoderm expansion (especially in crinoids; figures 26.1, 26.2) is probably the most striking feature shown by the echinoderm diversity record during the Ordovician. Many factors may have contributed to this high point in echinoderm diversification, and it is not easy to identify a single dominant cause (Sprinkle and Guensburg 2003). Environmental conditions were very favorable for echinoderms during the Mohawkian, with warm, stable, Greenhouse conditions (Sheehan 2001b), "Calcite Seas" with rapid carbonate cementation (Sandberg 1983; Stanley and Hardie 1998), and wide tropical and temperate climatic zones with only small areas of polar climate (Sheehan 2001b). High sea levels produced extensive shallow epeiric seas on the continents and wide continental shelves on the margins; thin-bedded carbonates interbedded with shales (often storm-generated) were deposited in both of these areas. In shallow-water areas with slow deposition, hardgrounds and large-brachiopod shell pavements formed, and bryozoan thickets, sponge and bryozoan mounds, and some larger reefs built up, all prime living areas for attached, suspension-feeding echinoderms. Orogenic activity was relatively subdued worldwide with the Taconic Orogeny beginning in eastern Laurentia during the mid Mohawkian (Holland and Patzkowsky 1997), and in many areas there was relatively little delta building supplying medium to coarse clastics. A moderate number of volcanic arcs produced scattered major eruptions with widespread bentonites (Kolata et al. 1996), and drifting continents and island arcs were still mostly isolated from each other within the slowly shrinking Iapetus Ocean, so that several faunal provinces still had their own distinctive echinoderm faunas (Paul 1976). Echinoderms had developed new life modes with the appearance of the first herbivores (ophiocistiods, echinoids; Kolata 1975) and carnivores (asteroids, ophiuroids, echinoids; Blake and Guensburg 1994). Suspension-feeding crinoids had also developed advanced arms with closely spaced pinnules for more efficient small-particle capture (Ausich 1980) and new types of distal stem attachments that could be vertically rooted in soft sediment, draped along the sea floor with cirri for rooting or support, or wrapped around other organisms, such as bryozoans and other stems (Brett 1981; Guensburg 1984).

These favorable conditions led to many different echinoderms living under a variety of conditions in the Mohawkian (or mid Caradoc) seas. Many echinoderms were then smothered and rapidly buried intact by storm-generated sediment. This has produced numerous, well-preserved, complete specimens that are relatively easy to collect and prepare from weathered shales or from thin limestone bedding surfaces. In shales, siltstones, or fine sandstones, molds can be recovered to produce excellent casts showing the exterior plating. This abundant echinoderm material from this time interval in varying depositional settings on different continents has resulted in a number of large monographs and numerous smaller papers to name and describe this wealth of nicely preserved echinoderms. Our experience collecting in Laurentia and Baltic Russia indicates that the echinoderms from this particular time interval are more abundant, easier to collect and prepare, more complete and better preserved, and more diverse than echinoderms from most other parts of the Ordovician. All these favorable environmental and biotic factors have apparently combined to produce the very high diversity of echinoderms in the Mohawkian (mid Caradoc) shown in figures 26.1, 26.2, 26.3, and 26.4 compared with earlier or later time intervals.

Toward the end of the Ordovician, a gradual decline in diversity was followed by a steep dropoff in diversity and abundance of echinoderms (figures 26.1, 26.2, 26.3, 26.4) in the latest Ordovician (Hirnantian), which was a major extinction interval apparently caused by glaciation (Sheehan 2001b). The large drop in sea level caused by the buildup of ice caps decreased the area for shallow marine sediments to be preserved, and thus not many echinoderm specimens have been found. Although diversity was low and many echinoderm groups were completely absent from Hirnantian *TS*.6c (figure 26.4), most echinoderm clades (23 out of 27 that existed in the Ashgill) survived this extinction interval and recovered during the Early Silurian (figures 26.1, 26.2, 26.3, 26.4; arrows at top signify which clades survived). Only two groups (caryocystitid rhombiferans in figure 26.3 and bothriocidarids in figure 26.4) became extinct at the Ordovician-Silurian boundary, but cornute stylophorans (figure 26.4) disappeared slightly earlier, at the end of *TS*.6b, and edrioasterids (figure 26.4) became extinct two time slices earlier, at the end of *TS*.6a. All four of these clades were rare by the earliest Ashgill, and so their demise during or shortly before a mass extinction event is not surprising.

Although many echinoderm clades originated or reappeared in the fossil record within a relatively short interval in the Early Ordovician, some later clades appear in an evolutionary sequence, from plesiomorphic ancestor, to partly changed intermediate group, to highly derived descendant group. We believe that the time of first appearance of a new clade in the fossil record often closely approximates the time of branching from an ancestor accompanied by rapid development of new morphological features. The best example of this pattern may be the sequence from edrioasterids that reappeared near the base of the Ordovician and continued to the mid Ashgill, to ophiocystioids that appeared at the beginning of the Mid Ordovician (*TS*.3a), to bothriocidarids, echinoids, and holothurians that appeared together early in the Late Ordovician (*TS*.5b; figure 26.4). Ophiocistioids have long been considered an intermediate group that might be related to echinoids or holothurians (Ubaghs 1966; Smith 1988b; Smith and Savill 2001), and the branching order and preserved time of first appearance closely agree with this origination model. Smith and Savill (2001:145) argued that "it is obvious that the origins of the group [echinoids] must lie deeper. There remains much hidden history of the group to be discovered in the Early and Middle Ordovician." This statement would agree with the downward range extensions predicted by Marshall (1990) for a small group such as echinoids using 95 percent confidence intervals. However, we do not agree with this Smith and Savill statement and predict that (1) after nearly 150 years of collecting Ordovician echinoids, the known fossil record for their origin and appearance is now close to the true record, even though not many species have been found, and (2) no bothriocidarid or echinoid older than *TS*.4c (latest Mid Ordovician) will be found in the future. Another example of sequential appearances of related clades might include the somasteroid, ophiuroid, asteroid sequence in star-shaped mobile echinoderms in Early Ordovician *TS*.1c, 2a, and 2c (figure 26.4).

A related pattern that shows up in several of these diversity plots is the "paired appearance" of closely related sister clades or a mostly formed precursor and fully formed descendant clade (Sprinkle 1995) at almost the same time in the fossil record. Examples of this pattern include (1) sister groups paracrinoids and rhipidocystids (figure 26.3) in *TS*.2c near the end of the Early Ordovician; (2) coronoids (the precursor?) and blastoids (the descendant?) (figure 26.3) in Late Ordovician *TS*.5b (see Brett et al. 1983; Broadhead 1984); and (3) precursor bothriocidarids and descendant echinoids (figure 26.4), also in Late Ordovician *TS*.5b. Sprinkle (1995) pointed out that these paired precursor-descendant groups would likely be sequential in time of origin, but the pattern found here indicates that the time of first appearance in the fossil record is almost identical. Also surprising in two of these three examples is that the less advanced precursor was more successful than the more advanced descendant early in these groups' history (coronoids > blastoids and bothriocidaroids > echinoids; see figures 26.3 and 26.4).

Finally, a few echinoderm groups, including diploporans and cornute stylophorans, had unusual diversity patterns that do not closely resemble those of most other groups. Diploporans (figure 26.3) were the most diverse group of blastozoans and ranged throughout almost the entire Ordovician, but they had an earlier peak in diversity in the Mid Ordovician *TS*.4c than other groups and had somewhat less diversity through the Mohawkian, when the diversity of other blastozoan and crinozoan groups peaked. Paul (1988) pointed out that diploporans were a heterogeneous and very likely polyphyletic grouping of blastozoans with similar but convergent pore structures. Consequently, they should probably be broken up into several monophyletic subgroups. Because many diploporans were stemless and either recumbent in soft muddy sediment or directly cemented to a hard substrate (Parsley 1988), they had a somewhat different life mode than most other stemmed blastozoans. Also possibly contributing to their different diversity pattern is that diploporans were apparently most common in cool-water (temperate) clastic environments.

Cornute stylophorans (figure 26.4) are a Cambrian group that continued into the Ordovician, branched to the more advanced and bilaterally symmetrical mitrate stylophorans apparently in *TS*.1b, and then reached their peak diversity in Early Ordovician *TS*.2b. This peak diversity is based almost entirely on the diverse cornutes in the St.-Chinian Formation in the Montagne Noire region of southern France (Vizcaïno and Lefebvre 1999). Although common earlier, cornutes had already disappeared from Laurentia by this time. As mitrate stylophorans slowly expanded during the Mid Ordovician, cornute stylophorans gradually declined, becoming rare in the Mohawkian (mid Caradoc), when most other echinoderm groups reached their peak diversity. They made their last appearance in the early and mid Ashgill before becoming extinct. This distinctive Ordovician diversity pattern for cornute stylophorans appears to mark a relict group that was replaced by more advanced mitrate stylophorans.

■ New Designs and Life Modes

Several divergent echinoderm body plans and new life modes appeared abruptly at various times during the Ordovician. The echinoderm groups having these new body plans are characterized by numerous morphological innovations as compared with their Cambrian predecessors. Few morphological intermediates link these new groups to possible ancestors, although some early members (called precursors by Sprinkle 1995) did not have the fully formed morphology characteristic of most later members.

Somasteroids (Spencer and Wright 1966) were a small precursor group to brittle stars and starfish (Blake and Guensburg 1993). These asterozoans were the first free-living echinoderms with a star-shaped body made up only of axial and perforate extraxial skeletal components (Mooi and David 2000). They had a downward-facing mouth, enlarged food grooves, and biserial ambulacral and adambulacral plates. All three asterozoan groups appeared sequentially in the Early Ordovician, and a Late Ordovician asteroid has been found with its arms wrapped in feeding position around a smaller bivalve, indicating a carnivorous life mode like that of many modern asteroids (Blake and Guensburg 1994).

Another new echinoderm design is shown by ophiocistioids that appeared in the Mid Ordovician, followed by bothriocidarids and echinoids in the early Late Ordovician. Ophiocistioids had a globular to biscuit-shaped test with partly organized plate columns, medium to long ambulacra bearing large pores for tube feet(?), a downward-facing mouth containing a lantern with teeth for rasping, and a lateral to upward-facing anus. In the two echinoidlike clades, the tube foot pores are smaller and usually paired, and articulated spines are mounted on pustules (later on bosses) on the columns of interambulacral and ambulacral test plates. The times of origin and possible phylogenetic relationships of ophiocistioides, bothriocidarids, echinoids, and sclerite-bearing holothurians have been discussed by Smith (1988b) and Smith and Savill (2001). The first three clades appear to have been mobile, epifaunal herbivores and scavengers; the earliest holothurians may have been epifaunal deposit feeders.

Surviving CEF blastozoan eocrinoids and newly appearing PEF crinoids and blastozoan groups such as glyptocystitid rhombiferans, hemicosmitid rhombiferans, and parablastoids developed similar body plans as stalked or stemmed, medium- to high-level, upright suspension feeders. All these clades standardized the thecal plating by reducing the numerous irregular plates and growing the remaining thecal plates larger (Sprinkle and Guensburg 2001). These larger thecal plates formed a few alternating circlets, developed better pentameral symmetry, and provided for stronger articulation at the stem facet (basals or infrabasals), better support of erect arms or recumbent ambulacra (radials), and protection of the central mouth (orals). Many Late Cambrian echinoderms respired through thin, nearly smooth thecal plates (Sprinkle 1973a; Sumrall et al. 1997). Most new PEF groups in the Early and Mid Ordovician changed this design to thicker and more ridged thecal plates that had specialized respiratory pores or folds shared between adjacent plates (many blastozoans, few crinoids). Ambulacral plating in blastozoans was standardized to an arrangement where one to two biserial floor plates supported each brachiole, and early crinoids developed erect arms having both an axial component and extraxial-derived brachials (Guensburg and Sprinkle 2001). These new crinoid arms converged on the "erect ambulacra" of Late Cambrian trachelocrinid eocrinoids, Mid to Late Ordovician hemicosmitid rhombiferans, and Late Ordovician coronoids (Bockelie 1979; Brett et al. 1983; Sumrall et al. 1997). However, crinoid arms were better organized and more efficient, especially when most camerate crinoids added closely spaced pinnules to the arms starting in the Mid Ordovician, soon followed by some disparids and cladids. The tegmens of the earliest crinoids were similar to the oral surfaces of ancestral edrioasteroids (Guensburg and Sprinkle 2001), but some camerate crinoids soon developed sunken ambulacra and thick cover plates that resembled adjacent tegmen plates, thus hiding the tegmen food grooves and providing more protection against predators or parasites.

The tiny-plated stalks of some early crinoids (Guensburg 1992) were initially organized into pentameres (earliest Ordovician) and later into one-piece columnals (latest Early Ordovician), convergent with those developed earlier in the Mid Cambrian by blastozoan eocrinoids (Sprinkle 1973a). Holdfasts at the distal tips of stems were differentiated for attachment to both hard and soft substrates (Brett 1981) or discarded in adults so the stem could be wrapped around other objects (camerate crinoids; Guensburg 1992) or to allow the adult echinoderm to become free living or recumbent on the seafloor (many glyptocystitid rhombiferans, some rhipidocystids, and a few crinoids; Paul 1984; Lewis et al. 1987).

■ Echinoderm Provincial Patterns

Another finding of the project, evident in the monographs and faunal lists that were used to assemble the database, is that the composition of the richest echinoderm faunas varied considerably from one occurrence to another. Figure 26.5 is a series of histograms plotting the number of species present in different echinoderm clades for five of the richest Ordovician echinoderm faunas surveyed for the project. All of these range from 40 to 79 echinoderm species, with the most diverse faunas in the Barrandian of Bohemia (figure 26.5C) spanning nearly the entire Ordovician (*TS*.1a–6c). The Fillmore and Wah Wah formations in western Laurentia (figure 26.5A) span most of the Early Ordovician Ibexian (*TS*.1b–2c), and these units have a balanced and moderately diverse fauna with about 15 percent each of disparids, glyptocystitid rhombiferans, and edrioasteroids but only one homalozoan (a mitrate) (Guensburg and Sprinkle 1994; Sprinkle and Guensburg 1995, 1997). The Montagne Noire region of southern France (figure 26.5B), on the northwestern edge of Gondwana in high to intermediate latitudes (chapter 5), has six formations spanning the Tremadocian and Arenig (*TS*.1a–4a), the lower half of the Ordovician, and this region has produced diverse echinoderm faunas dominated by cornute (41 percent) and to a lesser degree mitrate (16 percent) stylophorans but only two crinoids (Vizcaïno and Lefebvre 1999). The Barrandian fauna of Bohemia (figure 26.5C), a peri-Gondwanan terrane in high latitudes (chapter 5), has 12 formations spanning nearly the entire Ordovician and bearing several diverse echinoderm faunas dominated by diploporans (24 percent), mitrate stylophorans (17 percent), and edrioasteroids (11 percent) (Prokop and Petr 1999). The Lebanon Formation in Tennessee (central Laurentia) (figure 26.5D) has a moderately diverse early Mohawkian (early Late Ordovician; *TS*.5b) echinoderm fauna dominated by crinoids, especially diplobathrid camerates (30 percent) and disparids (16 percent), but lacking homalozoans (Guensburg 1984). Finally, the Lady Burn Starfish Bed in western Scotland (figure 26.5E), which was then an island belt perhaps adjacent to northeastern Laurentia, has produced a large and varied echinoderm fauna from a few thin sandstone beds in the Late Ordovician mid Ashgill (*TS*.6a) that has

FIGURE 26.5. Composition of five relatively diverse echinoderm faunas from different geographic areas and parts of the Ordovician. A, Ibexian echinoderms from the Fillmore and Wah Wah formations in the western United States; note the fairly even distribution of groups except for only one homalozoan (Sprinkle and Guensburg 1997). B, Tremadocian and Arenig echinoderms from the Montagne Noire region of southern France; note the dominance of cornute and mitrate stylophorans but only two crinoids (Vizcaïno and Lefebvre 1999). C, Early to Late Ordovician echinoderms from the Barrandian region of Bohemia (Czech Republic); note the dominance by diploporans and mitrate stylophorans and fairly even distribution of other echinoderms except for relatively few crinoids (Prokop and Petr 1999). D, Mohawkian echinoderms from the Lebanon Limestone in the central United States; note the dominance by diplobathrid camerates, relatively few blastozoans, and no homalozoans (Guensburg 1984). E, Middle Ashgill echinoderms from the Lady Burn Starfish Bed of Scotland; note the large numbers of ophiuroids, asteroids, bothriocidarids, and echinoids and fairly even distribution of other groups except for relatively few blastozoans (Donovan et al. 1996; Jefferies and Daley 1996).

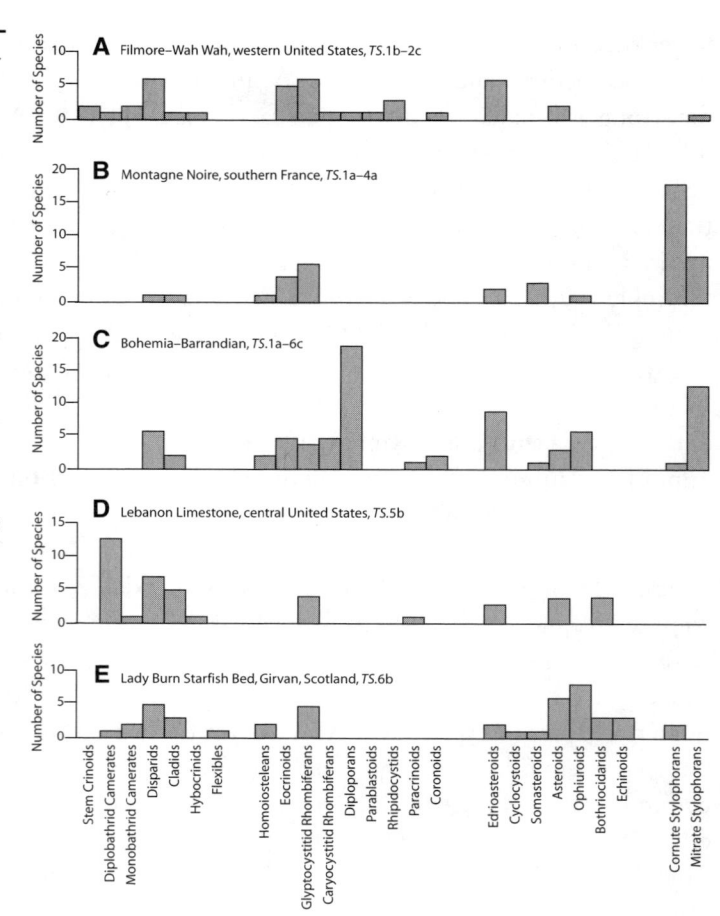

numerous ophiuroids (18 percent) and asteroids (13 percent), plus bothriocidarids and most of the echinoids (together 13 percent) known from the Ordovician (Donovan et al. 1996; Jefferies and Daley 1996).

Different echinoderm clades dominated these Ordovician faunas depending on their age, geographic location, climate zonation, and facies and environmental setting. Several authors have discussed the paleogeography, climatic zonation, and migration routes of Ordovician echinoderms including Paul (1976) for all echinoderms, Eckert (1988) for American and British crinoids, and Donovan (1989) for British crinoids. Paul (1976) identified three faunal provinces for Ordovician echinoderms, based on earlier work using trilobites and brachiopods: (1) North American (now commonly called Laurentian); (2) Baltic, including Scandinavia, Estonia, and western Russia; and (3) South European (now commonly called Northwestern Gondwanan), including Morocco, Spain, southern France, southern Britain, Bohemia, and perhaps some localities farther east in the "Tethyan" region. The Laurentian Province was tropical (within 0–30 degrees of the equator), the Baltic Province drifted from intermediate latitudes in the Early Ordovician to low latitudes in the Late Ordovician, and the Northwestern Gondwanan Province was temperate to polar (at 40–80 degrees south) (Paul 1976), although large ice caps were present on nearby land only in the latest Ordovician (Hirnantian).

During the Early Ordovician, these three faunal provinces were well differentiated and showed little mixing of echinoderms, many of which were endemic. In the Mid Ordovician, the provinces were still recognizable but showed more faunal migrations and a greater mixture of echinoderms. By the Late Ordovician, the faunal provinces were less distinct, and there were many cosmopolitan echinoderm families and genera (Paul 1976). This pattern probably represents both the geographic spread of genera and species as they diversified during the Ordovician and the gradual closure of part of the Iapetus Ocean by plate movements, bringing the faunal components of the Laurentian Province and at least one peri-Gondwanan terrane into contact with Northwestern Gondwanan provincal elements by the Late Ordovician.

In the Early and Mid Ordovician, the Laurentian Province had numerous crinoids, glyptocystitid rhombiferans, eocrinoids, paracrinoids, parablastoids, and edrioasteroids but relatively few or no diploporans, hemicosmitids, caryocystitids, coronoids, homoiosteleans, and cornute and mitrate stylophorans (figure 26.5A, D). In contrast, the Northwestern Gondwanan Province had numerous diploporans, glyptocystitid rhombiferans, caryocystitids, eocrinoids, edrioasteroids, somasteroids, and cornute and mitrate stylophorans but relatively few or no crinoids, paracrinoids, parablastoids, and homoiosteleans (figure 26.5B, C). By the latest Ordovician, the fauna from Scotland included a mixture of echinoderms from several faunal provinces, such as crinoids, glyptocystitid rhombiferans, asteroids, and ophiuroids, plus newly evolved groups such as bothriocidarids and echinoids and a few relict clades such as cornute stylophorans, minus relict groups that had already become extinct or were very rare, such as rhipidocystids, parablastoids, and paracrinoids (figure 26.5E).

■ Conclusions

1. Echinoderms diversified greatly during the Ordovician as new and more advanced members of the PEF added to and replaced older members of the CEF. Echinoderm diversity increased from 8–9 classes in the CEF to a peak of 17 classes and 29 distinctive clades in the PEF by the early Late Ordovician.

2. The echinoderm component changed from a small eocrinoid and stylophoran-dominated fauna during the Late Cambrian to much larger crinoid, rhombiferan, and diploporan-dominated faunas during the Ordovician. Crinoids then dominated the echinoderm fossil record from the Late Ordovician to the end of the Paleozoic.

3. In terms of the percentage of completeness of the echinoderm fossil record in the Ordovician, the 30 clades plotted over the 19 time slices had an average of 32 percent gaps (implying 68 percent occurrence). Most long-lived echinoderm clades had relatively complete occurrence records, but some closely related shorter-lived clades had very divergent occurrence patterns.

4. The echinoderm radiation began near the beginning of the Early Ordovician, during which 21 of the 30 echinoderm clades appeared, and continued until the early Late Ordovician, when all the echinoderm classes and Paleozoic clades had appeared. This Early Ordovician appearance of echinoderms preceded that of many other metazoan groups in the PEF, although the time of peak diversity was similar.

5. There was little change in species diversity in most echinoderm clades at the Early Ordovician–Mid Ordovician boundary, when many other metazoan groups diversified, and at the Mid Ordovician–Late Ordovician boundary. This implies that many echinoderm clades only slowly increased in diversity during the Early and Mid Ordovician.

6. However, there was a dramatic diversity increase in crinoids and many other echinoderm clades in the early Late Ordovician (Mohawkian or mid Caradoc), when echinoderms reached their peak species and clade diversity for the Ordovician and had developed nearly all their Paleozoic life modes. Although six new clades and several new life modes appeared at this time, they did not contribute much to overall diversity. Instead, there was a much larger diversity increase of clades (especially in crinoids) that had appeared much earlier.

7. Diversity then declined during the rest of the Late Ordovician as smaller and less advanced PEF echinoderm classes dropped out of the record. Echinoderms were scarce in the latest Ordovician (Hirnantian), even though 23 of 27 clades present in the Ashgill survived this glacially driven extinction interval.

8. Advanced echinoderm clades show a sequential appearance, asteroids during the Early Ordovician, echinoids and holothurians during the Late Ordovician, perhaps indicating their likely time of branching and rapid morphological change from ancestral stem group to intermediate precursor to standardized advanced clade.

9. A few groups show unusual diversity patterns, including diploporans, which have an earlier diversity peak than most other echinoderms, and cornute stylophorans, which also peaked early and were then replaced by more advanced mitrate stylophorans before becoming extinct in the Late Ordovician. Both diploporans and cornutes were most common in deeper-water clastics in temperate regions where other echinoderms were less common.

10. The composition of the richest echinoderm faunas varied considerably from one occurrence to another, based on time of deposition during the Ordovician, geographic location of the echinoderm fauna, type of facies and depositional setting, and climatic zonation.

ACKNOWLEDGMENTS

We thank Sergei V. Rozhnov (Paleontological Institute, Moscow, Russia), David L. Meyer (University of Cincinnati, Cincinnati, Ohio), Colin D. Sumrall (University of Tennessee, Knoxville, Tennessee), and Mark McKinzie (Grapevine, Texas) for assistance in compiling the echinoderm diversity of particular geographic areas for which they had detailed knowledge. Our part of the chapter is based on field work supported by the National Science Foundation under Grants BSR-8906568 (Sprinkle) and EAR-9304253 (Sprinkle, Guensburg), by CRDF Cooperative Research Grant RG1-242 (Rozhnov, Sprinkle, Guensburg), and by a Petroleum Research Fund, American Chemical Society Grant to Mark Wilson, College of Wooster (Guensburg). We also thank the University of Texas Geology Foundation, Austin, Texas, and Rock Valley College, Rockford, Illinois, for additional funds for travel and manuscript preparation. Chris Schneider, University of Texas at Austin, prepared the diversity plots in figures 26.1–26.5 from the spreadsheets and reviewed an early draft of the manuscript. Barry Webby (Macquarie University, Sydney, Australia), Peter Jell (Queensland Museum, Brisbane, Australia), and a third, anonymous reviewer read the completed manuscript and offered many helpful suggestions.

27 Graptolites: Patterns of Diversity Across Paleolatitudes

Roger A. Cooper, Jörg Maletz, Lindsey Taylor, and Jan A. Zalasiewicz

Graptolites (Graptoloidea or planktic Graptolithina) provide an ideal group for the study of biodiversity through the Ordovician because they were widely distributed around the globe and are well represented in numerous, relatively continuous black shale sequences. They originated at the base of the Ordovician and became nearly extinct at the top, thus providing a closed system and removing the problem of "edge effects" (Foote 2000a). In addition, because they are widely used for dating and correlation, the stratigraphic ranges of species are generally well known. The level of taxonomic and biostratigraphic investigation internationally ranges widely in quality. We use the complete species lists for Australasia and Avalonia, each of which regions have complete or relatively complete graptolite zonal suites through the Ordovician. In the Early to Middle Ordovician Avalonia occupied a relatively high paleolatitude (Scotese and McKerrow 1990; chapter 5), whereas Australasia lay in low paleolatitudes throughout the Ordovician. For comparison with a region that was, during the Early to Middle Ordovician, in middle paleolatitudes, the less complete graptolite sequence of Baltica is also used. In this study we attempt to distinguish between those features of diversity and evolutionary rates that were affected by latitude and those that were global. In a parallel study (Sadler and Cooper unpubl.) a global survey of Ordovician and Silurian graptolite successions is analyzed using the constrained optimization (CONOP) method to derive the global pattern of graptolite species diversity change.

■ Data Sets

The species lists and zonal range charts for the three regions have been compiled, updated, and revised by the authors. Species qualified by a query (?) in the source faunal lists are omitted. Species listed with a "cf." or "aff." are included only where it is clear that they are not synonymous with other species in the list for the region. Subspecies are omitted unless they represent a distinct stratigraphic horizon or are as morphologically distinctive as an average species. Thus the four subspecies of *Isograptus manubriatus* in the Australasian Yapeenian Ya1 and Ya2 zones (figure 27.1) are entered as a single taxon. *Dichograptus maccoyi maccoyi* (Bendigonian Be1 Zone) and *D. m. densus* (Chewtonian Ch1 Zone) are separate taxa. However, the five successional subspecies of *Isograptus victoriae* are combined into three taxa by grouping adjacent pairs. This is because Cooper (1973) found, in a multivariate analysis, that adjacent subspecies overlapped by up to 40 percent. The overlap reduced to zero or nearly zero (i.e., comparable with discrete species) when alternate subspecies were omitted. The

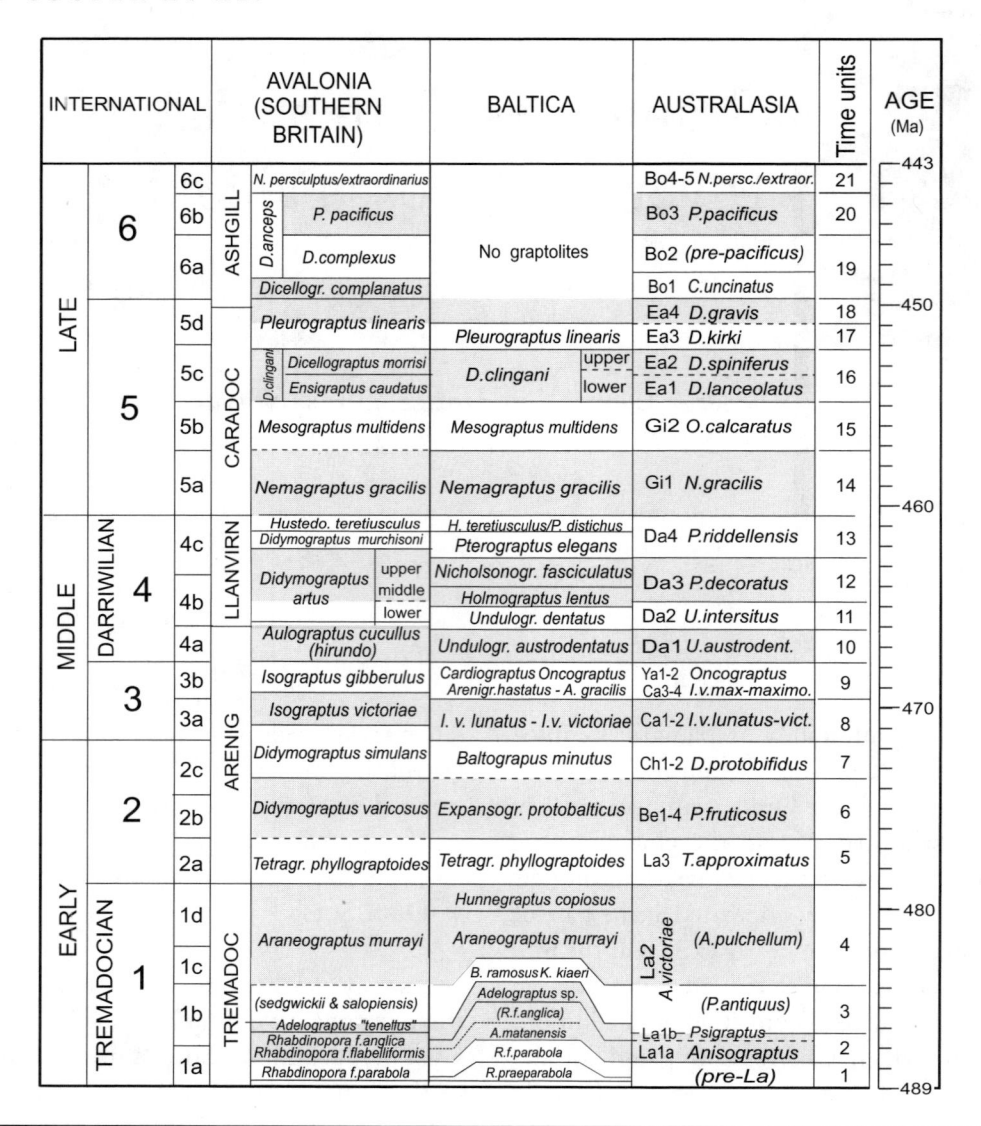

INTERNATIONAL			AVALONIA (SOUTHERN BRITAIN)			BALTICA		AUSTRALASIA		Time units	AGE (Ma)
LATE	6	6c		N. persculptus/extraordinarius				Bo4-5	N.persc./extraor.	21	443
		6b	ASHGILL	P. pacificus		D.anceps	No graptolites	Bo3	P.pacificus	20	
		6a		D. complexus				Bo2	(pre-pacificus)	19	
				Dicellogr. complanatus				Bo1	C.uncinatus		450
	5	5d	CARADOC	Pleurograptus linearis				Ea4	D.gravis	18	
							Pleurograptus linearis	Ea3	D.kirki	17	
		5c		Dicellograptus morrisi	D.clingani	upper	Ea2	D.spiniferus	16	
				Ensigraptus caudatus		lower	Ea1	D.lanceolatus		
		5b		Mesograptus multidens			Mesograptus multidens	Gi2	O.calcaratus	15	
		5a		Nemagraptus gracilis			Nemagraptus gracilis	Gi1	N.gracilis	14	460
MIDDLE	4	4c	LLANVIRN	Hustedo. teretiusculus			H. teretiusculus/P. distichus	Da4	P.riddellensis	13	
	DARRIWILIAN			Didymograptus murchisoni			Pterograptus elegans				
		4b		Didymograptus artus	upper		Nicholsonogr. fasciculatus	Da3	P.decoratus	12	
					middle		Holmograptus lentus				
					lower		Undulogr. dentatus	Da2	U.intersitus	11	
		4a		Aulograptus cucullus (hirundo)			Undulogr. austrodentatus	Da1	U.austrodent.	10	
	3	3b	ARENIG	Isograptus gibberulus			Cardiograptus Oncograptus Arenigr.hastatus - A. gracilis	Ya1-2 Ca3-4	Oncograptus I.v.max-maximo.	9	470
		3a		Isograptus victoriae			I. v. lunatus - I.v. victoriae	Ca1-2	I.v.lunatus-vict.	8	
	2	2c		Didymograptus simulans			Baltograpus minutus	Ch1-2	D.protobifidus	7	
		2b		Didymograptus varicosus			Expansogr. protobalticus	Be1-4	P.fruticosus	6	
		2a		Tetragr. phyllograptoides			Tetragr. phyllograptoides	La3	T.approximatus	5	480
EARLY	1	1d	TREMADOC	Araneograptus murrayi			Hunnegraptus copiosus		(A.pulchellum)	4	
	TREMADOCIAN						Araneograptus murrayi	La2 A.victoriae			
		1c					B. ramosus K. kiaeri				
		1b		(sedgwickii & salopiensis)			Adelograptus sp. (R.f.anglica)		(P.antiquus)	3	
				Adelograptus "teneltus"			A. matanensis				
				Rhabdinopora f.anglica				La1b	Psigraptus	2	
				Rhabdinopora f.flabelliformis			R.f.parabola	La1a	Anisograptus		
		1a		Rhabdinopora f.parabola			R.praeparabola		(pre-La)	1	489

FIGURE 27.1. Correlation of Ordovician graptolite zones (sources given in text) and time units used in this study. Alternate units are shaded. Age calibration of Australasian zones from Sadler and Cooper (chapter 3). Dashed lines in Britain and Baltica columns = correlation with Australasia uncertain. Dashed line in Australasian column = time calibration of boundary uncertain. Short-ranging zones are combined, and long ones split, in order to achieve time units of more even duration. The time units used here, numbered at right, are preferred to time slices recommended by Webby et al. (chapter 2, listed at left) because their boundaries are more readily located in the graptolite zonal succession.

taxa used here are therefore all loosely of "species-equivalent" rank and are referred to as "species."

In all three regions the stratigraphic ranges of several species within the time units used here are known, making it possible to refine the age range of taxa. It also enables some regional zones to be split to accord with the time units adopted here. The first and last appearances of species in the regional data sets should be regarded as regional events rather than global events.

As noted by Cooper et al. (1991), an equivalent of the deep-water graptolite biofacies of Australasia and other low paleolatitude regions, representing slope and base of slope depositional environments, is lacking from Avalonia and other intermediate to high paleolatitude regions, also from Baltica. It is unclear at present whether this lack is because the facies simply was not developed in these regions or whether it was likely to have been developed but is not now preserved. If the latter, our estimate of diversity in these regions is likely to be too low.

Responsibility for the regional data sets is as follows: Australasia, R. A. Cooper and A. H. M. Vanden-

Berg; Baltica, J. Maletz; Avalonia, J. A. Zalasiewicz and L. Taylor.

Australasia

The Australasian region, comprising Australia and New Zealand, includes the classic graptolitic sections described by Hall, Harris, Thomas, Benson, and Keble (see VandenBerg and Cooper 1992 for references), particularly in the Lachlan Fold Belt of Australia and its equivalent in New Zealand. The sedimentary succession comprises turbidites, cherts, and shales throughout the Ordovician, and the black shale facies is developed in all zones except for the Bolindian Bo5 Zone. The earliest Tremadocian international graptolite chronozones, *R. praeparabola* and *R. f. parabola,* however, are not represented by graptolites, and there are likely to be faunal gaps at the zonal or subzonal level higher in the Tremadocian (Cooper 1999a; Maletz and Egenhoff 2001). Diversity through the early Tremadocian is therefore possibly underestimated.

The sequence of zones, particularly the post-Tremadocian zones, has become a standard of reference within the Pacific faunal realm. The graptolite sequence has been fully reviewed by VandenBerg and Cooper, who list 30 zones, 2 of which are subdivided into subzones. The zones are grouped here into 26 time units. The very long Lancefieldian La2 Zone, *A. victoriae,* is informally split into upper and lower parts, using published (Cooper 1979; Cooper and Stewart 1979) and unpublished information on the stratigraphic ranges of species within the zone, in order to better match with the time unit divisions.

VandenBerg and Cooper (1992) listed 313 named species and subspecies, reduced here to 283 taxa of "species-equivalent" rank. Graptolitic shales are well developed in all time units (not to be confused with the 19 global time slices also employed in this volume; see figure 27.1).

Baltica

A total of 213 taxa is recorded for Baltica, which is here taken to include Sweden, Denmark, Estonia, and Norway and to exclude the richly graptolitic Bogo Shale of the Trondheim region, which was not part of the Baltica paleoplate (Dewey et al. 1970; Schmidt-Gündel 1994). Graptolites of the northern German boreholes in the Rügen area are also omitted, as it is not certain to which paleoplate these belong. A suite of 31 zones is used for plotting species ranges (Maletz 1995, 1997, unpubl.), and these are grouped here into 18 time units (figure 27.1). One zone, *D. clingani,* is informally split into upper and lower parts, using knowledge of the range of species within the zone.

A problem in sampling the graptolite fauna of Baltica for diversity studies is the limited number of stratigraphic sections and the unfavorable facies in some parts of the column. The graptolites in some stratigraphic intervals have been sampled and studied in detail, for example, the "Dictyonema Shales" and "Lower" and "Upper Didymograptus Shales" of earlier authors. On the other hand, other parts of the column, such as the Upper Ordovician zones, are little studied. The regrouping of zones into longer time units reduces the effects of unevenness in sampling intensity and biostratigraphic study to some extent. However, in time units 9, 10, and 11 (the *Cardiograptus, U. austrodentatus,* and *U. dentatus* zones of the late Arenig) the graptolite facies in most of the Oslo-Scania shale belt is replaced by a carbonate facies, the Komstad Limestone Formation and its equivalents, that is less favorable for preservation of graptolites. This interval is yet to be fully investigated for graptolites, and its apparently depressed graptolite diversity may therefore be largely an artifact of nonpreservation or of sampling deficiency. For these reasons, we give the Baltica data set less weight than the other two sets in interpretation.

On the other hand, the preservation quality of Baltic graptolites is exceptionally high, and the Tremadocian succession of the Oslo region is, in addition, among the stratigraphically best controlled in the world. Stratigraphic ranges of graptolites in the classic and richly diverse Töyen Shale (formerly known as the Lower Didymograptus Shale) used here are those revised by Maletz (unpubl.). Graptolites of late Caradocian and Ashgillian age are not known in Scandinavia, and time units 22 to 26 are therefore unrepresented.

Avalonia

This region includes England and Wales and contains the classic graptolite-bearing strata studied by

Lapworth, Elles, and Wood. Two hundred and twelve species are included in the present analysis. The graptolite census of Strachan (1996, 1997) has been revised and updated in the light of new work and the zonal ranges replotted in terms of the currently used zonal scheme (figure 27.1).

The zonal scheme largely follows that of Fortey et al. (2000). In the middle Tremadocian (*salopiensis* and *sedgwickii* trilobite zones), equivalent to the *B. ramosus* and *K. kiaeri* zones of Baltica, a widespread facies change and unconformity greatly reduce graptolite field occurrence, affecting the diversity pattern. Similarly, no graptolites are recorded in the late Caradoc equivalent of the *D. complanatus* Zone of Scotland and of the Ea4 Zone, *D. gravis,* of Australasia. Otherwise, the graptolitic facies is developed within (but not necessarily throughout) all time units in Avalonia. The *D. artus* Zone is split informally into two parts in order to align better with the time unit divisions used here (figure 27.1).

Reliability and Comprehensiveness

Of the three regions, Australasia and Britain have graptolite localities that number in the thousands, and they have numerous sections spanning successions of zones. Although exact numbers are unknown, it is unlikely that Baltica contains a similar abundance of collecting localities throughout the Ordovician. In this respect Australasia and Britain are likely to be more representative than Baltica of the original graptolite faunas that lived in the regions. This may be the reason for the mean duration of species in Baltica being appreciably shorter than in both Australasia and Britain (see later in this chapter). A relatively smaller number of collecting localities makes it possible that sampling and recorded stratigraphic ranges are less complete than in the other two regions. An artificially shortened mean species duration would affect diversity and evolutionary rate patterns.

■ Ordovician Timescale

A reliable and precise timescale for the Ordovician is necessary for the accurate representation of evolutionary rates. Graptolite stratigraphic ranges are expressed in zones in each region and the zonal schemes are correlated in figure 27.1. The zones must be calibrated in millions of years to derive rates. The graphic correlation-based composite timescale of Cooper (1999b) was an attempt to do this. It was based on 10 relatively long ranging, Early to Mid Ordovician, high-quality, deep-water graptolitic measured sections, representing two paleoplates, and on the conodont-based graphic correlation of 61 Late Ordovician, carbonate sections in North America (Sweet 1984, 1988a).

A much larger database, comprising 1,136 taxa in almost 200 deep-water shale measured sections from around the world, has been used in a novel application of the CONOP procedure by Sadler and Cooper (unpubl.) to derive a composite sequence that stands as a proxy relative timescale for the Ordovician and Silurian periods. In the biostratigraphic database are graptolite assemblages that are reliably tied to 22 high-resolution zircon dates. These are used, first, to test the linearity of the relative timescale and, second, to calibrate the scale. The method is outlined by Sadler and Cooper (chapter 3 and unpubl.). It is the scale used in this chapter.

Of the six new international time divisions for the Ordovician (figure 27.1) only the Darriwilian and the Tremadocian have been formally named. The Caradoc and Ashgill can be used informally, with their British definitions (Fortey et al. 1995), for the Late Ordovician. The gap between the end of the Tremadocian and start of the Darriwilian is here informally referred to as Arenig, so that we can refer to a complete set of global Ordovician divisions.

Time Units

Ideally, evolutionary rates and diversity trends should be measured against a scale with time units of equal duration to avoid the distorting effects of uneven duration (Sepkoski and Koch 1996). This is generally not practical for most paleontological data, which are recorded in zones or stages that are of unknown or uncertain relative duration or are of known but uneven duration. Although the relative duration of zones in each of the three regions studied here is known with unusually good precision, the zones range widely in duration (figure 27.1). Therefore, zones of short duration (such as the four zones of the Australasian Bendigonian) are grouped into a single time unit, and long zones (such as the La2 Zone of Australasia and *D. artus* Zone of Avalonia) are split in two.

The correlation of the three regional graptolitic zonal successions (figure 27.1) follows Webby et al. (chapter 2). Time unit boundaries are made to coincide with zone boundaries (or subzone boundaries), and therefore some time units vary slightly from region to region.

The result is a scale of 21 time units that, in duration, are more even than zones in any of the zonal schemes, thereby minimizing any distortion of measured rates. Time units average 2.195 million years (m.y.) in duration ($\sigma = 0.89$ m.y.) and range in duration from 1.3 m.y. to 5.1 m.y.

Mean Species Duration

Using the CONOP timescale (chapter 3), the mean duration of a species in each of the three regions, with standard deviation and total number of species recorded from the region, is as follows:

Australasia	2.382 m.y.	$\sigma = 1.894$	283 total
Baltica	1.43 m.y.	$\sigma = 1.09$	210 total
Avalonia	2.377 m.y.	$\sigma = 1.821$	212 total

■ Measures of Diversity and Taxonomic Rates

Possible Sources of Bias

The present species lists have been compiled by the authors and their colleagues from published and unpublished sources and checked for taxonomic consistency, and the stratigraphic ranges of species revised in the light of the current zonal schemes. Although inconsistencies and errors are certain to remain in our data, we believe that they are unlikely to affect significantly the patterns emerging from the analysis, at least for Avalonia and Australasia. As mentioned already, the patterns for Baltica are interpreted with more caution.

Other sources of bias arise from uneven outcrop area and sampling opportunity and uneven or incomplete preservation quality and collecting completeness. As discussed earlier, these are believed to be minimal in our data sets, especially those of Avalonia and Australasia. The possible effects of inaccurate or uneven duration of time units are discussed in the next section.

Effects of Time Unit Duration

Uncertainty or pronounced unevenness (or both) in time unit duration can have a distorting effect on the pattern of diversity change through time (Sepkoski and Koch 1996; Foote 2000a). Sepkoski and Koch recommend minimizing this problem by combining zones of short duration and subdividing zones of long duration in order to even out the differences in time unit durations. With a relatively precise timescale and a well-constrained zonal correlation of sequences, we are able to plot the stratigraphic ranges of species for the three regions in a common scale of time units that are of relatively uniform duration, as discussed earlier. We therefore believe that the distorting effects of uncertain and uneven time units are minimized in the present study. Consistent with this view, we note that in the total diversity and total faunal turnover curves (see the next section), the main peaks do not coincide with the longest time units; nor do the main troughs coincide with the shortest time units. Furthermore, by using artificial data sets, Cooper (chapter 4) was able to directly test and compare the performance of three alternative measures of diversity on time units of uneven duration and thereby compare them with "true" mean standing diversity (MSD).

Foote (2000a) has modeled the effects on measures of diversity, and origination and extinction rates, of interval length, preservation quality, "edge effects," and the taxonomic rates themselves. He concludes that single-interval taxa produce many undesirable distortions and recommends that diversity and rate measures be adopted that do not count single-interval taxa (taxa confined to a single time unit). However, to ignore these taxa in our data sets would be to ignore a large proportion of the database. Some of the concerns raised by Foote may be minimized in the present study. As mentioned earlier, "edge effects" are largely removed, and we are using a precise timescale with subequal time units, thereby reducing the distorting effects of time interval durations. Variable preservation quality and collection completeness will undoubtedly affect the data. Incomplete collecting will tend to shorten true ranges and reduce diversity, as is possibly the case in Baltica. It will also be unlikely to detect the rare species, affecting diversity as well as origination and extinction rates. The zonal ranges of taxa, especially in Avalonia and Australasia, however, are well "filled in" (there are few gaps throughout the range; VandenBerg and Cooper 1992; Taylor and Zalasiewicz unpubl.), testifying to the completeness of collecting. In any case, it is the trend of the diversity curve, rather than the absolute value of the diversity measure, that is important here.

Appropriate Diversity Measures

Cooper (chapter 4) constructed six model data sets in which the durations of time units and of species life spans are comparable with those found in our graptolite data sets. He then measured diversity in the data sets utilizing three commonly used estimates of diversity: total diversity, species per m.y., and normalized diversity. *Total diversity* (d_{tot}) is the total number of species that are recorded from the time unit; it is the simplest and most commonly used measure for estimating MSD. *Species per m.y.* (d_i) is simply total diversity divided by the duration of the time unit (i) in m.y.; it allows for the probability that longer time units will capture more species. *Normalized diversity* (d_{norm}) is a measure that normalizes for variability in time unit duration to the extent that the longer a time unit is, the more species will begin or end in it or are confined to it. It is the sum of species that range from the unit below to the unit above, plus half the number of species that range beyond the time unit but originate or become extinct within it, plus half those that are confined to the time unit itself (Sepkoski 1975). The measure also compensates for another bias. Species ranges in regional data sets are generally expressed as zonal ranges. These will overestimate true ranges because few species ranges will completely span the zones in which they first appear or last appear or to which they are confined.

Because the "true" time ranges of species in the data sets are known, the "true" MSD of successive time units can be calculated. This was then compared with each of the three estimates of diversity in turn. The results show that the total diversity (d_{tot}) measure consistently and substantially overestimates true MSD, deviating from it by 25 percent on average, and the species per m.y. (d_i) measure generally underestimates it, deviating by 29 percent on average. The normalized diversity (d_{norm}) measure is consistently the closest approximation of true MSD, deviating by 10 percent on average.

In our regional analyses we give the three measures of diversity just described (d_{tot}, d_i, and d_{norm}), but, unless otherwise stated, comparisons and conclusions are based on the normalized measure (d_{norm}), the best estimator of MSD.

Origination and Extinction Measures

Origination and extinction intensity and rate measures are included in this study of diversity because, apart from the effects of regional migration, diversity change is a function of the dynamics of origination and extinction. Three measures of extinction are given: the total number of extinction events in the time unit; the percentage of extinction, which is the number of extinction events divided by the total number of taxa present; and the per capita extinction rate, which is the percentage of extinction divided by the duration of the time unit, in m.y. (Sepkoski and Koch 1996). The equivalent measures for origination are also given:

Number of originations	$= o$
Percentage of origination	$= o_d \times 100$
Per capita rate of origination	$= o_{di}$
Number of extinctions	$= e$
Percentage of extinction	$= e_d \times 100$
Per capita rate of extinction	$= e_{di}$

Faunal Turnover

This is the faunal turnover that takes place in a time unit. Three measures are given. The first, total faunal turnover, is expressed as the sum of the number of extinctions and originations in a time unit. The second is the proportion or percentage of faunal turnover; it is total faunal turnover divided by twice the total diversity for the time unit. The third is the per capita rate of faunal turnover, which is the percentage of faunal turnover divided by the duration of the time unit, in m.y.

Faunal turnover	$= (o + e)$
Percentage of faunal turnover	$= (o + e)_{2d} \times 100$
Per capita rate of faunal turnover (in m.y.)	$= (o + e)_{2di}$

■ Diversity Patterns

Diversity patterns for the three regions are shown in figures 27.2, 27.3, and 27.4, and the basic statistics for diversity and evolutionary change are given in table 27.1. Features common to all three regions are likely to reflect global patterns, whereas contrasts between Avalonia and Australasia may reflect latitudinal effects.

Although the graptolite clade appeared with a burst at the base of the Tremadocian, rapidly invaded a range of biotopes in the oceans, and spread widely around the globe (Bulman 1971; Cooper 1999a), it was not rich in species for a long time. Graptolites originated as planktic organisms at the beginning of Ordovician time. The earliest species are found in facies representing continental slope water depths, and within 2 m.y. the clade had expanded its habitat range into farther offshore (and possibly deeper) biotopes, as well as into the epipelagic zone, in which it ranges from inner shelf to outer slope (Cooper 1998, 1999a). Yet diversity was slow to expand.

For the first 12 m.y. (25 percent) of Ordovician time, normalized mean standing species diversity remains relatively low, at no more than 9 species. Then, within the space of 3 m.y. (time unit 6 in the early Arenig), there is a dramatic expansion, as if some threshold had been crossed, and diversity reaches a maximum or near maximum for the entire Ordovician. This is most marked in Australasia, where there is a dramatic expansion in the Bendigonian stage, and normalized diversity reaches 37.5 species (total species diversity reaches 73). The sudden expansion is also seen in the species per m.y. curve (figure 27.2A). It is clearly not an effect of time unit duration. In Avalonia, the effect is less strong, but there is a rapid expansion in the early Arenig, and normalized species diversity reaches 22.3 (total diversity 33). The low

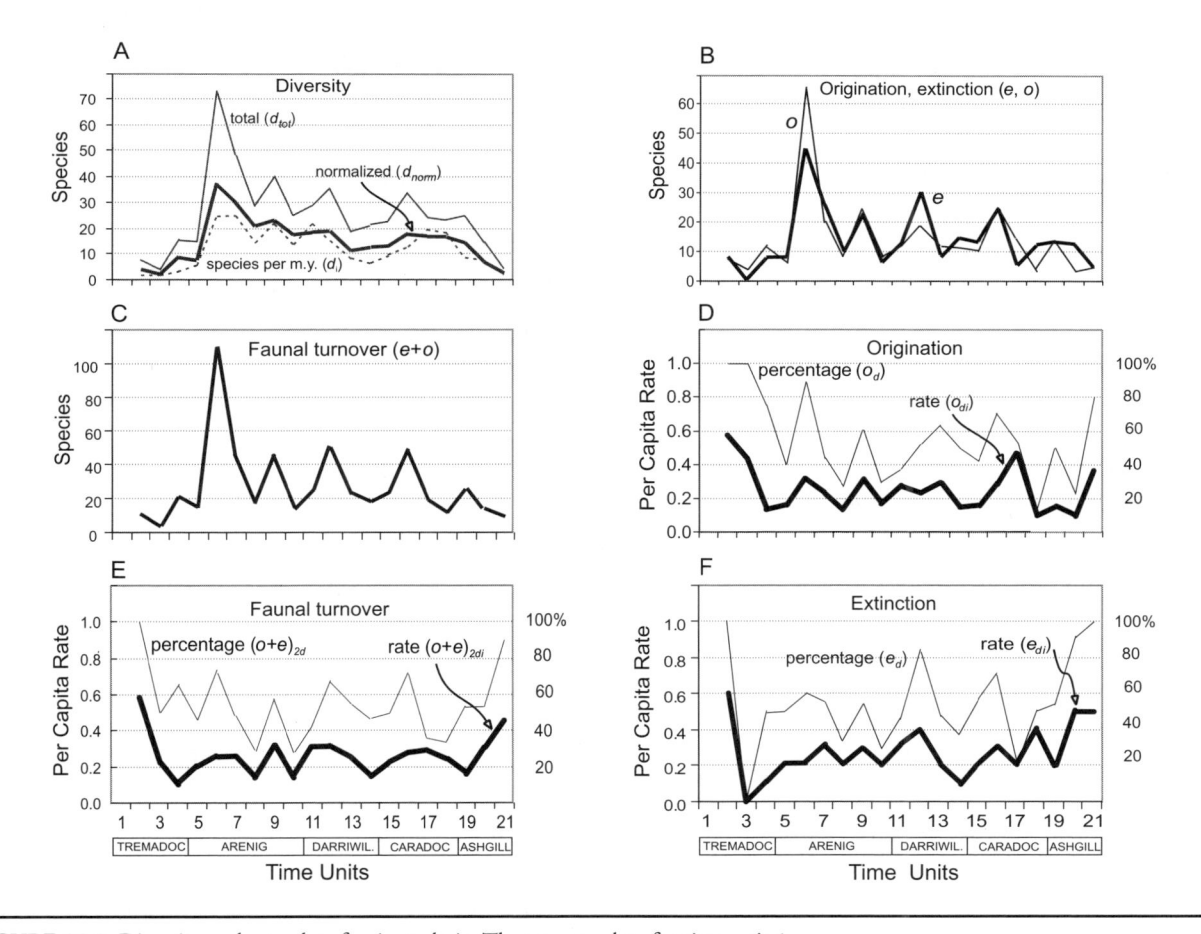

FIGURE 27.2. Diversity and rate plots for Australasia. There are no data for time unit 1.

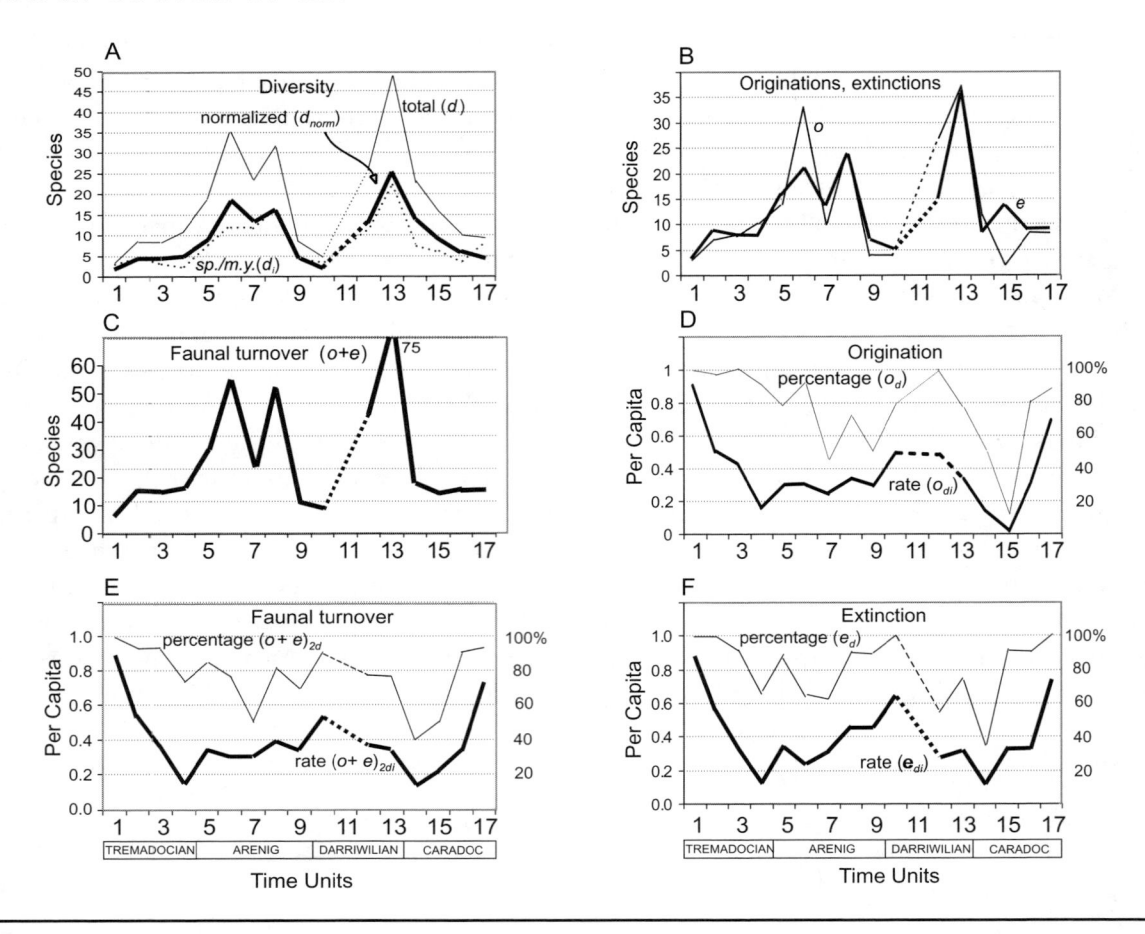

FIGURE 27.3. Diversity and rate plots for Baltica. There is little, or no, information in time units 9–11 inclusive.

diversity throughout the Tremadocian and rapid increase in the early Arenig are features common to all regions. They are likely to be global and not significantly influenced by latitude. The more extreme expansion in diversity in Australasia, however, may have been a latitudinal effect.

On the other hand, the reduction in diversity during the Late Ordovician, which led eventually to the near extinction of the group in the late Ashgill, appears to have begun much earlier in Avalonia, and possibly Baltica, than in Australasia. In Avalonia, a steep decline begins in the mid Caradoc, and diversity never recovers before the end of the Ordovician. In Australasia, the decline does not begin until the mid Ashgill. During the mid Caradoc to Ashgill, therefore, MSD in Avalonia is significantly lower than in Australasia (figure 27.5). This contributes to the lower overall total Ordovician diversity in Avalonia (212 species) being significantly lower than in Australasia (289 species). If latitude is the cause of this difference, it presumably operates either directly through surface water temperature (Skevington 1974) or indirectly through the earlier breakdown of ocean density structure in high paleolatitudes associated with global cooling, leading to decay of high productivity zones such as developed in the oxygen minimum zone and along continental margins due to upwelling, both regarded as favorable habitats for graptolites (Berry et al. 1987; Finney and Berry 1997; Cooper 1998).

However, according to the plate tectonic model of Cocks and Torsvik (chapter 5), the northward drift of Baltica and Avalonia carried both regions into lower latitudes by Late Ordovician time. Although they lay in opposite hemispheres, Australasia and Avalonia would have lain a similar distance from the Ordovician equator by the late Caradoc. In terms of this model, latitude may have had little effect, and the reason for the earlier decline in diversity in Avalonia remains uncertain.

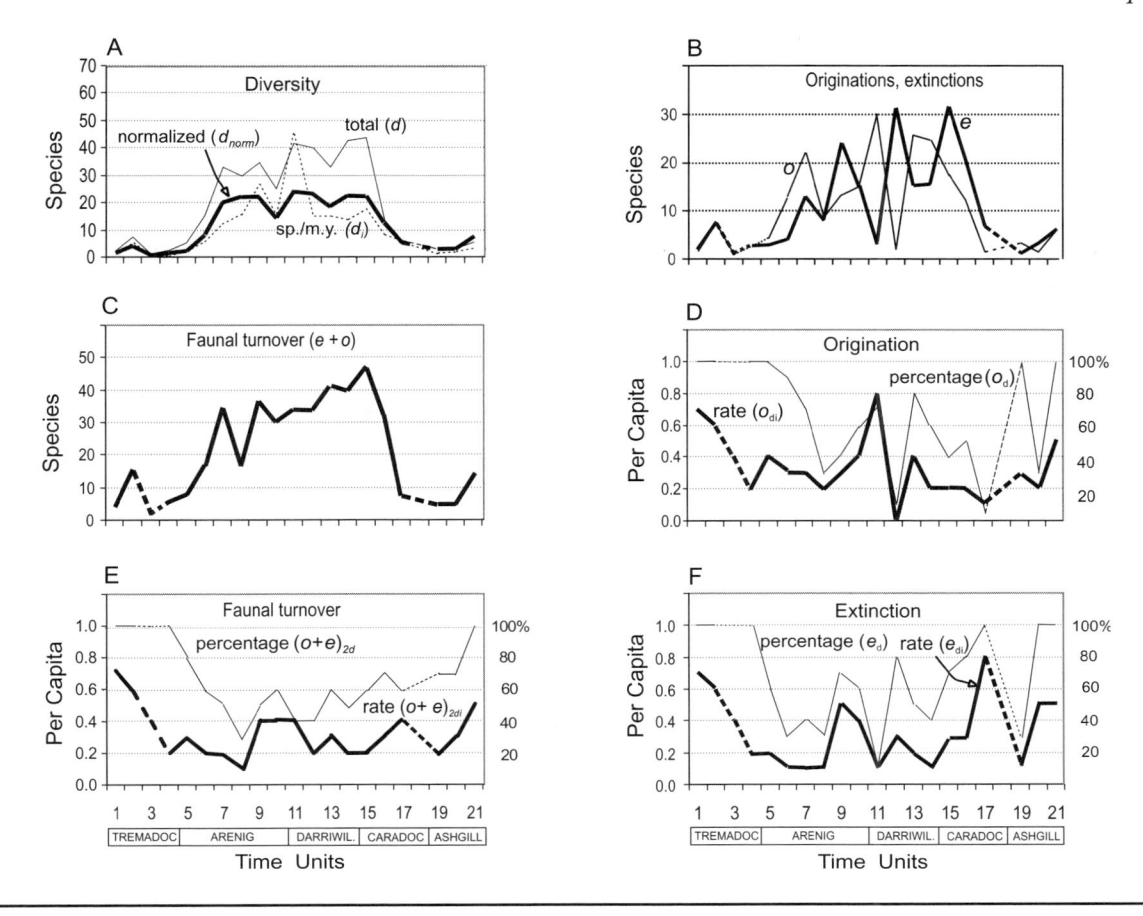

FIGURE 27.4. Diversity and rate plots for Avalonia. There are no data in time unit 4 and the *D. complanatus* Zone (time unit 18).

A comparison of the normalized diversity curves for Australasia and Avalonia (figure 27.5) reveals that the commonly held view, that standing diversity through the Ordovician was significantly lower in high paleolatitudes (Skevington 1974; Berry 1979; Cooper et al. 1991), is only partly true. In the early to mid Arenig, when latitudinal difference was most marked, the difference in diversity is indeed marked, being appreciably higher in Australasia (figure 27.5). From the mid Caradoc time on, the paleolatitudinal difference between the two regions may have been insignificant, and the observed difference between them in normalized diversity could not, therefore, be related to latitude. For most of the remainder of the Ordovician, normalized diversity is as high as, or higher than, that in Australasia. This is all the more significant in view of the lack, in Avalonia, of the deep-water graptolite biofacies (Cooper et al. 1991).

The lower overall diversity of Baltica possibly reflects its smaller sampling base and therefore may not be significant.

■ Evolutionary Rates

The curves for numbers of originations (o) and for numbers of extinctions (e) in Australasia (figure 27.2B) fluctuate markedly and are strongly positively correlated ($r^2 = 0.74$). As a result, originations and extinctions act in concert to drive the total faunal turnover curve, which fluctuates strongly. Faunal turnover seldom drops below 20 species in any time unit and reaches more than 100 species in the early Arenig (time unit 6), the highest level recorded globally. The high level of faunal change is what has enabled the fine subdivision of the Australasian graptolite succession into 32 zones and subzones (VandenBerg and Cooper 1992). The same trends can be seen in Baltica (figure 27.3), where the total faunal change curve has a similarly strong fluctuation. In Avalonia (figure 27.4), the origination (o) and extinction (e) curves also fluctuate strongly. They are positively correlated except for where diversity is high. In the late Arenig to mid Caradoc they are negatively correlated

TABLE 27.1. Graptolite Diversity, Origination, Extinction, and Faunal Turnover for Three Data Sets, Australasia, Baltica, and Avalonia

		Tremadocian				Arenig					Darriwilian				Caradoc				Ashgill		
Time Unit	1	2	3	4	5	6	7	8	9	10	11	12	13	14	15	16	17	18	19	20	21
Age of Base	489.3	488.0	486.3	483.9	478.8	476.4	473.6	471.7	469.7	467.9	466.1	464.8	462.6	460.4	457.3	454.9	452.3	451.0	449.7	446.7	444.8
Duration	0.8	1.7	2.4	5.1	2.4	2.9	1.9	2.0	1.8	1.8	1.3	2.2	2.2	3.1	2.4	2.6	1.3	1.3	3.0	1.9	2.1
AUSTRALASIA																					
Total diversity (d)		8.00	4.00	16.00	15.00	73.00	48.00	29.00	40.00	25.00	29.00	36.00	19.00	21.00	23.00	34.00	25.00	24.00	25.00	14.00	5.00
Species/m.y. (d_t)		1.90	1.68	3.14	6.30	25.44	25.95	14.50	22.22	13.89	21.97	16.51	8.64	6.77	9.58	12.88	19.84	18.46	8.33	7.37	2.38
Normalized diversity (d_{norm})		4.00	2.00	9.50	8.50	37.50	30.00	20.50	23.50	18.00	19.00	19.50	10.50	12.50	13.00	18.00	17.00	16.50	14.50	7.50	3.00
Originations (o)		8	4	12	6	66	20	8	24	8	11	19	12	11	10	24	14	4	13	3	4
Proportion of originations (o_d)		1.00	1.00	0.75	0.40	0.90	0.42	0.28	0.60	0.32	0.38	0.53	0.63	0.52	0.43	0.71	0.56	0.17	0.52	0.21	0.80
Std error of o_d		0.00	0.00	0.11	0.13	0.03	0.07	0.08	0.08	0.09	0.09	0.08	0.11	0.11	0.10	0.08	0.10	0.08	0.10	0.11	0.18
Per capita rate of origination (o_{di})		0.42	0.42	0.15	0.17	0.32	0.23	0.14	0.33	0.18	0.29	0.24	0.29	0.17	0.18	0.27	0.44	0.13	0.17	0.11	0.38
Extinctions (e)		8	0	8	8	44	26	10	22	7	13	30	9	8	13	24	5	12	14	13	5
Proportion of extinctions (e_d)		1.00	0.00	0.50	0.53	0.60	0.54	0.34	0.55	0.28	0.45	0.83	0.47	0.38	0.57	0.71	0.20	0.50	0.56	0.93	1.00
Std error of e_d		0.00	0.00	0.13	0.13	0.06	0.07	0.09	0.08	0.09	0.09	0.06	0.11	0.11	0.10	0.08	0.08	0.10	0.10	0.07	0.00
Faunal turnover ($o + e$)		16	4	20	14	110	46	18	46	15	24	49	21	19	23	48	19	16	27	16	9
Proportion of turnover $(o + e)_{2d}$		1.00	0.50	0.63	0.47	0.75	0.48	0.31	0.58	0.30	0.41	0.68	0.55	0.45	0.50	0.71	0.38	0.33	0.54	0.57	0.90
Std error of $(o + e)_{2d}$		0.00	0.18	0.09	0.09	0.04	0.05	0.06	0.06	0.06	0.06	0.05	0.08	0.08	0.07	0.06	0.07	0.07	0.07	0.09	0.09
Per capita rate of turnover $(o + e)_{2di}$		0.58	0.21	0.12	0.20	0.26	0.26	0.16	0.32	0.17	0.31	0.31	0.25	0.15	0.21	0.27	0.30	0.26	0.18	0.30	0.43
BALTICA																					
Total diversity (d)	3.00	8.00	8.00	11.00	18.00	35.00	23.00	32.00	8.00	5.00	0.00	27.00	49.00	23.00	16.00	10.00	9.00				
Species/m.y. (d_t)	2.73	4.65	3.36	2.16	7.50	12.50	12.11	16.00	4.00	3.13	0.00	13.50	22.27	7.42	6.67	3.85	6.92				
Normalized diversity (d_{norm})	1.50	3.99	4.00	5.50	9.00	18.00	13.50	16.00	4.50	2.50	0.00	13.50	25.00	14.00	9.00	6.00	4.50				
Originations (o)	3.00	7.00	8.00	10.00	14.00	33.00	10.00	23.00	4.00	4.00	0.00	27.00	38.00	11.00	2.00	8.00	8.00				
Proportion of originations (o_d)	1.00	0.88	1.00	0.91	0.78	0.94	0.43	0.72	0.50	0.80	0.00	1.00	0.78	0.48	0.13	0.80	0.89				
Std error of o_d	0.00	0.12	0.00	0.09	0.10	0.04	0.10	0.08	0.18	0.18	0.00	0.00	0.06	0.10	0.08	0.13	0.10				
Per capita rate of origination (o_{di})	0.91	0.51	0.42	0.18	0.32	0.34	0.23	0.36	0.25	0.50	0.00	0.50	0.35	0.15	0.05	0.31	0.68				
Extinctions (e)	3.00	8.00	7.00	7.00	16.00	22.00	14.00	28.00	7.00	5.00	0.00	15.00	37.00	8.00	14.00	9.00	9.00				
Proportion of extinctions (e_d)	1.00	1.00	0.88	0.64	0.89	0.63	0.61	0.88	0.88	1.00	0.00	0.56	0.76	0.35	0.88	0.90	1.00				
Std error of e_d	0.00	0.00	0.12	0.15	0.07	0.08	0.10	0.06	0.12	0.00	0.00	0.10	0.06	0.10	0.08	0.09	0.00				
Per capita rate of extinction (e_{di})	0.91	0.58	0.37	0.12	0.37	0.22	0.32	0.44	0.44	0.63	0.00	0.28	0.34	0.11	0.36	0.35	0.77				
Faunal turnover ($o + e$)	6.00	15.00	15.00	17.00	30.00	55.00	24.00	51.00	11.00	9.00	0.00	42.00	75.00	19.00	16.00	17.00	17.00				
Proportion of turnover $(o + e)_{2d}$	1.00	0.94	0.94	0.77	0.83	0.79	0.52	0.80	0.69	0.90	0.00	0.78	0.77	0.41	0.50	0.85	0.94				
Std error of $(o + e)_{2d}$	0.00	0.06	0.06	0.09	0.06	0.05	0.07	0.05	0.12	0.09	0.00	0.06	0.04	0.07	0.09	0.08	0.05				
Per capita rate of turnover $(o + e)_{2di}$	0.91	0.55	0.39	0.15	0.35	0.28	0.27	0.40	0.34	0.56	0.00	0.39	0.35	0.13	0.21	0.33	0.73				
AVALONIA																					
Total diversity (d)	2	8	1	3	5	15	33	30	35	26	41	40	33	42	43	24	7	0	3	3	7
Species/m.y. (d_t)	1.43	5.00	0.42	0.59	2.08	5.36	12.69	15.00	26.92	16.25	45.56	15.38	15.00	13.55	17.92	9.23	5.38	0.00	1.00	1.43	3.68
Normalized diversity (d_{norm})	1.00	4.00	0.50	1.50	2.50	8.50	20.00	22.33	21.50	13.50	24.50	22.82	18.00	23.00	23.50	13.00	5.64	0.00	3.00	3.00	8.21
Originations (o)	2	8	1	3	5	13	22	9	13	15	30	2	26	24	17	12	1	0	3	1	7
Proportion of originations (o_d)	1.00	1.00	1.00	1.00	1.00	0.87	0.67	0.30	0.37	0.58	0.73	0.05	0.79	0.57	0.40	0.50	0.14	0.00	1.00	0.33	1.00
Std error of o_d	0.00	0.00	0.00	0.00	0.00	0.09	0.08	0.08	0.08	0.10	0.07	0.03	0.07	0.08	0.07	0.10	0.13	0.00	0.00	0.27	0.00
Per capita rate of origination (o_{di})	0.71	0.63	0.42	0.20	0.42	0.31	0.26	0.15	0.29	0.36	0.81	0.02	0.36	0.18	0.16	0.19	0.11	0.00	0.33	0.16	0.53
Extinctions (e)	2	8	1	3	3	4	12	8	24	15	3	31	15	16	31	20	7	0	1	3	7
Proportion of extinctions (e_d)	1.00	1.00	1.00	1.00	0.60	0.27	0.36	0.27	0.69	0.58	0.07	0.78	0.45	0.38	0.72	0.83	1.00	0.00	0.33	1.00	1.00
Std error of e_d	0.00	0.00	0.00	0.00	0.22	0.11	0.08	0.08	0.08	0.10	0.04	0.07	0.09	0.07	0.07	0.08	0.00	0.00	0.27	0.00	0.00
Per capita rate of extinction (e_{di})	0.71	0.63	0.42	0.20	0.25	0.10	0.14	0.13	0.53	0.36	0.08	0.30	0.21	0.12	0.30	0.32	0.77	0.00	0.11	0.48	0.53
Faunal turnover ($o + e$)	4	16	2	6	8	17	34	17	37	30	33	33	41	40	48	32	8	0	4	4	14
Proportion of turnover $(o + e)_{2d}$	1.00	1.00	1.00	1.00	0.80	0.57	0.52	0.28	0.53	0.58	0.40	0.41	0.62	0.48	0.56	0.67	0.57	0.00	0.67	0.67	1.00
Std error of $(o + e)_{2d}$	0.00	0.00	0.00	0.00	0.13	0.09	0.06	0.06	0.06	0.07	0.05	0.06	0.06	0.05	0.05	0.07	0.13	0.00	0.19	0.19	0.00
Per capita rate of turnover $(o + e)_{2di}$	0.71	0.63	0.42	0.20	0.33	0.20	0.20	0.14	0.41	0.36	0.45	0.16	0.28	0.15	0.23	0.26	0.44	0.00	0.22	0.32	0.53

Note: Time units are defined in Figure 27.1.

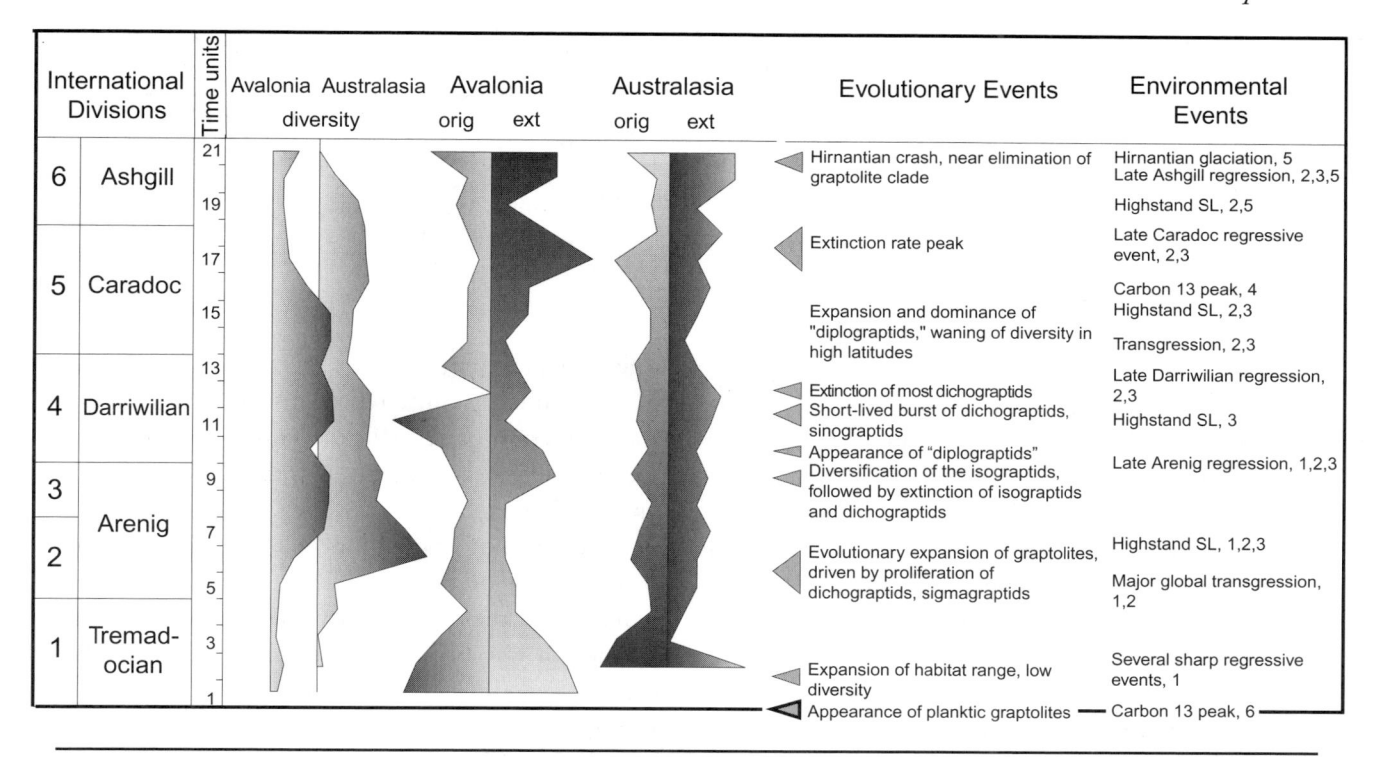

FIGURE 27.5. Comparison of normalized diversity plots, and per capita rates of origination (orig) and extinction (ext), for Australasia and Avalonia, showing main graptolite macroevolutionary events and events affecting the marine environment. Key to references: 1, Nielsen (1992a); 2, Ross and Ross (1995); 3, Fortey (1984); 4, Ainsaar et al. (1999); 5, Brenchley et al. (1994); 6, Ripperdan and Miller (1995).

($r^2 = 0.31$). The faunal turnover curve is much smoother than in the other two regions as a result. Faunal turnover is high (>30 species) from late Arenig to early Caradoc, reaching a high of 45 species in the mid Caradoc (*Mesograptus multidens* Zone).

When the numbers of extinctions and originations are normalized for total diversity, a similar but less marked pattern is seen in the Australasian data (figures 27.2D, F). The origination percentage (o_d) and extinction percentage (e_d) curves fluctuate strongly, driving a strongly fluctuating faunal turnover percentage curve $(o + e)_{2d}$. Ignoring the "end effects" in the earliest and latest Ordovician resulting from low total diversity, faunal turnover (figure 27.2E) averages about 50 percent of the species in a time unit and peaks at about 70 percent. Although overall correlation between origination and extinction is now very weak ($r^2 = 0.1$), there is an accordance of peaks in time units 6 (early Arenig), 9 (late Arenig), and 16 (mid Caradoc). In Baltica, the pattern is less clear, but faunal turnover percentage remains high throughout. However, in both Baltica and Avalonia, the marked accordance of peaks and troughs seen in the Australasian data set is lacking, and the faunal turnover percentage curves are smoother.

Fluctuations in origination and extinction numbers are less strong in Australasia when normalized for both diversity and time unit duration. These curves are the per capita rates of origination (o_{di}), extinction (e_{di}), and faunal turnover (($o + e)_{2di}$), shown in figure 27.2D–F. They suggest that the stronger fluctuations in the percentage and total numbers curves may, at least in part, be due to the effects of variation in time unit duration. Most conspicuously, the big peak in evolutionary activity in the early Arenig (time unit 6, Bendigonian stage), coinciding with the sharp rise in normalized diversity (figure 27.2A), is all but invisible. This suggests that although there are many extinctions and originations in this time unit, when expressed as a percentage of the total number present and normalized for duration of the time unit, it is nothing remarkable. However, the Bendigonian stage contains a high proportion of species with short stratigraphic ranges, spanning no more than one or two of the four zones (VandenBerg and Cooper 1992). We suspect that when the pattern is analyzed at a scale finer than the time units used here, it may change.

In Baltica, the per capita rates of origination and extinction are highest in the late Arenig–early Darriwilian (figure 27.3D–F), but they lie to either side of a sampling gap in the data. The per capita rate of faunal turnover peaks at 0.56, in the early Darriwilian (time unit 10). In Avalonia, both the percentage and per capita rate curves for originations and extinctions fluctuate strongly. The per capita rate of origination peaks in the lower *D. artus* Zone (Darriwilian) at 0.8, the highest level reached at any time in any region. Similarly, the per capita rate of extinction peaks in the lower *P. linearis* Zone (late Caradoc) at 0.8, higher than in any other region. In time units where diversity is high (mid Arenig to mid Caradoc), there is weak negative correlation between the percentage rates of origination and extinction ($r^2 = 0.34$), and the per capita rates of origination and extinction are uncorrelated ($r^2 = 0.01$).

Numbers of originations and extinctions are, of course, highly dependent on total diversity. Their decline in the earliest and latest Ordovician in all regions is due to the dropoff in total diversity at these times. However, the strong fluctuations in rates through the Ordovician in the two evolutionary processes are seen even after allowing for variation in total diversity and time unit duration and are likely to be significant. If the negative correlation of origination and extinction observed in the mid Arenig to mid Caradoc in Avalonia is significant, it contrasts with the other regions where positive correlation dominates. However, it is unlikely to be related to latitudinal differences, which diminished during this time.

Most noticeable is the dissimilarity between Australasia and Avalonia when either their extinction rate or their origination rate curves are compared (figure 27.5, table 27.2). Although the per capita rates of origination in the two regions are similar in having strong fluctuations, they are only weakly correlated ($r^2 = 0.24$). The per capita rates of extinction in the two regions also fluctuate strongly and are uncorrelated ($r^2 = 0.07$). Apart from the latest Arenig (time unit 9) and mid Darriwilian (time unit 12), where extinction peaks are present in both regions, there is little obvious accordance in either curves in the two regions. Unless there is an undetected bias in our data, this dissimilarity must reflect dissimilar origination and extinction patterns in the two regions. Because the dissimilarity persists through the Ordovician, it is unlikely to be related to latitude.

TABLE 27.2. Correlation of Origination and Extinction Rates (r^2 values)

	Per Capita Rate of Origination		
	Australasia	Baltica	Avalonia
Australasia	-	0.24	0.24
Baltica	0.01	-	0.02
Avalonia	0.07	0.72	-

Per Capita Rate of Extinction

When rate of origination in Baltica is compared with rates of origination in the other two regions (table 27.2), there is similarly no strong correlation. Extinction rate in Baltica is uncorrelated with that in Australasia but, interestingly, is strongly correlated with that in Avalonia ($r^2 = 0.72$), with which region it is most closely geographically located.

If the correlation coefficients can be taken at face value, the pattern suggests that origination rates are not related to either latitude or geographic proximity. Extinction rates are not strongly related to latitude but are related to geographic proximity, as might be expected. If the rates are influenced by environmental parameters such as nutrient supply and eutrophication (Allmon and Ross 2001), then these parameters vary, at least to a moderate extent, independently in Australasia and Avalonia.

■ Environmental Events

The normalized diversity curves and per capita rate curves for origination and extinction for Avalonia and Australasia are compared in figure 27.5. The main macroevolutionary events are compared in time with some major events that affect the ocean environment. The most spectacular evolutionary event (common to all three regions) is the rapid diversification in the early Arenig, marked by highs in the origination percentage curves for each region and with a diversity maximum for the entire Ordovician in Australasia. This coincides with a global transgression and highstand sea level (Nielsen 1992a), conditions that would suit the widespread development of the zone in the water column of oxygen depletion and nitrogen enrichment, recognized as a favored habitat for graptolites (Berry et al. 1987; Cooper et al. 1991; Cooper 1998) along with warm saline bottom waters in a greenhouse world (Wilde and Berry 1986). The expansion was driven by a proliferation of the Dichograptidae and Sigmagraptidae, following the decline of the Anisograptidae.

Highstand sea level has also been inferred in the mid Darriwilian (Fortey 1984), mid Caradoc (Ross and Ross 1995), and early Ashgill (Brenchley et al. 1994), all times with a high in the percentage origination curves. In the mid Darriwilian, a diverse dichograptacean and glossograptacean fauna, including didymograptids, sigmagraptids, sinograptids, isograptids, and glossograptids, makes a short-lived appearance. In the mid Caradoc, many diplograptids and orthograptids appear, and corynoidids become locally abundant. Ironically, the high productivity and oceanographic conditions associated with at least one of these highstands (mid Caradoc) is likely to have produced high extinction rates among the marine benthos on the continental shelves (Patzkowsky et al. 1997).

On the other hand, regressive events, such as those recognized in the late Arenig, late Caradoc, and possibly late Darriwilian, are matched by prominent extinction events, although not simultaneously in the three regions. The late Arenig marks the extinction of many isograptids and dichograptids. In the late Caradoc, many species of diplograptids, orthograptids, and Dicranograptidae became extinct. The major regression in the late Ashgill, associated with the Hirnantian glaciation (Brenchley et al. 1994), brought about the near extinction of the entire graptolite clade (Koren 1991; Melchin and Mitchell 1991). Only three or four lineages survived to give rise to Silurian stock.

During times of regression associated with climatic minima, density stratification and structure of the oceans break down, ocean circulation and turnover increase, the oceans become ventilated, and the nitrogen-rich, oxygen minimum zone decays (Wilde and Berry 1984). This would reduce the extent of the habitat zone for graptolites, consistent with the observed extinction pattern (Melchin and Mitchell 1991). Its effect on the distribution and extent of marginal upwelling zones, favored by Finney and Berry (1997) as the preferred habitat for graptolites, has yet to be assessed, however.

■ Conclusions

1. The mean duration of a species in Avalonia (2.377 m.y.) is not significantly different from that in Australasia (2.382 m.y.). As compared with Australasia, MSD in Avalonia is lower in the early to mid Arenig and late Caradoc–Ashgill and higher for much of the remainder of the Ordovician.

2. Origination and extinction rates fluctuate strongly through the Ordovician in all three regions. In Avalonia they appear to be negatively correlated during the time of high diversity (mid Arenig to mid Caradoc), whereas in the other regions positive correlation prevails.

3. Many of the major macroevolutionary extinction, origination, and diversity change events are matched by oceanic environmental events; transgressions and highstand sea levels are generally times of increased origination, and regressions are times of increased extinction. These effects are likely to be global.

4. The only feature detected that is possibly related to latitude is the rate of increase in diversity in the Early Ordovician; it is more rapid and reaches a higher level in Australasia (low latitude) than in Avalonia (high latitude). In terms of the plate tectonic model of Cocks and Torsvik (chapter 5), differences between the two regions in the Late Ordovician, such as the earlier decline in diversity and negative correlation of origination and extinction rates in Avalonia, are unlikely to be related to latitude.

ACKNOWLEDGMENTS

We thank Dan Goldman and Fons VandenBerg for their reviews of the manuscript. This project was supported in part by the Foundation for Research, Science and Technology, in New Zealand.

28 Chitinozoans

Florentin Paris, Aïcha Achab, Esther Asselin, Chen Xiao-hong, Yngve Grahn, Jaak Nõlvak, Olga Obut, Joakim Samuelsson, Nikolai Sennikov, Marco Vecoli, Jacques Verniers, Wang Xiao-feng, and Theresa Winchester-Seeto

Chitinozoans are an extinct group of organic-walled microfossils. The earliest known species appeared in the Early Ordovician (Tremadocian), and the group became extinct at the end of the Devonian. Chitinozoans have been reported from most types of Ordovician marine sediments. Major constraints affecting their occurrence and preservation are high-energy hydrodynamic regimes, weathering, and medium- to high-grade metamorphism. Because of their minimal dependence on lithology, chitinozoans are usually continuously present in most Ordovician marine successions, and the first-appearance datum (FAD) and last-appearance datum (LAD) of species can be precisely controlled.

Paris and Nõlvak (1999) postulated that chitinozoans are the eggs of soft-bodied marine metazoans and consequently can be used to document the biodiversification of their unknown producers. Chitinozoans diversified rapidly throughout the Ordovician Period. Of the 56 accepted chitinozoan genera (Paris et al. 1999), 63 percent first appeared in the Ordovician (28 percent in the Early Ordovician, 24 percent in the Mid Ordovician, and 11 percent in the Late Ordovician). Only 2 genera became extinct at the end of the Early Ordovician, 5 during the Mid Ordovician, and a total of 19 at the end of the Ordovician. The Ordovician biodiversification is also clearly reflected by the introduction of the major morphological innovations: 65 percent of the most important ones appeared before the end of the Mid Ordovician and 77 percent before the end of the Ordovician (Paris et al. 1999).

Chitinozoan generic and specific diversity is low. It normally ranges from a few to rarely more than 10 species representing only a few genera per sample. The abundance of chitinozoans, however, ranges from several hundred to a few thousand specimens per gram of rock for regions occupying high latitudes (e.g., North Gondwana). In general there are no more than a few hundred specimens per gram of rock in regions of low latitude (e.g., Laurentia, East Gondwana, Baltica); however, lithological and other environmental factors also influence the distribution of chitinozoans. Despite their inferred pelagic mode of distribution, chitinozoans display a definite provincialism during the Ordovician (Paris 1981; Achab 1991), when the major continental plates were far apart and the latitudinal contrast was most significant. Individual biodiversity curves have been established for all major paleocontinents where sufficiently abundant and reliable data were available.

■ Data and Counting Methods

Ordovician chitinozoans have been reported from all continents (figure 28.1), except Antarctica (see references in Miller 1996; Paris 1996). By the end of 2000, the global literature devoted to Ordovician

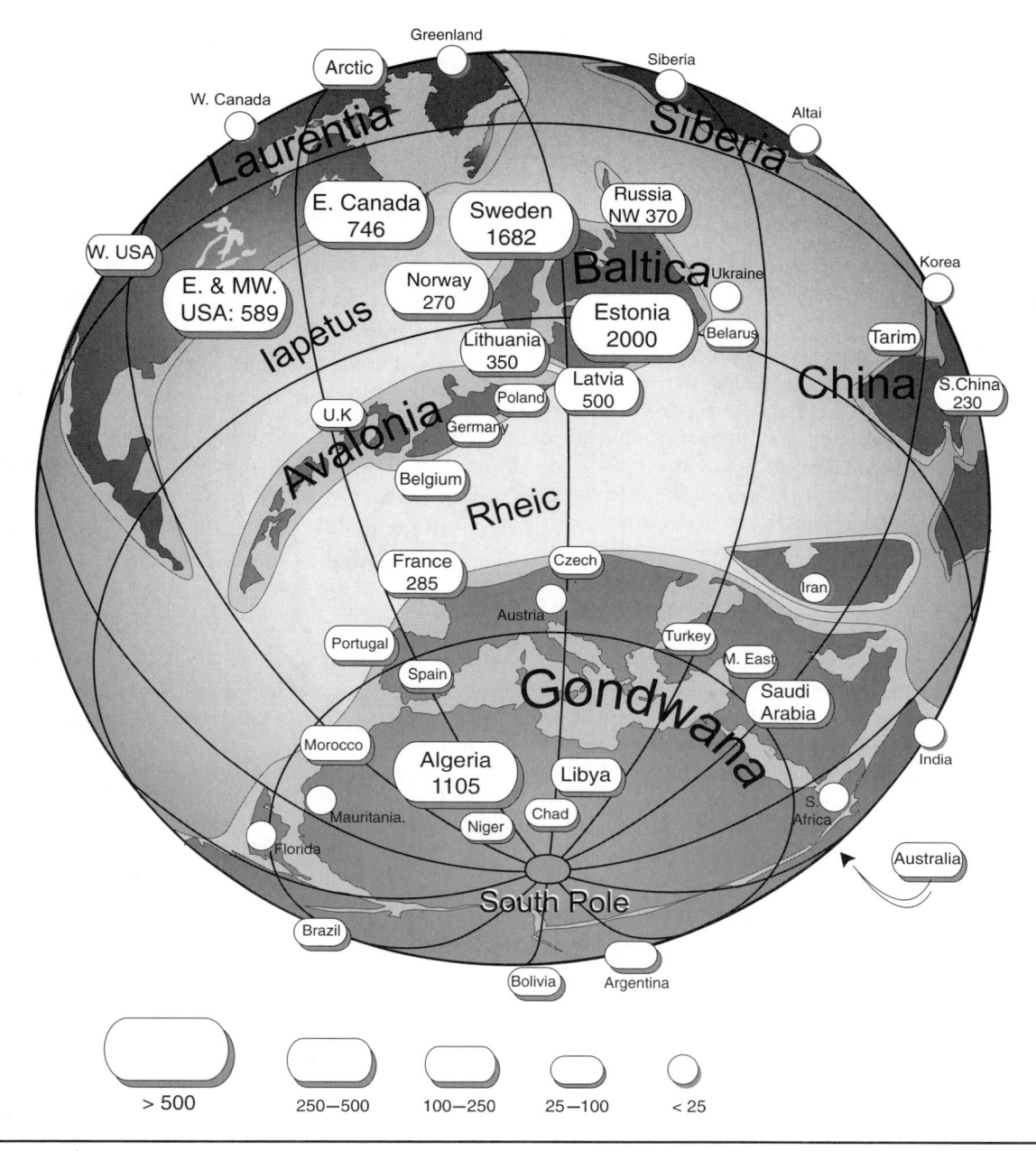

FIGURE 28.1. Regional distribution and density of the Ordovician samples used for documenting the chitinozoan diversity (the main references can be found in Achab 1989; Nõlvak and Grahn 1993; Paris 1996, 1999; Nõlvak 1999; Samuelsson and Verniers 1999; Wang and Chen 1999; Servais and Paris 2000). The paleogeographic reconstruction is based on faunal and sedimentologic evidence. It corresponds to Mid to Late Ordovician time (about 450–455 Ma). Data compiled and map drawn by F. Paris 2002.

chitinozoans included 380 publications (excluding abstracts and theses or other unpublished reports) and contained descriptions of some 397 Ordovician species (including subspecies and forms raised to the specific level). It is possible the Ordovician chitinozoan diversity may reach a total near 500 species. Regional databases have been built for each major paleoplate, using the normalized time slicing proposed by Webby et al. (chapter 2) and calibrated in million years (m.y.) (Cooper 1999b). Altogether, the data compiled by the "Chitinozoan Clade Team" of IGCP project no. 410 is based on more than 10,000 productive samples.

The respective position of FADs and LADs of most of the chitinozoan species can be precisely documented through fairly continuous sequences. A calculation method close to that suggested by Cooper (chapter 4) for his normalized diversity measure (d_{norm}) has been adopted. However, in our diversity measure, here referred to as the balanced total diversity measure (BTD), a full score is given to species confined to the time slice instead of a half score as adopted in Cooper's d_{norm}. For the sake of homogeneity, the following rules have been used in counting at species level.

- For the global analysis, only validly published species have been used. For regional analyses, validly published species, as well as published specimens left in open nomenclature (e.g., from theses and unpublished reports), have been taken into account.
- A full score (i.e., equal to 1) is given to species ranging through a whole time slice, or, in case of very short ranging species, to species restricted to part of a time slice.
- A full score (i.e., equal to 1) is also given for a species in each time slice between its FAD and LAD, even if not continuously recorded. This approach may introduce an overrepresentation of some long-ranging species resulting from poor diagnostic morphology (e.g., *Conochitina chydaea* and *Rhabdochitina magna*) or of poorly defined taxa (e.g., *Lagenochitina esthonica*).
- A half score (i.e., equal to 0.5) is given to species having their FAD or their LAD within a time slice but extending into the succeeding or preceding time slice.

Turnover events may be documented in different ways, either by taking into account or by ignoring the duration of the time slices in m.y.

- The origination (o_d) and extinction (e_d) rates are calculated for each time slice by dividing the number of originations or extinctions by the total number of species recorded in the corresponding time slice.
- The rate of species origination (o_{di}) or extinction (e_{di}) per m.y. are calculated by dividing the origination or extinction rates by the duration of the corresponding time slice expressed in m.y. These calculations correspond to the per capita rate of origination or of extinction suggested by Cooper (chapter 4).
- The turnover ratio (TR) has been calculated for some paleoplates; it corresponds to the number of originations, minus the number of extinctions, that is, net increase/decrease ($o - e$), divided by the BTD recorded in a given time slice. Extinction events have negative values, whereas radiation events correspond to positive values. A value close to 0 indicates that the number of originations and the number of extinctions are more or less identical. This is frequently the case for short-ranging taxa having their LAD and FAD within

the same time slice. The TR emphasizes the major events but is less significant in the case of reduced diversity (e.g., in the Early Ordovician).

■ Regional Biodiversity

North Gondwana (FP)

North Gondwana includes northern Africa, the Middle East, and the southern part of Europe. Most of the Ordovician sequences investigated in these regions (Paris 1998 and references therein) correspond to nearshore to outer-shelf environments at high or very high latitudes during most of the Ordovician. However, because of the greenhouse condition (Berner 1990) that prevailed from the Tremadocian to the middle part of the Ashgill, no permanent ice was reported in these areas until the late Ashgill (time slice *TS*.6c), when the major Hirnantian glaciation deeply affected all North Gondwanan environments (Brenchley and Marshall 1999 and references therein). Because of its erosive action (discontinuous record and recycling of older taxa), this glaciation greatly affected the records of Ordovician chitinozoans in North Gondwana (Paris et al. 1995; Paris et al. 2000a).

Database

The available data are from 139 published papers dealing with 1,415 productive samples. Additional information exists in numerous unpublished sources. More than 10,000 samples have yielded chitinozoans in North Gondwana. The data are contained mainly in oil company reports from different parts of southern Europe, North Africa, and Arabia (figure 28.1). The many unpublished literature sources are listed by Paris (1981, 1996) for southern Europe, Elaouad-Debbaj (1988a) for Morocco, Oulebsir and Paris (1995) for Algeria, Paris (1988) for Libya, and Paris et al. (2000b) for Saudi Arabia. Chitinozoan data have also been compiled by Paris et al. (1998) for Mauritania. Recent investigations have concentrated on the Upper Ordovician in order to document the Hirnantian glaciation (Oulebsir and Paris 1995; Paris et al. 2000a; Bourahrouh unpubl. data) and the Middle Ordovician, where the lithology (offshore shaley facies) is usually favorable for chitinozoan preservation and recovery.

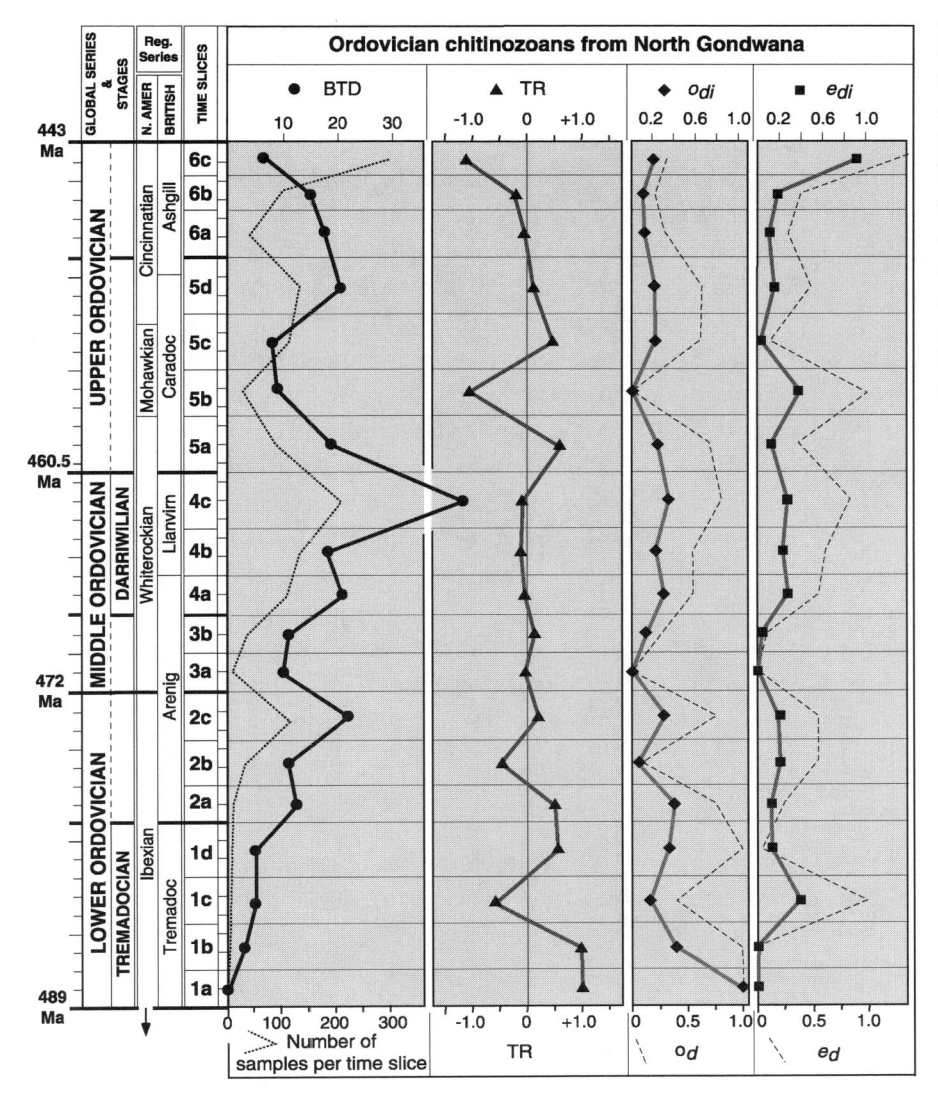

FIGURE 28.2. Biodiversity of Ordovician chitinozoan species (excluding taxa in open nomenclature) from North Gondwana (compiled by F. Paris). BTD curve: balanced total diversity; TR curve: turnover ratio; o_d curve: species origination rate; o_{di}: species origination rate per m.y.; e_d curve: species extinction rate; e_{di}: species extinction rate per m.y. For time slices and correlations adopted herein, see chapter 2.

The stratigraphic range of all specimens in the database is first documented in terms of the North Gondwanan chitinozoan biozones defined by Paris (1990, 1999) and then correlated with the 19 time slices of the Ordovician timescale following the calibration of Webby et al. (chapter 2).

Taxonomic Diversity

The BTD obtained for North Gondwanan chitinozoans (figure 28.2) shows average values ranging between 10 and 20 species per time slice (excluding specimens in open nomenclature). The BTD starts with very low values in the Tremadocian, where the clade originates, but ends with a dramatic drop in *TS*.6c (BTD = 6.5). The curve shows three main positive peaks during the late Early Ordovician (*TS*.2c), the late Mid Ordovician (*TS*.4c), and the middle part of the Late Ordovician (*TS*.5d), respectively. The most significant peak is in the upper part of the Darriwilian (BTD of 43 in *TS*.4c). Conversely, in addition to the major fall of diversity in *TS*.6c, two other significant falls of diversity are documented in the early Mid Ordovician (*TS*.3a–b) and in the Late Ordovician (*TS*.5b–c).

It must be stressed that the diversity curve (BTD) of the North Gondwanan chitinozoans mirrors the curve of the available data, except in the Ashgill (figure 28.2). The highest diversity (43 species) in the late Darriwilian (*TS*.4c) corresponds to one of the most intensively investigated intervals (more than 200 samples). This peak can be explained by several cumulative factors: (1) the duration of the time slices (3 m.y.); (2) the large variety of environments and localities investigated; (3) the thickness of the shaley sequences; and (4) the great chitinozoan diversity recorded in the Aquitaine Basin and Turkey. Both regions were located at more intermediate latitude than the Saharan localities and probably benefited from faunal transfers from Baltica (e.g., occurrence of *Armoricochitina granulata* in Turkey).

The Late Ordovician peak in *TS*.5d (more than 20 species) is mainly due to the diverse assemblages recorded in Saudi Arabia (Al-Hajri 1995; Paris et al. 2000b) and in Morocco (Elaouad-Debbaj 1984, 1986; Bourahrouh unpubl. data) that filled the data gap of previous reviews (Paris 1996).

Of the three recorded decreases in diversity (*TS*.3a–b, 5b–c, and 6b–c), the third one (BTD of 6.5 in *TS*.6c) is the most dramatic but not the most abrupt, as it began in *TS*.6a. This first-order event is of particular interest because it is contemporaneous with the Hirnantian glaciation (Brenchley and Marshall 1999) and cannot be related to the number of yielding samples (more than 300) (figure 28.2). The occurrence of reworked taxa in the Hirnantian diamictites (*TS*.6c) attributed to glacial processes (Oulebsir and Paris 1995) indicates that the actual diversity in the late Ashgill is even lower. The Hirnantian glaciation was finally responsible for the much reduced chitinozoan population (Paris et al. 2000a) after the steady decline from the late Caradoc diversity peak (*TS*.5d).

Turnover

The greatest diversity peak seen in the BTD curve at *TS*.4c is not matched in the TR curve (figure 28.2). Several positive values indicating a diversification trend (i.e., originations supplanting the extinctions) are, however, registered. The positive TR values recorded in the Early Ordovician document the radiation event. On the other hand, the positive values recorded in *TS*.5a and *TS*.5c are probably exaggerated because of a stratigraphic gap in the upper part of *TS*.5b and the lower part of *TS*.5c. The negative values for *TS*.1c and *TS*.2b may be biased because of the limited available data. The signal registered in the latest Ordovician is based on numerous assemblages from localities representing various environmental settings (i.e., deep-water deposits as well as nearshore environments from a large range of paleolatitudes). It corresponds to a first-order event related to the Hirnantian glaciation.

The origination (o_d) and extinction (e_d) rates, illustrated by the number of FADs or LADs per time slice and by the o_{di} and e_{di} curves (figure 28.2), respectively, provide additional information. With the exception of the Early Ordovician, the curve of the origination rate (o_d) is more or less similar to the BTD curve (peaks in *TS*.2c, 4c, and 5d). The o_{di} curve, however, gives a better filtered signal with a clear Tremadocian radiation event (first-order event), a flourishing Mid Ordovician interval, and a slight amelioration in *TS*.6c (postglacial recovery and setting of the next Rhuddanian morphotypes with *Spinachitina* species). The e_{di} curve perfectly demonstrates the dramatic extinction event that occurred during the late Ashgill glaciation (*TS*.6b–c pro parte). It also highlights an extinction peak registered within *TS*.5b.

Discussion

Once the chitinozoans had initially dispersed within the Ordovician oceans, and independently from the number of species recorded, originations in the North Gondwanan regions roughly counterbalanced extinctions, at least through the mid Arenig to the latest Darriwilian interval (*TS*.2c to 4c). The specific diversity (BTD) of the North Gondwanan Ordovician chitinozoans is fairly low (10 to 20 per time slice) when compared with those of other paleoplates. In contrast, their abundance is much higher than on other paleoplates. Significantly chitinozoans, or rather the population of "chitinozoan animals," display distribution patterns similar to those of modern pelagic marine fauna, that is, low diversity but with high productivity in areas located at high latitudes (cold environments).

Transgressions also seem to have had a positive impact on the diversity of chitinozoans as documented by the early Caradoc event (*TS*.5a–b). This is probably due to more open communication with Baltica, as shown by the occurrence of several typical Baltic species (e.g., *Laufeldochitina stentor*, *Lagenochitina dalbyensis*) in the North Gondwanan regions. The main transgressions registered in most of these regions, for example, in the mid Arenig (*TS*.2c), the early

Darriwilian (*TS*.4a–b), and the early Late Ordovician (basal *Nemagraptus gracilis* transgression; *TS*.5a), are associated with increased chitinozoan originations (second-order feature).

East Gondwana (TWS)

East Gondwana (i.e., Australia, Antarctica) was located at low latitudes during the Ordovician (figure 28.1). Data from East Gondwana include the Lower and Middle Ordovician chitinozoan assemblages from Australia, based on limited information from the Canning Basin (Western Australia: Combaz and Peniguel 1972; Winchester-Seeto et al. 2000; Winchester-Seeto unpubl.), Georgina Basin (western Queensland: Playford and Miller 1988; Winchester-Seeto unpubl.), and spot samples from the Tabita Formation (western New South Wales: Winchester-Seeto unpubl. data).

The calculated diversity (BTD) of chitinozoans for each of the time slices is 3 for *TS*.2b, 9.5 for *TS*.2c–4a, 11 for *TS*.4b, and 16 for *TS*.4c. There is a progressive increase in the number of species from the Lower Ordovician to the upper part of the Darriwilian, which is consistent with the general trend identified by Paris (1999) to a global peak within the North Gondwanan chitinozoan-based *jenkinsi* Zone, that is, within *TS*.4c.

The oldest samples employed in this compilation come from the upper Nambeet Formation (Canning Basin) representing *TS*.2b. The beginning of *TS*.2c is interpreted as coinciding with the first appearance of *Conochitina langei* (Combaz and Peniguel 1972; Winchester-Seeto unpubl.). The North American chitinozoan zonation has been adopted in this study because Australian chitinozoan faunas from the Lower and Middle Ordovician show a close resemblance to those from North America (Achab 1991; Winchester-Seeto et al. 2000). Except for very general information, there are not yet enough detailed biostratigraphic data for separation of *TS*.2c to 4a inclusive; thus they have been considered as one unit. *TS*.4b is indicated by the presence of *Conochitina subcylindrica;* however, the exact position of the beginning of this time slice is unknown. For this compilation, the boundary between *TS*.4b and *TS*.4c is taken at the beginning of Zone 04b established by Combaz and Peniguel (1972) and indicated by the first appearance of *Cyathochitina hunderumensis* and *Belonechitina vibrissa* (Winchester-Seeto et al. 2000 and unpubl. data). The youngest time slice for which there is adequate information is *TS*.4c. It contains the conodont *E. suecicus,* which correlates with the chitinozoan-based *jenkinsi* and *turgida/subcylindrica* zones from North America. However, Winchester-Seeto et al. (2000) found that uppermost Goldwyer and Nita formations can be correlated with the currently undefined chitinozoan zone above the *Cyathochitina jenkinsi* Zone in the North American zonation (Achab 1989), that is, still within *TS*.4c.

West Gondwana (YG)

South America was mainly located in medium to high latitudes during the Ordovician. It represents the western part of Gondwana (figure 28.1). Ordovician chitinozoans have been described from northwestern Argentina (Volkheimer et al. 1980; Ottone et al. 1982, 2001), southern Bolivia (Heuse et al. 1999), and northern Brazil (Grahn 1992; Grahn and Paris 1992). The chitinozoan occurrences in Argentina and Bolivia are calibrated with the graptolite zonation and in Argentina also partly with conodont zones. In Brazil, only poorly known acritarchs are available for independent stratigraphic control. Except for the Arenig, chitinozoans from Argentina display a Laurentian affinity (Achab 1989); those from Bolivia and Brazil show a Gondwanan affinity (Paris 1990), although Laurentian elements are common in the Ashgill.

Ordovician chitinozoan biodiversity of South America is low with respect to the data provided in the global curve presented by Paris (1999). The oldest faunas are from the late Tremadocian (*TS*.1b–d) of southern Bolivia and contain only one species, *Desmochitina* sp. gr. *minor*. An early Arenig fauna (*TS*.2a) comprises three species, including *Conochitina decipiens*. A somewhat younger chitinozoan assemblage with three species, including the index species *Eremochitina brevis,* has been described from *TS*.2c in northwestern Argentina. The Darriwilian is rather well represented. Four species, including *Calpichitina* cf. *C. lata* and *Lagenochitina langei*, are reported in northwestern Argentina. In northern Brazil, seven species, including *Conochitina* aff. *C. havliceki, Lagenochitina obeligis,* and *Conochitina decipiens,* occur in equivalent *TS*.4a–b. The youngest Darriwilian (*TS*.4c) has yielded eight species, among them *Calpichitina* cf. *C. lata*, *Cyathochitina jenkinsi*, and *Lagenochitina langei*. Increasing chitinozoan diversity in the Darriwilian reflects a major sea level rise. A main flooding event in the early Caradoc is evident in the *TS*.5a–b interval, where eight species are known, notably *Kalochitina multispinata*, *Belonechitina cactacea*, and *B. robusta*. The youngest Ordovician chitinozoan assemblage from South America has been described in a stratigraphic level corresponding to a worldwide rise in sea level during *TS*.6b (Ashgill). Of the seven chitinozoan species that are represented in *TS*.6a and lower *TS*.6b, the most important are *Armoricochitina nigerica*, *Lagenochitina prussica*, and *Tanuchitina anticostiensis*.

Baltica (JN)

Baltica drifted from high-intermediate to low latitudes during the Ordovician. Detailed investigations of chitinozoan assemblages show a rapid evolution in different parts of the Ordovician basin of the East European Platform. These investigations have been successful despite complications due to the presence of five main composite environmental belts in the Baltic regions (Nõlvak and Grahn 1993 and references therein). The belts are characterized by rather constant litho- and biofacies that have facilitated the construction of a detailed zonation. The difficulties encountered are due to secondary dolomitization; the occasional presence of barren marine red beds, carbonate mounds, or calcareous sandstone beds; and the limited outcrop area of Ordovician rocks in northern Estonia, Sweden (Grahn 1980, 1981), and Norway (Grahn et al. 1994). Most of the data were derived from core material in Ukraine, Belarus, Poland, Lithuania, Latvia, and Russia (St. Petersburg region), Gotland (Grahn 1982), and western and southern Estonia. The preservation of the chitinozoans varies from excellent to good in the dominantly bedded limestones (overall thicknesses commonly less than 250 m) but only acceptable to poor in graptolite shales (in Scania) or in other beds affected by thermal heating (e.g., Oslo area of Norway and central Sweden). The chitinozoan abundance usually ranges from a few to several tens of specimens per gram of rock.

Database

A total of 5,362 samples have been used to construct the biodiversity curve. About 5 percent were unproductive, with some exceptions, for example, in Öland, where 78 percent of the samples from *TS*.1b–4b (Grahn 1980) proved to be unproductive. In general, there is a great similarity in the regional faunal logs and in the origination and extinction rates. Regional differences do not exceed 10 percent of all taxa within the same time slice interval. This constancy allowed the construction of a very useful zonation (Nõlvak and Grahn 1993; Nõlvak 1999). As explained in the introduction of this chapter, the biodiversity record was established using the LAD and the FAD of each species even in the case of some species of *Cyathochitina* and *Conochitina*, which were interpreted as Lazarus taxa in East Baltic sections (e.g., with discontinuous occurrences due to changing environments).

The Ordovician chitinozoan list comprises 157 species belonging to 26 genera. The number of species per genus varies greatly. Five genera (*Lagenochitina*, *Cyathochitina*, *Desmochitina*, *Conochitina*, and *Belonechitina*) are represented by a relatively high number of species (15 to 24), whereas only 9 species are recorded in the genus *Spinachitina* and fewer than 6 species in each of the remaining genera.

A number of taxa identified at generic level only have been omitted. However, new undescribed species have been included in the local database. In some Upper Ordovician portions of the East Baltic sections it will be possible to arrive at more detailed subdivision of the suggested time slices (e.g., see Kaljo et al. 1996 for *TS*.4c and *TS*.5c). In the Lower Ordovician the situation is quite the opposite: because of well-known stratigraphic gaps in all investigated regions, the *TS*.1c–2a and *TS*.2b–c intervals are difficult to separate.

Taxonomic Diversity and TR

After the appearance of chitinozoans (*Lagenochitina destombesi*, *L. esthonica*) in the Tremadocian (*TS*.1b) of northern Estonia and Latvia, the most intensive origination and diversification took place at the beginning of the Darriwilian (*TS*.4a: Baltoscandian late Volkhov time; see figure 28.3). Below this level in the Lower Ordovician (*TS*.1b–3b), chitinozoans were

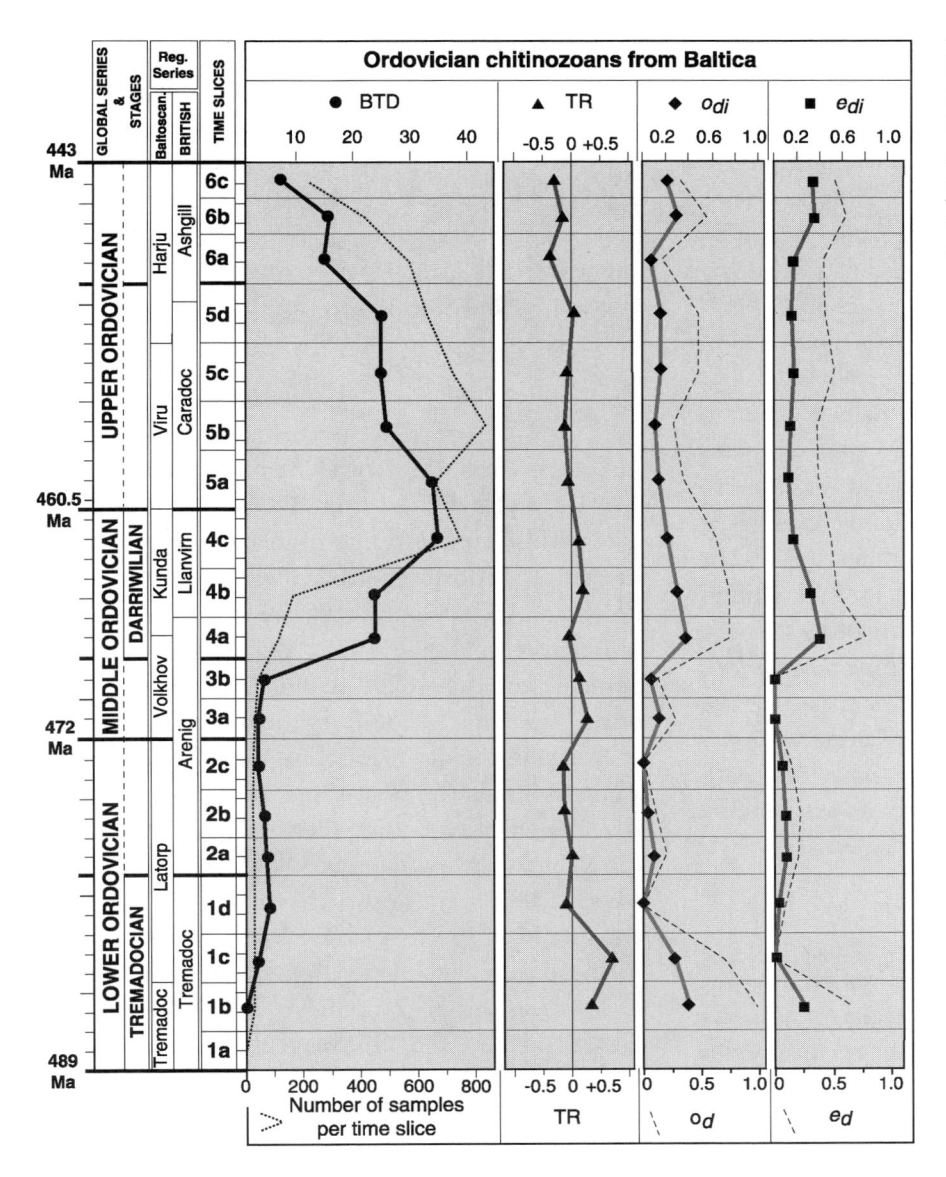

FIGURE 28.3. Biodiversity of Ordovician chitinozoan species (excluding taxa in open nomenclature) from Baltica (compiled by J. Nõlvak). For diversity measure abbreviations and time-slice correlations, see figure 28.2.

recovered, with rare exceptions, from the condensed sections of northern Estonia. Chitinozoans here are short-ranging forms with *Lagenochitina* being dominant. In the other regions, condensed beds corresponding to this interval have turned out to be barren because of preservational factors.

The BTD curve is simple and regular with low values in the Early Ordovician, followed by a noticeable rise in the Mid Ordovician and drop in the Late Ordovician. The number of investigated samples (figure 28.3) has no direct bearing on the sudden rise of the diversity curves in *TS*.4a, where all curves display positive peaks. The beginning of the Darriwilian was a time of rapid changes in chitinozoan assemblages, which is characterized by a relatively large number of short-ranging species and the extinction rate reaching its maximum value. The higher BTD curve shows stable (*TS*.4b) and rising values reaching a maximum during *TS*.4c to 5a intervals. The highest diversity is recorded during the latest Darriwilian in *TS*.4c with 35 species, including a high number of short-ranging species, yet the origination rate curve shows a slight decline. Origination and specific diversity was relatively stable in the mid Caradoc (*TS*.5c, or late Viru). However, in sections studied in detail, such as the Rapla drill core in northern Estonia, it has been clearly possible to subdivide the *TS*.5c into three local stages (Keila, Oandu, Rakvere). The middle part of this interval exhibits a minimum diversity level of chitinozoan assemblages, which was caused by a substantial extinction event in latest Keila time (63 percent of the species; see Kaljo et al. 1996). It is too early to determine whether this event in the northern Estonian sections is due to local environmental conditions (see discussion in Meidla et al. 1999). It is not registered in the more generalized curves presented herein because it is a short event counterbalanced by the higher diversity recorded in the remaining part of TS.5c.

The decline in diversity and the rate of originations (o_d) in the lower Ashgill (TS.6a) is attributed in part to the widely distributed red beds in Baltica, except for northern Estonia, northern Gotland, and Lithuania. In TS.6b, origination underwent a short intensive period followed by a sharp drop (latest Ordovician crisis), though the decline had already started near the Caradoc-Ashgill boundary in TS.5d. In TS.6c chitinozoans suffered major extinction (first-order event) at the upper boundary of the *Spinachitina taugourdeaui* Zone (see chapter 2), where one new species (*Conochitina scabra*) appears to continue, along with only three other species, into the Silurian. This level marks the onset of the environmental changes attributed to the Hirnantian glaciation (Kaljo et al. 2001).

Interpretation

The TR curve shows two main radiation events (TS.3a–b and 4b–c). The earlier event is somewhat overestimated because of the scarcity of chitinozoans in this interval. The causes are the paucity of localities suitable for chitinozoan preservation, long-ranging species, and the lack of definite extinctions. The TR negative values in TS.2b–c are biased by the lack of data. Comparison of the specific diversity curve with the number of samples curve shows no direct correlation: chitinozoan diversity in the Darriwilian and the Caradoc is relatively independent of the number of investigated samples. The origination (o_{di}) and extinction (e_{di}) curves show a good correlation, with regular changes in chitinozoan associations. They form the basis for the more detailed Darriwilian and Upper Ordovician biostratigraphy of Baltica.

Laurentia (AA, EA)

Following the breakup of Rodinia and the formation of Gondwana, Laurentia (the ancestral North American continent) developed as the second largest continental paleoplate. During the Ordovician, Laurentia roughly straddled the equator, and almost all the North American continent was submerged. Ordovician chitinozoans are known from shelf and slope deposits located at the Laurentian margin now approximately corresponding to eastern, southeastern, and western North America, the Canadian Arctic, Greenland, and Spitsbergen (see Achab 1989 and references therein).

Database

Laurentian chitinozoans have been described from 1,391 samples derived from 113 outcrops and drill cores. Most samples came from localities in the St. Lawrence Platform (Quebec and Ontario) and the northern Appalachians regions of eastern Canada (references in Achab 1989; Williams et al. 1999; Soufiane and Achab 2000a; and Albani et al. 2001); others are from 11 localities in the Arctic Platform (Achab and Asselin 1995; Soufiane and Achab 2000b), 22 localities in Tennessee and Alabama, the Black River Valley of New York, the Cincinnati region, and the Arbuckle Mountains of Oklahoma (Grahn and Miller 1986; Hart 1986; references in Achab 1989 and Grahn and Nöhr-Hansen 1989). A few localities have also been documented from the Interior Platform of western Canada (2), central Nevada (2), north of Greenland (1), and Spitsbergen (1) (references in Grahn and Nöhr-Hansen 1989; Martin 1992; Soufiane and Achab 2000b).

As shown in figure 28.4, the Upper Ordovician is the most investigated part of the system, with more than 83 percent of the samples belonging to the Caradoc-Ashgill (TS.5a–6c) interval. More than 130 papers and abstracts record Laurentian chitinozoans, but only 30 percent provide useful information on chitinozoan assemblages and their occurrence. The database contains 303 species (137 in open nomenclature) assigned to 26 genera.

Methodology

All published and documented Laurentian chitinozoan species were entered in the database under their original names and their stratigraphic occurrence expressed in terms of graptolite, conodont, or chitinozoan zones, as reported by the original authors. Although only original data have been used, the fol-

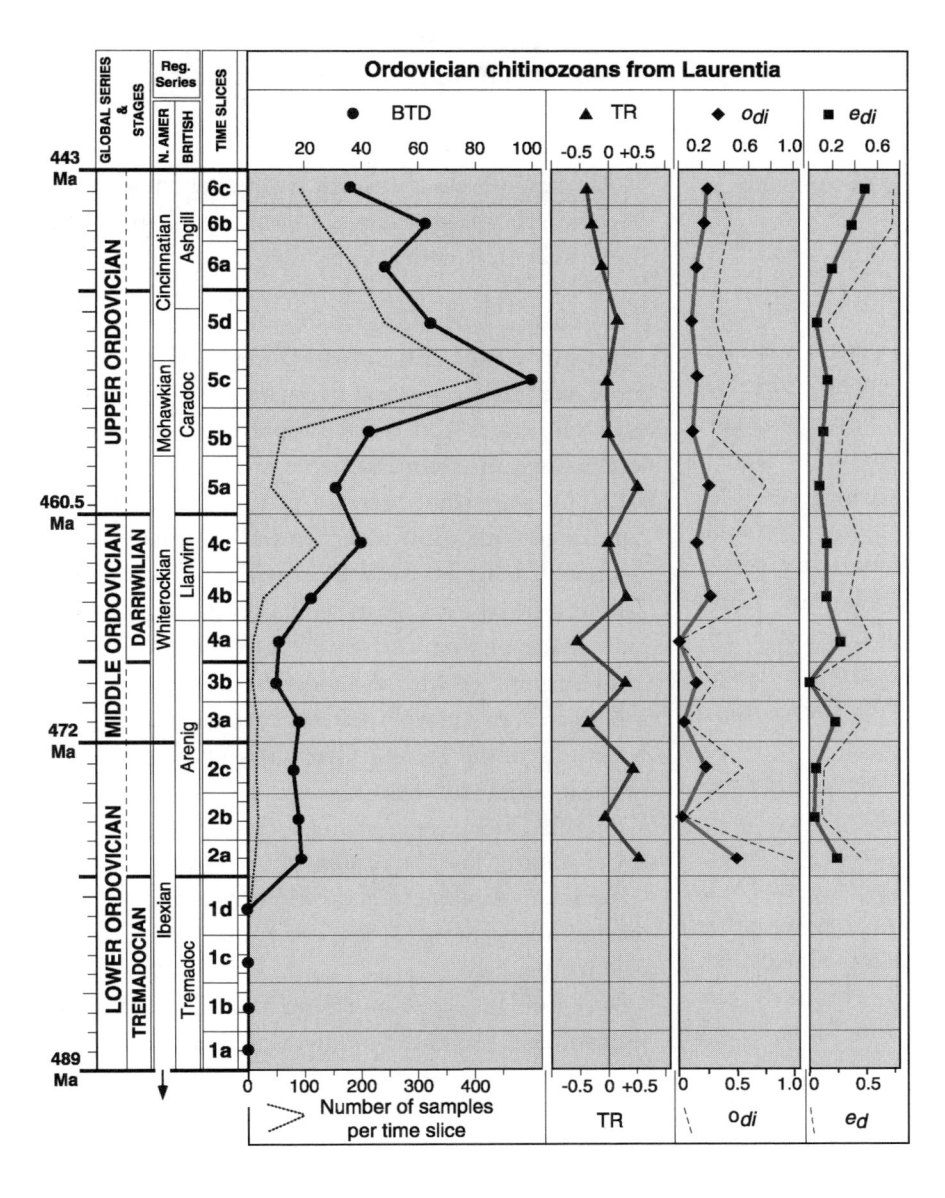

FIGURE 28.4. Biodiversity of Ordovician chitinozoan species (including taxa in open nomenclature) from Laurentia (compiled by A. Achab and E. Asselin). For diversity measure abbreviations and time-slice correlations, see figure 28.2.

lowing taxonomic modifications have been made: (1) the subspecies and varieties have been raised to species rank, and (2) the known occurrences or stratigraphic ranges of the species in the database were correlated with the 19 time slices of the timescale used in the IGCP 410 project (see chapter 2).

A total stratigraphic range has been established for each species. We assumed that long-ranging species were continuously present between their FAD and LAD. This approach appears to be valid when the gap is small, but where larger gaps occur, for example, in species such as *Conochitina chydaea, Conochitina subcylindrica, Cyathochitina hyalophrys, Lagenochitina baltica, Desmochitina complanata,* and *Tanuchitina bergstroemi,* it is possible that misidentifications of these species have been made.

Taxonomic Diversity

The BTD curve (figure 28.4) clearly shows the first documented occurrence of Laurentian chitinozoans in *TS*.2a at the base of the Arenig. It also shows that the diversity remains more or less constant during most of the Arenig (*TS*.2a–3a interval), with values ranging from 15 to 20 species. A slight decrease (about 10 species) is observed in the late Arenig (*TS*.3b–4a) followed by a continuous increase (11 to 40 species) during the Darriwilian (*TS*.4a–c). After a second decrease (31 species) in the early Caradoc

(*TS*.5a), the curve expands abruptly in the late Caradoc (*TS*.5c), where the highest diversity (BTD = 100) is observed. From this point, except in *TS*.6b, it continuously declines through the latest Caradoc and the Ashgill (*TS*.5d to 6c) but retains overall higher values than in the Arenig.

Turnover

Chitinozoan species origination or extinction rates (o_{di} or e_{di}) for the various Ordovician time slices in Laurentia are shown in figure 28.4. The first occurrence of chitinozoans in the early Arenig (*TS*.2a) is reflected in the origination rate curve (o_{di} = 0.5), which then fluctuates with values ranging from 0.00 to 0.23 through the Arenig (*TS*.2b–4a). Two pulses occur in *TS*.4b (Darriwilian) and *TS*.5a (early Caradoc), with values close to 0.30. The fluctuation appears to be less important during the Caradoc-Ashgill interval, with the rate increasing slightly from 0.12 in *TS*.5b to 0.24 in *TS*.6c.

The extinction rate curve (e_{di}) shows three distinct intervals. The first corresponds to the Arenig (*TS*.2a–4a), where the extinction rate fluctuates between time slices with distinct pulses in *TS*.3a (0.22) and *TS*.4a (0.27). The e_{di} values remain rather constant (0.09 to 0.16) through the late Darriwilian and most of the Caradoc (*TS*.4b–5c). From the latest Caradoc (*TS*.5d) onward, the extinction rate (e_{di}) increases continuously (from 0.06 to 0.49), paralleling the declining biodiversity at the end of the Ordovician.

The TR shows no obvious correlation with the number of samples and the BTD curves. It emphasizes the events documented by the o_{di} and e_{di} analyses by pointing to the levels where originations or extinctions are predominant.

Discussion

As shown in figure 28.4, the number of fossiliferous samples and the total diversity curves are very similar, suggesting that the latter has to be corrected to eliminate the bias caused by a large number of samples in some intervals. Nevertheless, despite the small number of fossiliferous samples in the Early and Mid Ordovician (*TS*.2a–4a), a significant biodiversity record is evident. The highest diversity is observed in the most studied time slice (*TS*.5c), and a brief increase in the diversity is evident during the Ashgill (*TS*.6b) while, at the same time, the number of fossiliferous samples is declining and the extinction rate increasing.

All these observations can be integrated with and related to the global events and to the geologic evolution of North America, as follows:

1. In the Early Ordovician, the passive margin sediments yield a chitinozoan fauna characterized by moderate biodiversity and fluctuating origination and extinction rates, reflecting the prevalent eustatic sea level variations (second-order feature).

2. An important increase in the extinction rate, combined with a low diversity and origination rate, is observed in the early Darriwilian (*TS*.4a). It is followed by an increase in the biodiversity (*TS*.4b–c) and in the origination rate (*TS*.4b). These changes can be correlated with the general change from a passive to an active margin setting, corresponding to the beginning of the Taconic Orogeny and the formation of the associated foreland basin.

3. The Taconic regime prevailed through the entire Caradoc. The deepening of the basin is marked by an increase in the total diversity of the chitinozoan fauna and by more or less constant extinction and origination rates.

4. From the late Caradoc (*TS*.5d) onward, a significant decrease in the biodiversity and a corresponding steady increase in the extinction rate may reflect the Late Ordovician ice cap development in Gondwana and the associated sea level fall (first-order feature). Changes in environmental conditions seem to have had an impact on Laurentian chitinozoan diversity since the late Caradoc.

Avalonia (JS, JV, MV)

Detailed analyses of the fossil faunas of Europe and North America led Cocks and Fortey (1982) to suggest the existence of an independent microcontinent, Avalonia, during the Ordovician, and the Tornquist Sea, separating Avalonia from the paleocontinent Baltica (figure 28.1). Avalonia had a short independent existence from the time it started to drift off North or West Gondwana in the late Early Ordovician to its collision with Baltica in the Ashgill. It follows that evaluation of the origination/extinction rates of Avalonian chitinozoan faunas is limited to the Mid and Late Ordovician.

The boundaries of Avalonia extended from Cape Cod in Massachusetts to the Atlantic Provinces of Canada, southern Ireland and Britain, Belgium, and northern Germany. Most strata deposited on Avalonia are now deeply buried and inaccessible. Avalonian faunas are generally poorly preserved in comparison with coeval ones from Baltoscandia and North Gondwana. Accordingly, few studies deal with Avalonian chitinozoans, and all of them have concentrated on the European part of Avalonia (i.e., eastern Avalonia).

Some problems have been encountered in compiling the biostratigraphic data, such as poorly figured and described taxa and the lack of well-constrained dating of the associated sediments. In addition, some chitinozoans are poorly preserved, and a limited number of samples were available for study (316 samples for the whole Ordovician).

Database

For the diversity curve (BTD), we have used both published and unpublished data. All data were critically evaluated, and only the taxa determined with certainty to the specific level were included (figure 28.5). The data came from Belgium (references in Samuelsson and Verniers 2000), Wales, Shropshire (Jenkins 1967), and the Ebbe Anticline in western Germany (Samuelsson et al. 2002a). Also included are

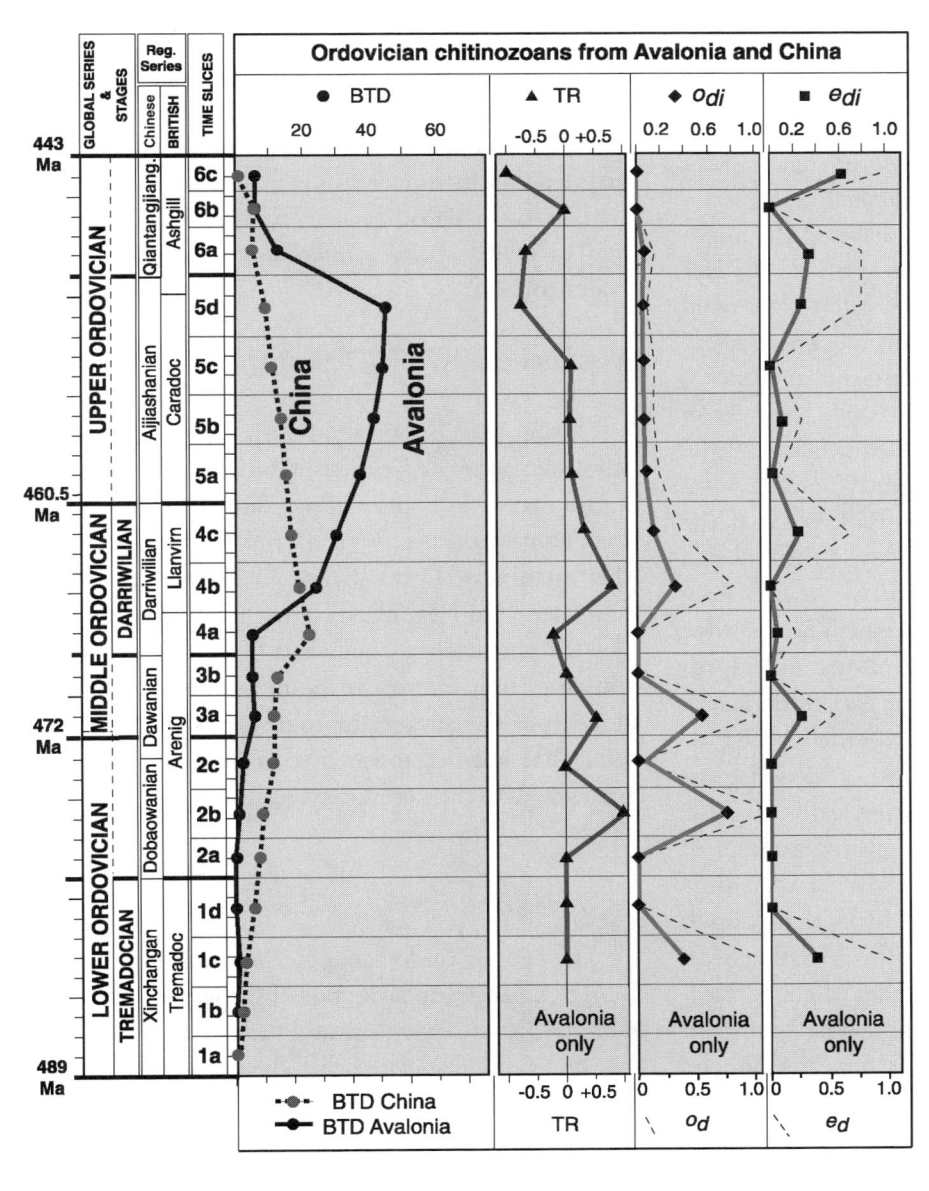

FIGURE 28.5. Biodiversity of Ordovician chitinozoan species from Avalonia (compiled by J. Samuelsson and including taxa in open nomenclature) and from China (compiled by X.-F. Wang and X.-H. Chen). For diversity measure abbreviations and time-slice correlations, see figure 28.2.

sections where recent work has demonstrated a probable Avalonian affinity of the sediments, that is, deep drillings at Rügen, northern Germany (Samuelsson et al. 2000; Samuelsson and Servais 2001; Vecoli and Samuelsson 2001), and Pomerania, northwestern Poland (Samuelsson et al. 2002b; Samuelsson unpubl.).

Taxonomic Diversity

The number of species recorded in Avalonian successions is low. In all, 81 species were counted, 75 of which were recorded from the time that Avalonia existed as a separate microcontinent (i.e., from Arenig to early Ashgill). For completeness, we have also included data from the remaining Ordovician interval (*TS*.6b–c), which were recorded in Avalonian areas after docking, when the microcontinent ceased to exist as a separate entity.

As with other North Gondwanan successions, low-diversity chitinozoan assemblages characterize the oldest units of what was to become Avalonia (figure 28.5). Among these, *Lagenochitina destombesi* and *Eremochitina brevis* are the most distinctive taxa (Samuelsson et al. 2000). When Avalonia began its existence as an independent microcontinent, a relatively rapid diversification took place during the late Darriwilian. In general, the assemblage consists of ubiquitous, North Gondwanan taxa (Jenkins 1967; Samuelsson and Verniers 2000; Vecoli and Samuelsson 2001). Most taxa extended to the end of the Caradoc. A less conspicuous diversification event took place at the base of the Caradoc, and the majority of these taxa (a notable exception being *Lagenochitina stentor*) were still present in upper Caradoc rocks. The Caradoc chitinozoans are still of the ubiquitous and North Gondwanan types, but in the later part of the Caradoc a few taxa, which can be regarded as Baltoscandian, make their appearance. No extinction event follows the two late Darriwilian and early Caradoc diversification events, although a limited number of species disappear at the end of early Caradoc time (*TS*.5b) (figure 28.5). In the pre-Hirnantian Ashgill, Avalonia docked with Baltica (Vecoli and Samuelsson 2001), and from the beginning of the Ashgill (*TS*.6a), diversity declined very rapidly (e_{di} close to 0.6 in *TS*.6c). One or two taxa, however, appeared for the first time in *TS*.6c.

Discussion

The observed high chitinozoan diversity in Avalonia during the Caradoc (*TS*.5a–d) can be explained in terms of the quality of the sedimentary record or the paleobiogeographic evolution, especially given that most Avalonian sections hitherto investigated are of Caradoc age. Thus, the high diversity might be due to overrepresentation of sampled Caradoc units. It is also during the Darriwilian and the Caradoc that Avalonia was an independent microcontinent in the temperate Southern Hemisphere (Cocks 2001). This position might have positively contributed to the higher diversity, since the taxa occurring on North Gondwana and in Baltoscandia have also been recovered from Avalonia. The diversity patterns can be explained as the result of a free interchange of species as Avalonia drifted northward between Gondwana and Baltica. There is no firm evidence that any chitinozoans are exclusively Avalonian. In the Ashgill, although few successions have been investigated, a major faunal turnover seems to have occurred with characteristic Baltic (*Conochitina scabra*) and Laurentian (*Hercochitina gamachiana*) taxa playing the major roles.

China (XFW, XHC)

During the Ordovician, South China was located at low to moderate latitudes. Chitinozoans have been found mainly in South China (Yangtze craton and its shelf-slope biofacies) (Wang et al. 1994) and in the shelf margin of Tarim (figure 28.1). Integrated bio-, sequence-, and event-stratigraphic studies suggest that 17 chitinozoan zones and 5 major chitinozoan diversification events can be recognized. Their relationship with relevant graptolite zones and conodont zones has been established to correlate these events precisely with the time slices adopted herein (see chapter 2).

Taxonomic Diversity

The earliest Ordovician chitinozoans appeared in *TS*.1b. This event is marked by the appearance of *Lagenochitina destombesi* and *Conochitina?* sp. (figure 28.5). It is associated with the transgression at the base of the Nanjinguan Formation (Wang et al.

1996). The chitinozoan diversity increases (BTD = 3) in the lower Fenxiang Formation (*TS*.1c) with the appearance of *Conochitina* cf. *decipiens*, *C.* cf. *pervulgata*, *C. poumoti*, and *Cyathochitina*? cf. *clepsydra* (Wang et al. 1996).

The appearance of the *Conochitina symmetrica*, the worldwide index species of the *symmetrica* Zone, heralds an initial chitinozoan radiation in *TS*.1d–2a (BTD increasing from 5.5 to 7 in this interval). *C. symmetrica* is found with *Lagenochitina* cf. *obeligis*, *L. obeligis*, *Euconochitina vulgaris*, and *Eisenackitina tongziensis* (Hou and Wang 1982; Grahn and Geng 1990; Wang and Chen 1992, 1994; Chen 1994; Chen et al. 1996). These species coexist with conodonts of the *Serratognathus-Paroistodus proteus* Zone, indicating that the *C. symmetrica* Zone is correlative with the latest Tremadocian to earliest Arenig. The overlying *Eremochitina baculata* Zone (*TS*.2b) corresponds to most of the graptolite-based *approximatus* and *fruticosus* zones.

A slight increase in the chitinozoan diversity is recorded in *TS*.2b to *TS*.3a (BTD increasing from 8 to 11.5). The abundance and diversity of *Conochitina* and the change in chitinozoan assemblages characterize this interval. At least 12 species of chitinozoans (e.g., *C. raymondi*, *C. brevis*, *C. baculata*, *L. esthonica*) occur simultaneously or successively in this interval. The assemblage of the *Eremochitina brevis* Zone (*TS*.2c) is quite similar to that of the underlying *raymondi* Zone (*TS*.2b), except for the first appearance of the index species with graptolites of part of the *suecicus* Zone. *Conochitina langei* has its FAD in the *suecicus* Zone (*TS*.2c–3a) and ranges up to the graptolite-based *austrodentatus* Zone in association with *C. poumoti* and other taxa (Chen et al. 1996). The mixture of chitinozoans from different paleobiogeographic realms is believed to be related to warm-water currents associated with transgression and the clockwise rotation and movement of the South China plate from moderate to low southern latitudes along with the northward movement of Gondwana during this time—Dobaowanian (*TS*.2a–c) to early Dawanian (*TS*.3a) (Wang and Chen 1999).

The maximum chitinozoan diversity (BTD = 22) is recorded in *TS*.4a with no fewer than 17 chitinozoan species referred to the *Cyathochitina protocalix* and *C. calix* zones (Geng 1984; Wang and Chen 1994 and references therein; Chen et al. 1996). This is a new stage in the evolution and diversity of the species including the chitinozoan genera *Cyathochitina* and *Rhabdochitina*. The base of the *C. protocalix* Zone is defined by the FAD of the index species in association with *Sagenachitina oblonga*, *Conochitina pirum*, *Conochitina havliceki*, and graptolites of the *sinodentatus* Zone (= *clavus* Zone). The association of *S. oblonga*, *C. pirum*, and *C. havliceki* is also documented from the graptolite-based *clavus* Zone to the lower part of the *austrodentatus* Zone (Chen et al. 1996). The *C. calix* Zone characterizes the interval from the graptolite-based upper part of the *clavus* to the lower part of the *austrodentatus* zones. Most of the chitinozoans (e.g., *R. usitata*, *R. turgida*) occurring in the *C. calix* Zone range up to the *C. jenkinsi* Zone, except the 5 species originating in the underlying *protocalix* Zone. The *jenkinsi* Zone has a diversified chitinozoan assemblage with 15 documented species (Wang and Chen 1994). The base of the zone is defined by the first appearance of the index species in association with graptolites of the upper part of the *austrodentatus* Zone. The FAD of *C. jenkinsi* therefore occurs earlier in China. Its top is below the base of the *Lagenochitina shizipuensis* Zone (Chen and Wang 1996). In the Yangtze platform graptolites from the *sinodentatus* to the *austrodentatus* zones (*TS*.3b–4a, upper Dawan Formation) are considered to be contained in Transgressive System Track (TST) deposits of Darriwilian age (Wang et al. 1996). The overlying Guniutan Formation comprises highstand regressive deposits and yields conodonts of the upper part of the *variabilis* Zone to the lower part of the *anserinus* Zone and chitinozoans of the *jenkinsi* Zone. The abundance and diversity of the chitinozoans in this zone decrease owing to the influence of sea level changes.

In *TS*.5a, after a period of extinction (e.g., six taxa disappeared in *TS*.4c) with very few originations (e.g., one FAD in *TS*.4c), new chitinozoan species appeared during *TS*.5a (BTD = 15). The event is characterized by the occurrence of chitinozoans in deep-water settings during the earliest Caradoc associated with the largest Ordovician transgressive episode across the platform areas of South China and Tarim. Chitinozoans of the *Eisenackitina uter* and *Lagenochitina deunffi* zones occur in association with graptolites of the *gracilis* Zone in condensed sections—the deposits associated with maximum flooding (Wang et al. 1996). A similar chitinozoan assemblage has been reported from the upper Sargan Formation of the Tarim Basin (Cai 1991; Zeng et al. 1996).

Following the regression, mid to late Caradoc (*TS*.5b–c) chitinozoan assemblages belonging to the *hirsuta, tanvillensis,* and *robusta* zones are characterized by high abundance and low diversity (BTD respectively of 13.5 and 11 in *TS*.5b and *TS*.5c). These assemblages are known from the Tarim Basin (Cai 1991; Zeng et al. 1996) and are correlated to the graptolite-based *americanus* and *quadrimucronatus* zones (Wang et al. 1996).

The progressive lowering of the chitinozoan diversity registered from the Darriwilian continued up to the Ashgill (e.g., BTD = 5 in *TS*.6a) (figure 28.5). During the Ashgill the Yangtze region was transformed into a semi-isolated basin following a sea level drop caused by the glaciation in North and Central Gondwana. The deposition of large amounts of organic matter generated an anoxic environment, and the accompanying change of salinity brought about a decrease in the abundance and diversity of chitinozoans in the Yangtze Basin. Only a few species belonging to the *Ancyrochitina merga* and *Tanuchitina fusiformis* zones have been found in the black shales of western Hubei, Jiangsu, and Guizhou (Qian and Geng 1989). Associated graptolites suggest that they belong to *TS*.6b. Deposits equivalent of the Wufeng Formation are absent or, at best, incomplete in Tarim with no Ashgill chitinozoans identified.

Siberia (NS, OO)

The Siberian paleoplate was located at low latitudes during the Ordovician (figure 28.1). The Ordovician chitinozoan assemblages briefly reported here are from the Siberian Platform and the Altai Mountains.

Siberian Platform

Chitinozoan assemblages (15 productive samples) are known from six stratigraphic levels in terrigenous sediments of Arenig to Ashgill age (Obut and Zaslavskaya 1980; Zaslavskaya 1982, 1984). Almost all the recorded species are long-ranging. At the present stage of investigation, no regional chitinozoan zonation is available. The recorded species are difficult to correlate with the chronostratigraphic subdivisions adopted for this volume, as the Ordovician rocks from the Siberian Platform have yielded very few graptolites and most reported conodont assemblages are endemic.

Gorny Altai

The Altai Mountains (Gorny Altai) belong to the western part of the Altai-Sayan folded area and form the southwestern margin of the Siberian Platform. Samples were collected from terrigenous strata in the northern, northeastern, and central parts of the Gorny Altai. Several chitinozoan assemblages (17 productive samples) were described (Zaslavskaya et al. 1978; Obut and Zaslavskaya 1980; Zaslavskaya and Obut 1984). Revision of the graptolites and chitinozoans allowed recognition of four regional zones from the Arenig to the early Ashgill: (1) *Conochitina raymondi,* (2) *Conochitina parvicolla,* (3) *Cyathochitina calix,* and (4) *Lagenochitina dalbyensis–Desmochitina lecaniella.* Only the chitinozoan assemblages occurring with graptolites are mentioned here. Two additional chitinozoans are reported from the Ordovician of the Gorny Altai: (1) *Laufeldochitina* aff. *stentor* from Arenig strata in association with graptolites of the *densus* Zone (Zaslavskaya et al. 1978; Petrunina et al. 1984); and (2) *Lagenochitina* aff. *deunffi* from upper Arenig strata of the graptolite-based *hirundo* Zone (Zaslavskaya and Obut 1984). Because *Laufeldochitina stentor* and *Lagenochitina deunffi* are respectively the index species of a late Darriwilian–early Caradoc zone in Baltica (Nõlvak and Grahn 1993) and of an early Caradoc zone in North Gondwana, the two Siberian forms might yet prove to be new species. The data concerning the Ordovician chitinozoans from Siberia are still too sparse to develop a meaningful biodiversity curve.

■ Global Biodiversity

A global database was assembled from the different regional data sets. The same generic names were adopted, but the taxa in open nomenclature were omitted. Additional adjustments were necessary in documenting the total range of some species (e.g., the

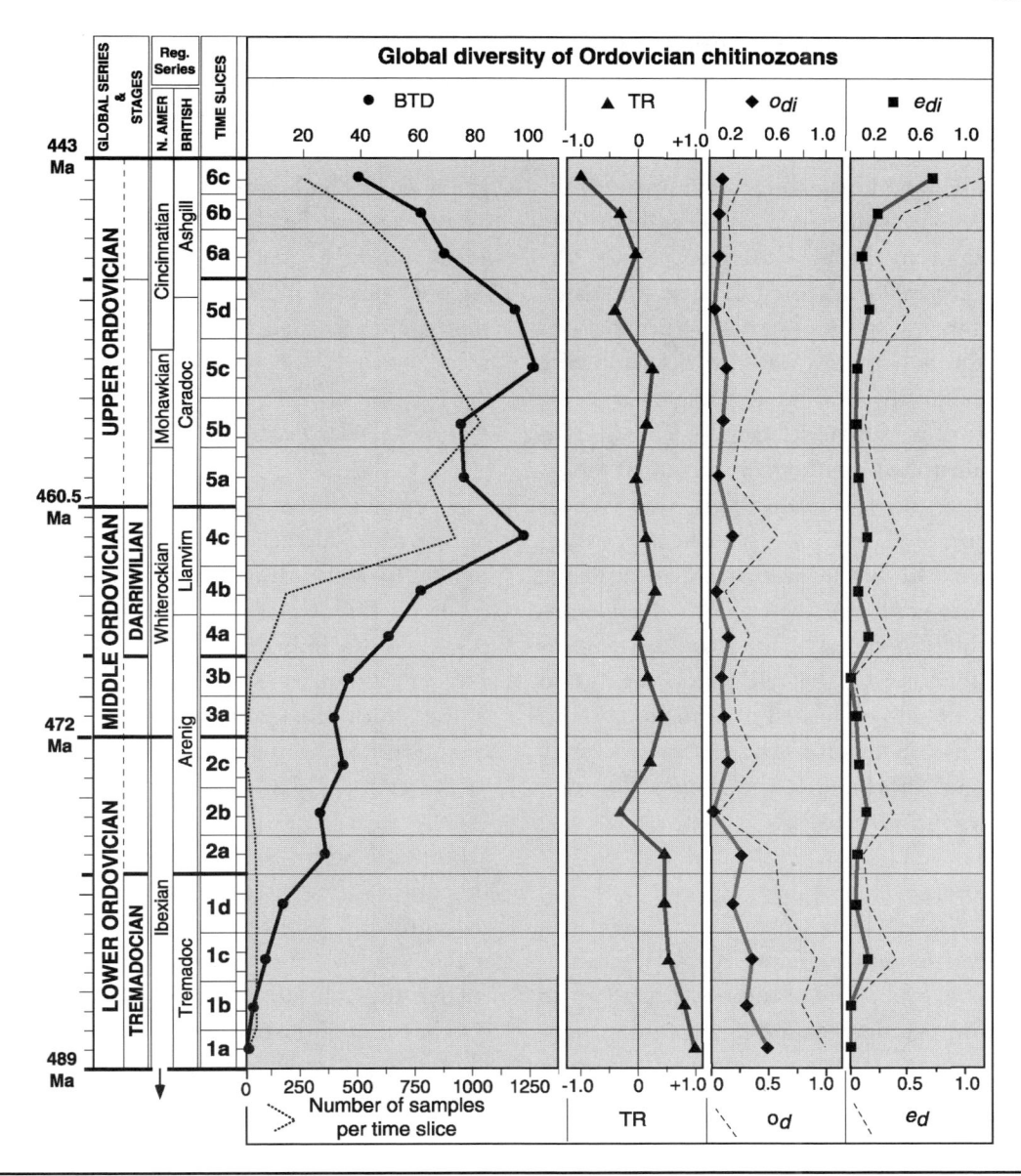

FIGURE 28.6. Global biodiversity of Ordovician chitinozoan species (excluding species in open nomenclature) and major turnover. For diversity measure abbreviations and time-slice correlations, see figure 28.2.

earliest FAD and the latest LAD were used to show the total range of each species).

Of the 287 species compiled in the global database, only 2 percent are distributed across all five paleocontinents, 6 percent are distributed across four paleoplates, 9 percent are represented in three paleoplates, and 19.5 percent are found in two paleoplates. The remaining species (62.7 percent) are reported only from one paleoplate. It is apparent that any temporary input from another paleoplate increased the diversity (e.g., Caradoc assemblages from Avalonia). As with the regional chitinozoan diversity data, several curves have been plotted at a global scale (figure 28.6), that is, BTD graph, TR graph, and o_{di} and e_{di} curves. These global curves differ from their regional counterparts, suggesting that regional factors may also have been important in the diversification of chitinozoans.

The global BTD curve is depicted as a continuous diversification of the group from the early Tremadocian to a first peak in the late Darriwilian (BTD = 96 in TS.4c) and then a short-lived lowering of diversity with a loss of about 25 percent of the BTD in TS.5a and 5b. A second maximum of diversification followed in the late Caradoc (BTD = 101 in *TS*.5c) and then decreased progressively until the end of the Ordovician (BTD = 38 in *TS*.6c).

The TR provides additional information suggesting brief pulses of extinction in the early Arenig (*TS*.2b) and again in the late Caradoc (*TS*.5d). The latter pulse is better documented because it is based on a larger number of species records (11 originations against 48 extinctions). However, the most dramatic extinction event (only 7 originations against 46 extinctions) occurs in the latest Ashgill (*TS*.6c). This event is well illustrated by the steep decline of the e_{di} curve. In contrast, no significant origination event is registered by the o_{di} curve on a global scale, except during the initial Early Ordovician radiation of the group. The various plots in figure 28.6 clearly confirm that the high biodiversity levels are not systematically synchronous with the origination events and low diversity is not always related to extinction events. Each of these parameters must be shown separately in order to depict more clearly the biodiversification patterns of the chitinozoans.

■ Conclusions

Biodiversity appraisal requires reliable identifications, and this is probably the most difficult step to achieve because taxonomic studies of chitinozoans have been so neglected recently. Previous attempts to evaluate the chitinozoan diversification (Paris 1999; Paris and Nõlvak 1999) have shown that the size of the data set for each time slice and the duration of the time slices control the shape of the biodiversity curves. Large data sets increase the probability of recording rare taxa, and long time intervals increase the cumulative effect of very short ranging species. In order to be aware of the size of the data set per time slice, a curve illustrating the number of available samples is drawn parallel with the regional diversity curve (figures 28.2, 28.3, 28.4, 28.6). Because they have roughly the same duration (i.e., 2–3 m.y.), the adopted time slices (chapter 2) avoid this duration bias. As a consequence, the o_{di} and e_{di} curves integrating the duration of the time slices produce curves that are less irregular than those of the origination (o_d) and extinction (e_d) rates (figures 28.2, 28.3, 28.4, 28.6). Long-ranging taxa, even if they are suspect on taxonomic grounds, and even if they require, at times, the introduction of virtual occurrences between their FADs and LADs, do not affect the graphs significantly.

The differences observed between the regional and global curves highlight the existence of various levels of changes in the diversity of Ordovician chitinozoans. The first-order trends are registered in both the regional and the global curves, whereas second-order trends are recorded only regionally, in one or two paleoplates.

First-Order Features

The major feature of the chitinozoan diversification during the Ordovician is the radiation beginning in the Tremadocian and developing rather regularly to the latest Darriwilian (figures 28.2–28.6). This radiation was probably driven by such intrinsic factors as a high evolutionary potential (plasticity of the genome of the "chitinozoan animals"), as revealed by the numerous innovations recorded during the early history of the group (Paris et al. 1999). The Ordovician is usually regarded as predominantly represented by a greenhouse state (Berner 1990). The radiation was probably favored by stable climatic conditions during the Early and Mid Ordovician and a large variety of niches available for occupation of pelagic organisms such as the "chitinozoan animals."

The onset of the chitinozoan diversity decline during the Late Ordovician (mainly in the Ashgill), associated with the dramatic extinction event in the Hirnantian (*TS*.6c), also represents a first-order change. The Hirnantian crisis is contemporaneous with a glaciation of the first magnitude (chapter 9; Brenchley and Marshall 1999 and references therein) and thus is most probably linked to a drastic fall in temperature and drop in sea level. This relationship is confirmed by the first evidence of chitinozoan recovery in the topmost Hirnantian, just after the melting of the major part of the African ice cap (Paris et al. 2000a).

The explanation for the progressive global lowering of chitinozoan diversity through the Late Ordovician (from *TS*.5d onward) is less clear, but it could be linked to increased volcanic activity or raised carbon dioxide levels as possible causes for this global change. It was not an intrinsic factor for the chitinozoans because the group flourished again in the Silurian, especially in the Pridoli, reaching an even higher diversity than in the Mid or Late Ordovician (Paris and Nõlvak 1999).

Second-Order Features

These features have been reported previously from different paleoplates. In eastern Laurentia, the increase in the chitinozoan diversity culminated in the middle part of the Caradoc (*TS*.5c) and is related to tectonic activity (e.g., the Taconic Orogeny, which was responsible for the deepening of the Appalachian foreland basin). Large marine transgressions are also believed to have increased chitinozoan diversity, probably because they favor argillaceous sedimentation in the more distal environment (i.e., better preservation and fuller registration of the chitinozoans). This is shown by the Darriwilian transgression in China and, to a lesser extent, by the mid Darriwilian (base of the Llanvirn) transgression in North Gondwana. However, the largest Ordovician transgression at the beginning of the graptolite-based *gracilis* Zone is not accompanied by numerous originations in China. In North Gondwana, the positive balance between origination and extinction in *TS*.5a is probably linked to external factors. For example, from *TS*.5b to 5d, Baltoscandian taxa probably drifted in the anticlockwise oceanic gyre of the southern oceans to higher latitudes, which were not yet disrupted or modified by the docking of Avalonia with Baltica during the Ashgill. The few meteorite effects documented in the Ordovician are not likely to have had any significant effect on patterns of chitinozoan diversity (Grahn et al. 1996).

ACKNOWLEDGMENTS

The authors are grateful to John Riva, Barry Webby, and Theresa Winchester-Seeto for the improvement of the text and to Merrell Miller, who reviewed the manuscript. We thank Kathleen Lauzière for her support in database management and Denis Lavoie for his constructive advice (Laurentia). We also acknowledge the help of A. Ancilletta, B. Billiaert, P. De Geest, J. Vanmeirhaeghe (Avalonia), M. Mechin (Laurentia), S. Al-Hajri, A. Bourahrouh, and L. Oulebsir (Gondwana) for use of their unpublished data. The project was supported by IGCP 410, NSER Canada (AA and EA), STINT and Uppsala University (JS), Estonian Science Foundation grant no. 4674 (JN), and the "CRISEVOLE" project of the French CNRS (FP).

29 Conodonts: Lower to Middle Ordovician Record

Guillermo L. Albanesi and Stig M. Bergström

Conodonts, eel-shaped animals that were common inhabitants of Paleozoic and Triassic seas, are now interpreted to be chordates (Aldridge et al. 1993). Recent phylogenetic analyses based on morphological, biochemical, and physiological characters suggest conodonts to be the most plesiomorphic member of the group Gnathostomata (Donoghue et al. 2000). Their apatitic feeding microelements are usually well preserved, and conodonts have a fossil record whose completeness competes with that of any other animal group (Foote and Sepkoski 1999). The excellent fossil record of conodonts and their rapid evolution make them key tools for establishing high-resolution biostratigraphy from the Middle Cambrian through the Triassic.

Application of advances in systematics methodology to conodonts shows them to exhibit important attributes of vertebrate phylogenetic paleontology. Coupled with the virtues of micropaleontology (with a strong tradition in stratophenetics; see Gingerich 1990), this particular fossil group has now become involved in the discussions of the importance of morphology and stratigraphy in the resolution of relationships in order to work out superior phylogenetic models (Dzik 1999; Donoghue 2001). With recent advances in the understanding of the phylogeny of the group and the access to a database that in its detail is now unrivaled among Paleozoic and Triassic fossil groups, we are now in the position to achieve a much improved understanding of many aspects of conodonts and the environments in which they thrived. This will assist us in a variety of phylogenetic, paleoecologic, paleobiogeographic, paleogeographic, and paleoceanographic interpretations (Purnell 2001; Sweet and Donoghue 2001).

Analyses of biodiversity, viewed as the number and variability of genes, species, and communities in space and time along with resulting processes and patterns of diversification in marine ecosystems, have always been problematic and uncertain given the incomplete preservation of fossils, inadequate sampling, and arbitrary taxonomy (Sepkoski 1997; Conway Morris 1998a). Conodonts are not an exception to this generalization; however, after the pioneer attempts to reveal general patterns of diversification (Clark et al. 1981; Aldridge 1988; Sweet 1988b; Barnes et al. 1996), modern taxonomy and much more extensive occurrence records now permit detailed studies of diversity changes in particular chronostratigraphic intervals, as exemplified by the present analysis of the Lower and lower Middle Ordovician conodont diversity.

■ Stratigraphic Framework and Taxonomic Database

We analyze a database that consists of the stratigraphic ranges of Lower and lower Middle Ordovician conodonts from well-documented localities in

Laurentia, Baltica, the Gondwanan margin in South America, China, and Australia. Sections at selected localities record a continuous succession of biostratigraphic units, which represent diverse platform and slope environments of different paleobiogeographic nature. Most published and unpublished conodont data from these localities represent recent studies. Older significant contributions whose conodont systematics needed revision were updated using conodont multielement taxonomy. Despite our efforts, many taxa previously classified *sensu formae* or in open nomenclature had to be excluded from the conodont database and remain as residual taxa. The database also is constrained by the number of selected localities and by the observed taxon ranges as a confident estimate of real duration of a taxon (Hayek and Bura 2001). It is important to note that, except for a few localities with biostratigraphically well-controlled information, Ordovician conodont data from Asia and Australia are incomplete and some faunas could not be revised based on the published information. Consequently, our database represents mainly regions surrounding the Iapetus Ocean (Williams et al. 1995). However, because of the broad range of environments represented by the large number of study localities in both the Midcontinent and Atlantic realms (see tables 29.1, 29.2), we believe that the present database is extensive enough to cover all the major paleoecologic environments and to serve as a reasonably reliable basis for a first assessment of the global diversity patterns exhibited by Lower and lower Middle Ordovician conodont faunas.

Our database includes 430 species representing 106 genera that are present in the stratigraphic interval from the *Paltodus deltifer* through the *Eoplacognathus suecicus* zones of the Atlantic realm, that is, from the middle part of the Tremadocian Stage of the Lower Ordovician to the upper-middle part of the Middle Ordovician. The conodonts of the lowest part of the Lower Ordovician are not considered because of the incompleteness of the information available on a global scale and because of the poorly understood relations between euconodonts and paraconodonts. We start this study at the onset of a major diversification of conodont lineages that occurred just after the existence of the lowest Ordovician *Iapetognathus-Cordylodus* faunas (Nicoll 1992; Nicoll et al. 1999; Dubinina 2000), that is, from the *manitouensis-deltifer* interval, where a stock of taxa gave rise to a main plexus of adaptive Ordovician conodont communities (Miller 1984; Sweet 1988b). This diversification of conodont genera is an early phase of the Ordovician Radiation (cf. Miller 1997c), associated with biogeographic differentiation into the Midcontinent and Atlantic conodont realms (Barnes et al. 1973; Bergström 1990; Pohler and Barnes 1990).

Stratigraphic ranges of taxa in this study span parts of Boucot's (1983) Ecologic-Evolutionary Units (EEU) 3 and 4, whose limit coincides with the North American Sauk/Tippecanoe sequence boundary, which has been proposed as a level for the base of the global Middle Ordovician Series (Webby 1998; Finney and Ethington 2000). The time intervals for global analysis of conodont faunas used herein correspond to those of previously defined biozones within these EEUs (figures 29.1, 29.2). Analyzed intervals represent six biostratigraphic intervals, which constitute single biozones in the Lower Ordovician of the Midcontinent and Atlantic faunal realms, or combined two to three successive biozones in the lower Middle Ordovician of these realms. Combinations of these global biostratigraphic intervals of 3–5 million years' (m.y.) duration are considered equivalent to Ecologic-Evolutionary Subunits (EESs), biomeres, or community groups. They include a large number of time-successive communities that are unique at the species level (Boucot 1983, 1990). Because of correlation difficulties and the way the conodont succession has been described, it proved impractical to use the time slice (*TS*) subdivisions referred to elsewhere in this book directly for our analysis, but the approximate correlation between the time slices and our biostratigraphic intervals is as follows (figures 29.1, 29.2): *manitouensis-deltifer* interval, late *TS*.1a through *TS*.1b to early *TS*.1c; *deltatus-proteus* interval, late *TS*.1c through *TS*.1d to early *TS*.2a; *communis-elegans* interval, middle to late *TS*.2a and middle to late *TS*.2b; *andinus-evae* interval, late *TS*.2b through *TS*.2c; *laevis-norrlandicus* interval, *TS*.3a–b and early *TS*.4a; and *sinuosa-suecicus* interval, late *TS*.4a through *TS*.4b to early *TS*.4c.

Biodiversity can be evaluated by describing patterns of stability and change in marine ecosystems as they are structured on long (about 10 m.y.) timescales and defined as "coordinated stasis" (Brett and Baird 1995). These patterns involve intervals of stability

TABLE 29.1. Recent Contributions on Lower Ordovician Conodont Records from Selected Localities Worldwide

Locality	Lithostratigraphic Unit/Section	Reference	Biostratigraphic Interval	Zone
Precordillera (Cuyania Terrane), western Argentina	San Jorge Formation	Bergström, unpublished collection	M-D	*Rossodus manitouensis*
	La Silla Formation	Lehnert 1995a; Albanesi et al. 1998	M-D D-P	*R. manitouensis, Paltodus deltifer* *Paroistodus proteus*
	San Juan Formation Ponón Trehué Formation	Serpagli 1974; Lehnert 1995a; Albanesi et al. 1998; Lehnert et al. 1998	D-P C-E A-E	*P. proteus* *Priondious elegans* *Oepikodus evae, O. intermedius*
Northwestern Argentina	Volcancito Formation, Famatina	Albanesi et al. 2000	M-D	*Paltodus deltifer*
	Suri Formation, Famatina	Albanesi and Astini 2000a	A-E	*O. evae, O. intermedius*
	Santa Victoria Group, Eastern Cordillera	Rao et al. 1994; Rao and Flores 1997; Albanesi et al. 2001	M-D D-P A-E	*R. manitouensis* *Acodus deltatus, P. proteus* *O. evae*
Baltoscandian Region	Lava River Section, western Russia	Tolmacheva 2001	M-D D-P C-E A-E	*P. deltifer* *P. proteus* *P. elegans* *O. evae*
	Talubäcken Section, Sweden	Bergström 1988	D-P C-E A-E	*P. proteus* *P. elegans* *O. evae*
	Furuhäll Section, central Öland, Sweden	Bagnoli et al. 1988	M-D D-P C-E A-E	*P. deltifer* *P. proteus* *P. elegans* *O. evae*
	Horns Udde sections, northern Öland, Sweden	Bagnoli and Stouge 1997	A-E	*O. evae, Trapezognathus diprion, Microzarkodina* n. sp. A
	Brattefors, Västergötland, Sweden	Löfgren 1997	M-D	*P. deltifer*
	Hunneberg, Västergötland, Sweden	Löfgren 1993a	M-D D-P C-E	*P. deltifer* *P. proteus* *P. elegans*
	Siljan District, Sweden	Löfgren 1994	M-D D-P C-E A-E	*P. deltifer* *P. proteus* *P. elegans* *O. evae*
North America	Marathon Limestone, western Texas	Izold 1993	M-D D-P C-E A-E	*R. manitouensis* *A. deltatus* *P. elegans* *O. evae*
	El Paso Formation, western Texas, New Mexico	Repetski 1982; Izold 1993	M-D D-P C-E A-E	*R. manitouensis, Colaptoconus quadraplicatus* *A. deltatus* *Fahraeusodus marathonensis, Oepikodus communis* *Jumudontus gananda/Reutterodus andinus*
	Ouachita Mountains, Arkansas and Oklahoma	Repetski and Ethington 1977; Repetski and Ethington in Stone et al. 1994	M-D D-P A-E	*R. manitouensis* *P. proteus* *O. evae*
	Green Point Formation, Cow Head Group, St. Pauls Inlet Section, Newfoundland	Johnston and Barnes 1999, 2000	D-P C-E A-E	*Paracordylodus gracilis* *P. elegans* *O. evae*
	The Ledge Point of Head Section, Cow Head Group, Newfoundland	Stouge and Bagnoli 1988	M-D D-P C-E A-E	*Prioniodus gilberti* *Prioniodus adami, P. oepiki,* *P. elegans* *O. evae*
	Watts Bight, Boat Harbour, Catoche formations, St. George Group, Newfoundland	Ji and Barnes 1994b	M-D D-P C-E A-E	*Cordylodus angulatus* *Drepanoistodus nowlani, Macerodus dianae* *A. deltatus, Acodus? primus* *O. communis, Protoprioniodus simplicissimus*

Locality	Lithostratigraphic Unit/Section	Reference	Biostratigraphic Interval	Zone
North America	The Ibexian Series Composite Stratotype Section and adjacent strata, House-Confusion Range Area, west-central Utah	Ross et al. 1997	M-D D-P C-E A-E	*R. manitouensis* *Macerodus dianae, A. deltatus – Stultodontus costatus,* *O. communis* *Reutterodus andinus*
	Ninemile Formation, Whiterock Narrows Section, Monitor Range, Nevada	Finney and Ethington 2000	A-E	*R. andinus*
	Little Falls and Tribes Hill formations, eastern New York	Landing et al. 1996	M-D	*R. manitouensis*
	Cape Weber Formation, Ella Section, East Greenland	Smith 1991	M-D D-P C-E	*R. manitouensis* Fauna D (middle-upper part) *O. communis*
	Manitou Formation, Colorado	Seo and Ethington 1993	M-D	*R. manitouensis*
Australia	Allochthonous limestones Hensleigh Siltstone, New South Wales	Webby et al. 2000	C-E	*P. elegans*

Note: The database used for calculated diversity values and measures was updated following most recent revisions on multielement taxonomy for every taxon.

when species associations maintain their relative abundance with little change in composition at the species level. These intervals may be bounded by short periods of ecologic and evolutionary turnover (<10 percent of the duration of the stable times) that are characterized by high levels of extinction, speciation, and immigration (less than 20 percent of indigenous species persisted from one EES to the next). Considerable attention has been paid to "coordinated stasis" in regional studies by paleoecologists because it involves widespread patterns across time and space, but the concept has been mostly applied to level-bottom organisms (for instance, Ivany and Schopf 1996; Patzkowsky and Holland 1997). Nevertheless, doubts have been expressed regarding its validity on a global scale, the changes of EESs probably being related to changes in the basinwide lithologic suite (Miller 1997d; Budd and Johnson 2001).

The present study on conodont diversity applies commonly used measures and rates of taxonomic diversity for Lower and lower Middle Ordovician biostratigraphic intervals at the species and genus levels. It currently is not possible to test our results on the development of nektobenthic-pelagic conodont communities using direct comparisons with published analyses—not only because of the absence of previous analyses of this type at the species level for this time span but also because published case studies deal with benthic fossil groups, whose ecologic-evolutionary mechanisms may differ significantly from those of conodonts. Our attempt to illustrate conodont diversity patterns on a global scale is viewed by us as an initiation to future, more extensive studies on the pattern of diversification in space and time of this important fossil group.

■ Diversity Patterns and Turnover Rates

Patterns of conodont global diversity throughout the *manitouensis-deltifer* (M-D), *deltatus-proteus* (D-P), *communis-elegans* (C-E), *andinus-evae* (A-E), *laevis-norrlandicus* (L-N), and *sinuosa-suecicus* (S-S) biostratigraphic intervals (spanning a total of about 25 m.y.) are here interpreted on the basis of assessment of total diversity (d_{tot}), normalized diversity (d_{norm}), and turnover rates—that is, species originations-appearances (o_i) and extinctions-disappearances (e_i) per m.y., per taxon rates of origination (o_{di}) and extinction (e_{di}) per m.y., percentage of species and genera carryover (C), percentage of species and genera holdover (H), and species-to-genus ratios (sp./gen.) (figures 29.1, 29.2, 29.3; table 29.3). All these metrics have been widely used in paleoecologic studies, and their advantages or particular problems have been discussed elsewhere (see, for instance, Sepkoski and Koch 1996; Cooper 1999c; Foote 2000a, 2001).

TABLE 29.2. Recent Contributions on Middle Ordovician Conodont Records from Selected Localities Worldwide

Locality	Lithostratigraphic Unit/Section	Reference	Biostratigraphic Interval	Zone
Precordillera (Cuyania Terrane), western Argentina	San Juan, Gualcamayo, Los Azules, Yerba Loca, La Invernada, Las Aguaditas, and Las Chacritas formations	Lehnert 1995a, Albanesi et al. 1995b, 1998, Ottone et al. 1999, Albanesi and Astini 2000b	L-N S-S	*Tripodus laevis, Baltoniodus navis, Microzarkodina parva* *Lenodus variabilis, Eoplacognathus suecicus*
Famatina, western Argentina	Suri Formation, Famatina	Albanesi and Vaccari 1994	L-N	*B. navis*
Baltoscandian Region	Mäekalda, North Estonia Lava River S., eastern Russia	Viira et al. 2001 Tolmacheva 2001	L-N L-N	*B. navis, B. norrlandicus* *B. triangularis, B. norrlandicus*
	Sjurberg, Siljan District, central Sweden	Löfgren 1994	L-N	*B. triangularis, B. navis*
	Jämtland, central Sweden	Löfgren 1978, 1993b	L-N S-S	*B. triangularis, B. navis* *L. variabilis, E. suecicus*
	Gillberga, northern Öland, Sweden	Löfgren 2000	L-N S-S	*B. navis, B. norrlandicus* *L. variabilis, Yangtzeplacognathus crassus, Eoplacognathus pseudoplanus*
	Hagudden, northern Öland, Sweden	Stouge and Bagnoli 1990	L-N S-S	*B. navis, M. parva* *L. variabilis*
	Horns Udde sections, northern Öland, Sweden	Bagnoli and Stouge 1997	L-N	*B. triangularis, B. norrlandicus*
	Huk Formation, Slemmestad, southern Norway	Rasmussen 1991	L-N S-S	*B. navis, M. parva* *L. variabilis*
	Scandinavian Caledonides, Norway-Sweden	Rasmussen 2001	L-N	*Microzarkodina flabellum, B. norrlandicus*
	Mójcza Limestone, Poland	Dzik 1994a	S-S	*L. variabilis*
North America	Saint George Group, western Newfoundland	Ji and Barnes 1994b	L-N	*Pteracontiodus cryptodens*
	Saint Pauls Inlet Section, western Newfoundland	Johnston and Barnes 1999	L-N	*T. laevis*
	Table Head Formation, western Newfoundland	Stouge 1984	S-S	*L. variabilis, E. suecicus*
	Fort Peña Formation, Marathon Area, Texas	Bradshaw 1969	S-S	*L. variabilis, E. suecicus*
	Whiterock Narrows Section, Monitor Range, Roberts Mountains, Nevada	Finney and Ethington 2000	L-N S-S	*T. laevis* *Histiodella sinuosa*
	Antelope Valley Limestone, Eagan Range, Martin Ridge–Monitor Range, Copper Mountain–March Spring–Ikes Canyon, Groom Range, Test Site, Meiklejohn Peak sections, Nevada	Harris et al. 1979	S-S	*L. variabilis, E. suecicus*
	Ibex Area, Utah	Ethington and Clark 1981, Ross et al. 1997	L-N S-S	*T. laevis* *H. sinuosa, H. holodentata*
Australia	Amadeus Basin, central Australia	Cooper 1981	L-N	*B. triangularis, B. navis*
	Canning Basin, western Australia	Watson 1988	S-S	*L. variabilis, E. suecicus*
China	East-central China	Wang and Bergström 1999	L-N	*Paroistodus originalis, B. norrlandicus*
	South-central China	Zhang 1998	S-S	*L. variabilis, E. suecicus*

Note: The database used for calculated diversity values and measures was updated following most recent revisions on multielement taxonomy for every taxon.

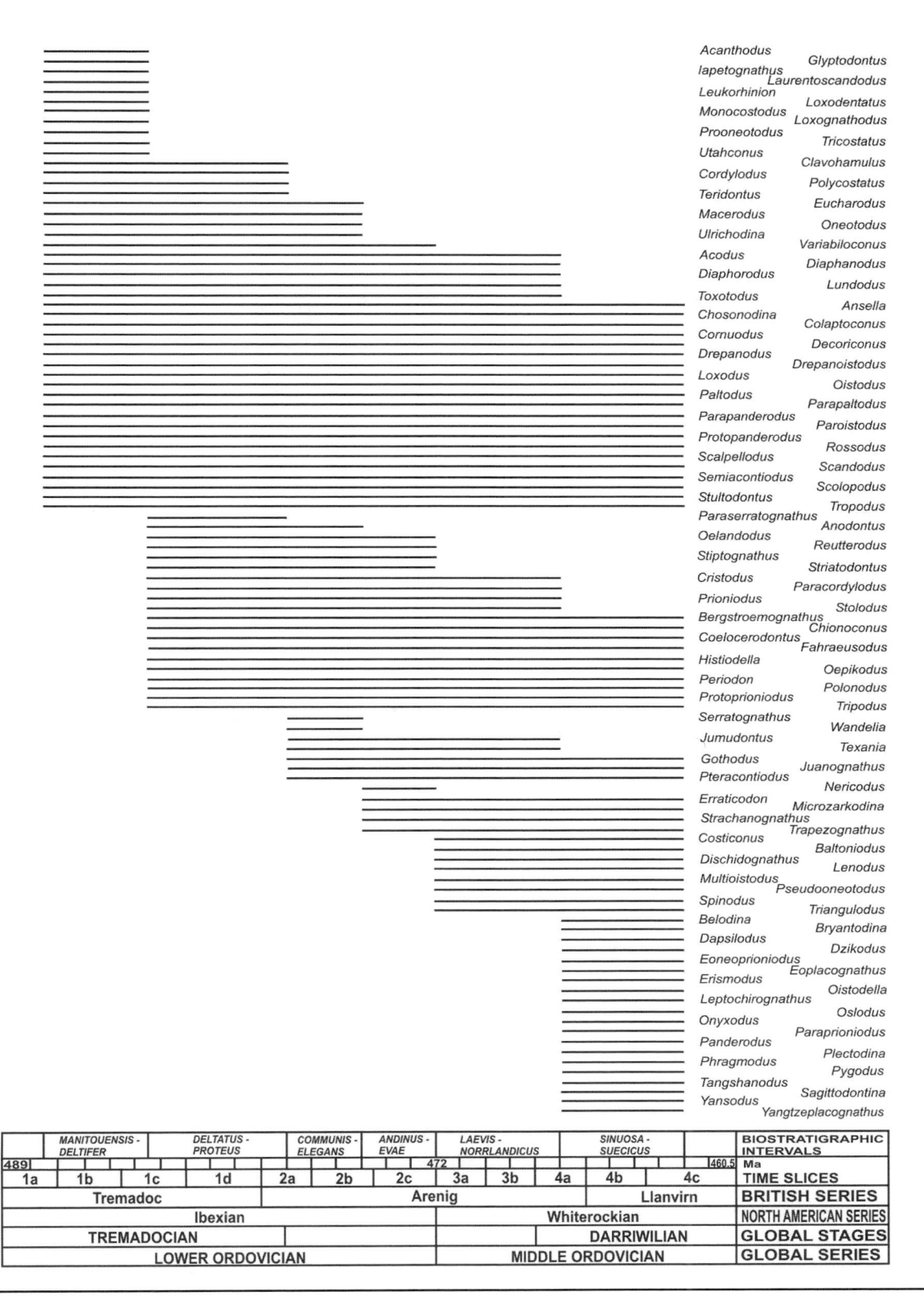

FIGURE 29.1. Stratigraphic ranges of 106 conodont genera based on the occurrence of 430 species within the study interval. Gaps in ranges of genera at particular localities due to the absence of a particular facies are not illustrated, and the ranges shown are based on occurrences at diverse localities with appropriate facies. However, because of the imprecise nature of much published distribution information combined with some regional correlation problems, we have chosen to show first and last appearances of these genera as coinciding with boundaries of biostratigraphic intervals, thereby unfortunately creating sharp, partly artificial faunal breaks at these levels.

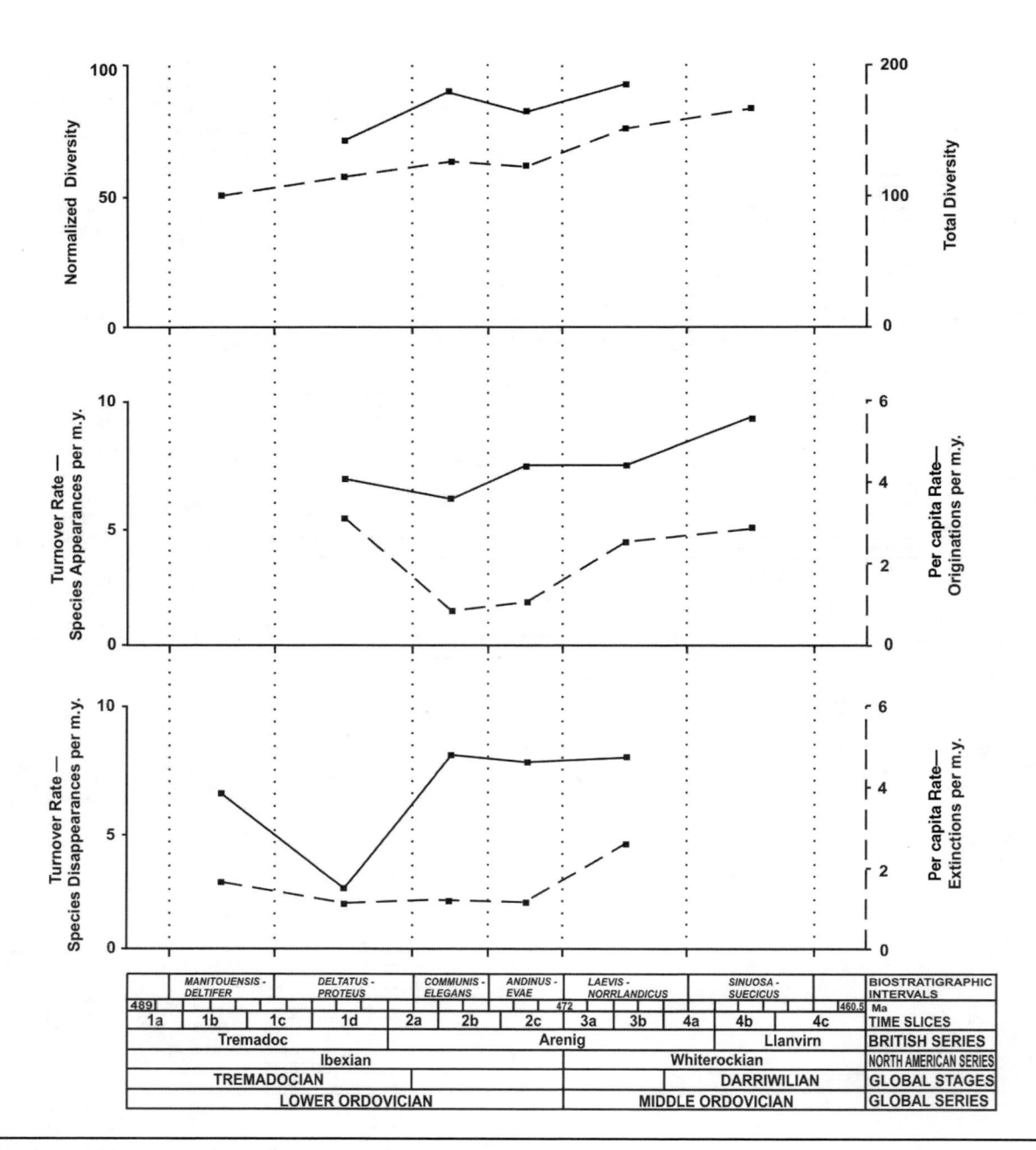

FIGURE 29.2. Three plots of species diversity and rate measures for each biostratigraphic interval (values are shown in table 29.1). In the upper plot the total diversity (d_{tot}) refers to the number of species per m.y., and the normalized diversity (d_{norm}) is the number of species ranging through a time unit (from the preceding to the following time unit) plus half the number of species ranging beyond the time unit but originating or ending within it, plus half the number of species confined to the time unit—that is: ($d_{norm} = n_r + 0.5\ n_{ex} + 0.5\ n_t$). In the middle and lower plots, the turnover rates represent the number of appearances (o_i) and disappearances (e_i) of species per m.y. through the biostratigraphic interval, and the per capita rates are expressed as (1) origination (o_{di}): number of species confined to the interval + number of species that cross the top boundary only / total number of species / duration of interval—that is: $(n_{fl} + n_{ft})/(n_{tot})/i$; and (2) extinction ($e_{di}$): number of species confined to the interval + number of species that cross the bottom boundary only / total number of species / duration of interval—that is: $(n_{fl} + n_{bl})/(n_{tot})/i$. No estimates of errors in turnover rates and interval duration have been incorporated.

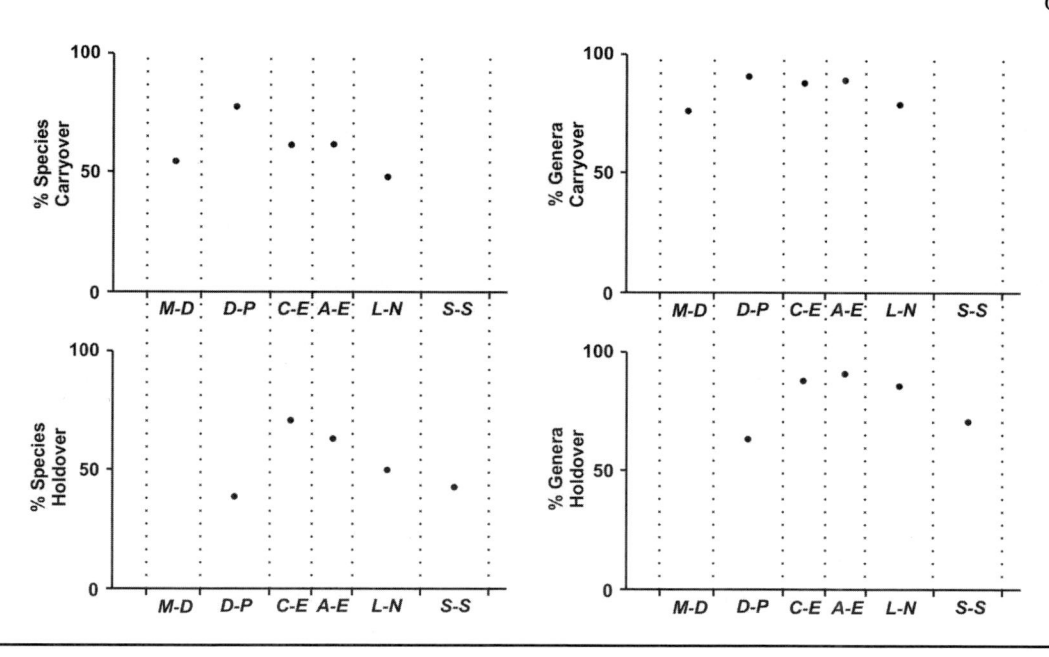

FIGURE 29.3. Plots of holdover (lower row) and carryover (upper row) measurements for species- and genus-level data for each biostratigraphic interval. Percentage of taxa (species or genera) holdover (H): percentage of all taxa within a biostratigraphic interval that are also present in the preceding biostratigraphic interval. Percentage of taxa (species or genera) carryover (C): percentage of all taxa within a biostratigraphic interval that are also present in the subsequent biostratigraphic interval. No error estimates for calculated measurements have been incorporated. Abbreviations of biostratigraphic intervals correspond to those units discussed in the text.

Genus and Species Diversity

The overall diversity pattern of conodont genera (figure 29.1, table 29.3) is similar to the pattern of total diversity (d_{tot}) (figure 29.2 and see later in this chapter), suggesting that taxonomic practice has not significantly distorted the species-level trend. The trajectory of diversity at the genus level, where taxonomic assignments are more stable, may provide a test of potential taxonomic oversplitting. Only the L-N interval does not show a proportional increase in the number of genera, which could be attributed to a bias in taxonomic treatment in successive biostratigraphic intervals. Generic total diversity ranges from a low of 46 genera in the M-D interval to a high of 65 genera in the S-S interval. Intermediate biostratigraphic intervals display an almost constant number of 56–58 genera.

Conodont species diversity (figure 29.2, table 29.3) is relatively high already in the M-D interval and constantly increases through the studied time interval. Total diversity (d_{tot}) exhibits a first peak in the C-E interval with 126 species and reaches a maximum of 167 species in the S-S interval. At the end of the Early Ordovician total diversity (d_{tot}) is 122 in the A-E interval. The diversity fluctuates within the Early Ordovician between a low of 101 and a high of 126 species. A notable rapid increase of diversity occurs in

TABLE 29.3. Calculated Generic and Specific Diversity Values and Measures Derived from the Compiled Database of Tables 29.1, 29.2

Biostratigraphic Interval	Duration m.y.	d_{tot}	d_{norm}	o_i	e_i	o_{di}	e_{di}	C	H	d_{tot}	C	H	Species/Genus Rate
M-D	4.0	101	-	-	6.4	-	1.60	55	-	46	76	-	2.20
D-P	5.5	115	72.5	7.0	2.5	3.04	1.10	78	39	56	91	63	2.05
C-E	3.0	126	92	6.2	8.0	0.90	1.14	62	71	58	88	88	2.17
A-E	3.0	122	85	7.5	7.7	1.11	1.13	62	63	56	89	91	2.18
L-N	5.0	151	95.5	7.5	7.9	2.50	2.60	48	50	58	79	86	2.60
S-S	5.0	167	-	9.4	-	2.80	-	-	43	65	-	71	2.57

Note: Abbreviations of biostratigraphic intervals correspond to those units discussed in the text. Abbreviations of rate measures are shown in figure 29.2. Additional abbreviations: C, carryover; H, holdover. Calculations of total diversity refer to the number of species and genera, respectively, per biostratigraphic interval.

the early Mid Ordovician (L-N interval), since there is a diversity difference of more than 30 species in that interval compared with that of the latest Early Ordovician (A-E interval).

The formula used for normalized diversity (d_{norm}) significantly reduces the values of total diversity (d_{tot}) from a low of 72.5 species in the D-P interval to a high of 95.5 in the L-N interval. These sensitive lower values probably represent a more meaningful estimate of global diversity patterns (cf. Cooper 1999c). However, there are several potential artifacts obscuring the diversity patterns, such as the effects of taxonomic oversplitting or lumping, of sibling species, of Lazarus taxa (cf. Hughes and Labandeira 1995), and of the Penelope effect (see Stouge and Bagnoli 1988). Estimates of these effects are beyond the scope of the present study.

The diversity pattern exhibits a constant increase in the number of species, as might be expected in the global Ordovician radiation event, though with a noticeable drop at the end of the Early Ordovician. This was followed by a recovery of the faunas from the beginning of the Mid Ordovician onward.

Species Turnover Rates

The value of species appearances per m.y. (o_i) is relatively high in the D-P interval, with 7 species, and except for the slight decline in the next C-E interval to 6.2 species, it continues to increase up to a maximum of 9.4 species in the S-S interval. A similar pattern is displayed by the per capita species origination rate (o_{di}), although lower numbers of taxa per m.y. are shown by the curve. In particular, drops of the trajectory within the C-E and A-E intervals are even more marked than those in the appearance curve. Species disappearance (e_i) shows a considerable value (6.4) in the M-D interval, but the curve is punctuated by a low trough (2.5) in the D-P interval that coincides with the high appearance value in same interval. This curve follows a trajectory of high values (8, 7.7, 7.9) of disappearances through the C-E, A-E, and L-N intervals. The curve of per capita rate of species extinction per m.y. (e_{di}) displays a lower value in the D-P interval compared with a high level for the per capita origination rate. Extinction trends through the next three intervals maintain values exceeding those for originations. This tendency changes in the S-S interval, as expected based on the high values shown by the appearance-origination rates.

The general trends of both sets of curves reflect the significant origination pulse in the D-P interval, which is followed by a decline of the conodont faunas in the late Early Ordovician (but also the major and relatively abrupt disappearance of faunas at the end of the M-D interval, which is masked by the averaging techniques that were used). The emergence of a new cycle of conodont evolution is shown by the marked radiation of taxa in the early Mid Ordovician (Sweet 1988b).

Our assembled species data show that origination and extinction peaks are closely linked in time (L-N interval), indicating that there are long intervals of relatively low background turnover that are interrupted by pulses of relatively high turnover. This pattern is apparently showing how identified biostratigraphic intervals are assembled together to behave as the EESs of EEUs 3 and 4 of Boucot (1983, 1990). The M-D interval could be distinguished as the first EES 3.1, followed by the second EES 3.2 that consists of the D-P, C-E, and A-E intervals. The third EES 4.1 is composed of the L-N and S-S intervals (figure 29.2). Species/genus ratios assessed for each biostratigraphic interval (table 29.3) show two extreme values, 2.05 and 2.60 at D-P and L-N intervals, respectively, which differ notably from the other intervals exhibiting values that lie between these limits.

In order to quantify faunal stability and amount of turnover between Silurian-Devonian EESs, Brett and Baird (1995) used "coordinated stasis" measures. Percentage of holdover (proportion of species and genera in an EES that also occur in the previous EES) and percentage of carryover (proportion of taxa within an EES that also occur in the subsequent EES) are measures applied here for each biostratigraphic interval. Relatively low values (<40 percent) of these measures characterize, and assist to test, "coordinated stasis" (Brett et al. 1996; Patzkowsky and Holland 1997). Our calculations of H and C in the case of both species and genera (figure 29.3, table 29.3), are persistently high (>40 percent) for every biostratigraphic interval. This shows a pattern for selected units that is not consistent with "coordinated stasis." Instead, trends of turnover rates suggest the possibility of defining EES as species assemblages in biostratigraphic intervals, such as those identified in the previous paragraph.

Biofacies Patterns and Faunal Change

Lower Ordovician Record

Manitouensis-Deltifer *(M-D) Interval (Late* TS.*1a Through* TS.*1b to Early* TS.*1c)*

The M-D interval (lower to middle Tremadocian) is characterized by a considerable diversity (figure 29.2), and a few forms usually dominate in different lithofacies. The *Rossodus manitouensis* and *Paltodus deltifer* zones represent this interval in the Midcontinent and Atlantic faunal realms, respectively. In addition to the zonal denominators, several other short-ranging taxa are found in particular biofacies within both realms. An example of this is *Variabiloconus transiapeticus*, which may be used to establish a reliable correlation between Laurentia and Baltica for this interval (Löfgren et al. 1999). *Colaptoconus* and *Parapanderodus* are typical genera in the warm and shallow, and eventually hypersaline, waters of the Midcontinent realm. In the Argentine Precordillera these genera are present in the restricted carbonate platform environments of the La Silla Formation (Lehnert 1995a; Albanesi 1998a), which were similar to those in modern shelf lagoons. It is a shallow subtidal, open-platform lithofacies without land-derived siliciclastics (Cañas 1999). Comparable lithofacies are common along the eastern margin of Laurentia. Hence, the peritidal-subtidal lithofacies of the Boat Harbour Formation and the lower part of the Catoche Formation in western Newfoundland, contain the *Glyptoconus* (= *Colaptoconus*)–*Stultodontus* community (Ji and Barnes 1994a). Further, *Parapanderodus* and *Colaptoconus* dominate the *Scolopodus* (= *Parapanderodus*) community recognized by Fortey and Barnes (1977) in the Nordporten Member of the Kirtonryggen Formation in Spitsbergen. The lower part of Fauna D by Ethington and Clark (1971, 1981) and the "Low Diversity Interval" of Ross et al. (1997), which includes components of these communities, are particularly well represented in the intertidal-subtidal deposits of western North America (Ethington and Repetski 1984; Izold 1993). In the less restricted environments of Baltoscandia, the M-D faunas can be quite diverse (cf. Löfgren 1996, 1997).

Third-order eustatic cyclicity overprinted on the second-order eustatic oscillation that represents the Cambrian-Ordovician sea level rise (Vail et al. 1977) is well recognized in diverse environments surrounding the Iapetus Ocean (Bagnoli 1994). The discussed M-D interval is part of a first cycle up to the boundary between the *Paltodus deltifer* and *Paroistodus proteus* zones, where a global regression occurs (Fortey 1984; Erdtmann 1986). It is represented in deeper-water facies by oxidized sediments, for instance, in the Cow Head Group of western Newfoundland (Stouge and Bagnoli 1988; Williams et al. 1994), and by a hiatus at numerous other localities in North America (Ross and Ross 1995). In Australia, Nicoll et al. (1992) recognized the hiatus as the "Kelly Creek Eustatic Event," which correlates with the "Ceratopyge Regressive Event" (CRE) defined by Erdtmann (1986) in Scandinavia (see also Nielsen, figure 10.2). The boundary has also been identified on the Russian platform (Dronov and Holmer 1999; Dubinina 2000) and in South America (Erdtmann 1995; Albanesi 1998a; Moya et al. 1998). Severe extinctions and faunal turnover in the middle Tremadocian, which are particularly well documented in the conodont record (Ethington et al. 1987; Ji and Barnes 1993, 1994a), are apparently related to this eustatic event. Incidentally, this important marine regression has been used as a criterion for the definition of a major intrasystemic chronostratigraphic boundary (Erdtmann and Maletz 1995). However, it currently is preferred practice to use index fossils showing significant morphological innovations promoted by the subsequent transgressive cycle (Maletz et al. 1996) for the definition of significant chronostratigraphic boundaries.

Deltatus-Proteus *(D-P) Interval (Late* TS.*1c Through* TS.*1d to Early* TS.*2a)*

A second eustatic cycle followed in the late Tremadocian with a major transgression, but it commonly is not represented by preserved sediments on the Laurentian and Baltoscandian platforms, where an extensive hiatus is developed over large areas (Bagnoli 1994). This eustatic cycle 2 is characterized in slope facies by the acme of *Prioniodus adami* in western Newfoundland (Stouge and Bagnoli 1988); *Acodus deltatus* and *Paroistodus proteus* are often associated in the platform facies of Laurentia, Baltica, the Gondwanan margin, and the Argentine Precordillera. This relationship allows for long-distance intercontinental correlation (Ethington and Clark 1971; Lindström 1971; Lehnert 1995b; Albanesi et al. 2001).

A significant change in the Midcontinent conodont faunas was recognized by Ethington and Clark (1971, 1981) and involved an abrupt shift of communities, which was apparently induced by eustatic pulses (see Ethington et al. 1987). The dominant Fauna C (*Rossodus manitouensis* Zone of Landing et al. 1986 and of Ross et al. 1997), which is composed of *R. manitouensis, Acanthodus lineatus, Polycostatus oneotensis,* and *Variabiloconus bassleri,* is replaced, after an interval of a low diversity fauna, by the radiation of the upper Fauna D. This new fauna consists of several characteristic species, such as *Colaptoconus bolites, C. floweri,* and *Macerodus dianae.* Ji and Barnes (1993, 1994a) carried out a detailed evaluation of this extinction-radiation event in the Boat Harbour Formation (St. George Group) in western Newfoundland. There, about 30 species of major lineages disappear, providing space for a subsequent radiation of hyaline coniform species of the Drepanoistodontidae and Oistodontidae, and taxa with complex apparatus plans developed in new lineages, such as the Prioniodontidae.

A less extensive faunal change has been documented in Baltoscandia, where the association of conodonts of the *Paltodus deltifer* Zone (Löfgren 1997) is replaced by a conodont fauna that includes widespread species of deep/cold environments, such as those belonging to *Oelandodus* and *Paracordylodus* (Bagnoli et al. 1988; Löfgren 1993a; Tolmacheva and Löfgren 2001; Tolmacheva et al. 2001b).

In the Argentine Precordillera, the transgressive event after the "Ceratopyge Regressive Event" starts the deposition of the open-platform lithotopes of the San Juan Formation (Keller et al. 1994; Cañas 1999). The uppermost strata of the La Silla Formation gradually incorporates new taxa up to the level of the sudden appearance of a complex biofacies at the base of the San Juan Formation. This includes pandemic forms such as species of *Paroistodus, Acodus, Tropodus, Drepanodus, Paltodus, Scolopodus, Protopanderodus,* and *Drepanoistodus,* as well as typical representatives of the deep/cold water region, for instance, species of *Paracordylodus, Oelandodus, Prioniodus, Periodon,* and temperate ecophenotypic forms such as species of *Bergstroemognathus, Reutterodus, Diaphorodus,* and *Juanognathus* (sensu Bagnoli and Stouge 1991), which characterize the lower portion of this formation.

Communis-Elegans *(C-E) Interval (Middle to Late TS.2a and Middle to Late TS.2b)*

Fauna E in North America was defined by Ethington and Clark (1971) as an association of coniform species along with ramiform elements assigned to *Oepikodus communis,* which apparently inhabited open-platform environments. Records of diverse species associations that include *Acodus deltatus, Bergstroemognathus extensus, Juanognathus variabilis, Oepikodus evae, Paracordylodus gracilis, Paroistodus parallelus, Protopanderodus gradatus, P. leonardii, "Polonodus" corbatoi, Prioniodus elegans, Reutterodus andinus, Rossodus barnesi,* and *Tropodus australis* were documented from recurrent shallow subtidal facies or relatively deeper platforms (Ethington and Clark 1981; Repetski 1982; Izold 1993; Ross et al. 1997). Pohler (1994) distinguished the biofacies represented by *O. communis* and *Prioniodus-Texania* as characteristic of open-platform and platform-margin environments, respectively, for the time span corresponding to the *P. elegans* Zone in the Cow Head Group of western Newfoundland. Equivalent biofacies occur in the San Juan Formation of the Argentine Precordillera, although significantly more endemic species are present (Lehnert 1995a; Albanesi 1998a). The litho-skeletal limestone facies indicates a well-oxygenated, open subtidal, low-energy environment within the photic zone (Cañas 1999). Deeper-water conodont communities thriving apparently below the permanent thermocline were common in diverse types of basins in central Asia (Dubinina 1998). These faunas are exclusively composed of Baltic taxa (Bergström 1988) and show close relations to those of the Atlantic realm. However, apparent broad dispersal of such pelagic taxa could reflect the presence of cryptic species and hidden ecologic and genetic diversity that may actually involve numerous biologic species. This may have significant macroevolutionary implications (cf. Norris 2000).

Andinus-Evae *(A-E) Interval (Late TS.2b and TS.2c)*

The upper part of the *Oepikodus communis* Zone (late Ibexian in North American) as defined in particular areas of Laurentia (e.g., Greenland; Smith 1991) spans the range of the *Reutterodus andinus* and *Oepikodus evae* zones of more widely used biostratigraphic schemes (for instance, Lindström 1971; Pohler and Orchard 1990; Ross et al. 1997; Albanesi et al. 1998; Tolmacheva 2001; Pyle and Barnes 2002). These include communities inhabiting shallow-subtidal to relatively deep-water environments (Ethington and Repetski 1984; Izold 1993). The biofacies described by Pohler (1994) and Johnston and Barnes (1999) from the *O. evae* Zone of the Cow Head Group, western Newfoundland, are composed of conodont species associations that are similar to those present in the San Juan Formation, Argentine Precordillera (Serpagli 1974), in terms of both the species association and relative frequency of taxa. Generic associations, which are defined as the *Oepikodus-Periodon* and *Oepikodus* biofacies, characterize deeper-water environments of the platform margin or upper slope there, as in many localities in Baltoscandia (Lindström 1971; van Wamel 1974; Bagnoli et al. 1988). These biofacies, when compared with the taxonomic structure of the *O. evae* Zone in the relatively shallower environment of the Precordillera, are mainly represented by different numerical proportions of the same genera. The biofacies composition is also shown by the *Prioniodus* (= *Oepikodus*) Community defined by Fortey and Barnes (1977), based on collections from the lower part of the Olenidsletta Member, Valhallfonna Formation, Spitsbergen. This represents an inner platform lithofacies. *Periodon*-rich faunas dominate in the eastern part of Baltica (Tolmacheva 2001; Viira et al. 2001). The biofacies that characterizes the *O. evae* Zone in the Precordillera is typical of open-platform environments, where a transgressive systems tract ends and a highstand systems tract is established, as is also the case in other parts of the world (Nicoll et al. 1992; Nielsen 1992a; Ross and Ross 1995; Dronov and Holmer 1999; Cañas 1999; see also chapter 10).

The *Periodon-Texania* biofacies in the lower part of the Aguathuna Formation, St. George Group, western Newfoundland, represents the shelf margin environment (Pohler 1994). This biofacies includes an association of *Periodon*, *Protoprioniodus*, and *Texania* that also occurs in the *Oepikodus intermedius* Zone of the Precordilleran San Juan Formation. Pohler (1994) interprets the faunal change in the upper part of Bed 11 of the Cow Head Group, which represents the establishment of this biofacies, as a possible response to a significant regression on the carbonate platform. During this event, *O. intermedius* migrated from the platform edge to the upper-slope biofacies, while *Periodon* species occupied the vacant ecospace. In the latest Ibexian, several taxa that were common in older communities disappeared, such as *Bergstroemognathus, Reutterodus,* and *Tropodus,* as is also the case in the Argentine Precordillera (Albanesi 1998a) and the Gondwanan margin (Rao et al. 1994; Albanesi and Astini 2000a), where siliciclastic and volcaniclastic facies were introduced locally.

Middle Ordovician Record

Laevis-Norrlandicus *(L-N) Interval (TS.3a–b and Early TS.4a)*

The interval of the *Tripodus laevis* through *Baltoniodus norrlandicus* zones occurs in the lower part of the nodular limestone facies of the San Juan Formation in the Argentine Precordillera. At the base of this interval a transitional sequence of gray-green glauconitic-skeletal limestones is interpreted as representing the beginning of a new transgressive systems tract. At this stage a slow rise of sea level allowed the establishment of reefal structures in the central-eastern part of the basin (Carrera and Cañas 1996; Keller 1999). Under this eustatic influence, typical Laurentian conodonts, such as *Pteracontiodus cryptodens* and *Histiodella* species, invaded the Precordilleran sea. At the same time, taxa of deep/cold environments, for example, *Baltoniodus* and *Trapezognathus,* also began to appear in the deepest environments of the northern Precordillera (Albanesi et al. 1999).

In the North American Ibexian-Whiterockian boundary interval, important bioevents (Berry 1995; Sepkoski 1995; Droser and Sheehan 1997b) are related to a pronounced global marine regression (Barnes 1984; Fortey 1984; Erdtmann 1986; Nielsen 1992a; Ross and Ross 1995). Profound extinction of numerous marine invertebrates was followed by complex faunal shifts beginning in the early Whiterockian (Benedetto et al. 1999; Li and Droser 1999). The appearance of the *Isograptus* fauna is recognized as a new cosmopolitan graptolitic biofacies (Cocks and Fortey

1990), and the conodonts exhibit a new cycle of species evolution (Sweet 1988b).

A great similarity exists in the species replacement patterns of conodont faunas across the Ibexian-Whiterockian boundary in western North America (Ethington and Repetski 1984; Finney and Ethington 2000) and in the Precordillera (Lehnert 1995a; Albanesi et al. 1999). Many widespread conodont species became extinct just below or near the boundary, for instance, *Bergstroemognathus extensus, Juanognathus variabilis, Oepikodus intermedius, Reutterodus andinus, Scolopodus krummi, Stolodus stola,* and *Tropodus australis.* Several long-ranging species, such as *Cornuodus longibasis, Drepanodus arcuatus, Paroistodus originalis, Periodon flabellum,* and *Protopanderodus gradatus,* are represented as continuing across the boundary of both areas. Additionally, some characteristic species, such as *Erraticodon balticus, Juanognathus jaanussoni, Jumudontus gananda, Histiodella altifrons, Microzarkodina flabellum, Parapaltodus simplicissimus, Periodon gladysi, Protoprioniodus aranda,* and *Spinodus spinatus,* appeared nearly simultaneously in Laurentia and in the Precordillera. Deep/cold water taxa of the Atlantic realm (e.g., species of *Baltoniodus, Gothodus,* and *Trapezognathus*) like those represented in the Gondwanan margin of northwestern Argentina also appeared in the Precordillera. Although specimens of *Baltoniodus* and *Microzarkodina* are scarce in the latter regions, in Baltoscandia they often occur in mud rocks or grainstones (Lindström 1984), which suggest particular ecologic conditions. Löfgren (1995, 1996, 2000), Rasmussen and Stouge (1995), Bagnoli and Stouge (1997), Tolmacheva and Federov (2001), and Viira et al. (2001) have described and quantified the diverse epicontinental-shelf faunas from Baltica in this interval.

Sinuosa-Suecicus *(S-S) Interval (Late* TS.*4a Through* TS.*4b to Early* TS.*4c)*

The most common taxa in the *Lenodus variabilis* Zone of the Argentine Precordillera are species of *Periodon, Paroistodus,* and *Protopanderodus.* They are best known in the Atlantic realm (Lindström 1971, 1984; Stouge and Bagnoli 1990; Rasmussen and Stouge 1995; Bagnoli and Stouge 1996; Löfgren 2000) or within platform-margin deeper-water environments elsewhere (Rasmussen 1998). A species association that represents the typical distal carbonate-terrigenous facies in Laurentia and includes *Fahraeusodus, Microzarkodina,* and *Pteracontiodus* (Ethington and Repetski 1984) comprises about 7 percent of the components of this biofacies in the Precordillera (Albanesi 1998a). The biofacies is present in the lower member of the Gualcamayo Formation within a transgressive systems tract that includes depositional environments ranging from distal carbonate ramp to black shale basin (Astini 1994) associated with a global sea level rise (Barnes 1984; Fortey 1984; Erdtmann 1986), eruption of volcanic ashes (Huff et al. 1998), and local active tectonism (Keller 1999). Such distal platform to proximal slope facies of ancient continental margins, which are not commonly preserved, are critical for the correlation of coeval platform-slope sequences and for interpreting the interplay between different faunal regions and the thermocline near the shelf edge (Pohler et al. 1987). A microevolutionary event involving the *Paroistodus* lineage is registered with unusual detail in this particular environment, in the Argentine Precordillera (Albanesi and Barnes 2000). Complex paleoecologic relationships, such as that observed by Stouge (1984) for the *Periodon-Cordylodus?* (= *Paroistodus*) biofacies, where the frequency of *Periodon* significantly increases in association with *Paroistodus,* could be explained by differences in population dynamics in relation to sea level changes within these transitional environments (Bagnoli and Stouge 1996).

Occurrences of *Periodon, Paroistodus, Protopanderodus, Drepanoistodus, Ansella,* and *Walliserodus* (= *Costiconus*) in the lower member of the Gualcamayo Formation (Albanesi 1998a) constitute a typical biofacies that represents a deep-water upper-slope environment according to the model proposed by Stouge (1984) based on the Table Head succession in western Newfoundland. Other abundant species of *Drepanodus, Rossodus, Pteracontiodus,* and *Fahraeusodus* also preferred deeper-water environments in the Precordillera. *Histiodella, Erraticodon,* and *Lenodus* are also significant components of this biofacies. Some species of *Histiodella* are abundant in the deep subtidal facies of Laurentia (Ethington and Repetski 1984) and not uncommon in deeper facies in Baltica (Rasmussen 2001). *Erraticodon* is a common taxon, but it is not restricted to particular biotopes of both realms (Cooper 1981; Albanesi and Vaccari 1994; Wang and

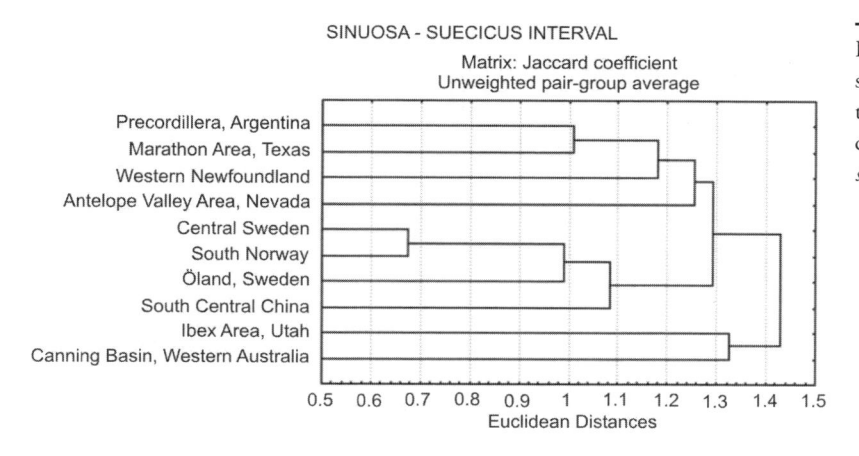

FIGURE 29.4. Dendrograph of cluster analysis showing conodont paleobiogeographic relationships during the highest peak of conodont diversity in the Ordovician Period (*Histiodella sinuosa–Eoplacognathus suecicus* Interval).

Bergström 1999; Rasmussen 2001). *Lenodus* is typical of deeper/cold environments in mixed carbonate-terrigenous facies (Barnes and Fåhraeus 1975; Sarmiento 1985), but it is also present in rather pure limestones (Dzik 1994a; Löfgren 2000).

In the Argentine Precordillera, the upper part of the S-S interval represents different biofacies of deep-ramp or slope environments, as in the carbonate sequence of the Las Chacritas Formation, or the black shales of the Gualcamayo and Los Azules formations (Ottone et al. 1999; Peralta et al. 1999; Albanesi and Astini 2000b; Ortega and Albanesi 2000). The *Periodon, Polonodus, Pygodus,* and *Dzikodus* association inhabiting a lower-slope to open-ocean zone has a low faunal diversity. These taxa are typical representatives of the Atlantic realm, being common in the Baltoscandian region (Löfgren 1978, 2000; Bergström 1983; Stouge and Bagnoli 1990; Bagnoli and Stouge 1997; Rasmussen 2001), in deeper-water continental margins facing the Iapetus Ocean (Bradshaw 1969; Bergström and Carnes 1976; Stouge 1984; Rasmussen 1998), in marginal areas of the Ordos basin of North China (An and Zheng 1990), and in the Yangtze region of south-central China (Zhang 1998). In the same S-S interval, the Midcontinent realm was characterized by endemic taxa in shallow-water facies, such as those from Laurentia (Ethington and Clark 1981), Australia (Watson 1988), and Korea (Lee and Lee 1986) (figure 29.4).

■ Conclusions

The refined taxonomy and the extensive occurrence records now available have allowed a detailed analysis of conodont diversity changes at the species level during a major part of the Ordovician Period. Because of the wide variety of environments covered by biostratigraphic studies and the particularly extensive record of conodont faunas available from areas surrounding the Iapetus Ocean, the present study has been centered on this region, but the results are believed to apply adequately to more general diversity patterns of conodonts worldwide during Early Ordovician to early Mid Ordovician time.

Diversity patterns at the species level indicate constant additions of conodont species through time as may be expected in the global Ordovician radiation event. A conspicuous drop in conodont diversity in the late Early Ordovician was followed by a significant radiation of conodont lineages, as is the case in other fossil groups, from the beginning of the Mid Ordovician.

An important origination pulse occurred in the late Tremadocian (D-P interval), just after the "low diversity interval" coinciding with the global "Ceratopyge Regressive Event." The initiation of a new and very important cycle in conodont evolution, after the late Ibexian demise and subsequent turnover in the conodont faunas, is shown by the marked radiation of taxa in the early Mid Ordovician (L-N and S-S intervals).

Despite the fact that some errors in diversity calculations cannot be avoided owing to the incompleteness of the fossil record and to different taxonomic interpretations, the pattern of diversity changes appears to be related directly to major eustatic sea level oscillations, which might explain why these principal changes seem to be on a global scale. Overprinting of regional trends on the replacement of conodont communities may have been due to multiple factors, including the effects of paleogeographic, paleoceanographic, paleoclimatic, and/or local paleoecologic changes.

ACKNOWLEDGMENTS

G. L. Albanesi acknowledges continued support for conodont reseach by CONICET and, in particular, thanks the Fulbright Commission (United States) and Fundación Antorchas (Argentina) for grants that enabled him to stay for several months in 2001–2002 at the the Ohio State University, Columbus, Ohio, where most of the present study was carried out as a part of a major conodont research project. He also thanks Dr. G. Ortega for her assistance during the final preparation of this work. F. Zeballo helped with graphic designs. S. M. Bergström is indebted to the Department of Geological Sciences, the Ohio State University, for making its laboratories and other facilities available for this research. Helpful comments on the manuscript were provided by Drs. R. L. Ethington, A. Löfgren, and Yongyi Zhen.

30 Vertebrates (Agnathans and Gnathostomes)

Susan Turner, Alain Blieck, and Godfrey S. Nowlan

Although Cambrian-Ordovician vertebrate occurrences have been repeatedly claimed, confirmed taxa bearing mineralized tissues with such a histomorphology are still relatively rare. Cambrian records are few and controversial. Shu et al. (1999a), for example, identified possible Lower Cambrian vertebrates from South China, and Young et al. (1996) have described Late Cambrian remains from Australia that might represent the oldest definite vertebrate with hard tissues. The term "vertebrate," as used here, designates one of a group of animals, the Vertebrata, characterized by the occurrence of a longitudinal, dorsal skeletal rachis transformed into vertebrae, sometimes cartilaginous, bearing certain phosphatic hard tissues, and having a cranium or equivalent. First we shall clarify what we understand as "vertebrates," excluding the basal chordates reviewed by Blieck (1992) and Smith et al. (2001), which are not vertebrates. In addition, despite claims over the past decade, conodonts do not form key elements in Cambro-Ordovician vertebrate evolution (contra Sansom et al. 2001), especially when considering homology of early vertebrate mineralized tissues (see later in this chapter) with conodont tissues (e.g., Schultze 1996; Kemp 2002; Reif 2002). In the phylogenetic analysis of Donoghue et al. (2000), conodonts appear as basal vertebrates within the topology: [Petromyzontida (Conodonta (Pteraspidomorphi (Anaspida (Thelodonti (Gnathostomata (Osteostraci-Pituriaspida-Galeaspida))))))] (simplified after their figure 7A). However, especially in respect to the interpretation of their hard tissues, we consider the problem to be unsolved and assign conodonts to unresolved or questionable basal chordates.

By Late Cambrian to Ordovician time, vertebrates had developed a mineralized exoskeleton made of multilayered tissues. These hard tissues include the following:

1. *Bone* is formed by cells called osteoblasts and includes various mineralized tissues of internal, mesenchyme origin; sometimes it is acellular, but often it has enclosed bone cells; and aspidine is a particular type of bone without traces of bone cells.
2. *Dentine* is a mineralized tissue formed by cells called odontoblasts, which leave behind fine, simple, or complex tubuli, thought to be a derivative of the neural crest of embryos and typical (an apomorphic character) for vertebrates.
3. *Lamelline* is a special dentine or aspidinelike tissue with a lamellar structure.
4. *Enameloid* is a dense hypermineralized tissue found external to dentine that represents the initial cap of formation. True *enamel* is a hypermineralized, external tissue in the exoskeleton of higher vertebrates, characterized by transverse crystals under crossed nicols of a polarized microscope, and is still controversially claimed as observable in early vertebrates.

As Walcott (1892) concluded, true vertebrates (as well as the related conodont clade) probably had a Precambrian (Neoproterozoic) origin. This hypothesis has been supported by molecular analysis of amelogenin, the major protein of enamel (Delgado et al. 2001).

We divide the vertebrates into *agnathans,* or jawless vertebrates, an early, basal paraphyletic group; and *gnathostomes,* or jawed vertebrates, which are regarded as a monophyletic group (clade). The Ordovician vertebrate clades, though some are still poorly defined, comprise the agnathan groups of basal Pteraspidomorphi, basal Osteostraci?, and basal Thelodonti, and the gnathostome groups include the basal Placodermi?, Mongolepida?, basal Chondrichthyes, and Acanthodii. Many remains are fish microfossils, and generally they are less than 5 mm long. These include head bones or platelets, teeth, body scales, spines, branchial denticles, and lepidotrichia (synonyms: vertebrate microremains, "ichthyoliths," or "microvertebrates"). In the Cambro-Ordovician most are scales: single units (odontodes) or complex (odontodium) body denticles that form the covering (squamation) of the trunk, tail, and fins and sometimes internal surfaces of the mouth and branchial region. Another term used is "tessera," originally a small element of a mosaic, which is a bony dermal element of the head or body. The tesserae are not fused together at their bases.

No consensus has been reached on the phylogenetic relationships of early vertebrates (e.g., Gagnier 1995; Janvier 1996a, 1996b, 1998; Donoghue et al. 2000), and so it is preferable to separate the known taxa as follows: (1) the unnamed Late Cambrian species from Australia; (2) various pteraspidomorph taxa, *Astraspis* (including *Pycnaspis* Ørvig), *Eriptychius,* "*Tesakoviaspis*"; (3) Arandaspidiformes, *Arandaspis,* and *Sacabambaspis;* (4) thelodonts; (5) basal gnathostomes, that is, *Skiichthys, Areyongalepis, Apedolepis,* basal chondrichthyans, mongolepids, acanthodians; and (6) a purported anaspid from South Africa.

One further "problem" taxon is *Anatolepis* (*A. heintzi* Bockelie and Fortey 1976), which was first determined as "fragments of the earliest heterostracan" (Bockelie and Fortey 1976:38) based on its mineralogy (hydroxy?-apatite), its morphology (plates covered with small "scales"), and its histology (three-layered carapace with dentine? and "aspidine"?). Later, other Cambrian-Ordovician phosphatic remains were attributed to *Anatolepis* (Elliott et al. 1991; Blieck 1992), but the vertebrate affinity of *Anatolepis* was not considered well established because the macro- and microstructure seemed different from that of other Paleozoic vertebrates.

The debate has continued until recently (e.g., Ørvig 1989; Smith and Hall 1990). For example, Smith and Sansom (1995) argued that the external scutes of some *Anatolepis* consist of a form of "tubular dentine," although the nature of the basal lamellar tissue remains enigmatic (either similar to hyaloine of extant teleosts, or acellular bone—either aspidine? or lamelline of Karatajute-Talimaa and Predtechenskyj 1995). If, as traditionally accepted, dentine is a vertebrate characteristic (e.g., Smith and Hall 1990) and the tubular dentine tissue is verified, then *Anatolepis* should be a vertebrate. Recently, Smith et al. (1996) showed the presence of dentine tubules 0.5–1µm thick. Sansom et al. (2001) also press vertebrate claims for the type species (*A. heintzi*) from the late Early and early Mid Ordovician of Ny Friesland, Spitsbergen, and other material determined as *Anatolepis* sp. from Cambrian and Ordovician localities from around margins of Laurentia. Nevertheless, not all remains attributed to *Anatolepis,* including some of the type series of *A. heintzi,* belong to this taxon (Turner, Blieck, and Nowlan unpubl.).

In summary, the taxon "*Anatolepis*" seems to be represented by three different organisms: (1) a group reminiscent of vertebrates, for example, Nitecki et al. (1975: figure 3), Repetski (1980: figures e, f), and Smith et al. (1996: figures 1–3); (2) a group that is not vertebrate, including original *A. heintzi* Bockelie and Fortey (1976) and ?*Anatolepis* of Smith and Hall (1990); and (3) a group of doubtful affinity, for example, *Milaculum* of Nitecki et al. (1975: figure 5) or the "Ordovician vertebrate" of Wang and Zhu (1996, 1997) from the Darriwilian of Inner Mongolia (China), which has been compared to *Anatolepis* by Janvier (1998), but reattributed to arthropods by Sansom et al. (2000). All *Anatolepis* material is in need of revision. Only members of the first group are included in this biodiversity survey.

■ Fossil Record

Only limited biostratigraphic study of Cambrian and Ordovician vertebrate and presumed vertebrate remains has been undertaken. Owing to an upsurge of interest in microvertebrate work particularly in the past decade (e.g., Young 1997; Sansom et al. 2001), there has been a marked increase in claimed Ordovician records. Talimaa (2000) outlined a series of Late Ordovician to Early Silurian microvertebrates with biostratigraphic potential in the northern Russian territories. Young (in Shergold et al. 1991) was the first to recognize a sequence of microvertebrate assemblages in the Ordovician of central Australia. No detailed sequencing of North American deposits, despite their important status, has been realized. Work in other parts of the world has barely begun, but through the auspices of IGCP projects 406 and 410, more Ordovician to Early Silurian vertebrate finds have been made in the Circum-Arctic.

Discovery of soft-bodied vertebrates in the Lower Cambrian of South China has extended the range of vertebrates and their diversity (e.g., Shu et al. 1999a). However, the Chengjiang vertebrates do not throw much light as yet on relationships. Despite earlier claims by Repetski (1978), other verified Cambrian records based on phosphatic microremains are rare. Smith et al. (2001) reviewed the data including *Anatolepis* and conodonts, which they regard as vertebrates. However, for us, the discoveries of G. C. Young (1995) and Young et al. (1996) are the only confirmed Cambrian remains to help in assessing relationships with the Ordovician biodiversity patterns. The phosphatic remains from the Gola Beds of western Queensland (~500 Ma) show the complex multilayered structure containing canals and bearing external tubercles typical of later pteraspidomorphs; Karatajute-Talimaa (1997) regarded it as a predentinous tissue.

Vertebrates were first discovered in the Upper Ordovician (Caradoc) Harding Sandstone (Colorado) in the late nineteenth century, with continuing finds through the twentieth century, but often these remains were misunderstood, commonly relegated to the Problematica. This paradoxical situation has continued, with many finds provoking disagreement over taxonomic placement (e.g., Erdtmann et al. 2000; Sansom et al. 2001). This is due in part to relatively poor preservation and also lack of understanding of what constitutes an early vertebrate. Until recently, the Harding Sandstone provided the best fauna and virtually the only accepted vertebrate remains (e.g., Blieck et al. 1991; Elliott et al. 1991; Smith and Sansom 1997; Sansom et al. 2001). The formation is stratigraphically constrained within the conodont-based *Phragmodus undatus* Zone, time-slice boundary TS.5b–c, within the middle Mohawkian (chapter 2: figures 2.1, 2.2). Apart from this American fauna, purported vertebrates have been recorded from a variety of Ordovician deposits around the world (e.g., Smith et al. 2001; see also database in Turner, Blieck, and Nowlan unpubl.). Articulated remains are scarce and the histological framework even less well understood. Some work on the nature and evolution of vertebrate hard tissues is controversial (e.g., Schultze 1996). Significant paleontological discoveries of recent decades include articulated agnathans from North America (Elliott 1987; Sansom et al. 1997), South America (Gagnier et al. 1986; Gagnier 1987), and Australia (Ritchie and Gilbert-Tomlinson 1977; Ritchie 1985) and microremains from several countries (e.g., Smith and Sansom 1997; Young 1997; Karatajute-Talimaa 1998). Here, we review the published record and, for the first time, attempt to place all accepted vertebrate records into the highly resolved timescale with 19 time slices (see chapters 2 and 3) in order to assess the biodiversity trends (figure 30.1).

In some cases, we have not been able to plot all data using the scheme. This is due to various factors, such as a lack of precise biostratigraphic data for some vertebrate records because they derive from shallow-water facies, or in cases in which a close correlation with the pelagic conodont/graptolite/chitinozoan standard zonation is difficult to achieve. So we have used the data available, based on relevant IUGS stratigraphic charts for the countries involved. Where there is doubt on the position of a formation within a time slice, a thin vertical line is shown to identify the uncertainty (see figure 30.1). This provides for the first time a framework from which to discuss vertebrate diversity in the Ordovician. Later in this chapter we emphasize the originations and disappearances of taxa and clades, where resolved, in relation to the time slices, and we propose to distinguish an early Gondwanan Endemic Assemblage (GEA) from a later Laurentia-Baltica-Siberia Assemblage (LBSA) (figure 30.1).

Lower Ordovician: Tremadocian (TS.1a–d)

No good evidence of vertebrates is known from this time span, but records of *Anatolepis* are suspected; therefore they are plotted on figure 30.1 for the five different geographic locations clustered in

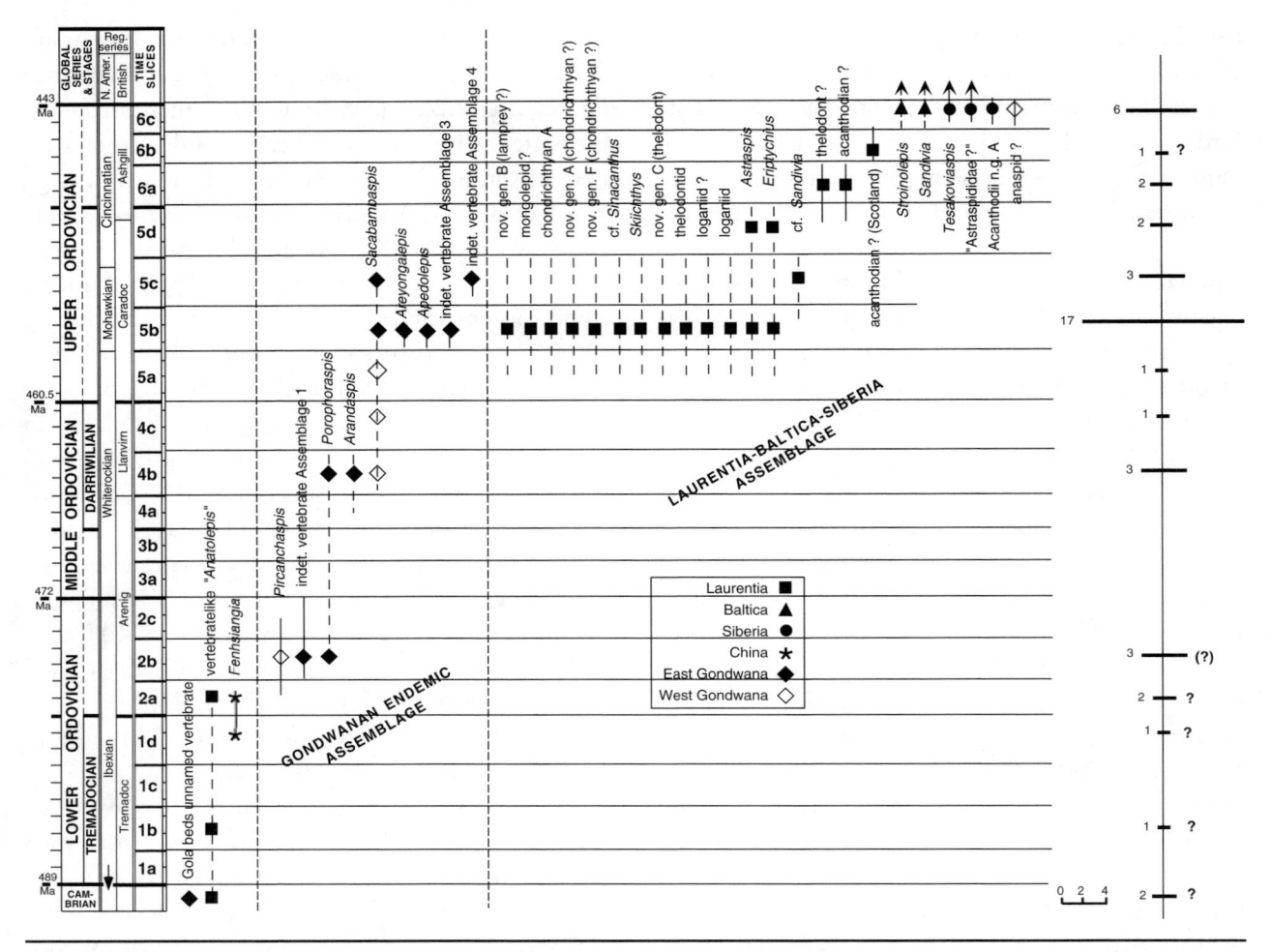

FIGURE 30.1. Stratigraphic range chart of latest Cambrian and Ordovician vertebrates and supposed vertebrates (identified by "?"). The thin vertical lines indicate uncertainty about stratigraphic distribution. The two main biogeographic assemblages are delineated. The horizontal bars on the right illustrate the total diversity of taxa (d_{tot} sensu Cooper, chapter 4) at generic level (some of them, e.g., *Astraspis* and *Sandivia*, corresponding to at least two different species). Note that taxa whose names have been kept in open nomenclature are considered at generic level for the calculation of diversity measures of table 30.1.

North America (see earlier discussion in this chapter). Long and Burrett's (1989) *Fenhsiangia,* which first appears in the Fenxiang Formation (upper Tremadocian *TS.*1d), Hupeh Province, Yangtze Platform, China, is the only other candidate, given its vertebratelike stellate tubercles.

Lower Ordovician (TS.2a–c)

Again, no certain vertebrate is found in *TS.*2a, but we note here *Fenhsiangia* from the Hunghuayuan Formation (*TS.*2a–b). Schultze, in Erdtmann et al. (2000), described an enigmatic form, *Pircanchaspis rinconensis,* from the Pircancha Formation (probable *TS.*2b: early Arenig [Moridunian], graptolite-based

"*Didymograptus deflexus*" Zone) of the Tarija district, southern Bolivia, but, in fact, little evidence of its vertebrate affinity is presented. Young (1997), and Young in Webby et al. (2000), referred to isolated material in the Amadeus Basin, central Australia; his Bendigonian Assemblage 1 (basal Horn Valley Siltstone Member, *TS.*2b) comprises *Porophoraspis* sp. indet. and another indeterminate vertebrate.

Middle Ordovician (TS.3a–4c), Including Darriwilian (TS.4a–c)

In central Australia, Ritchie and Gilbert-Tomlinson's (1977) taxa *Arandaspis prionotolepis* and *Porophoraspis crenulata,* plus further forms intimated by

Ritchie (1985, 1991) from the Stairway Sandstone (*TS*.4b), along with Young's (1997) Assemblage 2 containing the same genera, belong here. Other records from South America appear to straddle the Middle Ordovician–Upper Ordovician boundary. In the late Darriwilian to earliest Caradoc (*TS*.4c?–5a), *Sacabambaspis janvieri* is known principally from articulated specimens collected from the Anzaldo and Cuchupunata formations (Cochabamba Group) of Bolivia (e.g., Gagnier et al. 1996), along with fragmentary material from the La Cantera Formation of Argentina (Albanesi et al. 1995a) and an unnamed Darriwilian formation (*TS*.4b?–c) from Teoponte in Bolivia (Janvier in Ramirez et al. 1992).

Upper Ordovician: Caradoc (TS.5a–d)

Young's (1997) Assemblage 3 from the basal Stokes Formation, central Australia (with late Gisbornian conodonts, *TS*.5b: Zhang et al. 2000), yielded four taxa, including some of the oldest possible gnathostomes: *Sacabambaspis* sp., *Apedolepis tomlinsonae* (a polymerolepid?), *Areyongalepis oervigi*, indet. ?chondrichthyan, and an indet. vertebrate "Assemblage 3" (an early nonchondrichthyan gnathostome).

Renewed research (e.g., by Sansom et al. 2001) in North America has produced a fourfold increase in taxa during this time span. Vertebrate material similar to that from the Harding Sandstone and contemporaneous deposits (*TS*.5a–c) in the Cordilleran region of Colorado has been detected in Ordovician strata from Alberta, British Columbia (not yet verified), Ontario, Quebec, Michigan, Minnesota (?), Montana, Oklahoma, South Dakota, and Wyoming. The placement of these records in a definite time slice is still imprecise, as some equate with *TS*.5a based on graptolites while others correlate with the *TS*.5b–c boundary using conodonts. The principal pteraspidomorphs in these faunas are *Astraspis* and *Eriptychius* (Sansom et al. 2001). Typical isolated material from the Braeside section in the Lowville Formation (*TS*.5b) of the Ottawa Embayment is illustrated here (figure 30.2A–D).

Stetson (1931) and Spjeldnaes (1967) noted further taxa in the Harding Sandstone. Sansom et al. (1995, 1996) and Smith et al. (1995) now recognize at least 12 different histomorphological types, comprising *Astraspis, Eriptychius, Skiichthys* (a possible osteostracan [Vertebrate indet. A of Denison 1967]), a "scale morphology A" attributed to chondrichthyan placoid scales, a "scale morphology B" identified as loganiid thelodont scales, and six other undescribed taxa (among them "possible gnathostomes," Sansom et al. 1995; and/or "possible acanthodians," Sansom et al. 1996) comprising new genera A–F (Sansom et al. 2001: figure 10.4c–i). One is an unnamed spine cf. *Sinacanthus* (Sansom et al. 2001: figure 10.4i), which may be acanthodian or chondrichthyan (see, e.g., Zhu 1998; Burrow 2001, 2003).

One purported thelodont (Sansom et al. 2001: figure 10.3b) lacks typical characters of loganiid or other known Late Ordovician or Early Silurian forms (*Sandivia, Zuegelepis,* or *Valyalepis;* Turner 1991; Karatajute-Talimaa 1997; Turner et al. 1999) and may be broken astraspidiform tubercles. However, with the discovery of "thelodontidid"-like scales from residues, Sansom et al. (2001: figure 10.3a; Sansom and Elliott 2002) have verified the presence of thelodonts in the Caradoc Gull River Formation of Ontario (Darby 1982).

In Canada, Barnes (1964), Darby (1982), and unpublished theses have noted true astraspids (figure 30.2A–C). One of us (GSN) has also found a teardrop-shaped thelodontlike scale (figure 30.2E) associated with conodonts and possible conulariid "bars" in rocks of the Red River Formation in the Bruce Well, subsurface Alberta (mid Maysvillian to Richmondian, latest Caradoc–earliest Ashgill, top *TS*.5d–6a). Other problematic forms also come from subsurface Alberta (stratigraphic data of Nowlan 2002). Nitecki et al. (1975: figure 1) illustrated possible water-worn astraspid bone from the Elgin Member, Scales Formation, basal Maquoketa Group (Maysvillian) of Indiana. Also illustrated here (figure 30.2F–I) are new thelodont scales reminiscent of *Sandivia* (see later in this chapter) from the Pont Rouge Formation (top *TS*.5b–c), Trenton Group of Cap-à-l'Aigle, Quebec.

Young's (1997) Eastonian Assemblage 4, from the Carmichael Sandstone of central Australia, with fragmentary remains of *Sacabambaspis* sp. indet. and another indeterminate vertebrate fits in the *TS*.5c–d as the last Ordovician record from Gondwana.

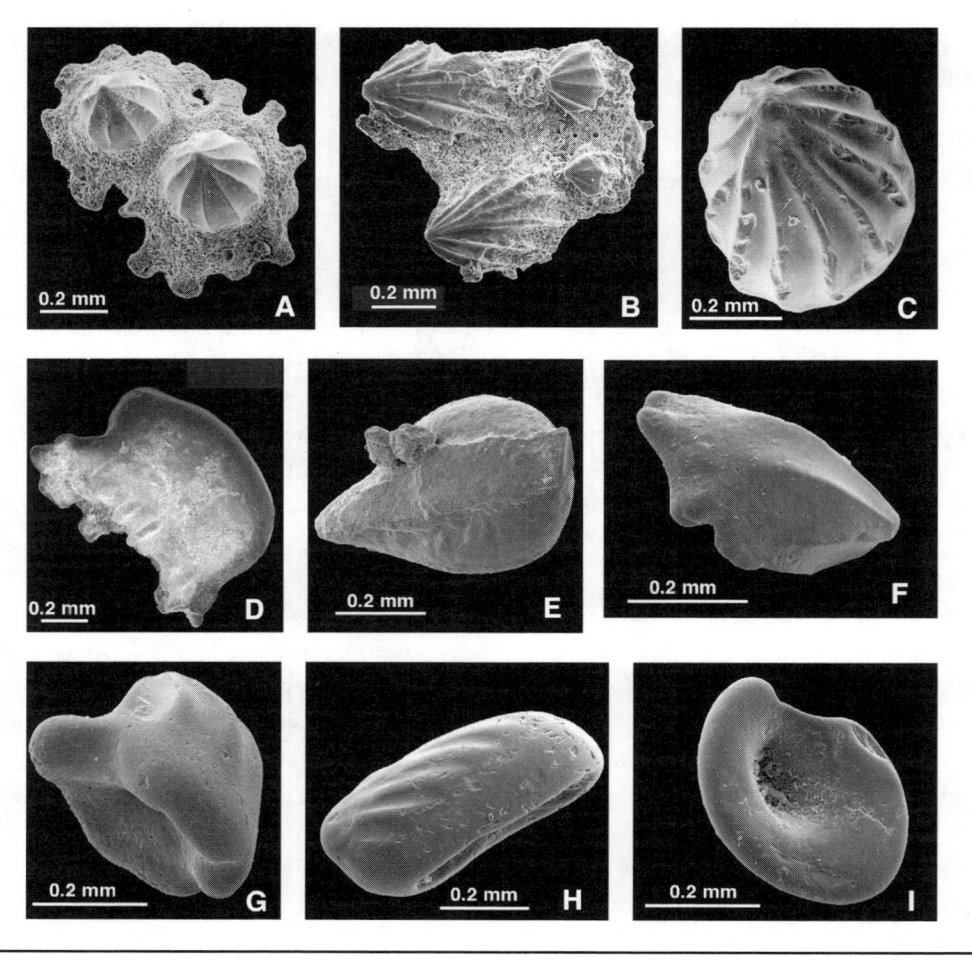

FIGURE 30.2. New vertebrate microremains from the Ordovician of Canada (Geological Survey of Canada collections). A–D, specimens from the Lowville Formation (LF), Ottawa Embayment, St. Lawrence Platform, road cut section southwest of Sand Point, Ontario; lower Upper Ordovician (Turinian; equivalent TS.5b); A–C, *Astraspis* sp.: (A) tessera GSC 121707 from 25 cm above base of LF, (B) broken platelet GSC 121708 from the basal 10 cm of LF, (C) isolated tubercle GSC 121709 from the same locality as A; D, isolated tubercle of an indet. vertebrate (astraspid or eriptychiid) GSC 121710 from the same locality as A. E, a probable scale of an indet. vertebrate (thelodont or acanthodian) GSC 121711 from the Red River Formation, at –1,632.2 m in Suncor Bruce 8-2-46-14W4 well, subsurface Alberta (GSC loc. C-252303); Upper Ordovician (middle Maysvillian to Richmondian; equivalent top TS.5d–6a). F–I, specimens from the uppermost 7 cm of Pont Rouge Formation at Cap-à-l'Aigle, Quebec (GSC loc. O-099533); lower Upper Ordovician (just below beds with a conodont fauna of the Chatfieldian; equivalent top TS.5b–c); water-worn scales of thelodont cf. *Sandivia*: (F) GSC 121712 in lateral view, (G) GSC 121713 in basal-lateral view, (H) GSC 121714 in lateral view, (I) GSC 121715 in basal view.

Upper Ordovician: Ashgill (TS.6a–c)

Although Ørvig (1989) and Janvier (1996b, 1998) reported early vertebrates (?thelodont scales) from the Ordovician of the Siberian Platform, we support Moskalenko's (1970, 1973) alternative identification of these remains as true conodonts. Nevertheless, Karatajute-Talimaa and Predtechenskyj (1995) and Märss and Karatajute-Talimaa (2002) have presented good evidence of Ashgill remains, namely, questionable and undescribed pteraspidomorphs (originally called Propataraspidida and "astraspids") including "*Tesakoviaspis*" Karatajute-Talimaa (1978, 1998) from the Khondelen Beds of Tuva, in the Khoreiver depression and Ijemsk region in Timan-Pechora, and the Viljui Basin.

Ashgill thelodonts *Sandivia melnikovi* and *S. angusta* are described by Karatajute-Talimaa (1997) and Märss and Karatajute-Talimaa (2002) from the Late Ordovician Ust'Zyb and Khoreiver "stages" and correlative Van'ju Formation of Timan-Pechora, as well as *Stroinolepis maenniki*, from the uppermost Ozernaya and Strojnaya formations of October Revolution Island, Severnaya Zemlya, Russia. Karatajute-Talimaa and Predtechenskyj (1995: figure 3) also noted the only definite gnathostomes in *TS*.6c, "supposedly Ordovician" scales of Acanthodii indet. in the Balturino Formation of Tchuna-Biriussa, Siberia. This discovery and the presence of an acanthodian-like spine in the Harding Sandstone suggest the need for a reevaluation of the supposed spines from the Lady Burn Starfish Beds, Drummuck Group, Craighead inlier, Midland Valley of Scotland (Harper 1979).

Finally, a purported, undescribed Late Ordovician "anaspid" has been mentioned from the Soom Shale, Cedarberg Formation, Table Mountain Group, South Africa (Anderson et al. 1999; Sansom et al. 2001).

■ Turnover

Preliminary analysis of vertebrate diversity patterns by Morrow et al. (1996:54) showed no Ordovician turnover because the scale of analysis was not detailed enough and their database dealt only with "terrestrial" vertebrates and was out of date. Purnell (2001:190–191, figure 12.3) considered the topic superficially but biased his sample by using pre-1990s agnathan data with stratigraphic ranges that were imprecise. His diversity plots using families might have worked in theory because almost all families then defined equated to a monotypic genus, but he skewed the data by assuming that conodonts were vertebrates. In addition, Purnell's analysis implied that the Late Ordovician extinction event had almost no effect on turnover. Long's (1993: figure 3.1) analysis, also using pre-1982 data but omitting conodonts, instead showed that a small number of families (five or fewer) first existed, subsequently decreased in end Ordovician times, and then expanded, accompanying the strong radiation event that followed in the Early Silurian. Long (1993) showed agnathan decline paralleling gnathostome increase in diversity. However, Long united in "Agnatha" taxa that are here kept as separate clades, for example, Arandaspidiformes, Astraspidiformes (*Astraspis*), and Eriptychiiformes (*Eriptychius*); and he ignored other taxa, for example, Vertebrate indet. A of Denison (1967; now *Skiichthys*) and Russian vertebrates. Purnell's (2001: figure 12.4) preliminary analysis was therefore a slightly better approach in that he surveyed more or less separate clades.

Our data show slow turnover rates (table 30.1). From the Late Cambrian starting point there is a gap of about 15 million years (m.y.) (excluding the problematic forms, *Anatolepis, Fenhsiangia,* and *Pircanchaspis*), then a first appearance of the GEA in the Early Ordovician *TS*.2b (with a mean per capita origination rate o_{di} of 0.4), and strong incoming of Arandaspidiformes in the Darriwilian (o_{di} = 0.26 in *TS*.4b). Originations of the Astraspidiformes and

TABLE 30.1. Diversity of Ordovician Vertebrates

Time Slices	d_{tot}	d_i	d_{norm}	o	o_i	o_d	o_{di}	e	e_i	e_d	e_{di}
6c	6	4	3	6	4	100	0.66	2	1.33	33.33	0.22
6b	1	0.5	0.5	1	0.5	100	0.5	1	0.5	100	0.5
6a	2	0.8	1	2	0.8	100	0.4	2	0.8	100	0.4
5d	2	0.66	1	0	0	0	0	2	0.66	100	0.33
5c	3	1	1.5	2	0.66	66.66	0.22	3	1	100	0.33
5b	17	6.8	9	16	6.4	94.11	0.37	14	5.6	82.35	0.32
5a	1	0.33	1	0	0	0	0	0	0	0	0
4c	1	0.33	1	0	0	0	0	0	0	0	0
4b	3	1.2	1.5	2	0.8	66.66	0.26	2	0.8	66.66	0.26
4a											
3b											
3a											
2c											
2b	3	1.2	1.5	3	1.2	100	0.4	2	0.8	66.66	0.26
2a	2	1	1	0	0	0	0	2	1	100	0.5
1d	1	0.33	0.5	1	0.33	100	0.33	0	0	0	0
1c											
1b	1	0.4	1	0	0	0	0	0	0	0	0
1a											

Note: This table uses the standardized diversity measures and symbols of Cooper (chapter 4) for the data shown in Figure 30.1. Symbols d_{tot}, total diversity at generic level; d_i, taxa per m.y.; d_{norm}, normalized diversity at generic level; o_d/e_d, percentage of originations/extinctions; o_{di}/e_{di}, mean per capita origination/extinction rate.

Eriptychiiformes occur in the Late Ordovician (early Caradoc) of Laurentia along with the appearance of the first thelodonts and stem-group gnathostomes ($o_{di} = 0.37$ in *TS*.5b). The appearances of GEA taxa *Sacabambaspis* and allied gnathostomelike vertebrates of Assemblages 3–4 of central Australia overlap with the Laurentian assemblages in the Caradoc (*TS*.5b–c). Caradoc taxa all seem to disappear by the end Caradoc, with a new fauna emerging in the late Ashgill (*TS*.6c; Hirnantian; $o_{di} = 0.66$) in Siberia and Baltica (Timan-Pechora). Did an extinction event (possibly related to glacial changes) affect the earlier forms and spur on a Late Ordovician radiation? We need more data to assess this for the vertebrates. The highest mean per capita extinction rate ($e_{di} = 0.50$) occurred both in the late Early (*TS*.2a) and late Late Ordovician (*TS*.6b) (table 30.1).

■ Conclusions

Taxonomic richness of vertebrates is low in the Ordovician. In the Early Ordovician time span of 18 m.y., the first vertebrate clade appeared in the Arenig, with a possible precursor in the Late Cambrian. However, there is still a dearth of knowledge for the Tremadocian and early Mid Ordovician intervals. The Ordovician vertebrate clades include basal Pteraspidomophi, Osteostraci?, Thelodonti, Placodermi?, Mongolepida?, basal Chondrichthyes?, and Acanthodii. Some "*Anatolepis*" specimens are also considered as possible vertebrates, whereas others are related to the problematic *Mongolitubulus*.

Vertebrates are known from North America, Russia, South America, and Australia, the last two regions producing the Gondwanan clade assemblage defined here as the GEA comprising Late Cambrian, Early Ordovician to early Late Ordovician (Caradoc) taxa spanning some 20 m.y. Origination of the GEA with its typical arandaspidiform component occurs in the Arenig (by *TS*.2b) at least. Most GEA forms disappear completely before the end of the Darriwilian, with only *Sacabambaspis* continuing into the Late Ordovician (mid to late Caradoc *TS*.5c–d); vertebrates do not reappear in Gondwana until mid Late Silurian.

From the end Darriwilian onward, however, the record becomes increasingly firm, and diversity rises rapidly especially in the Laurentia-Baltica-Siberia blocks. By the Caradoc all major vertebrate clades appear to be present. Not only is the Ordovician a major evolutionary period for fishes, but it is now clear that the Darriwilian to Caradoc is the time of major expansion for the clade. The disappearance of the bulk of the GEA around 458 Ma slightly overlaps the incoming of vertebrates in North America, but with only one taxon, *Sacabambaspis* sp., persisting for a brief time. Most vertebrates are known from the last 17.5 m.y. of the Late Ordovician. Maximum diversity occurred during the mid Caradoc (*TS*.5b). By about 455 Ma (or earlier), several taxa had appeared in Laurentia (especially the western United States) as follows: (1) pteraspidomorphs; (2) putative loganiid and thelodontidid thelodonts; and (3) gnathostomes including chondrichthyan-, placoderm-, and acanthodianlike remains. Late Ordovician vertebrates (astraspid? *Tesakoviaspis,* thelodonts) are known from Arctic Canada, Wisconsin, Quebec, Timan-Pechora, the Severnaya Zemlya archipelago, and Siberia (including western Mongolia). We define these faunas of Caradoc to Ashgill age as comprising the LBSA.

The most significant changeover is in the latest Ordovician. The last 3 m.y. show an upsurge of new taxa and possible new clades. In the top Ashgill (*TS*.6c), there is a post-Hirnantian recovery indicating major innovation in the thelodonts, pteraspidomorphs, and possibly the acanthodians. Almost everywhere a latest Ordovician to earliest Silurian (Rhuddanian) gap of around 3 m.y. in the vertebrate record exists, which we here name Talimaa's Gap.

The Tremadocian of the Southern Hemisphere and the Caradoc and Ashgill of the Northern Hemisphere pinpoint times and places needing greater effort in collecting, especially for microfossils, in order to fill the gaps in our knowledge of vertebrate paleobiodiversity. Detailed analysis of the existing vertebrate database will be published elsewhere (Turner, Blieck, and Nowlan unpubl.).

ACKNOWLEDGMENTS

This is a contribution to IGCP 410: The Great Ordovician Biodiversification Event; to IGCP 406: Circum-Arctic Palaeozoic Vertebrates; and to IGCP

421: North Gondwana Mid Palaeozoic Biodynamics. We thank G. Arratia, H.-P. Schultze, P. Janvier, and I. Sansom for valuable information and help provided. S. Turner thanks the CNRS for a three-month term as Directeur de Recherche Associé at USTL in late 2001. Our gratitude goes to L. Stricanne (USTL) and R. Netter (CNRS) for help in creating the figures. Finally, the critical and constructive remarks of our editor and referees B. D. Webby, D. K. Elliott, and G. C. Young helped to improve the chapter.

31 Receptaculitids and Algae

*Matthew H. Nitecki, Barry D. Webby,
Nils Spjeldnaes, and Zhen Yong-Yi*

The meaning of the term *algae* is still debatable among neoalgologists; they disagree on a number of higher-level algal taxa; they continuously revise their classification based on the ultrastructural, reproductive, and biochemical analyses; and they place many algal phyla (divisions) among protists, among plant metaphytes, or, with bacteria, among Monera. While the concept of the word *algae* is thus changing (or even losing its meaning), paleoalgologists accept blue-greens, greens, reds, and occasionally browns as the major groups of Paleozoic calcareous algae and readily admit that the systematics of most Paleozoic algal groups is in a very uncertain state (Riding 1977; Gnilovskaya 1980).

The literature on the Ordovician algae is less extensive than that on the Precambrian and Cambrian algae and the problematic acritarchs. With the exception of a few of the Ordovician calcareous algae, no survey of the entire Ordovician algal flora is available, and no specialists on Ordovician algae have emerged.

Morphological similarities and dissimilarities are insufficient features to identify algae, which display greater ultrastructural diversities than all metazoan groups, and hence it is not surprising that the editors of this volume have been unable to secure a contribution on the general overview of the Ordovician algal diversity (except acritarchs and cyclocrinitids). This chapter includes, first, a species-level biodiversity survey of two autonomous Paleozoic groups, the problematic, possibly metazoan, receptaculitids and the algal cyclocrinitids, and, second, a discussion of the timing of the diversification of selected groups of algae that predominantly belong to Chuvashov and Riding's 1984 "Ordovician Flora." Only a few calcified taxa of the "Ordovician Flora" are here discussed in the context of the timing of their appearances in the major evolutionary radiation. It is hoped that this will provide a basis for comparisons between the timing of earliest appearances of Ordovician algal floras and the great radiation events of metazoans discussed elsewhere in this volume.

■ Receptaculitids and Cyclocrinitids (MHN, BDW, NS)

Paleozoic receptaculitids and cyclocrinitids (now vernacular terms in English) are often considered to be related to each other and have been classified among many high-level taxa, particularly sponges and algae. They are here treated together for convenience and to provide the basis for constructing charts of their Ordovician distributional diversity. The latest, but surely not the last, consensus is that whereas receptaculitids are not sponges or algae, cyclocrinitids are a problematic order of algae, possibly a sister group of Dasycladales (Nitecki et al. 1999). However, the morphological architecture of receptaculitids and cyclocrinitids is comparable.

Only the monographic review of North American cyclocrinitids by Nitecki (1970) has been published, hence there is no agreement on the classification of both groups. Many receptaculitid and cyclocrinitid species, particularly *Receptaculites* sp. and *Cyclocrinites* sp., are listed from numerous localities and from stratigraphic settings; however, because of their poor preservation and many doubtful specific and even generic identifications, the majority of these listings cannot be included here. In addition, although operational taxonomic units are retained, it is necessary to exclude a few species that lack adequate stratigraphic information. Given the sporadic occurrences of receptaculitids and cyclocrinitids in the geologic column and the still scanty knowledge of their internal morphology, we choose to retain the two independent orders Receptaculitida and Cyclocrinales. First-published authors of the named class, orders, families, and genera used here are listed in Nitecki and Toomey (1979b), Spjeldnaes and Nitecki (1990a, 1990b), Nitecki and Spjeldnaes (1992), and Nitecki et al. (1999).

Both receptaculitids and cyclocrinitids are characterized by (1) a distinct vertical cylindrical, conical, or ampullar central axis paralleling the axis of the radially symmetrical body; (2) a generally uncalcified, simple axis (dichotomously branched, in one receptaculitid family) often with an ampulla (ampulla is the distal end of the axis of inverse ampullar shape); (3) globose bodies varying from perfectly ovoid to pyriform or claviform, rarely flattened at one or both ends and rarely asymmetrical; (4) occasional elongation of the lower part of the body to form a corniculum (in receptaculitids the central conical projection in the nuclear region) or pedunculus (in cyclocrinitids the central conical or stemlike projection in the nuclear region); (5) skeletons consisting of verticillate branches (laterals in cyclocrinitids, meroms in receptaculitids), originating from the axis and consisting of heads, shafts, and frequently feet (in receptaculitids); (6) heads forming a relatively compact outer surface; (7) sinistral and dextral, often disrupted (less distinct in cyclocrinitids), spiral alignment of branches running from a point at or above the abapical nucleus to the apical lacuna; and (8) the presence of unique morphological structures of unrecognized function— the quadribrachia, stellata, and cribella (see below)— that are unknown in any other organisms. Many of these morphological features are discussed in detail in Fisher and Nitecki (1982) for receptaculitids and briefly in Nitecki and Spjeldnaes (1992) for cyclocrinitids.

The differences between receptaculitids and cyclocrinitids are as follows: (1) receptaculitid bodies generally attained a larger size; (2) body shapes are more diverse in receptaculitids; (3) axes may be dichotomous in receptaculitids but never in cyclocrinitids; (4) receptaculitid axes are generally broad and rarely ampullar—cyclocrinitid axes are thin and mostly ampullar; (5) nuclei (the first formed whorls of branches in the base of body) in receptaculitids are always well defined—less so in cyclocrinitids; (6) lacunae (the apical openings encircled by the last-formed whorls of branches) are apparently always present in receptaculitids—in cyclocrinitids seemingly only in one family; (7) receptaculitid branches (meroms) are distributed along the entire length of axis, rarely on ampullae—in cyclocrinitids (laterals), on ampullae, rarely along the entire length of the axis; (8) all receptaculitid branches are morphologically similar— cyclocrinitid branches are occasionally of two morphologies; (9) receptaculitid branches are undivided— in cyclocrinitids, they may divide to a fifth order; (10) receptaculitid shafts may be thick or thin and may have feet—in cyclocrinitids, shafts are generally thin or clavate, and feet are absent; (11) receptaculitid heads terminate with a calcified, generally rhomboidal, rarely hexagonal plate—in cyclocrinitids the comparable structure (rarely or weakly calcified) is hexagonal, rarely rhomboidal; (12) receptaculitid quadribrachia consist of four (or multiples of four) interlocking ribs—comparable structures in cyclocrinitids (stellata) vary from three to five, or (cribella) six or multiples of six noninterlocking ribs; and (13) structures interpreted as gametophores are present in some cyclocrinitid heads but not in receptaculitids.

Distribution and Diversity Patterns

Order Receptaculitida

The order Receptaculitida, composed of 5 families, 17 genera, and 120 species, ranges from the Ordovician to the Carboniferous and possibly to the Permian. They have been described from all continents except Antarctica and are found in pure carbonates

FIGURE 31.1. Composite diagram showing receptaculitid families and genera with the species represented by their stratigraphic ranges (thickened horizontal lines) during Ordovician time (lower half), and plots illustrating normalized species diversity and turnover (appearance and disappearance rates per m.y.) for these receptaculitid species (upper half). The Ordovician 19 time-slice (*TS*) subdivisions and million-year subdivisions in the timescale at the bottom of the diagram form the basis for calculating the species diversity totals and turnover rates, and the arrows on the diagram show the genera with extended ranges beyond the Ordovician.

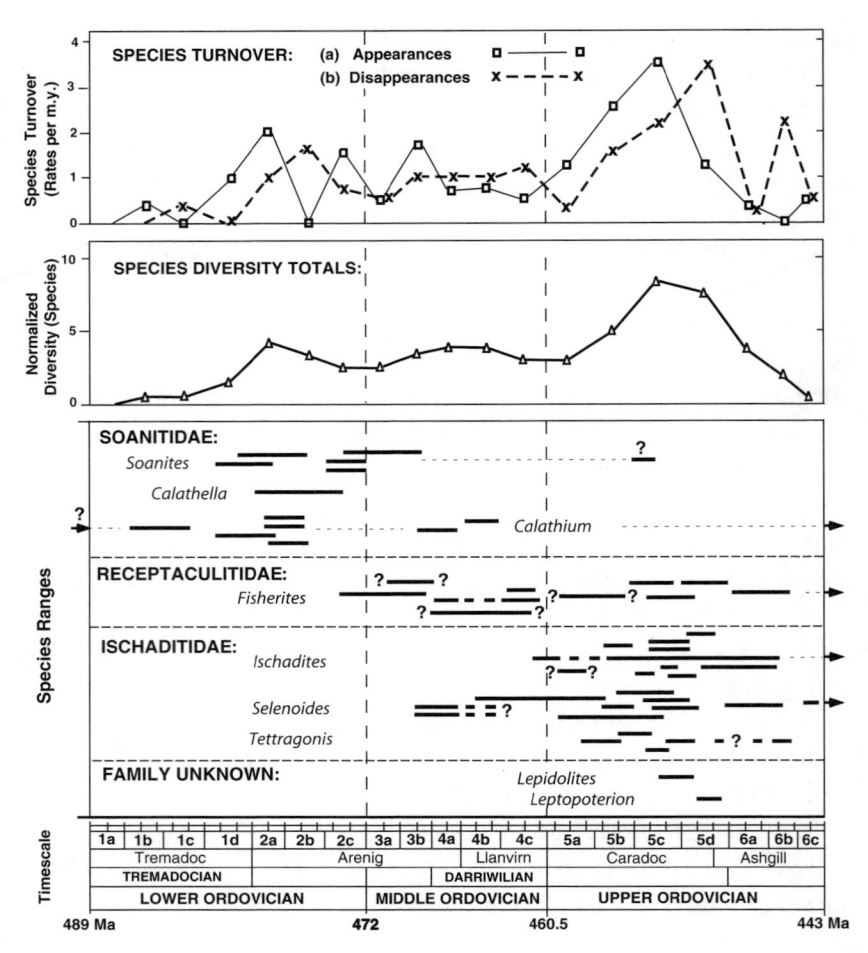

and in sandy, siliciclastic sediments (Nitecki 1972). In the Ordovician, the order contains 4 families, 9 genera, and 48 species.

Uncertainties in reconstructing diversity patterns are caused by the incompleteness of the fossil record, the paucity of monographic studies, and the imprecision of stratigraphic data. Thus, the graphs and their interpretations (figure 31.1) are preliminary and tentative. The receptaculitids have a comparatively low diversity record through the Ordovician, showing only three discernible pulses of diversity increase. The first two are represented by modest rises across the late Tremadocian–Arenig boundary and during the early Mid Ordovician, and the third involves a more significant event during late Caradoc time. This last may in part be biased by the more complete European and North American documentation and better knowledge of Caradoc rocks, but overall the signals seem to be robust, showing a definite pattern of staggered diversification of the three main receptaculitid families through Ordovician time. The Early Ordovician diversification involved mainly soanitids, the second minor pulse included mainly the Receptaculitidae (*Fisherites*), and the third radiation event also included mainly one family, the Ischaditidae, which rapidly became the most diversified receptaculitid group. A marked decline in species diversity occurred during the Ashgill and was probably related, as with most other clade groups, to events associated with the end Ordovician glaciation. At the generic level no taxa seem to have disappeared entirely, and so they remained available for the discernible increase in the diversity of Silurian receptaculitids that followed.

The first family to appear is the Ordovician-Silurian Soanitidae, which in the Ordovician ranges from *TS*.1b to 4b (see time slices in chapter 2) and consists of three genera: *Soanites* with three species, *Calathella* with one species, and *Calathium* with eight species. One of these species (*Calathium? pannosum*) comes from allochthonous deposits of the Pointe-de-Lévy thrust slice in the St. Lawrence Valley, and so it may be derived from an older horizon of Late Cambrian age (Bassler 1915). The family is found in shallow carbonates, and in North America, Siberia, and China it is frequently a major component of Lower to lower Middle Ordovician reefs (Webby 2002). It is distributed in a wide belt including Argentina, North America (Texas, the St. Lawrence Valley, Newfoundland), Scotland, Svalbard (Spitzbergen, Bear Island), the Siberian Platform and China (Liaoning, Jilin, Hubei and Sichuan provinces), Vietnam, Malaya, and possibly Tasmania (Nitecki and Toomey 1979a).

The best-known receptaculitids are the Ordovician to Carboniferous family Receptaculitidae, which in the Ordovician consists of a single genus, *Fisherites,* with nine species ranging from TS.2c through 6b. They are the largest receptaculitids known and are widely distributed in limestones and dolomites. Their wide geographic distribution is comparable to that of soanitids, except that the concentration of their distribution is in central North America. They are also found in the Canadian Arctic, Greenland, Baltoscandia, Burma, North Korea, Thailand, and the Argentine Precordillera but have not been reported from China.

The Ordovician-Silurian family Ischaditidae comprises 3 genera in the Ordovician: *Ischadites* with 12 species, *Selenoides* with 9 species, and *Tettragonis* with 4 species. These range from TS.3b to 6b. In the Ordovician they are found in shallow carbonates and siliciclastic sediments. Their distribution is again similar to that of Soanitidae and Receptaculitidae—from Argentina to the Canadian Arctic—but they are also known from Great Britain, Baltoscandia (and shores of the Baltic Sea), Sardinia (Italy), and India but have not been described from the rest of Asia. The genus *Ischadites* was recorded by Hammann and Serpagli (2003) from the upper Caradoc to lower Ashgill Portixeddu Formation of Sardinia.

Ordovician to Devonian receptaculitids that are not here attributed to a described family include, in the Ordovician, two genera—*Lepidolites,* with one species, and *Leptopoterion,* also with one species—both based on a single locality, one in Kentucky, the other in Ohio. These genera have a short range (TS.5c–d) and are found in siliciclastic and skeletal carbonate lithofacies.

The members of the family Ischaditidae, representatives of *Lepidolites* and *Leptopoterion,* and a few additional species of *Fisherites* were largely responsible for producing the normalized diversity peak through the TS.5c–d interval of on average 7 species and the maximum species turnover during TS.5c of up to 3.3 species appearances per million years (m.y.) and during TS.5d of up to 3.2 species disappearances per m.y. Species/genus ratios remained at very low levels through most of the Ordovician; highest levels were attained during the diversification of the soanitids during TS.2a, with 2.3 species per genus, and during the radiation of the ischaditids in TS.5c, with 2.1 species per genus.

Order Cyclocrinales

The order Cyclocrinales is composed of 5 families, 12 genera, and 68 species. The order ranges from the upper Middle Cambrian (trilobite-based *Cedaria* to *Crepicephalus* zones) to Lower Devonian (Lochkovian) and is found in various lithofacies and bottom conditions of pure carbonates or siliciclastic sediments. In the Middle to Upper Ordovician, the order is represented by 5 families, 7 genera, and 50 species. Difficulties with the interpretation of cyclocrinitid species diversity and turnover are similar to those of the receptaculitids. Their rate of occurrence is highest in the Caradoc, but well-defined and unquestioned cyclocrinitids are known from the Upper Cambrian and Lower Silurian. Therefore, figure 31.2 shows an incomplete distributional pattern, representing only the known Ordovician taxa.

The Ordovician-Silurian family Cyclocrinaceae consists of 3 genera in the Ordovician: *Cyclocrinites* with 20 species, *Mastopora* with 5 species, and *Nidulites* with 3 species. It ranges from TS.4c to 5d and is found predominantly in carbonates, often in high-purity lithographic limestones, but also may be preserved in mudstones and siltstones with moderate to no carbonate. One species of *Mastopora* (*M. parva*) from Girvan, Scotland, even occurs in a completely uncalcified form (Currie and Edwards 1943). The family ranges from California-Nevada, across the North American midcontinent, to the St. Lawrence Valley, Newfoundland, Arctic Canada, southern Appalachia, Wales, Scotland, Baltoscandia (where it is best known and common), Sardinia (Italy), to Kazakhstan and China (Tibet). The Sardinian occurrences of *Cyclocrinites* are from the upper Caradoc to lower

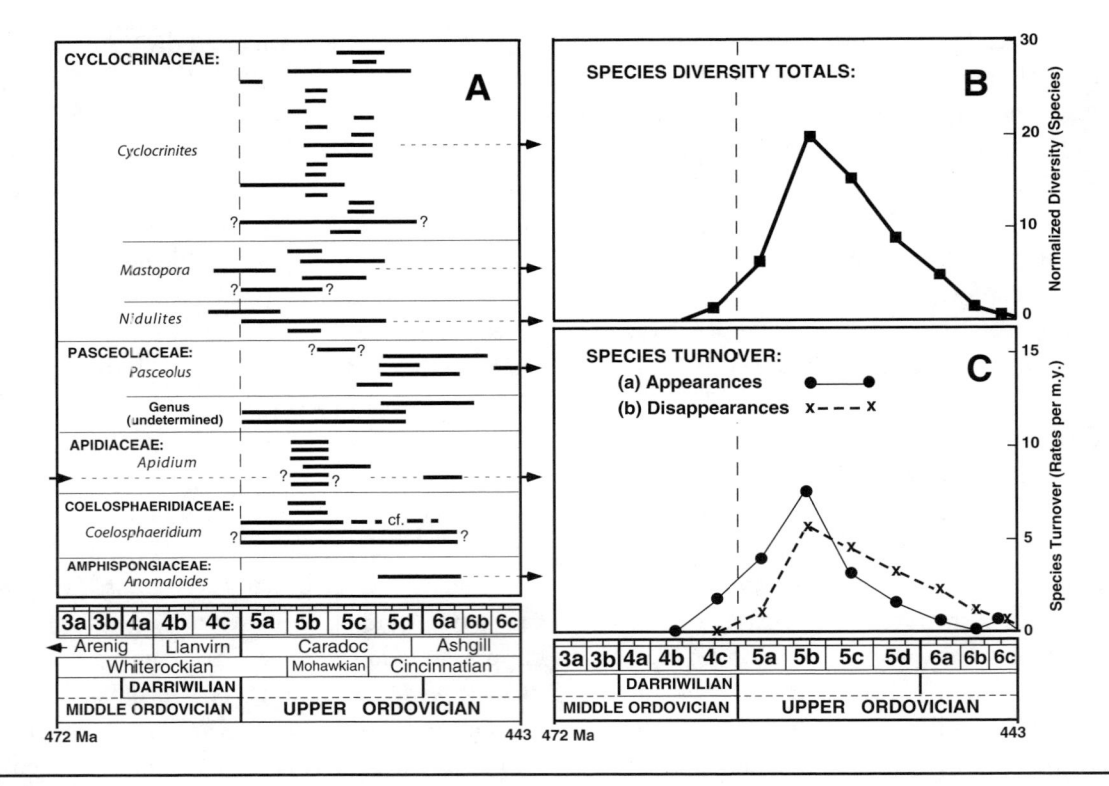

FIGURE 31.2. Composite diagram showing cyclocrinitid families and genera with the species shown by their stratigraphic ranges (thickened horizontal lines) during the Middle to Upper Ordovician (A), and the plots depicting normalized species diversity totals (B) and species turnover (appearance and disappearance rates per m.y.) for these cyclocrinitid species (C). The Ordovician time-slice (*TS*) subdivisions and million-year subdivisions are shown in the timescales at the bottom of the diagram, and the arrows in the left-hand diagram represent the genera with extended ranges beyond the Middle to Upper Ordovician.

Ashgill (*TS*.5d–6b) Portixeddu Formation (Hammann and Serpagli 2003).

The Ordovician to Devonian family Pasceolaceae is represented in the Ordovician by two genera: *Pasceolus* with six species and an undetermined genus with three species and a range from *TS*.5a to 6b. The family is found in siliciclastics as well as carbonates; when in mudstones, the fossils are frequently compressed. They are found in the Ohio Valley region, upper Mississippi Valley, south-central and Arctic Canada, New York, St. Lawrence Valley, New Brunswick, Greenland, Wales, and India.

The Cambrian to Devonian family Apidiaceae is represented in the Ordovician by a single genus, *Apidium*, with seven species ranging from *TS*.5b to 6a. They are found in limestones and are best known from Baltoscandia and Norway; they have also been described from Kazakhstan. The previously reported *Apidium* from India (Reed 1912) does not belong with cyclocrinitids. Only one unnamed apidiacean genus is currently known from sequences older than Mid Ordovician (early Darriwilian) age. This taxon occurs in Upper Cambrian deposits—the Pilgrim Formation of Montana and the Bonneterre Dolomite of Missouri (Lochman and Duncan 1944).

The family Coelosphaeridiaceae is restricted to the Ordovician and is based on a single genus, *Coelosphaeridium*, with five species and a number of subspecies. They are found in shallow-water carbonates and silty mudstones, range from *TS*.5a (?4c) to 6b, are known mainly from Baltoscandia (including the St. Petersburg area), and have been reported from Wales, India, Tadzhikistan, and China. Across Baltoscandia the associated lithofacies vary from dominant carbonates in Sweden and the eastern Baltic, to mudstones and siltstones in Norway, these latter derived from sources of terrigenous supply to the northwest that were uplifted parts of the Caledonide fold belt.

The Ordovician-Silurian family Amphispongiaceae is represented in the Ordovician by a single genus, *Anomaloides*, with one(?) species ranging from *TS*.5d to 6a. The genus is found in a variety of lithofacies from mudstones to skeletal carbonates and is recorded from Kentucky, Ohio, Sweden, Austria, and New South Wales (Australia).

The cyclocrinitids appear to have diversified rapidly from the late Darriwilian to early Caradoc with species attaining a normalized diversity maximum of 20 by the mid Caradoc (*TS*.5b). A total of five families and seven genera were present during the Caradoc interval (*TS*.5a–d). From the sharply pointed peak of diversity in *TS*.5b the normalized species totals decline rather rapidly, and this continued until early in the Ashgill, when comparatively low levels were attained (figure 31.2). The species/genus ratios show similar patterns of progressive rise from *TS*.4c to 5b, when 4.4 species per genus were diversifying; then the ratios decline steadily to background levels by the Ashgill.

The turnover patterns show rather similar trends—single, major turnover spikes for both appearances and disappearances—with a maximum of 7.2 species appearing per m.y. during the *TS*.5b interval and 5.8 species and 4.5 species, respectively, disappearing per m.y. during successive *TS*.5b and 5c intervals. It is noticeable that appearance rates decline more rapidly than disappearance rates at the beginning of this sharp decline. The end Ordovician glaciation seems to have had some impact on cyclocrinitid species; only one species is recorded as being present immediately after the basal Hirnantian "glacial" (*TS*.6b–c boundary) event. The last undetermined species of *Cyclocrinites* is reported from the Early Devonian (Lochkovian) of New York (Nitecki 1970).

■ Other Groups of Algae (MHN, BDW, YYZ)

The systematics of algae is a story of unresolved and existing conflicts of what to include, or exclude, in each group of algae. There is still no consensus among neoalgologists of how many divisions (or phyla) of algae, or classes within divisions, should be recognized. It is perhaps safe for paleontologists to refer to all "algae" as thallophytes, within which are a number of divisions of which the blue-green, brown, red, and green algae are our concern here.

Almost 20 years ago, Chuvashov and Riding (1984) tentatively identified and drew the outlines of the major Paleozoic algal floras, which they differentiated into the Cambrian, Ordovician, and Carboniferous. It was impossible then, as it is now, to delineate the algal groups exactly; hence, Chuvashov and Riding did not identify the major groups of algae definitively but referred to them with such modifiers as containing or dominated by a small group of genera, sometimes only two. Excluding the groups of doubtful affinities, the main components of their Early Paleozoic floras were the blue-greens in the Cambrian and the dasyclads and the codiaceans/udoteaceans in the Ordovician.

Blue-Green Algae (Cyanophytes)

The cyanophytes are the only algae that are prokaryotes, and although they are now considered by bacteriologists to be phylogenetically closer to bacteria than to eucaryotic algae, many neoalgologists and most paleontologists consider them together with "algae" because they have a chlorophyll *a* and a typical thallus without roots, stems, and leaves. Such a definition of algae is preferred by paleontologists but cannot apply to bacteria (Lee 1980). The unicellular blue-green microorganisms encrusting or forming filaments or mats are common from the Lower to the Upper Ordovician but are, however, poorly known. The most prevalent, cabbagelike stromatolitic structures, considered to have been built by an "algal" problematic organism *Cryptozoon,* with, for example, 11 species, are described from a wide range of Cambro-Ordovician successions across North America. There are also a number of distinctive forms such as Precambrian to Devonian *Renalcis* (North America, Kazakhstan); Cambro-Devonian *Epiphyton,* the Precambrian-Permian? boring and encrusting *Girvanella* with six species in North America, Lower Ordovician (Little Metis, Quebec; Chazyan, Crown Point, New York, Vermont; Lenoir, Tennessee; Richmond, Whitewater, Indiana) and two species in Kazakhstan.

Other Ordovician blue-green algae listed by Luchinina 1987 are *Chabakovia* (Cambro-Devonian), *Shuguria* (Cambro-Carboniferous), *Proaulopora* (Cambro-Ordovician), *Tubophyllum* (Cambro-Devonian), *Batenevia* (Cambro-Ordovician), *Subtifloria* (Precambrian-Carboniferous), *Obruchevella* (Cambro-Devonian), and *Halysis, Ortonella* (Precambrian-Jurassic). Gnilovskaya (1974) described *Setula* from the Kazakhstan middle Caradoc,

Anderski horizon, and Webby (1983) recorded *Cliefdenia* from the middle Caradoc Fossil Hill Limestone of New South Wales, Australia. Chuvashov and Riding's (1984) list included *Hedstroemia-Ortonella* (Cambro-Mesozoic) and, more doubtfully, *Rothpletzella* (Ordovician-Devonian) and *Wetheredella* (Ordovician-Permian). For a more complete differentiation of the blue-greens, see Riding (1991) and, in addition, Riding and Fan (2001), who described *Botomaella* (originally *Hedstroemia*), *Girvanella,* and ?*Subtifloria* from Tarim, China.

Most of these "blue-green" components were members of Chuvashov and Riding's (1984) Cambrian flora. In the reef habitats, for example, many of these forms took an active role in the development of Mid to Late Cambrian microbial-dominated reefs, especially the *Renalcis* group, *Epiphyton* group, and *Girvanella* forms (Rowland and Shapiro 2002), but these "holdover" groups became less conspicuous in reef building from the Mid Ordovician onward as new reef-building metazoans and the mainly Ordovician-style algae became more important (Webby 2002).

Brown Algae (Phaenophytes)

Brown algae, generally by comparison with other thallophytes, have giant branched, complex, and uncalcified thalli and occur mainly in shaley to more dolomitic intervals of well-bedded carbonate sequences. A few have been described from the Ordovician, of which the best known is the *Callithamniopsis* (originally *Oldhamia*) with four species—the first from the Lowville Limestone at Glen Falls, New York (Ruedemann 1909), the second from the Platteville Formation at Platteville, Wisconsin (both from the *TS*.5b interval), the third from the Glen Falls Limestone ("Trentonian") (*TS*.5c) of New York (Whitfield 1894), and the fourth (described as possibly a brown by Gnilovskaya 1980) from the Upper Ordovician, Chokparsk horizon (*TS*.6a–b) of Kazakhstan. In addition, a large, comparatively diversely branched and abundant, uncalicified flora of *Manitobia* (= *Chondrites, Bythotrepsis*) (one species), *Winnepegia* (= *Chondrites, Bythotrepsis*) (one species), *Dowlingia* (= *Chondrites*) (three species), *Enfieldia* (one species), *Inmostia* (one species), *Whiteavesia* (one species), *Westonia* (one species), *Amia* (one species), *Kinwowia* (one species), and a problematic brown occur in the Red River Formation (Cats Head Member; *TS*.5d) of Lake Winnipeg, Manitoba (Fry 1983).

Red Algae (Rhodophytes)

Reds are lime-encrusting, filamentous, often nodular, thallophytes, many of which have been described from the Ordovician in North America (e.g., those of Ruedemann and Hall) but have not been studied in the past 100 years. Pia (1927) recognized the family Solenoporaceae with seven genera as the best-known fossil taxon of red algae. Subsequent revisions by Korde (1965, 1973) were based mostly on the Russian fossils.

Gnilovskaya (1972) recorded the following red algae from eastern Kazakhstan: *Moniliporella* (two species), *Contexta* (four species), *Ansoporella, Furcatoporella, Plexa, Texturata, Villosoporella,* and *Solenopora* (two species) spanning the Late Ordovician, from early Caradoc (*TS*.5b) to Ashgill (*TS*.6b–c).

Chuvashov (1987), in his extensive review of red algae, added to Gnilovskaya's list the following Ordovician reds: *Parachaetetes* (Ordovician-Cretaceous), *Pseudochaetetes* (Ordovician-Silurian), and *Tubomorphophyton* (Lower Cambrian-Upper Devonian).

The red algae *Solenopora* sp. has also been reported in association with blue-green *Girvanella* and problematic *Nuia* and *Sphaerocodium* from the Chazy Group reefs (*TS*.4c–5a) in northeastern New York and western Vermont (Toomey and Nitecki 1979). The patch reefs of the Carters Limestone (*TS*.5b) and Chickamauga Group (*TS*.5b–c), the near-reef settings of the Lenoir Limestone (*TS*.4c–5a), and shelf lagoonal tidal channels of the Wardell Formation (*TS*.5b) in Tennessee include records of *Solenopora, Parachaetetes,* and *Petrophyton* (Alberstadt et al. 1974; Moore 1977). The slightly younger Mjøsa Limestone (*TS*.5c) of southern Norway contains large nodular thalli of *Solenopora* that form in closely packed shoal or reef complexes up to 14 m thick (Harland 1981; Opalinski and Harland 1981; Spjeldnaes 1982; Webby 2002). Solenoporans from the Karlstad Limestone near Trondheim (Høeg 1961) are probably of similar age (Bruton and Bockelie 1982). *Parachaetetes* cf. *compacta* has also been identified from the Late Ordovician (*TS*.6) in the carbonate and volcaniclastic deposits of Cabo Peñas, Asturia, northern Spain (N. Spjeldnaes, unpubl. data).

Riding (2000) and Riding and Fan (2001) clarified the relationships and the distribution of some groups of thallophytes of questionable affinities and removed the reds from the Cambrian but maintained them in the Ordovician as members of the "Ordovician Flora." They also tentatively transferred the moniliporellaceans from the reds (Chuvashov and Riding 1984) to the dasyclads (Riding and Fan 2001).

The corallinelike alga *Arenigiphyllum* Riding (in Riding et al. 1998), type species, *A. crustosum,* from the early Arenig (*TS*.2a–b) of southwestern Wales remains a problematic taxon. It is based on a single, tiny, fragmentary, sheetlike specimen, and although it shows morphological similarities to extant coralline vegetative parts (as described in detail by Riding et al. 1998), it exhibits little close resemblance to other confirmed (as well as the more questionable) red algae recorded later in the Ordovician. In addition, it greatly predates the records of these confirmed calcified red algae, such as *Petrophyton kiaeri* Høeg from the mid Caradoc (*TS*.5c) of Norway and ?*Graticula* sp. from the early Caradoc of the Tarim Basin, northwestern China (Riding and Fan 2001; see further discussion of age later in this chapter).

"Solenoporaceans"

Solenopora in North America and elsewhere has been described under many names from the Middle and Upper Ordovician and from many localities. The Ordovician record of the solenoporan group (as well as many other algal taxa) remains poorly understood, and its placement within the reds also remains uncertain and in urgent need of further studies (Riding 1977; Roux 1991b; Brooke and Riding 1998). Riding (2000) argued that the fossils conventionally assigned to the solenoporans are a heterogeneous group, containing metazoans (presumably chaetetid sponges and including the type species of *Solenopora*), blue-greens (e.g., *Solenopora compacta*), and true reds (e.g., *Solenopora gotlandica*).

However, the continuity in the temporal record of occurrences of loosely defined *Solenopora*-like forms within Ordovician successions seems clear. The first significant occurrences of solenoporans appear to be related to the initial development of late Mid Ordovician (*TS*.4b–c) reefs (Webby 2002). Riding (2000) noted that in the Cambrian *Solenopora* has been "confused" with *Epiphyton* and with *Bija. Solenopora* has not been confirmed in Lower Ordovician successions, but it has been listed by Klappa and James (1980) as a common constituent of an assemblage with the blue-greens, *Girvanella, Hedstroemia,* and *Halysis,* in the matrix of small reefs in the Table Point Formation (mid Darriwilian age; *TS*.4b) of western Newfoundland. In addition, Toomey and LeMone (1977) reported the presence of *Solenopora* in the upper Middle Ordovician part of the Simpson Group (McLish Formation, *TS*.4c) of Oklahoma, in an occurrence that broadly equates with the reef-bearing Chazy Group successions of New York, Vermont, and Quebec. Common nodular and less common sheetlike thalli of solenoporans occur in the Chazy successions, and in places these accumulations appear to have formed substrates that facilitated early reefal growth (Pitcher 1964; Desrochers and James 1989).

Solenoporans of early Late Ordovician age (*TS*.5a–b) are more widely distributed, occurring not only in parts of eastern North America (e.g., Tennessee) and Norway but also in Estonia; the northern Urals; northwestern (Tarim), southern, and southeastern China; and eastern Australia.

A solenoporan from the Tarim Basin of northwestern China has been described as ?*Graticula* sp. by Riding and Fan (2001). It comes from sample LN108a about 6,160 m below the surface in the Lunnan-46 borehole (Zhao and Zhang 1992; Fan et al. 1996) and is closely associated with age-diagnostic conodonts. These comprise a distinctive early Caradoc (*TS*.5b) conodont assemblage of *Belodina compressa, B. monitorensis, Pseudobelodina dispansa, Yaoxianognathus lijiapoensis,* and others (Zhao et al. 2000). Stratigraphically about 40 m lower in this borehole are samples with *Pygodus anserinus,* the key conodont zonal index species for the globally established Middle Ordovician–Upper Ordovician boundary interval (Bergström et al. 2000). These conodont occurrences indicate that ?*Graticula* sp. has an early Caradoc (TS.5b), rather than a Mid Ordovician, age as previously claimed by Riding and Fan.

Unfortunately, Riding and Fan's (2001: see text figure 2) "provisional" Ordovician age assignments are in error. The junction between their "Early Ordovician" and "Mid Ordovician" successions in the three Tarim boreholes, as shown in their text figure 2, is actually very close to the level of the globally established Middle Ordovician–Upper Ordovician boundary, now ratified by the International Subcommission on Ordovician Stratigraphy (chapter 2). Consequently, the statement in Riding and Fan (2001:797) that the highest percentages of samples (20.5 percent) of undifferentiated "solenoporaceans" from borehole samples belong to the Mid Ordovician interval also needs revision. These samples virtually all come from the early Late (not Mid) Ordovician interval. Only 1.4 percent of samples belong to the Mid Ordovician interval (i.e., those assigned by Riding and Fan to the Early Ordovician).

The solenoporans appeared in the Mid Ordovician (with initial diversification mainly in the Darriwilian), and by early Late Ordovician (early to mid Caradoc) the group was flourishing, especially in reefal habitats.

Green Algae (Chlorophytes)

Green algae are the most complex and the most diverse group of all thallophytes, which, with the advent of ultrastructural studies, became a complex of unicellular or multicellular or colonial, uninucleate or coenocytic, flagellate or filamentous, coccoid or parenchymatous organisms. For the purpose of this discussion we recognize the green as a division Chlorophyta, with the class Ulvophyceae, containing orders Dasycladales, Cyclocrinales, and Caulerpales (previously known as Codiaceae).

The following Ordovician green algae have been listed by Shuyskiy (1987): *Ajakmalaisoria, Aphroporella* (Middle Ordovician), *Apidium* (Middle Ordovician), *Bogutschanophycus, Callisphenus, Callithamniopsis, Chaetocladus, Coelosphaeridium, Cyclocrinus, Dasyporella, Dimorphosiphon* (Middle Ordovician), *Dimorphosiphonoides, Diversoporella* (Middle Ordovician), *Intermurella, Kazakhstanelia, Lowvillia, Mastopora, Novantiella, Nuia* [sic], *Palaeoporella, Primicorallina, Rhabdoporella* (Ordovician-Lower Carboniferous), *Sinuatoporella* (Middle Ordovician), *Uralella,* and *Vermiporella.*

Gnilovskaya in 1974 described *Cyclocrinites, Mastopora,* and *Striola* from the Middle Ordovician, middle Caradoc, Anderski horizon in Kazakhstan. In 1972 she described the following dasyclads from the early Caradoc (*TS*.5b) to Ashgill (*TS*.6b–c) of eastern Kazakhstan: *Vermiporella* (11 species), *Dasyporella, Apidium, Mastopora;* problematic dasyclad: *Aphroporella* (2 species); codiacean: *Dimorphosiphon* (3 species), *Palaeoporella* (2 species), *Diversoporella, Sinuatoporella* (2 species), and *Doliporella.* The distribution of the cyclocrinitids is presented earlier in this chapter.

Order Dasycladales

The dasyclads belong to the group of calcareous coenocytic thallophytes composed of a central siphonous tubular structure containing cytoplasm, from which radiate branches or laterals or perhaps segments of a generally uniform plan.

The more or less widely accepted classification of dasyclads is that of Pia (1927), who based his systematics on thallus morphology, arrangements of laterals on the axis, and assumed position of gametangia. His view was deeply rooted in the works of other German algologists, particularly Hermann Solms-Laubach and Ernst Stolley. He believed, among other things, that *Rhabdoporella* is the oldest dasyclad and that it has spirally arranged, isolated undivided laterals with gametangia in the main axis. He also theorized that in the more advanced cyclocrinitids the cysts were formed within unmodified branches. Contrary to Pia, cyclocrinitids have a well-developed globellum, and their axis is rarely capitate. Pia's interpretations have been falsified by Høeg 1932 (who is frequently quoted but misunderstood—his clear photographs of undoubted terminal gametangia within the secondary or tertiary laterals of the Ordovician *Apidium* are never mentioned). Although Pia's contributions to paleoalgology are second to none, his phylogenetic schemes and his assessment of primitiveness in the random arrangement of laterals should be modified or abandoned (see Høeg 1932; Spjeldnaes and Nitecki 1990b; and the discussion of cyclocrinitids earlier in this chapter).

Riding and Fan (2001), in their revision of dasyclads, removed *Vermiporella* from dasyclads to the order Ulotrichales and transferred many moniliporellacean genera to the Dasycladales. The distribution of *Vermiporella* and the dasyclad *Rhabdoporella* has been reviewed previously by Roux (1991a).

Vermiporella is shown by Roux (1991a) as having its early distribution in "Llanvirn-Llandeilo" horizons of eastern North America (New York, Quebec, Ontario) and in Scotland. The four species Roux recorded from eastern North America, however, come from horizons in the Black River (Lowville to Leray formations) and Trenton groups, rather than the older "Llanvirn-Llandeilo" equivalent Chazyan; thus these occurrences are in fact of early to mid Caradoc age (*TS*.5b–c). One of these species also occurs in the Stinchar Limestone of Scotland (Elliott 1972), and this is an earlier occurrence, close to the base of the graptolite-based *Nemagraptus gracilis* Zone (Ingham 2000), which equates with the base of the Upper Ordovician (or *TS*.4c–5a boundary). The genus exhibits rapid diversification and spread worldwide during the early to mid Caradoc to other parts of North America (Tennessee, Canadian Arctic, and Hudson Bay), Norway, Sweden, Estonia, Kazakhstan, North China, Tarim, and New South Wales (Australia).

In the Mjøsa Limestone (*TS*.5c) of southern Norway, *Vermiporella* is recognized as an important contributor to interreef and back reef sites (Opalinski and Harland 1981; Webby 2002). In Estonia *Vermiporella* ranges from mid Late Ordovician to latest Late Ordovician (*TS*.5c–6c—Oandu to Porkuni regional stages; Körts and Mark-Kurik 1997), and there are good records of the genus in the late Late Ordovician (Ashgill) on Anticosti Island, in southeastern China, and in other regions.

Rhabdopleurella has a more restricted but continuous Ordovician record, for example, in Estonia, where it first appears in the late Caradoc Rakvere stage (upper *TS*.5c) and has a continuous record to the basal Silurian Juuru stage (Körts and Mark-Kurik 1997). *Rhabdoporella* in the latest Ordovician Porkuni stage, together with *Vermiporella*, formed extensive algal mats across the shallow Baltic epicontinental sea (Körts and Mark-Kurik 1997). The genus is also recorded from the Ashgill of Sweden and from Anticosti Island (Roux 1991a).

Plexa (a junior synonym of *Intermurella;* Riding and Fan 2001) occurs in the mid Caradoc of Scotland and in the late Caradoc of eastern Kazakhstan (Gnilovskaya 1972). In Tarim (Riding and Fan 2001) cf. *Plexa* is recorded from sample LN 120a-b in the Lunnan-46 borehole, from above a horizon with the Mid Ordovician–Late Ordovician boundary conodont zonal index, *Pygodus anserinus,* and just below the next age-diagnostic assemblage yielding *Belodina compressa, B. monitorensis, Pseudobelodina dispansa,* and *Yaoxianognathus lijiapoensis* (Zhao et al. 2000) of early Caradoc age (probably *TS*.5b). Consequently cf. *Plexa* is probably of earliest Caradoc (*TS*.5a) age. It is not a Mid Ordovician occurrence as suggested by Riding and Fan (2001).

The genus *Dasyporella* (including its junior synonyms *Contexta, Texturata,* and *Villosoporella;* Riding and Fan 2001) occurs in the Late Ordovician of eastern Kazakhstan (Gnilovskaya 1972), from early to late Caradoc and Ashgill horizons (*TS*.5b–6b). In Tarim, *Dasyporella* is identified in three horizons (samples LN 7, 14, and 22) from the upper part of the Ordovician section in the Lunnan-46 borehole (Riding and Fan 2001). Age-diagnostic conodonts include *Taoqupognathus blandus, Phragmodus undatus,* and *Belodina confluens* of mid to late Caradoc (*TS*.5c) aspect (Zhao et al. 2000). *Dasyporella* is also recorded from the Rakvere and Nabala stages (late Caradoc *TS* 5c–d) in Estonia (Körts and Mark-Kurik 1997), and it is well represented in patch reefs and lagoon and bank deposits of the early Caradoc (*TS*.5b) Chickamauga Group, Lenoir and Wardell formations, and related units in Tennessee (Moore 1977) and in Late Ordovician (Cincinnatian) Fish Haven and Ely Springs dolomites of Utah and Nevada (Johnson and Sheehan 1985).

The genus *Moniliporella* has a more restricted Ordovician distribution in Kazakhstan, Tarim, Baltoscandia, and possibly Tennessee. In eastern Kazakhstan it is restricted to the late Caradoc (*TS*.5d) (Gnilovskaya 1972) and in the Baltic area presumably to the late Caradoc–mid Ashgill (*TS*.5d–6b). Riding and Fan's (2001) illustration of the Tarim *Moniliporella* is based on two samples (Y2-48c and Y2-78a) from the Yingmai-2 borehole, claimed to be of Early and Mid Ordovician ages. However, Zhao et al. (2000) have recorded an age-diagnostic Darriwilian conodont assemblage that includes, among others, *Eoplacognathus suecicus* and *Erraticodon balticus,* from the upper part of the borehole section that Riding and Fan assigned to the "Early Ordovician" and seemingly including the Y2-48c sample. The stratigraphically much higher Y2-78a sample is from the top of the Ordovician section in the Yingmai-2 borehole and includes another associated age-diagnostic conodont assemblage—species such as *Taoqupognathus blandus, Phragmodus undatus, Yaoxianognathus? tunguskaensis, Y.* sp., and *Belodina confluens,* suggestive of a mid to late Caradoc (*TS.*5c) age. The global Mid Ordovician–Late Ordovician boundary in this Yingmai-2 borehole is, as in the Lunnan-46 borehole, close to the junction between Riding and Fan's (2001) "Early Ordovician" and "Mid-Ordovician" subdivisions (see their text figure 2).

Moniliporella, based on the two samples from the Yingmai-2 borehole of the Tarim Basin and on the revised globally based correlations, ranges from Darriwilian to late Caradoc in age. According to the revised stratigraphic assignments, the Tarim *Moniliporella* first appeared in the late Mid Ordovician (Darriwilian), and its distribution is consistent with Chuvashov and Riding's (1984) "the Mid-Ordovician radiation of green calcified algae" (Riding and Fan 2001:807).

Riding and Fan (2001:794) referred to relative proportions of the "Undifferentiated Dasyporelleae" in all 170 samples of calcified algae and cyanobacteria in the boreholes from Tarim as being 1.4 percent, 16.4 percent, and 37.5 percent of the Early, Mid, and Late Ordovician samples, respectively. However, when the globally based stratigraphic constraints are applied (Zhao et al. 2000), the relative dasyporellean proportions are quite different, with values of 1.4 percent for the Early to Mid Ordovician and 53.9 percent for the Late Ordovician.

A number of other "Blackriveran" to "Trentonian" (i.e., Mohawkian) dasyclads were described by Whitfield (1894): for example, the problematic dasyclad *Chaetocladus* in the Platteville Formation of Wisconsin and Minnesota, *Chaetomorpha* also in the Platteville Formation of Wisconsin, and *Primicorallina* in the "Trenton" limestones of Middleville, New York. The dasyclads *Dasyporella, Vermiporella,* and *Rhabdoporella,* with cyclocrinitid *Cyclocrinites,* also occur in abundance in the Upper Ordovician (Cincinnatian) shallow-water, dasyclad-colonial coral community in the Great Basin of western North America (Johnson and Sheehan 1985).

Gnilovskaya (1972) recorded some unrelated, problematic algae from eastern Kazakhstan—*Guttoporella densa, Crinitella radiata, Palmatoporella lata, P. stena*—that span the Late Ordovician, from early Caradoc (*TS.*5b) to Ashgill (*TS.*6b–c).

Class Caulerpales

The class Caulerpales is characterized by the well-known, sediment-producing extant *Halimeda.* The group is defined by its ultrastructures, which are distinctly different from all other green thallophytes (O'Kelly and Floyd 1984). It contains a single order, Udoteaceae. The fossil members of the order were first recognized by Elliott (1984) and the Ordovician udoteaceans by Roux (1991a, 1991b), with at least four genera—*Dimorphosiphon, Dimorphosiphonoides, Lowvillia,* and *Palaeoporella.* The first three genera appear in the early Late Ordovician—*Dimorphosiphonoides* and *Lowvillia* in the Lowville Limestone (*TS.*5b) and *Dimorphosiphon* in the Ouareau Formation (*TS.*5c) of the St. Lawrence Lowlands region (Guilbault and Mamet 1976). *Dimorphosiphon* also occurs in the Craighead Limestone (mid Caradoc, *TS.*5c) of Scotland (Elliott 1972) and Norway (Karlstad and Mjøsa limestones) and in the late Caradoc of Kazakhstan (Gnilovskaya 1972). Estonian species of *Dimorphosiphon* appear later, in the Ashgill (Pirgu stage). *Palaeoporella* is less widely distributed, first appearing in the late Caradoc of eastern Kazakhstan and also occurring in the Ashgill of Tadjikistan, Sweden (Boda Limestone), and southern Norway (Sørbakken Formation).

The udoteaceans, as noted by Roux (1991a), show rather distinctive biogeographic patterns: (1) *Dimorphosiphonoides* and *Lowvillia* with restricted North American provincialism in the early Late Ordovician (early Caradoc); (2) *Dimorphosiphon,* with a wider but still geographically limited spread across eastern North America, Scotland, Baltica, and Kazakhstan through the Late Ordovician (mid Caradoc onward); and (3) *Palaeoporella,* with a marked, rather restricted Kazakhstan-Tadjikistan to Baltic connection in the latter part of the Late Ordovician. In contrast other dascyclads exhibit broader, equatorially disposed, cosmopolitan floras (Riding and Fan 2001), and the "solenoporaceans" also appear to show a similar, mainly equatorially contained distribution.

Concluding Remarks

In contrast to the staggered diversification of the non-algal problematic receptaculitids through most of the Ordovician, the cyclocrinitids exhibit only one major rapid diversification that spanned only late Darriwilian to mid Caradoc and then a progressive decline to the end of the Ordovician. The timing of the initial diversification of other algae ranges from expansion that commenced during the mid to late Darriwilian or a little later through to the mid Caradoc, like that of the cyclocrinitids. Only the dasyclad genus *Moniliporella* and the solenoporans apparently originated a little earlier in the Darriwilian than most other algal genera, which mainly appeared during the Caradoc (especially through the early to mid Caradoc interval). Two other genera, the dasyclad *Rhabdopleurella* and udoteacean *Palaeoporella*, commenced their diversification a little later, during the late Caradoc.

Consequently, the major radiation of calcified marine green and red algae of Chuvashov and Riding's (1984) Ordovician Flora extended through about 10 m.y. of the Mid Ordovician–Late Ordovician boundary interval (*TS*.4c–5b or 5c inclusive), more or less coinciding with the major diversification of certain animal groups (e.g., bryozoans, corals) and the great expansion of reef habitats, which included the establishment of an array of new and complex community interrelationships (Webby 2002). The uncalcified marine brown algae also diversified significantly through Late Ordovician time.

The blue-greens, on the other hand, declined in importance through the Ordovician. In the reef habitat this may have been related in part to the great expansion of metazoan reef-building activity and the displacement of the microbialite-forming blue-greens to more restricted depositional regimes (Webby 2002), for example: (1) the shallow basins of continental interiors with a restricted circulation; (2) and mud mounds in deeper platform and slope habitats and across paleolatitudes. Another factor that may have been influential has been suggested by Riding (1992), namely, that blue-greens, like the *Epiphyton* and *Renalcis* group taxa, lost their ability to calcify their sheaths as global temperatures declined, and this may have happened in the approach to the end Ordovician glaciation.

32 Acritarchs

Thomas Servais, Jun Li, Ludovic Stricanne, Marco Vecoli, and Reed Wicander

The organisms that produced the acritarchs are thought to have been the major component of the organic-walled microphytoplankton in Proterozoic and Paleozoic oceans. As the primary producers, these organisms represent the base of the food chain in the Proterozoic and Paleozoic marine ecosystem. In order to understand the functioning and evolution of the Paleozoic marine ecosystem, it is essential to understand the marine phytoplankton. An interesting question discussed by such authors as Strother et al. (1996b) is whether Phanerozoic phytoplankton diversity and invertebrate diversity are linked or decoupled from each other.

Is biodiversification of the major Ordovician fossil groups related to the evolution of the microphytoplankton, for example, acritarchs? This chapter discusses some of the problems in producing an Ordovician phytoplankton (acritarch) diversity curve and whether biodiversification, speciation, and radiation events can be identified, as they seemingly have for the major macrofossil groups.

■ The Group "Acritarcha" (TS, RW)

Definition and Interpretation

During the early 1960s, palynologists were able to attribute most Mesozoic "hystrichospheres" (organic-walled microfossils with a "spherical" body and "spines") to the dinoflagellates (a class of the Pyrrhophyta). However, there remained a large number of predominantly Proterozoic and Paleozoic organic-walled microfossils whose biologic affinities were still uncertain. Evitt (1963) thus proposed the term "Acritarcha" to include all organic-walled microfossils of variable shape and size and of unknown biologic affinity. Etymologically, *acritarch* means "of uncertain origin" (from the Greek: *akritos,* uncertain, confused; and *arche,* origin). Acritarchs are, therefore, by definition, an informal polyphyletic group of organic-walled *incertae sedis* microfossils without any nomenclatural rank or status and must be considered a palynological catchall category. For a detailed review of subsequent definitions and a discussion on the value and meaning of the term *acritarch,* see Servais et al. (1997).

The interpretation of what acritarchs are tries to answer the question about their biological affinities, that is, to which biologic entities can the different acritarch morphotypes be assigned? A consensus exists today that most acritarchs are probably the cysts of diverse marine microphytoplankton groups and that most of them were probably "pre–dinoflagellates." Although microfossils displaying the morphological criteria of dinoflagellate cysts (e.g., the presence of an "archeopyle" excystment opening and paratabulation) first appear in the Middle Triassic, there is now sufficient ultrastructural and molecular phylogenetic evidence to indicate that the dinoflagellates, together with other groups of the alveolate lineage, such as the ciliates and apicomplexans, possibly diverged as early as the Proterozoic (Fensome et al. 1996). Additionally, geochemical (biomarker) studies (Moldowan et al. 1996) suggest that some Early Paleozoic acritarchs, particularly acanthomorphic acritarchs, may belong to the dinoflagellates (Talyzina et al. 2000).

Many acritarchs display openings that can be compared with excystment structures, thus strengthening the belief that they probably represent cysts of unicellular algae belonging to the Dinophyceae. Some morphotypes are, however, undoubtedly phycomata of unicellular prasinophycean green algae (Colbath and Grenfell 1995), while others can likely be attributed to chlorophycean algae (Brenner and Foster 1994), the Chlorococcales (families Hydrodictyacea and Scenedesmaceae), and Zygnematales (family Zygnemataceae) (Wood and Miller 1996).

There are also those acritarch morphotypes that probably do not belong to the microphytoplankton but are related to miospores (reproductive elements of plants), the exoskeletal remains of higher organisms (Fatka and Konzalova 1995), or even egg cases of meroplanktonic zooplankton (Van Waveren and Marcus 1993). However, the number of Ordovician acritarch species that belong to these latter groups is probably quite limited.

As stated earlier, acritarchs are generally interpreted as marine organisms. Most taxa certainly lived in marine environments based on their association with various known marine fossils. A few taxa, however, indicate probable brackish or even freshwater environments, and a small number of nonmarine acritarchs are known from both modern and ancient oceans (Riding and Duxbury 1993).

Despite the problems of attributing biologic affinity, most authors today agree that acritarch morphotypes represent, in most cases, cysts of various marine phytoplanktonic organisms. Therefore, acritarch diversity changes during the Paleozoic do not reflect the evolution of a single biologic group but rather the whole organic-walled phytoplankton (even if one includes the very limited number of taxa that may possibly be related to nonalgal groups). An analysis of Ordovician acritarch diversity thus attempts to understand the biodiversification, including radiations and extinctions, of the cyst-forming organic-walled microorganisms that formed the major part of the marine microphytoplankton and thus the evolution of the principal primary producers of the Ordovician oceans.

History and Methods

Alfred Eisenack, a pioneer worker on Paleozoic organic-walled microfossils, first published on Ordovician acritarchs, as well as chitinozoans and melanosclerites, during the early 1930s. However, stratigraphic palynological research did not seriously develop until the 1950s, when the oil industry recognized the value of palynomorphs for dating and correlation. Since that time, acritarch investigations have greatly increased in number. A review of the literature on Ordovician acritarchs (Servais 1998) shows that since the early 1960s between 10 and 20 papers have been published every year, versus less than 43 papers per year before 1960. Compared with most Ordovician macrofossil groups, the study of acritarchs is still in its infancy.

In analyzing acritarchs, the enclosing rock matrix often determines the method of study. In most cases, standard palynological preparation techniques involving dissolution of the rock with hydrochloric (HCl) and hydrofluoric acid (HF) are employed (Wood et al. 1996). The liberated specimens can then be studied by light or scanning electron microscopy.

In low-grade metamorphic rocks, where they are often the only fossils still present and typically are poorly preserved, acritarchs can be studied only by means of petrographic thin sections. This is because palynological extraction using acids yields only fragments of organic matter that can no longer be identified. The technique of polished sections (Munnecke and Servais 1996) is very useful for the observation of specimens in situ, that is, inside the rock matrix. However, neither of these techniques is suitable for routine biostratigraphic studies.

Acritarchs, as silt-sized particles, normally occur in mudstones, but they may also occur in low numbers in limestones and fine-grained sandstones. They are less frequent and generally absent in coarse sandstones because they have been washed out or destroyed by taphonomic processes. Acritarchs are typically completely flattened in clastic sediments, but they may, in some limestones, be three-dimensionally preserved.

Acritarchs generally range in size between 20 and 100 micrometers (microns). Some forms, however, are smaller (less than 10 microns) but remain almost totally undescribed because the filters used in palynological preparation are typically 10-micron mesh size and usually only the fraction greater than this size is analyzed. Although some Proterozoic acritarchs are as large as several hundred microns in diameter, most Ordovician acritarchs described are between 20 and 50 microns, while morphotypes in most post-Devonian rocks are small (typically less than 20 microns) and display a very simple morphology (Molyneux et al. 1996).

The number of acritarch specimens in Lower Paleozoic sediments is commonly between 20 and 100 individuals per gram of rock but can easily reach 10,000 specimens in well-preserved material. Dorning (1999) noted that the Tremadocian Shineton Shales of the Welsh Borderland contained many samples with more than 100,000 acritarchs per gram. The large number of individuals available for study constitutes both advantages and problems for the group. Highly diverse acritarch assemblages with abundant specimens provide a large database with a high potential for meaningful biostratigraphic, paleoecologic, and paleobiogeographic interpretations. However, the complete description of all constituents in a large set of samples can easily take many years to complete.

Biostratigraphic, Paleoecologic, and Paleobiogeographic Distribution

Just as for the dinoflagellates in Mesozoic and Cenozoic sediments, acritarchs have a high biostratigraphic potential that is not yet realized. Published Ordovician stratigraphic schemes integrate acritarch zones with a duration of about 2 million years (Cooper et al. 1995). Because of a pronounced provincialism that is also observed in most other Ordovician fossil groups, the publication of a global acritarch zonation appears at this time to be unrealistic. Although the exact stratigraphic range of many acritarchs remains to be determined, the first-appearance datum (FAD) of some selected species with a restricted stratigraphic range is now known (Brocke et al. 1995) and should be very useful for the definition of biohorizons.

It is noteworthy that the last-appearance datum (LAD) is difficult to establish, as small palynomorphs can commonly be recycled into younger rocks. In the peri-Gondwanan area, which includes not only the British Isles but also southern China, the formal definition of a succession of biohorizons, based on the FAD of selected taxa, should soon be possible for the Tremadocian (Dorning 1999), as well as other parts of the Lower and Middle Ordovician (Li et al. 2002a). The duration of these zones should be about 2 million years, perhaps even shorter. For some regions and some age intervals it is reasonable to expect zones of about a 1-million-year duration.

In terms of paleogeography, two widely accepted "provinces" have been defined on the basis of the occurrence of certain taxa. Volkova (1997) defined a "warm-water" province that is characterized by the presence of the genera *Athabascaella*, *Aryballomorpha*, and *Lua* in the Tremadocian of Baltica, Laurentia, Australia, and North China. Li (1989) redefined the "Mediterranean" province on the presence of such taxa as *Arbusculidium filamentosum*, *Coryphidium*, and *Striatotheca*. This latter geographic assemblage, also named "peri-Gondwanan" province, occurs from high to low latitudes in many localities along the border of Gondwana (Li and Servais 2002) during the Early and Mid Ordovician. Although taxa from both provinces are present on Baltica, a clear distinction between the assemblages of peri-Gondwana and Baltica can be drawn (Servais and Fatka 1997).

Modern and fossil dinoflagellates reflect not only a latitudinal (climatic) signal but also a coastal-neritic signal, that is, a nearshore-offshore trend, as well as a salinity signal. They are therefore very useful for paleoceanographic reconstructions (Dale 1996). Similar to the dinoflagellates, the composition of an acritarch assemblage results from a combination of environmental parameters. These include such factors as temperature, light, nutrient supply, water chemistry, salinity, pH, water depth, surface currents, and turbidity.

The potential of acritarchs for Paleozoic paleoecologic interpretations has yet to be realized. Various studies have demonstrated nearshore to offshore trends (Staplin 1961; Dorning 1981; Wicander and Wood 1997; Vecoli 2000) and that the highest diversity of acritarchs occurs at intermediate depths. Future detailed studies are needed to better understand the influence of fluctuating sea levels on the composition of acritarch assemblages. Thus far, only a few such studies have been done, and these concern the

Late Ordovician of Laurentia (Jacobson 1979; Colbath 1980).

As with all fossil groups, taphonomy must be considered when reconstructing the phytoplankton biocoenosis. In terms of fossilization potential, it is interesting to note that only 17 percent of modern dinoflagellate species produce resting cysts (Head 1996). Furthermore, it is well known that oxygenation, burial diagenesis, and metamorphism can damage or destroy the organic matter contained in the sediments. This is the reason why many originally palyniferous sediments have proved to be barren after laboratory processing. Nevertheless, it is important to remember that the extremely resistant acritarchs have often been the only fossils found in otherwise unfossiliferous sediments. Furthermore, acritarchs have proved to be extremely useful for understanding the geology of ancient basins and orogenesis as well as being biostratigraphically useful in mildly metamorphosed sediments, such as phyllites (Montenari et al. 2000).

Previous Compilations

Some 40 years after Eisenack's first investigations, Tappan and Loeblich (1973) summarized the status of acritarch research. They counted more than 400 species described from the Ordovician, making it the system with the highest recorded number of species. Six years later, Cramer and Díez (1979) compiled an index listing the diagnoses and stratigraphic ranges of all known Paleozoic acritarch genera.

During the ensuing years, the data set of acritarchs increased, and thanks to a consortium of oil companies that brought together their databases, Fensome et al. (1990) published the first complete catalog of acritarch species for all geologic epochs, including a species index listing almost all species described to 1989.

The important data set of Fensome et al. (1990) was used by Strother (1996) to establish distribution curves of genera through time. The latter author incorporated the age information provided by Fensome et al. (1990) for the holotype of each taxon to plot the number of all validly published genera over time by normalizing the values per million years. This confirmed the general temporal trend inferred by Tappan and Loeblich (1973). After a continuous increase in acritarch diversity from the Precambrian to the Devonian, a rapid decrease occurred near the end of the Devonian. This decline led to a "phytoplankton-blackout" during the Carboniferous and Permian (Riegel 1996:133) and was followed by the replacement of acritarchs by dinoflagellates in the Middle Triassic.

Strother (1996) also used Downie's (1984) review of Paleozoic British acritarchs to compile summary charts showing the stratigraphic distribution of selected acritarchs from the Cambrian through the Devonian. Strother (1996), in applying the data sets of Downie (1984) and Fensome et al. (1990), was the first to plot the distribution of acritarchs for the Paleozoic since Tappan and Loeblich (1973).

Downie's (1984) data set was based on his own investigations during the previous 25 years and on the Ph.D. research of seven of his students, four of whom investigated the British Ordovician. In Downie's (1984) compilation, the ranges of 57 Cambrian, 131 Ordovician, 122 Silurian, and 104 Devonian species were given for the British series and stages. The indicated ranges included both the distribution of the aforementioned species in the British Isles and their ranges elsewhere in the world.

We also use Downie's (1984) data set in this study (figure 32.1). However, we also include the data available from the Llanvirn with reference to British stratigraphic nomenclature (Fortey et al. 2000). The plotted results (figure 32.1) provide a biodiversity curve for the Cambrian through Devonian for the British Isles, that is, the eastern part of Avalonia. Because the British sequence is by far the best studied of all global sections, and because only a few new species have been described since the mid-1980s, the distribution curve given in figure 32.1 provides the most up-to-date information about the biodiversity of Paleozoic acritarchs, at least for Britain and other parts of Avalonia. According to these data, acritarch diversity increased from the Late Cambrian to the Tremadocian and Arenig (more than 40 species present), reaching a peak in the Llanvirn (about 70 species). A rapid decrease in the number of species can be observed in the Caradoc, and this trend is even more accentuated in the Ashgill. This does not necessarily reflect a decrease in diversity but the fact that, at this time, almost no acritarch studies had been conducted on Caradoc and Ashgill assemblages of the British Isles and other areas.

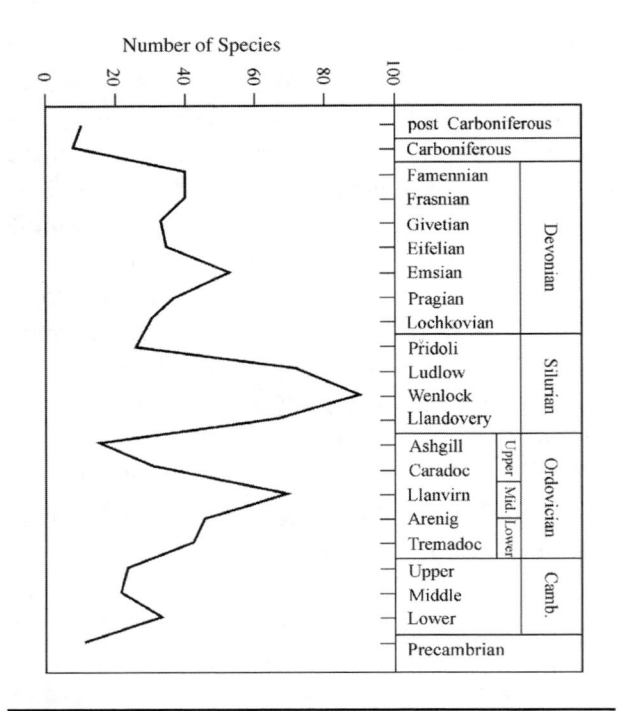

FIGURE 32.1. Number of acritarch species from the Precambrian to immediately post Carboniferous, based on Downie's (1984) data set from the British Isles (following Strother 1996).

■ The Data Set (TS, LS)

A marked contrast exists between the Proterozoic and Cambrian acritarch biodiversity patterns of speciation and extinctions, and the total number of species never exceeds 100, according to Vidal and Moczydłowska-Vidal (1997). The situation in the Ordovician appears to be more complex, with a developing provincialism and much greater diversity, making a general overview more difficult.

Following a suggestion from the International Geological Correlation Programme (IGCP) no. 410, "The Great Ordovician Biodiversification Event," Ordovician acritarch workers formed an acritarch clade group with the objective of assembling all data to establish a global biodiversity curve for Ordovician acritarchs. Accordingly, all the known Ordovician acritarch literature was first compiled (Servais 1998) into a complete bibliographic index with more than 700 references, including theses and abstracts, and listings of more than 250 acritarch genera from the Ordovician.

The compilation of published papers updated to 2001 is plotted on a paleogeographic map (figure 32.2) to document the current status of Ordovician acritarch research. Most studies (several hundred papers) have been concentrated in Europe (Baltica, Avalonia, and peri-Gondwana) and North Africa. The documentation from other regions comprises 50 papers from China (about 40 from the Yangtze Platform alone), 25 papers from North America, about 10 articles from South America, and 4 papers from Australia (2 that only included detailed descriptions of taxa). Some paleocontinents still have no coverage.

The next step for the acritarch clade team was to establish a complete database of Ordovician species (Servais and Stricanne 2001). The team compiled files that included for each species (1) its current generic attribution, (2) the year of its description, (3) the author(s), (4) the age of the type material, and (5), where possible, the stratigraphic range of the species. Compilation is now in the final stage and includes almost all species (more than 1,300). Together with the other taxa that have been described from older or younger sediments, the number of species known to have existed during the Ordovician currently exceeds 1,500.

The exact stratigraphic range for most species is not known. Whereas many Ordovician taxa have been described in monographic works in which the biostratigraphy was not the first objective, others have come from samples of sediments lacking independent age control. Many acritarch investigations had as their prime objective the dating of otherwise unfossiliferous rocks, commonly from areas with a complex geologic history, involving orogenesis and metamorphism (Albani 1989). Thus, for many of the 1,300 known species from the Ordovician, the age of the holotype has not been satisfactorily established.

Another problem is that for many species the holotype was neither well illustrated nor described, leading to numerous potential synonymies. Thus, the real number of Ordovician species is probably lower than documented. Nevertheless, many species from well-investigated sequences are currently well known and are extremely useful in the biostratigraphy of some areas (especially in the Lower and Middle Ordovician of Baltica and peri-Gondwana, including South China).

The compilation of a catalog of Ordovician acritarch species indicates that for most species a taxonomic revision is needed and precise stratigraphic

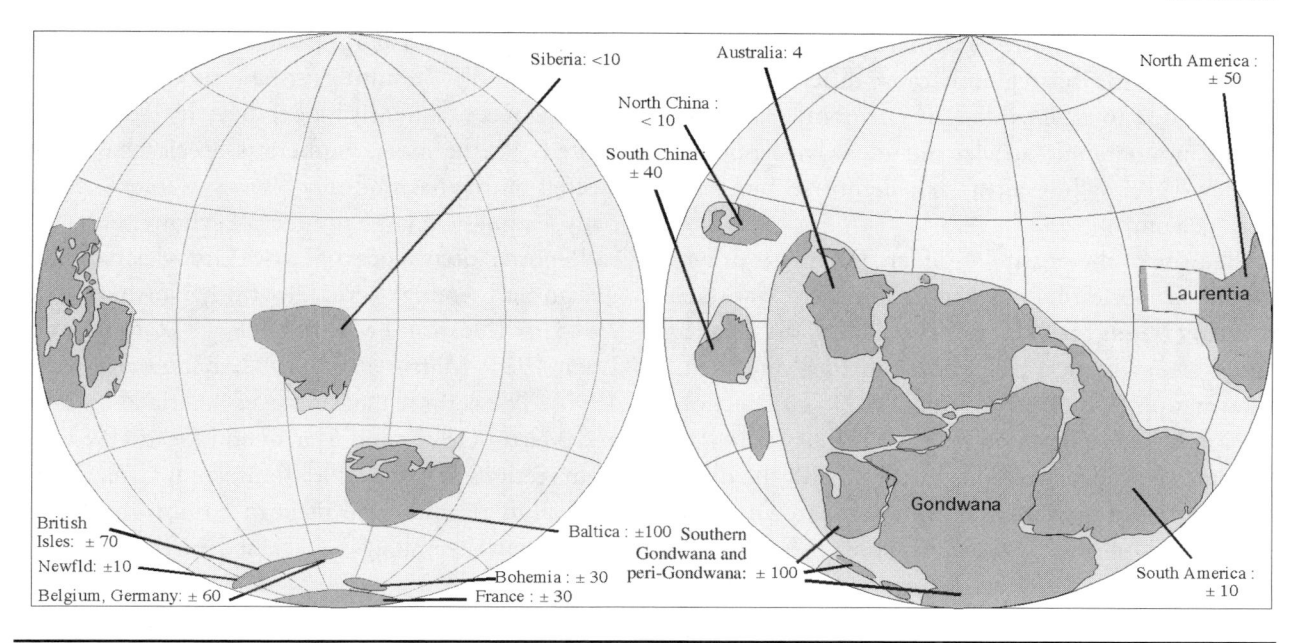

FIGURE 32.2. Number of articles on Ordovician acritarch research plotted on a paleogeographic reconstruction of Li and Servais (2002) for the Arenig (~ 480 Ma), based on the literature compilation of Servais (1998). Data from Avalonia are from Newfoundland, the British Isles, Belgium, and Germany. The area of "southern Gondwana and peri-Gondwana" includes the data from Iran, Pakistan, Saudi Arabia, Jordan, and northern Africa.

ranges must be established before attempting to draw diversity curves.

The counting of all new Ordovician species indicates that many species were described from the Baltica continent, including parts of the former Soviet Union, with 150 to 200 species for each of the British series. In the investigations of Ordovician acritarchs from southern Europe and North Africa, more than 100 new species were described in both the Tremadocian and Arenig, more than 80 species in the Llanvirn, but fewer than 40 species in both the Caradoc and Ashgill. New species from the Lower and Middle Ordovician of Laurentia are few in number. However, more than 100 species are described in the Upper Ordovician of this continent. Only a few species are described from Australia, and Chinese Ordovician acritarchs are mostly described from the Arenig (more than 70 species for this interval).

The data set also provides the first estimation of the total number of species in the Ordovician. Although this number never exceeds 100 and generally lies between 30 and 60, compared with the Cambrian (Vidal and Moczydłowska-Vidal 1997), the Ordovician appears to have a much higher diversity, with possibly more than 200 species being present worldwide during all time intervals of this period.

In view of this information, it is premature to propose a global biodiversity curve. However, the revision of regional diversity trends of areas that have been investigated in more detail provides some useful insights, as shown in the next section.

■ Local and Regional Diversity Trends

Some research projects have included investigations of acritarch assemblages from continuous sections with excellent independent biostratigraphic control. These include monographs that depict local or regional biodiversity trends. Summarized in the following paragraphs are results from Baltica and a reevaluation of publications from three other areas, the aim being to document those regional biodiversity trends possibly in relation to geologic events.

Baltica

The acritarch data set from Baltica is the largest compared with other areas in terms of the number of described taxa. However, it is difficult to provide an overview for this region because different authors have tended to adopt different taxonomic concepts. Many monographic articles include a large number of new species, but most lack definitive biostratigraphic control.

Because of the possibility of large numbers of synonymous species, it is difficult to establish acritarch diversity trends for Baltica. Nevertheless, the detailed study of one section in Estonia provides very useful information (Uutela and Tynni 1991). This section, the Rapla borehole (some 60 km south of Tallinn), is biostratigraphically well documented, and the diversity patterns of the acritarchs were compared not only with those of the chitinozoans but also with the changing lithofacies (Kaljo et al. 1996). A major result was recognition of regularly alternating acritarch and chitinozoan peaks (high acritarch diversity corresponding to low chitinozoan diversity and vice versa), except at the beginning of the Llanvirn, when both groups radiated strongly. The extinction curves of both groups are similar, with major extinctions occurring during the late Caradoc (Keila) and late Ashgill (Pirgu), the latter reflecting the Late Ordovician extinction event.

North America (RW)

The current Ordovician acritarch data set indicates that 103 genera and 337 species that are reported in the literature occur in North American Ordovician strata. Of these, 26 genera and 153 species were originally described as new.

Despite what appears to be a reasonably large data set for North American Ordovician acritarchs, many of the papers only describe one or a few acritarch species and do not include the entire acritarch assemblage. Furthermore, several of the described assemblages are based on only one or a few samples from a single locality and do not represent an entire formation. Most of the reported occurrences are not from continuous sections. Moreover, the sampling did not include several formations and lithotopes; nor did it cross series or stage boundaries. This makes it difficult to determine precise stratigraphic ranges for many species and to compare assemblages. Additionally, some of the illustrations and original descriptions of species are inadequate for positive attribution in subsequent studies, resulting in new species being created and a consequently higher diversity than actually existed. Furthermore, numerous species have been erected on the basis of only a few specimens, potentially leading to a large number of synonymies.

Reported occurrences of Lower Ordovician (Tremadocian and Arenig) acritarchs from North America are sparse (Dean and Martin 1978, 1982; Martin and Dean 1981; Martin 1984, 1992; Barker and Miller 1990). The acritarch assemblages described by Dean and Martin (1978) and Martin and Dean (1981) are from sections in eastern Newfoundland, Canada (part of Avalonia), and have little in common with the Laurentian assemblages discussed in the following paragraphs.

Lower Ordovician Laurentian acritarch assemblages have been described by Dean and Martin (1982), Martin (1984, 1992), and Barker and Miller (1990). Dean and Martin (1982) and Martin (1984, 1992) reported on Tremadocian and Arenig acritarch assemblages from Wilcox Pass, Alberta, Canada. At this locality, assemblages were reported from the Survey Peak Formation (Tremadocian) and the Outram and Skoki formations (Arenig). Twenty-four species were recorded (six new) from these three formations, the majority from the Survey Peak Formation. Important taxa include *Aryballomorpha grootaertii*, *Athabascaella playfordii*, *A. rossi*, and *Lua erdaopuziana*, which are also found in northeastern China (Martin and Yin 1988).

Barker and Miller (1990) reported a Tremadocian acritarch assemblage recovered from a stratigraphic test well in Terrell County, Texas. The assemblage was dominated by sphaeromorphic and acanthomorphic acritarchs, including representatives of the genera *Aryballomorpha*, *Athabascaella*, and *Lua*.

The genera *Aryballomorpha*, *Athabascaella*, and *Lua* have been reported from Laurentia, northeastern China, Australia, and Öland, Sweden. This association is considered by some authors to be characteristic of a warm-water province (Barker and Miller 1990; Servais and Fatka 1997; Volkova 1997). Except for Öland, the aforementioned locations all were located at low latitudes during the Tremadocian, and Öland (part of Baltica) was located in intermediate latitudes, between the warm- and cold-water assemblages.

There are no Middle Ordovician acritarch assemblages reported from North America. Loeblich and Tappan (1978 and authors quoted therein) did, however, describe numerous new acritarch species from the Mountain Lake Member of the "Middle Ordovician" Bromide Formation of Oklahoma. But, as reported by Bauer (1994), conodonts indicate that the Mountain Lake Member is not Mid Ordovician but late Whiterockian to early Mohawkian in age, which corresponds to the early Caradoc.

The Upper Ordovician has received the greatest attention, and most reported occurrences come from locations with well-documented stratigraphic and paleontological control. Upper Ordovician localities from North America include Indiana (Loeblich and Tappan 1978; Colbath 1979), Kansas (Wright and Meyers 1981), Kentucky (Jacobson 1978; Loeblich and Tappan 1978), Missouri (Miller 1991; Wicander et al. 1999), Ohio (Loeblich and Tappan 1978), Oklahoma (Loeblich and Tappan 1978), Anticosti Island, Quebec, Canada (Staplin et al. 1965; Jacobson and Achab 1985; Jacobson 1987), St. Lawrence Lowland in the Ottawa, Montreal, and Quebec City areas, Ontario and Quebec, Canada (Martin 1983), and Gaspé, Quebec, Canada (Martin 1980). These assemblages are very diverse with representation of the following genera: *Baltisphaeridium, Dorsennidium, Eupoikilofusa, Leiofusa, Lophosphaeridium, Multiplicisphaeridium, Orthosphaeridium, Peteinosphaeridium, Polygonium,* and *Veryhachium.* Many of the species that occur in several North American localities are also found outside North America and are thus useful for correlating Upper Ordovician strata. These include *Dorsennidium hamii, Excultibrachium concinnum, Multiplicisphaeridium irregulare, Orthosphaeridium insculptum, O. rectangulare,* and *Villosacapsula setosapellicula.*

An understanding of biodiversification trends in North American (Laurentian) Ordovician acritarch assemblages remains still in its infancy. Only a few areas have been studied in the Lower Ordovician of Laurentia, making conclusions for this time interval speculative. Additionally, no North American Middle Ordovician acritarch assemblages have been reported in the literature. Even the best-known Upper Ordovician North American acritarch assemblages require further studies before any meaningful trends can be ascertained.

South China (LJ)

During the Ordovician, China was probably composed of a series of paleocontinents. Three of these major blocks are the North China (Sino-Korean), Tarim, and South China plates. More than 50 publications have dealt with Ordovician acritarchs of China, and some of these include detailed taxonomic analyses.

Only a few publications concern North China. This area has so far been investigated only for the biostratigraphy of acritarchs across the Cambro-Ordovician boundary, especially in the Dayangcha section (Yin 1985). Although the diversity trend in this section is not precisely established, a gradual increase in the number of species can be seen from the latest Cambrian to the early Tremadocian. The only studies from the Tarim Plate concern the Qilang Formation (e.g., Li 1995). The acritarchs are from the mid–Caradoc graptolite-based *Dicranograptus clingani* Zone. Diversity trends are not yet determinable because acritarch studies of the underlying and overlying strata are still in progress.

Most investigations have focused on the Lower and Middle Ordovician of South China, that is, on the Yangtze Platform and the Jiangshan-Changshan-Yushan (JCY) area, from levels between the late Tremadocian and early Darriwilian. The acritarchs of the Yangtze Platform have been detailed by several groups of workers but are principally based on the Ph.D. theses of Li (1991), Xu (1995), and Brocke (1998). Thirty-nine publications covering this region have been published to date and are listed in Li et al. (2002b).

The paleogeography and the paleoecology of the Yangtze Platform are fairly well understood, as this area has been better studied than any other region of China (Chen et al. 2001). The dominance of carbonates and marls indicates a relatively stable carbonate platform persisting from the late Tremadocian throughout the remainder of the Ordovician. Southeastward, these carbonate rocks gradually change into the graptolitic shale facies of the Jiangnan Belt, which includes the JCY area and locally comprises some intercalated, lenticular beds of limestone. Local lithofacies changes can be observed within the Yangtze Platform, and these are reflected by biofacies changes from shelly to graptolite-dominated faunas. Sea level on the platform generally rose through earliest Bendigonian (early Arenig) to Darriwilian time.

Acritarchs have been recovered from many sections of the Yangtze Platform. The assemblages clearly reflect a gradual lithofacies change from the northwest to the southeast as indicated by changing abundance and diversity patterns that can be observed in the assemblages recovered from samples in the graptolite-based *Azygograptus suecicus* Zone (earliest Mid Ordovician) (Li et al. 2002a). In the shallow-water part of the western Yangtze Platform in eastern Yunnan, the acritarch assemblages are not very rich, with low abundance and diversity. Acritarch assemblages from deeper-water conditions to the east show much higher diversity, but with variable composition, depending on their location on the platform. Within the deeper parts of the basin, that is, toward the JCY area, the diversity again decreases (albeit based on poorly preserved acritarch assemblages). These compositional differences probably reflect, in part, a nearshore to offshore trend (Li et al. 2002a).

In some continuous sections, acritarchs have been investigated from a succession of graptolite zones. Together with shorter stratigraphic sections representing shorter time intervals, the data set of acritarch samples is today quite large, providing a fairly complete spatial and temporal picture of acritarch distribution for this area.

The highest number of taxa recorded are from samples within the graptolite-based *A. suecicus* Zone. A review of the acritarchs from this zone indicates a total of 111 acritarch species, 26 of which were newly described with 35 species remaining in open nomenclature. This number is probably too high and reflects "splitting" of taxa by some authors. Tongiorgi et al. (1995), for example, named more than 60 taxa from a single sample of this interval in the Daping section, and Yin (1995) provided a different list of 60 species from a sample of the same interval from the nearby Huanghuachang section. Taxonomic reconsiderations including the analysis of synonymies are necessary in order to understand the real acritarch diversity in southern China. In the present study, we use an internally consistent acritarch taxonomy based on Brocke and his co-workers (Brocke et al. 2000; Li et al. 2002a) in order to establish the diversity trend in the Yangtze Platform, as shown in figure 32.3.

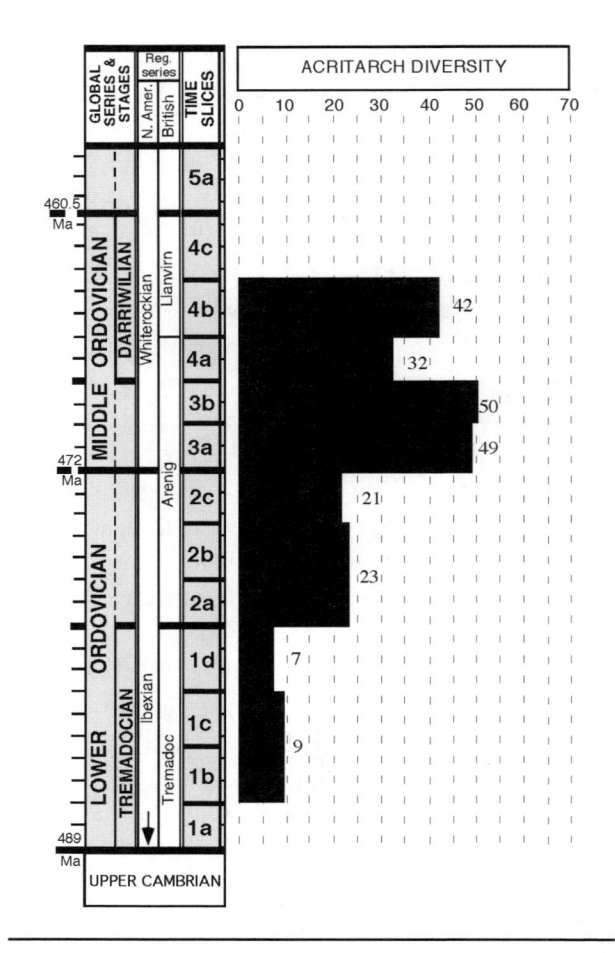

FIGURE 32.3. Diversity trends in the Yangtze Platform, South China (numbers refer to the species described from intervals between the middle Tremadocian and the lower Darriwilian). Time slices (*TS*) are after Webby et al., chapter 2.

Eight acritarch assemblages have been distinguished (Li 1987; Brocke 1997; Brocke et al. 2000; Li et al. 2002a). The first three acritarch assemblages have been documented by Brocke (1997). The oldest association appears in the early Tremadocian (conodont-based *Cordylodus angulatus* Zone) and continues into the mid Tremadocian, showing a low diversity of 9 species. This assemblage is here referred to *TS*.1b–c (figure 32.3; see time slices in chapter 2). The second assemblage of late Tremadocian (conodont-based *Paltodus deltifer* Zone, *TS*.1d) age also displays low diversity, with only 7 species recorded by Brocke (1997). The next assemblage, identified in the Yangtze Platform, is of latest Tremadocian and early Arenig age and is here attributed to *TS*.2a–b. It is moderately well preserved, with upwardly increasing diversity, and comprises 23 species (Brocke 1997). The assemblage recorded in the graptolite-based *Didymograptus (Corymbograptus) deflexus* Zone (Li 1987, unpubl. data) and here assigned to *TS*.2c shows a similar diversity (21 species recorded).

The other four assemblages are documented in Brocke et al. (2000). A single acritarch assemblage is recorded from graptolite zones—the *A. suecicus* Zone and in the lower part of the *Exigraptus clavus* (formerly *sinodentatus/nexus*) Zone (*TS*.3a)—with a maximum of 49 species recorded (Brocke et al. 2000; Li et al. 2002a). In the lower part of this assemblage, that is, in the *A. suecicus* Zone, the diversity is higher than in the upper part, that is, in the lower *E. clavus* Zone; the middle part contains almost no acritarchs. This interval with no acritarchs corresponds to the reddish carbonates and nodular limestones of the middle parts of the Dawan and Meitan formations and is considered to represent a regressive marine phase. In the middle and upper parts of the *E. clavus* Zone (*TS*.3b), the diversity is highly variable and reflective of various sedimentologic facies (Brocke et al. 2000). The diversity in the samples from this interval is generally lower than in the previous assemblage, but the total number of species, recovered from different facies types, is closely consonant with the *A. suecicus* Zone (50 species). The next acritarch assemblage, from the base of the Darriwilian (graptolite-based *Undulograptus austrodentatus* Zone), is assigned to *TS*.4a and is only moderately diversified, with 32 species recorded by Brocke et al. (2000). The youngest assemblage so far recorded in the Yangtze Platform is placed approximately in *TS*.4b. Its diversity is still lower than that in the *A. suecicus* Zone, with some 42 species identified.

The Yangtze Platform studies demonstrate the difficulty of establishing diversity trends without knowledge of the biostratigraphy and lithofacies conditions of the wider area. Although the influences of lithofacies (nearshore-offshore trends) are not yet completely understood, and despite some remaining biostratigraphic problems, the following conclusions can be drawn:

1. There is evidence for a continuous increase in acritarch diversity from the lower Tremadocian up to the lowermost Middle Ordovician (*TS*.3a). This increase most probably corresponds to the general transgression observed in southern China.

2. The highest diversity is observed in the lower part of the Middle Ordovician (*A. suecicus* Zone). Following a brief regression, coinciding with the local occurrence of reddish carbonates and barren samples, the acritarch diversity remains high in the *E. clavus* Zone (*TS*.3b) and can possibly be attributed to renewed transgression.

3. At the base of the Darriwilian, a further decline in diversity is possibly related to the regional regression during this interval.

Further data from continuous acritarch successions in southern China, together with a more consistent taxonomic approach by all involved workers and greater biostratigraphic resolution, are needed to achieve a more comprehensive analysis.

North Africa (MV)

The analysis of Ordovician acritarch diversity of North Africa is based principally on borehole sections in Algeria, Libya, and Tunisia investigated by the present author (Vecoli 1996, 1999) and colleagues (Vecoli et al. in press). Other sources are published and unpublished data from palynologists working for the French oil industry during the 1960s and 1970s (e.g., Deunff 1961; Combaz 1967; Jardiné et al. 1974). More recently published work (e.g., Molyneux 1988; Oulebsir and Paris 1995; Paris et al. 2000a) was used mainly for calibrating the biostratigraphic correlations. Palynological data from other sources (e.g., the Ordovician of Morocco; Cramer and Díez 1977; Elaouad-Debbaj 1988a, 1988b, and references therein) were taken into account but not used directly to derive the current diversity curve, which is based only on the author's database.

Compared with other areas, and despite some problems of stratigraphic continuity as discussed later, the North African Ordovician acritarch record presents some advantages. These include accurate taxonomic studies of the acritarch floras (Vecoli 1996, 1999), typically excellent preservation, and very high palynomorph abundance, as well as independent biostratigraphic control provided by co-occurring macrofauna (graptolites; not very abundant but important at particular levels) and microfauna (chitinozoans; of great importance in all post-Tremadocian sedimentary strata).

Therefore, for the dating and correlation of Ordovician strata in the Sahara Platform subsurface, acritarchs can now be considered indispensable tools, complementary to the chitinozoans (Vecoli et al. in press). Recent acritarch studies in North Africa have permitted placement of the Cambrian-Ordovician boundary (Vecoli 1996), as well as paleobiogeographic (Colbath 1990) and paleoenvironmental reconstructions (Paris et al. 2000a; Vecoli 2000). Based on analyses of approximately 500 samples from several subsurface sections in Algeria (boreholes Uc101 and Nl2), southern Tunisia (boreholes Tt1, St1, Sn1), and Libya (boreholes A1-70, A1-23, B1-23, B2-34, C1-34, A1-70, A2-70, and A3-70), a first attempt is made here to assess semiquantitatively the diversity dynamics and the main evolutionary trends of North African Ordovician acritarchs. Acritarch diversity has been calculated as simply the number of species per time slice (*TS*), thus representing the "standing diversity" for each time interval (figure 32.4). Mostly direct or indirect correlation of acritarch biostratigraphy with chitinozoan or macrofossil (i.e., graptolites) formal zonations permits a stratigraphic resolution comparable to the duration of a time slice (or a chitinozoan zone). In some specific chronostratigraphic intervals (i.e., the lower Tremadocian), it is even possible to propose a finer resolution, as discussed later.

The present study is only preliminary and, in part, incomplete because of the presence of two major stratigraphic hiatuses in the studied area (spanning the middle Tremadocian through the lower Arenig and the middle Caradoc through middle Ashgill; see figure 32.4). The author is currently completing a more comprehensive study of the area. Moreover, although the data used herein were derived from a relatively restricted area (i.e., approximately equivalent to the extension of the Ghadamis Basin sensu lato) and therefore might not be totally representative for the entire North Saharan Platform, they at least represent a workable base for future studies. Some important features of acritarch diversification dynamics can thus be outlined as follows.

1. **Cambro-Ordovician radiation.** Uppermost Cambrian acritarch associations of North Africa are fairly diversified. Together with some typical Cambrian genera that became extinct at or about the systemic transition (e.g., *Timofeevia, Trunculumarium, Ooidium, Cristallinium, Phenacoon, Ladogella*), Late Cambrian acritarchs already include taxa that became well established during the Early Ordovician (e.g., *Cymatiogalea, Stelliferidium, Acanthodiacrodium, Vulcanisphaera*). Overall, the Cambro-Ordovician transition is characterized by increased diversity.

2. **Early Tremadocian differentiation.** The early Tremadocian is characterized by a conspicuous increase in morphological variety among the diacromorph and galeate (herkomorph) acritarchs. The increase in diversity occurs entirely within *TS*.1a; according to Vecoli (1996), a lowermost assemblage is recognizable (HM/B of Vecoli 1996) that is here tentatively correlated with the lowermost part of *TS*.1a (figure 32.4). In general, *Stelliferidium, Cymatiogalea,* and *Acanthodiacrodium* tend to dominate the early Tremadocian assemblages, and innovative morphologies such those of *Arbusculidium* (probably related to

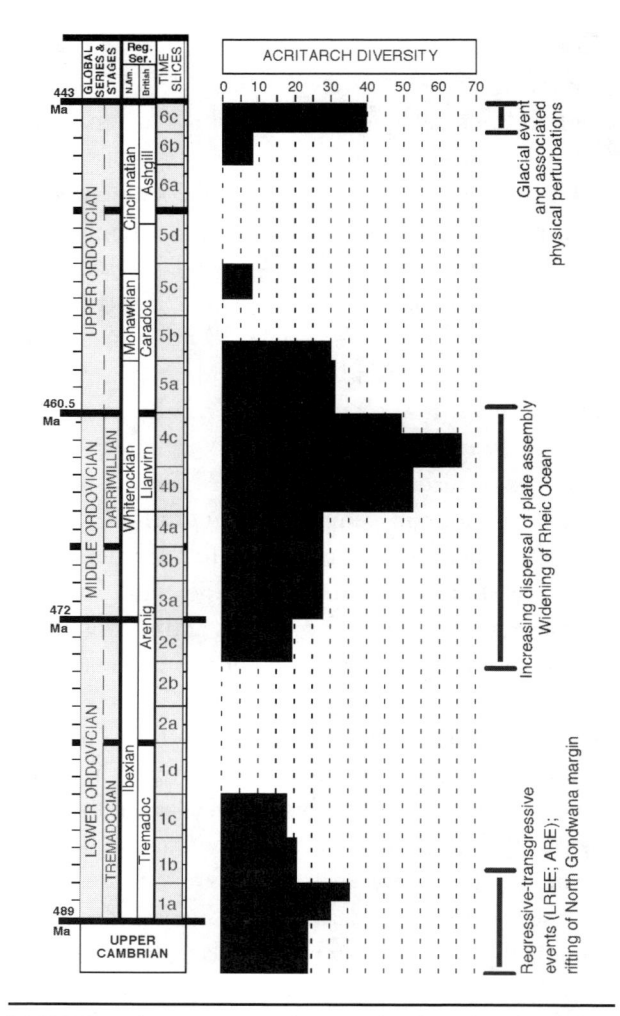

FIGURE 32.4. Diversity trends in North Africa (data from Algeria, Tunisia, and Libya). LREE = Lange Ranch Eustatic Event; ARE = Acerocare Regressive Event—see also chapter 10; and time slices (*TS*) after Webby et al., chapter 2.

the Upper Cambrian genus *Ladogella*) evolved. Numerous species of *Vulcanisphaera* also first appear during the early Tremadocian, as Rasul (1976) observed in the Tremadocian Shineton Shales of Shropshire, England. The FAD of the morphologically distinctive *Acanthodiacrodium angustum* can be considered as defining a horizon approximating the Cambrian-Ordovician boundary.

3. Middle to upper Arenig record. Arenig strata, only partly represented in the investigated North African sections, include relatively well-diversified acritarch assemblages, with first appearances of such genera showing innovative morphologies as *Coryphidium*, *Vavrdovella*, *Dicrodiacrodium*, and *Frankea*. The Tremadocian-Arenig transition corresponds to a stratigraphic hiatus in the study area, and therefore the acritarch dynamics of the transition from Tremadocian to typical Arenig are not observable.

4. Llanvirn radiation and differentiation. The basal Llanvirn exhibits a marked increase in acritarch diversity. The assemblages commonly contain up to 60–70 species, many with complex morphologies such as strongly sculptured, polyhedral vesicles bearing simple or, more commonly, complexly branched processes (e.g., species of *Arkonia*, *Frankea*, *Dicrodiacrodium*, *Coryphidium*, *Aureotesta*, *Striatoteca*, *Multiplicisphaeridium*, and *Vogtlandia*). Some genera that originated during Tremadocian-Arenig times show characteristic patterns of diversification, such as *Acanthodiacrodium* (*A. costatum*, *A. uniforme*), *Stelliferidium* (*S. striatulum*, *S. stelligerum*), *Cymatiogalea* (*C. granulata*), *Peteinosphaeridium* (*P. velatum*), and *Liliosphaeridium* (*L. pennatum*). The following genera were introduced in the North African region during Llanvirn times: *Ericanthea*, *Orthosphaeridium*, *Ordovicidium*, *Stellechinatum*, and *Villosacapsula*.

5. Lower Caradoc diversity decrease. Only data on acritarchs from the basal part of the Caradoc, which have a sufficient degree of taxonomic accuracy, are available. Correlation with the chitinozoan biozonation of Oulebsir and Paris (1995) and results from unpublished investigations by the present author shows that probably only *TS*.5a and 5b of the Caradoc are represented in northern Algeria, as well as the Ghadamis Basin of Libya. Upper Caradoc acritarchs have been recovered from the Illizi Basin in southern Algeria, but taxonomic information is not available. In general, Caradoc acritarch assemblages are of relatively low diversity and include species ranging from Llanvirn strata; they are dominated by both triangular and quadrangular forms of *Veryhachium*, netromorphic acritarchs, *Multiplicisphaeridium*, and less commonly species of *Ordovicidium*. Although incomplete, these data indicate that the Caradoc was most probably characterized by very low inception rates among microphytoplankton biotas.

6. Upper Ashgill record. In the upper Ashgill, the acritarch floras appear almost completely renewed and show relatively low generic yet high specific diversity. The most represented morphotypes are veryhachiid (both triangular and rectangular and including the genus *Villosacapsula*) and netromorphic acritarchs (e.g., *Dactylofusa*, *Poikilofusa*). Uncommon or rare occurrences of characteristic forms such as *Baltisphaeridium*, *Ordovicidium*, and *Multiplicisphaeridium* are noted. Just as important, the Ashgill record includes the first occurrence of distinct genera having a "Silurian" affinity, such as *Diexallophasis* (= *Evittia*) and *Oppilatala*.

The observed pattern of fossil microphytoplankton diversification in the Ordovician of North Africa is tentatively correlated with the main geologic and paleoenvironmental changes affecting the region during this time (figure 32.4). Given the complexity of factors influencing microphytoplankton dynamics, no direct causative relationships are implied in figure 32.4. The observed evidence suggests that the most abundant and diverse acritarch assemblages correspond to periods of intense tectonic activity that took place in the more peripheral parts of North Gondwana and also to periods of sea level highstand in the Gondwanan platforms. Latest Cambrian–early Tremadocian and Arenig-Llanvirn times were, in fact, periods of plume-induced rifting of the North Gondwanan margin (Winchester et al. 2002) that caused the separations of various terranes (including Avalonia) and perhaps were primarily responsible for sea level fluctuations. During late Arenig through Llanvirn times, the dispersal of the plate assembly in the region of the Gondwana-Baltica interface was at its maximum, and this paleogeographic and paleoceanographic situation was probably reflected in acritarchs by strong provincial differentiation.

The late Ashgill acritarch assemblages of the North Sahara Platform certainly reflect the response of microphytoplankton communities to the paleoenvironmental perturbations linked to the end Ordovician glaciation. It has been recently demonstrated (Paris et al. 2000a) that the upper Ashgill sediments of the Algerian basins actually correspond to the final stages of glaciation and the beginning of the postglacial climatic amelioration. Although of relatively low generic diversity, the late Ashgill North African acritarch floras are characterized by extreme polymorphism within the veryhachiid and netromorphic groups and by the first occurrences of numerous new morphotypes such as *Dicommopalla, Diexallophasis, Leprotolypa, Cheleutochroa,* and *Oppilatala.* It is interesting to note that most members of the latter group of acritarch genera have a definite "Silurian affinity" and survive the end Ordovician biotic crisis. The acritarch signal of the "end Ordovician mass extinction" is difficult to detect; actually, our data suggest that during the end Ordovician and the Ordovician-Silurian transitions, origination rates were unusually high, and hence it would be more suitable to use the term "end Ordovician turnover" rather than "mass extinction" for the acritarch flora.

■ Summary

It is impossible at this time to compile a global diversity curve, as the taxonomy of many species remains obscure and the exact stratigraphic ranges of most species are still unknown. However, regional diversity trends can be observed. The plotting of the information of the data set of British acritarchs by Downie (1984) indicates an ongoing diversification in the Early and Mid Ordovician with a peak of diversity in the mid to late Darriwilian (Llanvirn) before a decrease of the diversity in the Late Ordovician (figure 32.1).

Following detailed investigations in regional areas such as China (figure 32.3) and North Africa (figure 32.4), it appears that acritarch diversity basically mirrors transgressive-regressive trends. During marine transgressions, acritarch diversity increases, and during regressive phases the reverse applies. Similar trends can be seen among the Mesozoic-Cenozoic dinoflagellate floras in which diversity changes are closely linked to relative sea level fluctuations (Stover et al. 1996: figure 5). The present data set is not precise enough, however, to provide information about the nature of specific/generic diversifications/extinctions or turnover rates of the phytoplanktonic organisms that produced the resting cysts that were preserved in the fossil record. It is thus today still difficult to establish relations between the evolution of the marine phytoplankton and the invertebrate diversity in the Ordovician.

ACKNOWLEDGMENTS

As this chapter on acritarchs is basically a review of published data, we are grateful to numerous colleagues (including almost all Ordovician acritarch workers) for valuable information and discussions. We especially thank those who contributed to the database: A. Wallin (Stockholm), M. Masiak and M. Stempien (Warsaw), O. Fatka (Prague), T. Heuse (Weimar), K. Dorning (Sheffield), G. Playford (Brisbane), C. Rubinstein (Mendoza), A. Di Milia (Pisa), and M. Vanguestaine (Liège). We are particularly grateful to the reviewers of this chapter, G. Playford and an anonymous reviewer, as well as B. D. Webby (Sydney) and F. Paris (Rennes) for their editorial corrections and comments. Special thanks must also go to the principal organizers of IGCP project no. 410, who initiated the work and encouraged us to make the database, which otherwise may not have been completed.

33 Miospores and the Emergence of Land Plants

Philippe Steemans and Charles H. Wellman

During the past 30 years it has become increasingly clear that for much of the Ordovician a terrestrial vegetation existed. Evidence is in the form of dispersed microfossils (miospores and phytodebris) that are interpreted as deriving from land plants (embryophytes). As yet no megafossils have been recovered, probably owing to the low preservation potential of the plants. In the first part of this chapter the dispersed miospore fossil record is summarized, along with evidence it provides for the biodiversity and paleophytogeography of the parent plants. In the second part the affinities and nature of the vegetation are assessed based on evidence from the microfossil record and other sources.

■ Miospores (PS)

Cryptospores (Richardson et al. 1984 emend. Steemans 2000) are alete miospores (nonpollen grains) produced by primitive embryophytes. They are by far the most abundant miospores during the Ordovician, in contrast to trilete spores, which are very rare. Cryptospores have no haptotypic feature such as trilete or monolete marks, although some forms have a well-differentiated contact face (hilate monads). Cryptospores are preserved in tetrad, dyad, or monad configurations. Different basic forms of permanent tetrads and dyads are based on the nature of contact between the spores that constitute the tetrads or dyads. All these forms may be naked or enclosed within an outer envelope. Cryptospores with or without an outer envelope may be smooth or ornamented. Usually the ornamentation is discrete, constituting grana or convoluted folds, but sometimes it is structured into a reticulate pattern (see Richardson 1996). Examples of cryptospores are illustrated in figure 33.1.

Palynomorphs showing an imperfect trilete structure, resulting from tearing on breakup of the tetrad, are considered as cryptospores (*Imperfectotriletes*). According to Steemans et al. (2000), the trilete spore group should include only specimens with a complete trilete structure and without parts of the tetrad still adhering. The trilete spores observed in the Ordovician belong to the morphon *Ambitisporites avitus* (Hoffmeister 1959) Steemans et al. 1996. They are small (around 25 μm), very simple, without ornamentation, and with a narrow equatorial thickening.

Some enigmatic palynomorphs such as *Attritasporites*, *Virgatasporites*, and *Moyeria* have been considered as either acritarchs or miospores. However, there is no evidence that these palynomorphs are produced by embryophytes (land plants). Most of them are known only from marine sediments. Therefore, these palynomophs of enigmatic origin should be considered as acritarchs (Richardson 1996; Steemans 2000) and are not discussed here.

FIGURE 33.1. Examples of cryptospores and a spore mass from the Caradoc of the Arabian Peninsula. All figures at magnification ×700, except for M and O, which are at ×350, and P and Q, which are at ×155 and ×420, respectively. A–D, naked permanent tetrads. E–G, envelope-enclosed permanent tetrads. The envelope is murornate in E and smooth in F and G. H–K, naked permanent dyads. L–O, envelope-enclosed permanent dyads. The envelope is murornate in L and M and smooth in N and O. P and Q, SEM image of a spore mass comprising naked permanent tetrads.

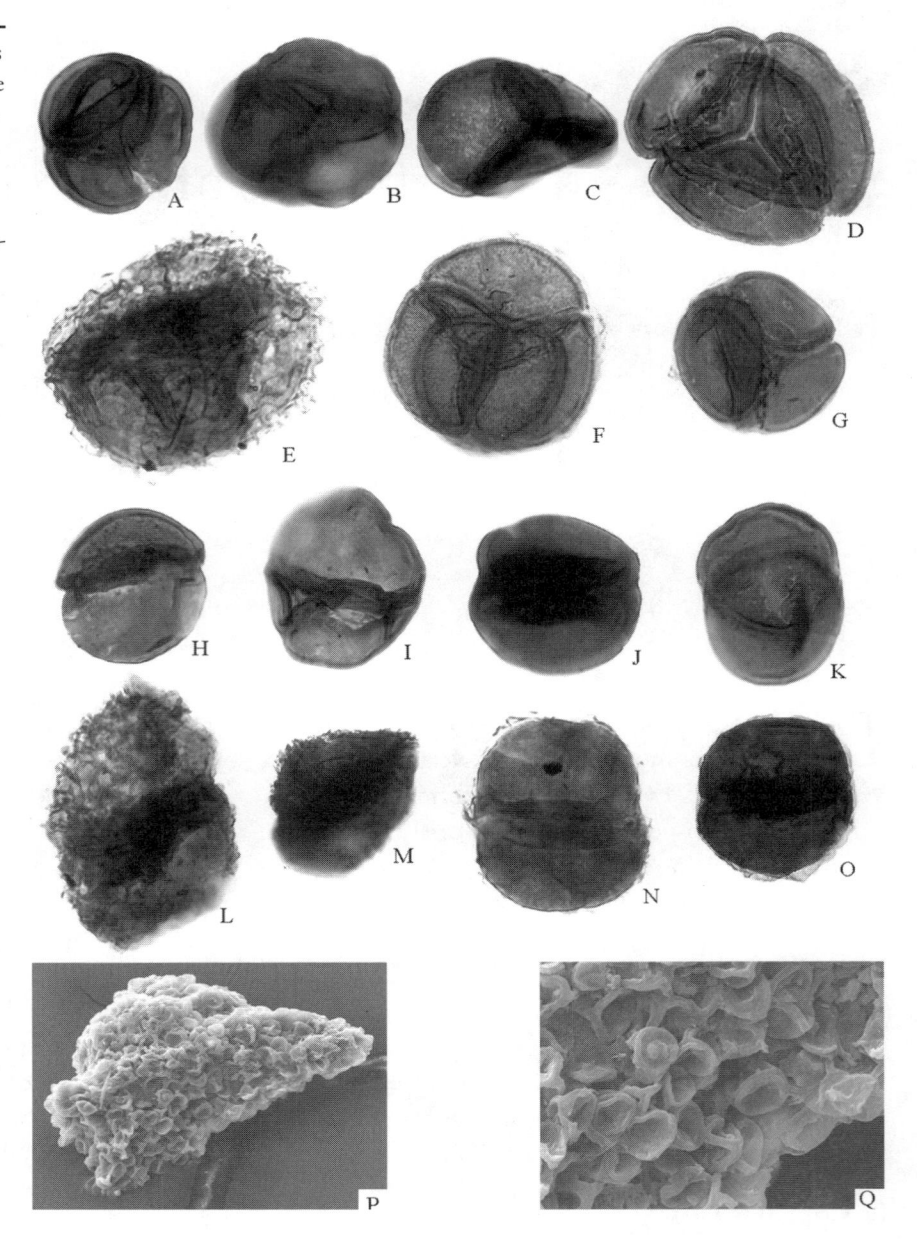

Stratigraphic Occurrence

Only the richest assemblages are reported here. For a more complete review, see Richardson (1996), Wellman (1996), and Steemans (1999a, 1999b, 2000).

Sporelike microfossils have been described from the Middle Cambrian strata (Strother and Beck 2000), but their classification as cryptospores is controversial because the spores lack the rigidity and symmetry of genuine embryophyte spores (Wellman 2003). The oldest indisputable cryptospores have been observed from the Hanadir Shales of Saudi Arabia (Strother et al. 1996a). Graptolites of the *Didymograptus murchisoni* Zone indicate a Llanvirn age. Unfortunately, the specimens are poorly described, and no systematic treatment has been provided. The assemblage contains monads, dyads, and tetrads. Some of these are enclosed within a laevigate membrane. Some tetrads exhibit disorganized attachment configuration. Except for the ornamented outer envelopes, most of the cryptospore morphologies known from younger Ordovician strata are present in the Hanadir Shales. Despite the absence of more detailed systematic information, at least nine genera and nine species can be distinguished in the assemblage on the basis of the illustrations.

Tetrads and monads (*Imperfectotriletes* type) have been reported in Bohemia, Czech Republic (Vavrdová 1984), from the Šárka Formation of Llanvirn age.

Millward and Molyneux (1992) have reported tetrads and dyads from the Over Water Formation of the Lake district, northern England. The age of these sediments ranges from the Llanvirn to Caradoc.

Wellman (1996) described in detail a cryptospore assemblage from the Caradoc type area in the Welsh Borderland of southern Britain, previously reported by Richardson (1988). The assemblage contains 17 species and 13 genera. Species enclosed within an ornamented envelope appear for the first time in this Caradoc sequence.

Gray et al. (1982) reported tetrads and single spores physically broken out of permanent tetrads from the Melez Chograne Formation in the Murzuck and Southern Rhadames basins of Libya. This formation is considered to be Caradoc in age, although evidence for this is limited.

There are numerous observations of miospores from the Ashgill. The richest assemblages have been found by Richardson (1988) in northeastern Libya, by Steemans et al. (1996) in southeastern Turkey, by Burgess (1991) in southwestern Wales, by Vavrdová (1982, 1984, 1988, 1989) in the Czech Republic, by Steemans (2001) in Belgium, and by Wang et al. (1997) in China. All these assemblages are very similar in composition, dominated by tetrads, dyads, and monads, naked or envelope-enclosed. Usually dyads and tetrads remain firmly attached, and the rare monads such as *Laevolancis* and *Imperfectotriletes* are physically dissociated from polyads. The most important event is the first appearance of true trilete spores in the Hirnantian assemblage from Turkey (Steemans et al. 1996).

Early Llandovery miospore assemblages are very similar to those from the Ashgill in their gross composition (Gray 1985; Richardson 1996; Wellman 1996; Steemans 1999a, 1999b, 2000). Early Llandovery assemblages are known from southwestern Wales (Burgess 1991), Libya (Hill et al. 1985; Richardson 1988), Saudi Arabia (Steemans et al. 2000; Wellman et al. 2000); Brazil (Melo 1997; Melo and Steemans 1997; Mizusaki et al. 2002); Paraguay (Steemans and Pereira 2002); and the United States (Strother and Traverse 1979). Similar assemblages are known from regions in paleolatitudes close to the equator and from regions close to the South Pole. All these observations indicate that the parent plants were able to survive in a wide range of climates.

The Hirnantian glaciation is believed to have caused mass extinction in many fossil groups. However, no impoverishment in spore biodiversity is observed during the Hirnantian glaciation (figure 33.2). The cosmopolitan nature of the earliest land plants in respect to climate could explain why climatic changes did not affect the biodiversity. Figure 33.2 illustrates the evolution of cryptospore biodiversity from the Llanvirn to the Telychian. There is a major biodiversity event in the upper part of the Llandovery with significant decrease in the number of cryptospore species (Steemans 1999a, 1999b, 2000).

■ Evidence for Embryophyte (Land Plant) Affinities (CHW)

The Fossil Record

The earliest fossil evidence for land plants is in the form of dispersed microfossils: spores and phytodebris (fragmentary cuticles and tubular structures). Dispersed spores first appear in the Llanvirn, and a more or less continuous record exists throughout the remainder of the Ordovician (summarized in the previous section of this chapter). The fossil record of phytodebris is patchy. Fragmentary cuticles are known from at least the Caradoc but have rarely been reported from the Ordovician, although they are common in younger sediments. There are sporadic reports of fragmentary tubular structures from the Ordovician. These consist entirely of forms that are internally smooth but externally are either smooth or ornamented. Those with an internal ornament of annular/spiral thickenings have not been reported from the Ordovician.

The earliest convincing land plant megafossils (i.e., relatively complete fossil plants) are from the Silurian but are rare, with only about 25 assemblages currently reported from this period (Edwards and Wellman 2001). There are no convincing land plant megafossils from the Ordovician. It is likely that the earliest land plants lacked recalcitrant tissues (e.g., those containing lignin) and consequently had very low fossilization potential and left no megafossil record. This is concordant with the hypothesis that the earliest land plants were bryophytelike (see later in this

FIGURE 33.2. Stratigraphic ranges of the most common miospore morphologies. Evolution of miospore biodiversity (number of species per stratigraphic level) through the Ordovician and Lower Silurian–Llandovery, with its stage subdivisions up to and including the Telychian. *TS*.3a–b, 4a–c, 5a–d, and 6a–c represent the time slices for the Middle to Upper Ordovician Series (*TS*.4a–c encompass the Darriwilian Stage). The middle column shows the British regional series subdivisions.

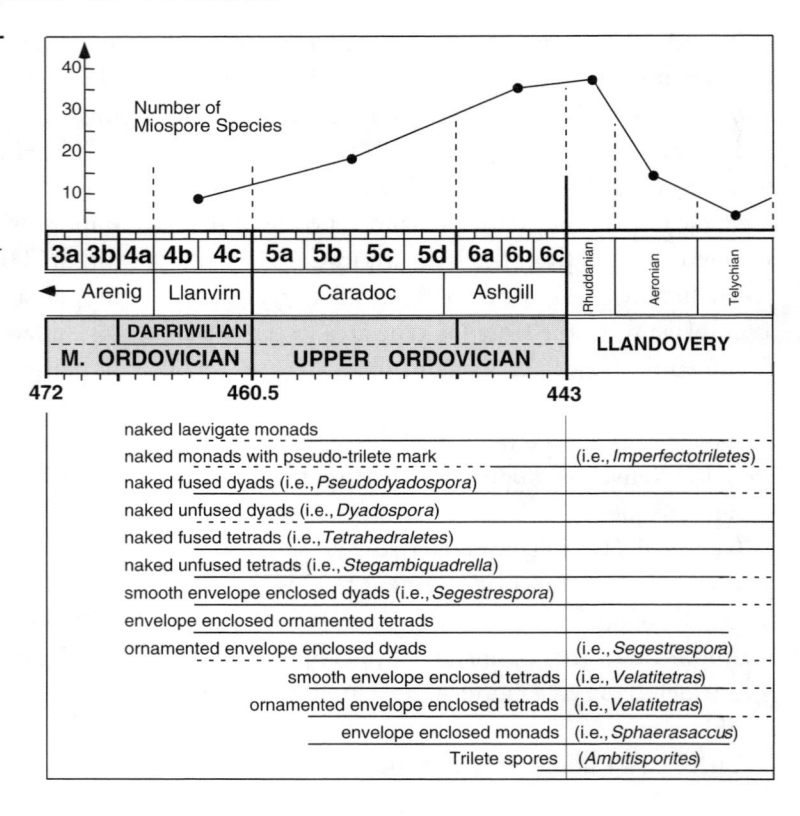

chapter), as extant bryophytes lack lignin. The land plant megafossil record probably coincides with the evolution of recalcitrant lignified tissues (e.g., stereome and conducting tissues), which dramatically increased fossilization potential. Some plants from the later Silurian (rhyniophytes and lycopsids) are clearly vascular plants with these attributes.

Affinities of Ordovician Land Plants

Phylogenetic analyses provide strong evidence that the embryophytes are monophyletic, with extant charophycean green algae their sister group. This suggests that the earliest land plants evolved from aquatic green algal ancestors. However, evolutionary relationships among the embryophytes are poorly resolved. There is some disagreement as to the exact relationship between the three extant bryophyte groups (liverworts, hornworts, and mosses) and the vascular plants. Most recent analyses indicate that the bryophytes are paraphyletic with respect to the vascular plants, with a moss/vascular plant sister group relationship. It is most likely that either the liverworts or hornworts are the most primitive, early divergent extant land plants.

There are several lines of evidence indicating that the Ordovician dispersed spores derive from land plants. Distribution of the Ordovician spores is similar to that of extant spores/pollen (i.e., they occur in nonmarine sediments, and when they are found in marine sediments, their abundance declines offshore). They are similar in size and morphology to the spores of land plants. In fact, some of the permanent tetrads, including envelope-enclosed forms, resemble the spores of extant liverworts (Gray 1985, 1991). Trilete spores, appearing later in the Ordovician, are more closely similar to the spores of vascular plants. Wall ultrastructure in Ordovician spores is varied but in some morphotypes is similar to that in extant liverworts (W. A. Taylor 2000). Intriguingly, some of the earliest known land plant megafossils, from the latest Silurian–earliest Devonian, contain spores similar in morphology to the Ordovician forms (e.g., cryptospore dyads/tetrads and trilete spores) (Edwards and Wellman 2001). It is possible that some of these plants represent a relic flora. The cryptospore producers are of uncertain affinity, although some possess tantalizing bryophytelike characters. The trilete spore producers include true vascular plants (rhyniophytes). In summary, the evidence suggests that the cryptospore producers and trilete spore producers most likely have bryophyte and tracheophyte affinities, respectively.

The dispersed cuticles are interpreted as deriving from land plants owing to their morphology and distribution (similar forms have been reported from nonmarine deposits of Early Silurian age). The type of plant from which they derive is unknown, but they lack stomata. It has been suggested that some may represent the covering of nematophytes. The nematophytes are an enigmatic group of organisms with tubular anatomy. They are of unknown affinity, although it has recently been suggested that they might represent either pathogens or decomposers and may belong with the fungi or lichens (e.g., Wellman and Gray 2000).

The dispersed tubular structures are linked to land plants because similar forms occur in nonmarine deposits of Early Silurian age, and they resemble the tubes of nematophytes (particularly those with internal annular/spiral thickenings). However, their affinities are far from clear, although a nematophyte origin seems most likely for at least some of them (e.g., Wellman and Gray 2000). Recently it has been suggested that some of the cuticles and tubes may represent the fragmentary remains of sporangia of bryophytes (Graham and Gray 2001), although the size and symmetry of the fossils do not support such an origin.

Nature and Ecology of Ordovician Land Plants

Interpretation of the nature and ecology of Ordovician land plants is based on (1) interpretation of the fossil evidence; (2) comparisons with extant analogues (based on reputed affinities and phylogenetic relationships); and (3) theoretical consideration of the attributes required by the first plants to colonize the land. Recent exhaustive discussion of evidence for the nature and ecology of the earliest land plants is provided by Gray (1984, 1985) and Graham and Gray (2001). Gray (1985:167) suggests that the cryptospore-producing plants were at a nonvascular (bryophytelike) grade of organization with "rapid colonization by founder populations with limited genetic diversity and with life-history strategies that included an ecophysiological tolerance to desiccation and a short vegetative life cycle" and that the trilete spore-producing plants were at a vascular grade of organization with "major establishment of large populations of genetically diverse plants exploiting a broad spectrum of ecological sites."

The actual fossils provide several lines of evidence that support Gray's hypotheses. Ordovician cryptospores display a number of characteristics expected of the reproductive propagules of founder populations. Their small size and abundance suggest that they were produced in vast numbers, subaerially dispersed and capable of long-distance dispersal. Furthermore, they were almost certainly produced by homosporous plants, a useful attribute for founder populations in that it permits selfing, although it restricts reproduction to environments that were at least periodically damp. However, it seems likely that Ordovician land plants could tolerate varying climatic conditions, since the spores are cosmopolitan, occurring from the paleoequator to near the paleo-pole. If the plants could reproduce only in damp environments and tolerated varying climatic conditions, an ecophysiological tolerance to desiccation and a short vegetative life cycle would have been desirable. Limited genetic diversity is suggested by the minimal evolutionary change (among the spores) during the Ordovician, although it is possible that the elusive gametophytes/sporophytes were evolving rapidly.

In summary, the earliest land plants appear to have been bryophytelike. They were abundant and cosmopolitan (based on the cryptospore fossil record) but probably confined to environments that were at least periodically damp. It has been proposed that they had a bryophytelike grade of organization and exhibited life-history strategies similar to those of extant liverworts (Gray 1984, 1985). Late in the Ordovician vascular plants may have appeared, as suggested by the first occurrence of trilete spores. There is little evidence as to the nature of the parent plants, but based on studies of later trilete spore-producing plants, they may have been similar to Late Silurian–Early Devonian rhyniophytes. The paucity of trilete spores in latest Ordovician spore assemblages suggests that the producers may initially have been a minor element of the flora. Furthermore, although data are scarce, Ordovician trilete spores have to date been reported only from a confined paleogeographic area, suggesting that the producers may not have been widespread at this time.

■ Conclusions

Dispersed microfossils, particularly spores, provide convincing evidence for a terrestrial vegetation in the Ordovician. The nature and affinities of the plants comprising this vegetation are interpreted based on limited evidence from the microfossils. There are currently no convincing megafossils reported from the Ordovician, but these will be critical in testing inferences based on the microfossils and in providing an insight into the actual gametophyte/sporophyte generation of the plants.

ACKNOWLEDGMENTS

C. H. Wellman's work was supported by NERC grant NER/B/S/2001/00211.

PART IV

Aspects of the Ordovician Radiation

34 The Ichnologic Record of the Ordovician Radiation

M. Gabriela Mángano and Mary L. Droser

The picture of the Ordovician Radiation comes primarily from the record of body fossils, in particular, patterns of diversity increase (Sepkoski 1981a, 1995; Sheehan 2001a; and elsewhere in this volume). Biodiversity in earth history has been accommodated via the colonization of new or previously underused ecospace (Bambach 1983; Orr 2001). The ichnologic record contributes significantly to our understanding of paleoecologic breakthroughs associated with the Ordovician Radiation, including an increase in complexity of shallow marine endobenthic communities, invasion of continental environments, and establishment of a modern aspect deep marine ecosystem (e.g., Droser et al. 1994; Buatois et al. 1998; Orr 2001; MacNaughton et al. 2002). Trace fossil evidence demonstrates much greater ecologic change than that revealed by body fossils alone. In particular, our knowledge of the ecology and biodiversity of deep marine Ordovician communities depends largely on trace fossil evidence.

Paleoecologic levels allow for the ranking of ecologic changes through the Phanerozoic (Droser et al. 1997). First-level changes, the highest level, indicate colonization of a new ecosystem (e.g., life on land, in the sky), and fourth-level changes, at the other end, indicate turnover at the community level. Whereas body fossil evidence reveals second- to fourth-level changes associated with the Ordovician biodiversification event, trace fossil data point to first- and second-level changes, such as the colonization of terrestrial environments and the establishment of deep marine ecosystems of modern aspect. The aim of this chapter is to discuss how trace fossil analysis can illuminate ecologic and evolutionary aspects of the Ordovician Radiation. To do so, we discuss ichnofaunas from different depositional environments, including shallow marine siliciclastic, shallow marine carbonates, deep marine, volcanic terrains, marginal marine, and continental. A combination of both available conceptual and methodological tools, ichnofacies, and ichnofabrics, are used in the present analysis. An outline of the significance of Ordovician trace fossil assemblages in biostratigraphy is also presented.

■ The Conceptual Framework of Ichnology

Ichnology comprises the analysis of postdepositional biologic effects on sedimentary deposits, dealing with the study of trackways, trails, burrows, borings, and other biogenic structures (Frey 1973; Ekdale et al. 1984; Bromley 1996). The aim of ichnology is the study of all aspects involved in organism-substrate interrelationships that focus on how animals and plants leave a record of their activity in or on the substrate, including the processes of bioturbation and bioerosion and the production of biogenic structures and fabrics (Ekdale et al. 1984; Pemberton 1992).

In contrast to body fossils, trace fossils are often characterized by long temporal ranges and narrow facies ranges. In addition, they are produced by both skeletonized and soft-bodied biotas and are rarely reworked. Significantly, they also occur in otherwise unfossiliferous rocks and represent evidence of behavior (Frey 1975; Pemberton et al. 1990). As a result of these characteristics, trace fossils are extremely useful in paleoenvironmental analysis and less so in biostratigraphic studies.

Paleoenvironmental, paleobiologic, and paleoecologic studies using trace fossil evidence have been based on ichnofacies or ichnofabrics or both. Ichnofacies are trace fossil associations that typically recur through long intervals of geologic time and are characteristic of a given set of environmental conditions (Frey 1973; Pemberton et al. 1992). At present, three groups of ichnofacies are currently recognized: (1) marine softground ichnofacies, including the *Psilonichnus*, *Skolithos*, *Cruziana*, *Zoophycos*, and *Nereites* ichnofacies, (2) substrate-controlled ichnofacies, namely, the *Teredolites*, *Glossifungites*, and *Trypanites* (subdivided into *Entobia* and *Gnathichnus*) ichnofacies, and (3) continental ichnofacies, including the *Coprinisphaera*, *Scoyenia*, and *Mermia* ichnofacies (see Buatois et al. 2002 for review). Ichnofabrics include all aspects of biogenic sedimentary structures as viewed in cross section. Thus, they include the record of discrete trace fossils as well as burrow mottling. The ichnofabric approach particularly emphasizes ecologic aspects of the bioturbation process, such as degree of bioturbation, tiering structure, and ichnoguilds.

■ Significance of Trace Fossils for Ordovician Biostratigraphy

Although most ichnogenera display long temporal ranges, when viewed within trace fossil assemblages, they commonly exhibit notable changes through geologic time. In addition, there are some trace fossils that register the activity of a particular kind of animal as its body morphology and behavior underwent closely related evolutionary transformations through time (Seilacher 2000). The more complex (in terms of fine morphological detail) a structure is, the more direct its biologic relationship is and the more distinctive its behavioral program is and hence the larger its stratigraphic significance is. This particular subset of trace fossils displays stratigraphic significance. A prime example is the so-called *Cruziana* stratigraphy, which uses as stratigraphic tools the ichnogenera *Cruziana* and *Rusophycus*. Early attempts to use *Cruziana* for biostratigraphic work were performed in rocks of the Upper Cambrian–Lower Ordovician (e.g., Crimes 1970, 1975; Crimes and Marcos 1976), which still remains the stratigraphic interval with maximum utility. The *Cruziana* stratigraphy has been further developed by Seilacher (1970, 1990, 1992, 1994), who extended the stratigraphic scheme from the Lower Cambrian to the Lower Carboniferous, proposing more than 30 *Cruziana* and *Rusophycus* ichnospecies with stratigraphic significance (Seilacher 1992: figure 2). The *Cruziana* stratigraphy uses the ribbonlike bilobate structures (*Cruziana*, sensu strictum) and the coffee-bean-shaped structures (*Rusophycus*) identified to ichnospecies level. *Cruziana* and *Rusophycus* ichnospecies are based on fine morphological features, particularly the "claw formula" (i.e., the fingerprint left by the tips of the endites displaying groupings of claws or setae) and, secondarily, on the presence of exopodal brushings, pleural or genal spine impressions, and cephalic and coxal marks, reflecting burrowing behavior. Seilacher (1990, 1992: figure 2) recognized 16 *Cruziana* ichnospecies, clustered in 7 groups (e.g., *Cruziana rugosa* group) relevant to the Ordovician stratigraphy (note that in Seilacher's scheme the Tremadocian is included in the Upper Cambrian). Seilacher (1985) presented comprehensive evidence to support the trilobite authorship of *Cruziana* (both cruzianiform and rusophyciform expressions), although some *Cruziana* may be produced by other arthropods such as aglaspids, limulids, and phyllopods (cf. Seilacher 1992, 1994; Bromley 1996). The *Cruziana* stratigraphy is a promising but still incomplete ichnostratigraphic device that provides only a low level of stratigraphic resolution (e.g., Lower, Middle, and Upper Ordovician; see Seilacher 1990, 1992). Even so, in the absence of body fossils it can play a crucial role in relative dating of sedimentary sequences (Seilacher and Alidou 1988). A similar proposal for the Ordovician and Silurian stratigraphy, based on arthrophycid structures (e.g., *Arthrophycus*, *Daedalus*, and *Phycodes*) produced by worms of unknown taxonomic affinity, has been presented recently by Seilacher (2000) and awaits further testing.

Trends in Colonization and Diversity Patterns

Paleontologists typically rely on taxonomic diversity data to track changes in skeletonized metazoans. Although we apply generic and, less commonly, specific names to trace fossils, trace fossil diversity does not equate with body fossil diversity. Biotaxa are essentially different from ichnotaxa. Most ichnologists accept that morphological traits holding major behavioral significance should rank at ichnogenus level and that more peripheral but nevertheless distinctive features should be used for subdivision as ichnospecies (Fürsich 1974; Bromley 1996). A simple trace fossil not reflecting a clear taxonomic relationship (e.g., *Skolithos*) may be produced by different animals at different times in geologic history. Its significance in terms of reconstructing biodiversity is certainly limited, but the expansion of this structure into different environments may still be of evolutionary significance and shed light on the colonization history of ancient ecosystems.

Shallow Marine Siliciclastic Environments

Analysis of ichnodiversity changes through the Ordovician (figure 34.1) does not support the widely accepted belief that shallow-water ichnofaunas were fully diversified by Cambrian times with no subsequent ichnodiversity increase (Seilacher 1974, 1977; Crimes 2001). Additionally, substantial changes in biofabrics (Kidwell and Brenchley 1994; Li and Droser 1999; Droser and Li 2000) and compositional turnovers by the dominant bioturbators of shallow-water environments occurred through the Phanerozoic. Lower Ordovician distal shoreface and offshore siliciclastic deposits display abundant trilobite-produced biogenic structures (e.g., *Cruziana, Rusophycus, Dimorphichnus, Monomorphichnus*) (figure 34.2). Examples of Lower Ordovician ichnofaunas dominated by trilobite trace fossils are widely known from many areas, including Europe (Baldwin 1977; Pickerill et al. 1984), North America (Bergström 1976), North Africa (Seilacher 1992), the Middle East (El-Khayal and Romano 1988), Australia (Webby 1983; Droser et al. 1994), and South America (Aceñolaza and Durand 1978; Mángano et al. 1996a, 2001a). In peri-Gondwanan settings, the most significant trace fossil turnover event is recorded by the *Cruziana* ichnostratigraphy. Elements of the *Cruziana semiplicata* group (Upper Cambrian–Tremadocian) are replaced by elements of the *Cruziana rugosa* group (upper Tremadocian to Darriwilian [= uppermost Arenig-Llanvirn]) (Crimes 1975; Seilacher 1992). Other common components of the *Cruziana* ichnofacies are vermiform structures such as *Planolites, Palaeophycus, Teichichnus, Phycodes,* and *Helminthopsis* commonly associated with trilobite structures (e.g., Baldwin 1977; Mángano et al. 1996a).

Mid to Late Ordovician shallow marine ichnofaunas generally show more varied behavioral patterns (figure 34.2). Trilobite traces are rarely the dominant component in wave-dominated open marine clastics, possibly reflecting the development of a deeper infauna (i.e., taphonomic bias). The dominant patterns

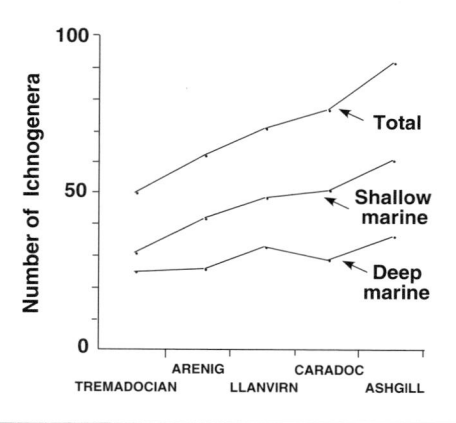

FIGURE 34.1. The changing ichnogeneric diversity numbers through Ordovician time based on a compilation of 65 published ichnologic studies and personal data. Environmental assignments were made on the basis of sedimentologic data obtained from the literature and, in the case of our own field data, on the basis of detailed facies analysis. The ichnodiversity curves were compiled at the ichnogenus level because the taxonomy is more firmly established than for ichnospecies. Synonymies have been checked. The ichnogeneric compilation was plotted as "range-through" data, that is, for each ichnogenus, recording its lower and upper appearances, then extrapolating its presence through any intervening gap in the continuity of its record. A continuous increase in ichnogeneric diversity is reflected through the Ordovician. The number of shallow marine ichnogenera doubled from the Tremadocian to the Ashgill. The Late Ordovician ichnodiversity peak may reflect, at least in part, a monographic effect owing to the comprehensive studies by Osgood (1970) and Stanley and Pickerill (1998). Total curve includes not only shallow and deep marine ichnofossils but also continental trace fossils and boring ichnotaxa. The shallow marine curve does not include borings.

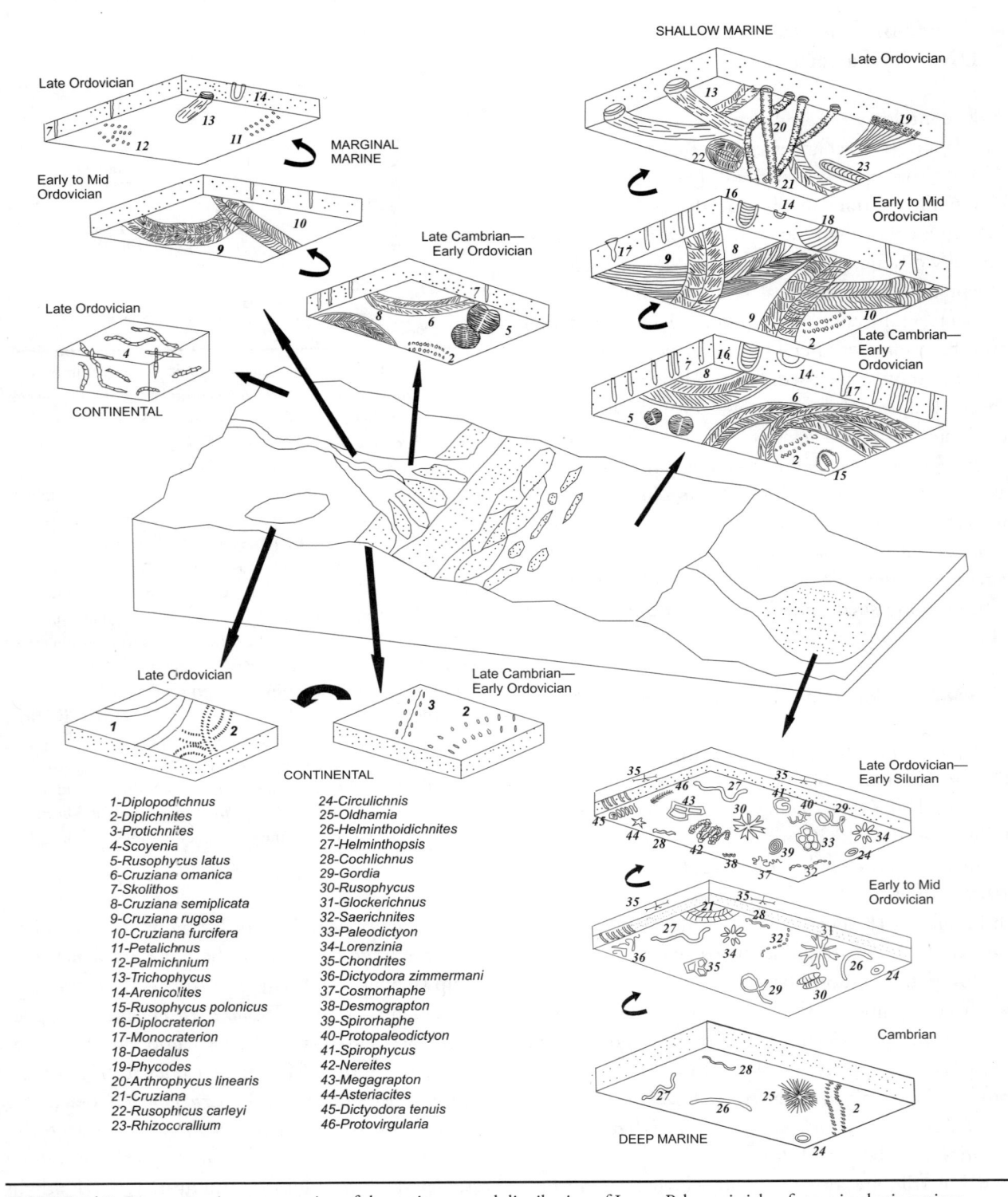

FIGURE 34.2. Diagrammatic representation of the environmental distribution of Lower Palaeozoic ichnofaunas in clastic settings. Shallow and marginal marine environments record the turnover from the *Cruziana semiplicata* and *C. omanica* groups (Upper Cambrian–Tremadocian) to the *C. rugosa* group (upper Tremadocian–Darriwilian) in Gondwana. Upper Ordovician trace fossil assemblages are best known from Laurentia; trilobite trace fossils are commonly present but do not represent the dominant component.

include branched, spreiten burrow systems (e.g., *Phycodes, Trichophycus*), branched, constricted burrow systems (e.g., *Arthrophycus*), branched burrow mazes and boxworks (e.g., *Thalassinoides*), dumbbell-shaped traces (e.g., *Arthraria*), and chevronate trails (e.g., *Protovirgularia*). Most of these behavioral architectures were present in the Cambrian and Early Ordovician but generally were subordinate in abundance and diversity to trilobite trace fossils. Examples of Middle to Upper Ordovician terrigenous, shallow marine ichnofaunas with diverse morphological patterns are known from several areas, including Ohio (Osgood 1970), central Nevada (Chamberlain 1977), eastern Canada (Hofmann 1979; Stanley and Pickerill 1998), the Czech Republic (Mikuláš 1988, 1990, 1992), and Norway (Stanistreet 1989).

In contrast to Cambrian faunas, shallow marine Ordovician biotas display more complex community structures. Ichnologic data suggest an increase in the complexity of tiering structure of infaunal resident and opportunistic communities. Droser et al. (1994) documented an ichnofauna from the Cambro-Ordovician Bynguano Formation of Australia, recognizing pre- and postdepositional ichnocoenoses. The predepositional ichnocoenosis almost exclusively consists of *Rusophycus latus* (see also Webby 1983), whereas the more diverse postdepositional ichnocoenosis, including *Skolithos, Arenicolites, Thalassinoides, Trichichnus,* and *Monocraterion,* is partitioned into three distinct tiers. Mángano et al. (2001b) recognized a complex tiering structure recording the activity of the resident community in upper Tremadocian deposits of the upper-offshore to offshore transition in the Santa Rosita Formation of northwestern Argentina. Shallow tiers are represented by trilobite feeding and locomotion traces (*Cruziana semiplicata*) and by small *Rusophycus* and *Palaeophycus,* preserved at the base of distal tempestites. A relatively deep tier (up to 12 cm below the base of the overlying tempestite) exhibits an ichnofabric dominated by *Trichophycus*. Trilobite trace fossils are crosscut by vertical *Skolithos*. Whereas the *Cruziana-Trichophycus* predepositional suite records the activity of the resident benthic fauna, the *Skolithos* postdepositional suite reflects opportunistic colonization of thin storm layers. All these ichnogenera were present in the Cambrian but apparently did not occur in complex multiple tiered communities at relatively deep levels within the sediment. This increase in tiering complexity is a feature of Ordovician endobenthic communities.

Shallow Marine Carbonate Environments

Unlike siliciclastic shallow marine settings, softgrounds in carbonate environments do not show a significant increase in trace fossil diversity per se through the Ordovician but rather reveal increased ecospace utilization and tiering. Ichnofabric analyses of inner shelfal carbonate deposits of the Great Basin (western United States) reveal two major increases in the extent and depth of bioturbation during the Early Paleozoic: the first between pretrilobite and trilobite-bearing Cambrian rocks, and the second between the Mid and Late Ordovician (Droser and Bottjer 1989). The Ordovician increase in bioturbation results in part from an increase in the size of discrete structures and in the architecture of *Thalassinoides* burrows from networks to mazes (Droser and Bottjer 1989). *Thalassinoides* burrow systems up to 4 cm in diameter displaying classic T and Y branching reach up to 1 m in depth in dolomites of the Upper Ordovician Ely Springs and Fish Haven formations of the eastern Great Basin (Sheehan and Schiefelbein 1984). Although *Thalassinoides* does occur in Cambrian and Lower Ordovician rocks, examples typically are less than 10 mm in burrow diameter and architecturally more simple and form two-dimensional networks (cf. Myrow 1995). In contrast, Upper Ordovician *Thalassinoides* resemble modern structures produced by decapod crustaceans recording extensive reworking with severe obliteration of primary structures (Sheehan and Schiefelbein 1984; Droser and Bottjer 1989). Specimens of *Thalassinoides* reported by Cañas (1995) from Upper Cambrian-Tremadocian lagoonal carbonates of Argentine Precordillera display unquestionable three-dimensional morphology, suggesting an earlier origin of boxwork architecture (Mángano and Buatois 2003). Furthermore specimens from Arenig siliciclastic deposits of northwestern Argentina provide an early occurrence of *Thalassinoides* of modern aspect compared with the typically smaller specimens that occur in Lower Paleozoic strata (Mángano and Buatois 2003). Ichnofabric evidence also indicates an onshore-offshore pattern because extensive bioturbation first developed in shallow-water settings and only later developed in more offshore settings

(Droser and Bottjer 1989). Additional ichnotaxa common in Ordovician carbonates include *Planolites, Chondrites,* and *Arenicolites.* Monospecific suites of *Chondrites* are traditionally thought to be characteristic of low-oxygen environments. *Chondrites* appears, however, to be very common in some carbonates from what were clearly fully oxygenated settings and yet uncommon in carbonate successions thought to have accumulated under low-oxygen conditions. Trace fossil studies in carbonates are hampered by diagenesis, compaction, and the lack of heterolithic bedding (which is ideal for trace fossil preservation). Thus, diversity lists from carbonates are likely to underestimate real diversity.

Significant changes in the evolution of macroboring organisms occurred during the Ordovician (Kobluk et al. 1978; Ekdale and Bromley 2001; Wilson and Palmer 2001a). This significant rise in bioeroders recently has been referred to as "the Ordovician bioerosion revolution" by Wilson and Palmer (2001a:248) and probably occurred by the end of the Mid Ordovician or the beginning of the Late Ordovician. The ability to excavate into lithified material is evolutionarily significant because it requires physiological and anatomical adaptations on the part of the bioeroding organism (Bromley 1996; Ekdale and Bromley 2001).

The oldest record of a macroboring biota comes from Early Cambrian hardgrounds and archaeocyathid reefs containing high densities of *Trypanites* (James et al. 1977). The same ichnotaxon occurs in Lower Ordovician hardgrounds (Jaanusson 1961; Hecker 1970). The first record of *Gastrochaenolites* (*G. oelandicus,* Arenig of Sweden) predates the next earliest occurrence of this ichnogenus by at least 160 million years. In the context of the Lower Ordovician major niche diversification event, the occurrence of *G. oelandicus* is interpreted as recording a bioerosional innovation produced by an organism of uncertain taxonomic affinities (Ekdale and Bromley 2001). An increase in diversity and abundance is detected in the Mid to early Late Ordovician (Chazyan-Mohawkian), with hardgrounds and coral reefs containing not only *Trypanites* but also *Myzostomites* and *Palaeosabella* (Kobluk et al. 1978; Wilson and Palmer 2001a). The bivalve-boring *Petroxestes* and the bryozoan-boring *Ropalonaria* were added to this list by the Late Ordovician (Wilson and Palmer 2001a).

The history of predation has been the subject of considerable debate (Kowalewski et al. 1998, 1999; E. M. Harper et al. 1999; Kaplan and Baumiller 2000, 2001; Wilson and Palmer 2001b). Although some small Vendian and Cambrian borings have been interpreted as predatory in origin (Bengtson and Zhao 1992; Conway Morris and Bengtson 1994), unequivocal predatory borings are first known in the Devonian, with a scant and conflictive record during the Ordovician and Silurian. Recently, the presence of presumed predatory borings in Upper Ordovician shells (Kaplan and Baumiller 2000) has been challenged by Wilson and Palmer (2001b). Borings on *Onniella meeki* from the Waynesville Formation (Richmondian) of the Cincinnatian Series were originally interpreted as gastropod predatory structures by Fenton and Fenton (1931) and Bucher (1938). These structures were later examined by Carriker and Yochelson (1968), who presented compelling evidence for dwelling structures rather than predation borings. Recently, Kaplan and Baumiller (2000, 2001) proposed a mixed-motive boring hypothesis for the holes in these brachiopod shells, suggesting that predators were likely included within the borers. However, evidence presented by Wilson and Palmer (2001b) is best explained by a domicile boring hypothesis. According to these authors, *Onniella* shell beds are storm lags, cemented by early diagenesis (i.e., hardgrounds) and colonized by domicile borers. Further work is needed to clarify the habit of some Upper Ordovician boring producers. Kobluk et al. (1978) mentioned that the increase in macroboring diversity through the Ordovician parallels a rapid development of skeletal reefs. Wilson and Palmer (2001a) noted that boring communities that appeared in the Ordovician do not significantly change until the Mesozoic marine revolution.

Deep Marine Environments

The traditional view of deep marine settings—that of a virtually empty terminal Proterozoic and Cambrian environment and an Early Ordovician colonization event of the deep marine biotope (Crimes 1974; Seilacher 1974)—is being replaced by the idea of a protracted process initiated during the terminal Proterozoic (MacNaughton et al. 2000; Crimes 2000; Orr 2001). Incipient colonization of slope to deep-sea environments took place during the terminal Proterozoic and Cambrian (Crimes and Fedonkin 1994; Crimes 2000; Orr 2001). At the end of the Cambrian increased competition for ecospace or resources (or both) within shallow marine ecosystems forced soft-bodied animals and shallow-water skeletal animals into deeper settings (Crimes et al. 1992; Crimes 2000; Orr 2001). During the Early Ordovician the main lineages of deep marine traces (i.e., rosette, meandering, patterned, spiral) were established in deep-sea sediments (figure 34.2).

Spiral, meandering, rosette, and network structures that typify post-Cambrian *Nereites* ichnofacies were present in shallow-water environments during the terminal Proterozoic and Cambrian, suggesting migration of sophisticated "pascichnia" (patterned grazers) and "agrichnia" (entrapment meshes) in the Ordovician (see Crimes and Fedonkin 1994; Orr 2001). The history of some lineages, however, may be much more complicated. Stanley and Pickerill (1993, 1998) documented *Paleodictyon* ichnopecies (ispp.) in middle- to outer-shelf settings of the Upper Ordovician Georgian Bay Formation of eastern Canada. This record suggests an initial phase of expansion to deep marine settings and then post-Ordovician retreat (Stanley and Pickerill 1993, 1998).

The classic idea that Early Paleozoic deep marine ichnofaunas are simple and of low diversity (Seilacher 1974, 1977) has been challenged by findings of relatively diverse ichnofaunas in Ordovician deep marine successions in England, Ireland, Canada, and Nevada (Chamberlain 1977; Pickerill 1980; Pickerill et al. 1987; McCann 1990; Crimes and Crossley 1991; Orr 1996; Orr and Howe 1999). Thirty-six distinct ichnogenera have been identified in clastic turbidites of the Middle to Upper Ordovician Grog Brook Group and calcareous turbidites of the Upper Ordovician to Llandovery Matapedia Group of eastern Canada (Pickerill 1980; Pickerill et al. 1987; Orr 2001). Individual sites can yield up to 12 ichnogenera, although 3 or 4 are more typical (Pickerill 1980). Crimes et al. (1992) documented 14 ichnogenera in Lower Ordovician (Arenig) flysch deposits of the Ribband Group in southeastern Ireland. Crimes and Crossley (1991) presented histograms of ichnogeneric diversity through time showing high levels of diversification by the Late Ordovician–Early Silurian. Orr (1996) also documented a moderately diverse ichnofauna (9 described ichnogenera and 3 ichnotaxa left in open nomenclature) from the Lower Ordovician Skiddaw Group of the Lake District, northern England. As pointed out by all these authors, the diversity of Ordovician assemblages is inconsistent with a model that predicts only 4 to 8 ichnotaxa for Ordovician flysch facies (Seilacher 1974). However, as noted by several authors (Orr 1996; Buatois et al. 2001a), these differences could reflect variations in the methods and approaches among authors. Whereas Seilacher (1974) recorded species diversity of individual ichnofaunas within a specific time interval, other authors (e.g., Crimes and Crossley 1991) reported the diversity as the total number of ichnotaxa within a time interval (Buatois et al. 2001a). Although alpha diversity of Lower Paleozoic deep marine ichnofaunas has been underestimated, Late Mesozoic and Cenozoic flysch ichnofaunas record a qualitative shift in community complexity and partitioning of energy resources, as reflected by tiering structure (e.g., Leszczyński 1991) and event-bed colonization history (Wetzel and Uchman 2001).

In contrast, Cambrian deep marine ichnofaunas are composed mostly of ethologically simple, facies-crossing ichnogenera (e.g., *Palaeophycus, Planolites, Helminthoidichnites, Taenidium*) and shallow-water components, such as "presumed trilobite traces" (e.g., *Cruziana, Diplichnites, Dimorphichnus, Monomorphichnus*) (Crimes and Crossley 1968; Aceñolaza and Durand 1973; Crimes et al. 1992; Hofmann et al. 1994; Pickerill 1995; Orr 2001; Buatois and Mángano 2003) (figure 34.2). This first exploratory incursion of shallow-water migrants was not ultimately successful, based on the rarity of typical shallow-water trace fossils in post–Lower Paleozoic flysch sequences (see Orr 2001: figure 3); this, however, could be preservational (cf. Droser et al. 2001). *Oldhamia*, often forming monospecific or low-diversity assemblages, recurs in deep marine Cambrian deposits (Crimes and Crossley 1968; Aceñolaza and Durand 1973; Lindholm and Casey 1990; Hofmann et al. 1994; Orr 2001; Buatois and Mángano 2003). The mode of life of the producer of *Oldhamia* is a relict strategy of undermat mining that persisted in the deep sea after the agronomic revolution that arose in shallow water (Seilacher 1999; Buatois and Mángano 2002, 2003). The ichnogenus is uncommon, if present at all, in Ordovician and younger flysch associations and may reflect the extension of the agronomic revolution to deep-water settings during the Ordovician.

Most of the Lower Paleozoic deep marine trace fossils represent the activity of shallow tier inhabitants. However, enigmatic full relief structures up to 40 cm deep have been reported from levee deposits of an outer submarine fan complex of the Meguma Group (Cambrian–Early Ordovician of Nova Scotia), suggesting attempts of colonization of the deeper infaunal ecospace in deep marine environments (Pickerill and Williams 1989). These burrows may have been produced by doomed pioneers transported from shallow marine settings to deep marine environments via turbidity currents (Waldron 1992; Allison and Briggs 1994). Orr (written commun.) analyzed Middle Ordovician deep marine ichnofabrics from Arkansas that reflect the activity of a climax suite that may have penetrated at least 40 cm into the substrate.

Lower Ordovician flysch ichnofaunas seem to be moderately diverse, and "fodinichnia" (feeding traces) usually dominate (figure 34.2). In contrast, Upper Ordovician–Lower Silurian ichnofaunas are considerably more diverse, and both "pascichnia" and "agrichnia" are well represented (Orr 1996, 2001) (figure 34.2). Benton (1982) recognized the *Nereites* and *Paleodictyon* subichnofacies in Upper Ordovician–Lower Silurian deep marine deposits of the Southern Uplands of Scotland—at least three compositionally different assemblages can be identified within the *Nereites* subichnofacies. These differences suggest environmental controls, such as oxygen level (Benton 1982) or nutrients (cf. Wetzel and Uchman 1998). *Dictyodora* is a peculiar meandering to spiral complex form. It is very abundant in Paleozoic flysch ichnofaunas starting in the Upper Ordovician (Benton and Trewin 1980; Benton 1982; Pickerill et al. 1987; Orr 1994). *Dictyodora* records a clear evolution from the Upper Ordovician to the Carboniferous, the key features being an increase in the height of the wall and an improvement in feeding efficiency (Seilacher 1967; Benton 1982). The architectural plan of *Dictyodora* suggests a strategy of underground mining progressively deeper into the sediment through time (Seilacher 1967; Benton and Trewin 1980; Benton 1982; Seilacher-Drexler and Seilacher 1999). However, maximum depth of the wall structure in the Ordovician *Dictyodora zimmermanni* is only up to 25 mm.

Trace fossil evidence records that the advent of a deep marine ecosystem of modern aspect originated during the Ordovician, representing a second-level change in terms of evolutionary paleoecology. Compared with late Mesozoic to Cenozoic assemblages, however, Ordovician assemblages are significantly less diverse and display less complex ecologic structures.

Volcanic Environments

In volcanic terranes, biotic activity is strongly controlled by volcanic processes (e.g., ash falls, lava flows) as well as sedimentary processes linked to volcanism (e.g., volcaniclastic debris flows). Volcanic activity has a strong impact on faunal composition and diversity, paleocommunity structure, and thus the trace fossil record. A series of studies in the Arenig Suri Formation of the Famatina region of northwestern Argentina by Mángano and Buatois (1995, 1996, 1997) and Mángano et al. (1996b) have documented the ichnologic content and associated facies of volcanic arc-related successions. Slope-apron deposits of the Vuelta de Las Tolas Member of the Suri Formation host a depauperate ichnofauna, comprising *Planolites montanus, Palaeophycus tubularis, Helminthopsis abeli,* and simple, unbranched, horizontal traces. Ichnofabric indices are typically low and tiering patterns very simple. Trace fossil distribution seems to be controlled by turbidity current-induced oxygen fluctuations within an overall oxygen-depleted setting. Presence of trace fossil–bearing beds within unbioturbated successions indicates short-term increases in oxygen content. Oxygen depletion was probably linked to limited deep-water circulation in topographically restricted depressions within an arc setting. Shelf deposits of the Loma del Kilómetro Member of the Suri Formation contain trilobite traces (*Cruziana furcifera*) as well as very simple forms, such as *Palaeophycus tubularis, Phycodes* isp., *Planolites beverleyensis* and *Helminthopsis* isp. (Mángano and Buatois 1996; Mángano et al. 1996b). As in slope-apron deposits, ichnofabrics are simple and ichnofabric indices usually low. Ichnofaunas from the Suri Formation record the ability of benthic organisms to colonize stressful and unstable ecosystems affected by active volcanism. The low diversity and complexity of these volcanic arc ichnofaunas contrast with coeval ichnofaunas in siliciclastic and carbonate marine settings.

Marginal Marine Environments

Relatively little is known about the colonization history of marginal marine, brackish water environments (see Buatois et al. 2001b). However, marginal marine ichnofaunas have been documented from a number of localities in Argentina (Mángano et al. 1996a, 2001a), Jordan (Selley 1970), the Czech Republic (Mikuláš 1995), South Africa (Braddy and Almond 1999), and the United States (Martin 1993). Ichnologic studies in the Paseky Shale of the Czech Republic (Mikuláš 1995) and in the Disi Group of Jordan (Selley 1970; Amireh et al. 1994) indicate that arthropods were able to foray into brackish water environments by the Early Cambrian.

This trend was accentuated during the Late Cambrian to Ordovician and is recorded in peri-Gondwanan marginal marine clastic successions. Whereas Cambrian brackish water ichnofaunas are restricted to the outermost regions of estuaries and bays, Ordovician assemblages seem to reflect a landward expansion into inner areas of marginal marine systems (figure 34.2). An Upper Cambrian ichnofauna from tide-dominated estuarine deposits of the Santa Rosita Formation of northwestern Argentina consists of trilobite traces of the *Cruziana semiplicata* and *C. omanica* groups (*C. semiplicata, C. omanica, Rusophus latus*), vertical dwelling and resting traces (*Skolithos, Conostichus*), and horizontal feeding and grazing traces (*Helminthopsis, Planolites*) (Mángano et al. 1996a; Buatois and Mángano 2001). The well-known West Gondwanan *Cruziana semiplicata* group and the East Gondwanan *C. omanica* group co-occur in northwestern Argentina. According to Seilacher (1992), *C. semiplicata* is probably replaced by *C. omanica* east of Poland, both ichnospecies displaying similar stratigraphic ranges but contrasting paleogeographic distributions. However, the *C. omanica* group apparently reached some West Gondwanan areas, which probably functioned as a melting pot of complex paleogeographic affinities (Mángano et al. 1996a). Ecologically similar trace fossil assemblages occur in tide-dominated marginal marine deposits of the Arenig-Llanvirn Mojotoro Formation of northwestern Argentina (Mángano et al. 2001a). Trilobite trace fossils of the *Cruziana rugosa* group are abundant in intertidal flat and interbar deposits, while *Skolithos* is dominant in colonization surfaces within the sand-wave complex. Restricted ichnofaunas of the basal units of the Santa Rosita Formation and the Mojotoro Formation record the *Cruziana* turnover, from the *C. semiplicata* group to the *C. rugosa* group (figure 34.2). Eurypterid trackways (*Palmichnium* and *Petalichnus*) occur in paralic deposits of the Lower to Upper Ordovician Table Mountain Group of South Africa (Braddy and Almond 1999). An impoverished mixed *Cruziana* and *Skolithos* ichnofacies in estuarine deposits of the Upper Ordovician Sequatchie Formation of Georgia and Tennessee includes *Skolithos, Arenicolites, Monocraterion, Planolites,* and *Trichophycus* (Martin 1993).

A low-diversity trace fossil assemblage occurs in delta front facies of a fluvio-dominated delta system of the Darriwilian-Caradoc Jbel Gaiz succession (First Bani Group) in Morocco. It is dominated by distinctive vermiform structures similar to *Parataenidium monoliformis* (= *Eione monoliformis* and *Margaritichnus reptilis*) (figure 34.3A). Arthropod trackways (*Diplichnites,* figure 34.3C), cleft-foot bivalve structures (*Protovirgularia,* figure 34.3C), simple subsurface locomotion and feeding structures (*Planolites*), and rare trilobite trace fossils (*Rusophycus,* figure 34.3B) also occur. The ichnofauna represents an example of a depauperate *Cruziana* ichnofacies with no representatives of the *Skolithos* ichnofacies. The establishment of a suspension-feeding fauna was probably inhibited by highly suspended sediment load at the delta front (cf. Gingras et al. 1998). Although the name-bearing ichnotaxon is not present, the overall features of this assemblage warrant its inclusion in the *Curvolithus* association (Lockley et al. 1987; Bromley 1996), characteristic of sandy substrates, generally in marginal marine settings influenced by freshwater discharge or by an energy regime that favored deposition over physical reworking (Lockley et al. 1987).

Analysis of Ordovician marginal marine ichnofaunas demonstrates that some trilobites and soft-bodied animals were able to invade estuarine and other marginal marine settings (Buatois et al. 2001b). Trilobite-produced traces are concentrated in low-energy settings, such as tidal flat and interbar areas, and vertical burrows in moderate- to high-energy settings, mostly sand-wave complexes. Mixed *Skolithos-Cruziana* ichnofacies in brackish water deposits thus can be traced back in time to the Early Paleozoic. As

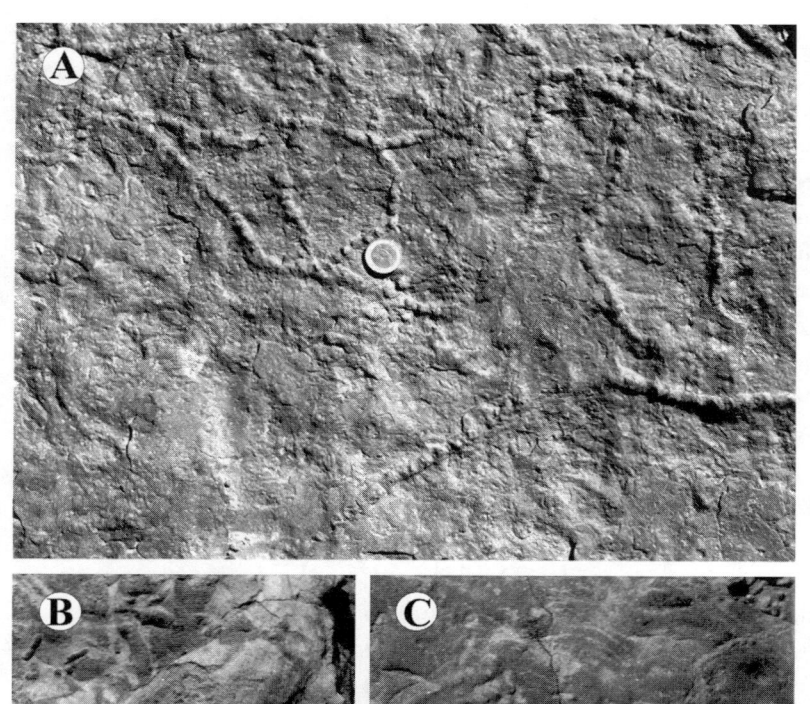

FIGURE 34.3. The *Curvolithus* association from Darriwilian-Caradoc deltaic prograding clinoforms at Jbel Gaiz (First Bani Group, Eastern Anti-Atlas, Morocco). A, general bedding plane view of *Parataenidium monoliformis* (= *Margaritichnus,* = *Eione*). This vermiform structure is characterized by imbricated sediment pads forming protruding ridges on the upper bedding plane. Coin is 21 mm in diameter. B, two superimposed specimens of *Rusophycus* n. isp., most likely produced by an indet. phacopid producer, preserved as positive hyporeliefs. Scale bar is 10 mm. C, chevroned structures, *Protovirgularia* isp., produced by small nuculoid bivalves (upper right) and a wide arthropod locomotion structure, *Diplichnites* isp. (center). Preserved as positive hyporeliefs. Scale bar is 10 mm.

in the case of open marine ecosystems, trilobites were the dominant trace makers in Cambrian to Lower Ordovician estuarine ichnofaunas, but they are apparently less important in Upper Ordovician assemblages (figure 34.2).

Continental Environments

Ichnofossil occurrences predate data from the body fossil record (Rolfe 1985) and demonstrate that the terrestrial invasion had already started by the Ordovician (Buatois et al. 1998). Recent research suggests that arthropods walked on land as early as the Late Cambrian to Early Ordovician, as indicated by trackways produced by an amphibious animal in coastal eolian dunes of the Nepean Formation in Canada (MacNaughton et al. 2002) (figure 33.2). Retallack and Feakes (1987) and Retallack (2001) discussed the presence of backfilled traces attributed to millipedes (figure 34.2). Johnson et al. (1994) documented arthropod trackways (*Diplichnites, Diplopodichnus*) undoubtedly produced by myriapodlike invertebrates in pond deposits that were desiccated periodically, suggesting the presence of an incipient *Scoyenia* ichnofacies (figure 34.2). Although myriapods are typically considered terrestrial, Early Ordovician to Late Silurian representatives have been interpreted as aquatic or possibly amphibious (Almond 1985).

■ Discussion

Trace fossils are key elements in reconstructing trends in colonization and ecology of benthic communities. Ecologic breakthroughs in deep marine, shallow marine, marginal marine, and continental environments rank at different paleoecologic levels (sensu Droser et al. 1997). The invasion of continents by invertebrates ranks as a first-level change; the establishment of a deep marine ecosystem of modern aspect represents a second-level change.

Although there is an initial colonization of deep marine settings during the Vendian (first level), the establishment of typical deep marine behavioral patterns ("agrichnia" and sophisticated "pascichnia") in slope and abyssal depths during the Early Ordovician is unquestionably a unique event (second level). Mid to Late Ordovician changes in community structure and increases in species richness within the deep marine community generally rank at lower levels, probably third- to fourth-level changes.

At the other end of the environmental spectrum, short-term incursions to terrestrial environments probably by amphibious marine invertebrates took place in the Late Cambrian–Early Ordovician (MacNaughton et al. 2002). Arthropod continental ichnofaunas are known since the Late Ordovician in paleosols (Retallack and Feakes 1987; Retallack 2001) and ponds (Johnson et al. 1994).

The misconception that infaunal activity was minimal previous to the Late Paleozoic radiation of crustaceans (see Thayer 1979) has been challenged in the past decades. Sediments were bioturbated to significant depths in shallow marine settings during the Cambrian (Droser 1991; Droser and Li 2000; Mángano and Buatois 2001), when the Bambachian megaguild of deep passive suspension feeders was established. The structure of infaunal communities remained simple, and partitioning of infaunal ecospace was meager. In contrast, Lower Ordovician shallow-water siliciclastic communities record middle-tier depth of bioturbation by active deposit feeders (Mángano et al. 2001b). The passive suspension feeder Bambachian megaguild became increasingly complex, resulting in a significant partitioning of the infaunal ecospace (Droser at el. 1994). Shallow-water carbonate settings record second-level changes, such as the transition to Paleozoic-type hardgrounds (Droser et al. 1997) and the appearance of burrow systems constructed by deep suspension feeders (i.e., deep *Thalassinoides* boxworks) (Sheehan and Schiefelbein 1984; Droser and Bottjer 1989). In shallow marine peri-Gondwanan siliciclastic environments, major trace fossil turnovers (e.g., the *Cruziana semiplicata* group being replaced by the *C. rugosa* group) are most likely related to intraclade taxonomic changes in the components of a community, representing fourth-level paleoecologic changes. Some of these changes may reflect onshore-offshore evolutionary trends, such as the retreat of olenids to deeper-water settings during the Early Ordovician (cf. Fortey and Owens 1990b), where biogenic structure preservation is biased by the scarcity of sandstone-mudstone interfaces.

Although marginal marine settings were first invaded in the Early Cambrian, the *Curvolithus* assemblage is first recorded in Ordovician rocks. This ichnocoenosis records the establishment of a distinctive type of community that may represent a third-level paleoecologic change. In marginal marine settings the incursions of trilobites were related to feeding and nesting behavior (Mángano et al. 2001a), providing evidence that trilobites were able, at least temporarily, to survive in brackish water conditions and were significant bioturbators in estuarine and other marginal marine settings.

ACKNOWLEDGMENTS

For financial support, M. Gabriela Mángano thanks Grants-in-Aid-of-Research by Sigma Delta Epsilon, the Antorchas Foundation, the Percy Sladen Memorial Fund, the National Agency of Science and Technology, and the Argentinean Research Council (CONICET). The National Science Foundation provided funding to M. Droser (EAR-9219731). M. Gabriela Mángano also thanks Naima Hamoumi for the opportunity to see the interesting Ordovician outcrops in the Anti Atlas of Morocco. Luis Buatois and Sören Jensen are thanked for valuable feedback during the preparation of this chapter. Richard Bromley and Mark Wilson provided valuable comments on Ordovician bioerosion. Roland Goldring and Patrick Orr undertook detailed reviews of the manuscript. Patrick Orr is also thanked for providing unpublished data on deep marine ichnofabrics. Noelia Carmona assisted us with the construction of a database, and Rodolfo Aredes helped with the compilation of the reference list. This manuscript is a contribution to IGCP 410.

35　The Ordovician Radiation: Toward a New Global Synthesis

Arnold I. Miller

By any measure, the Ordovician Period was a remarkable interval in the evolution of marine biodiversity. As described in a series of papers by the late J. John Sepkoski Jr. (e.g., Sepkoski 1979, 1981a, 1984, 1995), the period was marked by the most broadly based global radiation, at taxonomic levels ranging from species to order, in the history of life. In some ways, the preceding Cambrian explosion may have been more profound, given that it marked the first major proliferation of basic body plans expressed at the phylum and class levels (e.g., Erwin et al. 1987). However, it was during the Ordovician Radiation that marine ecosystems, and the taxa that comprised them, became established in ways that continue to resonate to the present day. In part, this involved the establishment of taxa and morphotypes that would come to dominate marine benthos for the rest of the Paleozoic Era and, in some cases, beyond (e.g., see chapter 17 for a detailed description of the advent and diversification of major brachiopod morphologies and chapter 16 for a similar description of bryozoans).

Even among taxa that did not become broadly dominant in marine settings until the post-Paleozoic, the Ordovician was a crucial time. To cite but one example, most of the major shell forms exhibited by bivalve mollusks, which were major contributors to Sepkoski's (1981a) Modern Evolutionary Fauna, became established during the Ordovician (see Stanley 1968, 1972; Pojeta 1971; chapter 20). Even among taxa that have been perceived as exhibiting their heyday prior to the Ordovician, significant components radiated anew during the Ordovician (e.g., trilobites; see Adrain et al. 1998 and chapter 24). Thus, on a global scale, Ordovician marine biotas exhibited a remarkable breadth of taxonomic representation.

■ The Two Facets of Why

In the past several years, there has been an intellectual radiation in the paleobiologic investigation of Ordovician strata and biotas, with a concerted effort to develop a more synthetic approach to the study of the diversification. This approach is exemplified in part by the chapters in this volume, in which global diversity trajectories are presented and evaluated for nearly all significant contributors to Ordovician marine diversity. More broadly, researchers publishing in a variety of outlets have endeavored to understand the relationships among Ordovician diversity trends at hierarchical levels ranging from local communities (typically referred to as the *alpha* level) to the global biota. There are two primary motivations for this approach:

1. From a theoretical standpoint, we can learn much about how global diversification proceeded, during the Ordovician and at other times, by understanding whether global diversity trajectories tran-

scended hierarchical scales. In essence, it can be asked whether the Ordovician Radiation displayed parallel diversity trajectories at local, regional, and global scales. The impression emerging thus far from investigations of hierarchical patterns during the Ordovician (e.g., Patzkowsky 1995c; Miller and Mao 1998; Adrain et al. 2000) is that different scales of diversification were, in fact, characterized by unique signatures that may be indicative of unique processes not necessarily linked to one another.

2. From a practical standpoint, the assessment of regional diversity patterns provides an opportunity to compare diversification among different areas, characterized by unique environmental attributes. Thus, hypotheses can be tested concerning the importance of particular physical and biologic attributes of environments in promoting diversity transitions. Opportunities abound to play off different environmental factors (e.g., substrate type and climate) against one another in various permutations and combinations, thereby permitting the definitive isolation of factors that govern regional proliferations or declines of taxa in given instances (e.g., Patzkowsky and Holland 1993, 1999; Waisfeld and Sánchez 1996; Waisfeld et al. 1999; McCormick and Owen 2001; Sánchez et al. 2002).

Of course, underlying much of this collective effort is a desire to understand *why* the Ordovician Radiation took place. In framing this issue, however, we would do well to remember that there are actually two facets to this perplexing question. First, regardless of the taxa involved, we can ask why there was such an extensive, worldwide increase in total biodiversity through the period. The Ordovician Radiation was obviously contingent at a fundamental level on the Cambrian explosion; nearly all the phyla and most of the classes that radiated during the Ordovician originated during the Cambrian (Erwin et al. 1987). However, this does not necessarily imply that the Ordovician Radiation was a direct by-product of the Cambrian explosion, with morphological innovations of the Ordovician representing the inevitable release of some sort of pent-up, collective genetic potential that became realized in the occupation of previously unexploited ecospace. It is possible, instead, that physical or biologic triggers unique to the Ordovician permitted a diversification that would otherwise have been far more limited in scope, had it occurred at all.

Second, we can ask why certain taxa radiated more appreciably than others during the period. Put another way, can we explain the taxonomic selectivity of the radiation? For example, among trilobites, why did Ibex Fauna I give way to Ibex Fauna II and, ultimately, to the Whiterock Fauna (chapter 24)? Among articulated brachiopods and bivalve mollusks, why did the former radiate much more appreciably during the Ordovician than the latter (see later in this chapter and the discussion in chapter 17)? To date, explanations for Ordovician biotic transitions and the selective radiations of some taxa have been remarkably diverse in their own right, encompassing a spectrum of contrasting views, including fundamental variations in the characteristic turnover dynamics of higher taxa (e.g., Sepkoski 1979, 1984); physical transitions among major geographic regions that favored the diversification of some taxa and, perhaps, promoted the demise of others (e.g., Miller and Mao 1995; Miller and Connolly 2001); and secular changes in, but continued limitations to, food availability (Bambach 1993; chapter 17).

In the end, we may discover that the "two facets of why" were related intimately, but this did not have to be the case. On the one hand, it seems reasonable to suppose that the same factors responsible for promoting an overall diversification also favored the diversification of some taxa more than others. On the other hand, it is plausible that, at some level, there were broad triggers to diversification that were nonselective and that, independent of this trigger, unique attributes of extant biotas or physical settings promoted the diversification of some taxa and not others. In this regard, it is useful to consider the major Phanerozoic mass extinctions, the physical triggering of which suggests that they would have occurred regardless of the composition of the biota that existed at the times that they occurred. Of course, analyses have suggested that some mass extinctions were biologically selective in various ways and that the nature of the selectivity, if any, was contingent on the biota that was extant at the time. But the actual occurrences of the events, at a basic level, were probably inevitable, given the physical mechanisms that caused them. Similarly, it is possible that an overarching mechanism(s) acted to produce the Ordovician Radiation, independent of the factors responsible for its taxonomic selectivity.

The purpose of this chapter is to consider the Ordovician Radiation from the perspective of the "two facets of why," with the hope of illuminating what we should expect to learn from comparing diversification among higher taxa and at different ecologic or geographic scales.

■ The Uniqueness of the Global Scale

To construct his seminal graphs that depict the Phanerozoic history of marine global biodiversity, Jack Sepkoski compiled databases that catalog the known *global* stratigraphic ranges of families (Sepkoski 1982, 1992a) and genera (Sepkoski 2002). There is no geographic information or charting of individual occurrences of taxa in Sepkoski's compendia. In recent years, however, in conjunction with a new generation of research into the history of marine biodiversity in the Ordovician (e.g., Miller and Foote 1996; Miller 1997a, 1997b) and the rest of the Phanerozoic (e.g., Alroy et al. 2001), the unit of data compilation has shifted from the *taxon* to the *taxonomic occurrence*. An emerging goal of data compilation is to catalog individual occurrences of taxa from around the world regardless of whether (1) the taxon has been encountered previously in the compilation of data or (2) cataloging the occurrence extends the known global stratigraphic range of the taxon. In addition, newer compilations generally record a variety of subsidiary data, when available, for each occurrence, including geographic, stratigraphic, and environmental data.

In cases in which data compilation in one of the newer investigations has progressed sufficiently to yield a reasonably "complete," globally representative sample, the expectation is that an aggregate, global-scale diversity curve built from the new database for some set of taxa should yield a pattern comparable to a curve constructed for the same taxa based on one of the earlier, global compendia (e.g., Miller and Foote 1996). The reason for this expectation is straightforward: if a sufficient number of occurrences for each taxon has been cataloged, the database should effectively capture the oldest and youngest known global occurrences of the taxon, the requisite data for constructing global diversity curves under the standard assumption that a taxon ranges through the entire interval between its first and last global occurrences.

The unit of data compilation in newer databases, the taxonomic occurrence, is, by its very nature, a *local* property, whereas the unit in older compendia, the taxon, is a *global* property. Thus, data in the two kinds of compilations are being captured at two fundamentally different levels (the only exceptions are true singletons; i.e., taxa known from just a single occurrence). Because we can construct an "accurate" global diversity curve from a database of local occurrences, we might conclude that there is a direct linkage between local and global diversification. In an operational sense, this is certainly the case: global patterns must be born of local occurrences, in aggregate. However, this is by no means tantamount to saying that diversification exhibits a parallel trajectory at local, regional, and global scales, despite some previous suggestions that we should expect diversity trajectories at the alpha and global levels to be similar to each other (see Bambach 1977; Sepkoski et al. 1981). In fact, during the Ordovician Radiation, clear differences have been documented between diversity trajectories at different levels, as noted earlier. These range from significant differences in diversification exhibited by different paleocontinents in the Ordovician world (e.g., Miller 1997a, 1997b; chapter 28), to the decoupling of alpha diversity from *beta* diversity (the degree of differentiation between communities; e.g., Patzkowsky 1995c; Miller and Mao 1998), to a lack of parallel between alpha and global diversity trajectories (e.g., Adrain et al. 2000 and chapter 24).

These differences suggest strongly that different factors operate, perhaps independently, to govern diversification at different ecologic or geographic scales. Given the operational linkage between the data used to reconstruct diversity trajectories at all hierarchical levels, we might expect that, aside from the globally mediated major mass extinctions, global diversity trends simply represent the epiphenomenal accumulations of processes operating locally and regionally. However, there may be two compelling reasons to suggest that processes operating at the global level, not just the local and regional levels, led to the Ordovician Radiation:

1. Although we have yet to discover a definitive link between the Cambrian explosion and the subsequent unfolding of the Ordovician Radiation, it is

plausible that such a link exists. In effect, Sepkoski's coupled logistic modeling of global diversification (Sepkoski 1978, 1979, 1984) advocated this view; diversification in the simulations unfolded in a way that effectively captured the trajectory of the Ordovician Radiation as an outcome of kinetics set in motion earlier, at the start of the simulation. The overall pattern, and the transitions among Sepkoski's evolutionary faunas (the Cambrian, Paleozoic, and Modern faunas), were governed in part by measured differences in the turnover rates of higher taxa belonging to each of the three faunas (Sepkoski 1981a, 1984). The broad, explanatory power of the coupled logistic model has been a matter of some contention (see Miller 1998; Benton 2001), but its success in modeling the Early Paleozoic pattern of diversification, including the distinct sequence of transitions from the Cambrian to the Ordovician, remains worthy of note. Alternatively, as suggested earlier, there may have been some global, physical trigger to Ordovician diversification, independent of the Cambrian. The specific pattern of diversification that unfolded may have been governed by collective biotic responses to environmental factors expressed regionally or locally at the same time that the fundamental impetus for the radiation was global.

2. Sepkoski's logistic modeling is also predicated on a second factor for which there is growing support in analyses of the Ordovician (Connolly and Miller 2002): the existence of an upper limit to diversity (an *equilibrium* level), the approach to which inhibits diversification. To the extent that global equilibrium was a reality—a possibility, even if Sepkoski's coupled system is viewed as questionable (see Courtillot and Gaudemer 1996)—it suggests that there was a global agent shaping the history of diversity, regardless of what transpired at other hierarchical levels.

We still are not in a position to determine definitively whether truly global processes played major roles in mediating Ordovician diversity trends. My purpose in articulating these views is to make the case that, in attempting to understand the Ordovician Radiation, we must maintain a global-scale outlook at the same time that we delve ever more deeply into the study of local and regional patterns. There is still much to be learned about the Ordovician Radiation by thinking globally.

■ Comparisons Among Higher Taxa: Geographic and Environmental Dissection

Regardless of whether one accepts the contention, implicit in the coupled logistic model, that Sepkoski's three evolutionary faunas constitute meaningful macroevolutionary units, Ordovician higher taxa can be classified broadly into groups characterized by comparable global diversity trends. Moreover, these similarities were paralleled, to at least some extent, by similarities in the rates of turnover exhibited by their constituent taxa (e.g., rates of genus origination and extinction within a class). Sepkoski (1981a, 1984) observed that higher taxa comprising the Cambrian evolutionary fauna exhibited relatively high turnover rates, whereas the Paleozoic Fauna was characterized by moderate rates and the Modern Fauna by low rates, although the distinctions among evolutionary faunas were not as "clean" as once envisaged (see Sepkoski 1998 and several chapters in this volume).

One of the interesting sidelights to the calibration of turnover rates is that we do not really understand, from an evolutionary perspective, what *causes* the dramatic differences observed in some cases among higher taxa; this remains as one of the enduring mysteries of paleobiology. It does not appear that they are merely artifacts of variations in taxonomic practices among specialists working with different higher taxa.

Despite the enigmatic underpinning of turnover rates, it has proved worthwhile to compare the diversification trajectories and characteristic turnover rates of higher taxa that were major contributors to Ordovician diversity. This is true because it helps us to address both of the "facets of why." As discussed earlier, these patterns may be direct reflections of overarching, global-scale processes that were at least partly responsible for the Ordovician Radiation. More directly, however, these comparisons get to the heart of the second facet: if we can calibrate accurately common trajectories of increasing or declining diversity among higher taxa, this should help us to better understand why Ordovician biotic *transitions* took place, *provided that we can understand what it was that taxa with similar diversity trajectories had in common*, perhaps from an autecologic standpoint (e.g., see Bambach 1983, 1985). Of course, if a definitive biologic explanation cannot be determined for parallel patterns of global diversification among two or more higher taxa, we cannot rule out the possibility that the similarities were simply a matter of chance.

In this context, any inference that the macroevolutionary histories of two or more taxa were governed by similar processes should be evaluated in the light of analyses that determine whether the patterns transcend geography and environment. Thus, comparisons among higher taxa of diversification trajectories have literally taken on new dimensions, with the definitive addition of geographic and environmental data to the mix. From some of the early quantitative studies of the relationship between global diversification and spatial transitions among evolutionary faunas (e.g., Sepkoski and Sheehan 1983; Sepkoski and Miller 1985), a sense emerged that the two were linked intimately: radiation was associated with spatial expansion, declining diversity with spatial contraction. Although this general observation has largely been borne out in subsequent research, it also appears that the early studies overestimated the extent to which evolutionary faunas functioned as distinct macroevolutionary units during the Ordovician, as well as through the rest of the Paleozoic. There was not a straightforward environmental segregation of Cambrian, Paleozoic, and Modern faunal elements on Ordovician seafloors (e.g., Westrop and Adrain 1998), and at least some higher taxa responded independently to local environmental conditions (e.g., Miller 1988). Even among higher taxa with highly comparable Ordovician global diversity trajectories, there clearly were cases in which their geographic and/or environmental distributions were rather different from one another (see later in this chapter). Collectively, these observations suggest the need for caution when assessing relationships among higher taxa that share similar patterns of global diversification.

In an exemplary series of studies, Babin (1993, 1995), Cope and Babin (1999), and Cope (chapter 20) analyzed in detail the paleogeographic history of the Ordovician radiation of bivalves. They demonstrated convincingly that the initial, Early Ordovician bivalve radiation took place in siliciclastic environments associated with Gondwana. The regions in which bivalves thrived included areas marginal to Gondwana at high southern latitudes and locations in East Gondwana (present-day central Australia) that straddled the equator. It was not until the Mid and Late Ordovician epochs that bivalves diversified appreciably at low latitudes outside Gondwana, as best exemplified in Laurentia. Interestingly, perhaps because of the position of Laurentia in the tropics, Cope and Babin argued that this Late Ordovician radiation was focused in carbonate environments, in direct contrast to the earlier proliferation elsewhere of bivalves in siliciclastics. However, a perusal of bivalve-rich, Upper Ordovician strata in eastern and midwestern North America (i.e., eastern Laurentia), including the type Cincinnatian, shows clearly that bivalves were most diverse and abundant in siliciclastic-rich sediments produced in association with the Taconic Orogeny (Bretsky 1970a, 1970b; Frey 1987; Miller 1989). In fact, in much of the siliciclastic, storm-deposited Upper Ordovician record in New York and Pennsylvania, bivalves are the dominant biotic elements. Thus, a common theme in the Ordovician radiation of bivalves appears to be the availability of siliciclastic sediments; significant diversification in eastern Laurentia may have been inhibited until siliciclastic sediments became available more readily in association with orogenic activity during the Mid and, especially, Late Ordovician.

This clarification about bivalves becomes relevant when considering the comparative diversity histories of bivalves and gastropods, the two major components of Sepkoski's Modern Evolutionary Fauna. Because they exhibited highly comparable global diversity trajectories during the Ordovician and thereafter, as well as similarly sluggish turnover rates relative to taxa comprising the Cambrian and Paleozoic Faunas (see Sepkoski 1998: figure 3), Sepkoski grouped bivalves and gastropods as major members of the Modern Fauna. This classification bolsters the impression that the two clades shared highly comparable evolutionary histories. However, a detailed dissection of their geographic and environmental distributions throughout the Ordovician shows that they exhibited rather different distributional patterns, with their primary centers of diversification focused in different environmental and geographic regimes (Novack-Gottshall 1999; Novack-Gottshall and Miller 2003a,b). Throughout the Ordovician world, gastropods were more diverse and abundant in comparatively shallow, carbonate-dominated settings and thus were more dominant on paleocontinents that were more carbonate-rich overall; bivalves were more dominant in siliciclastic settings and, correspondingly, were more important on paleocontinents with more extensive siliciclastic sediments. Locally, as exemplified in the mixed siliciclastic-carbonate regime of the type Cincinnatian, this dichotomy was manifested by marked differences in the diversities and abundances of bivalves and gastropods in siliciclastic- versus carbonate-rich strata.

In a sense, then, alpha-level differences in the environmental preferences of bivalves and gastropods can be seen as scale-transcendent, in that they were responsible not only for ecologic differences observed at local and regional levels such as the type Cincinnatian but also for differences among regions throughout Laurentia (e.g., the Appalachian Basin versus the Great Basin) and among continents throughout the world (e.g., low- versus high-latitude continents). Because of these differences, the Ordovician diversity trajectories of bivalves and gastropods were rather different from each other when evaluated at any level below the global scale; some continents, for example, exhibited bivalve radiations that were more substantial than those of gastropods, whereas the reverse is true elsewhere. Returning to a theme highlighted earlier, these observations suggest that, on a global scale, some set of characteristics that likely unite both of these molluscan groups were responsible for imparting the comparability observed in their global diversity trajectories and global turnover rates. At the same time, clear differences in their environmental preferences governed the ways in which local and regional patterns combined to produce significantly different local and regional diversity trajectories (see Miller 1997a, 1997b) and highly variable local rates of origination/immigration and extinction/emigration.

Similarly scaled comparisons of other taxa would almost certainly improve our understanding of the Ordovician Radiation at all ecologic scales, with respect to the "two facets of why." The seeds of such comparisons can be recognized, at least in a preliminary way, by considering collectively the chapters in the present volume. In addition to the preliminary comparisons among subclades presented in many chapters, it would be particularly valuable to compare different clades with one another. In some instances, we might discover—as we did for bivalves and gastropods—that closely related clades with comparable global trajectories exhibited significantly different patterns when evaluated in geographic or environmental contexts. In other cases, we might discover the inverse: seemingly unrelated clades that exhibit rather comparable geographic or environmental patterns of diversification. There are some tantalizing hints of this. For example, as with bivalves, Popov et al. (in chapter 17) suggest that the global proliferation of Ordovician shallow-shelf lingulide assemblages began initially in locations marginal to Gondwana, with a subsequent spread elsewhere (see also Harper's discussion of late Tremadocian Orthidina). In addition, Paris et al. (chapter 28) showed that, as with bivalves, there was a major pulse of chitinozoan diversification in Laurentia during the Late Ordovician, although chitinozoan diversity in Laurentia was not as limited as that of bivalves prior to that time; paralleling bivalves, the Ordovician acme of chitinozoan diversity in Gondwana occurred earlier. (It almost goes without saying that a careful geographic mapping of the diversification of all or most major Ordovician taxa would enhance significantly the prospects of definitive comparisons with spatio-temporal patterns exhibited by chitinozoans. These comparisons, in turn, would certainly aid in determining the biologic affinities of these enigmatic fossils.)

Just as it would be valuable to recognize similar geographic patterns of diversification among disparate taxa, it would be useful to delineate patterns that appear to be very nearly the opposite for different groups. In his discussion of plectambonitoid brachiopods, for example, Cocks (in chapter 17) noted that they originated at low latitudes during the Tremadocian and migrated to higher latitudes thereafter. This pattern stands in marked contrast to that of various groups described earlier.

Collectively, these geographic overprints on diversification should motivate us to evaluate further what it was that taxa with similar geographic or environmental trajectories may have had in common and what may have been a key difference among taxa with very nearly the opposite trajectories. Given that similarities in geographic trajectories likely transcended phylogenetic groupings in many instances, we would do well to look beyond morphological features that are diagnostic of phylogeny, perhaps paying greater attention to "ecomorphological" features, in evaluating the significance of these patterns. We should also recognize, as implied earlier, that taxonomic rank should in no way govern our comparisons among higher taxa with respect to assessments of diversification; we should not restrict ourselves, say, to class-versus-class or order-versus-order comparisons. It may be just as meaningful—perhaps more so—to compare a taxon with class rank in one phylum to one of ordinal rank in another phylum.

Geographic and environmental data can also help us to solve enduring questions about the comparative global trajectories of higher taxa that are thought—or perhaps assumed—to have been in direct competition with one another. As exemplified by the long-standing debate about the Phanerozoic diversity trajectories of bivalves and brachiopods (e.g., Gould and Calloway 1980), when the global radiation of one taxon is approximately mirrored through time by the demise of another taxon, there has been a tendency to suggest that the former taxon outcompeted the latter (see Benton 1987, 1991, 1996; Miller 2000). However, except perhaps in rare instances (e.g., Sepkoski et al. 2000), little evidence has been offered in such cases to demonstrate directly that the taxa in question actually competed with one another, let alone that one taxon was competitively superior. As a start, we would do well at least to document that the taxa in question were living in the same places! In considering the Ordovician global diversity trajectories of bivalves and rostroconchs, Cope (chapter 20) addressed a previous suggestion that a decline among rostroconchs was *caused* by the radiation of bivalves. As Cope pointed out, even if one ignores paleogeography, the scenario seems unlikely on autecologic grounds because rostroconchs were infaunal, while a large percentage of Ordovician bivalves were sedentary, epi-, and endobyssate forms. However, when one brings paleogeography into the mix, the competition scenario can be ruled out definitively. Cope noted that most of the significant Ordovician diversity increases or decreases in each of the two groups took place in regions from which the other group was largely absent at the time. The only documented instance in which there was significant regional representation of bivalves *and* rostroconchs, the Late Ordovician of Laurentia, was marked by diversity increases in *both* groups.

Finally, geographic and environmental data can provide unexpected insights into the macroevolutionary transitions that accompany diversification. For example, Taylor and Ernst (chapter 16) showed that, from the Mid to Late Ordovician, there was a substantial increase in the proportion of bryozoan genera that occurred in two or more provinces; that is, on average, individual bryozoan genera were becoming more widespread. This finding is similar to one that emerged from an earlier analysis that I conducted (Miller 1997c) on a genus-level database of Ordovician trilobite, brachiopod, and molluscan occurrences from several paleocontinents. I found that, paralleling the pattern documented by Taylor and Ernst with bryozoans, there was a significant increase during the Late Ordovician in the proportion of genera that occurred in two or more paleocontinents (there was also an expansion in the *environmental* ranges of genera). Moreover, I was able to show that this was correlated with an overall aging of the genus pool: as a direct consequence of the Ordovician Radiation, significantly more old genera were extant during the Late Ordovician than the Early Ordovician. Genera that persisted for long durations tended to become more widespread as they aged. This result was demonstrably not an artifact of the growing proximity of continents to one another during the Late Ordovician and, in fact, suggests the need for caution in using biogeographic data to delineate the changing proximities of continents to one another through the period.

■ An Ordovician Wish List

As exemplified by the chapters in this volume, there have been significant advances in documenting definitively the global diversity trajectories of nearly the entire suite of preserved Ordovician taxa. This progress has resulted from confluence of several factors, including the improved understanding, from a phylogenetic perspective, of several important clades; the inclusion of new data from parts of the world that had previously been undersampled; and improvements to the global Ordovician timescale. As much as anything, however, this advancement is testimony to the unprecedented willingness of Ordovician workers to become more synthetic in their treatments of the data they collect. The examples highlighted in this chapter provide only a small sample of what has been—or could be—accomplished by the integration of new kinds of data and analyses into investigations of Ordovician diversity. To keep the momentum going and especially to get at the "two facets of why," future efforts should probably be targeted on the following topics:

1. The incorporation of sampling standardization and related analytical procedures to overcome global variations in sample size from interval to interval that may well impart artifactual overprints on global or regional diversity trajectories (see Miller and Foote 1996; Alroy et al. 2001). The adoption, throughout this volume, of stratigraphic intervals for diversity compilation that represent roughly equivalent lengths of time can help to minimize these effects, but it is nevertheless probable that the number of fossil occurrences collected and cataloged (or published) varies significantly, in some cases from interval to interval. This variation can and should be overcome in the development of diversity curves.

2. The continued dissection of global diversity patterns at finer geographic scales. In the present volume the comparison of chitinozoan diversification among several paleocontinents (chapter 28), using consistent techniques, is particularly exemplary and can serve as a model for how this can be done by a unified group of researchers for other taxa. Beyond these analyses of individual clades, it will be important to compare patterns among clades, as discussed earlier.

3. The *definitive* integration of physical data (e.g., lithology or geochemical markers) into analyses of biodiversity. Given the recognition that physical transitions may have been important factors in diversity transitions at all hierarchical levels, these assertions must be analyzed more rigorously than has been the case heretofore. This research may be impeded by our inability to envision ways of quantifying the lithological/paleoenvironmental preferences of taxa or the logistical difficulty of sampling extensively the appropriate geochemical proxies contained in fossils or the sediments that encase them. However, recent accomplishments on both fronts for the Ordovician (e.g., Patzkowsky et al. 1997; Pancost et al. 1999; Miller and Connolly 2001) demonstrates that significant progress is attainable.

4. The continuation and expansion of field-based regional studies of biotic transitions in key regions. As discussed earlier, these analyses provide important opportunities to compare, at high stratigraphic resolution, diversity transitions in the face of unique combinations of suspected causal agents. Beyond that, local and regional studies permit the analysis of an important question that has emerged in recent years: the potential decoupling of diversity trends and abundance trends. In some cases, secular changes in taxonomic richness for major taxa have not been paralleled by changes in their relative abundances within sampled fossil assemblages (see McKinney et al. 1998 for a compelling example in association with the Cretaceous/Tertiary boundary). In recent years, transitions at the base of the Middle Ordovician in the Great Basin have been the subjects of an important discussion related to this topic, in this case focused on local transitions from trilobite- to brachiopod-dominated assemblages (e.g., Li and Droser 1999). It would be valuable for these kinds of analyses to expand to key transition intervals in other regions.

5. The expanded integration of measurements of morphological diversity into analyses of the Ordovician Radiation. Pioneering studies of morphological diversification through the Ordovician (e.g., Foote 1993 for trilobites and Wagner 1995a for gastropods) have demonstrated that taxonomic richness is not always a dependable proxy for morphological variation within a clade and that, in some instances, morphological diversification is decoupled from taxonomic diversification. Moreover, by comparing internal versus external morphological features of gastropods, Wagner (1995a) was able to distinguish morphological transitions among characters that were likely governed by phylogeny (internal characters) from those that were likely governed ecologically (external features). Certainly, these approaches would be welcome for other taxa.

As discussed at the start of this chapter, Jack Sepkoski raised our consciousness about the significance of the Ordovician Radiation in the first place through synthetic studies that incorporated *all* taxa together on the same analytical playing field. Sepkoski put us in a position to ask about the "two facets of why" by sorting his taxonomic compendia into evolutionary faunas and, subsequently, by modeling the evolutionary faunas quantitatively. These pivotal studies enabled us to see clearly the profound global biotic transitions that characterized the Ordovician and the apparent decoupling of the Ordovician Radiation from the Cambrian explosion. More than anything, this success should convince us of the need to maintain a synthetic approach as we seek in the future to meld together information from a disparate set of taxa at unprecedented levels of stratigraphic and geographic resolution.

ACKNOWLEDGMENTS

I thank Mary Droser, Michael Foote, Steven Holland, and Barry Webby for their comments on an earlier draft of this chapter. My research on the Ordovician Radiation has been supported by grants from NASA's Program in Exobiology (Grants NAGW-3307, NAG5-6946, and NAG5-9418), and NSF's Program in Biocomplexity (DEB-0083983).

FIGURES AND TABLES

Figures

1.1. Phanerozoic taxonomic diversity of marine animal families and Middle-Upper Cambrian, Ordovician and Silurian taxonomic diversity of marine animal genera   2
2.1. Ordovician stratigraphic chart   42
2.2. Ordovician stratigraphic chart illustrating correlations between the main conodont and chitinozoan zonal sequences   44
3.1. A projection of the six numbered chronostratigraphic time-slice boundaries   50
4.1. The four ways in which a species can be presented as ranging within and/or through a given time interval   53
4.2. Model data set comparing the three measures of diversity with mean standing diversity   54
4.3. Comparison of alternative diversity measures in five model data sets   56
5.1. Southern Hemisphere Ordovician terrane disposition in latest Tremadocian and base Arenig time   62
5.2. Southern Hemisphere Ordovician terrane disposition in early Caradoc time   63
5.3. Southern Hemisphere Ordovician terrane disposition in latest Ordovician–earliest Silurian time   65
5.4. Global reconstruction for latest Ordovician–earliest Silurian time   65
6.1. Strontium, carbon, and oxygen isotopic trends during the Ordovician   70
7.1. Paleogeographic reconstructions for the Early and Late Ordovician   74
7.2. Summary plot of neodymium isotope values from Ordovician conodonts   75
8.1. Plot of relative reversal rate calculated as a function of age for the Phanerozoic Global Paleomagnetic Database   78
9.1. Environmental changes associated with the two phases of the Late Ordovician mass extinction   82
9.2. Paleogeographic reconstruction of Late Ordovician Gondwana   82
10.1. Generalized facies belts of Baltoscandia during the Mid and Late Ordovician   85
10.2. Sea-level curve for the Ordovician of Baltoscandia plotted against the standard stratigraphic framework   86
10.3. Comparison of North American and Baltoscandian sea level curves   91
11.1. Range of Ordovician radiolarian clades and the contained genera   100
12.1. Normalized diversity curve for the Ordovician sponges at species level   103
12.2. Normalized diversity curve for the Ordovician sponges at species level and contribution of major sponge groups to total diversity   103
12.3. Patterns of geographical distribution, diversification, and major migration routes of Early to Mid Ordovician sponges   108

12.4. Patterns of geographic distribution, diversification, and major migration routes of Mid to Late Ordovician sponges   109
13.1. Range chart showing the temporal distribution of Ordovician stromatoporoids worldwide   113
13.2. Diversity curves for the Stromatoporoidea, both at generic and species levels, through Ordovician time   115
14.1. Approximate ranges of the 19 conulariid species in the Ordovician of the Bohemian Massif and species of the nine Ordovician genera known to occur in cratonic North America   121
15.1. Diagram showing the global record of stratigraphic ranges for species of tetradiid coral families and genera through Mid to Late Ordovician time   130
15.2. Cincinnatian coral biogeography and geologic features in Laurentia   133
15.3. Cincinnatian coral diversity in the Williston Basin outcrop belt, southern Manitoba   135
15.4. Distribution and diversity dynamics of the Baltoscandian Late Ordovician rugose corals   139
15.5. The temporal distribution of Ordovician coral genera from Australasia   142
15.6. Ordovician diversity curve for Australasian coral generic data   143
16.1. Cladogram summarizing inferred relationships between main bryozoan groups present in the Ordovician   149
16.2. Bryozoan species diversity changes through 19 Ordovician time slices   150
16.3. Logarithmic plot showing bryozoan species diversification through 19 Ordovician time slices   152
16.4. Bryozoan generic diversification pattern among major taxonomic groups through 19 Ordovician time slices   152
16.5. Changes in per capita generic turnover rates through Ordovician time slices   153
16.6. Geographic aspects of Ordovician bryozoan radiation   153
16.7. Proportion of Lazarus genera of bryozoans for *TS.*3a to 6c   154
17.1. Ordovician standing diversity patterns of the linguliformeans, craniiformeans, and rhynchonelliformeans   159
17.2. Extinction and origination rates of the linguliformeans, lingulides, siphonotretides, and acrotretides   160
17.3. Absolute generic abundances of the linguliformean and craniiformean orders   163
17.4. Strophomenide diversity   165
17.5. Comparison of standing diversity and corrected diversity of strophomenide brachiopods   166
17.6. Orthide diversity   168
17.7. Comparison of standing diversity and corrected diversity of orthide brachiopods   169
17.8. Pentameride, atrypide, and rhynchonellide biodiversity   170
17.9. Increase in size of Plectorthoidea and Orthoidea brachiopods through the Ordovician Radiation   172
17.10. Time environment diagram of the onshore-offshore pattern of the Ordovician Radiation in the Prague Basin   173
17.11. Standing diversities of all major rhynchonelliformean groups compared in terms of numbers of genera   177
18.1. Range chart for genera of Ordovician polyplacophorans   181
18.2. Plot illustrating generic biodiversity and turnover rates of Amphigastropoda (bellerophontiform mollusks)   183
19.1. Stratigraphic ranges of main gastropod groups based on protoconch morphology   186
19.2. Examples of main groups of Ordovician gastropods and bellerophontiform mollusks   188
19.3. Stratigraphic ranges of the main Ordovician gastropod groups   189
19.4. Generic diversity of the Ordovician gastropods and their rates of originations and extinctions   192
20.1. Classification of the Bivalvia showing the probable phylogenetic links between the major bivalve groups   197

20.2. Range chart of the genera of Ordovician Bivalvia belonging to the Nuculoida, Solemyoida, Cardiolarioidea, and Trigonioida   199
20.3. Range chart of the genera of Ordovician Bivalvia belonging to the Heteroconchia, Anomalodesmata, and Pteriomorphia   201
20.4. Range chart of the genera of Ordovician Bivalvia belonging to the Pteriomorphia   203
20.5. Range chart of the genera of Ordovician rostroconchs   205
20.6. Bivalve diversity measures   207
20.7. Rostroconch diversity measures   208
21.1. Genus-level nautiloid diversity through Ordovician time   210
21.2. Originations and extinctions of nautiloid genera through Ordovician time   210
22.1. Stratigraphic distribution of Ordovician hyolith genera, and numbers of species for each genus   216
22.2. Number of Ordovician hyolith species through time   217
22.3. Tubes and disk-like attachments of bryoniids   220
23.1. Genus-level diversity pattern of Ordovician jawed polychaetes   224
23.2. A selection of different Ordovician machaeridian types   226
23.3. Stratigraphic ranges of Ordovician palaeoscolecidans and chaetognaths   229
24.1. Ordovician and Early Silurian trilobite genus diversity   232
24.2. Cluster analysis of Ordovician trilobite families, with plots of their diversity through time   234
24.3. Four trilobite-based biogeographic realms during the time of the Ordovician Radiation   235
24.4. Definition of Laurentian trilobite biofacies during the onset of the Ordovician Radiation   238
24.5. Percentage of total species occurrence in each early Whiterockian Laurentian biofacies that is contributed by the Whiterock Fauna   239
24.6. Representative Laurentian faunas through time and along an environmental gradient   240
24.7. Generic turnover and normalized generic diversity curves for Ordovician trilobites of Australia and New Zealand   241
24.8. Normalized diversity of trilobite species in South America   243
24.9. Sampled and range-through trilobite normalized biodiversity curves for the Anglo-Welsh sector of Avalonia   245
24.10. Range-through, normalized trilobite diversity curves for Baltoscandia, representing Baltica   246
24.11. Sampled and range-through normalized generic biodiversity curves for Ordovician trilobites of the South China Block   248
24.12. Proportion and diversity of Ordovician trilobite genera belonging to faunas of the South China Block   249
25.1. Vertical range of South American Ordovician phyllocarids   260
25.2. Diversity of Ordovician ostracodes in Baltica and the Barrandian area of the Bohemian Massif   262
26.1. Species-level total diversity throughout the Ordovician of the five echinoderm subphyla   268
26.2. Species-level crinozoan total diversity for the Ordovician   269
26.3. Species-level blastozoan total diversity for the Ordovician   270
26.4. Species-level echinozoan, asterozoan, and homalozoan total diversity for the Ordovician   271
26.5. Composition of five relatively diverse echinoderm faunas from different geographic areas and parts of the Ordovician   278
27.1. Correlation of Ordovician graptolite zones and time units   282
27.2. Graptolite species diversity and rate plots for Australasia   287
27.3. Graptolite species diversity and rate plots for Baltica   288
27.4. Graptolite species diversity and rate plots for Avalonia   289
27.5. Comparison of normalized diversity plots, and rates of origination and extinction for graptolite species in Australasia and Avalonia   291

28.1. Regional distribution and density of Ordovician samples used for documenting the diversity of chitinozoans   295
28.2. Biodiversity of Ordovician chitinozoan species from North Gondwana   297
28.3. Biodiversity of Ordovician chitinozoan species from Baltica   301
28.4. Biodiversity of Ordovician chitinozoan species from Laurentia   303
28.5. Biodiversity of Ordovician chitinozoan species from Avalonia and from China   305
28.6. Global biodiversity of Ordovician chitinozoan species and major turnover   309
29.1. Stratigraphic ranges of 106 conodont genera   317
29.2. Conodont species diversity and rate measures for each biostratigraphic interval   318
29.3. Holdover and carryover measurements for species- and genus-level conodont data for each biostratigraphic interval   319
29.4. Dendrograph of cluster analysis showing conodont paleobiogeographic relationships during the highest peak of conodont diversity in the Ordovician Period   325
30.1. Stratigraphic range chart of latest Cambrian and Ordovician vertebrates and supposed vertebrates   330
30.2. New vertebrate microremains from the Ordovician of Ontario, Quebec, and Alberta   332

31.1. Receptaculitid families and genera with the species represented by their stratigraphic ranges during the Ordovician   338
31.2. Cyclocrinitid families and genera with the species represented by their stratigraphic ranges during the Middle-Upper Ordovician   340
32.1. Number of acritarch species versus time   352
32.2. Number of Ordovician acritarch research publications plotted on a paleogeographic reconstruction for the Arenig   353
32.3. Acritarch diversity trends for the Yangtze Platform, South China   356
32.4. Acritarch diversity trends for North Africa   358
33.1. Examples of cryptospores and a spore mass from the Caradoc of the Arabian Peninsula   362
33.2. Stratigraphic ranges of the most common miospore morphologies and diversity change of miospore species through the Ordovician and Lower Silurian (Llandovery)   364
34.1. Ichnogeneric diversity numbers through Ordovician time   371
34.2. Environmental distribution of Lower Paleozoic ichnofaunas in clastic settings   372
34.3. The *Curvolithus* association from Darriwilian-Caradoc deltaic prograding clinoforms in the First Bani Group at Jbel Gaiz, eastern Anti-Atlas, Morocco   378

Tables

1.1. A preliminary listing of genus- and species-level totals for the Ordovician fossil groups   4
1.2. Comparative generic lists of Ordovician animal diversity data   5
1.3. Summary of the global subdivisions of the Ordovician system/period   7
3.1. Radiometric control points   49
4.1. Properties of the six trial data sets and comparison of trial diversity measures with mean standing diversity (MSD)   55
11.1. Number of radiolarian species and genera per assemblage   98

11.2. Radiolarian taxon diversity measures   99
15.1. Occurrences of coral genera in the Cincinnatian of Laurentia   134
15.2. Baltoscandian rugose coral-bearing stratigraphic units and localities assigned to Upper Ordovician time slices   138
16.1. Summary of the Ordovician Bryozoan taxic database   151
24.1. Distribution by faunal realm of Whiterock Fauna families and subfamilies of trilobites during the onset of the Ordovician radiation   235

24.2. Latitudinal distribution of Whiterock Fauna families and subfamilies of trilobites during the time of radiation contrasted with their End Ordovician fate   236
25.1. Reliable and unverified occurrences of Ordovician Eurypterida   256
25.2. Annotated list of described Ordovician Phyllocarid taxa   258
27.1. Graptolite diversity, origination, extinction, and faunal turnover for three data sets, Australasia, Baltica, and Avalonia   290
27.2. Correlation of graptolite origination and extinction rates   292
29.1. Lower Ordovician conodont records from selected localities worldwide   314
29.2. Middle Ordovician conodont records from selected localities worldwide   316
29.3. Calculated generic and specific diversity values and measures derived from the compiled conodont database of tables 29.1, 29.2   319
30.1. Diversity of Ordovician vertebrates   333

REFERENCES

Abushik, A., and L. Sarv. 1983. Ostracodes from the Molodovo Stage of Podolia; pp. 101–134 *in* E. Klaamann (ed.), Paleontologiya drevnego paleozoya Pribaltiki i Podolii. Tallinn, Estonia.

Aceñolaza, F. G., and F. R. Durand. 1973. Trazas fósiles del basamento cristalino del Noroeste argentino. Boletín de la Asociación Geológica de Córdoba 2:45–55.

Aceñolaza, F. G., and F. R. Durand. 1978. Trazas de trilobites en los estratos del Ordovícico basal de la Puna Argentina. Acta Geológica Lilloana 15:5–12.

Aceñolaza, F. G., and S. Esteban. 1996. Filocáridos (Crustacea) en el Tremadociano del Sistema de Famatina, Provincia de La Rioja, Argentina. Memorias del XII Congreso Geológico de Bolivia, Tarija, 1:281–288.

Achab, A. 1989. Ordovician chitinozoan zonation of Quebec and western Newfoundland. Journal of Paleontology 63:14–24.

Achab, A. 1991. Biogeography of Ordovician Chitinozoa; pp. 135–142 *in* C. R. Barnes and S. H. Williams (eds.), Advances in Ordovician Geology. Geological Survey of Canada, Paper 90-9.

Achab, A., and E. Asselin. 1995. Ordovician chitinozoans from the Arctic Platform and the Franklinian miogeosyncline in northern Canada. Review of Palaeobotany and Palynology 86:69–90.

Adrain, J. M., and R. A. Fortey. 1997. Ordovician trilobites from the Tourmakeady Limestone, western Ireland. Bulletin of the Natural History Museum, London, Geology Series 53:79–115.

Adrain, J. M., and S. R. Westrop. 2000. An empirical assessment of taxic paleobiology. Science 289:110–112.

Adrain, J. M., B. D. E. Chatterton, and R. B. Blodgett. 1995. Silurian trilobites from southwestern Alaska. Journal of Paleontology 69:723–736.

Adrain, J. M., R. A. Fortey, and S. R. Westrop. 1998. Post-Cambrian trilobite diversity and evolutionary faunas. Science 280:1922–1925.

Adrain, J. M., S. R. Westrop, B. D. E. Chatterton, and L. Ramsköld. 2000. Silurian trilobite alpha diversity and the end-Ordovician mass extinction. Paleobiology 26:625–646.

Ainsaar, L., and T. Meidla. 2001. Facies and stratigraphy of the middle Caradoc mixed siliciclastic-carbonate sediments in eastern Baltoscandia. Proceedings of the Estonian Academy of Sciences. Geology 50:5–23.

Ainsaar, L., T. Meidla, and T. Martma. 1999. Evidence for a widespread carbon isotopic event associated with late Middle Ordovician sedimentological and faunal changes in Estonia. Geological Magazine 136:49–62.

Aitchison, J. C. 1998. A new Lower Ordovician (Arenigian) radiolarian fauna from the Ballantrae Complex Scotland. Scottish Journal of Geology 34:73–81.

Aitchison, J. C., P. G. Flood, and J. Malpas. 1998. Lowermost Ordovician (basal Tremadoc) radiolarians from the Little Port Complex, western Newfoundland. Geological Magazine 135:413–419.

Albanesi, G. L. 1998a. Biofacies de conodontes de las secuencias ordovícicas del cerro Potrerillo, Precordillera Central de San Juan, R. Argentina. Actas Academia Nacional de Ciencias, Córdoba, 12:75–98.

Albanesi, G. L. 1998b. Taxonomía de conodontes de las secuencias ordovícicas del cerro Potrerillo, Precordillera Central de San Juan, R. Argentina. Actas Academia Nacional de Ciencias, Córdoba, 12:101–253.

Albanesi, G. L., and R. A. Astini. 2000a. Nueva fauna de conodontes de la Formación Suri (Ordovícico Inferior-Medio), Sistema de Famatina, Argentina. Ameghiniana 37:68Ra.

Albanesi, G. L., and R. A. Astini. 2000b. Bioestratigrafía de conodontes de la Formación Las Chacritas, Precordillera de San Juan, Argentina. Ameghiniana 37:68Rb.

Albanesi, G. L., and C. R. Barnes. 2000. Subspeciation within a puctuated equilibrium evolutionary event: Phylogenetic history of the Lower-Middle Ordovician *Paroistodus originalis–P. horridus* complex (Conodonta). Journal of Paleontology 74:492–502.

Albanesi, G. L., and N. E. Vaccari. 1994. Conodontos del Arenig en la Formación Suri, Sistema del Famatina,

Argentina. Revista Española de Micropaleontología 26: 125–146.

Albanesi, G. L., J. L. Benedetto, and P.-Y. Gagnier. 1995a. *Sacabambaspis janvieri* (Vertebrata) y conodontes del Llandeiliano temprano en la Formación la Cantera, Precordillera de San Juan, Argentina; pp. 519–543 *in* M. A. Hünicken (ed.), IGCP Project N° 271: South American Paleozoic Conodontology—Proceedings, Latin American Conodont Symposium LACON I (1990, Cochabamba [Bolivia], Córdoba y San Juan [Argentina]), LACON II (1992, Córdoba, Argentina), and the III International Meeting (1991, Porto Alegre, Brazil). Boletín de la Academia Nacional de Ciencias, Córdoba, 60(3–4).

Albanesi, G. L., G. Ortega, and M. A. Hünicken. 1995b. Conodontes y graptolitos de la Formación Yerba Loca (Arenigiano-Llandeiliano) en las quebradas de Ancaucha y El Divisadero, Precordillera de San Juan, Argentina. Boletín de la Academia Nacional de Ciencias, Córdoba, 60:365–400.

Albanesi, G. L., M. A. Hünicken, and C. R. Barnes. 1998. Bioestratigrafía de conodontes de las secuencias ordovícicas del cerro Potrerillo, Precordillera Central de San Juan, R. Argentina. Actas Academia Nacional de Ciencias, Córdoba, 12:7–72.

Albanesi, G. L., G. Ortega, C. R. Barnes, and M. A. Hünicken. 1999. Conodont-graptolite biostratigraphy of the Gualcamayo Formation (Middle Ordovician) in the Gualcamayo-Guandacol rivers area, Argentina Precordillera; pp. 45–48 *in* P. Kraft and O. Fatka (eds.), *Quo vadis* Ordovician? Short papers of the 8th International Symposium on the Ordovician System. Acta Universitatis Carolinae, Geologica 43(1–2).

Albanesi, G. L., S. B. Esteban, M. A. Hünicken, and C. R. Barnes. 2000. Las biozonas de conodontes de la Formación Volcancito (Cámbrico tardío–Ordovícico temprano), Sistema de Famatina, Noroeste de Argentina. Ameghiniana 37:5R.

Albanesi, G. L., G. Ortega, and F. Zeballo. 2001. Late Tremadocian conodont-graptolite biostratigraphy from NW Argentine basins; pp. 125–127 *in* C. Minjin (comp.), The Guide Book, Mongolian Ordovician and Silurian Stratigraphy and Abstracts for the Joint Field Meeting of IGCP 410 and IGCP 421 in Mongolia. Tempus-Tacis Project (Ulaanbaatar, Mongolia) 20091-98.

Albani, R. 1989. Ordovician (Arenigian) acritarchs from the Solanas Sandstone Formation, Central Sardinia, Italy. Bollettino della Società Paleontologica Italiana 28:3–37.

Albani, R., G. Bagnoli, J. Maletz, and S. Stouge. 2001. Integrated chitinozoan, conodont, and graptolite biostratigraphy from the upper part of the Cape Cormorant Formation (Middle Ordovician), western Newfoundland. Canadian Journal of Earth Sciences 38:387–409.

Alberstadt, L. P., and J. E. Repetski. 1989. A Lower Ordovician sponge/algal facies in the southern United States and its counterparts elsewhere in North America. Palaios 4:225–242.

Alberstadt, L. P., K. R. Walker, and R. P. Zurawski. 1974. Patch reefs in the Carters Limestone (Middle Ordovician) in Tennessee, and vertical zonation in Ordovician reefs. Geological Society of America, Bulletin 85:1171–1182.

Aldridge, R. J. 1988. Extinction and survival in the Conodonta; pp. 231–256 *in* G. P. Larwood (ed.), Extinction and Survival in the Fossil Record. Systematics Association Special Volume 34. Clarendon Press, Oxford.

Aldridge, R. J., D. E. G. Briggs, M. P. Smith, E. N. K. Clarkson, and N. D. L. Clark. 1993. The anatomy of conodonts. Philosophical Transactions of the Royal Society of London, B 340:405–421.

Alexander, R. R. 1986. Resistance to and repair of shell breakage induced by durophages in Late Ordovician brachiopods. Journal of Paleontology 60:273–285.

Alexander, R. R. 2001. Functional morphology and biomechanics of articulate brachiopod shells; pp. 145–169 *in* S. J. Carlson and M. R. Sandy (eds.), Brachiopods Ancient and Modern: A Tribute to G. Arthur Cooper. The Paleontological Society (New Haven, Connecticut), Paper 7.

Al-Hajri, S. 1995. Biostratigraphy of the Ordovician Chitinozoa of northwestern Saudi Arabia. Review of Palaeobotany and Palynology 89:27–68.

Allison, P. A., and D. E. G. Briggs. 1994. Exceptional fossil record: Distribution of soft-issue preservation through the Phanerozoic: Reply. Geology 22:184.

Allmon, W. D., and R. M. Ross. 2001. Nutrients and evolution in the marine realm; pp. 105–148 *in* W. D. Allmon and D. J. Bottjer (eds.), Evolutionary Paleoecology: The Ecological Context of Macroevolutionary Change. Columbia University Press, New York.

Almond, J. E. 1985. The Silurian-Devonian fossil record of the Myriapoda. Philosophical Transactions of the Royal Society of London, B 309:227–237.

Alroy, J. 1992. Conjunction among taxonomic distributions and the Miocene mammalian biochronology of the Great Plains. Paleobiology 18:326–343.

Alroy, J. 2000. New methods for quantifying macroevolutionary patterns and processes. Paleobiology 26:707–733.

Alroy, J., C. R. Marshall, R. K. Bambach, K. Bezusko, M. Foote, F. T. Fursich, T. A. Hansen, S. M. Holland, L. C. Ivany, D. Jablonski, D. K. Jacobs, D. C. Jones, M. A. Kosnik, S. Lidgard, S. Low, A. I. Miller, P. M. Novack-Gottshall, T. D. Olszewski, M. E. Patzkowsky, D. M. Raup, K. Roy, J. J. Sepkoski, M. G. Sommers, P. J. Wagner, and A. Webber. 2001. Effects of sampling standardization on estimates of Phanerozoic marine diversification. Proceedings of the National Academy of Sciences (U.S.) 98:6261–6266.

Alvarez, F., J.-Y. Rong, and A. J. Boucot. 1998. The classification of athyridid brachiopods. Journal of Paleontology 72:827–855.

Amireh, B. S., W. Schneider, and A. M. Abed. 1994. Evolving fluvial-transitional-marine deposition through the Cambrian sequence of Jordan. Sedimentary Geology 89:65–90.

An, T.-X., and Z. Zheng. 1990. The conodonts of the marginal areas around the Ordos basin, north China. Science Press, Beijing, 101 pp. (in Chinese with English abstract).

Anderson, M. E., J. E. Almond, F. J. Evans, and J. A. Long. 1999. Devonian (Emsian-Eifelian) fish from the Lower Bokkeveld Group (Ceres Subgroup), South Africa. Journal of African Earth Sciences 29:179–193.

Anstey, R. L. 1986. Bryozoan provinces and patterns of generic evolution and extinction in the Late Ordovician of North America. Lethaia 19:33–51.

Anstey, R. L. 1990. Bryozoans; pp. 232–252 in K. J. McNamara (ed.), Evolutionary Trends. Belhaven Press, London.

Anstey, R. L., and J. F. Pachut. 1995. Phylogeny, diversity history, and speciation in Paleozoic bryozoans; pp. 239–284 in D. H. Erwin and R. L. Anstey (eds.), New Approaches to Speciation in the Fossil Record. Columbia University Press, New York.

Apollonov, M. K. 1974. Ashgill Trilobites of Kazakhstan. Nauka, Alma-Ata, SSSR, 136 pp.

Apollonov, M. K. 1975. Ordovician trilobite assemblages of Kazakhstan. Fossils and Strata 4:375–380.

Archibald, J. D. 1993. The importance of phylogenetic analysis for the assessment of species turnover: A case history of Paleocene mammals in North America. Paleobiology 19:1–27.

Armstrong, H. A. 1996. Biotic recovery after mass extinction: The role of climate and ocean-state in the postglacial (Late Ordovician–Early Silurian) recovery of the conodonts; pp. 105–117 in M. B. Hart (ed.), Biotic Recovery from Mass Extinction Events. The Geological Society, London, Special Publication 102.

Armstrong, H. A. 1997. Conodonts from the Ordovician Shinnel Formation, South Uplands, Scotland. Palaeontology 40:763–797.

Armstrong, H. A., and A. W. Owen. 2002a. Euconodont diversity changes in a cooling and closing Iapetus Ocean; pp. 85–98 in J. A. Crame and A. W. Owen (eds.), Palaeobiogeography and Biodiversity Change: The Ordovician and Mesozoic-Cenozoic Radiation. The Geological Society, London, Special Publication 194.

Armstrong, H. A., and A. W. Owen. 2002b. Euconodont paleobiogeography and the closure of the Iapetus Ocean. Geology 30:1091–1094.

Astini, R. A. 1994. Geología e interpretación de la Formación Gualcamayo en su localidad clásica (suroeste de Guandacol y cordón de Perico-Potrerillo), Precordillera septentrional. Revista de la Asociación Geológica Argentina 49:55–70.

Astini, R. A. 1995. Sedimentología de la Formación Las Aguaditas (talud carbonático) y evolución de la cuenca precordillerana durante el Ordovícico medio. Revista de la Asociación Geológica Argentina 50:143–164.

Astini, R. A. 1998. Stratigraphic evidence supporting the rifting, drifting, and collision of the Laurentian Precordilleran terrane of western Argentina; pp. 11–33 in R. J. Pankhurst and C. W. Rapela (eds.), The Proto-Andean Margin of Gondwana. The Geological Society, London, Special Publication 142.

Astini, R. A. 1999a. Sedimentological constraints on the Middle-Upper Ordovician extension in the exotic-to-Gondwana Precordillera terrane; pp. 119–122 in P. Kraft and O. Fatka (eds.), Quo vadis Ordovician? Short papers of the 8th International Symposium on the Ordovician System. Acta Universitatis Carolinae, Geologica 43(1–2).

Astini, R. A. 1999b. The Late Ordovician glaciation in the Proto-Andean margin of Gondwana revisited: Geodynamic implications; pp. 171–173 in P. Kraft and O. Fatka (eds.), Quo vadis Ordovician? Short papers of the 8th International Symposium on the Ordovician System. Acta Universitatis Carolinae, Geologica 43(1–2).

Astini, R. A., and J. L. Benedetto. 1996. Paleoenvironmental features and basin evolution of a complex volcanic-arc region in the Pre-Andean western Gondwana: The Famatina belt. Third International Symposium on Andean Geodynamics, St. Malo, France, Extended Abstracts, 755–758.

Astini, R. A., J. L. Benedetto, and N. E. Vaccari. 1995. The Early Paleozoic evolution of the Argentine Precordillera as a Laurentian rifted, drifted, and collided terrane: A geodynamic model. Geological Society of America, Bulletin 107:253–273.

Astrova, G. G. 1978. The history of development, system, and phylogeny of the Bryozoa: Order Trepostomata. Akademia Nauk SSSR, Trudy Paleontologicheskogo Instituta 169:1–240.

Ausich, W. I. 1980. Synecology—Niche differentiation; pp. 59–72 in T. W. Broadhead and J. A. Waters (eds.), Echinoderms: Notes for a Short Course. University of Tennessee (Knoxville), Department of Geological Sciences, Studies in Geology 3.

Ausich, W. I. 1998. Phylogeny of Arenig to Caradoc crinoids (Phylum Echinodermata) and suprageneric classification of the Crinoidea. University of Kansas Paleontological Contributions, n.s., 9:1–36.

Ausich, W. I., and D. J. Bottjer. 1982. Tiering in suspension-feeding communities on soft substrata throughout the Phanerozoic. Science 216:173–174.

Ausich, W. I., and P. Copper. 2002. New latest Ordovician (Rawtheyan and Hirnantian) crinoid faunas from Anticosti Island, Quebec, Canada. Geological Society of America, Abstracts with Programs 34(6):428.

Babcock, L. E., R. M. Feldmann, and M. T. Wilson. 1987. Teratology and pathology of some Paleozoic conulariids. Lethaia 20:93–105.

Babin, C. 1966. Mollusques bivalves et céphalopodes du Paléozoïque armoricain. Imprimerie Commerciale et Administrative, Brest, France, 470 pp.

Babin, C. 1982. Mollusqes bivalves et rostroconches; pp. 37–49 in C. Babin, R. Courtessole, M. Melou, J. Pillet, D. Vizcaino, and E. L. Yochelson. Brachiopodes (articulés) et mollusques (bivalves, rostroconches, monoplacophores, gastropodes) de l'Ordovicien inférieur (Trémadocien-Arenigien) de la Montagne Noire (France méridionale). Mémoire de la Société des Études Scientifiques de l'Aude, Sival, Carcassonne.

Babin, C. 1993. Rôle des plates-formes gondwaniennes dans les diversifications des mollusques bivalves durant l'Ordovicien. Bulletin de la Société Géologique de France 164:141–153.

Babin, C. 1995. The initial Ordovician bivalve mollusc radiations on the western Gondwanan shelves; pp. 491–498 in J. D. Cooper, M. L. Droser, and S. C. Finney (eds.), Ordovician Odyssey: Short Papers, 7th International Symposium on the Ordovician System. Book 77, Pacific Section Society for Sedimentary Geology (SEPM), Fullerton, California.

Babin, C. 2000. Ordovician to Devonian diversification of the Bivalvia. American Malacological Bulletin 15:167–178.

Babin, C., and J. C. Gutiérrez-Marco. 1991. Middle Ordovician bivalves from Spain and their phyletic and palaeogeographic significance. Palaeontology 34:109–147.

Bagnoli, G. 1994. Sea level changes and conodont correlation across the Iapetus Ocean during late Tremadocian-early Arenigian. Palaeopelagos 4:61–71.

Bagnoli, G., and S. Stouge. 1991. Paleogeographic distribution of Arenigian (Lower Ordovician) conodonts. Anais Academia Brasileira de Ciencias 63:171–183.

Bagnoli, G., and S. Stouge. 1996. Changes in conodont provincialism and biofacies during the lower Ordovician in Öland, Sweden. Palaeopelagos 6:19–29.

Bagnoli, G., and S. Stouge. 1997. Lower Ordovician (Billingenian-Kunda) conodont zonation and provinces based on sections from Horns Udde, north Öland, Sweden. Bollettino della Società Paleontologica Italiana 35:109–163.

Bagnoli, G., S. Stouge, and M. Tongiorgi. 1988. Acritarchs and conodonts from the Cambro-Ordovician Furuhäll (Köpingsklint) section (Öland, Sweden). Rivista Italiana di Paleontologia e Stratigrafia 94:163–248.

Balashov, Z. G. 1962. Nautiloidei ordovika sibirskoy platformy [Ordovician nautiloids of the Siberian Platform]. Izdatelstvo, Leningrad University, Leningrad, 206 pp.

Baldwin, C. T. 1977. The stratigraphy and facies associations of trace fossils in some Cambrian and Ordovician rocks of north western Spain; pp. 9–40 in T. P. Crimes and J. C. Harper (eds.), Trace Fossils 2. Geological Journal, Special Issue 9. Seel House Press, Liverpool.

Bambach, R. K. 1977. Species richness in marine benthic habitats through the Phanerozoic. Paleobiology 3:152–167.

Bambach, R. K. 1983. Ecospace utilization and guilds in marine communities through the Phanerozoic; pp. 719–746 in M. J. S. Tevesz and P. L. McCall (eds.), Biotic Interactions in Recent and Fossil Benthic Communities. Plenum Press, New York.

Bambach, R. K. 1985. Classes and adaptive variety: The ecology of diversification in marine faunas through the Phanerozoic; pp. 191–253 in J. W. Valentine (ed.), Phanerozoic Diversity Patterns: Profiles in Macroevolution. Princeton University Press, Princeton.

Bambach, R. K. 1993. Seafood through time: Changes in biomass, energetics, and productivity in the marine ecosystem. Paleobiology 19:372–397.

Bandel, K. 1982. Morphologie und Bildung der frühontogenetischen Gehäuse bei conchiferen Mollusken. Facies 7:1–198.

Bandel, K. 1997. Higher classification and pattern of evolution of the Gastropoda. Courier Forschungsinstitut Senckenberg 201:57–81.

Bandel, K., and J. Frýda. 1996. Balbinipleura, a new slit bearing archaeogastropod (Vetigastropoda) from the Lower Devonian of Bohemia and the Lower Carboniferous of Belgium. Neues Jahrbuch für Geologie und Paläontologie 6:325–344.

Bandel, K., and J. Frýda. 1998. Position of Euomphalidae in the system of the Gastropoda. Senckenbergiana Lethaea 78(1/2):103–131.

Bandel, K., and J. Frýda. 1999. Notes on the evolution and higher classification of the subclass Neritimorpha (Gastropoda) with the description of some new taxa. Geologica et Palaeontologica 33:219–235.

Bandel, K., and W. Geldmacher. 1996. The structure of the shell of Patella crenata connected with suggestions to the classification and evolution of the Archaeogastropoda. Freiberger Forschungsheft, C 464:1–71.

Banks, M. R. 1988. The base of the Silurian System in Tasmania. Bulletin of the British Museum (Natural History), Geology, 43:191–194.

Banks, M. R., and C. F. Burrett. 1980. A preliminary Ordovician biostratigraphy of Tasmania. Journal of the Geological Society of Australia 26:363–376.

Barker, G. W., and M. M. Miller. 1990. Tremadocian (Lower Ordovician) acritarchs from the subsurface of West Texas. Palynology 14:209.

Barnes, C. R. 1964. Conodont biofacies analysis of some Wilderness (Middle Ordovician) limestones, Ottawa Valley, Ontario. Doctoral thesis, Ottawa University, Ottawa.

Barnes, C. R. 1984. Early Ordovician eustatic events in Canada; pp. 51–63 in D. L. Bruton (ed.), Aspects of the Ordovician System. Palaeontological Contributions from the University of Oslo 295.

Barnes, C. R. 2001. Modeling atmospheric O_2 over Phanerozoic time. Geochemica et Cosmochimica Acta 65:685–694.

Barnes, C. R., and L. E. Fåhraeus. 1975. Provinces, communities, and the proposed nektobenthic habit of Ordovician conodontophorids. Lethaia 8:133–149.

Barnes, C. R., C. B. Rexroad, and J. F. Miller. 1973. Lower Paleozoic conodont provincialism. Geological Society of America, Special Paper 141:157–190.

Barnes, C. R., R. A. Fortey, and S. H. Williams. 1996 [dated 1995]. The pattern of global bio-events during the Ordovician Period; pp. 139–172 in O. H. Walliser (ed.), Global Events and Event Stratigraphy in the Phanerozoic. Springer-Verlag, New York.

Barrande, J. 1867. Système Silurien du centre de la Bohême. 1ère partie, Recherches paléontologiques. Vol. 3, Classe des Mollusques, Ordre des Ptéropodes, 1–179. Prague.

Barrande, J. 1872. Système Silurien du centre de la Bohême. 1ère partie, Recherches paléontologiques. Supplément au Vol. 1, Trilobites, Crustacés divers et Poissons, texte et planches, 565–577, pls. 20 and 35. Prague.

Barrois, C. 1891. Mémoire sur la faune du grès armoricain. Annales de la Société Géologique du Nord 19:134–351.

Bassett, M. G., and V. Berg-Madsen. 1993. *Protocimex*: A phyllocarid crustacean, not an Ordovician insect. Journal of Paleontology 67:144–147.

Bassett, M. G., M. Dastanpour, and L. E. Popov. 1999a. New data on Ordovician faunas and stratigraphy of the Kerman and Tabas regions, east-central Iran; pp. 483–486 in P. Kraft and O. Fatka (eds.), *Quo vadis* Ordovician? Short papers of the 8th International Symposium on the Ordovician System. Acta Universitatis Carolinae, Geologica 43(1–2).

Bassett, M. G., L. E. Popov, and L. E. Holmer. 1999b. Organophosphatic brachiopods: Patterns of biodiversification and extinction in the Early Palaeozoic. Geobios 32:145–163.

Bassett, M. G., L. E. Popov, and E. V. Sokiran. 1999c. Patterns of diversification in Ordovician cyrtomatodont rhynchonellate brachiopods; pp. 329–332 in P. Kraft and O. Fatka (eds.), *Quo vadis* Ordovician? Short papers of the 8th International Symposium on the Ordovician System. Acta Universitatis Carolinae, Geologica 43(1–2).

Bassler, R. S. 1911. The Early Paleozoic Bryozoa of the Baltic Provinces. United States National Museum, Bulletin 77:1–382.

Bassler, R. S. 1915. Bibliographic index of American Ordovician and Silurian fossils. United States National Museum, Bulletin 92 (2 vols.):1–1521.

Bassler, R. S. 1941. The Nevada Early Ordovician (Pogonip) sponge fauna. United States Natural History Museum, Proceedings 91:91–102.

Bassler, R. S. 1950. Faunal lists and descriptions of Paleozoic corals. Geological Society of America, Memoir 44:1–315.

Bassler, R. S., and M. W. Moodey. 1943. Bibliographic and faunal index of Paleozoic pelmatozoan echinoderms. Geological Society of America, Special Paper 45:1–734.

Bauer, J. A. 1994. Conodonts from the Bromide Formation (Middle Ordovician) south-central Oklahoma. Journal of Paleontology 68:358–376.

Bell, B. M. 1976. A Study of North American Edrioasteroidea. New York State Museum, Memoir 21:1–446.

Bender, F. 1974. Geology of Jordan: Contributions to the Regional Geology of the Earth. Supplementary Edition, Vol. 7. Gebrüder Borntraeger, Berlin, 196 pp.

Benedetto, J. L. 1998. Early Palaeozoic brachiopods and associated shelly faunas from western Gondwana: Their bearing on the geodynamic history of the pre-Andean margin; pp. 57–83 in R. J. Pankhurst and C. W. Rapela (eds.), The Proto-Andean Margin of Gondwana. The Geological Society, London, Special Publication 142.

Benedetto, J. L., T. M. Sánchez, M. G. Carrera, E. D. Brussa, and M. J. Salas. 1999. Paleontological constraints on successive paleogeographic positions of Precordillera terrane during the Early Paleozoic; pp. 21–42 in D. Keppie and V. A. Ramos (eds.), Laurentia-Gondwana Connections before Pangea. Geological Society of America, Special Paper 336.

Bengtson, S., and Y. Zhao. 1992. Predatorial boring in late Precambrian mineralized exoskeletons. Science 257:367–369.

Benton, M. J. 1982. *Dictyodora* and associated trace fossils from the Palaeozoic of Thuringia. Lethaia 15:115–132.

Benton, M. J. 1987. Progress and competition in macroevolution. Biological Reviews 62:305–338.

Benton, M. J. 1991. Extinction, biotic replacements, and clade interactions; pp. 89–102 in E. C. Dudley (ed.), The Unity of Evolutionary Biology. Dioscorides Press, Portland, Oregon.

Benton, M. J. 1996. On the nonprevalance of competitive replacement in the evolution of tetrapods; pp. 185–210 in D. Jablonski, D. H. Erwin, and J. H. Lipps (eds.), Evolutionary Paleobiology. University of Chicago Press, Chicago.

Benton, M. J. 1999. The history of life: large databases in palaeontology; pp. 249–283 in D. A. T. Harper (ed.), Numerical Palaeobiology. Wiley, Chichester, England.

Benton, M. J. 2001. Biodiversity on land and in the sea. Geological Journal 36:211–230.

Benton, M. J., and N. H. Trewin. 1980. *Dictyodora* from the Silurian of Peeblesshire, Scotland. Palaeontology 23:501–513.

Berdan, J. M. 1976. Middle Ordovician leperditicopid ostracodes from the Ibex area, Millard County, western Utah. Brigham Young University Geology Studies 23:37–65.

Berdan, J. M. 1984. Leperditicopid ostracodes from Ordovician rocks of Kentucky and nearby states and characteristic features of the Order Leperditicopida. United

States Geological Survey, Professional Paper 1066J: 1–40.

Berdan, J. M. 1988. Middle Ordovician (Whiterockian) palaeocopid and podocopid ostracodes from the Ibex area, Millard County, western Utah. New Mexico Bureau of Mines and Mineral Resources, Memoir 44: 273–301.

Bergenhayn, J. R. M. 1955. Die fossilen Schwedischen Loricaten nebst einer vorläufigen Revision des Systems der ganzen Klasse Loricata. Lunds Universitets Arsskrift, Nya Förhandlingar, Avdelningen 2, 51(8):1–46.

Bergenhayn, J. R. M. 1960. Cambrian and Ordovician loricates from North America. Journal of Paleontology 34:168–178.

Berg-Madsen, V. 1987. *Tuarangia* from Bornholm (Denmark) and similarities in Baltoscandian and Australasian late Middle Cambrian faunas. Alcheringa 11:245–259.

Berg-Madsen, V., and J. M. Malinky. 1999. A revision of Holm's late Mid and Late Cambrian hyoliths of Sweden. Palaeontology 42:841–885.

Bergman, C. F. 1998. Reversal in some fossil polychaete jaws. Journal of Paleontology 72:632–638.

Bergström, J. 1976. Lower Palaeozoic trace fossils from eastern Newfoundland. Canadian Journal of Earth Sciences 13:1613–1633.

Bergström, S. M. 1983. Biogeography, evolutionary relationships, and biostratigraphic significance of Ordovician platform conodonts. Fossils and Strata 15:35–58.

Bergström, S. M. 1986. Biostratigraphic integration of Ordovician graptolite and conodont zones—a regional review; pp. 61–78 *in* C. P. Hughes and R. B. Rickards (eds.), Palaeoecology and Biostratigraphy of Graptolites. The Geological Society, London, Special Publication 20.

Bergström, S. M. 1988. On Pander's Ordovician conodonts: Distribution and significance of the *Prioniodus elegans* fauna in Baltoscandia. Senckenbergiana Lethaea 69:217–251.

Bergström, S. M. 1990. Relations between conodont provincialism and changing palaeogeography during the Early Palaeozoic; pp. 105–121 *in* W. S. McKerrow and C. R. Scotese (eds.), Palaeozoic Palaeogeography and Biogeography. The Geological Society, London, Memoir 12.

Bergström, S. M. 1995. The search for global biostratigraphic reference levels in the Ordovician System: Regional correlation potential of the base of the North American Whiterockian Stage; pp. 149–152 *in* J. D. Cooper, M. L. Droser, and S. C. Finney (eds.), Ordovician Odyssey: Short Papers, 7th International Symposium on the Ordovician System. Book 77. Pacific Section Society for Sedimentary Geology (SEPM), Fullerton, California.

Bergström, S. M. 1996. Tentaculitoids; pp. 282–287 *in* R. M. Feldmann and M. Hackathorn (eds.), Fossils of Ohio. Department of Natural Resources, Division of Geology (Columbus, Ohio), Bulletin 70.

Bergström, S. M. 1997. The oldtimers were right: New data on the relations between the type Cincinnatian and the Upper Ordovician in the upper Mississippi Valley, Oklahoma, Texas, New Mexico, and the western interior. Geological Society of America, Abstracts with Programs 29(4):5.

Bergström, S. M., and J. B. Carnes. 1976. Conodont biostratigraphy and paleoecology of the Holston Formation (Middle Ordovician) and associated strata in eastern Tennessee. Geological Association of Canada Special Paper 15:27–57.

Bergström, S. M., and C. E. Mitchell. 1994. Regional relationships between late Middle and early Late Ordovician standard successions in New York and Quebec and the Cincinnati region in Ohio, Indiana, and Kentucky; pp. 5–20 *in* E. Landing (ed.), Studies in Stratigraphy and Paleontology in Honor of Donald W. Fisher. New York State Museum/Geological Survey, Bulletin 461.

Bergström, S. M., S. C. Finney, X. Chen, C. Pålsson, Z.-H. Wang, and Y. Grahn. 2000. A proposed global boundary stratotype for the base of the Upper Series of the Ordovician System: The Fågelsång section, Scania, southern Sweden. Episodes 23:102–109.

Bergström, S. M., W. Huff, D. R. Kolata, and H. Bauert. 1995. Nomenclature, stratigraphy, chemical fingerprinting, and areal distribution of some Middle Ordovician K-bentonites in Baltoscandia. GFF [Geologiska Föreningens i Stockhom Förhandlingar] 117:1–13.

Berner, R. A. 1990. Atmospheric carbon dioxide levels over Phanerozoic time. Science 249:1382–1386.

Berner, R. A. 1994. GEOCARB II: A revised model of atmospheric CO_2 over Phanerozoic time. American Journal of Science 294:56–91.

Berner, R. A. 2001. Modelling atmospheric O_2 over Phanerozoic time. Geochimica et Cosmochimica Acta 65:685–694.

Berner, R. A., D. J. Beeling, R. Dudley, J. M. Robinson, and R. A. Wildman Jr. 2003. Phanerozoic atmospheric oxygen. Annual Reviews of Earth and Planetary Science 31:105–134.

Berry, W. B. N. 1979. Graptolite biogeography: A biogeography of some Lower Paleozoic plankton; pp. 105–115 *in* J. Gray and A. J. Boucot (eds.), Historical Biogeography, Plate Tectonics, and the Changing Environment. Oregon State University Press, Corvallis.

Berry, W. B. N. 1995. Ibexian: A unique interval in the Ordovician; pp. 37–40 *in* J. D. Cooper, M. L. Droser, and S. C. Finney (eds.), Ordovician Odyssey: Short Papers, 7th International Symposium on the Ordovician System. Book 77, Pacific Section Society for Sedimentary Geology (SEPM), Fullerton, California.

Berry, W. B. N., and A. J. Boucot. 1973. Glacio-eustatic control of late Ordovician-early Silurian platform sedimentation and faunal changes. Geological Society of America, Bulletin 84:275–284.

Berry, W. B. N., and P. Wilde. 1978. Progressive ventilation of the oceans—an explanation for the distribution of the Lower Paleozoic black shales. American Journal of Science 278:257–275.

Berry, W. B. N., P. Wilde, and M. S. Quinby-Hunt. 1987. The oceanic non-sulfidic oxygen minimum zone: A habitat for graptolites? Geological Society of Denmark, Bulletin 35:103–114.

Beuf, S., B. Biju-Duval, O. De Charpal, P. Rognon, O. Gariel, and A. Bennacef. 1971. Les Grés du Paléozoique Inférieur au Sahara. Sédimentation et Discontinuités, Structurale d'un Craton. Technip, Paris, 464 pp.

Bieler, R. 1992. Gastropod phylogeny and systematics. Annual Review of Ecology and Systematics 23:311–338.

Biggelaar, J. A. M., and G. Haszprunar. 1996. Cleavage patterns and mesentoblast formation in the Gastropoda: An evolutionary perspective. Evolution 50:1520–1540.

Billings, E. 1861–1865. Palaeozoic Fossils, Vol. 1. Canadian Geological Survey. Dawson Brothers, Montreal, 426 pp.

Bischoff, G. C. O. 1978. Internal structures of conulariid tests and their functional significance, with special reference to the Circonulariina n. suborder (Cnidaria, Scyphozoa). Senckenbergiana Lethaea 59:275–327.

Bischoff, G. C. O. 1981. *Cobcrephora* n.g., representative of a new polyplacophora order Phosphatoloricata with chitinophosphatic shells. Senckenbergiana Lethaea 61: 173–215.

Bischoff, G. C. O. 1989. Byroniida, new order, from early Palaeozoic strata of eastern Australia (Cnidaria, thecate scyphopolyps). Senckenbergiana Lethaea 69:467–521.

Blake, D. B., and T. E. Guensburg. 1993. New Lower and Middle Ordovician stelleroids (Echinodermata) and their bearing on the origins and early history of the stelleroid echinoderms. Journal of Paleontology 67:103–113.

Blake, D. B., and T. E. Guensburg. 1994. Predation by the Ordovician asteroid *Promopalaester* on a pelecypod. Lethaia 27:235–239.

Blieck, A. 1992. At the origin of chordates. Geobios 25: 101–113.

Blieck, A., D. K. Elliott, and P.-Y. Gagnier. 1991. Some questions concerning the phylogenetic relationships of heterostracans, Ordovician to Devonian jawless vertebrates; pp. 1–17 *in* M.-M. Chang, Y.-H. Liu, and G.-R. Zhang (eds.), Early Vertebrates and Related Problems of Evolutionary Biology. Science Press, Beijing.

Blodgett, R. B., K. L. Wheeler, D. M. Rohr, A. G. Harris, and F. R. Weber. 1987. A Late Ordovician age reappraisal for the upper Fossil Creek Volcanics, and possible significance for glacio-eustasy; pp. 54–58 *in* T. D. Hamilton and J. P. Galloway (eds.), Geologic Studies in Alaska by the United States Geological Survey during 1986. U.S. Geological Survey, Circular 998, 54–58.

Blome, C. D., and K. M. Reed. 1993. Acid processing of pre-Tertiary radiolarian cherts and its impact on faunal content and biozonal correlation. Geology 20:177–180.

Blumenstengel, H. 1965. Zur Ostracodenfauna eines Kalkgerölls aus dem Thüringer Lederschiefer (Ordovizium). Freiberger Forschungshefte, C 182:63–78.

Bockelie, J. F. 1974. Ordovician echinoderms from the Trondheim Region, Norway. Norsk Geologisk Tidskrift 54:221–226.

Bockelie, J. F. 1979. Taxonomy, functional morphology and palaeoecology of the Ordovician cystoid family Hemicosmitidae. Palaeontology 22:363–406.

Bockelie, J. F. 1981a. The Middle Ordovician of the Oslo region, Norway, 30: The eocrinoid genera *Cryptocrinites, Rhipidocystis* and *Bockia*. Norsk Geologisk Tidskrift 61: 123–147.

Bockelie, J. F. 1981b. A re-evaluation of the Ordovician cystoid *Stichocystis* Jaekel and the taxonomic implications. GFF [Geologiska Föreningens i Stockhom Förhandlingar] 103:51–59.

Bockelie, J. F. 1981c. Functional morphology and evolution of the cystoid *Echinosphaerites*. Lethaia 14:189–202.

Bockelie, J. F. 1982a. Symmetry and ambulacral pattern of the rhombiferan superfamily Caryocystitida and the relationship to other Blastozoa. GFF [Geologiska Föreningens i Stockhom Förhandlingar] 103:491–498.

Bockelie, J. F. 1982b. Morphology, growth and texonomy of the Ordovician rhombiferan *Caryocystites*. GFF [Geologiska Föreningens i Stockhom Förhandlingar] 103: 499–513.

Bockelie, J. F. 1984. The Diploporita of the Oslo region, Norway. Palaeontology 27:1–68.

Bockelie, T. G., and R. A. Fortey. 1976. An early Ordovician vertebrate. Nature 260(5546):36–38.

Bockelie, T. G., and E. L. Yochelson. 1979. Variation in a species of "worm" from the Ordovician of Spitsbergen. Saertrykk av Norsk Polarinstitutt 167:225–237.

Bodenbender, B. E., M. A. Wilson, and T. J. Palmer. 1989. Paleoecology of *Sphenothallus* on an Upper Ordovician hardground. Lethaia 22:217–225.

Bogolepova, O. K. 1999. Ordovician cephalopods and lingulate brachiopods from the Southern Alps: Remarks on palaeogeography; pp. 409–411 *in* P. Kraft and O. Fatka (eds.), *Quo vadis* Ordovician? Short papers of the 8th International Symposium on the Ordovician System. Acta Universitatis Carolinae, Geologica 43(1–2).

Bolton, T. E. 1980. Colonial coral assemblages and associated fossils from the Late Ordovician Honorat Group and White Head Formation, Gaspé Peninsula, Québec. Geological Survey of Canada, Paper 80-1C:13–28.

Bolton, T. E. 1988. Stromatoporoidea from the Ordovician rocks of central and eastern Canada. Geological Survey of Canada, Bulletin 379:17–45.

Bolton, T. E. 1994. *Sphenothallus angustifolius* Hall, 1847 from the Lower Upper Ordovician of Ontario and Quebec. Geological Survey of Canada, Bulletin 479:1–11.

Bottjer, D. J., and W. I. Ausich. 1986. Phanerozoic development of tiering in soft substrata suspension-feeding communities. Paleobiology 12:400–420.

Bottjer, D. J., M. L. Droser, and D. Jablonski. 1988. Palaeoenvironmental trends in the history of trace fossils. Nature 333:252–255.

Bottjer, D. J., M. L. Droser, P. M. Sheehan, and G. R. McGhee Jr. 2001. The ecological architecture of major events in the Phanerozoic history of marine invertebrate life; pp. 35–61 in W. D. Allmon and D. J. Bottjer (eds.), Evolutionary Paleoecology. Columbia University Press, New York.

Bouček, B. 1928. Revision des conulaires Paléozoïques de la Bohême. Paleontographica Bohemiae 11:60–108.

Bouček, B. 1939. Conularida; pp. A113–A131 in O. H. Schindewolf (ed.), Handbuch der Paläozoologie, Band 2A. Gebrüder Borntraeger, Berlin.

Bouček, B. 1964. The Tentaculites of Bohemia: Their morphology, taxonomy, ecology, phylogeny and biostratigraphy. Publishing House of the Czechoslovak Academy of Science, Prague, 215 pp.

Boucot, A. J. 1975. Evolution and Extinction Rate Controls. Elsevier, Amsterdam, 427 pp.

Boucot, A. J. 1983. Does evolution take place in an ecological vacuum? Journal of Paleontology 57:1–30.

Boucot, A. J. 1990. Modern paleontology: Using biostratigraphy to the utmost. Revista Española de Paleontología 5:63–70.

Bova, J. A., and J. F. Read. 1987. Incipiently drowned facies within a cyclic peritidal ramp sequence, Early Ordovician Chepultepec interval, Virginia Appalachians. Geological Society of America, Bulletin 98:714–727.

Bowring, S. A., and D. H. Erwin. 1998. A new look at evolutionary rates in deep time: Uniting paleontology and high-precision geochronology. GSA Today 8(9):1–8.

Brabcová, Z. 1999. Ordovician conulariids of the Prague Basin (Czech Republic); pp. 433–435 in P. Kraft and O. Fatka (eds.), Quo vadis Ordovician? Short papers of the 8th International Symposium on the Ordovician System. Acta Universitatis Carolinae, Geologica 43(1–2).

Brabcová, Z. 2000. Vybrané druhy konulárií spodního a středního ordoviku Barrandienu [Study on selected conulariids from the Lower and Middle Ordovician of the Barrandian]. Master's thesis, Charles University, Prague.

Braddy, S. J. 2001. Eurypterid palaeoecology: Palaeobiological, ichnological and comparative evidence for a "mass-moult-mate" hypothesis. Palaeogeography, Palaeoclimatology, Palaeoecology 172:115–132.

Braddy, S. J., and J. E. Almond. 1999. Eurypterid trackways from the Table Mountain Group (Lower Ordovician) of South Africa. Journal of African Earth Sciences 29:165–177.

Braddy, S. J., R. J. Aldridge, and J. N. Theron. 1995. A new eurypterid from the Late Ordovician Table Mountain Group, South Africa. Palaeontology 38:563–581.

Bradshaw, L. E. 1969. Conodonts from the Fort Peña Formation (Middle Ordovician), Marathon basin, Texas. Journal of Paleontology 43:1137–1168.

Branson, C. C., A. LaRocque, and N. D. Newell. 1969. Order Conocardioida; pp. N859–N860 in R. C. Moore (ed.), Treatise on Invertebrate Paleontology. Part N, Mollusca 6, Bivalvia. Geological Society of America, Boulder, Colorado, and University of Kansas Press, Lawrence.

Brenchley, P. J., and L. R. M. Cocks. 1982. Ecological associations in a regressive sequence: The latest Ordovician of the Oslo-Asker district. Norway. Palaeontology 24:783–815.

Brenchley, P. J., and D. A. T. Harper. 1998. Palaeoecology: Ecosystems, Environments and Evolution. Blackwell Science, Oxford, 402 pp.

Brenchley, P. J., and J. D. Marshall. 1999. Relative timing of critical events during the late Ordovician mass extinction—new data from Oslo; pp. 187–190 in P. Kraft and O. Fatka (eds.), Quo vadis Ordovician? Short papers of the 8th International Symposium on the Ordovician System. Acta Universitatis Carolinae, Geologica 43(1–2).

Brenchley, P. J., and G. Newall. 1980. A facies analysis of Upper Ordovician regressive sequences in the Oslo Region, Norway: A record of glacio-eustatic changes. Palaeogeography, Palaeoclimatology, Palaeoecology 31:1–38.

Brenchley, P. J., J. D. Marshall, G. A. F. Carden, D. B. R. Robertson, D. G. F. Long, T. Meidla, L. Hints, and T. F. Anderson. 1994. Bathymetric and isotopic evidence for a short-lived Late Ordovician glaciation in a greenhouse period. Geology 22:295–298.

Brenchley, P. J., G. A. F. Carden, and J. D. Marshall. 1995. Environmental changes associated with the "first strike" of the late Ordovician mass extinction. Modern Geology 20:69–72.

Brenchley, P. J., J. D. Marshall, and C. J. Underwood. 2001. Do all mass extinctions represent an ecological crisis? Evidence from the Late Ordovician. Geological Journal 36:329–340.

Brenner, W., and C. B. Foster. 1994. Chlorophycean algae from the Triassic of Australia. Review of Palaeobotany and Palynology 80:209–234.

Bretsky, P. W., Jr. 1970a. Late Ordovician benthic marine communities in north-central New York. New York State Museum and Science Service, Bulletin 414:1–34.

Bretsky, P. W., Jr. 1970b. Upper Ordovician ecology of the central Appalachians. Peabody Museum of Natural History, Bulletin 34:1–150.

Brett, C. E. 1981. Terminology and functional morphology of attachment structures in pelmatozoan echinoderms. Lethaia 14:343–370.

Brett, C. E. 1995. Stasis: Life in the balance. Geotimes 40(3):18–20.

Brett, C. E., and G. C. Baird. 1995. Coordinated stasis and evolutionary ecology of Silurian to Middle Devonian faunas in Appalachian Basin; pp. 285–315 in D. H. Erwin and R. L. Anstey (eds.), New Approaches to Speciation in the Fossil Record. Columbia University Press, New York.

Brett, C. E., T. J. Frest, J. Sprinkle, and C. R. Clement. 1983. Coronoidea, a new class of blastozoan echinoderms based on taxonomic reevaluation of *Stephanocrinus*. Journal of Paleontology 57:627–651.

Brett, C. E., L. C. Ivany, and K. M. Schopf. 1996. Coordinated stasis: An overview. Palaeogeography, Palaeoclimatology, Palaeoecology 127:1–20.

Briggs, D. E. G., D. H. Erwin, and F. J. Collier. 1994. The Fossils of the Burgess Shale. Smithsonian Institution Press, Washington, D.C., 238 pp.

Broadhead, T. W. 1984. *Macurdablastus*, a Middle Ordovician blastoid from the southern Appalachians. University of Kansas Paleontological Contributions, Paper 110:1–9.

Brocke, R. 1997. First results of Tremadoc to lower Arenig acritarchs from the Yangtze Platform, South China. Acta Universitatis Carolinae, Geologica 40:337–356.

Brocke, R. 1998. Palynomorpha (Acritarchen, Prasinophyceae, Chlorophyceae) aus dem Ordovizium der Yangtze-Plattform, Südwest-China. Doctoral thesis, Technische Universität Berlin, Germany.

Brocke, R., O. Fatka, S. G. Molyneux, and T. Servais. 1995. First appearance of selected early Ordovician acritarch taxa from peri-Gondwana; pp. 473–476 *in* J. D. Cooper, M. L. Droser, and S. C. Finney (eds.), Ordovician Odyssey: Short Papers, 7th International Symposium on the Ordovician System. Book 77, Pacific Section Society for Sedimentary Geology (SEPM), Fullerton, California.

Brocke, R., J. Li, and Y. Wang. 2000. Upper Arenigian to lower Llanvirnian acritarch assemblages from South China: A preliminary evaluation. Review of Palaeobotany and Palynology 113:27–40.

Bromley, R. G. 1996. Trace Fossils: Biology, Taphonomy and Applications. Second Edition. Chapman and Hall, London, 361 pp.

Bronn, H. G. 1835. Lethaea Geognostica, Vol. 1. E. Schweizerbart, Stuttgart, 672 pp.

Brooke, C., and R. Riding. 1998. Ordovician and Silurian coralline algae. Lethaia 31:185–195.

Brower, J. C. 1973. Crinoids from the Girardeau Limestone (Ordovician). Palaeontographica Americana 7(46):259–499.

Brower, J. C. 1992a. Cupulocrinid crinoids from the Middle Ordovician (Galena Group, Dunleith Formation) of northern Iowa and southern Minnesota. Journal of Paleontology 66:99–128.

Brower, J. C. 1992b. Hybocrinid and disparid crinoids from the Middle Ordovician (Galena Group, Dunleith Formation) of northern Iowa and southern Minnesota. Journal of Paleontology 66:973–993.

Brower, J. C. 1994. Camerate crinoids from the Middle Ordovician (Galena Group, Dunleith Formation) of northern Iowa and southern Minnesota. Journal of Paleontology 68:570–599.

Brower, J. C. 1995a. Eoparisocrinid crinoids from the Middle Ordovician (Galena Group, Dunleith Formation) of northern Iowa and southern Minnesota. Journal of Paleontology 69:351–366.

Brower, J. C. 1995b. Dendrocrinid crinoids from the Ordovician of northern Iowa and southern Minnesota. Journal of Paleontology 69:939–960.

Brower, J. C. 1996. Carabocrinid crinoids from the Ordovician of northern Iowa and southern Minnesota. Journal of Paleontology 70:614–631.

Brower, J. C. 1997. Homocrinid crinoids from the Upper Ordovician of northern Iowa and southern Minnesota. Journal of Paleontology 71:442–458.

Brower, J. C. 2001. Flexible crinoids from the Upper Ordovician Maquoketa Formation of the northern midcontinent and the evolution of early flexible crinoids. Journal of Paleontology 75:370–382.

Brower, J. C., and H. L. Strimple. 1983. Ordovician calceocrinids from northern Iowa and southern Minnesota. Journal of Paleontology 57:1261–1281.

Brower, J. C., and J. Veinus. 1974. Middle Ordovician crinoids from southwestern Virginia and eastern Tennessee. Bulletins of American Paleontology 66(283):1–125.

Brower, J. C., and J. Veinus. 1978. Middle Ordovician crinoids from the Twin Cities area of Minnesota. Bulletins of American Paleontology 74(304):372–506.

Browne, R. G. 1965. Some upper Cincinnatian (Ordovician) colonial corals of north-central Kentucky. Journal of Paleontology 39:1177–1191.

Brunton, C. H. C., and L. R. M. Cocks. 1996. The classification of the brachiopod Order Strophomenida; pp. 47–52 *in* P. Copper and J. Jin (eds.), Brachiopods, Proceedings of the Third International Brachiopod Congress, Sudbury, Ontario. Balkema, Rotterdam.

Brunton, F. R., and O. A. Dixon. 1994. Siliceous sponge-microbe biotic associations and their recurrence through the Phanerozoic as reef mound constructors. Palaios 9:370–387.

Bruton, D. L., and J. F. Bockelie. 1982. The Løkken-Hølonda-Støren areas, and Road Log. Palaeontological Contributions from the University of Oslo 279:77–91.

Bruton, D. L., O. A. Hoel, L. T. Beyene, and A. Y. U. Ivantsov. 1997. Catalogue of the trilobites figured in Friedrich Schmidt's "Revision der ostbaltischen silurischen" (1881–1907). Contributions from the Palaeontological Museum, University of Oslo, 403:1–117.

Buatois, L. A., and M. G. Mángano. 2001. Ichnology, sedimentology and sequence stratigraphy of the Upper Cambrian to Tremadoc Santa Rosita Formation in northwest Argentina; pp. 17–25 *in* L. A. Buatois and M. G. Mángano (eds.), Ichnology, Sedimentology and Sequence Stratigraphy of Selected Lower Paleozoic, Mesozoic and Cenozoic Units of Northwest Argentina. Fourth Argentinian Ichnologic Meeting and Second

Ichnologic Meeting of Mercosur. Field Guide. San Miguel de Tucumán, Argentina.

Buatois, L. A., and M. G. Mángano. 2002. Ichnology of the Puncoviscana Formation in northwest Argentina: Anactualistic ecosystems and the Precambrian-Cambrian transition. First International Palaeontological Congress (IPC2002) Geological Society of Australia, Abstracts 68:25–26.

Buatois, L. A., and M. G. Mángano. 2003. La icnofauna de la Formación Puncoviscana en el noroeste argentino: Implicancias en la colonización de fondos oceánicos y reconstrucción de paleoambientes y paleoecosistemas de la transición precámbrica-cámbrica. Ameghiniana 40:103–117.

Buatois, L. A., M. G. Mángano, J. F. Genise, and T. N. Taylor. 1998. The ichnologic record of the invertebrate invasion of nonmarine ecosystems: Evolutionary trends in ecospace utilization, environmental expansion, and behavioral complexity. Palaios 13:217–240.

Buatois, L. A., M. G. Mángano, and Z. Sylvester. 2001a. A diverse deep marine ichnofauna from the Eocene Tarcau Sandstone of the Eastern Carpathians, Romania. Ichnos 8:23–62.

Buatois, L. A., M. K. Gingras, J. MacEachern, M. G. Mángano, A. J. Martin, R. G. Netto, S. G. Pemberton, and J.-P. Zonneveld. 2001b. Colonization of brackish-water environments through time: Evidence from the trace-fossil record. VI International Ichnofabric Workshop, Isla Margarita and Puerto La Cruz, Venezuela, Abstracts, 24–25.

Buatois, L. A., M. G. Mángano, and F. G. Aceñolaza. 2002. Trazas fósiles: Señales de comportamiento en el registro estratigráfico. Publicación 2. Museo Paleontológico Egidio Feruglio, Trelew, 382 pp.

Bucher, W. H. 1938. A shell-boring gastropod in Dalmanella bed of upper Cincinnatian age. American Journal of Science 36:1–7.

Budd, A. E., and K. G. Johnson. 2001. Contrasting patterns in rare and abundant species during evolutionary turnover; pp. 295–325 in J. B. C Jackson, S. Lidgard, and F. K. McKinney (eds.), Evolutionary Patterns: Growth, Form, and Tempo in the Fossil Record. University of Chicago Press, Chicago.

Bullard, E. C., J. E. Everett, and A. G. Smith. 1965. The fit of the continents around the Atlantic. Philosophical Transactions of the Royal Society of London, A 258: 41–51.

Bulman, O. M. B. 1971. Graptolite faunal distribution; pp. 47–60 in F. A. Middlemiss, P. F. Rawson, and G. Newall (eds.), Faunal Provinces in Space and Time. Geological Journal, Special Issue 4.

Burgess, N. D. 1991. Silurian cryptospores and miospores from the type Llandovery area, south-west Wales. Palaeontology 34:575–599.

Burrett, C. F., B. Stait, and J. R. Laurie. 1983. Trilobites and microfossils from the Middle Ordovician of Surprise Bay, southern Tasmania, Australia. Association of Australasian Palaeontologists, Memoir 1:177–193.

Burrow, C. J. 2001. Late Silurian to Middle Devonian acanthodians of eastern Australia. Doctoral thesis, University of Queensland, Australia.

Burrow, C. J. 2003. Redescription of the gnathostome fish fauna from the Mid-Palaeozoic Silverband Formation, the Grampians, Victoria. Alcheringa 27:37–49.

Butterfield, N. J. 1997. Plankton ecology and the Proterozoic-Phanerozoic transition. Paleobiology 23:247–262.

Buttler, C. J. 1989. New information on the ecology and skeletal ultrastructure of the Ordovician cyclostome bryozoan *Kukersella* Toots, 1952. Paläontologishe Zeitschrift 63:215–227.

Cai, X.-R. 1991. Chitinozoa; pp. 150–157 in S.-B. Zhang and Q.-Q. Kao (eds.), Sinian to Permian Stratigraphy and Paleontology of the Tarim Basin, Xinjiang, (2) Kalping-Bachu Region. Petroleum Industry Press, Beijing.

Caldeira, K., and M. R. Rampino. 1991. The mid-Cretaceous super-plume, carbon dioxide, and global warming. Geophysical Research Letters 18:987–990.

Cameron, D., and P. Copper. 1994. Paleoecology of giant Late Ordovician cylindrical sponges from Anticosti Island, eastern Canada; pp. 13–21 in R. W. M. van Soest, Th. M. G. van Kempen, and J. C. Braekmman (eds.), Sponges in Time and Space. Balkema, Rotterdam.

Cañas, F. 1995. Early Ordovician carbonate platform facies of the Argentine Precordillera: Restricted shelf to open platform evolution; pp. 221–224 in J. D. Cooper, M. L. Droser and S. C. Finney (eds.), Ordovician Odyssey: Short Papers, 7th International Symposium on the Ordovician System. Book 77, Pacific Section Society for Sedimentary Geology (SEPM), Fullerton, California.

Cañas, F. L. 1999. Facies and sequences of the Late Cambrian–Early Ordovician carbonates of the Argentine Precordillera: A stratigraphic comparison with Laurentian platform; pp. 43–62 in D. Keppie and V. A. Ramos (eds.), Laurentia-Gondwana Connections before Pangea. Geological Society of America, Special Paper 336.

Cañas, F. L., and M. G. Carrera. 1993. Early Ordovician microbial-sponge-receptaculitid bioherms of the Precordillera basin, western Argentina. Facies 29:169–178.

Cardinale, B. J., M. A. Palmer, and S. L. Collins. 2002. Species diversity enhances ecosystem functioning through interspecific facilitation. Nature 415:426–429.

Carlson, S. J. 1996. Revision and review of the Order Pentamerida; pp. 53–58 in P. Copper and J. Jin (eds.), Brachiopods, Proceedings of the Third International Brachiopod Congress, Sudbury, Ontario. Balkema, Rotterdam.

Carlson, S. J., and M. R. Sandy (eds.). 2001. Brachiopods Ancient and Modern: A Tribute to G. Arthur Cooper. The Paleontological Society (New Haven, Connecticut), Paper 7, 261 pp.

Carrera, M. G. 1994. An Ordovician sponge fauna from San Juan Formation, Precordillera Basin, western Argentina. Neues Jahrbuch für Geologie und Paläontologie, Abhandlungen, 191:201–220.

Carrera, M. G. 1996. Ordovician megamorinid demosponges from San Juan Formation, Precordillera, western Argentina. Géobios 29:643–650.

Carrera, M. G. 1998. First Ordovician sponge from the Puna region, northwestern Argentina. Ameghiniana 35: 393–412.

Carrera, M. G., and F. L. Cañas. 1996. Los biohermos de la Formación San Juan (Ordovícico temprano, Precordillera Argentina): Paleoecología y comparaciones. Revista Asociación Argentina de Sedimentología 3:85–104.

Carrera, M. G., and J. K. Rigby. 1999. Biogeography of the Ordovician sponges. Journal of Paleontology 73: 26–37.

Carriker, M. R., and E. C. Yochelson. 1968. Recent gastropod boreholes and Ordovician cylindrical borings. United States Geological Survey, Professional Paper 593: B1–B26.

Carter, J. G., and R. Seed. 1998. Thermal potentiation and mineralogical evolution in *Mytilus* (Mollusca; Bivalvia); pp. 87–117 *in* P. A. Johnston and J. W. Haggart (eds.), Bivalves: An Eon of Evolution—Paleobiological Studies Honoring Norman D. Newell. University of Calgary Press, Calgary.

Carter, J. G., D. C. Campbell, and M. R. Campbell. 2000. Cladistic perspectives on early bivalve evolution; pp. 47–79 *in* E. M. Harper, J. D. Taylor, and J. A. Crame (eds.), The Evolutionary Biology of the Bivalvia. The Geological Society, London, Special Publication 177.

Carter, J. L., J. G. Johnson, R. Gourvennec, and H.-F. Hou. 1994. A revised classification of the spiriferid brachiopods. Annals of the Carnegie Museum 63: 327–374.

Caster, K. K., and E. N. Kjellesvig-Waering. 1955. [Untitled contributions]; *in* R. C. Moore (ed.), Treatise on Invertebrate Paleontology. Part P, Arthropoda 2. Geological Society of America, New York, and University of Kansas Press, Lawrence.

Caster, K. K., and E. N. Kjellesvig-Waering. 1964. Upper Ordovician eurypterids of Ohio. Palaeontographica Americana 4(32):297–358.

Chamberlain, C. K. 1977. Ordovician and Devonian trace fossils from Nevada. Nevada Bureau of Mines and Geology, Bulletin 90:1–24.

Chatterton, B. D. E., and S. E. Speyer. 1989. Larval ecology, life history strategies, and patterns of extinction and survivorship among Ordovician trilobites. Paleobiology 15:118–132.

Chauvel, J. 1966. Échinodermes de l'Ordovicien du Maroc. Cahiers de Paléontologie, Éditions du Centre National de la Recherche Scientifique, Paris, 120 pp.

Chauvel, J. 1967. Sur quelques representants du genre *Plumulites* Barrande (*Machairidiés*) provenant de l'ordovicien du Maroc et du Massif Armoricain. Bulletin de la Societé Géologique et Minéralogique de Bretagne 1966:73–85.

Chauvel, J. 1969. Les échinodermes Macrocystellidés de l'Anti-Atlas marocain. Bulletin de la Societé Géologique et Minéralogique de Bretagne, (C) 1:23–32.

Chauvel, J. 1977. Note complémentaire sur les Cystoïdes Rhombiferes (Echinodermes) de l'Ordovicien marocain. Notes du Service Géologique du Maroc 38(268):115–139.

Chauvel, J. 1978. Compléments sur les Echinodermes du Paléozoïque marocain (Diploporites, Eocrinoïdes, Edrioastéroïdés). Notes du Service Géologique du Maroc 39(272):27–78.

Chauvel, J. 1981. Etude critique de quelques échinodermes stylophores du Massif Armoricain. Bulletin de la Societé Géologique et Minéralogique de Bretagne, (C) 13:67–101.

Chauvel, J., and J. Le Menn. 1973. Echinodermes de l'Ordovicien Superieur de Coat-Carrec, Argol (Finistere). Bulletin de la Societé Géologique et Minéralogique de Bretagne, (C) 4:39–61.

Chauvel, J., and J. Le Menn. 1979. Sur quelques Echinodermes (Cystoides et Crinoides) de l'Ashgill d'Aragon (Espagne). Geobios 12:549–587.

Chauvel, J., and B. Melendez. 1978. Les Echinodermes (Cystoïdes, Asterozoaires, Homalozoaires) de l'Ordovicien moyen des Monts de Tolede (Espagne). Estudios Geologico 34:75–87.

Chauvel, J., B. Melendez, and J. Le Menn. 1975. Les Echinodermes (Cystoïdes et Crinoïdes) de l'Ordovicien Supérieur de Luesma (Sud de l'Aragon, Espagne). Estudios Geologico 31:351–364.

Chen, J.-Y. 1991. Bathymetric biosignals and Ordovician chronology of eustatic variations; pp. 299–311 *in* C. R. Barnes and S. H. Williams (eds.), Advances in Ordovician Geology. Geological Survey of Canada, Paper 90-9.

Chen L., H. Luo, S. Hu, J. Yin, Z. Jiang, Z. Wu, F. Li, and A. Chen. 2002. Early Cambrian Chengjiang Fauna in Eastern Yunnan, China. Yunnan Science and Technology Press, Kunming, 199 pp.

Chen, T.-E. 1995. Nautiloid assemblages; pp. 15–17 *in* X. Chen, J.-Y. Rong, X.-F. Wang, Z.-H. Wang, Y.-D. Zhang, and R.-B. Zhan, Correlation of the Ordovician Rocks of China. Charts and Explanatory Text. International Union of Geological Sciences, Publication 31.

Chen, X., and S. M. Bergström. 1995. The base of the *austrodentatus* Zone as a level for global subdivision of the Ordovician System. Palaeoworld 5:1–117.

Chen, X., Z.-Y. Zhou, J.-Y. Rong, and J. Li. 2001. Ordovician series and stages in Chinese stratigraphy: Steps toward a global usage. Alcheringa 25:131–141.

Chen, X.-H. 1994. Early Ordovician chitinozoan at Huaqiao of Changyang, Hubei province. Geoscience 8:259–263.

Chen, X.-H., and X.-F. Wang. 1996. Llanvirnian and Llandeilian chitinozoan biostratigraphy in central Yangtze Platform. Acta Micropalaeontologica Sinica 13:75–83 (in Chinese with English abstract).

Chen, X.-H., X.-F. Wang, and Z.-H. Li. 1996. Arenigian chitinozoan biostratigraphy and palaeobiogeography in South China. Geological Review 43:200–208.

Cherns, L. 1998a. *Chelodes* and closely related Polyplacophora (Mollusca) from the Silurian of Gotland, Sweden. Palaeontology 41:545–573.

Cherns, L. 1998b. Silurian Polyplacophora (Mollusca) from Gotland, Sweden. Palaeontology 41:939–974.

Cherns, L. 1999. Silurian chitons as indicators of rocky shores and lowstand on Gotland, Sweden. Palaios 14:172–179.

Chinese Committee of Stratigraphy. 2001. Stratigraphic Guidebook of China with a Synopsis. Geological Publishing House, Beijing, 65 pp.

Chlupáč, I. 1970. Phyllocarid crustaceans of the Bohemian Ordovician. Sborník Geologických Věd, Paleontologie 12:41–75.

Chlupáč, I. 1999. Some problematical arthropods from the Upper Ordovician Letná Formation of Bohemia. Journal of the Czech Geological Society 44:79–92.

Chlupáč, I., V. Havlíček, J. Kříž, Z. Kukal, and P. Štorch. 1998. Palaeozoic of the Barrandian (Cambrian to Devonian). Czech Geological Survey, Prague, 183 pp.

Choi, D. K. 1990. *Sphenothallus* ("Vermes") from the Tremadocian Dumugol Formation, Korea. Journal of Paleontology 64:403–408.

Choi, D. K., and Kim, K. H. 1989. Problematic fossils from the Dumugol Formation (Lower Ordovician), Dongjeom area, Korea. Journal of the Geological Society of Korea 25:405–412.

Churkin, M., Jr. 1966. Morphology and stratigraphic range of the phyllocarid crustacean *Caryocaris* from Alaska and the Great Basin. Palaeontology 9:371–380.

Chuvashov, B. I. 1987. Krasnyie vodorosli (Rhodophyta); pp. 109–134, 171–172 *in* V. N. Dubatopov, Iskopaemye Isvstkove Vodorosli. Akademia Nauk SSSR, Sibirskoe Otdelenie, Trudy Instituta Geologi i Geofiziki 674.

Chuvashov, B. I., and R. Riding. 1984. Principal floras of Palaeozoic marine calcareous algae. Palaeontology 27:487–500.

Clark, D. L., W. C. Sweet, S. M. Bergström, G. Klapper, R. L. Austin, F. H. T. Rhodes, K. J. Müller, W. Ziegler, M. Lindström, J. F. Miller, and A. G. Harris. 1981. Conodonta; pp. W1–W202 *in* R. A. Robison (ed.), Treatise on Invertebrate Paleontology. Part W, Miscellanea, Supplement 2. Geological Society of America, Boulder, Colorado, and University of Kansas Press, Lawrence.

Clark, T. H. 1924. The paleontology of the Beekmantown Series at Lévis, Quebec. Bulletins of American Paleontology 10(41):1–136.

Clark, T. H. 1925. On the nature of *Salterella*. Royal Society of Canada Proceedings and Transactions, Sec. 4, Series 3, 19:29–41.

Clarke, J. M., and R. Ruedemann. 1912. The Eurypterida of New York. New York State Museum, Memoir 14:1–439.

Claypool, G. E., W. T. Holser, I. R. Kaplan, H. Sakai, and I. Zak. 1980. The age curves of sulfur and oxygen isotopes in marine sulfate and their mutual interpretation. Chemical Geology 28:190–260.

Cobbold, E. S. 1921. The Cambrian horizons of Comley (Shropshire) and their Brachiopoda, Pteropoda, Gasteropoda. Quarterly Journal of the Geological Society of London 76:325–386.

Cobbold, E. S. 1931. Additional fossils from the Cambrian rocks of Comley, Shropshire. Quarterly Journal of the Geological Society of London 87:459–512.

Cobbold, E. S., and R. W. Pocock. 1934. The Cambrian area of Rushton (Shropshire). Philosophical Transactions of the Royal Society of London, B 223:305–409.

Cocks, L. R. M. 1988. The Ordovician-Silurian boundary and its working group; pp. 9–15 *in* L. R. M. Cocks and R. B. Rickards (eds.), A Global Analysis of the Ordovician-Silurian Boundary. Bulletin of the British Museum (Natural History), Geology 43.

Cocks, L. R. M. 2000. The Early Palaeozoic geography of Europe. Journal of the Geological Society of London 157:1–10.

Cocks, L. R. M. 2001. Ordovician and Silurian global geography. Journal of the Geological Society of London 158:197–210.

Cocks, L. R. M. 2002. Key Lower Palaeozoic faunas from near the Trans-European suture Zone; pp. 37–46 in J. A. Winchester, T. C. Pharoah, and J. Verniers (eds.), Palaeozoic Amalgamation of Central Europe. The Geological Society, London, Special Publication 201.

Cocks, L. R. M., and R. A. Fortey. 1982. Faunal evidence for oceanic separations in the Palaeozoic of Britain. Journal of the Geological Society of London 139:465–478.

Cocks, L. R. M., and R. A. Fortey. 1988. Lower Palaeozoic facies and faunas around Gondwana; pp. 183–200 *in* M. G. Audley-Charles and A. Hallam (eds.), Gondwana and Tethys. The Geological Society, London, Special Publication 37.

Cocks, L. R. M., and R. A. Fortey. 1990. Biogeography of Ordovician and Silurian faunas; pp. 97–104 *in* W. S. McKerrow and C. R. Scotese (eds.), Palaeozoic Palaeogeography and Biogeography. The Geological Society, London, Memoir 12.

Cocks, L. R. M., and R. A. Fortey. 1998. The Lower Palaeozoic margins of Baltica. GFF [Geologiska Föreningens i Stockhom Förhandlingar] 120:173–179.

Cocks, L. R. M., and J.-Y. Rong. 1988. A review of the Late Ordovician *Foliomena* brachiopod fauna with new data from China, Wales, and Poland. Palaeontology 31: 53–67.

Cocks, L. R. M., and J.-Y. Rong. 2000. Order Strophomenida; pp. 216–348 in R. L. Kaesler (ed.), Treatise on Invertebrate Paleontology. Part H, Brachiopoda, rev., Vol. 2. Geological Society of America, Boulder, Colorado, and University of Kansas Press, Lawrence.

Cocks, L. R. M., and T. H. Torsvik. 2002. Earth geography from 500 to 400 million years ago: A faunal and palaeomagnetic review. Journal of the Geological Society of London 159:631–644.

Cocks, L. R. M., W. S. McKerrow, and C. R. van Staal. 1997. The margins of Avalonia. Geological Magazine 134:627–636.

Colbath, G. K. 1979. Organic-walled microplankton from the Eden Shale (Upper Ordovician), Indiana, U.S.A. Palaeontographica, Abteilung B 171:1–38.

Colbath, G. K. 1980. Abundance fluctuations in Upper Ordovician organic-walled microplankton from Indiana. Micropaleontology 26:97–102.

Colbath, G. K. 1990. Palaeobiogeography of Middle Palaeozoic organic-walled phytoplankton; pp. 207–213 in W. S. McKerrow and C. R. Scotese (eds.), Palaeozoic Palaeogeography and Biogeography. The Geological Society, London, Memoir 12.

Colbath, G. K., and H. R. Grenfell. 1995. Review of biological affinities of Paleozoic acid-resistant, organic-walled eukaryotic algal microfossils (including "acritarchs"). Review of Palaeobotany and Palynology 86: 287–314.

Colgan, D. J., W. F. Ponder, and P. E. Eggler. 2000. Gastropod evolutionary rates and phylogenetic relationships assessed using partial 28S rDNA and histone H3 sequences. Zoologica Scripta 29:29–63.

Collins, A. G., and J. W. Valentine. 2001. Defining phyla: Evolutionary pathways to metazoan body plans. Evolution and Development 3:432–442.

Collins, A. G., A. C. Marques, and M. G. Simoes. 2000. The phylogenetic placement of Conulatae within Cnidaria. Geological Society of America, Annual Meeting, Abstracts with Programs 23:A443.

Combaz, A. 1967. Un microbios du Trémadocien dans un sondage d'Hassi-Messaoud. Actes de la Société Linnéenne de Bordeaux, B 104(29):1–26.

Combaz, A., and G. Peniguel. 1972. Etude palynostratigraphique de l'Ordovicien dans quelques sondages du Bassin de Canning (Australie Occidentale). Bulletin du Centre de Recherches de Pau–SNPA 6:121–165.

Compston, W., and I. S. Williams. 1992. Ion probe ages for the British Ordovician and Silurian stratotypes; pp. 59–67 in B. D. Webby and J. R. Laurie (eds.), Global Perspectives on Ordovician Geology. Balkema, Rotterdam.

Condie, K. C. 2001. Mantle Plumes and Their Record in Earth History. Cambridge University Press, Cambridge, 306 pp.

Connolly, S. R., and A. I. Miller. 2002. Global Ordovician faunal transitions in the marine benthos: Ultimate causes. Paleobiology 28:26–40.

Conway Morris, S. 1977. Fossil priapulid worms. Special Papers in Palaeontology 20:1–95.

Conway Morris, S. 1986. The community structure of the Middle Cambrian Phyllopod Bed (Burgess Shale). Palaeontology 29:423–467.

Conway Morris, S. 1997. The cuticular structure of the 495-Myr-old type species of the fossil worm *Palaeoscolex*, *P. piscatorum* (?Priapulida). Zoological Journal of the Linnean Society 119:69–82.

Conway Morris, S. 1998a. The evolution of diversity in ancient ecosystems: A review. Philosophical Transactions of the Royal Society of London, B 353:327–345.

Conway Morris, S. 1998b. The Crucible of Creation: The Burgess Shale and the Rise of Animals. Oxford University Press, Oxford, 242 pp.

Conway Morris, S., and S. Bengtson. 1994. Cambrian predators: Possible evidence from boreholes. Journal of Paleontology 68:1–23.

Conway Morris, S., R. K. Pickerill, and T. L. Harland. 1982. A possible annelid from the Trenton Limestone (Ordovician) of Quebec, with a review of fossil oligochaetes and other annulate worms. Canadian Journal of Earth Sciences 19:2150–2157.

Cooper, A. H., A. W. A. Rushton, S. G. Molyneux, R. A. Hughes, R. M. Moore, and B. C. Webb. 1995. The stratigraphy, correlation, provenance and palaeogeography of the Skiddaw Group (Ordovician) in the English Lake District. Geological Magazine 132:185–211.

Cooper, B. J. 1981. Early Ordovician conodonts from the Horn Valley Siltstone, central Australia. Palaeontology 24:147–183.

Cooper, R. A. 1968. Lower and Middle Palaeozoic fossil localities of North-west Nelson, New Zealand. Transactions of the Royal Society of New Zealand, Geology 6:75–89.

Cooper, R. A. 1973. Taxonomy and evolution of *Isograptus* Moberg in Australasia. Palaeontology 16:45–115.

Cooper, R. A. 1979. Ordovician geology and graptolite faunas of the Aorangi Mine area, north-west Nelson, New Zealand. New Zealand Geological Survey Paleontological Bulletin 47:1–127.

Cooper, R. A. 1998. Towards a general model for the depth ecology of graptolites; pp. 161–163 in J. C. Gutiérrez-Marco and I. Rábano (eds.), Proceedings of the Sixth International Graptolite Conference of the GWG (IPA) and the 1998 Field Meeting of the International Subcommission on Silurian Stratigraphy (ICS-IUGS). Instituto Technológico Geominero de España, Temas Geológico-Mineros 23.

Cooper, R. A. 1999a. Ecostratigraphy, zonation and global correlation of earliest planktic graptolites. Lethaia 32: 1–16.

Cooper, R. A. 1999b. The Ordovician time scale—calibration of graptolite and conodont zones; pp. 1–4 in P. Kraft and O. Fatka (eds.), Quo vadis Ordovician? Short papers of the 8th International Symposium on the Ordovician System. Acta Universitatis Carolinae, Geologica 43(1–2).

Cooper, R. A. 1999c. Graptolites and the great Ordovician biodiversification event; pp. 441–442 in P. Kraft and O. Fatka (eds.), Quo vadis Ordovician? Short papers of the 8th International Symposium on the Ordovician System. Acta Universitatis Carolinae, Geologica 43(1–2).

Cooper, R. A., and K. Lindholm. 1990. A precise worldwide correlation of early Ordovician graptolite sequences. Geological Magazine 127:497–525.

Cooper, R. A., and P. M. Sadler. 2002. Optimised biostratigraphy and its applications in basin analysis, timescale development and macroevolution. First International Palaeontological Congress (IPC 2002), Geological Society of Australia, Abstracts 68:38–39.

Cooper, R. A., and I. R. Stewart. 1979. The Tremadoc graptolite sequence of Lancefield, Victoria. Palaeontology 22:767–797.

Cooper, R. A., R. A. Fortey, and K. Lindholm. 1991. Latitudinal and depth zonation of Early Ordovician graptolites. Lethaia 24:199–218.

Cooper, R. A., G. S. Nowlan, and S. H. Williams. 2001. Global stratotype and point for base of the Ordovician System. Episodes 24:19–28.

Cope, J. C. W. 1995 [dated 1996]. The early evolution of the Bivalvia; pp. 361–370 in J. D. Taylor (ed.), Origin and Evolutionary Radiation of the Mollusca. Oxford University Press, Oxford.

Cope, J. C. W. 1996a. Bivalves; pp. 95–115 in D. A. T. Harper and A. W. Owen (eds.), Fossils of the Upper Ordovician. Palaeontontological Association (London), Field Guide to Fossils 7.

Cope, J. C. W. 1996b. Early Ordovician (Arenig) bivalves from the Llangynog Inlier, South Wales. Palaeontology 39:979–1025.

Cope, J. C. W. 1997a. Affinities of the early Ordovician bivalve *Catamarcaia* Sánchez and Babin, 1993 and its role in bivalve evolution. Geobios, Mémoire Spécial 20:126–131.

Cope, J. C. W. 1997b. The early phylogeny of the class Bivalvia. Palaeontology 40:713–746.

Cope, J. C. W. 1999. Middle Ordovician bivalves from mid-Wales and the Welsh Borderland. Palaeontology 42:467–499.

Cope, J. C. W. 2000. A new look at early bivalve phylogeny; pp. 81–95 in E. M. Harper, J. D. Taylor, and J. A. Crame (eds.), The Evolutionary Biology of the Bivalvia. The Geological Society, London, Special Publication 177.

Cope, J. C. W. 2002. Diversification and biogeography of bivalves during the Ordovician Period; pp. 35–52 in J. A. Crame and A. W. Owen (eds.), Palaebiogeography and Biodiversity Change: A Comparison of the Ordovician and Mesozoic-Cenozoic Radiations. The Geological Society, London, Special Publication 194.

Cope, J. C. W., and C. Babin. 1999. Diversification of bivalves in the Ordovician. Geobios 32:175–185.

Copeland, M. J. 1965. Ordovician Ostracoda from Lake Timiskaming, Ontario. Geological Survey of Canada, Bulletin 127:1–54.

Copeland, M. J. 1970. Ostracoda from the Vauréal Formation (Upper Ordovician) of Anticosti Island, Quebec. Geological Survey of Canada, Bulletin 187:15–29.

Copeland, M. J. 1973. Ostracoda from the Ellis Bay Formation (Ordovician), Anticosti Island, Quebec. Geological Survey of Canada, Paper 72-43:1–49.

Copeland, M. J. 1974. Middle Ordovician Ostracoda from southwestern District of Mackenzie. Geological Survey of Canada, Bulletin 244:1–55.

Copeland, M. J. 1977. Ordovician Ostracoda, southeastern District of Franklin. Geological Survey of Canada Bulletin 269:77–97.

Copeland, M. J. 1982. Bathymetry of Early Middle Ordovician (Chazy) Ostracodes, Lower Esbataottine Formation, District of Mackenzie. Geological Survey of Canada Bulletin 347:1–39.

Copeland, M. J. 1989. Silicified Upper Ordovician–Lower Silurian ostracodes from the Avalanche Lake Area, southwestern District of Mackenzie. Geological Survey of Canada Bulletin 341:1–100.

Copper, P. 1999. Brachiopods during and after the Late Ordovician mass extinctions on Anticosti Island, Eastern Canada; pp. 207–209 in P. Kraft and O. Fatka (eds.), Quo vadis Ordovician? Short papers of the 8th International Symposium on the Ordovician System. Acta Universitatis Carolinae, Geologica 43(1–2).

Copper, P. 2001a. Originations and extinctions in brachiopods; pp. 249–261 in S. J. Carlson and M. R. Sandy (eds.), Brachiopods Ancient and Modern: A Tribute to G. Arthur Cooper. The Paleontological Society (New Haven, Connecticut), Paper 7.

Copper, P. 2001b. Radiations and extinctions of atrypide brachiopods; pp. 201–211 in C. H. C. Brunton, L. R. M. Cocks, and S. L. Long (eds.), Brachiopods Past and Present. Taylor and Francis, London.

Copper, P. 2001c. Reefs during the multiple crises towards the Ordovician-Silurian boundary: Anticosti Island, eastern Canada, and worldwide. Canadian Journal of Earth Sciences 38:153–171.

Copper, P., and R. Gourvennec. 1996. Evolution of spire-bearing brachiopods (Ordovician-Jurassic); pp. 81–88 in P. Copper and J. Jin (eds.), Brachiopods. Proceedings of the Third International Brachiopod Congress, Sudbury, Ontario. Balkema, Rotterdam.

Copper, P., and R. Morrison. 1978. Morphology and paleoecology of Ordovician tetradiid corals from the Manitoulin district, northern Ontario. Canadian Journal of Earth Sciences 15:2006–2020.

Corbett, K. D., and M. R. Banks. 1974. Ordovician stratigraphy of the Florentine Synclinorium, south-west Tasmania. Papers and Proceedings of the Royal Society of Tasmania 107:207–238.

Crame, J. A., and A. W. Owen (eds.). 2002. Palaeobiogeography and Biodiversity Change: The Ordovician and Mesozoic-Cenozoic Radiations. The Geological Society, London, Special Publication 194:1–206.

Cramer, F. H., and M. d. C. R. Díez. 1977. Late Arenigian (Ordovician) acritarchs from Cis-Saharan Morocco. Micropaleontology 23:339–360.

Cramer, F. H., and M. d. C. R. Díez. 1979. Lower Palaeozoic acritarchs. Palinología, Número Extraordinario 1: 17–160.

Crick, R. E. 1980. Integration of paleobiogeography and paleogeography: Evidence from Arenigian nautiloid biogeography. Journal of Paleontology 54:1218–1236.

Crick, R. E. 1981. Diversity and evolutionary rates of Cambro-Ordovician nautiloids. Paleobiology 7:216–229.

Crick, R. E. 1988. Buoyancy regulation and macroevolution in nautiloid cephalopods. Senckenbergiana Lethaea 69:13–42.

Crick, R. E. 1990. Cambro-Devonian biogeography of nautiloid cephalopods; pp. 147–161 in W. S. McKerrow and C. R. Scotese (eds.), Palaeozoic Palaeogeography and Biogeography. The Geological Society, London, Memoir 12.

Crimes, T. P. 1970. Trilobite tracks and other trace fossils from the Upper Cambrian of North Wales. Geological Journal 7:47–68.

Crimes, T. P. 1974. Colonisation of the early ocean floor. Nature 248:328–330.

Crimes, T. P. 1975. The stratigraphical significance of trace fossils; pp. 109–130 in R. W. Frey (ed.), The Study of Trace Fossils: A Synthesis of Principles, Problems, and Procedures in Ichnology. Springer-Verlag, New York.

Crimes, T. P. 1992. The record of trace fossils across the Proterozoic-Cambrian boundary; pp. 177–199 in J. H. Lipps and P. W. Signor (eds.), Origin and Early Evolution of the Metazoa. Plenum Press, New York.

Crimes, T. P. 2000 [dated 2001]. Evolution of the deepwater benthic community; pp. 275–290 in A. Y. Zhuravlev and R. Riding (eds.), The Ecology of the Cambrian Radiation. Columbia University Press, New York.

Crimes, T. P., and M. M. Anderson. 1985. Trace fossils from Late Precambrian–Early Cambrian strata of southeastern Newfoundland (Canada): Temporal and environmental implications. Journal of Paleontology 59: 310–343.

Crimes, T. P., and J. D. Crossley. 1968. The stratigraphy, sedimentology, ichnology and structure of the Lower Paleozoic rocks of part of northeastern Co. Wexford. Proceedings of the Royal Irish Academy 67B:185–215.

Crimes, T. P., and J. D. Crossley. 1991. A diverse ichnofauna from Silurian flysch of the Aberystwyth Grits Formation, Wales. Geological Journal 26:27–64.

Crimes, T. P., and M. A. Fedonkin. 1994. Evolution and dispersal of deepsea traces. Palaios 9:74–83.

Crimes, T. P., and A. Marcos. 1976. Trilobite traces and the age of the lowest part of the Ordovician reference section for NW Spain. Geological Magazine 113:349–356.

Crimes, T. P., J. F. García Hidalgo, and D. G. Poiré. 1992. Trace fossils from Arenig flysch sediments of Eire and their bearing on the early colonisation of the deep seas. Ichnos 2:61–77.

Crowley, T. J., and S. K. Baum. 1991. Toward reconciliation of Late Ordovician (~440 Ma) glaciation with very high CO_2 levels. Journal of Geophysical Research 96: 22597–22610.

Crowley, T. J., and S. K. Baum. 1995. Reconciling Late Ordovician (440 Ma) glaciation with very high (14×) CO_2 levels. Journal of Geophysical Research 100 (D1): 1093–1101.

Cuffey, R. J., and D. B. Blake. 1991. Cladistic analysis of the Phylum Bryozoa. Bulletin de la Societe des Sciences Naturelles de l'Ouest de la France, Mémoire Hors Serie 1:97–108.

Currie, E. D., and Edwards, W. N. 1943. Dasycladaceous algae from the Girvan area. Quarterly Journal of the Geological Society of London 98:235–260.

Dale, B. 1996. Dinoflagellate cyst ecology: Modeling and geological applications; pp. 1249–1275 in J. Jansonius and D. C. McGregor (eds.), Palynology: Principles and Applications, Vol. 3. American Association of Stratigraphic Palynologists Foundation. Publishers Press, Salt Lake City, Utah.

Dana, J. D. 1846. Structure and Classification of Zoophytes: U.S. Exploring Expedition during the Years 1838, 1839, 1840, 1841, 1842 under the command of Captain Wilkes, U.S.N., Vol. 7, Lea and Blanchard, Philadelphia, x + 740 pp., atlas, 61 pls.

Danelian, T. 1999. Taxonomic study of some Ordovician (Llanvirn-Caradoc) Radiolaria from the Southern Uplands (Scotland). Geodiversitas 21:625–635.

Danelian, T., and J. Floyd. 2001. Progress in describing Ordovician siliceous biodiversity from the Southern Uplands (Scotland). Transactions of the Royal Society of Edinburgh, Earth Science 91:489–498.

Danelian, T., and L. Popov. 2003. Ordovician radiolarian diversity increase: Insights based on new and revised data from Kazakhstan. Bulletin de la Société Géologique de France 174 (4):325–335.

Darby, D. G. 1982. The early vertebrate *Astraspis* habitat based on a lithologic association. Journal of Paleontology 56:1187–1196.

David, B., B. Lefebvre, R. Mooi, and R. Parsley. 2000. Are homolozoans echinoderms? An answer from the axial-extraxial theory. Paleobiology 26:529–555.

Davidek, K., E. Landing, S. A. Bowring, S. R. Westrop, A. W. A. Rushton, R. A. Fortey, and J. M. Adrain. 1998. New uppermost Cambrian U-Pb date from Avalonian Wales and the age of the Cambrian-Ordovician boundary. Geological Magazine 135:305–309.

Dawson, J. W. 1889. New species of fossil sponges from the Siluro-Cambrian at Little Métis on the Lower St. Lawrence. Royal Society of Canada, Transactions 7(4):31–55.

Dean, W. T. 1974. Trilobites of the Chair of Kildare Limestone (Upper Ordovician) of eastern Ireland, Part 2. Palaeontographical Society Monograph 128(539):61–98.

Dean, W. T., and F. Martin. 1978. Lower Ordovician acritarchs and trilobites from Bell Island, Eastern Newfoundland. Geological Survey of Canada, Bulletin 284:1–35.

Dean, W. T., and F. Martin. 1982. The sequence of trilobite faunas and acritarch microfloras at the Cambrian-Ordovician boundary, Wilcox Pass, Alberta, Canada; pp. 131–140 in M. G. Bassett and W. T. Dean (eds.), The Cambrian-Ordovician Boundary: Sections, Fossil Distributions, and Correlations. National Museum of Wales, Geological Series 3.

Dean, W. T., O. Monod, R. B. Rickards, D. Osman, and P. Bultynck. 2000. Lower Palaeozoic stratigraphy and palaeontology, Karadere-Zirze area, Pontus Mountains, northern Turkey. Geological Magazine 137:555–582.

Debrenne, F., and J. Reitner. 2000 [dated 2001]. Sponges, Cnidarians and Ctenophores; pp. 301–325 in A. Zhuravlev and R. Riding (eds.), The Ecology of the Cambrian Radiation. Columbia University Press, New York.

de Freitas, T., and U. Mayr. 1995. Kilometer-scale microbial buildups in a rimmed carbonate platform succession, Arctic Canada: New insight on Lower Ordovician reef facies. Bulletin of Canadian Petroleum Geology 43:407–432.

Delgado, S., D. Casane, L. Bonnaud, M. Laurin, J.-Y. Sire, and M. Girondot. 2001. Molecular evidence for Precambrian origin of amelogenin, the major protein of vertebrate enamel. Molecular Biology and Evolution 18:2146–2153.

Denison, R. E., R. B. Koepnik, W. H. Burke, and E. A. Hetherington. 1998. Construction of the Cambrian and Ordovician seawater $^{87}Sr/^{86}Sr$ curve. Chemical Geology 152:325–340.

Denison, R. H. 1967. Ordovician vertebrates from western United States. Fieldiana: Geology 16:131–192.

Depitout, A. 1962. Etude des Gigantostraces Siluriens du Sahara Central. Publications du Centre de Recherches Sahariennes, Série Géologie 2:1–141.

Desrochers, A., and N. P. James. 1989. Middle Ordovician (Chazyan) bioherms and biostromes of the Mingan Islands, Québec; pp. 183–191 in H. H. J. Geldetzer, N. P. James, and G. E. Tebbutt (eds.), Reefs, Canada and Adjacent Areas. Canadian Society of Petroleum Geologists, Memoir 13.

Deunff, J. 1961. Un microplancton a hystrichosphères dans le Tremadoc du Sahara. Revue de Micropaléontologie 4:37–52.

Dewey, J. F., R. B. Rickards, and D. Skevington. 1970. New light on the age of Dalradian deformation and metamorphism in western Ireland. Norsk Geologisk Tidsskrift 50:19–44.

Dixon, O. A. 1974. Late Ordovician *Propora* (Coelenterata: Heliolitidae) from Anticosti Island, Quebec, Canada. Journal of Paleontology 48:568–585.

Donoghue, P. C. J. 2001. Conodonts meet cladistics: Recovering relationships and assessing the completeness of the conodont fossil record. Palaeontology 44:65–93.

Donoghue, P. C. J., P. L. Forey, and R. J. Aldridge. 2000. Conodont affinity and chordate phylogeny. Biological Review 75:191–251.

Donovan, S. K. 1989. The significance of the British Ordovician crinoid fauna. Modern Geology 13:243–255.

Donovan, S. K., C. R. C. Paul, and D. N. Lewis. 1996. Echinoderms; pp. 202–267 in D. A. T. Harper and A. W. Owen (eds.), Fossils of the Upper Ordovician. Palaeontological Association (London), Field Guide to Fossils 7.

Dorning, K. J. 1981. Silurian acritarch distribution in the Ludlovian Shelf Sea of South Wales and the Welsh Borderland; pp. 31–36 in R. G. Neale and M. D. Brasier (eds.), Microfossils from Recent and Fossil Shelf Seas. Ellis Horwood, Chichester, England.

Dorning, K. J. 1999. Ordovician acritarch biohorizons, palaeoenvironmental interpretation and event stratigraphy; pp. 237–240 in P. Kraft and O. Fatka (eds.), *Quo vadis* Ordovician? Short papers of the 8th International Symposium on the Ordovician System. Acta Universitatis Carolinae, Geologica 43(1–2).

Downie, C. 1984. Acritarchs in British stratigraphy. The Geological Society, London, Special Report 17:1–26.

Downie, C., D. W. Fisher, R. Goldring, and F. H. T. Rhodes. 1967. Miscellanea; pp. 613–626 in W. B. Harland et al. (eds.), The Fossil Record: A Symposium with Documentation. The Geological Society, London.

Dronov, A., and L. E. Holmer. 1999. Depositional sequences in the Ordovician of Baltoscandia; pp. 133–136 in P. Kraft and O. Fatka (eds.), *Quo vadis* Ordovician? Short papers of the 8th International Symposium on the Ordovician System. Acta Universitatis Carolinae, Geologica 43(1–2).

Droser, M. L. 1991. Ichnofabric of the Paleozoic *Skolithos* ichnofacies and the nature and distribution of the *Skolithos* piperock. Palaios 6:316–325.

Droser, M. L., and D. J. Bottjer. 1988. Trends in depth and extent of bioturbation in Cambrian carbonate marine environments, western United States. Geology 16:233–236.

Droser, M. L., and D. J. Bottjer. 1989. Ordovician increase in extent and depth of bioturbation: Implications for understanding early Paleozoic ecospace utilization. Geology 17:850–852.

Droser, M. L., and X. Li. 2000 [dated 2001]. The Cambrian radiation and the diversification of sedimentary fabrics; pp. 137–164 in A. Y. Zhuravlev and R. Riding (eds.), The Ecology of the Cambrian Radiation. Columbia University Press, New York.

Droser, M. L., and P. M. Sheehan. 1995. Paleoecology of the Ordovician radiation and the Late Ordovician extinction event: Evidence from the Great Basin; pp. 63–106 in J. D. Cooper (ed.), Ordovician of the Great Basin: Field Trip Guidebook and Volume. Book 78, Pacific Section, SEPM, Fullerton, California.

Droser, M. L., and P. M. Sheehan. 1997a. Evaluating the ecological architecture of major events in the Phanerozoic history of marine invertebrate life. Geology 25:167–170.

Droser, M. L., and P. M. Sheehan. 1997b. Palaeoecology of the Ordovician radiation: Resolution of large-scale patterns with individual clade histories, palaeogeography and environments. Geobios Mémoire Spécial 20:221–229.

Droser, M. L., R. A. Fortey, and X. Li. 1996. The Ordovician Radiation. American Scientist 84:122–131.

Droser, M. L., D. J. Bottjer, and P. M. Sheehan. 1997. Evaluating the ecological architecture of major events in the Phanerozoic history of marine invertebrate life. Geology 25:167–170.

Droser, M. L., D. J. Bottjer, P. M. Sheehan, and G. R. McGhee, Jr. 2000. Decoupling of taxonomic and ecologic severity of Phanerozoic marine mass extinctions. Geology 28:675–678.

Droser, M. L., S. R. Jensen, and J. G. Gehling. 2001. Cambrian firm muddy substrates: Significance for the preservation of trace fossils. Geological Society of America, Abstracts with Programs 33:430.

Drygant, D. M. 1971. Nakhodka ostatkov konulyarii v Ordovike Volyni i Silure Podolii [Discovery of remains of conulariids in the Ordovician of Volyn and the Silurian of Podolia]. Paleontologicheskii Sbornik 1(8):19–22.

Dubinina, S. V. 1991. Upper Cambrian and Lower Ordovician conodont associations from open ocean paleoenvironments, illustrated by Batyrbay and Sarykum sections in Kazakhstan; pp. 107–124 in C. R. Barnes and S. H. Williams (eds.), Advances in Ordovician Geology. Geological Survey of Canada, Paper 90-9.

Dubinina, S. V. 1998. Conodonts from the early Ordovician (mid-Arenig) deep water deposits of central Asia paleobasins; pp. 79–86 in H. Szaniawski (ed.), Proceedings of the Sixth European Conodont Symposium (ECOS VI), Warzawa. Palaeontologia Polonica 58.

Dubinina, S. V. 2000. Conodonts and zonal stratigraphy on the Cambrian-Ordovician boundary deposits. Transactions of the Geological Institute of Moscow, Russian Academy of Sciences 517:1–239.

Duncan, H. 1956. Ordovician and Silurian coral faunas of western United States. United States Geological Survey, Bulletin 1021-F:209–235.

Dunham, J. B., and M. A. Murphy. 1976. An occurrence of well-preserved Radiolaria from the Upper Ordovician (Caradocian), Eureka County, Nevada. Journal of Paleontology 50:882–887.

Dzik, J. 1981. Evolutionary relationships of the early Palaeozoic cyclostomatous Bryozoa. Palaeontology 24:827–861.

Dzik, J. 1983. Larval development and relationships of *Mimospira;* a presumably hyperstrophic Ordovician gastropod. GFF [Geologiska Föreningen i Stockhom Förhandlingar] 104:231–239.

Dzik, J. 1994a. Conodonts of the Mójcza Limestone; pp. 43–128 in J. Dzik, E. Olempska, and A. Pisera, Ordovician Carbonate Platform Ecosystem of the Holy Cross Mountains. Palaeontologia Polonica 53.

Dzik, J. 1994b. Machaeridians, chitons, and conchiferan molluscs of the Mójcza Limestone; pp. 213–252 *in* J. Dzik, E. Olempska, and A. Pisera, Ordovician Carbonate Platform Ecosystem of the Holy Cross Mountains. Palaeontologia Polonica 53.

Dzik, J. 1994c. Evolution of "small shelly fossils" assemblages of the Early Paleozoic. Acta Palaeontologica Polonica 39:247–313.

Dzik, J. 1999. Relationship between rates of speciation and phyletic evolution: Stratophenetic data on pelagic conodont chordates and benthic ostracods. Geobios 32:205–221.

Ebbestad, J.-O. R., and A. E. S. Högström. 1999. Gastropods and machaeridians of the Baltic Late Ordovician; pp. 401–404 *in* P. Kraft and O. Fatka (eds.), *Quo vadis* Ordovician? Short papers of the 8th International Symposium on the Ordovician System. Acta Universitatis Carolinae, Geologica 43(1–2).

Ebneth, S., G. A. Shields, J. Veizer, J. F. Miller, and J. H. Shergold. 2001. High-resolution strontium isotope stratigraphy across the Cambrian-Ordovician transition. Geochimica et Cosmochimica Acta 65:2273–2292.

Eckert, J. 1988. Late Ordovician extinction of North American and British crinoids. Lethaia 21:147–167.

Edwards, D., and C. H. Wellman. 2001. Embryophytes on land: The Ordovician to Lochkovian (Lower Devonian) record; pp. 3–28 *in* P. G. Gensel and D. Edwards (eds.), Plants Invade the Land. Columbia University Press, New York.

Egerquist, E. 1999. Early Ordovician (Billingen—Volkhov stages) brachiopod faunas from NW Russia; pp. 341–343 *in* P. Kraft and O. Fatka (eds.), *Quo vadis* Ordovician? Short papers of the 8th International Symposium on the Ordovician System. Acta Universitatis Carolinae, Geologica 43(1–2).

Eisenack, A. 1978. Phosphatische und glaukonitische Mikrofossilien aus dem Vaginatenkalk von Hälluden,

Öland. Neues Jahrbuch für Geologie und Paläontologie, Monatshefte 1978:1–12.

Ekdale, A. A., and R. G. Bromley. 2001. Bioerosional innovation for living in carbonate hardgrounds in the Early Ordovician of Sweden. Lethaia 34:1–12.

Ekdale, A. A., R. G. Bromley, and S. G. Pemberton. 1984. Ichnology, Trace Fossils in Sedimentology and Stratigraphy. Society of Economic Paleontologists and Mineralogists Short Course 15. Society for Sedimentary Geology (SEPM), Tulsa, 317 pp.

Elaouad-Debbaj, Z. 1984. Chitinozoaires ashgilliens de l'Anti-Atlas (Maroc). Géobios 17:45–68.

Elaouad-Debbaj, Z. 1986. Chitinozoaires de la Formation du Ktaoua inférieur, Ordovicien supérieur de l'Anti-Atlas (Maroc). Hercynica 2:35–55.

Elaouad-Debbaj, Z. 1988a. Acritarches de l'Ordovicien supérieur (Caradoc-Ashgill) de l'Anti-Atlas, Maroc. Revue de Micropaléontologie 30:232–248.

Elaouad-Debbaj, Z. 1988b. Acritarches et chitinozoaires du Trémadoc de l'Anti-Atlas Central (Maroc). Revue de Micropaléontologie 31:85–128.

Elias, R. J. 1980. An Upper Ordovician eurypterid from Manitoba. Journal of Paleontology 54:262–263.

Elias, R. J. 1981. Solitary rugose corals of the Selkirk Member, Red River Formation (late Middle or Upper Ordovician), southern Manitoba. Geological Survey of Canada, Bulletin 344:1–53.

Elias, R. J. 1982. Latest Ordovician solitary rugose corals of eastern North America. Bulletins of American Paleontology 81(314):1–116.

Elias, R. J. 1985. Solitary rugose corals of the Upper Ordovician Montoya Group, southern New Mexico and westernmost Texas. The Paleontological Society, Memoir 16 (Journal of Paleontology, 59[5], supplement):1–58.

Elias, R. J. 1989. Extinctions and origins of solitary rugose corals, latest Ordovician to earliest Silurian in North America. Association of Australasian Palaeontologists, Memoir 8:319–326.

Elias, R. J. 1991. Environmental cycles and bioevents in the Upper Ordovician Red River–Stony Mountain solitary rugose coral province of North America; pp. 205–211 in C. R. Barnes and S. H. Williams (eds.), Advances in Ordovician Geology. Geological Survey of Canada, Paper 90-9.

Elias, R. J. 1995. Origin and relationship of the Late Ordovician Red River–Stony Mountain and Richmond solitary rugose coral provinces in North America; pp. 439–442 in J. D. Cooper, M. L. Droser, and S. C. Finney (eds.), Ordovician Odyssey: Short Papers, 7th International Symposium on the Ordovician System. Book 77, Pacific Section Society for Sedimentary Geology (SEPM), Fullerton, California.

Elias, R. J., and G. A. Young. 1998. Coral diversity, ecology, and provincial structure during a time of crisis: The latest Ordovician to earliest Silurian Edgewood Province in Laurentia. Palaios 13:98–112.

Elias, R. J., D. S. Brandt, and T. H. Clark. 1990. Late Ordovician solitary rugose corals of the St. Lawrence Lowland, Québec. Journal of Paleontology 64:340–352.

Elias, R. J., A. W. Potter, and R. Watkins. 1994. Late Ordovician rugose corals of the northern Sierra Nevada, California. Journal of Paleontology 68:164–168.

El-Khayal, A. A., and M. Romano. 1988. A revision of the upper part of the Saq Formation and Hanadir Shale (Lower Ordovician) of Saudi Arabia. Geological Magazine 125:161–174.

Eller, E. R. 1945. Scolecodonts from the Trenton Series (Ordovician) of Ontario, Quebec, and New York. Annals of the Carnegie Museum 30:119–212.

Eller, E. R. 1969. Scolecodonts from well cores of the Maquoketa Shale, Upper Ordovician, Ellsworth County, Kansas. Annals of the Carnegie Museum 41:1–17.

Elliott, D. K. 1987. A reassessment of *Astraspis desiderata*, the oldest north American vertebrate. Science 237:190–192.

Elliott, D. K., A. Blieck, and P.-Y. Gagnier. 1991. Ordovician vertebrates; pp. 93–106 in C. R. Barnes and S. H. Williams (eds.), Advances in Ordovician Geology. Geological Survey of Canada, Paper 90-9.

Elliott, G. F. 1972. Lower Palaeozoic green algae from southern Scotland, and their evolutionary significance. Bulletin of the British Museum (Natural History), Geology Series 22(4):356–376.

Elliott, G. F. 1984. Modern developments in the classification of some fossil green algae; pp. 297–302 in D. E. G. Irvine and D. M. John (eds.), Systematics of the Green Algae. Systematics Association Special Volume 27. Academic Press, London.

Endo, R. 1932. The Canadian and Ordovician formations and fossils of South Manchuria. United States National Museum Bulletin 164:1–115.

Engeser, T., and F. Riedel. 1996. The evolution of the Scaphopoda and its implications for the systematics of the Rostroconchia (Mollusca). Mitteilungen der Geologisches-Paläontologisches Institut der Universität Hamburg 79:117–138.

Erdtmann, B.-D. 1986. Early Ordovician eustatic cycles and their bearing on punctuations in early nematophorid (planktic) graptolite evolution; pp. 139–152 in O. H. Walliser (ed.), Global Bio-Events. Lecture Notes in Earth Sciences 8. Springer-Verlag, Berlin.

Erdtmann, B.-D. 1995. Tremadoc of the East European platform: Stratigraphy, confacies regions, correlation and basin dynamics; pp. 237–239 in J. D. Cooper, M. L. Droser, and S. C. Finney (eds.), Ordovician Odyssey: Short Papers, 7th International Symposium on the Ordovician System. Book 77, Pacific Section Society for Sedimentary Geology (SEPM), Fullerton, California.

Erdtmann, B.-D., and J. Maletz. 1995. The Lower Hunneberg Interval (LHI): A proposal to solve the Tremadoc/Arenig boundary crisis; pp. 137–138 in J. D. Cooper,

M. L. Droser, and S. C. Finney (eds.), Ordovician Odyssey: Short Papers, 7th International Symposium on the Ordovician System. Book 77, Pacific Section Society for Sedimentary Geology (SEPM), Fullerton, California.

Erdtmann, B.-D., B. Weber, H.-P. Schultze, and S. Egenhoff. 2000. A possible agnathan plate from the Lower Arenig (Lower Ordovician) of south Bolivia. Journal of Vertebrate Paleontology 20:394–399.

Eriksson, M. 1997. Lower Silurian polychaetaspid polychaetes from Gotland. GFF [Geologiska Föreningens i Stockhom Förhandlingar] 119:213–230.

Eriksson, M., and C. F. Bergman. 1998. Scolecodont systematics exemplified by the polychaete *Hadoprion cervicornis* (Hinde, 1879). Journal of Paleontology 72:477–485.

Eriksson, M., and C. F. Bergman. 2001. Upper Ordovician scolecodonts from the type Cincinnatian in North America. Geological Society of America, Abstracts with Programs 33:A-307.

Ernst, R. E., and K. L. Buchan. 2001. Mantle plumes: Their identification through time. Geological Society of America, Special Paper 352:1–598.

Ernst, R. E., and K. L. Buchan. 2003. Recognizing mantle plumes in the geological record. Annual Reviews of Earth and Planetary Sciences 31:469–523.

Erwin, D. H. 1993. The Great Paleozoic Crisis: Life and Death in the Permian. Columbia University Press, New York, 327 pp.

Erwin, D. H., J. W. Valentine, and J. J. Sepkoski, Jr. 1987. A comparative study of diversification events: The Early Paleozoic versus the Mesozoic. Evolution 41:1177–1186.

Etheridge, R., Jr. 1909. An organism allied to *Mitcheldeania* Wethered, of the Carboniferous Limestone, in the Upper Silurian of Malongulli. Records of the Geological Survey of New South Wales 8:308–311.

Ethington, R. L., and D. L. Clark. 1971. Lower Ordovician conodonts in North America. Geological Society of America, Memoir 127:63–82.

Ethington, R. L., and D. L. Clark. 1981. Lower and Middle Ordovician conodonts from the Ibex area, western Millard County, Utah. Brigham Young University Geology Studies 28:1–160.

Ethington, R. L., and J. E. Repetski. 1984. Paleobiogeographic distribution of Early Ordovician conodonts in central and western United States. Geological Society of America, Special Paper 196:89–101.

Ethington, R. L., K. M. Engel, and K. L. Elliott. 1987. An abrupt change in conodont faunas in the Lower Ordovician of the Midcontinent Province; pp. 111–127 in R. J. Aldridge (ed.), Palaeobiology of Conodonts. Ellis Horwood, Chichester, England.

Evans, D. H. 1993. The cephalopod fauna of the Killey Bridge Formation (Ordovician, Ashgill), Pomeroy, County Tyrone. Irish Journal of Earth Science 13:11–29.

Evitt, W. R. 1963. A discussion and proposals concerning fossil dinoflagellates, hystrichospheres, and acritarchs. Proceedings of the National Academy of Sciences (U.S.) 49:158–164, 298–302.

Fan, J., Y. Wang, and K. Hou. 1996. Studies of the stratigraphy, sedimentary facies, and oil and gas potential of Cambrian-Ordovician carbonate rocks in eastern Tarim; pp. 372–382 in X. Tong, D. Liang, and C. Jia (eds.), Collected works on oil and gas exploration in the Tarim Basin. Science Publishing House, Beijing (in Chinese).

Fang Z.-J., and J. C. W. Cope. In press. Early Ordovician bivalves of Dali, West Yunnan, China. Palaeontology.

Fang, Z.-J., and N. J. Morris. 1997. The genus *Pseudosanguinolites* and some modioliform bivalves (mainly Palaeozoic). Palaeoworld 7:50–74.

Fanton, K. C., C. Holmden, G. S. Nowlan, and F. M. Haidl. 2002. ^{143}Nd/^{144}Nd and Sm/Nd stratigraphy of Upper Ordovician epeiric sea carbonates. Geochimica et Cosmochimica Acta 66:241–255.

Fatka, O., and M. Konzalova. 1995. Microfossils of the Paseky Shales (Lower Cambrian, Czech Republic). Journal of the Czech Geological Society 40:55–66.

Fensome, R. A., G. L. Williams, M. S. Barss, J. M. Freeman, and J. M. Hill. 1990. Acritarchs and fossil prasinophytes: An index to genera, species and infraspecific taxa. American Association of Stratigraphic Palynologists, Contributions Series 25:1–771.

Fensome, R. A., R. A. MacRae, J. M. Moldowan, F. J. R. Taylor, and G. L. Williams. 1996. The early Mesozoic radiation of dinoflagellates. Paleobiology 22:329–338.

Fenton, C. L., and M. A. Fenton. 1931. Some snail borings of Paleozoic age. American Midland Naturalist 12:522–528.

Fergusson, C. L., and C. M. Fanning. 2002. Late Ordovician stratigraphy, zircon provenance and tectonics, Lachlan Fold Belt, southeastern Australia. Australian Journal of Earth Sciences 49:423–436.

Finks, R. 1983. Fossil Hexactinellida; pp. 101–115 in T. W. Broadhead (ed.), Sponges and Spongiomorphs: Notes for a Short Course. University of Tennessee (Knoxville), Department of Geological Sciences, Geology Study 7.

Finnegan, S., and M. L. Droser. 2001. The relationship between diversity and abundance: An example from the lower-middle Ordovician boundary of the Great Basin. Paleobios 21:51.

Finney, S. C., and W. B. N. Berry. 1997. New perspectives on graptolite distributions and their use as indicators of platform margin dynamics. Geology 25:919–922.

Finney, S. C., and Ethington, R. L. 1995. Base of Whiterock Series correlates with base of *Isograptus victoriae lunatus* Zone in Vinini Formation, Roberts Mountains, Nevada; pp. 153–156 in J. D. Cooper, M. L. Droser, and S. C. Finney (eds.), Ordovician Odyssey: Short Papers, 7th International Symposium on the Ordovician

System. Book 77, Pacific Section Society for Sedimentary Geology (SEPM), Fullerton, California.

Finney, S. C., and Ethington, R. L. 2000. Global Ordovician Series boundaries and global event biohorizons, Monitor Range and Robert Mountains, Nevada. Geological Society of America, Field Guide 2:301–318.

Finney, S. C., W. B. N. Berry, J. D. Cooper, R. L. Ripperdan, W. C. Sweet, S. R. Jacobson, A. Soufiane, A. Achab, and P. J. Noble. 1999. Late Ordovician mass extinction: A new perspective from stratigraphic sections in central Nevada. Geology 27:215–218.

Finney, S. C., J. Gleason, G. Gehrels, S. Peralta, and G. Aceñolaza. 2003. Early Gondwanan connection for the Argentine Precordillera terrane. Earth and Planetary Science Letters 205:349–359.

Fischer, A. G. 1984. The two Phanerozoic supercycles; pp. 129–150 *in* W. A. Berggren and J. A. Van Couvering (eds.), Catastrophes in Earth History. Princeton University Press, Princeton.

Fisher D. C., and M. H. Nitecki. 1982. Standardization of the anatomical orientation of receptaculitids. The Paleontological Society, Memoir 13 (Journal of Paleontology, 56[1], supplement):1–40.

Fisher, D. W. 1962. Small conoidal shells of uncertain affinities; pp. W98–W143 *in* R. C. Moore (ed.), Treatise on Invertebrate Paleontology. Part W, Miscellanea. Geological Society of America, New York, and University of Kansas Press, Lawrence.

Fisher, D. W., and R. S. Young. 1955. The oldest known tentaculitid from the Chepultepec Limestone (Canadian) of Virginia. Journal of Paleontology 29:871–875.

Flower, R. H. 1945. A new Deepkill eurypterid. American Midland Naturalist 34:717–719.

Flower, R. H. 1957. Nautiloids of the Paleozoic; pp. 829–852 *in* H. S Ladd (ed.), Treatise on Marine Ecology and Paleoecology. Vol. 2, Paleoecology. Geological Society of America, Memoir 67.

Flower, R. H. 1961. Montoya and related colonial corals. New Mexico Bureau of Mines and Mineral Resources, Memoir 7:1–97.

Flower, R. H. 1964. The Nautiloid Order Ellesmerocerida (Cephalopoda). New Mexico Bureau of Mines and Mineral Resources, Memoir 12:1–234.

Flower, R. H. 1970. Early Paleozoic of New Mexico and the El Paso Region, Revision 2. New Mexico Bureau of Mines and Mineral Resources, reprint series, 44 pp.

Flower, R. H. 1976. Ordovician cephalopod faunas and their role in correlation; pp. 523–552 *in* M. G. Bassett (ed.), The Ordovician System: Proceedings of a Palaeontological Association Symposium, Birmingham (England), September 1974. University of Wales Press and National Museum of Wales, Cardiff.

Flower, R. H. 1988. Progress and changing concepts in cephalopod and particularly nautiloid phylogeny and distribution; pp. 17–24 *in* J. Weidmann and J. Kullmann (eds.), Cephalopods Present and Past. Schweizerbart'sche Verlagsbuchhandlung, Stuttgart.

Flower, R. H., and H. M. Duncan. 1975. Some problems in coral phylogeny and classification. Bulletins of American Paleontology 67(287):175–192.

Floyd, J. D., M. Williams, and A. W. A. Rushton. 1999. Late Ordovician (Ashgill) ostracodes from the Drummuck Group, Craighead Inlier, Girvan district, SW Scotland. Scottish Journal of Geology 35:15–24.

Foerste, A. F. 1916. Upper Ordovician formations of Ontario and Quebec. Geological Survey of Canada, Memoir 83:1–297.

Foerste, A. F. 1928. American Arctic and related cephalopods. Journal of the Scientific Laboratories, Denison University, Bulletin, 19:175–224.

Foerste, A. F. 1929. The cephalopods of the Red River formation of southern Manitoba. Journal of the Science Laboratories, Denison University, Bulletin 24:129–235.

Foote, M. 1991. Morphological patterns of diversification: Examples from trilobites. Palaeontology 34:461–485.

Foote, M. 1993. Discordance and concordance between morphological and taxonomic diversity. Paleobiology 19:185–204.

Foote, M. 1997a. Sampling, taxonomic description, and our evolving knowledge of morphological diversity. Paleobiology 23:181–206.

Foote, M. 1997b. Estimating taxonomic durations and preservability. Paleobiology 23:278–300.

Foote, M. 2000a. Origination and extinction components of taxonomic diversity: General problems; pp. 74–102 *in* D. H. Erwin and S. L. Wing (eds.), Deep Time: Paleobiology's Perspective. Special volume (Supplement to Paleobiology 26), Paleontological Society, Lawrence, Kansas.

Foote, M. 2000b. Origination and extinction components of taxonomic diversity: Paleozoic and post Paleozoic dynamics. Paleobiology 26:578–605.

Foote, M. 2001. Evolutionary rates and the age distributions of living and extinct taxa; pp. 245–294 *in* J. B. C. Jackson, S. Lidgard, and F. K. McKinney (eds.), Evolutionary Patterns: Growth, Form, and Tempo in the Fossil Record. University of Chicago Press, Chicago.

Foote, M., and J. J. Sepkoski, Jr. 1999. Absolute measures of the completeness of the fossil record. Nature 398:415–417.

Fortey, R. A. 1975a. Early Ordovician trilobite communities; pp. 331–352 *in* A. Martinsson (ed.), Evolution and morphology of the Trilobita, Trilobitoidea and Merostomata. Fossils and Strata 4. Universitetsforlaget, Oslo.

Fortey, R. A. 1975b. The Ordovician trilobites of Spitsbergen. II. Asaphidae, Nileidae, Raphiophoridae and Telephinidae of the Valhallfonna Formation. Norsk Polarinstitut Skrifter 162:1–207.

Fortey, R. A. 1979. Early Ordovician trilobites from the Catoche Formation (St. George Group), western Newfoundland. Geological Survey of Canada, Bulletin 321: 61–114.

Fortey, R. A. 1980. The Ordovician trilobites of Spitsbergen. III. Remaining trilobites of the Valhallfonna Formation. Norsk Polarinstitut Skrifter 171:1–163.

Fortey, R. A. 1984. Global earlier Ordovician transgressions and regressions and their biological implications; pp. 37–50 in D. L. Bruton (ed.), Aspects of the Ordovician System. Palaeontological Contributions from the University of Oslo 295. Universitetsforlaget, Oslo.

Fortey, R. A. 1985. Pelagic trilobites as an example of deducing the life habits of extinct arthropods. Transactions of the Royal Society of Edinburgh 76:219–230.

Fortey, R. A. 2000. Olenid trilobites: The oldest known chemoautotrophic symbionts? Proceedings of the National Academy of Sciences (U.S.) 97:6574–6578.

Fortey, R. A., and C. R. Barnes. 1977. Early Ordovician conodont and trilobite communities of Spitsbergen: Influence on biogeography. Alcheringa 1:297–309.

Fortey, R. A., and L. R. M. Cocks. 1986. Marginal faunal belts and their structural implications, with examples from the Lower Palaeozoic. Journal of the Geological Society of London 143:151–160.

Fortey, R. A., and L. R. M. Cocks. 1992. The early Palaeozoic of the North Atlantic region as a test case for the use of fossils in continental reconstruction. Tectonophysics 206:147–158.

Fortey, R. A., and L. R. M. Cocks. 1998. Biogeography and palaeogeography of the Sibumasu terrane in the Ordovician: A review; pp. 43–56 in R. Hall and J. D. Holloway (eds.), Biogeography and Geological Evolution of SE Asia. Backhuys, Leiden.

Fortey, R. A., and L. R. M. Cocks. 2003. Palaeontological evidence bearing on global Ordovician-Silurian continental reconstructions. Earth Science Reviews 61:245–307.

Fortey, R. A., and B. K. Holdsworth. 1971. The oldest known well-preserved Radiolaria. Bollettino della Società Paleontologica Italiana 10:35–41.

Fortey, R. A., and C. J. T. Mellish. 1992. Are some fossils better than others for inferring palaeogeography? The early Ordovician of the North Atlantic region as an example. Terra Nova 4:210–216.

Fortey, R. A., and S. F. Morris. 1982. The Ordovician trilobite *Neseuretus* from Saudi Arabia and the palaeogeography of the *Neseuretus* fauna related to Gondwanaland in the earlier Ordovician. Bulletin of the British Museum (Natural History), Geology Series 36: 63–75.

Fortey, R. A., and R. M. Owens. 1978. Early Ordovician (Arenig) stratigraphy and faunas of the Carmarthen district, south-west Wales. Bulletin of the British Museum (Natural History), Geology Series 30:225–294.

Fortey, R. A., and R. M. Owens. 1987. The Arenig series in South Wales. Bulletin of the British Museum (Natural History), Geology Series 41:69–307.

Fortey, R. A., and R. M. Owens. 1990a. Trilobites; pp. 121–142 in K. J. McNamara (ed.), Evolutionary Trends. Belhaven Press, London.

Fortey, R. A., and R. M. Owens. 1990b. Evolutionary radiations in the Trilobita; pp. 139–164 in P. D. Taylor and G. P. Larwood (eds.), Major Evolutionary Radiations. Systematics Association Special Volume 42. Clarendon Press, Oxford.

Fortey, R. A., and R. M. Owens. 1999. Feeding habits in trilobites. Palaeontology 42:429–465.

Fortey, R. A., and J. H. Shergold. 1984. Early Ordovician trilobites, Nora Formation, central Australia. Palaeontology 27:315–366.

Fortey, R. A., D. E. G. Briggs, and M. A. Wills. 1996. The Cambrian evolutionary explosion—decoupling cladogenesis from morphological disparity. Biological Journal of the Linnean Society 57:13–33.

Fortey, R. A., R. M. Owens, and A. W. A. Rushton. 1989. The palaeogeographic position of the Lake District in the earlier Ordovician. Geological Magazine 126:9–17.

Fortey, R. A., D. A. T. Harper, J. K. Ingham, A. W. Owen, and A. W. A. Rushton. 1995. A revision of Ordovician series and stages from the historical type area. Geological Magazine 132:15–30.

Fortey, R. A., D. A. T. Harper, J. K. Ingham, A. W. Owen, M. A. Parkes, A. W. A. Rushton, and N. H. Woodcock. 2000. A revised correlation of Ordovician rocks in the British Isles. The Geological Society, London, Special Report 24:1–83.

Foster, C. B., R. Wicander, and J. D. Reed. 1990. *Gloeocapsomorpha prisca* Zalessky, 1917: A new study. Part 2, Origin of kukersite, a new interpretation. Geobios 23: 133–140.

Foster, C. B., T. M. Winchester-Seeto, and T. O'Leary. 1999. Preliminary study of environmental significance of Middle Ordovician (Darriwilian) acid resistant microfossils from the Canning Basin, Western Australia; pp. 311–314 in P. Kraft and O. Fatka (eds.), *Quo vadis* Ordovician? Short papers of the 8th International Symposium on the Ordovician System. Acta Universitatis Carolinae, Geologica 43(1–2).

Fox, J. S., and P. E. Videtich. 1997. Revised estimate of $\delta^{34}S$ for marine sulfates from the Upper Ordovician: Data from the Williston Basin, North Dakota, U.S.A. Applied Geochemistry 12:97–103.

Frakes, L. A., J. E. Francis, and J. I. Sykes. 1992. Climate Modes of the Phanerozoic. Cambridge University Press, Cambridge, 274 pp.

Frest, T. J., and H. L. Strimple. 1982. A new camarocystitid (Echinodermata: Paracrinoidea) from the Kimmswick Limestone (Middle Ordovician), Missouri. Journal of Paleontology 56:358–370.

Frest, T. J., H. L. Strimple, and C. C. Coney. 1979. Paracrinoids (Platycystitidae) from the Benbolt Formation (Blackriveran) of Virginia. Journal of Paleontology 53:380–398.

Frest, T. J., H. L. Strimple, and B. J. Witzke. 1980. New Comarocystitida (Echinodermata: Paracrinoidea) from the Silurian of Iowa and Ordovician of Oklahoma. Journal of Paleontology 54:217–228.

Frest, T. J., C. E. Brett, and B. J. Witzke. 1999. Caradocian-Gedinnian echinoderm associations of central and eastern North America; pp. 638–783 in A. J. Boucot and J. D. Lawson (eds.), Paleocommunity Analysis: A Silurian–Lower Devonian Example. Cambridge University Press, Cambridge.

Frey, R. C. 1987. The paleoecology of a Late Ordovician shale unit from southwest Ohio and southeastern Indiana. Journal of Paleontology 61:242–267.

Frey, R. C., and C. H. Holland. 1995. Nautiloid faunas; pp. 12–13 in B. S. Norford et al., Correlation Chart and Biostratigraphy of the Silurian Rocks of Canada. International Union of Geological Sciences, Publication 33.

Frey, R. W. 1973. Concepts in the study of biogenic sedimentary structures. Journal of Sedimentary Petrology 43:6–19.

Frey, R. W. 1975. The realm of ichnology, its strengths and limitations; pp. 13–38 in R. W. Frey (ed.), The Study of Trace Fossils: A Synthesis of Principles, Problems, and Procedures in Ichnology. Springer-Verlag, New York.

Fridley, J. D. 2001. The influence of species diversity on ecosystem productivity: How, where and why? Oikos 93:514–526.

Fritz, M. A. 1947. Cambrian Bryozoa. Journal of Paleontology 21:434–435.

Frogner, P., R. S. Gislason, and N. Oskarsson. 2001. Fertilizing potential of volcanic ash in ocean surface water. Geology 29:487–490.

Fry, W. L. 1983. An algal flora from the Upper Ordovician of the Lake Winnipeg region, Manitoba, Canada. Review of Palaeobotany and Palynology 39:313–341.

Frýda, J. 1992. Mode of life of a new onychochilid mollusc from the Lower Devonian of Bohemia. Journal of Paleontology 66:200–205.

Frýda, J. 1998. Higher classification of the Paleozoic gastropods inferred from their early shell ontogeny. World Congress of Malacology, Washington, D.C., Abstracts, 108.

Frýda, J. 1999a. Shape convergence in gastropod shells: An example from the Early Devonian *Plectonotus* (*Boucotonotus*)–*Palaeozygopleura* community of the Prague Basin (Bohemia). Mitteilungen aus dem Geologisch-Paläontologischen Institut der Universität Hamburg 83:179–190.

Frýda, J. 1999b. Higher classification of Paleozoic gastropods inferred from their early shell ontogeny. Journal of the Czech Geological Society 44:137–153.

Frýda, J. 1999c. Suggestions for polyphyletism of Paleozoic bellerophontiform molluscs inferred from their protoconch morphology. 65th annual meeting, American Malacological Society, Pittsburgh, Abstracts, 30.

Frýda, J. 2001. Discovery of a larval shell in Middle Paleozoic subulitoidean gastropods with description of two new species from the Early Devonian of Bohemia. Bulletin of the Czech Geological Survey 76:29–37.

Frýda, J., and K. Bandel. 1997. New Early Devonian gastropods from the *Plectonotus* (*Boucotonotus*)–*Palaeozygopleura* vommunity in the Prague Basin (Bohemia). Mitteilungen aus dem Geologisch-Paläontologischen Institut der Universität Hamburg 80:1–58.

Frýda, J., and R. B. Blodgett. 2001. The oldest known heterobranch gastropod, *Kuskokwimia* gen. nov., from the Early Devonian of west-central Alaska, with notes on the early phylogeny of higher gastropods. Bulletin of the Czech Geological Survey 76:39–53.

Frýda, J., and D. Heidelberger. 2003. Systematic position of Cyrtoneritimorpha within class Gastropoda with description of two new genera from Siluro-Devonian strata of central Europe. Bulletin of Geosciences 78:35–39.

Frýda, J., and S. Manda. 1997. A gastropod faunule from the *Monograptus uniformis* graptolite Biozone (Early Lochkovian, Early Devonian) in Bohemia. Mitteilungen aus dem Geologisch-Paläontologischen Institut der Universität Hamburg 80:59–122.

Frýda, J., and D. M. Rohr. 1999. Taxonomy and paleobiogeography of the Ordovician Clisospiridae and Onychochilidae (Mollusca); pp. 405–408 in P. Kraft and O. Fatka (eds.), *Quo vadis* Ordovician? Short papers of the 8th International Symposium on the Ordovician System. Acta Universitatis Carolinae, Geologica 43(1–2).

Fu, L.-P. 1982. Phylum Brachiopoda; pp. 95–178 in Xi'an Institute of Geology and Mineral Resources (ed.), Paleontological Atlas of Northwest China, Shaanxi-Gansu-Ningxia Volume. Part 1, Precambrian and Early Paleozoic. Geological Publishing House, Beijing.

Furey-Greig, T. 1999. Initial report on discovery of Ordovician scolecodonts from eastern Australia. Proceedings of the Linnean Society of New South Wales 121:85–88.

Furnish, W. M., and B. F. Glenister. 1964. Nautiloidea—Ascocerida; pp. K261–277 in R. C. Moore (ed.), Treatise on Invertebrate Paleontology. Part K, Mollusca 3, Cephalopoda. Geological Society of America, New York, and University of Kansas Press, Lawrence.

Fürsich, F. T. 1974. On *Diplocraterion* Torell 1870 and the significance of morphological features in vertical, spreiten-bearing, U-shaped trace fossils. Journal of Paleontology 48:952–962.

Gagnier, P.-Y. 1987. *Sacabambaspis janvieri*, un heterostraceo del Ordovicico superior de Bolivia. Acta del IV Congreso Latinoamericano de Paleontologia (Santa Cruz, Bolivia) 2:665–677.

Gagnier, P.-Y. 1995. Ordovician vertebrates and agnathan phylogeny; pp. 1–37 *in* M. Arsenault, H. Lelièvre, and P. Janvier (eds.), Etudes sur les Vertébrés inférieurs (VIIe Symposium International, Parc de Miguasha, Québec, 1991). Bulletin du Muséum national d'Histoire naturelle, 4e série, 17, C (1–4).

Gagnier, P.-Y., A. Blieck, and S. G. Rodrigo. 1986. First Ordovician vertebrate from South America. Geobios 19:629–634.

Gagnier, P.-Y., A. Blieck, C. C. Emig, T. Sempere, D. Vachard, and M. Vanguestaine. 1996. New paleontological and geological data on the Ordovician and Silurian of Bolivia. Journal of South American Earth Sciences 9:329–347.

Gao, L. 1980. Lower Ordovician chitinozoans from Wuding and Luquan, Yunnan. Fifth International Palynological Conference, Bedford College, London, Abstracts, 148.

Geng, L.-Y. 1984. Chitinozoa from the Fenxiang, Honghuayuan and Dawan Formations of Huanghuachang, Yichang, Hubei; pp. 509–516 *in* Nanjing Institute of Geology and Palaeontology (ed.), Stratigraphy and Palaeontology of Systemic Boundaries in China, Ordovician-Silurian, Part 1. Anhui Science and Technology Publishing House, Hefei.

Geyer, G. 1994. Middle Cambrian mollusks from Idaho and early conchiferan evolution. New York State Museum Bulletin 481:69–85.

Gibbs, M. T., E. J. Barron, and L. R. Kump. 1997. An atmospheric pCO_2 threshold for glaciation in the Late Ordovician. Geology 25:447–450.

Gingerich, P. D. 1990. Stratophenetics; pp. 437–442 *in* D. E. G. Briggs and P. R. Crowther (eds.), Palaeobiology, a Synthesis. Blackwell, Oxford.

Gingras, M. K., J. A. MacEachern, and S. G. Pemberton. 1998. A comparative analysis of the ichnology of wave- and river-dominated allomembers of the Upper Cretaceous Dunvegan Formation. Bulletin of Canadian Petroleum Geology 46:51–73.

Glaessner, M. F. 1971. The genus *Conomedusites* Glaessner and Wade and the diversification of the Cnidaria. Paläontologische Zeitschrift 45:7–17.

Glaessner, M. F. 1979. Lower Cambrian Crustacea and annelid worms from Kangaroo Island, South Australia. Alcheringa 3:21–31.

Glaessner, M. F. 1984. The Dawn of Animal Life: A Biohistorical Study. Cambridge University Press, Cambridge, 244 pp.

Glebovskaja, E. M. 1949. Ortryad Ostracoda. Rakovinchatye raki. Atlas rukovodyashchikh form iskopaemykh faun SSSR 2:261–268, 338, 367–368. Moscow.

Gnilovskaya, M. B. 1972. Izvestkovye vodorosli srednego i pozdnego ordovika vostochnogo Kazakhstana. Akademia Nauk SSSR, Institut Geologii i Geokhronologii Dokembria. Nauka, Leningrad, 196 pp.

Gnilovskaya, M. B. 1974. Opisanie vodoroslei; pp. 149–158 *in* I. F. Nikitin, M. B. Gnilovskaya, I. T. Zhuravleva, V. A. Luchinina, and E. I. Myagkova, Anderkenskaya biohermnaya gryada i istoriya ee obrazhovaniya; *in* O. A. Betekhtina and I. T. Zhuravleva (eds.), Sreda i zhizn v geologicheckom proshlom (paleoekologishskie problemy). Nauka, Novosibirsk, Siberia.

Gnilovskaya, M. B. 1980. Tip Phaecophycophyta? Burye vodorosli? pp. 177–178 *in* M. K. Apollonov, S. M. Bandaletov, and I. F. Nikitin (eds.), Granica Ordovika i Silura v Kazakhstane. Nauka, Alma-Ata, SSSR.

Golikov, A. N. 1988. Problems of phylogeny and system of the prosobranchiate gastropods. Trudy Zoologicheskogo Instituta AN SSSR 176:1–77.

Golikov, A. N., and Y. I. Starobogatov. 1975. Systematics of prosobranch gastropods. Malacologia 15:185–232.

Golonka, J., and W. Kiessling. 2002. Phanerozoic time scale and definition of time slices; pp. 11–20 *in* W. Kiessling, E. Flügel, and J. Golonka (eds.), Phanerozoic Reef Patterns. SEPM [Society for Sedimentary Geology] Special Publication 72.

Gorka, H. 1994. Late Caradoc and early Ludlow Radiolaria from Baltic erratic boulders. Acta Palaeontologica Polonica 39:169–178 (in Polish).

Goto, H., M. Umeda, and H. Ishiga. 1992. Late Ordovician radiolarians from the Lachlan Fold Belt, Southeastern Australia. Faculty of Science, Shimane University, Memoirs 26:145–170.

Gould, S. J. 1989. Wonderful Life. W. W. Norton and Co., New York, 347 pp.

Gould, S. J., and C. B. Calloway. 1980. Clams and brachiopods—ships that pass in the night. Paleobiology 6:383–396.

Graham, L. E., and J. Gray. 2001. The origin, morphology and ecophysiology of early embryophytes: Neontological and paleontological perspectives; pp. 140–158 *in* P. G. Gensel and D. Edwards (eds.), Plants Invade the Land. Columbia University Press, New York.

Grahn, Y. 1980. Early Ordovician Chitinozoa from Öland. Sveriges Geologiska Undersökning C 775:1–41.

Grahn, Y. 1981. Middle Ordovician Chitinozoa from Öland. Sveriges Geologiska Undersökning C 784:1–51.

Grahn, Y. 1982. Caradocian and Ashgillian Chitinozoa from the subsurface of Gotland. Sveriges Geologiska Undersökning C 788:1–66.

Grahn, Y. 1992. Ordovician Chitinozoa and biostratigraphy of Brazil. Geobios 25:703–723.

Grahn, Y., and L.-Y. Geng. 1990. Early Ordovician chitinozoa from Honghuayuan at Tongzi, northern Guizhou. Acta Micropalaeontologica Sinica 7:219–229.

Grahn, Y., and M. A. Miller. 1986. Chitinozoa from the Middle Ordovician Bromide Formation, Arbuckle Mountains, Oklahoma, U.S.A. Neues Jahrbuch für Geologie und Paläontologie 172:381–403.

Grahn, Y., and H. Nøhr-Hansen. 1989. Chitinozoans from Ordovician and Silurian shelf and slope sequences in North Greenland. Rapport Grønlands Geologiske Undersøgelse 144:35–41.

Grahn, Y., and F. Paris. 1992. Age and correlation of the Trombetas Group, Amazonas Basin, Brazil. Revue de Micropaléontologie 35:197–209.

Grahn, Y., S. Idil, and A. M. Ostvedt. 1994. Caradocian and Ashgillian chitinozoan biostratigraphy of the Oslo-Asker and Ringerike districts, Oslo Region, Norway. GFF [Geologiska Föreningens i Stockhom Förhandlingar] 116:147–160.

Grahn, Y., J. Nõlvak, and F. Paris. 1996. Precise chitinozoan dating of Ordovician impact events in Baltoscandia. Journal of Micropalaeontology 15:21–35.

Gray, J. 1984. Ordovician-Silurian land plants: The interdependence of ecology and evolution. Special Papers in Palaeontology 32:281–295.

Gray, J. 1985. The microfossil record of early land plants: Advances in understanding of early terrestrialization, 1970–1984. Philosophical Transactions of the Royal Society of London, B 309:167–195.

Gray, J. 1991. *Tetrahedraletes, Nodospora,* and the "cross" tetrad: An accretion of myth; pp. 49–87 *in* S. Blackmore and S. H. Barnes (eds.), Pollen and Spores: Patterns of Diversification. The Systematics Association, Special Volume. 44. Clarendon Press, Oxford.

Gray, J., D. Massa, and A. J. Boucot. 1982. Caradocian land plant microfossils from Libya. Geology 10:197–201.

Greife, J. L., and R. L. Langenheim. 1963. Sponges and brachiopods from the middle Ordovician Mazourka Formation, Independence Quadrangle, California. Journal of Paleontology 37:564–574.

Gubanov, A. P., and J. S. Peel. 2000. Cambrian monoplacophoran molluscs (Class Helcionelloida). American Malacological Bulletin 15:139–145.

Gubanov, A. P., and D. M. Rohr. 1995. Paleogeography of the Macluritidae (Ordovician-Gastropoda); pp. 461–464 *in* J. D. Cooper, M. L. Droser, and S. C. Finney (eds.), Ordovician Odyssey: Short Papers, 7th International Symposium on the Ordovician System. Book 77, Pacific Section Society for Sedimentary Geology (SEPM), Fullerton, California.

Gubanov, A. P., and J. A. Tait. 1998. *Maclurites* (Mollusca) and Ordovician palaeogeography. Schriften des Staatlichen Museums für Mineralogie und Geologie zu Dresden 9:140–142.

Guensburg, T. E. 1984. Echinodermata of the Middle Ordovician Lebanon Limestone, central Tennessee. Bulletins of American Paleontology 86:1–100.

Guensburg, T. E. 1992. Paleoecology of hardground encrusting and commensal crinoids, Middle Ordovician, Tennessee. Journal of Paleontology 66:129–147.

Guensburg, T. E., and J. Sprinkle. 1992. Rise of echinoderms in the Paleozoic Evolutionary Fauna: Significance of paleoenvironmental controls. Geology 20:407–410.

Guensburg, T. E., and J. Sprinkle. 1994. Revised phylogeny and functional interpretation of the Edrioasteroidea based on new taxa from the Early and Middle Ordovician of western Utah. Fieldana: Geology, n.s., 29:1–43.

Guensburg, T. E., and J. Sprinkle. 2000 [dated 2001]. Ecologic radiation of Cambro-Ordovician echinoderms; pp. 428–444 *in* A. Y. Zhuravlev and R. Riding (eds.), The Ecology of the Cambrian Radiation. Columbia University Press, New York.

Guensburg, T. E., and J. Sprinkle. 2001. Earliest crinoids: New evidence for the origin of the dominant Paleozoic echinoderms. Geology 29:131–134.

Guensburg, T. E., and J. Sprinkle. 2003. The oldest known crinoids (Early Ordovician, Utah) and a new crinoid plate homology system. Bulletins of American Paleontology, 364:1–43.

Guex, J. 1991. Biochronological Correlations. Springer-Verlag, Berlin, 252 pp.

Guilbault, J. P., and B. L. Mamet. 1976. Codiacées (Algues) ordoviciennes de Basses-Terres du Saint-Laurent. Canadian Journal of Earth Sciences 13:636–660.

Gurnis, M. 2001. Sculpting the earth from the inside out. Scientific American 284:40–47.

Gutiérrez-Marco, J. C., R. Albani, C. Aramburu, M. Arbizu, C. Babin, J. C. García-Ramos, I. Méndez-Bedia, I. Rábano, J. Truyols, J. Vannier, and E. Villas. 1996. Biostratigraphy of the Sueve Formation (Middle Ordovician) in the northern part of the Laviana-Sueve thrust-sheet (Cantabrian Zone, N. Spain). Revista Española de Paleontología 11:48–74.

Gutiérrez-Marco, J. C., R. Schallreuter, M. El Bourkhissi, and I. Hinz-Schallreuter. 1997. Identificación del género *Reuentalina* (ostrácodo Palaeocopa) en el Ordovícico Medio del Anti-Atlas central marroquí; pp. 80–83 *in* A. Grandal d'Anglade, J. C. Gutiérrez-Marco, and L. Santos Fidalgo (eds.), XIII Jornadas de Paleontología Reunión anual de la Sociedad Española de Paleontología "Fósiles de Galicia"/V Reunión Internacional Proyecto 351 Programa Internacional de Correlación Geológica Paleozoico Inferior del Noroeste de Gondwana A Coruña Libro de Resúmenes y Excursiones, Madrid (Sociedad Española de Paleontología).

Guo, F. 1988. New genera of fossil bivalves from Yunnan. Yunnan Geology 7:112–114 (in Chinese with English abstract).

Hacht, U. von, and F. Rhebergen. 1997. Ordovizische Geschiebespongien Europas; pp. 51–62 *in* M. Zwanzig and H. Löser (eds.), Berliner Beiträge zur Geschiebeforschung. Cpress, Dresden.

Hageman, S. J., P. E. Bock, Y. Bone, and B. McGowran. 1998. Bryozoan growth habits: Classification and analysis. Journal of Paleontology 72:418–436.

Hall, J. 1847. Palaeontology of New York, Vol. 1, containing descriptions of the organic remains of the lower division of the New-York system; *in* Natural History of New York, Part VI. C. Van Benthuysen, Albany, 338 pp.

Hall, J., and R. P. Whitfield. 1875. Description of invertebrate fossils, mainly from the Silurian System. Report of the Geological Survey of Ohio 2(2), Palaeontology, Sec. 1:65–161.

Hall, R. L. 1975. Late Ordovician coral faunas from northeastern New South Wales. Journal and Proceedings of the Royal Society of New South Wales 108:75–93.

Hallam, A. 1992. Phanerozoic Sea-Level Changes. Columbia University Press, New York, 266 pp.

Hambrey, M. J. 1985. The Late Ordovician–Early Silurian glaciation. Palaeogeography, Palaeoclimatology, Palaeoecology 51:273–289.

Hammann, W. 1974. Phacopina und Cheirurina (Trilobita) aus dem Ordovizium von Spanien. Senckenbergiana Lethaea 55:1–151.

Hammann, W. 1983. Calymenacea (Trilobita) aus dem Ordovizium von Spanien; ihre Biostratigraphie, Ökologie und Systematik. Abhandlungen der Senckenbergischen Naturforschenden Gesellschaft 542:1–177.

Hammann, W. 1985. Life habit and enrolment in Calymenacea (Trilobita) and their significance for classification. Transactions of the Royal Society of Edinburgh, Earth Sciences 76:307–318.

Hammann, W., and E. Serpagli. 2003. The algal genera *Ischadites* Murchison 1839 and *Cyclocrinites* Eichwald 1840 from the Upper Ordovician Portixeddu Formation of SW Sardinia. Bolletino della Paleontolgia Italiana 42:1–29.

Hammer, Ø. In press. Biodiversity curves for the Ordovician of Baltoscandia. Lethaia.

Hammer, Ø., D. A. T. Harper, and P. D. Ryan. 2001. PAST—Paleontological Statistics Software: Package for Education and Data Analysis. Paleontologia Electronica 4, 9 pp.

Hamoumi, N. 1999. Upper Ordovician glaciation spreading and its sedimentary record in Moroccan North Gondwana margin; pp. 111–114 in P. Kraft and O. Fatka (eds.), *Quo vadis* Ordovician? Short papers of the 8th International Symposium on the Ordovician System. Acta Universitatis Carolinae, Geologica 43(1–2).

Hanken, N.-M. 1979. The presence of *Rhabdotetradium* (tabulate coral) in the Upper Ashgillian of the Oslo region, Norway. Norsk Geologisk Tidsskrift 59:97–100.

Hannibal, J. T., and R. M. Feldmann. 1996. *Caryocaris* (Crustacea: Phyllocarida) from the Ordovician of the Cordillera Oriental of Souhern Bolivia. Kirtlandia 49:7–11.

Häntzschel, W. 1965. Vestigia invertebratorum et Problematica: Fossilium Catalogus. I. Animalia, Pars 108. W. Junk, s'Gravenhage, Netherlands, 142 pp.

Harland, T. L. 1981. Middle Ordovician reefs of southern Norway. Lethaia 14:169–188.

Harland, W. B., A. V. Cox, P. G. Llewellyn, C. A. G. Pickton, A. G. Smith, and R. Walters. 1982. A Geologic Time Scale. Cambridge University Press, Cambridge, 131 pp.

Harland, W. B., R. L. Armstrong, A. V. Cox, L. W. Craig, A. G. Smith, and D. G. Smith. 1990. A Geologic Time Scale 1989. Cambridge University Press, Cambridge, 263 pp.

Harper, C. W. 1975. Standing diversity of fossil groups in successive intervals of geologic time: A new measure. Journal of Paleontology 49:752–757.

Harper, C. W. 1996. Patterns of diversity, extinction and origination in the Ordovician-Devonian Stropheodontacea. Historical Biology 11:267–288.

Harper, D. A. T. 1979. Ordovician fish spines from Girvan, Scotland. Nature 278(5705):634–635.

Harper, D. A. T. 1989. Brachiopods from the Upper Ardmillan succession (Ordovician) of the Girvan District, Scotland, Part 2. Palaeontographical Society Monograph 142(579):79–128.

Harper, D. A. T. 2000. Suborder Dalmanellidina; pp. 782–844 in R. L. Kaesler (ed.), Treatise on Invertebrate Paleontology. Part H, Brachiopoda, rev., Vol. 3. Geological Society of America, Boulder, Colorado, and University of Kansas Press, Lawrence.

Harper, D. A. T., and E. Gallagher. 2001. Diversity, disparity and distributional patterns amongst the orthide brachiopod groups. Journal of the Czech Geological Society 46:87–93.

Harper, D. A. T., and C. Mac Niocaill. 2002. Early Ordovician rhynchonelliformean brachiopod diversity: Comparing some platforms, margins and intra-oceanic sites around the Iapetus Ocean; pp. 25–34 in J. A. Crame and A. W. Owen (eds.), Palaeobiogeography and Biodiversity Change: The Ordovician and Mesozoic-Cenozoic Radiations. The Geological Society, London, Special Publication 194.

Harper, D. A. T., and J.-Y. Rong. 1995. Patterns of change in the brachiopod faunas through the Ordovician-Silurian interface. Modern Geology 20:83–100.

Harper, D. A. T., and J.-Y. Rong. 2001. Palaeozoic brachiopod extinctions, survival and recovery: Patterns within the rhynchonelliformeans. Geological Journal 36:317–328.

Harper, D. A. T., and M. R. Sandy. 2001. Paleozoic brachiopod biogeography; pp. 207–222 in S. J. Carlson and M. R. Sandy (eds.), Brachiopods Ancient and Modern: A Tribute to G. Arthur Cooper. The Paleontological Society (New Haven, Connecticut), Paper 7.

Harper, D. A. T., and A. D. Wright. 1996. Brachiopods; pp. 63–94 in D. A. T. Harper and A. W. Owen (eds.), Fossils of the Upper Ordovician. Palaeontological Association (London), Field Guide to Fossils 7.

Harper, D. A. T., C. Mac Niocaill, and S. H. Williams. 1996. The palaeogeography of the early Ordovician Iapetus terranes: An integration of faunal and palaeomagnetic constraints. Palaeogeography, Palaeoclimatology, Palaeoecology 121:297–312.

Harper, D. A. T., J.-Y. Rong, and P. M. Sheehan. 1999a. Ordovician diversity patterns in early rhynchonelliform (protorthide, orthide and strophomenide) brachiopods; pp. 325–327 in P. Kraft and O. Fatka (eds.), *Quo vadis* Ordovician? Short papers of the 8th International Symposium on the Ordovician System. Acta Universitatis Carolinae, Geologica 43(1–2).

Harper, D. A. T., J.-Y. Rong, and R.-B. Zhan. 1999b. Late Ordovician development of deep-water brachiopod faunas; pp. 351–353 *in* P. Kraft and O. Fatka (eds.), *Quo vadis* Ordovician? Short papers of the 8th International Symposium on the Ordovician System. Acta Universitatis Carolinae, Geologica 43(1–2).

Harper, E. M., G. T. W. Forsythe, and T. J. Palmer. 1999. A fossil record full of holes: The Phanerozoic history of drilling predation. Comment. Geology 27:959–960.

Harper, J. A. 2000. The bellerophont controversy revisited. American Malacological Bulletin 15:147–156.

Harper, J. A., and H. B. Rollins. 1982. Recognition of Monoplacophora and Gastropoda in the fossil record: A functional morphological look at the bellerophont controversy. The Third North American Paleontological Convention, Montreal, Proceedings 1:227–232.

Harrington, H. J. 1938. Sobre las faunas del Ordoviciano inferior del Norte Argentino. Revista del Museo de La Plata, n.s., 1:109–289.

Harrington, H. J., and A. F. Leanza. 1957. Ordovician trilobites of Argentina. Department of Geology, University of Kansas, Special Publication 1:1–276.

Harris, A. G., S. M. Bergström, R. L. Ethington, and R. J. Ross, Jr. 1979. Aspects of Middle and Upper Ordovician conodont biostratigraphy of carbonate facies in Nevada and southeast California and comparison with some Appalachian successions. Brigham Young University Geology Studies 26:7–43.

Hart, C. P. 1986. Trenton Group chitinozoans from New York state: A brief review; pp. 17–33 *in* M. A. Miller (ed.), A Field Excursion to Trenton Group (Middle and Upper Ordovician) and Hamilton Group (Middle Devonian) Localities in New York, and a Survey of Their Chitinozoans. American Association of Stratigraphic Palynologists. Publishers Press, Salt Lake City, Utah.

Haszprunar, G. 1985. The fine morphology of the osphradial sense organs of the Mollusca. I. Gastropoda, Prosobranchia. Philosophical Transactions of the Royal Society of London, B 307:457–496.

Haszprunar, G. 1988. A preliminary phylogenetic analysis of the streptoneurous Gastropoda; pp. 7–16 *in* W. F. Ponder (ed.), Prosobranch Phylogeny, Malacological Review Supplement.

Haszprunar, G. 1993. The Archaeogastropoda: A clade, a grade or what else? American Malacological Bulletin 10:165–177.

Haszprunar, G., and K. Schaefer. 1997. Monoplacophora; pp. 415–457 *in* F. W. Harrison and A. J. Kohn (eds.), Microscopic Anatomy of Invertebrates, Mollusca II. Wiley, New York.

Hatch, J. R., S. R. Jacobson, B. J. Witzke, J. B. Risatti, D. E. Anders, W. L. Watney, K. D. Newell, and A. K. Vuletich. 1987. Possible Late Middle Ordovician organic carbon isotope excursion: Evidence from Ordovician oils and hydrocarbon source rocks, mid-continent and east-central USA. American Association of Petroleum Geologists Bulletin 71:1342–1354.

Havlíček, V. 1998. Prague Basin: Ordovician; pp. 39–79, pls. VIII–XXII *in* I. Chlupáč, V. Havlíček, J. Kříž, Z. Kukal, and P. Štorch. Palaeozoic of the Barrandian (Cambrian to Devonian). Czech Geological Survey, Prague.

Havlíček, V., and J. Vaněk. 1990. Ordovician invertebrate communities in black-shale lithofacies (Prague Basin, Bohemia). Věstník Ústředního ústavu geologického 65:223–236.

Havlíček, V., J. Vaněk, and O. Fatka. 1994. Perunica microcontinent in the Ordovician (its position within the Mediterranean Province, series division, benthic and pelagic associations). Sborník geologických Věd, Geologie 46:23–56.

Hayek, L. A. C., and E. Bura. 2001. On the ends of the taxon range problem; pp. 221–244 *in* J. B. C Jackson, S. Lidgard, and F. K. McKinney (eds.), Evolutionary Patterns: Growth, Form, and Tempo in the Fossil Record. University of Chicago Press, Chicago.

Head, M. 1996. Modern dinoflagellate cysts and their biological affinities; pp. 1197–1248 *in* J. Jansonius and D. C. McGregor (eds.), Palynology: Principles and Applications, Vol. 3. American Association of Stratigraphic Palynologists Foundation. Publishers Press, Salt Lake City, Utah.

Hecker, R. Th. 1970. Palaeoichnological research in the Palaeontological Institute of the Academy of Science of the USSR; pp. 215–226 *in* T. P. Crimes and J. C. Harper (eds.), Trace Fossils. Geological Journal, Special Issue 3. Seel House Press, Liverpool.

Hedegaard, C., D. R. Lindberg, and K. Bandel. 1997. Shell microstructure of a Triassic patellogastropod limpet. Lethaia 30:331–335.

Henningsmoen, G. 1953. The Middle Ordovician of the Oslo Region, Norway. 4. Ostracoda. Norsk Geologisk Tidsskrift 32:35–56.

Henningsmoen, G. 1954a. Lower Ordovician Ostracods from the Oslo Region, Norway. Norsk Geologisk Tidsskrift 33:41–68.

Henningsmoen, G. 1954b. Upper Ordovician Ostracods from the Oslo Region, Norway. Norsk Geologisk Tidsskrift 33:69–108.

Henningsmoen, G. 1957. The trilobite family Olenidae: Skrifter utgitt av der Norsk Videnskaps-Akademi i Oslo, I. Matematik-naturvidenskapelig Klasse 1:1–303.

Henry, J.-L. 1980. Trilobites ordoviciens du Massif Armoricain. Mémoires de la Société Géologique et Minéralogique de Bretagne 22:1–250.

Hergarten, B. 1985. Die Conularien des Rheinischen Devons. Senckenbergiana Lethaea 66:269–297.

Hergarten, B. 1988. Conularien in Deutschland. Aufschluss 39:321–356.

Hessland, I. 1949. A Lower Ordovician *Pseudoconularia* from the Siljan District: Investigations of the Lower

Ordovician of the Siljan District, Sweden III. Bulletin of the Geological Institute of Uppsala 33:429–435.

Heuse, T., O. Lehnert, and P. Kraft. 1996. Organic-walled microfossils from the Lower Ordovician of the Argentine Precordillera; pp. 425–439 in O. Fatka and T. Servais (eds.), Acritarcha in Praha. Acta Universitatis Carolinae, Geologica 40.

Heuse, T., Y. Grahn, and B.-D. Erdtmann. 1999. Early Ordovician chitinozoans from the east Cordillera of southern Bolivia. Revue de Micropaléontologie 42:43–55.

Hewitt, R. A., and B. Stait. 1985. Phosphatic connecting rings and ecology of an Ordovician ellesmerocerid nautiloid. Alcheringa 9:229–243.

Hickman, C. S. 1988. Archaeogastropod evolution, phylogeny and systematics: A re-evaluation; pp. 17–34 in W. F. Ponder (ed.), Prosobranch Phylogeny, Malacological Review Supplement.

Hickman, C. S., and J. H. McLean. 1990. Systematic Revision and Suprageneric Classification of Trochacean Gastropods. Natural History Museum of Los Angeles County, Los Angeles, 167 pp.

Hill, D. 1942. Some Tasmanian Palaeozoic corals. Royal Society of Tasmania, Papers and Proceedings for 1941:3–11.

Hill, D. 1955. Ordovician corals from Ida Bay, Queenstown and Zeehan, Tasmania. Royal Society of Tasmania, Papers and Proceedings 89:237–254.

Hill, D. 1957. Ordovician corals from New South Wales. Journal and Proceedings of the Royal Society of New South Wales 91:97–107.

Hill, D. 1981. Rugosa and Tabulata; pp. 1–762 in C. Teichert (ed.), Treatise on Invertebrate Paleontology, Part F (Supplement 1) in 2 vols. Geological Society of America, Boulder, Colorado, and University of Kansas Press, Lawrence.

Hill, D., and E. C. Stumm. 1956. Tabulata; pp. 444–477 in R. C. Moore (ed.), Treatise on Invertebrate Paleontology, Part F. Geological Society of America, New York, and University of Kansas Press, Lawrence.

Hill, D., G. Playford, and J. T. Woods. 1969. Ordovician and Silurian Fossils of Queensland. Queensland Palaeontographical Society, Brisbane, 32 pp.

Hill, J. P., F. Paris, and J. B. Richardson. 1985. Silurian palynomorphs; pp. 27–48 in B. Thusu and B. Owens (eds.), Palynostratigraphy of North-East Libya. Journal of Micropaleontology 4.

Hinde, G. J. 1879. On annelid jaws from the Cambro-Silurian, Silurian and Devonian formations in Canada and from Lower Carboniferous in Scotland. Quarterly Journal of the Geological Society of London 35:370–389.

Hinde, G. J. 1890. Specimens and microscopic sections of Radiolarian chert from the Ordovician strata (Llandeilo-Caradoc) of the Southern Uplands of Scotland. Journal of the Geological Society, Proceedings 111–112.

Hints, L. 1997. Aseri Stage; pp. 66–67 in A. Raukas and A. Teedumäe (eds.), Geology and Mineral Resources of Estonia. Estonian Academy Publishers, Tallinn.

Hints, O. 1998. Late Viruan (Ordovician) polychaete jaws from North Estonia and the St. Petersburg Region. Acta Palaeontologica Polonica 43:471–516.

Hints, O. 2000. Ordovician eunicid polychaetes of Estonia and surrounding areas: A review of their distribution and diversification. Review of Palaeobotany and Palynology 113:41–55.

Hints, O., M. Eriksson, and C. F. Bergman. 2000. Ordovician eunicid polychaete faunas of Baltica and Laurentia: Affinities and differences. 31st International Geological Congress, Rio de Janeiro, Brazil, August 6–17, 2000, Abstracts, on CD-ROM.

Hinz, I., P. Kraft, M. Mergl, and K. J. Müller. 1990. The problematic *Hadimopanella, Kaimenella, Milaculum* and *Utahphospha* identified as sclerites of Palaeoscolecida. Lethaia 23:217–221.

Hinz-Schallreuter, I. 2000. Middle Cambrian Bivalvia from Bornholm and a review of Cambrian bivalved Mollusca. Revista Española de Micropaleontología 32:225–242.

Hintze, L. F. 1953. Lower Ordovician trilobites from western Utah and eastern Nevada. Utah Geological and Mineralogical Survey Bulletin 48:1–249.

Hoare, R. D. 2000. Considerations on Paleozoic Polyplacophora, including the description of *Plasiochiton curiosus,* n. gen. and sp. American Malacological Bulletin 15:131–137.

Høeg, O. A. 1932. Ordovician algae from the Trondheim area. Skrifter utgitt av det Norske Videnskaps-Akademi i Oslo 1932(4):63–96.

Høeg, O. A. 1961. Ordovician algae in Norway; pp. 103–120 in J. H. Johnson (ed.), Studies of Ordovician algae, Part 2. Colorado School of Mines, Quarterly 56(2).

Hofmann, H. J. 1979. Chazy (Middle Ordovician) trace fossils in the Ottawa–St. Lawrence Lowlands. Geological Survey of Canada Bulletin 321:27–59.

Hofmann, H. J., M. P. Cecile, and L. S. Lane. 1994. New occurrences of *Oldhamia* and other trace fossils in the Cambrian of the Yukon and Ellesmere Island, Arctic Canada. Canadian Journal of Earth Sciences 31:767–782.

Högström, A. E. S. 2000. Aspects of machaeridian ecology. Geological Society of America, Abstracts with Programs 32(7):370.

Högström, A. E. S., and M. L. Droser. 2001. Machaeridians of the Al Rose Formation, Inyo Mountains, California. Geological Society of America, Cordilleran Section, Abstracts with Programs 33(3):73.

Högström, A. E. S., and W. L. Taylor. 2001. The machaeridian *Lepidocoleus sarlei* Clarke, 1896, from the Rochester Shale (Silurian) of New York State. Palaeontology 44:113–130.

Holland, C. H. 1976. Introduction; pp. 9–11 in M. G. Bassett (ed.), The Ordovician System: Proceedings of a

Palaeontological Association Symposium, Birmingham (England), September 1974. University of Wales Press and National Museum of Wales, Cardiff.

Holland, C. H. 1984. Form and function in Silurian Cephalopoda. Special Papers in Palaeontology 32:151–164.

Holland, C. H. 1987. The nautiloid cephalopods: A strange success. Journal of the Geological Society of London 144:1–15.

Holland, C. H., M. G. Audley-Charles, M. G. Bassett, J. W. Cowie, D. Curry, F. J. Fitch, J. M. Hancock, M. R. House, J. K. Ingham, P. E. Kent, N. Morton, W. H. C. Ramsbottom, P. F. Rawson, D. B. Smith, C. J. Stubblefield, H. S. Torrens, P. Wallace, and A. W. Woodland. 1978. A guide to stratigraphical proceedure. The Geological Society, London, Special Report 10:1–18.

Holland, S. M. 1995. The stratigraphic distribution of fossils. Paleobiology 21:92–109.

Holland, S. M. 1997. Using time/environment analysis to recognize faunal events in the Upper Ordovician of the Cincinnati Arch; pp. 309–334 in C. E. Brett and G. C. Baird (eds.), Paleontological Events: Stratigraphic, Ecological, and Evolutionary Implications. Columbia University Press, New York.

Holland, S. M. 2000. The quality of the fossil record and sequence stratigraphic perspective; pp. 148–168 in D. H. Erwin and S. L. Wing (eds.), Deep Time: Paleobiology's Perspective. Special volume (Supplement to Paleobiology 26). Paleontological Society, Lawrence, Kansas.

Holland, S. M., and M. E. Patzkowsky. 1997. Distal orogenic effects on peripheral bulge sedimentation: Middle and Upper Ordovician of the Nashville Dome. Journal of Sedimentary Research 67:250–263.

Holland, S. M., and M. E. Patzkowsky. 1999. Models for simulating the fossil record. Geology 27:491–494.

Holm, G. 1893. Sveriges Kambrisk-Siluriska Hyolithidæ och Conulariidæ. Sveriges Geologiska Undersökning, C 112:1–172.

Holmden C., R. A. Creaser, K. Muehlenbachs, S. M. Bergström, and S. A. Leslie. 1996. Isotopic and elemental systematics of Sr and Nd in 454 Ma biogenic apatites: Implications for paleoseawater studies. Earth and Planetary Science Letters 142:425–437.

Holmden C., R. A. Creaser, K. Muehlenbachs, S. A. Leslie, and S. M. Bergström. 1998. Isotopic evidence for geochemical decoupling between ancient epeiric seas and bordering oceans: Implications for secular curves. Geology 26:567–570.

Holmer, L. E. 1986. Inarticulate brachiopods around the Middle-Upper Ordovician boundary in Västergötland. GFF [Geologiska Föreningens i Stockhom Förhandlingar] 108:97–126.

Holmer, L. E. 1987. Ordovician mazuelloids and other microfossils from Västergötland (Sweden). GFF [Geologiska Föreningens i Stockhom Förhandlingar] 109:67–71.

Holmer, L. E. 1989. Middle Ordovician phosphatic inarticulate brachiopods from Västergötland and Dalarna, Sweden. Fossils and Strata 26:1–172.

Holmer, L. E., and L. E. Popov. 2000. Subphylum Linguliformea; pp. 30–146 in R. L. Kaesler (ed.), Treatise on Invertebrate Paleontology. Part H, Brachiopoda, rev., Vol. 2. Geological Society of America, Boulder, Colorado, and University of Kansas Press, Lawrence.

Holmer, L. E., S. P. Koneva, L. E. Popov, and A. M. Zhylkaidarov. 1996. Lingulate brachiopods and associated conodonts from the Middle Ordovician (Llanvirn) of the Malyi Karatau Range, Kazakhstan. Paläontologische Zeitschrift 70:481–495.

Holmer, L. E., L. E. Popov, and M. G. Bassett. 2000. Early Ordovician organophosphatic brachiopods with Baltoscandian affinities from the Alay Range, southern Kyrgyzstan. GFF [Geologiska Föreningens i Stockhom Förhandlingar] 122:367–375.

Holmer, L. E., L. E. Popov, S. P. Koneva, and M. G. Bassett. 2001. Cambrian–early Ordovician brachiopods from Malyi Karatau, the western Balkhash Region, and northern Tien Shan, Central Asia. Special Papers in Palaeontology 65:1–180.

Horný, R. 1963. O systematickém postavení cyrtoneloidních měkkýšů (Molusca). Časopis Národního Muzea, Řada přírodovědná 132:90–94.

Horowitz, A. S., and J. F. Pachut. 2000. The fossil record of bryozoan species diversity; pp. 245–248 in A. Herrera Cubilla and J. B. C. Jackson (eds.), Proceedings of the 11th International Bryozoology Association Conference, Smithsonian Tropical Research Institute, Balboa, Republic of Panama.

Hou, J.-P., and X.-F. Wang. 1982. Chitinozoa biostratigraphy in China. Bulletin de la Société géologique et minéralogique de Bretagne, Serié C:79–82.

Hou Xianguang, and J. Bergström. 1994. Palaeoscolecid worms may be nematomorphs rather than annelids. Lethaia 27:11–17.

Hou Xianguang, D. J. Siveter, M. Williams, D. Walossek, and J. Bergström. 1996. Appendages of the arthropod *Kunmingella* from the early Cambrian of China: Its bearing on the systematic position of the Bradoriida and the fossil record of the Ostracoda. Philosophical Transactions of the Royal Society of London, B 351:1131–1145.

Hou, Y.-T. 1953. Some Tremadocian Ostracods from Taitzeho Valley, Liaotung. Acta Palaeontologica Sinica 1:40–50.

House, M. R. 1988. Extinction and survival in the Cephalopoda; pp. 139–154 in G. P. Larwood (ed.), Extinction and Survival in the Fossil Record. Systematics Association Special Volume 34. Clarendon Press, Oxford.

Howell, B. F. 1952. New Carboniferous serpulid worm from Missouri. Bulletin of the Wagner Free Institute of Sciences of Philadelphia 27:37–40.

Hsü, S.-C., and C.-T. Ma. 1948. The Ichang Formation and the Ichangian fauna. National Research Institute of Geology, Academia Sinica, Contributions 8:1–51.

Hu, Z.-X., and N. Spjeldnaes. 1991. Early Ordovician bryozoans from China. Bulletin de la Societe des Sciences Naturelles de l'Ouest de la France, Mémoire Hors 1:179–185.

Huff, W. D., S. M. Bergström, and D. R. Kolata. 1992. Gigantic Ordovician ash fall in North America and Europe: Biological, tectonomagmatic, and event stratigraphic significance. Geology 20:875–878.

Huff, W. D., D. Davis, S. M. Bergström, M. P. S. Krekeler, D. R. Kolata, and C. Cingolani. 1997. A biostratigraphically well constrained K-bentonite U-Pb zircon age of the lowermost Darriwilian Stage (Middle Ordovician) from the Argentine Precordillera. Episodes 20:29–33.

Huff, W. D., S. M. Bergström, D. R. Kolata, C. A. Cingolani, and R. A. Astini. 1998. Ordovician K-bentonites in the Argentine Precordillera: Relations to Gondwana margin evolution; pp. 107–126 in R. J. Pankhurst and C. W. Rapela (eds.), The Proto-Andean Margin of Gondwana. Geological Society, London, Special Publication 142.

Hughes, C. P., R. B. Rickards, and A. Williams. 1980. The Ordovician fauna from the Contaya Formation of eastern Peru. Geological Magazine 117:1–21.

Hughes, N. C., and C. C. Labandeira. 1995. The stability of species in taxonomy. Paleobiology 21:401–403.

Hughes, N. C., G. O. Gunderson, and M. J. Weedon. 2000. Late Cambrian conulariids from Wisconsin and Minnesota. Journal of Paleontology 74:828–838.

Hynda, V. A. 1973. The tubes of some problematic organisms from the Middle Ordovician of Volynia. Paleontological Journal 7:250–253.

Hynda, V. A. 1986. Melkaja bentosnaja fauna ordovika jugo-zapada Vostocno-Evropejskoj platformy. Akademia Nauk USSR—Naukova Dumka, Kiev, Ukraine, 153 pp.

Ingham, J. K. 2000. Scotland: The Midland Valley Terrane–Girvan; pp. 43–47 in R. A. Fortey et al., A Revised Correlation of Ordovician Rocks in the British Isles. The Geological Society, London, Special Report 24.

Ingham, J. K., and A. D. Wright. 1970. A revised classification of the Ashgill Series. Lethaia 3:233–242.

Ivanovskii, A. B. 1972. The evolution of the Ordovician and Silurian Rugosa. Twenty-Third International Geological Congress. Proceedings of the International Palaeontological Union 2:69–78.

Ivany, L. C., and K. M. Schopf (eds.). 1996. New perspectives on faunal stability in the fossil record. Palaeogeography, Palaeoclimatology, Palaeoecology 127:1–361.

Izold, M. D. 1993. Early Ordovician shelf-slope conodont biostratigraphy and biofacies differentiation of western Iapetus in west Texas and New York. Master's thesis, Ohio State University, Columbus, Ohio.

Jaanusson, V. 1953a. Untersuchungen über baltoskandische Asaphiden. I. Revision der mittordovizischen Asaphiden des Siljan-Gebietes in Dalarna. Arkiv för Mineralogi och Geologi 1:377–464.

Jaanusson, V. 1953b. Untersuchungen über baltoskandische Asaphiden. II. Revision der *Asaphus* (*Neoasaphus*)-Arten aus dem Geschiebe des südbottnischen Gebietes. Arkiv för Mineralogi och Geologi 1:465–499.

Jaanusson, V. 1956. Untersuchungen über baltoskandische Asaphiden. III. Über die Gattungen *Megistaspis* n. nom. und *Homalopyge* n. gen. Bulletin of the Geological Institutions of the University of Uppsala 36:59–77.

Jaanusson, V. 1957. Middle Ordovician Ostracodes of Central and Southern Sweden. Bulletin of the Geological Institutions of the University of Uppsala 37:173–442.

Jaanusson, V. 1960. On the series of the Ordovician System. Report of the 21st International Geological Congress (Copenhagen) 7:70–81.

Jaanusson, V. 1961. Discontinuity surfaces in limestones. Bulletin of the Geological Institutions of the University of Uppsala 40:221–241.

Jaanusson, V. 1973. Aspects of carbonate sedimentation in the Ordovician of Baltoscandia. Lethaia 6:11–34.

Jaanusson, V. 1979. Ordovician; pp. A136–A166 in R. A. Robison and C. Teichert (eds.), Treatise on Invertebrate Paleontology. Part A, Introduction: Fossilization (Taphonomy), Biogeography and Biostratigraphy. Geological Society of America, Boulder, Colorado, and University of Kansas Press, Lawrence.

Jaanusson, V. 1995. Confacies differentiation and Upper Middle Ordovician correlation in the Baltoscandian Basin. Proceedings of the Estonian Academy of Sciences 44:73–86.

Jablonski, D. 1986. Larval ecology and macroevolution in marine invertebrates. Bulletin of Marine Science 39:565–587.

Jablonski, D. 1991. Extinctions: A paleontological perspective. Science 253:754–757.

Jablonski, D. 1995. Extinctions in the fossil record; pp. 25–44 in J. H. Lawton and R. M. May (eds.), Extinction Rates. Oxford University Press, Oxford.

Jacobson, S. R. 1978. Acritarchs from the Upper Ordovician Clays Ferry Formation, Kentucky, U.S.A. Palinología, Número Extraordinario 1:293–301.

Jacobson, S. R. 1979. Acritarchs as paleoenvironmental indicators in Middle and Upper Ordovician rocks from Kentucky, Ohio and New York. Journal of Paleontology 53:1197–1212.

Jacobson, S. R. 1987. "Middle Ordovician" acritarchs are guide fossils for the Upper Ordovician. Lethaia 20:91–92.

Jacobson, S. R., and A. Achab. 1985. Acritarch biostratigraphy of the *Dicellograptus complanatus* graptolite Zone from the Vaureal Formation (Ashgillian), Anticosti Island, Quebec, Canada. Palynology 9:165–198.

Jacobson, S. R., J. R. Hatch, S. C. Teerman, and R. A. Askin. 1988. Middle Ordovician organic matter assemblages and their effect on Ordovician-derived oils. Bulletin of the American Association of Petroleum Geologists 72:1090–1100.

James, N. P., D. R. Kobluk, and S. G. Pemberton. 1977. The oldest macroborers: Lower Cambrian of Labrador. Science 197:980–983.

James, N. P., R. K. Stevens, C. R. Barnes, and I. Knight. 1989. Evolution of a Lower Paleozoic continental margin carbonate platform, northern Canadian Appalachians; pp. 123–146 in P. D. Crevello, J. L. Wilson, J. F. Sarg, and J. F. Read (eds.), Controls on Carbonate Platform and Basin Development. Society of Economic Paleontologists and Mineralogists, Special Publication 44.

Janvier, P. 1996a. The dawn of the vertebrates: Characters versus common ascent in the rise of current vertebrate phylogenies. Palaeontology 39:259–287.

Janvier, P. 1996b. Early vertebrates. Clarendon Press, Oxford, 393 pp.

Janvier, P. 1998. Les Vertébrés avant le Silurien. Geobios 30:931–950.

Jardiné, S., A. Combaz, L. Magloire, G. Peniguel, and G. Vachey. 1974. Distribution stratigraphique des Acritarches dans le Paléozoïque du Sahara Algérien. Review of Palaeobotany and Palynology 18:99–129.

Jefferies, R. P. S., and P. E. J. Daley. 1996. Calcichordates; pp. 268–276 in D. A. T. Harper and A. W. Owen (eds.), Fossils of the Upper Ordovician. Palaeontological Association (London). Field Guide to Fossils 7.

Jell, P. A. 1979. *Plumulites* and the machaeridian problem. Alcheringa 3:253–259.

Jell, P. A. 1980. Two arthropods from the Lancefieldian (La 1) of central Victoria. Alcheringa 4:37–46.

Jell, P. A. 1985. Tremadoc trilobites of the Digger Island Formation, Waratah Bay, Victoria. Museum of Victoria, Memoir 46:53–88.

Jell, P. A., and B. Stait. 1985a. Tremadoc trilobites from the Florentine Valley Formation, Tim Shea area, Tasmania. Museum of Victoria, Memoir 46:1–34.

Jell, P. A., and B. Stait. 1985b. Revision of an early Arenig trilobite faunule from the Caroline Creek Sandstone, near Latrobe, Tasmania. Museum of Victoria, Memoir 46:35–51.

Jenkins, W. A. M. 1967. Ordovician Chitinozoa from Shropshire. Palaeontology 10:436–488.

Jeppson, L. 1990. An oceanic model for lithological and faunal changes tested on the Silurian record. Journal of the Geological Society of London 147:663–674.

Jerre, F. 1994a. Anatomy and phylogenetic significance of *Eoconularia loculata* (Wiman), a Silurian conulariid from Gotland. Lethaia 27:97–109.

Jerre, F. 1994b. Taxonomy and functional morphology of Silurian conulariids from Gotland. Lund Publications in Geology 117:1–33.

Ji, Z., and C. R. Barnes. 1993. A major conodont extinction event during the Early Ordovician within the Midcontinent Realm. Palaeogeography, Palaeoclimatology, Palaeoecology 104:37–47.

Ji, Z., and C. R. Barnes. 1994a. Conodont paleoecology of the Lower Ordovician St. George Group, Port au Port Peninsula, western Newfoundland. Journal of Paleontology 68:1368–1383.

Ji, Z., and C. R. Barnes. 1994b. Lower Ordovician conodonts of the St. George Group, Port au Port Peninsula, western Newfoundland, Canada. Palaeontographica Canadiana 11:1–149.

Ji, Z., and C. R. Barnes. 1996. Uppermost Cambrian and Lower Ordovician conodont biostratigraphy of the Survey Peak Formation (Ibexian/Tremadoc), Wilcox Pass, Alberta. Journal of Paleontology 70:871–890.

Jin, J. 2001. Evolution and extinction of the North American *Hiscobeccus* brachiopod fauna during the Late Ordovician. Canadian Journal of Earth Sciences 38:143–151.

Jin, J., and R.-B. Zhan. 2001. Late Ordovician Articulate Brachiopods from the Red River and Stony Mountain Formations, Southern Manitoba. NRC Research Press, Ottawa, 117 pp.

Johns, R. A. 1994. Ordovician lithistid sponges of the Great Basin. Nevada Bureau of Mines and Geology, Open-File Report 1994-1:1–160.

Johnson, E. W., D. E. G. Briggs, R. J. Suthren, J. L. Wright, and S. P. Tunncliff. 1994. Non-marine arthropod traces from the subaerial Ordovician Borrowdale Volcanic Group, English Lake District. Geological Magazine 131:395–406.

Johnson, H. P., D. Van Patten, M. Tivey, and W. W. Sager. 1995. Geomagnetic polarity reversal rate for the Phanerozoic. Geophysical Research Letters 22:231–234.

Johnson, R. E., and P. M. Sheehan. 1985. Late Ordovician dasyclad algae of the Eastern Great Basin; pp. 79–84 in D. F. Toomey and M. H. Nitecki (eds.), Paleoalgology: Contemporary Research and Applications. Springer-Verlag, New York.

Johnston, D. I., and C. R. Barnes. 1999. Early and middle Ordovician (Arenig) conodonts from St. Pauls Inlet and Martin Point, Cow Head Group, western Newfoundland, Canada. 1. Biostratigraphy and paleoecology. Geologica et Palaeontologica 33:21–70.

Johnston, D. I., and C. R. Barnes. 2000. Early and middle Ordovician (Arenig) conodonts from St. Pauls Inlet and Martin Point, Cow Head Group, western Newfoundland, Canada. 2. Systematic paleontology. Geologica et Palaeontologica 34:11–87.

Jones, C. R. 1986. Ordovician (Llandeilo and Caradoc) Beyrichiocope Ostracoda from England and Wales. Palaeontographical Society Monograph 138(569):1–76.

Jones, C. R. 1987. Ordovician (Llandeilo and Caradoc) Beyrichiocope Ostracoda from England and Wales, Part 2. Palaeontographical Society Monograph 139(571): 77–114.

Jull, R. K. 1976. Review of some species of *Favistina, Nyctopora,* and *Calapoecia* (Ordovician corals from North America). Geological Magazine 113:457–467.

Kaesler, R. L. (ed.). 2000. Treatise on Invertebrate Paleontology. Part H, Brachiopoda, rev., Vol. 1. Geological Society of America, Boulder, Colorado, and University of Kansas Press, Lawrence.

Kaljo, D. 1958. Some new and little-known Baltic tetracorals. Institute of Geology, Academy of Sciences, Estonian SSR, Uurimused 3:101–123 (in Russian with English summary).

Kaljo, D. 1961. Stratigraphical importance of the Ordovician and Llandoverian rugose corals of the Estonian SSR; pp. 49–56 *in* K. Orviku (ed.), Geoloogiline kogumik. Estonian Society of Naturalists, Tartu (in Estonian with English summary).

Kaljo, D. 1996. Diachronous recovery patterns in Early Silurian corals, graptolites and acritarchs; pp. 127–133 *in* M. B. Hart (ed.), Biotic Recovery from Mass Extinction Events. The Geological Society, London, Special Publication 102.

Kaljo, D., and E. Klaamann. 1973. Ordovician and Silurian corals; pp. 37–45 *in* A. Hallam (ed.), Atlas of Palaeobiogeography. Elsevier, Amsterdam.

Kaljo, D., J. Nõlvak, and A. Uutela. 1996. More about Ordovician microfossil diversity patterns in the Rapla section, northern Estonia. Proceedings of the Estonian Academy of Sciences, Geology 45:131–148.

Kaljo, D., L. Hints, T. Martma, and J. Nõlvak. 2001. Carbon isotope stratigraphy in the latest Ordovician of Estonia. Chemical Geology 175:49–59.

Kammer, T. W., C. E. Brett, D. R. Boardman, and R. H. Mapes. 1986. Ecologic stability of the dysaerobic biofacies during the Late Paleozoic. Lethaia 19:109–121.

Kampschulte, A. 2001. Schwefelisotopenuntersuchungen an strukturell substituierten Sulfaten in marinen Karbonaten des Phanerozoikums—Implikationen fur die geochemische Evolution des Meerwassers und die Korrelation verschiedener Stoffkreisläufe. Doctoral thesis, Ruhr-Universität Bochum, Germany.

Kampschulte, A., and H. Strauss. In press. The sulfur isotopic composition of Phanerozoic seawater based on the analysis of structurally substituted sulfates in carbonates. Chemical Geology.

Kanygin, A. V. 1967. Ostrakody ordovika gornoj sistema Cherskogo. Nauka, Moscow, 154 pp.

Kanygin, A. V. 1971. Ostrakody i biostratigrafiya ordovika khrebta Sette-Daban (Verkhoyanskaya gornaya sistema). Akademia Nauk SSSR, Sibirskoe Otdelenie, Trudy Instituta Geologii i Geofiziki 128:1–110.

Kanygin, A. V. 1977. Ostrakody ordovika Chukotskogo poluostrova. Akademia Nauk SSSR, Sibirskoe Otdelenie, Trudy Instituta Geologii i Geofiziki 351:73–86.

Kanygin, A. V. 2001. The Ordovician explosive divergence of the Earth's organic realm: Causes and effects of the biosphere evolution. Russian Geology and Geophysics 42:599–633.

Kaplan, P., and T. K. Baumiller. 2000. Taphonomic inferences on boring habit in the Richmondian *Onniella meeki* epibole. Palaios 15:499–510.

Kaplan, P., and T. K. Baumiller. 2001. A misuse of Occam's Razor that trims more than just the fat. Palaios 16:525–527.

Kapp, U. S., and C. W. Stearn. 1975. Stromatoporoids of the Chazy Group (Middle Ordovician) Lake Champlain, Vermont and New York. Journal of Paleontology 49:163–186.

Karatajute-Talimaa, V. N. 1978. Telodonty Silura i Devona SSSR i Shpitsbergena [Silurian and Devonian thelodonts of the USSR and Spitsbergen]. Mokslas, Vilnius, Lithuania, 334 pp.

Karatajute-Talimaa, V. N. 1997. Taxonomy of loganiid thelodonts; pp. 1–15 *in* S. Turner and A. Blieck (eds.), Gross Symposium, Vol. 2. Modern Geology 21(1–2).

Karatajute-Talimaa, V. N. 1998. Determination methods for the exoskeletal remains of early vertebrates. Mitteilungen aus dem Museum für Naturkunde in Berlin, Geowissenschaftliche Reihe 1:21–52.

Karatajute-Talimaa, V. N., and N. Predtechenskyj. 1995. The distribution of the vertebrates in the Late Ordovician and Early Silurian palaeobasins of the Siberian Platform; pp. 39–55 *in* M. Arsenault, H. Lelièvre, and P. Janvier (eds.), Etudes sur les Vertébrés inférieurs (VIIe Symposium International, Parc de Miguasha, Québec, 1991). Bulletin du Muséum national d'Histoire naturelle, 4e série, 17, C (1–4).

Karhu, J., and S. Epstein. 1986. The implication of the oxygen isotope records in coexisting cherts and phosphates. Geochimica et Cosmochimica Acta 50:1745–1756.

Keller, M. 1999. Argentine Precordillera: Sedimentary and plate tectonic history of a Laurentian crustal fragment in South America. Geological Society of America, Special Paper 341:1–131.

Keller, M., F. Cañas, O. Lehnert, and N. E. Vaccari. 1994. The Upper Cambrian and Lower Ordovician of the Precordillera (Western Argentina): Some stratigraphic reconsiderations. Newsletters on Stratigraphy 31:115–132.

Kemp, A. 2002. Amino acid residues in conodont elements. Journal of Paleontology 76:518–528.

Kemple, W. G., P. M. Sadler, and D. J. Strauss. 1995. Extending graphic correlation to many dimensions: Stratigraphic correlation as constrained optimization; pp. 65–82 *in* K. O. Mann and H. R. Lane (eds.), Graphic Correlation. Society for Sedimentary Geology (SEPM), Special Publication 53.

Keto, L. S., and S. B. Jacobsen. 1987. Nd and Sr isotopic variations of Early Paleozoic oceans. Earth and Planetary Science Letters 84:7–41.

Kidwell, S. M., and P. J. Brenchley. 1994. Patterns in bioclastic accumulation through the Phanerozoic: Changes in input or in destruction? Geology 22:1139–1143.

Kielan-Jaworowska, Z. 1966. Polychaete jaw apparatuses from the Ordovician and Silurian of Poland and comparison with modern forms. Palaeontologia Polonica 16:1–152.

King, A. H. 1993. Mollusca: Cephalopoda (Nautiloidea); pp. 169–188 in M. J. Benton (ed.), The Fossil Record 2. Chapman and Hall, London.

King, A. H. 1999. A review of Volkhovian and Kundan (Arenig-Llanvirn) nautiloids from Sweden; pp. 137–159 in F. Oloriz and F. J. Rodriguez-Tovar (eds.), Advancing Research on Living and Fossil Cephalopods. Kluwer Academic/Plenum Press, New York.

Kjellesvig-Waering, E. N., and C. A. Heubusch. 1962. Some eurypterids from the Ordovician and Silurian of New York. Journal of Paleontology 36:211–221.

Klaamann, E. R. 1966. Inkommunikatnye tabulyaty Estonii [Incommunicate Tabulata of Estonia]. Eesti NSV Teaduste Akademia, Geloogia Instituut, Tallinn, Estonia, 96 pp.

Klappa C. F., and N.P. James. 1980. Small lithistid sponge bioherms, early Middle Ordovician Table Head Group of western Newfoundland. Bulletin of Canadian Petroleum Geology 28:435–451.

Klenina, L. N. 1989. Pozdneordovikskiye khitinozoi i skolekodonty Vostachnogo Kazakhstana [Late Ordovician chitinozoans and scolecodonts from eastern Kazakhstan]. Ezhegodnik Vsesoyuznogo Paleontologitsheskogo Obshestva 32:232–249.

Knight, J. B., and E. L. Yochelson. 1958. A reconsideration of the relationships of the Monoplacophora and the primitive Gastropoda. Proceedings of the Malacological Society of London 33:37–48.

Knight, J. B., L. R. Cox, A. Myra, R. L. Batten, E. L. Yochelson, and R. Robertson. 1960. Systematic descriptions (Archaeogastropoda); pp. 310–324 in R. C. Moore (ed.), Treatise on Invertebrate Paleontology. Part 1, Mollusca 1. Geological Society of America, New York, and University of Kansas Press, Lawrence.

Knüpfer, J. 1968. Ostracoden aus dem Oberen Ordovizium Thüringens. Freiberger Forschungshefte C234:63–78.

Kobayashi, T. 1927. Ordovician Fossils from Corea and South Manchuria. Japanese Journal of Geology and Geography 5:173–212.

Kobayashi, T. 1933. Faunal study of the Wanwanian (basal Ordovician) series with special notes on the Ribeiridae and the ellesmeroceroids. Journal of the Faculty of Science, Tokyo Imperial University, Sec. 2, 3:249–328.

Kobayashi, T. 1934. The Cambro-Ordovician formations and faunas of South Chosen (Korea). Part 2, Lower Ordovician faunas. Journal of the Faculty of Science, Tokyo Imperial University, Sec. 2, 3:521–585.

Kobayashi, T. 1954. Fossil estherian and allied fossils. Journal of the Faculty of Science, Tokyo Imperial University, Sec. 2, 9:1–192.

Kobayashi, T., and T. Hamada. 1976. Occurrences of the Machaeridia in Japan and Malaysia. Proceedings of the Japan Academy 52:371–374.

Kobluk, D. R. 1984. *Archaeotrypa* Fritz, 1947 (Cambrian, Problematica) reinterpreted. Canadian Journal of Earth Sciences 21:1343–1348.

Kobluk, D. R., and I. Noor. 1990. Coral microatolls and a probable middle Ordovician example. Journal of Paleontology 64:39–43.

Kobluk, D. R., N. P. James, and S. G. Pemberton. 1978. Initial diversification of macroboring ichnofossils and exploitation of macroboring niche in the lower Paleozoic. Paleobiology 4:163–170.

Koeberl, C., W. U. Reimold, and S. P. Kelley. 2001. Petrography, geochemistry, and argon-40/argon-39 ages of impact-melt rocks and breccias from the Ames impact structure, Oklahoma: The Nicor Chestnut 18-4 drill core. Meteoritics and Planetary Science 36:651–669.

Kolata, D. R. 1973. *Sclenocystites strimplei,* a new Middle Ordovician belemnocystitid solute from Minnesota. Journal of Paleontology 47:969–974.

Kolata, D. R. 1975. Middle Ordovician echinoderms from northern Illinois and southern Wisconsin. The Paleontological Society, Memoir 7 (Journal of Paleontology, 48[5], supplement):1–74.

Kolata, D. R. 1976. Crinoids from the Upper Ordovician Bighorn Formation of Wyoming. Journal of Paleontology 50:444–453.

Kolata, D. R. 1982. Camerates; pp. 170–205 in J. Sprinkle (ed.), Echinoderm Faunas from the Bromide Formation (Middle Ordovician) of Oklahoma. University of Kansas (Lawrence) Paleontological Contributions, Monograph 1.

Kolata, D. R. 1983. *Cataraquicrinus elangatus,* a new disparid inadunate crinoid from the Middle Ordovician of Ontario. Canadian Journal of Earth Sciences 20:1609–1613.

Kolata, D. R. 1986. Crinoids of the Champainian (Middle Ordovician) Guttenberg Formation–upper Mississippi Valley region. Journal of Paleontology 60:711–718.

Kolata, D. R., and T. E. Guensburg. 1979. *Diamphidiocystis,* a new mitrate "carpoid" from the Cincinnatian (Upper Ordovician) Maquoketa Group in southern Illinois. Journal of Paleontology 53:1121–1135.

Kolata, D. R., and M. Jollie. 1982. Anomalocystitid mitrates (Stylophora-Echinodermata) from the Champlainian (Middle Ordovician) Guttenberg Formation of the upper Mississippi Valley region. Journal of Paleontology 56:631–653.

Kolata, D. R., J. C. Brower, and T. J. Frest. 1987. Upper Mississippi Valley Champlainian and Cincinnatian echinoderms; pp. 179–181 *in* R. E. Sloan (ed.), Middle and Late Ordovician Lithostratigraphy and Biostratigraphy of the Upper Mississippi Valley. Minnesota Geological Survey, Report of Investigations 35.

Kolata, D. R., W. D. Huff, and S. M. Bergström. 1996. Ordovician K-bentonites of eastern North America. Geological Society of America, Special Paper 313:1–84.

Kolosnitsyna, G. R. 1984. Ostrakody iz ordovika Ajchal'skogo rajona. Akademia Nauk SSSR, Sibirskoe Otdelenie, Trudy Instituta Geologii i Geofiziki 584: 25–32.

Korde, K. B. 1965. Algae. Typi Rhodophyta i Chlorophyta. Razvitie i zmiena morskih organizmov na rubezhe paleozoia i mezozoya. Akademia Nauk SSSR, Trudy Paleontologichieskogo Instituta (Moscow) 108:268–284.

Korde, K. B. 1973. Vodorosli kembria. Akademia Nauk SSSR, Trudy Paleontologichieskogo Instituta (Moscow) 139:1–349.

Koren, T. N. 1991. Evolutionary crisis of the Ashgill graptolites; pp. 157–164 *in* C. R. Barnes and S. H. Williams (eds.), Advances in Ordovician Geology. Geological Survey of Canada, Paper 90-9.

Körts, A. 1992. Ordovician oil shale of Estonia: Origin and palaeoecological characteristics; pp. 445–454 *in* B. D. Webby and J. R. Laurie (eds.), Global Perspectives on Ordovician Geology. Balkema, Rotterdam.

Körts, A., and E. Mark-Kurik. 1997. Algae and vascular plants; pp. 213–215 *in* A. Raukas and A. Teedumäe (eds.), Geology and Mineral Resources of Estonia. Institute of Geology, Estonian Academy Publishers, Tallinn.

Koslowski, R. 1967. Sur certains fossils ordoviciens a test organique. Acta Palaeontologica Polonica 12:99–132.

Koukharsky, M., R. Torres Claro, M. Echeverría, N. E. Vaccari, and B. G. Waisfeld. 1996. Episodios volcánicos del Tremadociano y Arenigiano en Vega Pinato, Puna salteña, Argentina. XIII Congreso Geológico Argentino (Buenos Aires), Actas 5:535–542.

Kowalewski, M., F. Dulai, and F. T. Fürsich. 1998. A fossil record full of holes: The Phanerozoic history of drilling predation. Geology 26:1091–1094.

Kowalewski, M., F. Dulai, and F. T. Fürsich. 1999. A fossil record full of holes: The Phanerozoic history of drilling predation. Reply. Geology 27:957–960.

Kozur, H. W., H. Mostler, and J. E. Repetski. 1996. Wellpreserved Tremadocian primitive Radiolaria from the Windfall Formation of the Antelope Range, Eureka County, Nevada, U.S.A. Geologisch-Palaontologische Mitteilunge Innsbruck 21:245–271.

Kraft, P., and O. Fatka (eds.). 1999. *Quo vadis* Ordovician? Short papers of the 8th International Symposium on the Ordovician System. Acta Universitatis Carolinae, Geologica 43(1–2), 534 pp.

Kraft, P., and M. Mergl. 1989. Worm-like fossils (Palaeoscolecida; ?Chaetognatha) from the Lower Ordovician of Bohemia. Sborník geologickych Věd, Paleontologie 30:9–36.

Kraft, P., O. Lehnert, and J. Frýda. 1999. *Titerina,* a living fossil in the Ordovician: A young protoconodont (?) and the oldest chaetognath; pp. 451–454 *in* P. Kraft and O. Fatka (eds.), *Quo vadis* Ordovician? Short papers of the 8th International Symposium on the Ordovician System. Acta Universitatis Carolinae, Geologica 43(1–2).

Krandievsky, V. S. 1969. Stratigraphic distribution of Ostracoda in the Ordovician deposits of the VolynPodolye. Dopovidi Akademia Nauk Ukrains'koi RSR (B) 1969:870–874.

Krause, F. F., and A. J. Rowell. 1975. Distribution and systematics of the inarticulate brachiopods of the Ordovician carbonate mud mound, Meiklejohn Peak, Nevada. University of Kansas Paleontological Contributions 61: 1–74.

Kruse, P. D. 1987. Further Australian Cambrian sphinctozoans. Geological Magazine 124:543–553.

Kruse, P. D. 1996. Update on the northern Australian Cambrian sponges *Rankenella, Jawonya* and *Wagina.* Alcheringa 20:161–178.

Kruse, P. D. 1997. Hyolith guts in the Cambrian of northern Australia—turning hyolithomorphs upside down. Lethaia 29:213–217.

Kump, L. R., M. A. Arthur, M. E. Patzkowsky, M. T. Gibbs, D. S. Pinkus, and P. M. Sheehan. 1999. A weathering hypothesis for glaciation at high atmospheric pCO_2 during the Late Ordovician. Palaeogeography, Palaeoclimatology, Palaeoecology 152:173–187.

Kunk, M. J., J. Sutter, J. D. Obradovitch, and M. A. Lanphere. 1985. Age of biostratigraphic horizons within the Ordovician and Silurian Systems; pp. 89–92 *in* N. J. Snelling (ed.), The Chronology of the Geological Record. The Geological Society, London, Memoir 10.

Land, L. S. 1995. Oxygen and carbon isotopic composition of Ordovician brachiopods: Implications for coeval seawater: Discussion. Geochimica et Cosmochimica Acta 59:2843–2844.

Landing, E., R. Ludvigsen, and P. H. Von-Bitter. 1980. Upper Cambrian to Lower Ordovician conodont biostratigraphy and biofacies, Rabbitkettle Formation, district of Mackenzie, Northwest Territories, Canada. Royal Ontario Museum, Life Sciences Contributions 126: 1–42.

Landing, E., C. R. Barnes, and R. K. Stevens. 1986. Tempo of early Ordovician graptolite faunal succession: Conodont-based correlations from the Tremadocian of Quebec. New York State Science Service, Journal Series, Paper 482:1928–1949.

Landing, E., S. R. Westrop, and L. A. Knox. 1996. Conodonts, stratigraphy, and relative sea level changes in the Tribes Hill Formation (Lower Ordovician, east-central New York). Journal of Paleontology 70:656–680.

Landing, E., S. A. Bowring, R. A. Fortey, and K. L. Davidek. 1997. U-Pb zircon date from Avalonian Cape Breton Island and geochronological calibration of the Early Ordovician. Canadian Journal of Earth Sciences 34:724–730.

Landing, E., S. A. Bowring, K. L. Davidek, A. W. A. Rushton, R. A. Fortey, and W. A. P. Wimbledon. 2000. Cambrian-Ordovician boundary age and duration of the lowest Tremadoc Series, based on U-Pb zircon dates from Avalonian Wales. Geological Magazine 137:485–494.

Lapworth, C. 1879. On the tripartite classification of the Lower Palaeozoic. Geological Magazine 26:1–15.

Lardeux, H. 1969. Les Tentaculites d'Europe occidentale et d'Afrique du Nord. Cahiers de Paléontologie du Centre National de la Recherche Scientifique (C.N.R.S.), Paris, 238 pp.

Larson, R. L. 1991. Latest pulse of Earth: Evidence for a mid-Cretaceous superplume. Geology 19:547–550.

Larsson, K. 1979. Silurian Tentaculitids from Gotland and Scania. Fossils and Strata 11:1–180.

Larwood, G. P., and P. D. Taylor. 1979. Early structural and ecological diversification in the Bryozoa; pp. 209–234 in M. R. House (ed.), The Origin of Major Invertebrate Groups. Academic Press, London.

Lasaga, A. C., and H. Ohmoto. 2002. The oxygen geochemical cycle: Dynamics and stablity. Geochimica et Cosmochimica Acta 66:361–381.

Laub, R. S. 1984. *Lichenaria* Winchell and Schuchert, 1895, *Lamottia* Raymond, 1924 and the early history of the tabulate corals. Palaeontographica Americana 54:159–163.

Laurie, J. R., and J. H. Shergold. 1996a. Early Ordovician trilobite taxonomy and biostratigraphy of the Emanuel Formation, Canning Basin, Western Australia, Part 1. Palaeontographica, Abteilung A 240:65–103.

Laurie, J. R., and J. H. Shergold. 1996b. Early Ordovician trilobite taxonomy and biostratigraphy of the Emanuel Formation, Canning Basin, Western Australia, Part 2. Palaeontographica, Abteilung A 240:105–144.

Law, R. H., and C. W. Thayer. 1990. Articulate fecundity in the Phanerozoic: Steady state or what? pp. 183–190 in D. I. MacKinnon, D. E. Lee, and J. D. Campbell (eds.), Brachiopods through Time. Balkema, Rotterdam.

Lawton, J. H., and R. M. May (eds.). 1995. Extinction Rates. Oxford University Press, Oxford, 233 pp.

Lee, D.-J., and R. J. Elias. 2000. Paleobiologic and evolutionary significance of corallite increase and associated features in *Saffordophyllum newcombae* (Tabulata, Late Ordovician, southern Manitoba). Journal of Paleontology 74:404–425.

Lee, H.-Y., and S.-J. Lee. 1986. Conodont biostratigraphy of the Jigunsan Shale and Duwibong Limestone in the Nokjeon-Sangdong area, Yeongweol-Gun, Kangweondo, Korea. Journal of the Geological Society of Korea 2:114–136.

Lee, R. E. 1980. Phycology. Cambridge University Press, Cambridge, 478 pp.

Lees, D. C., R. A. Fortey, and L. R. M. Cocks. 2002. Quantifying paleogeography using biogeography: A test case for the Ordovician and Silurian of Avalonia based on brachiopods and trilobites. Paleobiology 28:343–363.

Le Fèvre, J., C. R. Barnes, and M. Tixier. 1976. Paleoecology of Late Ordovician and Early Silurian conodontophorids, Hudson Bay Basin; pp. 69–89 in C. R. Barnes (ed.), Conodont Paleoecology. Geological Association of Canada, Special Paper 15.

Legall, R. D., C. R. Barnes, and R. W. Macqueen. 1982. Organic metamorphism, burial history and hotspot development, Paleozoic strata of southern Ontario-Quebec, from conodont and acritarch alteration studies. Bulletin of Canadian Petroleum Geology 29:492–539.

Legg, D. P. 1976. Ordovician trilobites and graptolites from the Canning Basin, Western Australia. Geologica et Palaeontologica 10:1–58.

Leggett, J. K. 1980. British Lower Palaeozoic black shales and their palaeo-oceanographic significance. Journal of the Geological Society of London 137:139–156.

Legrand, P. 1995. Evidence and concerns with regard to the late Ordovician glaciation in North Africa; pp. 165–169 in J. D. Cooper, M. L. Droser, and S. C. Finney (eds.), Ordovician Odyssey: Short Papers, 7th International Symposium on the Ordovician System. Book 77, Pacific Section Society for Sedimentary Geology (SEPM), Fullerton, California.

Lehmann, D., and J. K. Pope. 1987. Tidal flat and shallow subtidal fauna from the Upper Ordovician, Martinsburg Formation, Swatara Gap, Lebanon County, Pennsylvania. Geological Society of America, Abstracts with Programs 19(1):25.

Lehnert, O. 1995a. Ordovizische Conodonten aus der Präkordillere Westargentiniens: Ihre Bedeutung für Stratigraphie und Paläogeographie. Erlanger Geologische Abhandlungen 125:1–193.

Lehnert, O. 1995b. The Tremadoc/Arenig transition in the Argentine Precordillera; pp. 145–148 in J. D. Cooper, M. L. Droser, and S. C. Finney (eds.), Ordovician Odyssey: Short Papers, 7th International Symposium on the Ordovician System. Book 77, Pacific Section Society for Sedimentary Geology (SEPM), Fullerton, California.

Lehnert, O., M. Keller, and O. Bordonaro. 1998. Early Ordovician conodonts from the southern Cuyania terrane (Mendoza Province, Argentina). Palaeontologia Polonica 58:47–65.

Lehnert, O., I. Hinz-Schallreuter, and H.-H. Krueger. 1999. Paläozoische Conodontenfaunen aus eiszeitlichen Geschieben Norddeutschlands (I): Eine Mittelordoviz-Fauna aus Rügen. Natur und Mensch '98 (Jahresmitteilungen der Naturhistorischen Gesellschaft Nürnberg): 29–44.

Leighton, L. R. 2001. New directions in the paleoecology of Paleozoic brachiopods; pp. 185–205 in S. J. Carlson and M. R. Sandy (eds.), Brachiopods Ancient and Modern: A Tribute to G. Arthur Cooper. The Paleontological Society (New Haven, Connecticut), Paper 7.

Leighton, L. R., and M. Savarese. 1996. Functional and taphonomic implications of Ordovician strophomenid brachiopod morphology; pp. 161–168 in P. Copper and J. Jin (eds.), Brachiopods. Proceedings of the Third International Brachiopod Congress, Sudbury, Ontario, Canada. Balkema, Rotterdam.

Leslie, S. A. 2000. Mohawkian (Upper Ordovician) conodonts of eastern North America and Baltoscandia. Journal of Paleontology 74:1122–1147.

Leslie, S. A., and S. M. Bergström. 1995. Revision of the North American late Middle Ordovician standard stage classification and timing of the Trenton transgression based on K-bentonite bed correlation; pp. 49–54 in J. D. Cooper, M. L. Droser, and S. C. Finney (eds.), Ordovician Odyssey: Short Papers, 7th International Symposium on the Ordovician System. Book 77, Pacific Section Society for Sedimentary Geology (SEPM), Fullerton, California.

Leszczyński, S. 1991. Trace fossil tiering in flysch sediments: Examples from the Guipuzcoan flysch (Cretaceous-Paleogene), northern Spain. Palaeogeography, Palaeoclimatology, Palaeoecology 88:167–184.

Levin, L. A. 2002. Deep-ocean life where oxygen is scarce. American Scientist 90:436–444.

Lewis, R. D. 1982. Depositional environments and paleoecology of the Oil Creek Formation (Middle Ordovician), Arbuckle Mountains and Criner Hills, Oklahoma. Doctoral thesis, University of Texas at Austin.

Lewis, R. D., J. Sprinkle, J. B. Bailey, J. Moffit, and R. L. Parsley. 1987. *Mandalacystis,* a new rhipidocystid eocrinoid from the Whiterockian Stage (Ordovician) in Oklahoma and Nevada. Journal of Paleontology 61: 1222–1235.

Li, H. 1995. New genera and species of Middle Ordovician Nassellaria and Albaillellaria from Baijingsi, Qilian Mountains, China. Scientia Geologica Sinica 4:331–346 (in Chinese).

Li, J. 1987. Ordovician acritarchs from the Meitan Formation of Guizhou Province, south-west China. Palaeontology 30:613–634.

Li, J. 1989. Early Ordovician Mediterranean province acritarchs from Upper Yangtze Region, China; pp. 231–234 in Developments in Geoscience: Contribution to the 28th Geological Congress 1989, Washington, D.C., U.S.A. Chinese Academy of Science, Beijing.

Li, J. 1991. The Early Ordovician acritarchs from Southwest China. Doctoral thesis, Nanjing Institute of Geology and Palaeontology, Academia Sinica, Nanjing (in Chinese with English summary).

Li, J. 1995. Ordovician (Caradoc) acritarchs from Qilang Formation of Kalpin, Xinjiang, China. Acta Palaeontologica Sinica 34:454–467 (in Chinese with English abstract).

Li, J., and T. Servais. 2002. The Ordovician acritarchs of China and their implication in the global palaeobiogeography. Bulletin de la Société Géologique de France 173:399–406.

Li, J., R. Brocke, and T. Servais. 2002a. The acritarchs of the South Chinese *Azygograptus suecicus* graptolite Biozone and their bearing on the definition of the Lower/Middle Ordovician boundary. CR Palevol 1:75–81.

Li, J., T. Servais, and R. Brocke. 2002b. Chinese Palaeozoic acritarch research: Review and perspectives. Review of Palaeobotany and Palynology 118:181–193.

Li, X., and M. L. Droser. 1999. The nature and distribution of Ordovician shell beds: Evidence from the Great Basin of Nevada, Utah, and California. Palaios 14:215–233.

Lin, B.-Y. 1983. Ordovician tabulate corals of China. Acta Palaeontologica Sinica 22:487–491 (in Chinese with English abstract).

Lin, B.-Y., and B. D. Webby. 1989. Biogeographic relationships of Australian and Chinese Ordovician corals and stromatoporoids. Association of Australasian Palaeontologists, Memoir 8:207–217.

Lindholm, R. M., and J. F. Casey. 1990. The distribution and possible biostratigraphic significance of the ichnogenus *Oldhamia* in the shales of the Blow Me Down Brook Formation, western Newfoundland. Canadian Journal of Earth Sciences 27:1270–1287.

Lindström, M. 1971. Lower Ordovician conodonts of Europe. Geological Society of America, Memoir 127: 21–61.

Lindström, M. 1984. Baltoscandic conodont life environments in the Ordovician: Sedimentologic and paleogeographic evidence. Geological Society of America, Special Paper 196:33–42.

Linsley, R. M., and W. M. Kier. 1984. The Paragastropoda: A proposal for a new class of Paleozoic Mollusca. Malacologia 25:241–254.

Lipps, J. H. 1992a. Proterozoic and Cambrian skeletonized protists; pp. 237–240 in W. J. Schopf and C. Klein (eds.), The Proterozoic Biosphere. Cambridge University Press, Cambridge.

Lipps, J. H. 1992b. Origin and early evolution of Foraminifera; pp. 3–9, in T. Saito and T. Takayangi (eds.), Studies in Benthic Foraminifera. Proceedings of the Fourth International Symposium on Benthic Foraminifera "Benthos 90," Sendai, Japan. Tokai University Press, Shimizu.

Lipps, J. H., and P. W. Signor (eds.). 1992. Origin and Early Evolution of the Metazoa. Plenum Press, New York, 570 pp.

Little, H. P. 1936. Ordovician fossils from Labrador. Science 84(2177):268–269.

Liu, B.-L., J. K. Rigby, Y.-W. Jiang, and Z.-D. Zhu. 1997. Lower Ordovician lithistid sponges from the eastern Yangtze Gorge area, Hubei, China. Journal of Paleontology 71:194–207.

Liu, D.-Y., C.-Y. Zhu, and C.-T. Xue. 1985. Ordovician brachiopods from northwestern Xiao Hinggan Ling, northeast China. Bulletin of the Shenyang Institute of Geology and Mineral Resources 11:1–46 (in Chinese).

Lochman, C., and D. Duncan. 1944. Early Upper Cambrian faunas of central Montana. Geological Society of America, Special Paper 54:1–181.

Lockley, M. G. 1993. A review of brachiopod dominated palaeocommunities from the type Ordovician. Palaeontology 26:111–145.

Lockley, M. G., A. K. Rindsberg, and R. M. Zeiler. 1987. The paleoenvironmental significance of the nearshore *Curvolithus* ichnofacies. Palaios 2:255–262.

Loeblich, A. R., Jr., and H. Tappan. 1978. Some Middle and Late Ordovician microplankton from central North America. Journal of Paleontology 50:1233–1287.

Löfgren, A. 1978. Arenigian and Llanvirnian conodonts from Jämtland, northern Sweden. Fossils and Strata 13:1–129.

Löfgren, A. 1993a. Conodonts from the Lower Ordovician at Hunneberg, south-central Sweden. Geological Magazine 130:215–232.

Löfgren, A. 1993b. Arenig conodont successions from central Sweden. GFF [Geologiska Föreningens i Stockhom Förhandlingar] 115:193–207.

Löfgren, A. 1994. Arenig (Lower Ordovician) conodonts and biozonation in the Eastern Siljan District, Central Sweden. Journal of Paleontology 68:1350–1368.

Löfgren, A. 1995. The Middle Lanna/Volkhov Stage (Middle Arenig) in Sweden and its conodont fauna. Geological Magazine 132:693–711.

Löfgren, A. 1996. Lower Ordovician conodonts, reworking, and biostratigraphy of the Orreholmen quarry, Västergötland, south-central Sweden. GFF [Geologiska Föreningens i Stockhom Förhandlingar] 118:169–183.

Löfgren, A. 1997. Conodont faunas from the upper Tremadoc at Brattefors, south-central Sweden, and reconstruction of the *Paltodus* apparatus. GFF [Geologiska Föreningens i Stockhom Förhandlingar] 119:257–266.

Löfgren, A. 2000. Early to early middle Ordovician conodont biostratigraphy of the Gillberga quarry, northern Öland, Sweden. GFF [Geologiska Föreningens i Stockhom Förhandlingar] 122:321–338.

Löfgren, A., J. E. Repetski, and R. L. Ethington. 1999. Some trans-Iapetus faunal conections in the Tremadocian. Bollettino della Società Paleontologica Italiana 37:159–173.

Long, J. A. 1993. Early-Middle Palaeozoic vertebrate extinction events; pp. 54–63 in J. A. Long (ed.), Palaeozoic Vertebrate Biostratigraphy and Biogeography. Belhaven Press, London.

Long, J. A., and C. F. Burrett. 1989. Tubular phosphatic microproblematica from the Early Ordovician of China. Lethaia 22:439–446.

Lu, Y.-H. 1975. Ordovician trilobite faunas of central and southwestern China. Palaeontologia Sinica, n.s., B, 11:1–463 (in Chinese and English).

Luchinina, V. A. 1987. Sinezelenye vodorosli (Cyanobacteria); pp. 12–38 in V. N. Dubatopov, Iskopaemye Isvstkove Vodorosli. Akademia Nauk SSSR Sibirskoe Otdelenie, Trudy Instituta Geologii i Geofiziki, bypusk 674.

Ludvigsen, R. 1978. Middle Ordovician trilobite biofacies, southern Mackenzie Mountains; pp. 1–37 in C. R. Stelck and B. D. E. Chatterton (eds.), Western and Arctic Canadian Biostratigraphy. Geological Association of Canada, Special Paper 18.

Ludvigsen, R. 1982. Upper Cambrian and Lower Ordovician trilobite biostratigraphy of the Rabbitkettle Formation, Western District of Mackenzie. Royal Ontario Museum, Life Sciences Publications 134:1–188.

Ludvigsen, R., and S. R. Westrop. 1983. Trilobite biofacies in Cambrian-Ordovician boundary interval in northern North America. Alcheringa 7:301–319.

Ludvigsen, R., S. R. Westrop, and C. H. Kindle. 1989. Sunwaptan (Upper Cambrian) trilobites of the Cow Head Group, western Newfoundland. Palaeontographica Canadiana 6:1–175.

MacArthur, R. H., and E. O. Wilson. 1967. The Theory of Island Biogeography. Princeton University Press, Princeton, 203 pp.

MacNaughton, R. B., and G. M. Narbonne. 1999. Evolution and ecology of Neoproterozoic-Lower Cambrian trace fossils, NW Canada. Palaios 14:97–115.

MacNaughton, R. B., G. M. Narbonne, and R. W. Dalrymple. 2000. Neoproterozoic slope deposits, Mackenzie Mountains, northwestern Canada: Implications for passive-margin development and Ediacaran faunal ecology. Canadian Journal of Earth Sciences 37:997–1020.

MacNaughton, R. B., J. M. Cole, R. W. Dalrymple, S. J. Braddy, D. E. G. Briggs, and T. D. Lukie. 2002. First steps on land: Arthropod trackways in Cambrian-Ordovician eolian sandstone, southeastern Ontario, Canada. Geology 30:391–394.

Mac Niocaill, C. 1996. Secular increase of nutrient levels through the Phanerozoic: Implications for productivity, biomass, and diversity of the biosphere. Palaios 11:209–219.

Mac Niocaill, C. 2001. Palaeozoic paleogeography and the evolution of the Caledonian-Applachian Orogen. http://www.earth.ox.ac.uk/~conallm.

Mac Niocaill, C., B. A. van der Pluijm, and R. Van der Voo. 1997. Ordovician paleogeography and the evolution of the Iapetus ocean. Geology 25:159–162.

Maletz, J. 1995. The Middle Ordovician (Llanvirn) graptolite succession of the Albjära core (Scania, Sweden) and its implication for a revised biozonation. Zeitschrift für geologische Wissenschaften 23:249–259.

Maletz, J. 1997. Arenig biostratigraphy of the Pointe-de-Levy slice, Quebec Appalachians, Canada. Canadian Journal of Earth Sciences 34:731–752.

Maletz, J., and S. O. Egenhoff. 2001. Late Tremadoc to early Arenig graptolite faunas of southern Bolivia and their implications for a worldwide biozonation. Lethaia 34:47–62.

Maletz, J., A. Löfgren, and S. M. Bergström. 1996. Proposal for the adoption of the Diabasbrottet section (Hunneberg, Västergötland) as a Global Stratotype Section (GSSP) for the second Series of the Ordovician System. Newsletters on Stratigraphy 34:129–159.

Malinky, J. M. 1990. *Solenotheca*, new Hyolitha (Mollusca) from the Ordovician of North America. Proceedings of the Biological Society of Washington 103:265–278.

Malinky, J. M. 2002. A revision of Early to Middle Ordovician hyoliths from Sweden. Palaeontology 45:511–555.

Malinky, J. M. 2003. Ordovician and Silurian hyoliths and gastropods reassigned from the Hyolitha from the Girvan District, Scotland. Journal of Paleontology 77:625–645.

Malinky, J. M., and V. Berg-Madsen. 1999. A revision of Holm's Early and early Mid Cambrian hyoliths of Sweden. Palaeontology 42:25–65.

Malinky, J. M., and R. H. Mapes. 1983. First occurrences of Hyolitha (Mollusca) in the Pennsylvanian of North America. Journal of Paleontology 57:347–352.

Malinky, J. M., and S. Sixt. 1990. Early Mississippian Hyolitha from northern Iowa. Palaeontology 33:343–357.

Malvina, R. G., A. H. Knoll, and R. Siever. 1990. Secular change in chert distribution: A reflection of evolving biological participation in the silica cycle. Palaios 4:519–532.

Mángano, M. G., and L. A. Buatois. 1995. Biotic response to volcanic and sedimentologic processes in a Gondwanic active plate margin basin: The Arenig-Llanvirn Suri Formation, Famatina Basin, northwest Argentina; pp. 229–232 *in* J. D. Cooper, M. L. Droser, and S. C. Finney (eds.), Ordovician Odyssey: Short Papers, 7th International Symposium on the Ordovician System. Book 77, Pacific Section Society for Sedimentary Geology (SEPM), Fullerton, California.

Mángano, M. G., and L. A. Buatois. 1996. Shallow marine event sedimentation in a volcanic arc-related setting: The Ordovician Suri Formation, Famatina Range, northwest Argentina (Famatina System). Sedimentary Geology 105:63–90.

Mángano, M. G., and L. A. Buatois. 1997. Slope apron deposition in an Ordovician arc-related setting, Chaschuil area, Famatina Basin, Northwest Argentina. Sedimentary Geology 109:155–180.

Mángano, M. G., and L. A. Buatois. 2001. The *Syringomorpha* ichnofabric: Pervasive bioturbation and the Cambrian Explosion. VI International Ichnofabric Workshop, Isla Margarita and Puerto La Cruz, Venezuela, Abstracts, 31.

Mángano, M. G. and L. A. Buatois. 2003. Trace Fossils; pp. 507–553 *in* J. L. Benedetto (ed.), Ordovician Fossils of Argentina. Secretaría de Ciencia y Tecnología, Universidad Nacional de Córdoba, Córdoba.

Mángano, M. G., L. A. Buatois, and G. F. Aceñolaza. 1996a. Trace fossils and sedimentary facies from an Early Ordovician tide-dominated shelf (Santa Rosita Formation, northwest Argentina): Implications for ichnofacies models of shallow marine successions. Ichnos 5:53–88.

Mángano, M. G., L. A. Buatois, and F. G. Aceñolaza. 1996b. Icnología de ambientes marinos afectados por vulcanismo: La Formación Suri, Ordovícico del extremo norte de la sierra de Narváez, Sistema del Famatina. Asociación Paleontológica Argentina, Publicación Especial 4:69–88.

Mángano, M. G., L. A. Buatois, and M. C. Moya. 2001a. Trazas fósiles de trilobites de la Formación Mojotoro (Ordovícico inferior-medio de Salta, Argentina): Implicancias paleoecológicas, paleobiológicas y bioestratigráficas. Revista Española de Paleontología 16:9–28.

Mángano, M. G., L. A. Buatois, and F. Muñiz-Guinea. 2001b. Stop 2B: The Upper Cambrian to Tremadoc Santa Rosita Formation at Angosto de Chucalezna; pp. 50–54 *in* L. A. Buatois and M. G. Mángano (eds.), Ichnology, Sedimentology and Sequence Stratigraphy of Selected Lower Paleozoic, Mesozoic and Cenozoic Units of Northwest Argentina. Fourth Argentinean Ichnologic Meeting and Second Ichnologic Meeting of Mercosur, San Miguel de Tucumán, Argentina, Field Guide.

Männil, R. 1959. Voprosy stratigrafii i Mshanki Ordovika Estonii [Problems in the stratigraphy and Bryozoa of the Ordovician of Estonia]. Doklady Akademia Nauk Éstontkoi SSR 1959:1–39.

Männil, R., and T. Meidla. 1994. The Ordovician System of the East European Platform (Estonia, Latvia, Lithuania, Byelorussia, parts of Russia, Ukraine, Moldavia). International Union of Geological Sciences, Publication 28:1–52. Boulder, Colorado.

Marek, J. 1999. Ordovician cephalopods of the Prague Basin (Barrandian area, Czech Republic): A review; pp. 413–416 *in* P. Kraft and O. Fatka (eds.), *Quo vadis* Ordovician? Short papers of the 8th International Symposium on the Ordovician System. Acta Universitatis Carolinae, Geologica 43(1–2).

Marek, L. 1963. New knowledge on the morphology of *Hyolithes*. Sborník geologických Věd, Paleontologie 1:53–72.

Marek, L. 1966. New hyolithid genera from the Ordovician of Bohemia. Časopis národního muzea v Praze 135:89–92.

Marek, L. 1967. The class Hyolitha in the Caradoc of Bohemia. Sborník geologických Věd, Paleontologie 9:51–113.

Marek, L. 1976. The distribution of the Mediterranean Ordovician Hyolitha; pp. 491–499 *in* M. G. Bassett (ed.), The Ordovician System: Proceedings of a Palaeontological Association Symposium, Birmingham (England), September, 1974. University of Wales Press and National Museum of Wales, Cardiff.

Marek, L. 1983. The Ordovician hyoliths of Anti-Atlas (Morocco). Sborník národního muzea v Praze 39:1–36.

Marek, L. 1989. The hyoliths of the Králuv Dvur Formation (Bohemian Ordovician). Sborník geologických Věd, Paleontologie 30:37–59.

Marek, L., and E. L. Yochelson. 1976. Aspects of the biology of Hyolitha (Mollusca). Lethaia 9:65–81.

Marek, L., R. L. Parsley, and A. Galle. 1997. Functional morphology of hyoliths based on flume studies. Věstník Českého geologického ústavu 72:351–358.

Markov, A. V., E. B. Naimark, and R. V. Goryunova. 1998. Quantitative regularities in the evolution of Paleozoic bryozoans: Results of the system analysis. Paleontological Journal 32:493–502.

Marshall, C. R. 1990. Confidence intervals on stratigraphic ranges. Paleobiology 16:1–10.

Marshall, J. D., and J. D. Middleton. 1990. Changes in marine isotopic composition and the late Ordovician glaciation. Journal of the Geological Society of London 147:1–4.

Märss, T., and V. N. Karatajute-Talimaa. 2002. Ordovician and Lower Silurian thelodonts from Severnaya Zemlya Archipelago (Russia). Geodiversitas 24(2):381–404.

Martin, A. J. 1993. Semiquantitative and statistical analysis of bioturbate textures, Sequatchie Formation (Upper Ordovician), Georgia and Tennessee, USA. Ichnos 2:117–136.

Martin, F. 1980. Quelques chitinozoaires et acritarches ordoviciens supérieurs de la formation de White Head en Gaspésie, Québec. Canadian Journal of Earth Sciences 17:106–117.

Martin, F. 1983. Chitinozoaires et acritarches ordoviciens de la plate-forme du Saint-Laurent (Québec et sud-est de l'Ontario). Geological Survey of Canada, Bulletin 310:1–59.

Martin, F. 1984. New Ordovician (Tremadoc) acritarch taxa from the middle member of the Survey Peak Formation at Wilcox Pass, southern Canadian Rocky Mountains, Alberta. Geological Survey of Canada, Current Research, Part A, Paper 84-1A:441–448.

Martin, F. 1992. Uppermost Cambrian and Lower Ordovician acritarchs and Lower Ordovician chitinozoans from Wilcox Pass, Alberta. Geological Survey of Canada, Bulletin 420:1–57.

Martin, F., and W. T. Dean. 1981. Middle and Upper Cambrian and Lower Ordovician acritarchs from Random Island, Eastern Newfoundland. Geological Survey of Canada, Bulletin 343:1–43.

Martin, F., and Yin Leiming. 1988. Early Ordovician acritarchs from southern Jilin Province, north-east China. Palaeontology 31:109–127.

Martin, R. E. 1996. Secular increase of nutrient levels through the Phanerozoic: Implications for productivity, biomass, and diversity of the biosphere. Palaios 11:209–219.

Martin, R. E. 1998. Catastrophic Fluctuations in Nutrient Levels as an Agent of Mass Extinction: Upward Scaling of Ecological Processes? pp. 405–429 *in* M. L. McKinney and J. A. Drake (eds.), Biodiversity Dynamics: Turnover of Populations, Taxa, and Communities. Columbia University Press, New York.

Martinsson, A. 1956. Neue Funde kambrischer Gänge und ordovizischer Geschiebe im südwestlichen Finnland. Bulletin of the Geological Institutions of the University of Uppsala 36:79–105.

Mason, C., and E. L. Yochelson. 1985. Some tubular fossils (*Sphenothallus:* "Vermes") from the Middle and Late Paleozoic of the United States. Journal of Paleontology 59:85–95.

Mayoral, E., J. C. Gutiérrez Marco, and J. Martinell. 1994. Primeras evidencias de briozoos perforantes (Ctenostomata) en braquiópodos Ordovícicos de los Montes de Toledo (Zona Centroiibérica Meridional, España). Revista Española de Paleontología 9:185–194.

McAuley, R. J., and R. J. Elias. 1990. Latest Ordovician to earliest Silurian solitary rugose corals of the east-central United States. Bulletins of American Paleontology 98(333):1–82.

McCann, T. 1990. Distribution of Ordovician-Silurian ichnofossil assemblages in Wales—implications for Phanerozoic ichnofaunas. Lethaia 23:243–255.

McCormick, T., and R. A. Fortey. 1998. Independent testing of a paleobiological hypothesis: The optical design of two Ordovician pelagic trilobites reveals their relative paleobathymetry. Paleobiology 24:235–253.

McCormick, T., and R. A. Fortey. 1999. The most widely distributed trilobite species: Ordovician *Carolinites genacinaca.* Journal of Paleontology 73:202–218.

McCormick, T., and A. W. Owen. 2001. Assessing trilobite biodiversity change in the Ordovician of the British Isles. Geological Journal 36:279–290.

McCracken A. D., and G. S. Nowlan. 1989. Conodont paleontology and biostratigraphy of Ordovician carbonates and petroliferous carbonates on Southampton, Baffin, and Akpatok islands in the eastern Canadian Arctic. Canadian Journal of Earth Sciences 26:1880–1903.

McKerrow, W. S. (ed.). 1978. The Ecology of Fossils: An Illustrated Guide. MIT Press, Cambridge, Massachusetts, and Duckworth, London, 384 pp.

McKerrow, W. S. 1979. Ordovician and Silurian changes in sea level. Journal of the Geological Society of London 136:137–145.

McKerrow, W. S., and C. R. Scotese (eds.). 1990. Palaeozoic palaeogeography and biogeography. The Geological Society, London, Memoir 12:1–435.

McKinney, F. K. 2000. Phylloporinids and the phylogeny of the Fenestrida; pp. 54–65 in A. Herrera Cubilla and J. B. C. Jackson (eds.), Proceedings of the 11th International Bryozoology Association Conference, Smithsonian Tropical Research Institute, Balboa, Republic of Panama.

McKinney, F. K., and J. B. C. Jackson. 1989. Bryozoan Evolution. Unwin Hyman, Boston.

McKinney, F. K., S. Lidgard, J. J. Sepkoski, and P. D. Taylor. 1998. Decoupled temporal patterns of evolution and ecology in two post-Paleozoic clades. Science 281: 807–809.

McKinney, F. K., S. Lidgard, and P. D. Taylor. 2001. Macroevolutionary trends: Perception depends on the measure used; pp. 348–385 in J. B. C. Jackson, S. Lidgard, and F. K. McKinney (eds.), Evolutionary Patterns: Growth, Form and Tempo in the Fossil Record. University of Chicago Press, Chicago.

McLean, J. H. 1981. The Galapagos rift limpet *Neomphalus;* relevance to understanding the evolution of a major Paleozoic-Mesozoic radiation. Malacologia 21: 291–336.

McLean, J. H. 1984. A case for derivation of the Fissurellidae from the Bellerophontacea. Malacologia 25: 3–20.

McLean, R. A., and B. D. Webby. 1976. Upper Ordovician rugose corals of central New South Wales. Proceedings of the Linnean Society of New South Wales 100:231–244.

McLeod, J. D. 1979. A Lower Ordovician (Canadian) lichenarid coral from the Ozark Uplift area. Journal of Paleontology 53:505–506.

McNamara, A. K., C. Mac Niocaill, B. A. van der Pluijm, and R. Van der Voo. 2001. West African proximity of the Avalon terrane in the latest Precambrian. Geological Society of America, Bulletin 113:1161–1170.

Meert, J. G., and R. Van der Voo. 1997. The assembly of Gondwana 800–550 Ma. Journal of Geodynamics 23: 223–235.

Meidla, T. 1996. Late Ordovician Ostracodes of Estonia. Fossilia Baltica 2:1–222.

Meidla, T. 1997. Hunneberg and Billingen stages; pp. 58–61 in A. Raukas and A. Teedumäe (eds.), Geology and Mineral Resources of Estonia. Estonian Academy Publishers, Tallinn.

Meidla, T., L. Ainsaar, L. Hints, O. Hints, T. Martma, and J. Nõlvak. 1999. The mid-Caradocian biotic and isotopic event in the Ordovician of East Baltic; pp. 503–506 in P. Kraft and O. Fatka (eds.), *Quo vadis* Ordovician? Short papers of the 8th International Symposium on the Ordovician System. Acta Universitatis Carolinae, Geologica 43(1–2).

Melchin, M. J., and C. E. Mitchell. 1991. Late Ordovician extinction in the Graptoloidea; pp. 143–156 in C. R. Barnes and S. H. Williams (eds.), Advances in Ordovician Geology. Geological Survey of Canada, Paper 90-9.

Melnikova, L. M. 1978. Some Late Ordovician Ostracodes from Central Mongolia. Paleontological Journal 11:206–214.

Melnikova, L. M. 1980. Some Early Ordovician Ostracodes of the Southern Urals. Paleontological Journal 13:71–76.

Melnikova, L. M. 1986. Ordovikskie ostrakody Kazakhstana. Akademia Nauk SSSR, Trudy Paleontologicheskogo Instituta 218:1–104.

Melnikova, L. M. 1999. Ostracodes from the Billingen Horizon (Lower Ordovician) of the Leningrad Region. Paleontological Journal 33:147–152.

Melnikova, L. M., and Yu. E. Dmitrovskaja. 1997. Ostrakody i zamkovye brakhiopody ordovika Moskovskoj sineklizy. Stratigrafiya, Geologicheskaya Korrelyatsiya 5(5):10–23.

Melo, J. H. G. 1997. Nova dataçao palinologica da Formaçao Vila Maria (Siluriano, Bacia do Parana) em sua faixa de afloramentos no SW de Goias; pp. 1–15 in Comunicaçao tecnica SEBIPE 40/97 CENPES/MCT 650-18341. Petrobras, Rio de Janeiro.

Melo, J. H. G., and P. Steemans. 1997. Resultados de investigacoes palinoestratigraficas em amostras de superficie da regiao de Presidente Figueiredo (AM), Bacia do Amazonas; pp. 1–11 in Comunicaçao tecnica SEBIPE 048/97 CENPES/MCT 650-18478. Petrobras, Rio de Janeiro.

Menard, H. W. 1986. Islands. Scientific American Books, New York, 230 pp.

Mergl, M. 1997. Distribution of the lingulate brachiopod *Thysanotus* in Central Europe. Věstník Českého geologického ústavu 72:27–35.

Mergl, M. 1999. Inarticulated brachiopod communities in Tremadoc-Arenig of Prague Basin: A review; pp. 337–340 in P. Kraft and O. Fatka (eds.), *Quo vadis* Ordovician? Short papers of the 8th International Symposium on the Ordovician System. Acta Universitatis Carolinae, Geologica 43(1–2).

Mergl, M. 2001. Extinction of some lingulate brachiopod families: New stratigraphical data from the Silurian and Devonian of Central Bohemia; pp. 345–351 in C. H. C. Brunton, L. R. M. Cocks, and S. L. Long (eds.), Brachiopods Past and Present. Taylor and Francis, London.

Mikuláš, R. 1988. Trace fossils from the pelitic sediments of the Bohemian Upper Ordovician. Acta Universitatis Carolinae, Geologica 3:343–363.

Mikuláš, R. 1990. Trace fossils from the Zahořany Formation (Upper Ordovician, Bohemia). Acta Universitatis Carolinae, Geologica 1990/3:307–335.

Mikuláš, R. 1992. Trace fossils from the Kosov Formation of the Bohemian Upper Ordovician. Sborník geologických Věd, Paleontologie 32:9–54.

Mikuláš, R. 1995. Trace fossils from the Paseky Shale (Early Cambrian, Czech Republic). Journal of the Czech Geological Society 40:37–45.

Miller, A. I. 1988. Spatio-temporal transitions in Paleozoic Bivalvia: An analysis of North American fossil assemblages. Historical Biology 1:251–273.

Miller, A. I. 1989. Spatio-temporal transitions in Paleozoic Bivalvia: A field comparison of Upper Ordovician and upper Paleozoic bivalve-dominated fossil assemblages. Historical Biology 2:227–260.

Miller, A. I. 1997a. Comparative diversification dynamics among palaeocontinents during the Ordovician Radiation. Geobios, Mémoire Spécial 20:397–406.

Miller, A. I. 1997b. Dissecting global diversity trends: Examples from the Ordovician radiation. Annual Review of Ecology and Systematics 28:85–104.

Miller, A. I. 1997c. A new look at age and area: The geographic and environmental expansion of genera during the Ordovician radiation. Paleobiology 23:410–419.

Miller, A. I. 1997d. Coordinated stasis or coincident relative stability? Paleobiology 23:155–164.

Miller, A. I. 1998. Biotic transitions in global marine diversity. Science 281:1157–1160.

Miller, A. I. 2000. Conversations about Phanerozoic global diversity; pp. 53–73 in D. H. Erwin and S. L. Wing (eds.), Deep Time: Paleobiology's Perspective. Special Volume (Supplement to Paleobiology 26), Paleontological Society, Lawrence, Kansas.

Miller, A. I., and S. R. Connolly. 2001. Substrate affinities of higher taxa and the Ordovician Radiation. Paleobiology 27:768–778.

Miller, A. I., and M. Foote. 1996. Calibrating the Ordovician radiation of marine life: Implications for Phanerozoic diversity trends. Paleobiology 22:304–309.

Miller, A. I., and S. Mao. 1995. Association of orogenic activity with the Ordovician radiation of marine life. Geology 23:305–308.

Miller, A. I., and S. Mao. 1998. Scales of diversification and the Ordovician radiation; pp. 288–310 in M. L. McKinney and J. A. Drake (eds.), Biodiversity Dynamics: Turnover of Populations, Taxa, and Communities. Columbia University Press, New York.

Miller, J. F. 1984. Cambrian and earliest Ordovician conodont evolution, biofacies, and provincialism. Geological Society of America, Special Paper 196:43–68.

Miller, M. A. 1991. *Paniculaferum missouriensis* gen. et sp. nov., a new Upper Ordovician acritarch from Missouri, U.S.A. Review of Palaeobotany and Palynology 70: 217–223.

Miller, M. A. 1996. Chitinozoa; pp. 307–336 in J. Jansonius and D. C. McGregor (eds.), Palynology: Principles and Applications. American Association of Stratigraphic Palynologists Foundation (Dallas), Vol. 1. Publishers Press, Salt Lake City, Utah.

Miller, S. A. 1874. Genus *Megalograptus*. Cincinnati Quarterly Journal of Science 1:343–346.

Millward, D., and S. G. Molyneux. 1992. Field and biostratigraphic evidence for an unconformity at the base of the Eycott Volcanic Group in the English Lake District. Geological Magazine 129:77–92.

Min, K., P. R. Renne, and W. D. Huff. 2001. ^{40}Ar/^{39}Ar dating of Ordovician K-bentonites in Laurentia and Baltoscandia. Earth and Planetary Science Letters 185: 121–134.

Minjin, C. (comp.). 2001. The Guide Book, Mongolian Ordovician and Silurian Stratigraphy and Abstracts for the Joint Field Meeting of IGCP 410 and IGCP 421 in Mongolia. Tempus-Tacis Project (Ulaanbaatar, Mongolia) 20091-98, 131 pp.

Mitchell, C. E., E. D. Brussa, and R. A. Astini. 1998. A diverse Da2 fauna preserved within an altered volcanic ash fall, Eastern Precordillera, Argentina: Implications for graptolite paleoecology; pp. 222–223 in J. C. Gutiérrez and I. Rábano (eds.), Proceedings of the Sixth International Graptolite Conference of the GWG (IPA) and the SW Iberia Field Meeting 1998 of the International Subcommission on Silurian Stratigraphy (ICS-IUGS). Instituto Technológico Geominero de España, Temas Geológico-Mineros 23.

Mitchell, C. E., X. Chen, S. M. Bergström, Y.-D. Zhang, W.-H. Wang, B. D. Webby, and S. C. Finney. 1997. Definition of a global boundary stratotype for the Darriwilian stage of the Ordovician System. Episodes 20: 158–166.

Mizusaki, A. M., J. H. G. Melo, M. L. Lelarge, and P. Steemans. 2002. Vila Maria Formation, Parana Basin, Brazil—an example of integrated geochronological and palynological datings. Geological Magazine 139(2): 453–463.

Moberg, J. C. 1914. Om Svenska Silurcirripedier. Lunds Universitets Årsskrift, Ny Följd, Afdelning 2, 11:1–20.

Moberg, J. C., and C. O. Segerberg. 1906. Bidrag till kännedomen om Ceratopygeregionen med särskild hänsyn till dess utveckling i Fogelsångstrakten. Lunds Universitets Årsskrift, Ny Följd, Afdelning 2, 2(7): 1–116.

Modzalevskaya, E. A. 1953. Trepostomaty ordovika Pribaltiki i ikh stratigraficheskoje znachenije [Trepostomata of the Ordovician and East Baltic and their stratigraphic significance]. Trudy Vsesojuznogo Nefdtjanogo Nauchno-Issledovatelskogo Geologo-Razvedochnogo Instituta 78:91–167.

Moldowan, J. M., J. Dahl, S. R. Jacobson, B. J. Huizinga, F. J. Fago, R. Shetty, D. S. Watt, and K. E. Peters. 1996. Chemostratigraphic reconstruction of biofacies: Molecular evidence linking cyst-forming dinoflagellates with pre-Triassic ancestors. Geology 24:159–162.

Molyneux, S. G. 1988. Late Ordovician acritarchs from northeast Libya; pp. 45–59 *in* A. El-Arnauti, B. Owens, and B. Thusu (eds.), Subsurface Palynostratigraphy of Northeast Libya. Garyounis University Publications, Benghazi, Libya.

Molyneux, S. G., A. Le Hérissé, and R. Wicander. 1996. Paleozoic phytoplankton; pp. 493–529 *in* J. Jansonius and D. C. McGregor (eds.), Palynology: Principles and Applications, Vol. 2. American Association of Stratigraphic Palynologists Foundation. Publishers Press, Salt Lake City, Utah.

Montenari, M., T. Servais, and F. Paris. 2000. Palynological dating (acritarchs and chitinozoans) of Lower Paleozoic phyllites from the Black Forest/southwestern Germany. Compte Rendu de l'Académie des Sciences de Paris, Sciences de la Terre et des planètes 330:493–499.

Mooi, R., and B. David. 2000. What a new model of skeletal homologies tells us about asteroid evolution. American Zoologist 40:326–339.

Moore, N. K. 1977. Distribution of the benthic algal flora in the Middle Ordovician carbonate units of the southern Appalachians; pp. 18–33 *in* S. C. Ruppel and K. R. Walker (eds.), The Ecostratigraphy of the Middle Ordovician of the Southern Appalachians (Kentucky, Tennessee, and Virginia), U.S.A.: A Field Excursion. University of Tennessee (Knoxville), Department of Geological Sciences, Studies in Geology 77-1.

Moore, R. C. 1952. Evolution rates among crinoids; pp. 338–352 *in* L. G. Henbest (ed.), Distribution of Evolutionary Explosions in Geologic Time. Journal of Paleontology, 26(3).

Moore, R. C. (ed.). 1966. Treatise on Invertebrate Paleontology. Part U, Echinodermata 3(1–2). Geological Society of America, New York, and University of Kansas Press, Lawrence, 695 pp.

Moore, R. C. (ed.). 1968 [dated 1967]. Treatise on Invertebrate Paleontology. Part S, Echinodermata 1(1–2). Geological Society of America, New York, and University of Kansas Press, Lawrence, 650 pp.

Moore, R. C., and L. R. Laudon. 1943. Evolution and classification of Paleozoic crinoids. Geological Society of America, Special Paper 46:1–153.

Moore, R. C., and C. Teichert (eds.). 1978. Treatise on Invertebrate Paleontology. Part T, Echinodermata 2(1–3). Geological Society of America, Boulder, Colorado, and University of Kansas Press, Lawrence, 1,027 pp.

Moore, R. C., C. Teichert, R. A. Robison, and R. L. Kaesler (eds.). 1953–1992. Treatise on Invertebrate Paleontology. Geological Society of America, New York and Boulder, Colorado, and University of Kansas Press, Lawrence.

Morgan, W. J. 1981. Hotspot tracks and the opening of the Atlantic and Indian Oceans; pp. 443–487 *in* C. Emilini (ed.), The Sea. Wiley, New York.

Morris, N. J. 1967. Mollusca: Scaphopoda and Bivalvia; pp. 469–477 *in* W. B. Harland et al. (eds.), The Fossil Record. The Geological Society, London.

Morris, N. J., and R. J. Cleevely. 1981. *Phanerotinus cristatus* (Phillips) and the nature of euomphalacean gastropods. Bulletin of the British Museum (Natural History), Geology Series 35:195–212.

Morris, R. W., and S. H. Felton. 1993. Symbiotic association of crinoids, platyceratid gastropods, and *Cornulites* in the Upper Ordovician (Cincinnatian) of the Cincinnati, Ohio region. Palaios 8:465–476.

Morris, R. W., and H. B. Rollins. 1971. The distribution and paleoecological interpretation of *Cornulites* in the Waynesville Formation (Upper Ordovician) of southwestern Ohio. Ohio Journal of Science 71:159–170.

Morrow, J. R., E. Schindler, and O. H. Walliser. 1996. Phanerozoic developments of selected global environmental features; pp. 53–61 *in* O. H. Walliser (ed.), Global Events and Event Stratigraphy in the Phanerozoic. Springer, Berlin.

Mory, A. J., R. S. Nicoll, and J. D. Gorter. 1998. Lower Palaeozoic correlations and thermal maturity, Carnarvon Basin, WA; pp. 599–611 *in* Sedimentary Basins of Western Australia 2, Proceedings of Petroleum Exploration Society of Australia Symposium, Perth.

Moskalenko, T. A. 1970. Konodonty krivolutskogo jarusa (Sredniy Ordovik) sibirskoj platformy [Conodonts of the Krivaya Luka stage (Middle Ordovician) from the Siberian Platform]. Akademia Nauk SSSR, Sibirskoye Otdelenie, Trudy Instituta Geologii i Geofiziki 61:1–116.

Moskalenko, T. A. 1973. Konodonty Srednego i Verkhnego Ordovika Sibirskoy Platformy [Conodonts of the Middle and Upper Ordovician on the Siberian Platform]. Akademia Nauk SSSR, Sibirskoye Otdelenie, Trudy Instituta Geologii i Geofiziki 137:1–143.

Moya, M. C. 1988. Lower Ordovician in the southern part of the Argentine Eastern Cordillera; pp. 55–68 *in* H. Bahlburg, C. Beitkreuz, and P. Giese (eds.), The Southern Central Andes. Lecture Notes in Earth Sciences, Vol. 17. Springer-Verlag, Berlin.

Moya, M. C., J. A. Monteros, and C. R. Monaldi. 1998. Graptolite dating of Lower Ordovician unconformity in the Argentinian Andes. Colección Temas Geológico-Mineros 23:227–230.

Müller, K. J., and I. Hinz-Schallreuter. 1993. Palaeoscolecid worms from the Middle Cambrian of Australia. Palaeontology 36:549–592.

Müller, K. J., Y. Nogami, and H. Lenz. 1974. Phosphatische Ringe als Mikrofossilien im Altpalaeozoikum. Palaeontographica, Abteilung A 146(4–6):79–99.

Munnecke, A., and T. Servais. 1996. Scanning electron microscopy of polished, slightly etched rock surfaces: A method to observe palynomorphs *in situ*. Palynology 20:163–176.

Myrow, P. M. 1995. *Thalassinoides* and the enigma of early Paleozoic open-framework burrow systems. Palaios 10: 58–74.

Natal'in, B. A., J. M. Amato, J. Toro, and J. E. Wright. 1999. Paleozoic rocks of northern Chukotka Peninsula, Russian Far East: Implications for the tectonics of the Arctic region. Tectonics 18:977–1003.

Nazarov, B. B. 1975. Lower and Middle Paleozoic radiolarians of Kazakhstan (research methods, systematics, stratigraphic importance). Akademia Nauk SSSR, Trudy Geologicheskii Institut (Moscow) 275:1–203.

Nazarov, B. B. 1988. Paleozoic Radiolaria. Practical manual of microfauna of the USSR, Vol. 2. Nedra, Leningrad, 232 pp. (in Russian).

Nazarov, B. B., and J. Nõlvak. 1983. Radiolari iz verkhnego Ordovika Estonii. Eesti NSV Teaduste Akademia toimetised, Geoloogia, Izvestiia 32:1–8.

Nazarov, B. B., and A. R. Ormiston. 1984. Tentative system of Paleozoic Radiolaria; pp. 64–87 *in* M. Petrushevskaya (ed.), Morphology, ecology and evolution of Radiolaria. Zoological Institute, Nauka, Leningrad (in Russian).

Nazarov, B. B., and A. R. Ormiston. 1985. Evolution of Radiolaria in the Palaeozoic and its correlation with the development of other marine fossil groups. Senckenbergiana Lethaea 66:203–235.

Nazarov, B. B., and A. R. Ormiston. 1993. New biostratigraphically important Paleozoic Radiolaria of Eurasia and North America. Micropaleontology Special Publication 6:22–60.

Nazarov, B. B., and L. E. Popov. 1980. Stratigraphy and fauna of the siliceous-carbonate deposits of the Ordovician of Kazakhstan (radiolarians and inarticulate brachiopods). Akademia Nauk SSSR, Trudy Geologicheskii Institut (Moscow) 331:1–190.

Neal, M. L., and J. T. Hannibal. 2000. Paleoecologic and taxonomic implications of *Sphenothallus* and *Sphenothallus*-like specimens from Ohio and areas adjacent to Ohio. Journal of Paleontology 74:369–380.

Nelson, S. J. 1963. Ordovician paleontology of the northern Hudson Bay Lowland. Geological Society of America, Memoir 90:1–152.

Nelson, S. J. 1981. Solitary streptelasmatid corals, Ordovician of northern Hudson Bay Lowland, Manitoba, Canada. Palaeontographica, Abteilung A 172:1–71.

Nestor, H. 1997. Evolutionary history of the single-layered, laminate, clathrodictyid stromatoporoids. Boletín de la Real Sociedad Española de Historia Natural (Sección Geológia) 91:319–328.

Nestor, H. 1999. Community structure and succession of Baltoscandian Early Palaeozoic stromatoporoids. Proceedings of the Estonian Academy of Sciences, Geology 48:123–139.

Nestor, H., and R. Einasto. 1997. Ordovician and Silurian carbonate sedimentation basin; pp. 192–208 *in* A. Raukas and A. Teedumäe (eds.), Geology and Mineral Resources of Estonia. Estonian Academy Publishers, Tallinn.

Nestor, H., and C. W. Stock. 2001. Recovery of the stromatoporoid fauna after the Late Ordovician extinction. Bulletin of the Tohoku University Museum 1:333–341.

Neuman, B. E. E. 1968. Two new species of Upper Ordovician rugose corals from Sweden. GFF [Geologiska Föreningens i Stockhom Förhandlingar] 90:229–240.

Neuman, B. E. E. 1969. Upper Ordovician streptelasmatid corals from Scandinavia. Bulletin of the Geological Institutions of the University of Uppsala, n.s., 1: 1–73.

Neuman, B. E. E. 1975. New Lower Palaeozoic streptelasmatid corals from Scandinavia. Norsk Geologisk Tidsskrift 55:335–359.

Neuman, B. E. E. 1984. Origin and early evolution of rugose corals. Palaeontographica Americana 54:119–126.

Neuman, B. E. E. 1997. Evaluation of rugose coral potentials as index fossils. Boletín de la Real Sociedad Española de Historia Natural (Sección Geológia) 92: 303–309.

Neuman, B. E. E. 1998. The latest Ordovician rugose corals in Baltoscandia; pp. 93–99 *in* S. Stouge (ed.), WOGOGOB-94 Symposium. Danmarks og Grönlands Geologiske Undersökelse Rapport 1996/98.

Neuman, B. E. E. In press. The new early Palaeozoic rugose coral genera *Eurogrewingkia* gen.nov. and *Fosselasma* gen.nov. Proceedings of the Estonian Academy of Sciences, Geology 52.

Neuman, R. B. 1951. St Paul Group: A revision of the "Stones River" Group of Maryland and adjacent states. Geological Society of America, Bulletin 62:267–324.

Neuman, R. B. 1972. Brachiopods of early Ordovician volcanic islands. Proceedings of the 24th International Geological Congress, Montreal 7:297–302.

Neuman, R. B. 1984. Geology and paleobiology of islands in the Ordovician Iapetus Ocean: Review and implications. Geological Society of America, Bulletin 95: 1188–1201.

Neuman, R. B., and D. A. T. Harper. 1992. Paleogeographic significance of Arenig-Llanvirn Toquima-Table Head and Celtic brachiopod assemblages; pp. 241–254 *in* B. D. Webby and J. R. Laurie (eds.), Global Perspectives on Ordovician Geology. Balkema, Rotterdam.

Newell, N. D., and A. LaRocque. 1969. Superfamily Praecardiacea; pp. N243–N248 *in* R. C. Moore (ed.), Treatise on Invertebrate Paleontology. Part N, Mollusca 6, Bivalvia. Geological Society of America, Boulder, Colorado, and University of Kansas Press, Lawrence.

Nicholson, H. A. 1872. On the genera *Cornulites* and *Tentaculites* and on a new genus *Conchicolites*. American Journal of Science, ser. 3, 3:202–206.

Nicholson, H. A. 1879. On the Structure and Affinities of the "Tabulate Corals" of the Palaeozoic period. W. Blackwood and Sons, Edinburgh, xii + 342 pp.

Nicholson, H. A., and J. Murie. 1878. On the minute structure of *Stromatopora* and its allies. Zoological Journal of the Linnean Society 14:187–246.

Nicoll, R. S. 1992. Evolution of the conodont genus *Cordylodus* and the Cambrian-Ordovician boundary; pp. 105–113 in B. D. Webby and J. R. Laurie (eds.), Global Perspectives on Ordovician Geology. Balkema, Rotterdam.

Nicoll, R. S., A. T. Nielsen, J. R. Laurie, and J. H. Shergold. 1992. Preliminary correlation of latest Cambrian to Early Ordovician sea level events in Australia and Scandinavia; pp. 381–394 in B. D. Webby and J. R. Laurie (eds.), Global Perspectives on Ordovician Geology. Balkema, Rotterdam.

Nicoll, R. S., J. F. Miller, G. S. Nowlan, J. E. Repetski, and R. L. Ethington. 1999. *Iapetonudus* (n. gen.) and *Iapetognathus* Landing, unusual earliest Ordovician multielement conodont taxa and their utility for biostratigraphy. Brigham Young University Geology Studies 44:27–101.

Nielsen, A. T. 1992a. Ecostratigraphy and the recognition of Arenigian (Early Ordovician) sea-level changes; pp. 355–366 in B. D. Webby and J. R. Laurie (eds.), Global Perspectives on Ordovician Geology. Balkema, Rotterdam.

Nielsen, A. T. 1992b. Intercontinental correlation of the Arenigian (Early Ordovician) based on sequence and ecostratigraphy; pp. 367–379 in B. D. Webby and J. R. Laurie (eds.), Global Perspectives on Ordovician Geology. Balkema, Rotterdam.

Nielsen, A. T. 1995. Trilobite systematics, biostratigraphy and palaeoecology of the Lower Ordovician Komstad Limestone and Huk Formations, southern Scandinavia. Fossils and Strata 38:1–374.

Nielsen, C., and K. J. Pedersen. 1979. Cystid structure and protrusion of the polypide in *Crisia* (Bryozoa, Cyclostomata). Acta Zoologica (Stockholm) 60:65–88.

Nikitin, I. F., and L. E. Popov. 1996. Strophomenid and triplesiid brachiopods from an Upper Ordovician carbonate mound in central Kazakhstan. Alcheringa 20:1–20.

Nikitin, I. F., N. M. Frid, and V. S. Zvontsov. 1991. Palaeogeography and main features of vulcanicity in the Ordovician of Kazakhstan and north Tien-Shan; pp. 259–270 in C. R. Barnes and S. H. Williams (eds.), Advances in Ordovician Geology. Geological Survey of Canada, Paper 90-9.

Nikitin, I. F., L. E. Popov, and L. E. Holmer. 1996. Late Ordovician brachiopod assemblage of Hiberno-Salairian type from Central Kazakhstan. GFF [Geologiska Föreningens i Stockhom Förhandlingar] 117:83–96.

Nitecki, M. H. 1970. North American cyclocrinitid algae. Fieldiana: Geology 2:1–182.

Nitecki, M. H. 1972. The paleogeographic significance of receptaculitids. 24th Session International Geological Congress, Montréal, Canada. Paleontology, sect. 7:303–309.

Nitecki, M. H., and N. Spjeldnaes. 1992. *Cyclocrinites spaskii*, a model of cyclocrinitid morphology. Institutt for Geologi, Universitetet i Oslo, Intern Skriftserie 63: 1–69. Preprint.

Nitecki, M. H., and D. F. Toomey. 1979a. The nature and distribution of calathid algae; p. 159 in Resumes, Deuxième Symposium International sur les Algues Fossiles, Paris, Avril 1979. Université Pierre-et-Marie Curie, Paris.

Nitecki, M. H., and D. F. Toomey. 1979b. Nature and classification of receptaculitids. Bulletin des Centres de Recherches Exploration-Production Elf-Aquitaine 3(2): 725–732.

Nitecki, M. H., R. C. Gutschick, and J. E. Repetski. 1975. Phosphatic microfossils from the Ordovician of the United States. Fieldiana: Geology 35:1–9.

Nitecki, M. H., H. Mutvei, and D. V. Nitecki. 1999. Receptaculitids: A Phylogenetic Debate on a Problematic Fossil Taxon. Kluwer/Plenum, New York, 241 pp.

Noble, P. J. 2000. Revised stratigraphy and structural relationships in the Roberts Mountains allochthon of Nevada (USA) based on radiolarian cherts; pp. 439–449 in J. K. Cluer, J. G. Price, E. M. Struhsacker, R. F. Hardyman, and C. L. Morris (eds.), Geology and Ore Deposits 2000: The Great Basin and Beyond: Symposium Proceedings, Geological Society of Nevada, Reno, May 15–18, 2000.

Noble, P. J., and J. C. Aitchison. 2000. Early Paleozoic radiolarian biozonation. Geology 28:367–370.

Nõlvak, J. 1999. Ordovician chitinozoan biozonation of Baltoscandia; pp. 287–290 in P. Kraft and O. Fatka (eds.), *Quo vadis* Ordovician? Short papers of the 8th International Symposium on the Ordovician System. Acta Universitatis Carolinae, Geologica 43(1–2).

Nõlvak, J., and Y. Grahn. 1993. Ordovician chitinozoan zones from Baltoscandia. Review of Palaeobotany and Palynology 79:245–269.

Nõlvak, J., T. Meidla, and A. Uutela. 1995. Microfossils in the Ordovician erratic boulders from southwestern Finland. Bulletin of the Geological Society of Finland 67:3–26.

Norford, B. S., and H. M. Steele. 1969. The Ordovician trimerellid brachiopod *Eodinobolus* from south-east Ontario. Palaeontology 12:161–171.

Norford, B. S., G. S. Nowlan, F. M. Haidl, and R. K. Bezys. 1998. The Ordovician-Silurian boundary interval in Saskatchewan and Manitoba. Saskatchewan Geological Society, Special Publication 13:27–45.

Norris, R. D. 2000. Pelagic species diversity, biogeography, and evolution; pp. 236–258 in D. H. Erwin and S. L. Wing (eds.), Deep Time, Paleobiology's Perspective. Special volume (Supplement to Paleobiology 26). Paleontological Society, Lawrence, Kansas.

Novack-Gottshall, P. M. 1999. Comparative geographic and environmental diversity dynamics of gastropods and bivalves during the Ordovician Radiation. Master's thesis, University of Cincinnati, Cincinnati.

Novack-Gottshall, P. M., and A. I. Miller. 2003a. Comparative geographic and environmental diversity dynamics of gastropods and bivalves during the Ordovician Radiation. Paleobiology 29:576–604.

Novack-Gottshall, P. M., and A. I. Miller. 2003b. Comparative taxonomic richness and abundance of the Late Ordovician gastropods and bivalves in the mollusk-rich strata of the Cincinnati Arch. Palaios 18:559–571.

Nowlan, G. S. 2002. Stratigraphy and conodont biostratigraphy of Upper Ordovician strata in the subsurface of Alberta, Canada. Special Papers in Palaeontology 67:185–203.

Nowlan, G. S., and C. R. Barnes. 1987. Thermal maturation of Paleozoic strata in eastern Canada from conodont colour alteration index (C.A.I.) data with implications for burial history, tectonic evolution, hotspot tracks and mineral and hydrocarbon potential. Geological Survey of Canada, Bulletin 367:1–47.

Nowlan, G. S., A. D. McCracken, and B. D. E. Chatterton. 1988. Conodonts from Ordovician-Silurian boundary strata, Whittaker Formation, Mackenzie Mountains, Northwest Territories. Geological Survey of Canada, Bulletin 373:1–99.

Nützel, A. 1997. Über die Stammesgeschichte der Ptenoglossa (Gastropoda). Berliner geowissenschaftliche Abhandlungen Reihe E:1–229.

Nützel, A. 2002. An evaluation of the recently proposed Palaeozoic subclass Euomphalomorpha. Palaeontology 45:259–266.

Nützel, A., D. H. Erwin, and R. H. Mapes. 2000. Identity and phylogeny of the Late Paleozoic Subulitoidea (Gastropoda). Journal of Paleontology 74:575–598.

Obut, A. M., and N. M. Zaslavskaya. 1980. Chitinozoans and outlook for studies in the Asian part of the USSR; pp. 122–126 in Paleontology, stratigraphy. Akademia Nauk SSSR, Nauka, Moscow (in Russian).

Ogienko, L. V., V. I. Bjaly, and G. R. Kolosnitsyna. 1974. Biostratigrafiya kembrijskich i ordovikskich otlozhenij yuga Sibirskoj platformy. Nedra, Moscow, 208 pp.

O'Kelly, C. J., and G. L. Floyd. 1984. Correlations among patterns of sporangial structure and development, life histories, and ultrastructural features in the Ulvophyceae; pp. 121–156 in D. E. G. Irvine and D. M. John (eds.), Systematics of the Green Algae. Systematics Association Special Volume 27. Academic Press, London.

Okulitch, V. J. 1935a. Fauna of the Black River Group in the vicinity of Montreal. Canadian Field Naturalist 49: 96–107.

Okulitch, V. J. 1935b. Tetradiidae—a revision of the genus *Tetradium*. Royal Society of Canada, Transactions ser. 3, sec. 4 29:49–74.

Okulitch, V. J. 1936. On the genera *Heliolites*, *Tetradium* and *Chaetetes*. American Journal of Science 232(191): 361–379.

Olempska, E. 1994. Ostracods of the Mójcza Limestone; pp. 129–212 in J. Dzik, E. Olempska, and A. Pisera (eds.), Ordovician Carbonate Platform Ecosystem of the Holy Cross Mountains. Palaeontologia Polonica 53.

Oliver, W. A., Jr., 1996. Origins and relationships of Paleozoic coral groups and the origin of the Scleractinia; pp. 107–134 in G. D. Stanley, Jr. (ed.), Biology and Paleobiology of Corals. The Paleontological Society Paper 1.

Opalinski, P. R., and T. L. Harland. 1981. Stratigraphy of the Mjøsa Limestone in the Toten and Nes-Hamar areas. Norsk Geologisk Tidsskrift 61:59–78.

Öpik, A. A. 1930. Beiträge zur Kenntnis der Kukruse- (C_2-C_3-) Stufe in Eesti IV. Acta et Commentationes Tartuensis (Dorpatensis) A19:1–34.

Orr, P. 1994. Trace fossil tiering within event beds and preservation of frozen profiles: An example from the Lower Carboniferous of Menorca. Palaios 9:202–210.

Orr, P. 1996. The ichnofauna of the Skiddaw Group (Early Ordovician) of the Lake District, England. Geological Magazine 133:193–216.

Orr, P. 2001. Colonization of the deep-marine environment during the early Phanerozoic: The ichnofaunal record. Geological Journal 36:265–278.

Orr, P., and M. P. A. Howe. 1999. Macrofauna and ichnofauna of the Manx Group (early Ordovician), Isle of Man; pp. 33–44 in N. H. Woodcock, D. G. Quirk, W. R. Fitches, and R. P. Barnes (eds.), In Sight of the Suture: The Palaeozoic Geology of the Isle of Man in Its Iapetus Ocean Context. The Geological Society, London, Special Publication 160.

Ortega, G., and G. L. Albanesi. 2000. Graptolitos de la Formación Gualcamayo (Arenigiano-Llanvirniano) en el cerro Potrerillo, Precordillera Central de San Juan, Argentina. Boletín de la Academia Nacional de Ciencias, Córdoba 64:27–60.

Ørvig, T. 1989. Histologic studies of ostracoderms, placoderms and fossil elasmobranchs. Part 6, Hard tissues of Ordovician vertebrates. Zoologica Scripta 18(3):427–446.

Osgood, R. G., Jr. 1970. Trace fossils of the Cincinnati area. Palaeontographica Americana 6:281–444.

Ottone, E. G., and G. D. Holfeltz. 1992. Hallazgo de escolecodontes en la Formación Gualcamayo, Llanvirniano inferior, Argentina. Simposio Argentino de Paleobotánica y Palinologia, Buenos Aires. Asociación Paleontologica Argentina, Publicación Especial 2: 85–88.

Ottone, E. G., B. A. Toro, and B. G. Waisfeld. 1982. Lower Ordovician palynomorphs from the Acoite Formation, northwestern Argentina. Palynology 16:93–116.

Ottone, E. G., G. L. Albanesi, G. Ortega, and G. Holfeltz. 1999. Palynomorphs, conodonts and associated graptolites from the Ordovician Los Azules Formation, Central Precordillera, Argentina. Micropaleontology 45:225–250.

Ottone, E. G., G. D. Holfeltz., G. L. Albanesi, and G. Ortega. 2001. Chitinozoans from the Ordovician Los Azules Formation, Central Precordillera, Argentina. Micropaleontology 47:97–110.

Oulebsir, L., and F. Paris. 1995. Chitinozoaires ordoviciens du Sahara Algérien: Biostratigraphie et affinités paléogéographiques. Review of Palaeobotany and Palynology 86:49–68.

Owen, A. W. 1981. The Ashgill trilobites of the Oslo Region, Norway. Palaeontographica, Abteilung A 175:1–88.

Owen, A. W., and T. McCormick. 1999. Ordovician biodiversity change in the British Isles: A database approach. I. Rationale and database structure; pp. 523–526 in P. Kraft and O. Fatka (eds.), *Quo vadis* Ordovician? Short papers of the 8th International Symposium on the Ordovician System. Acta Universitatis Carolinae, Geologica 43(1–2).

Owen, A. W., and T. McCormick. 2003. Ordovician trilobite biodiversity change in the Anglo-Welsh sector of Avalonia—a comparison of regional and global patterns. Special Papers in Palaeontology 70:271–280.

Owen, A. W., and D. B. R. Robertson. 1995. Ecological changes during the end-Ordovician extinction. Modern Geology 20:21–39.

Owen, A. W., D. L. Bruton, J. F. Bockelie, and T. G. Bockelie. 1990. The Ordovician successions of the Oslo Region, Norway. Norges Geologiske Undersøkelse, Special Publication 4:1–54.

Palmer, A. R. 1979. Biomere boundaries re-examined. Alcheringa 3:33–41.

Palmieri, V. 1978. Late Ordovician conodonts from the Fork Lagoons Beds, Emerald area, central Queensland. Geological Survey of Queensland, Publication 369 (Palaeontological Paper 43):1–31.

Pancost, R. D., K. H. Freeman, and M. E. Patzowsky. 1999. Organic-matter source variation and the expression of a late Middle Ordovician carbon isotope excursion. Geology 27:1015–1018.

Parfrey, S. M. 1982. Palaeozoic conulariids from Tasmania. Alcheringa 6:69–75.

Paris, F. 1981. Les Chitinozoaires dans le Paléozoïque du sud-ouest de l'Europe (cadre géologique—étude systématique—biostratigraphie). Mémoire de la Société Géologique et Minéralogique de Bretagne 26:1–496.

Paris, F. 1988. Late Ordovician and Early Silurian chitinozoans from central and southern Cyrenaica; pp. 61–71 in A. El-Arnauti, B. Owens, and B. Thusu (eds.), Subsurface Palynostratigraphy of Northeast Libya. Garyounis University Publications, Benghazi, Libya.

Paris, F. 1990. The Ordovician chitinozoan biozones of the Northern Gondwana Domain. Review of Palaeobotany and Palynology 66:181–209.

Paris, F. 1996. Chitinozoan biostratigraphy and palaeoecology; pp. 531–552 in J. Jansonius and D. C. McGregor (eds.), Palynology: Principles and Applications, Vol. 2. American Association of Stratigraphic Palynologists Foundation. Publishers Press, Salt Lake City, Utah.

Paris, F. 1998. Early Palaeozoic palaeogeography of northern Gondwana regions. Acta Universitatis Carolinae, Geologica 42:473–483.

Paris, F. 1999. Palaeobiodiversification of Ordovician chitinozoans from northern Gondwana; pp. 283–286 in P. Kraft and O. Fatka (eds.), *Quo vadis* Ordovician? Short papers of the 8th International Symposium on the Ordovician System. Acta Universitatis Carolinae, Geologica 43(1–2).

Paris, F., and J. Nõlvak. 1999. Biological interpretation and paleobiodiversity of a cryptic fossil group: The "chitinozoan-animal." Geobios 32:315–324.

Paris, F., Z. Elaouad-Debbaj, J. C. Jaglin, D. Massa, and L. Oulebsir. 1995. Chitinozoans and Late Ordovician glacial events on Gondwana; pp. 171–176 in J. D. Cooper, M. L. Droser, and S. C. Finney (eds.), Ordovician Odyssey: Short Papers, 7th International Symposium on the Ordovician System. Book 77, Pacific Section Society for Sedimentary Geology (SEPM), Fullerton, California.

Paris, F., M. Deynoux, and J. F. Ghienne. 1998. Découverte de chitinozoaires à la limite Ordovicien-Silurien en Mauritanie: Implications paléogéographiques. Comptes Rendus de l'Académie des Sciences de Paris 326:499–504.

Paris, F., Y. Grahn, V. Nestor, and I. Lakova. 1999. Revised chitinozoan classification. Journal of Paleontology 73:547–568.

Paris, F., A. Bourahrouh, and A. Le Hérissé. 2000a. The effects of the final stages of the Late Ordovician glaciation on marine palynomorphs (chitinozoans, acritarchs, leiospheres) in well Nl-2 (NE Algerian Sahara). Review of Palaeobotany and Palynology 113:87–104.

Paris, F., J. Verniers, and S. Al-Hajri. 2000b. Ordovician chitinozoans from Central Saudi Arabia; pp. 42–56 in S. Al-Hajri and B. Owens (eds.), Stratigraphic Palynology of the Palaeozoic of Saudi Arabia. Special GeoArabia Publication 1.

Parrish, J. T. 1982. Upwelling and petroleum source beds, with reference to Paleozoic. American Association of Petroleum Geologists Bulletin 66:750–774.

Parsley, R. L. 1970. Revision of the North American Pleurocystitidae (Rhombifera-Cystoidea). Bulletins of American Paleontology 58(260):131–213.

Parsley, R. L. 1988. Probable feeding and respiratory mechanisms in *Aristocystites* (Diploporita, Middle-Upper Ordovician of Bohemia, CSSR); pp. 103–108 in R. D. Burke, P. V. Mladenov, P. Lambert, and R. L. Parsley (eds.), Echinoderm Biology. Balkema, Rotterdam.

Parsley, R. L. 1991. Review of selected North American mitrate stylophorans (Homalozoa: Echinodermata). Bulletins of American Paleontology 100(336):1–57.

Parsley, R. L. 1998. Taxonomic revision of the Stylophora; pp. 111–117 in R. Mooi and M. Telford (eds.), Echinoderms: San Francisco. Balkema, Rotterdam.

Parsley, R. L., and L. W. Mintz. 1975. North American Paracrinoidea (Ordovician: Paracrinozoa, new, Echinodermata). Bulletins of American Paleontology 68:1–115.

Patzkowsky, M. E. 1995a. A hierarchical branching model of evolutionary radiations. Paleobiology 21:440–460.

Patzkowsky, M. E. 1995b. Gradient analysis of Middle Ordovician brachiopod biofacies: Biostratigraphic, biogeographic, and macroevolutionary implications. Palaios 10:154–179.

Patzkowsky, M. E. 1995c. Ecologic aspects of the Ordovician radiation of articulate brachiopods; pp. 413–414 in J. D. Cooper, M. L. Droser, and S. C. Finney (eds.), Ordovician Odyssey: Short Papers, 7th International Symposium on the Ordovician System. Book 77, Pacific Section Society for Sedimentary Geology (SEPM), Fullerton, California.

Patzkowsky, M. E., and S. M. Holland. 1993. Biotic response to a Middle Ordovician paleoceanographic event in eastern North America. Geology 21:619–622.

Patzkowsky, M. E., and S. M. Holland. 1996. Extinction, invasion, and sequence stratigraphy: Patterns of faunal change in the Middle and Upper Ordovician of the eastern United States. Geological Society of America, Special Paper 306:131–142.

Patzkowsky, M. E., and S. M. Holland. 1997. Patterns of turnover in Middle and Upper Ordovician brachiopods of the eastern United States: A test of coordinated stasis. Paleobiology 23:420–443.

Patzkowsky, M. E., and S. M. Holland. 1999. Biofacies replacement in a sequence stratigraphic framework: Middle and Upper Ordovician of the Nashville Dome, Tennessee, USA. Palaios 14:301–323.

Patzkowsky, M. E., L. M. Slupik, M. A. Arthur, R. D. Pancost, and K. H. Freeman. 1997. Late Middle Ordovician environmental change and extinction: Harbinger of the Late Ordovician or continuation of Cambrian patterns? Geology 25:911–914.

Paul, C. R. C. 1972. *Cheirocystella antiqua* gen. et sp. nov. from the Lower Ordovician of western Utah, and its bearing on the evolution of the Cheirocrinidae (Rhombifera: Glyptocystitida). Brigham Young University Geology Studies 19:15–63.

Paul, C. R. C. 1973. British Ordovician Cystoids, Part 1. Palaeontographical Society Monograph 127(536):1–64.

Paul, C. R. C. 1976. Palaeogeography of primitive echinoderms in the Ordovician; pp. 553–574 in M. G. Bassett (ed.), The Ordovician System: Proceedings of a Palaeontological Association Symposium, Birmingham (England), September 1974. University of Wales Press and National Museum of Wales, Cardiff.

Paul, C. R. C. 1984. British Ordovician Cystoids, Part 2. Palaeontographical Society Monograph 136(563):65–152.

Paul, C. R. C. 1988. The phylogeny of the cystoids; pp. 199–213 in C. R. C. Paul and A. B. Smith (eds.), Echinoderm Phylogeny and Evolutionary Biology. Clarendon Press, Oxford.

Paul, C. R. C. 1997. British Ordovician Cystoids, Part 3. Palaeontographical Society Monograph 151(604):153–213.

Paul, C. R. C. 1998. Adequacy, completeness and the fossil record; pp. 1–22 in S. K. Donovan and C. R. C. Paul (eds.), The Adequacy of the Fossil Record. Wiley, Chichester, England.

Pchelintsev, V. F., and I. A. Korobkov. 1960. Mollusca—Gastropoda; pp. 1–360 in Yu. A. Orlov (ed.), Osnovy paleontologii, Vol. 4. Izdatel'stvo Akademia Nauk, Moscow.

Peck, L. S. 2001. Physiology; pp. 89–104 in S. J. Carlson and M. R. Sandy (eds.), Brachiopods Ancient and Modern: A Tribute to G. Arthur Cooper. The Paleontological Society (New Haven, Connecticut), Paper 7.

Peel, J. S. 1977. Systematics and palaeoecology of the Silurian gastropods of the Arisaig Group, Nova Scotia. Biologiske Skrifter 21:1–89.

Peel, J. S. 1978. Faunal succession and mode of life of Silurian gastropods in the Arisaig Group, Nova Scotia. Palaeontology 21:285–306.

Peel, J. S. 1984. Autecology of Silurian gastropods and monoplacophorans; pp. 165–182 in M. G. Bassett and J. D. Lawson (eds.), Autecology of Silurian Organisms. Palaeontological Association, London, Special Papers in Palaeontology 32.

Peel, J. S. 1991a. Functional morphology of the class Helcionelloida nov., and the early evolution of the Mollusca; pp. 157–177 in A. Simonetta and S. Conway Morris (eds.), The Early Evolution of Metazoa and the Significance of Problematic Taxa. Cambridge University Press, Cambridge.

Peel, J. S. 1991b. The classes Tergomya and Helcionelloida, and early molluscan evolution. Bulletin Grönlands Geologiske Undersoegelse 161:11–65.

Peel, J. S., and R. Horný. 1999. Muscle scars and systematic position of the Lower Palaeozoic limpets *Archinacella* and *Barrandicella* gen. n. (Mollusca). Journal of the Czech Geological Society 44:97–115.

Pemberton, S. G. (ed.). 1992. Applications of ichnology to petroleum exploration—a core workshop. Society of Economic Paleontologists and Mineralogists, Core Workshop 17:1–429.

Pemberton, S. G., R. W. Frey, and T. D. A. Saunders. 1990. Trace fossils; pp. 355–362 in D. E. G. Briggs and P. R. Crowther (eds.), Palaeobiology, a Synthesis. Blackwell Scientific Publications, Oxford.

Pemberton, S. G., J. A. MacEachern, and R. W. Frey. 1992. Trace fossil facies models: Environmental and allostratigraphic significance; pp. 47–72 in R. G. Walker and N. P. James (eds.), Facies Models and Sea Level

Changes. Geological Association of Canada, St. John's, Newfoundland.

Peralta, S. H., S. Heredia, and M. S. Beresi. 1999. Upper Arenigian–Lower Llanvirnian sequence of the Las Chacritas river, Central Precordillera, San Juan province, Argentina; pp. 123–126 in P. Kraft and O. Fatka (eds.), Quo vadis Ordovician? Short papers of the 8th International Symposium on the Ordovician System. Acta Universitatis Carolinae, Geologica 43(1–2).

Percival, I. G. 1995. Evolution of trimerellid brachiopods exemplified by *Eodinobolus* and its relatives from the Late Ordovician of New South Wales. Association of Australasian Paleontologists, Memoir 16:41–60.

Percival, I. G., and B. D. Webby. 1996. Island benthic assemblages: With examples from the Late Ordovician of Eastern Australia. Historical Biology 11:171–185.

Percival, I. G., and B. D. Webby. 1999. Ordovician biodiversity profiles in eastern Australia; pp. 463–466 in P. Kraft and O. Fatka (eds.), Quo vadis Ordovician? Short papers of the 8th International Symposium on the Ordovician System. Acta Universitatis Carolinae, Geologica 43(1–2).

Percival, I. G., B. D. Webby, and J. W. Pickett. 2001. Ordovician (Bendigonian, Darriwilian to Gisbornian) faunas from the northern Molong Volcanic Belt of central New South Wales. Alcheringa 25:211–250.

Petrunina, Z. E., N. V. Sennikov, V. D. Ermikov, L. L. Zeifert, A. V. Krivchikov, and A. A. Puzyrev. 1984. Lower Ordovician Stratigraphy of the Gorny Altai; pp. 3–33 in A. V. Kanygin (ed.), Lower Ordovician stratigraphy and fauna of the Altai Mountains. Nauka, Moscow (in Russian).

Pia, J. 1927. Thallophyta; pp. 31–136 in M. Hirmer (ed.), Handbuch der Paläbotanik, Band 1. Oldenbourg, Munich.

Pickerill, R. K. 1980. Phanerozoic flysch trace fossils diversity—observations based on an Ordovician flysch ichnofauna from the Aroostook-Matapedia Carbonate Belt of northern New Brunswick. Canadian Journal of Earth Sciences 17:1259–1270.

Pickerill, R. K. 1995. Deep-water marine *Rusophycus* and *Cruziana* from the Ordovician Lotbinière Formation of Quebec. Atlantic Geology 31:103–108.

Pickerill, R. K., and P. F. Williams. 1989. Deep burrowing in the early Palaeozoic deep sea: Examples from the Cambrian(?)–Early Ordovician Meguma Group of Nova Scotia. Canadian Journal of Earth Sciences 26:1061–1068.

Pickerill, R. K., M. Romano, and B. Meléndez. 1984. Arenig trace fossils from the Salamanca area, western Spain. Geological Journal 19:249–269.

Pickerill, R. K., L. R. Fyffe, and W. H. Forbes. 1987. Late Ordovician–Early Silurian trace fossils from the Matapedia Group, Tobique River, Western New Brunswick, Canada. Maritime Sediments and Atlantic Geology 23:77–88.

Pickett, J. W., and P. A. Jell. 1983. Middle Cambrian Sphinctozoa (Porifera) from New South Wales. Association of Australasian Paleontologists Memoir 1:85–92.

Pickett, J. W., and I. G. Percival. 2001. Ordovician faunas and biostratigraphy in the Gunningbland area, central New South Wales. Alcheringa 25:9–52.

Pickett, J. W., and B. D. Webby. 2000. Poriferans; pp. 68–69 in A. J. Wright, G. C. Young, J. A. Talent, and J. R. Laurie (eds.), Palaeobiogeography of Australasian Faunas and Floras. Association of Australasian Palaeontologists, Memoir 23.

Pitcher, M. 1964. Evolution of Chazyan (Ordovician) reefs of eastern United States and Canada. Bulletin of Canadian Petroleum Geology 12:632–691.

Playford, G., and M. A. Miller. 1988. Chitinozoa from lower Ordovician strata of the Georgina Basin, Queensland (Australia). Geobios 21:17–39.

Plotnick, R. E. 1983. Patterns in the Evolution of the Eurypterids. Doctoral thesis, University of Chicago.

Pohler, S. M. L. 1994. Conodont biofacies of Lower to lower Middle Ordovician mega-conglomerates, Cow Head Group, Western Newfoundland. Geological Survey of Canada, Bulletin 459:1–71.

Pohler, S. M. L., and C. R. Barnes. 1990. Conceptual models in conodont paleoecology. Courier Forschungsinstitut Senckenberg 118:409–440.

Pohler, S. M. L., and M. J. Orchard. 1990. Ordovician conodont biostratigraphy, western Canadian Cordillera. Geological Survey of Canada, Paper 90-15:1–37.

Pohler, S. M. L., C. R. Barnes, and N. P. James. 1987. Reconstructing a lost faunal realm: Conodonts from megaconglomerates of the Ordovician Cow Head Group, western Newfoundland; pp. 341–362 in R. L. Austin (ed.), Conodonts: Investigative Techniques and Applications. Ellis Horwood, Chichester, England.

Pohowsky, R. A. 1978. The boring ctenostomate Bryozoa: Taxonomy and paleobiology based on cavities in calcareous substrata. Bulletins of American Paleontology 73:1–192.

Pojeta, J., Jr. 1971. Review of Ordovician pelecypods. United States Geological Survey, Professional Paper 695:1–46.

Pojeta, J., Jr. 1978. The origin and early taxonomic diversification of pelecypods. Philosophical Transactions of the Royal Society of London, B 284:225–246.

Pojeta, J., Jr. 1979. Geographic distribution of Cambrian and Ordovician rostroconch mollusks; pp. 27–36 in J. Gray and A. J. Boucot (eds.), Historical Biogeography, Plate Tectonics, and the Changing Environment. Oregon State University Press, Corvallis.

Pojeta, J., Jr. 1980. Molluscan phylogeny. Tulane Studies in Geology and Paleontology 16:55–80.

Pojeta, J., Jr. 1988. The origin and Paleozoic diversification of solemyoid bivalves. New Mexico Bureau of Mines and Mineral Resources, Memoir 44:201–271.

Pojeta, J., Jr., and J. Gilbert-Tomlinson. 1977. Australian Ordovician pelecypod molluscs. Bureau of Mineral Resources, Geology and Geophysics, Australia, Bulletin 174:1–64.

Pojeta, J., Jr., and B. Runnegar. 1976. The paleontology of rostroconch mollusks and the early history of the phylum Mollusca. United States Geological Survey, Professional Paper 968:1–88.

Pojeta, J., Jr., B. Runnegar, N. J. Morris, and N. D. Newell. 1972. Rostroconchia: A new class of bivalved mollusks. Science 177:264–267.

Pojeta, J., Jr., J. Gilbert-Tomlinson, and J. H. Shergold. 1977. Cambrian and Ordovician rostroconch molluscs from Northern Australia. Bureau of Mineral Resources, Geology and Geophysics, Australia, Bulletin 171:1–54.

Ponder, W. F. 1997. Towards a phylogeny of gastropod molluscs: An analysis using morphological characters. Zoological Journal of the Linnean Society 119:83–265.

Ponder, W. F., and D. R. Lindberg. 1995 [dated 1996]. Gastropod phylogeny—challenges for the 90s; pp. 135–154 in J. D. Taylor (ed.), Origin and Evolutionary Radiation of the Mollusca. Oxford University Press, Oxford.

Ponder, W. F., and D. R. Lindberg. 1997. Towards a phylogeny of gastropod molluscs: An analysis using morphological characters. Zoological Journal of the Linnean Society 119:83–265.

Ponder, W. F., and A. Warén. 1988. Classification of the Caenogastropoda and Heterostropha—a list of the family-group names and higher taxa; pp. 288–326 in W. F. Ponder (ed.), Prosobranch Phylogeny, Malacological Review Supplement.

Popov, L. E. 2000a. Late Ordovician linguliformean microbrachiopods from north-central Kazakhstan. Alcheringa 24:257–275.

Popov, L. E. 2000b. Late Ordovician (Ashgill) linguliformean microbrachiopods from the Bestyube Formation, north-central Kazakhstan. Geobios 33:419–435.

Popov, L. E., and L. E. Holmer. 1994. Cambrian-Ordovician lingulate brachiopods from Scandinavia, Kazakhstan, and South Ural Mountains. Fossils and Strata 35:1–156.

Popov, L. E., and L. E. Holmer. 1995. Distribution of brachiopods across the Cambrian-Ordovician boundary on the East European Plate and adjacent areas; pp. 117–120 in J. D. Cooper, M. L. Droser, and S. C. Finney (eds.), Ordovician Odyssey: Short Papers, 7th International Symposium on the Ordovician System. Book 77, Pacific Section Society for Sedimentary Geology (SEPM), Fullerton, California.

Popov, L. E., and T. Tolmacheva. 1995. Conodont distribution in a deep-water Cambrian-Ordovician boundary sequence from South-Central Kazakhstan; pp. 121–124 in J. D. Cooper, M. L. Droser, and S. C. Finney (eds.), Ordovician Odyssey: Short Papers, 7th International Symposium on the Ordovician System. Book 77, Pacific Section Society for Sedimentary Geology (SEPM), Fullerton, California.

Popov, L. E., K. K. Kazanovitch, N. G. Borovko, S. P. Sergeeva, and R. F. Sobolevskaya. 1989. The key sections and stratigraphy of the Cambrian-Ordovician phosphate bearing *Obolus* beds on the northeastern Russian platform. Ministerstvo Geologii SSSR, Mezhvedomstvennyi Stratigraficheskij Komitet SSSR, Trudy 18:1–222 (in Russian).

Popov, L. E., L. E. Holmer., and V. Yu. Gorjansky. 1997. Late Ordovician and Early Silurian trimerellide brachiopods from Kazakhstan. Journal of Paleontology 71:584–598.

Popov, L. E., I. F. Nikitin, and E. V. Sokiran. 1999a. The earliest atrypides and athyridides (Brachiopoda) from the Ordovician of Kazakhstan. Palaeontology 42:625–661.

Popov, L. E., M. G. Bassett, L. E. Holmer, and V. Yu. Gorjansky. 1999b. Ordovician patterns of diversification in craniiformean brachiopods; pp. 321–324 in P. Kraft and O. Fatka (eds.), *Quo vadis* Ordovician? Short papers of the 8th International Symposium on the Ordovician System. Acta Universitatis Carolinae, Geologica 43(1–2).

Popov, L. E., O. Vinn, and O. I. Nikitina. 2001. Brachiopods of the redefined family Tritoechiidae from the Ordovician of Kazakhstan and South Urals. Geobios 32:131–155.

Popov, L. E., J. C. W. Cope, and I. F. Nikitin. 2003. A new Ordovician rostroconch mollusc from Kazakhstan. Alcheringa 27:173–179.

Poussart, P. F., A. J. Weaver, and C. R. Barnes. 1999. Late Ordovician glaciation under high atmospheric CO_2: A coupled model analysis. Paleoceanography 14:542–558.

Pratt, B. R., and N. P. James. 1982. Cryptalgal-metazoan bioherms of Early Ordovician age in the St. George Group, western Newfoundland. Sedimentology 29:543–569.

Přibyl, A. 1984. Ostracodes from the Ordovician and Silurian of Bolivia. Časopis pro mineralogii a geologii 29:353–368.

Přibyl, A. 1996. Ordovizische Ostrakoden Argentiniens II. Mitteilungen aus dem Geologisch-Paläontologischen Institut der Universität Hamburg 79:139–169.

Prokop, R. J., and V. Petr. 1999. Echinoderms in the Bohemian Ordovician. Journal of the Czech Geological Society 44:63–68.

Purnell, M. A. 2001. Scenarios, selection and the ecology of early vertebrates; pp. 187–208 in P. E. Ahlberg (ed.), Major Events in Early Vertebrate Evolution—Palaeontology, Phylogeny, Genetics and Development. Systematics Association Special Volume 61. Taylor and Francis, London.

Pushkin, V. I., and L. E. Popov. 1999. Early Ordovician bryozoans from north-western Russia. Palaeontology 42:171–189.

Pyle, L. J., and C. R. Barnes. 2002. Taxonomy, Evolution, and Biostratigraphy of Conodonts from the Kechika Formation, Skoki Formation, and Road River Group (Upper Cambrian to Lower Silurian), Northeastern British Columbia. National Research Council, Research Press, Ottawa, 227 pp.

Qian, Y., and S. Bengtson. 1989. Palaeontology and biostratigraphy of the Early Cambrian Meishucunian Stage in Yunnan Province, south China. Fossils and Strata 24:1–156.

Qian, Z.-S., and L.-Y. Geng. 1989. Chitinozoans from the Wufeng Formation (Ashgillian) of the Yangtze Region. Acta Micropalaeontologica Sinica 6:45–64 (in Chinese with English abstract).

Qing, H., C. R. Barnes, D. Buhl, and J. Veizer. 1998. The strontium isotopic composition of Ordovician and Silurian brachiopods and conodonts: Relationships to geological events and implications for coeval seawater. Geochimica et Cosmochimica Acta 62:1721–1733.

Qvale, G. 1980. New Caradocian ostracodes from the Oslo-Asker district, Norway. Norsk Geologisk Tidsskrift 60:93–116.

Rábano, I. 1990. Trilobites del Ordovícico Medio del sector meridional de la zona Centroibérica española. Publicaciones Especiales del Boletín Geológico y Minero. Instituto Technológico GeoMinero de España, Madrid, 233 pp.

Racheboeuf, P. R. 1994. Silurian and Devonian phyllocarid crustaceans from the Massif armoricain, NW France. Revue de Paléobiologie 13:281–305.

Racheboeuf, P. R., J. Vannier, and G. Ortega 2000. Ordovician phyllocarids (Arthropoda; Crustacea) from Argentina. Paläontologische Zeitschrift 74:317–333.

Railsback, L. B., S. C. Ackerly, T. F. Anderson, and J. L. Cisne. 1990. Paleontological and isotope evidence for warm saline deep waters in Ordovician oceans. Nature 343:156–159.

Ramirez, V., C. Thompson, and G. Viscarra. 1992. Fauna vertebrada ordovicica en la parte norte de la Cordillera Oriental de Bolivia. X Congreso Geologico Boliviano, La Paz, 26–30 October 1992, Abstracts, 127–128.

Ramos, V. A. 1984. Filocáridos (Crustacea) del Ordovícico Argentino. Actas del III° Congreso Argentino de Paleontología y Bioestratigrafía, Corrientes, 29–38.

Ramos, V. A., and G. Blasco. 1975. Sobre la presensia de un Eurypteride un la facies Graptolino de la Formación Yerba Loca, Departmento Jachal, Provincia de San Juan. Revista de las Asociación Geológica Argentina 30:287–289.

Ramsbottom, W. H. C. 1961. A monograph on British Ordovician Crinoidea. Palaeontographical Society Monograph 114(492):1–36.

Rao, R. I., and F. J. Flores. 1997. Conodontes ordovícicos (Tremadoc Superior) de la sierra de Aguilar, provincia de Jujuy, República Argentina. Bioestratigrafía y tafonomía. Revista Española de Micropaleontología 30:5–20.

Rao, R. I., M. A. Hünicken, and G. Ortega. 1994. Conodontes y graptolitos del Ordovícico Inferior (Tremadociano-Arenigiano) en el área de Purmamarca, provincia de Jujuy, Argentina. Anais Academia Brasileira de Ciências 66:59–83.

Rasmussen, J. A. 1991. Conodont stratigraphy of the Lower Ordovician Huk Formation at Slemmestad, southern Norway. Norsk Geologisk Tidsskrift 71:265–288.

Rasmussen, J. A. 1998. A reinterpretation of the conodont Atlantic Realm in the late early Ordovician (early Llanvirn); pp. 67–77 in H. Szaniawski (ed.), Proceedings of the Sixth European Conodont Symposium (ECOS VI), Warsaw. Palaeontologia Polonica 58.

Rasmussen, J. A. 2001. Conodont biostratigraphy and taxonomy of the Ordovician shelf margin deposits in the Scandinavian Caledonides. Fossils and Strata 48:1–179.

Rasmussen, J. A., and S. Stouge. 1995. Late Arenig–Early Llanvirn conodont biofacies across the Iapetus Ocean; pp. 443–447 in J. D. Cooper, M. L. Droser, and S. C. Finney (eds.), Ordovician Odyssey: Short Papers, 7th International Symposium on the Ordovician System. Book 77, Pacific Section Society for Sedimentary Geology (SEPM), Fullerton, California.

Rasul, S. 1976. New species of the genus *Vulcanisphaera* (Acritarcha) from the Tremadocian of England. Micropaleontology 22:479–484.

Ratter, V. A., and J. C. W. Cope. 1998. New neotaxodont bivalves from the Silurian of South Wales and their phylogenetic significance. Palaeontology 41:975–991.

Raup, D. 1985. Mathematical models of cladogenesis. Paleobiology 11:42–52.

Reed, F. R. C. 1912. Ordovician and Silurian fossils from the Central Himalayas. Memoirs of the Geological Survey of India, Palaeontologica Indica, Ser. 15, 7(2):1–168.

Regnéll, G. 1945. Non-crinoid Pelmatozoa from the Paleozoic of Sweden. Meddelanden fran Lunds Geologisk-Mineralogiska Institition 108:1–255.

Reich, M. 2001. Ordovician holothurians from the Baltic Sea area; pp. 93–96 in M. Barker (ed.), Echinoderms 2000. Balkema, Lisse, Netherlands.

Reif, W.-E. 2002. Evolution of the dermal skeleton of vertebrates: Concepts and methods. Neues Jahrbuch für Geologie und Paläontologie, Abhandlungen 223:53–78.

Renz, G. W. 1990. Late Ordovician (Caradocian) radiolarians from Nevada. Micropaleontology 36:367–377.

Repetski, J. E. 1978. A fish from the Upper Cambrian of North America. Science 200:529–531.

Repetski, J. E. 1980. Fossils; pp. 202–204 in McGraw-Hill Yearbook of Science and Technology (for 1979).

Repetski, J. E. 1982. Conodonts from the El Paso Group (Lower Ordovician) of westernmost Texas and southern New Mexico. New Mexico Bureau of Mines and Mineral Resources, Memoir 40:1–12.

Repetski, J. E., and R. L. Ethington. 1977. Conodonts from graptolite facies in the Ouachita Mountains, Arkansas and Oklahoma. Arkansas Geological Commission 1:92–106.

Retallack, G. J. 2001. *Scoyenia* burrows from Ordovician palaeosols of the Juniata Formation in Pennsylvania. Lethaia 44:209–235.

Retallack, G. J., and C. R. Feakes. 1987. Trace fossil evidence for Late Ordovician animals on land. Science 235:61–63.

Rhebergen, F., R. Eggink, T. Koops, and B. Rhebergen. 2001. Ordovicische zwerfsteensponzen. Staringia 9. 55(2):1–144 Nederlandse Geologische Vereniging, Maastricht (in Dutch, with English and German summaries).

Rhodes M. C., and R. J. Thompson. 1993. Comparative physiology of suspension feeding in living brachiopods and bivalves: Evolutionary implications. Paleobiology 19:322–334.

Richards, R. P. 1974. Ecology of the Cornulitidae. Journal of Paleontology 48:514–523.

Richardson, J. B. 1988. Late Ordovician and Early Silurian cryptospores and miospores from northeast Libya; pp. 89–109 *in* A. El-Arnauti, B. Owens, and B. Thusu (eds.), Subsurface Palynostratigraphy of Northeast Libya. Garyounis University Publications, Benghazi, Libya.

Richardson, J. B. 1996. Lower and Middle Palaeozoic records of terrestrial palynomorphs; pp. 555–574 *in* J. Jansonius and D. C. McGregor (eds.), Palynology: Principles and Applications, Vol. 2. American Association of Stratigraphic Palynologists Foundation. Publishers Press, Salt Lake City, Utah.

Richardson, J. B., J. H. Ford, and F. Parker. 1984. Miospores, correlation and age of some Scottish Lower Old Red Sandstone sediments from the Strathmore region (Fife and Angus). Journal of Micropalaeontology 3:109–124.

Richardson, J. G., and L. E. Babcock. 2002. Weird things from the Middle Ordovician of North America interpreted as conulariid fragments. Journal of Paleontology 76:391–399.

Rickards, R. B. 1975. Palaeoecology of the Graptolithina, an extinct class of the phylum Hemichordata. Biological Reviews 50:397–436.

Rickards, R. B. 1990. Plankton; pp. 49–52 *in* D. E. G. Briggs and P. R. Crowther (eds.), Palaeobiology: A Synthesis. Blackwell, Oxford.

Rickards, R. B. 2002. The graptolite age of the type Ashgill Series (Ordovician), Cumbria, UK. Proceedings of the Yorkshire Geological Society 54:1–16.

Riding, J., and S. Duxbury. 1993. A new non-marine acritarch from the Middle Jurassic of Britain. Special Papers in Palaeontology 48:57–66.

Riding, R. 1977. Problems of affinity in Palaeozoic calcareous algae; pp. 202–211 *in* E. Flügel (ed.), Fossil Algae: Recent Results and Developments. Springer-Verlag, Berlin.

Riding, R. 1991. Cambrian calcareous Cyanobacteria and Algae; pp. 305–334 *in* R. Riding (ed.), Calcareous Algae and Stromatolites. Springer-Verlag, Berlin.

Riding, R. 1992. Temporal variation in calcification in marine cyanobacteria. Journal of the Geological Society of London 149:979–989.

Riding, R. 2000 [dated 2001]. Calcified algae and bacteria; pp. 445–473 *in* A. Yu. Zhuravlev and R. Riding (eds.), The Ecology of the Cambrian Radiation. Columbia University Press, New York.

Riding, R., and J. Fan. 2001. Ordovician calcified algae and cyanobacteria, northern Tarim Basin subsurface, China. Palaeontology 44:783–810.

Riding, R., J. C. W. Cope, and P. D. Taylor. 1998. A coralline-like red alga from the Lower Ordovician of Wales. Palaeontology 41:1069–1076.

Riegel, W. 1996. The geologic significance of the Late Paleozoic phytoplankton blackout. Ninth International Palynological Congress, Houston, Texas, Program and Abstracts, 133.

Rigby, J. K. 1971. Sponges of the Ordovician Cat Head Member, Lake Winnipeg, Manitoba. Geological Survey of Canada, Bulletin 202:35–68.

Rigby, J. K. 1986. Sponges from the Burgess Shale (Middle Cambrian), British Columbia. Paleontographica Canadiana 2:1–105.

Rigby, J. K. 1991. Evolution of Paleozoic heteractinid calcareous sponges and demosponges: Patterns and records; pp. 83–101 *in* J. Reitner and H. Keupp (eds.), Fossils and Recent Sponges. Springer-Verlag, Berlin.

Rigby, J. K., and T. Bayer. 1976. Sponges of the Ordovician Maquoketa Formation in Minnesota and Iowa. Journal of Paleontology 45:608–627.

Rigby, J. K., and B. D. E. Chatterton. 1989. Middle Silurian Ludlovian and Wenlockian sponges from Baillie-Hamilton and Cornwallis Islands, Arctic Canada. Geological Survey of Canada, Bulletin 391:1–69.

Rigby, J. K., and A. Desrochers. 1995. Lower and Middle Ordovician lithistid demosponges from the Mingan Islands, Gulf of St. Lawrence, Quebec, Canada. The Paleontological Society, Memoir 41 (Journal of Paleontology, 69[4], supplement):1–35.

Rigby, J. K., and L. F. Hintze. 1977. Early Middle Ordovician corals from western Utah. Utah Geology 4:105–111.

Rigby, J. K., and X.-G. Hou. 1995. Lower Cambrian demosponges and hexactinellid sponges from Yunnan, China. Journal of Paleontology 69:1009–1019.

Rigby, J. K., and B. D. Webby. 1988. Late Ordovician sponges from the Malongulli Formation of central New South Wales. Palaeontographica Americana 56:1–147.

Rigby, J. K., A. W. Potter, and R. B. Blodgett. 1988. Ordovician sphinctozoan sponges from Alaska and Yukon Territory. Journal of Paleontology 62:731–746.

Rigby, J. K., G. Budd, R. Wood, and F. Debrenne. 1993. Porifera; pp. 71–99 *in* M. J. Benton (ed.), The Fossil Record 2. Chapman and Hall, London.

Rigby, S. 1997. A comparison of the colonization of the planktic realm and the land. Lethaia 30:11–17.

Ripperdan, R. L., and J. F. Miller. 1995. Carbon isotope ratios from the Cambrian-Ordovician boundary section at Lawson Cove, Ibex area, Utah; pp. 129–132 in J. D. Cooper, M. L. Droser, and S. C. Finney (eds.), Ordovician Odyssey: Short Papers, 7th International Symposium on the Ordovician System. Book 77, Pacific Section Society for Sedimentary Geology (SEPM), Fullerton, California.

Ripperdan, R. L., M. Magaritz, R. S. Nicoll, and J. H. Shergold. 1992. Simultaneous changes in carbon isotopes, sea level, and conodont biozones within the Cambrian-Ordovician boundary interval at Black Mountain, Australia. Geology 20:1039–1042.

Ripperdan, R. L., M. Magaritz, and J. L. Kirschvink. 1993. Carbon isotope and magnetic polarity evidence for non-depositional events within the Cambrian-Ordovician boundary section near Dayangcha, Jilin Province, China. Geological Magazine 130:442–452.

Ritchie, A. 1985. *Arandaspis prionotolepis.* The southern four-eyed fish; pp. 95–101 in P. V. Rich, G. F. Van Tets, and F. Knight (eds.), Kadimakara—Extinct Vertebrates of Australia. Pioneer Design Studio, Victoria, Australia.

Ritchie, A. 1991. New discoveries of Ordovician vertebrates from the Northern Territory, Australia. VIIe Symposium International: Etudes des Vertébrés inférieurs (Parc de Miguasha, Miguasha, Quebec, 1991), Résumés, 39.

Ritchie, A., and J. Gilbert-Tomlinson. 1977. First Ordovician vertebrates from the Southern Hemisphere. Alcheringa 1:351–368.

Robardet, M., and F. Doré. 1988. The Late Ordovician diamictic formations from south-western Europe: North Gondwana glaciomarine deposits. Palaeogeography, Palaeoclimatology, Palaeoecology 66:19–31.

Robison, R. A. 1969. Annelids from the Middle Cambrian Spence Shale of Utah. Journal of Paleontology 43:1169–1173.

Rogers, N., and C. R. van Staal. In press. Volcanology and tectonic setting of the northern Bathurst Mining Camp. Part 2, Mafic volcanic contraints on back-arc opening; in W. D. Goodfellow, S. R. McCutcheon, and J. M. Peter (eds.), Massive sulfide deposits of the Bathurst Mining Camp, New Brunswick and northern Maine. Economic Geology, Monograph 11. Economic Geology Publishing Co., New Haven, Connecticut.

Rohr, D. M. 1979. Geographic distribution of the Ordovician gastropod *Maclurites*; pp. 45–52 in J. Gray and A. J. Boucot (eds.), Historical Biogeography, Plate Tectonics, and the Changing Environment. Oregon State University Press, Corvallis.

Rohr, D. M. 1994. Ordovician (Whiterockian) gastropods of Nevada—Bellerophontoidea, Macluritoidea, and Euomphaloidea. Journal of Paleontology 68:473–486.

Rohr, D. M., and A. P. Gubanov. 1997. Macluritid opercula (Gastropoda) from the Middle Ordovician of Siberia and Alaska. Journal of Paleontology 71:394–400.

Rohr, D. M., R. B. Blodgett, and W. M. Furnish. 1992. *Maclurina manitobensis* (Whiteaves) (Ordovician Gastropoda), the largest known Paleozoic gastropod. Journal of Paleontology 66:880–884.

Rolfe, W. D. I. 1969. Phyllocarida; pp. R296–331 in R. C. Moore (ed.), Treatise on Invertebrate Paleontology. Part R, Arthropoda 4. Geological Society of America, Boulder, Colorado, and University of Kansas Press, Lawrence.

Rolfe, W. D. I. 1981. *Septemchiton*—a misnomer. Journal of Paleontology 55:675–678.

Rolfe, W. D. I. 1985. Early terrestrial arthropods: A fragmentary record; pp. 207–218 in W. G. Chaloner and J. D. Lawson (eds.), Evolution and Environment in the Late Silurian and Devonian. Philosophical Transactions of the Royal Society of London, B 309.

Rong, J.-Yu. 1979. The *Hirnantia* fauna of China with comments on the Ordovician Silurian boundary. Acta Stratigraphica Sinica 3:1–29.

Rong, J.-Yu, R.-B. Zhan, and N.-R. Han. 1994. The oldest known *Eospirifer* (Brachiopoda) in the Changwu Formation (Late Ordovician) of western Zhejiang, east China, with a review of the earliest spiriferoids. Journal of Paleontology 68:763–776.

Rong, J.-Yu, A. J. Boucot, Y.-Z. Su, and D. L. Strusz. 1995. Biogeographical analysis of late Silurian brachiopod faunas, chiefly from Asia and Australia. Lethaia 28:39–60.

Rong, J.-Yu, R.-B. Zhan, and D. A. T. Harper. 1999. Late Ordovician (Caradoc-Ashgill) brachiopod faunas with *Foliomena* based on data from China. Palaios 14:412–431.

Rõõmusoks, A. 1970. Stratigraphy of the Viruan Series (Middle Ordovician) in Northern Estonia. Valgus, Tallinn, 355 pp. (in Russian with English summary).

Ropot, V. F., and V. I. Pushkin. 1987. Ordovik Belorussii. Nauka i technika, Minsk, 234 pp.

Rosenzweig, M. L. 1995. Species Diversity in Space and Time. Cambridge University Press, Cambridge, 436 pp.

Ross, C. A., and J. R. P. Ross. 1995. North American depositional sequences and correlations; pp. 309–313 in J. D. Cooper, M. L. Droser, and S. C. Finney (eds.), Ordovician Odyssey: Short Papers, 7th International Symposium on the Ordovician System. Book 77, Pacific Section Society for Sedimentary Geology (SEPM), Fullerton, California.

Ross, J. R. P. 1984. Palaeoecology of Ordovician Bryozoa; pp. 141–148 in D. L. Bruton (ed.), Aspects of the Ordovician System. Palaeontological Contributions from the University of Oslo, 295. Universitetsforlaget, Oslo.

Ross, J. R. P. 1985. Biogeography of Ordovician ectoproct (bryozoan) faunas; pp. 265–271 in C. Nielsen and G. P. Larwood (eds.), Bryozoa: Ordovician to Recent. Olsen and Olsen, Fredensborg, Denmark.

Ross, J. R. P., and C. A. Ross. 1992. Ordovician sea-level fluctuations; pp. 327–335 in B. D. Webby and J. R. Laurie (eds.), Global Perspectives on Ordovician Geology. Balkema, Rotterdam.

Ross, J. R. P., and C. A. Ross. 1996. Bryozoan evolution and dispersal and Paleozoic sea-level fluctuations; pp. 243–258 in D. P. Gordon, A. M. Smith, and J. A. Grant-Mackie (eds.), Bryozoans in Space and Time. NIWA (National Institute of Water and Atmospheric Research), Wellington, New Zealand.

Ross, R. J., Jr. 1951. Stratigraphy of the Garden City Formation in northeastern Utah, and its trilobite faunas. Peabody Museum of Natural History, Yale University, Bulletin 6:1–161.

Ross, R. J., Jr. 1975. Early Paleozoic trilobites, sedimentary facies, lithospheric plates, and ocean currents. Fossils and Strata 4:307–329.

Ross, R. J., Jr., et al. 1982. The Ordovician System in the United States: Correlation Chart and Explanatory Notes. Publication 12. International Union of Geological Sciences, Paris, 73 pp.

Ross, R. J., Jr., L. F. Hintze, R. L. Ethington, J. F. Miller, M. E. Taylor, and J. E. Repetski. 1997. The Ibexian, lowermost Series in the North American Ordovician. United States Geological Survey, Professional Paper 1579A:1–50.

Roux, A. 1991a. Ordovician algae and global tectonics; pp. 335–348 in R. Riding (ed.), Calcareous Algae and Stromatolites. Springer-Verlag, Berlin.

Roux, A. 1991b. Ordovician to Devonian marine calcareous algae; pp. 349–369 in R. Riding (ed.), Calcareous Algae and Stromatolites. Springer-Verlag, Berlin.

Rowland, S. M., and R. S. Shapiro. 2002. Reef patterns and environmental influences in the Cambrian and earliest Ordovician; pp. 95–128 in W. Kiessling, E. Flügel and J. Golonka (eds.), Phanerozoic Reef Patterns. SEPM (Society for Sedimentary Geology), Special Publication 72.

Rozhnov, S. V. 1989. New data about rhipidocystids (Eocrinoidea); pp. 38–57 in Fossil and recent echinoderm researches. Academy of Sciences Estonian SSR, Tallinn (in Russian with English summary).

Rozhnov, S. V. 2000 [dated 2001]. Evolution of the hardground community; pp. 238–253 in A. Yu. Zhuravlev and R. Riding (eds.), The Ecology of the Cambrian Radiation. Columbia University Press, New York.

Rubel, M., and A. D. Wright. 2000. Suborder Clitambonitidina; pp. 692–708 in R. L. Kaesler (ed.), Treatise on Invertebrate Paleontology. Part H, Brachiopoda, rev., Vol. 3. Geological Society of America, Boulder, Colorado, and University of Kansas Press, Lawrence.

Rudkin, D. 2001. The first complete scleritome of *Plumulites canadensis* (Machaeridia) from the Ordovician (Late Caradoc) of Ontario. Canadian Paleontological Convention—2001, Program and Abstracts 11:43.

Rudwick, M. J. S. 1970. Living and Fossil Brachiopods. Hutchinson, London, 199 pp.

Ruedemann, R. 1909. Some marine algae from the Trenton limestone of New York. New York State Museum, Bulletin 133:194–210.

Ruedemann, R. 1916. *Spathiocaris* and the Discinocarina. New York State Museum, Bulletin 189:98–112.

Ruedemann, R. 1942. Some new eurypterids from New York. New York State Museum, Bulletin 327:24–29.

Ruedemann, R., and T. Y. Wilson. 1936. Eastern New York Ordovician cherts. Geological Society of America, Bulletin 74:1535–1586.

Rundqvist, D. V., and F. P. Mitrofanov (eds.). 1993. Precambrian Geology of the U.S.S.R. Elsevier, Amsterdam, 528 pp.

Runnegar, B., and P. A. Jell. 1976. Australian Middle Cambrian molluscs and their bearing on early molluscan evolution. Alcheringa 1:109–138.

Runnegar, B., and J. Pojeta, Jr. 1974. Molluscan phyllogeny: The paleontological viewpoint. Science 186:311–317.

Runnegar, B., J. Pojeta, Jr., N. J. Morris, J. D. Taylor, M. E. Taylor, and G. McClung. 1975. Biology of the Hyolitha. Lethaia 8:181–191.

Runnegar, B., J. Pojeta, Jr., M. E. Taylor, and D. Collins. 1979. New species of the Cambrian and Ordovician chitons *Matthevia* and *Chelodes* from Wisconsin and Queensland: Evidence for the early history of polyplacophoran mollusks. Journal of Paleontology 53:1374–1394.

Rushton, A. W. A. 1988. Tremadoc trilobites from the Skiddaw Group in the English Lake District. Palaeontology 31:677–698.

Rushton, A. W. A., and M. Williams. 1996. The tail-piece of the crustacean *Caryocaris wrightii* from the Arenig rocks of England and Ireland. Irish Journal of Earth Sciences 15:107–111.

Ryland, J. S. 1970. Bryozoans. Hutchinson, London, 175 pp.

Sadler, P. M. 2001. Constrained Optimization Approaches to the Stratigraphic Correlation and Seriation Problems: A Users' Guide to the CONOP Program Family, version 6.1. Copyright Peter M. Sadler, Riverside, California, 142 pp.

Salvador, A. 1994. International Stratigraphic Guide: A Guide to Stratigraphic Classification. Second Edition. International Union of Geological Sciences and Geological Society of America, Washington D.C., xix + 214 pp.

Salvini-Plawen, L. von. 1980. A reconsideration of systematics in the Mollusca (phylogeny and higher classification). Malacologia 19:249–278.

Salvini-Plawen, L. von. 1990. Origin, phylogeny and classification of the phylum Mollusca. Iberus 9:1–33.

Salvini-Plawen, L. von, and G. Haszprunar. 1987. The Vetigastropoda and the systematics of streptonerous

Gastropoda (Mollusca). Journal of Zoology 11:747–770.

Salvini-Plawen, L. von, and G. Steiner. 1996. Synapomorphies and plesiomorphies in higher classification of Mollusca; pp. 29–52 in J. D. Taylor (ed.), Origin and Evolutionary Radiation of the Mollusca. Oxford University Press, New York.

Samuelsson, J., and T. Servais. 2001. Chitinozoa biostratigraphy of subsurface Ordovician sediments from the Lohme 2/70 well, Island of Rügen (NE Germany). Neues Jahrbuch für Paläontologie und Geologie, Abhandlungen 222:73–90.

Samuelsson, J., and J. Verniers. 2000. Ordovician chitinozoan biozonation of the Brabant Massif, Belgium. Review of Palaeobotany and Palynology 113:105–129.

Samuelsson, J., J. Verniers, and M. Vecoli. 2000. Chitinozoan faunas from the Rügen Ordovician (Rügen 5/66 and Binz 1/73 wells), NE Germany. Review of Palaeobotany and Palynology 113:131–143.

Samuelsson, J., A. Gerdes, L. Koch, T. Servais, and J. Verniers. 2002a. Chitinozoa and Nd isotope stratigraphy of the Ordovician rocks in the Ebbe Anticline, NW Germany. The Geological Society, London, Special Publication 201:115–131.

Samuelsson, J., M. Vecoli, W. S. Bednarczyk, and J. Verniers. 2002b. Timing of the Avalonia–Baltica plate convergence as inferred from palaeogeographic and stratigraphic data of chitinozoan assemblages in west Pomerania, northern Poland. The Geological Society, London, Special Publication 201:95–103.

Sánchez, T. M. 1997. Additional Mollusca (Bivalvia and Rostroconchia) from the Suri Formation, Early Ordovician (Arenig), western Argentina. Journal of Paleontology 71:1046–1054.

Sánchez, T. M. 2001. Moluscos bivalvos de la Formación Molles (Arenigiano medio), sierra de Famatina, Argentina. Ameghiniana 38:185–193.

Sánchez, T. M., and C. Babin. 1993. Un insolite bivalve, *Catamarcaia* n.g., de l'Arenig (Ordovicien inférieur) d'Argentine. Comptes rendus de l'Académie des Sciences, Paris, Série 2, 316:265–271.

Sánchez, T. M., J. L. Benedetto, and E. Brussa. 1991. Late Ordovician stratigraphy, paleoecology, and sea level in the Argentine Precordillera; pp. 245–258 in C. R. Barnes and S. H. Williams (eds.), Advances in Ordovician Geology. Geological Survey of Canada, Paper 90-9.

Sánchez, T. M., M. G. Carrera, and B. G. Waisfeld. 2002. Hierarchy of factors controlling faunal distribution: A case study from the Ordovician of the Argentine Precordillera. Palaios 17:309–326.

Sandberg, P. A. 1983. An oscillating trend in Phanerozoic non-skeletal carbonate mineralogy. Nature 305:19–22.

Sansom, I. J., M. P. Smith, M. M. Smith, and P. Turner. 1995. The Harding Sandstone revisited—a new look at some old bones; pp. 57–59 in H. Lelièvre, S. Wenz, A. Blieck, and R. Cloutier (eds.), Premiers Vertébrés et Vertébrés inférieurs (VIIIe Congrès International, Paris, 1995). Geobios, Mémoire Spécial 19.

Sansom, I. J., M. M. Smith, and M. P. Smith. 1996. Scales of thelodont and shark-like fishes from the Ordovician of Colorado. Nature 379(6566):628–630.

Sansom, I. J., M. P. Smith, M. M. Smith, and P. Turner. 1997. *Astraspis*—the anatomy and histology of an Ordovician fish. Palaeontology 40:625–643.

Sansom, I. J., R. J. Aldridge, and M. M. Smith. 2000. A microvertebrate fauna from the Llandovery of South China. Transactions of the Royal Society of Edinburgh, Earth Sciences 90:255–272.

Sansom, I. J., M. M. Smith, and M. P. Smith. 2001. The Ordovician radiation of vertebrates; pp. 156–171 in P. Ahlberg (ed.), Major Events in Early Vertebrate Evolution: Palaeontology, Phylogeny, Genetics and Development. Systematics Association Special Volume, 61. Taylor and Francis, London.

Sarmiento, G. N. 1985. La Biozona de *Amorphognathus variabilis–Eoplacognathus suecicus* (Conodonta), Llanvirniano inferior en el flanco oriental de la sierra de Villicum. Primeras Jornadas sobre Geología de Precordillera, San Juan, Actas:119–123.

Sarmiento, J. 1992. Biogeochemical ocean models; pp. 519–564 in K. E. Trenberth (ed.), Climate System Modelling. Cambridge University Press, Cambridge.

Sarv, L. I. 1959. Ordovician Ostracods in the Estonian S.S.R. Eesti NSV Teaduste Akadeemia Geoloogia Instituudi Uurimused 4:1–211.

Sayar, C. 1964. Ordovician conulariids from the Bosphorus area, Turkey. Geological Magazine 101:193–197.

Sayar, C., and R. Schallreuter. 1989. Ordovician ostracodes from Turkey. Neues Jahrbuch für Geologie und Paläontologie, Monatshefte 1989:233–242.

Schallreuter, R. 1986 [dated 1987]. Ostrakoden aus Öjlemyrflint-Geschieben von Sylt; pp. 203–232 in U. von Hacht (ed.), Fossilien von Sylt II. von Hacht, Hamburg.

Schallreuter, R. 1987. Geschiebe-Ostrakoden II. Neues Jahrbuch für Geologie und Paläontologie, Abhandlungen 174:23–53.

Schallreuter, R. 1990. Ordovician ostracodes and echinoids from the Carnic Alps and their relations to Bohemia and Baltoscandia. Neues Jahrbuch für Geologie und Paläontologie, Monatshefte 1990:120–128.

Schallreuter, R. 1993a. Beiträge zur Geschiebekunde Westfalens II Ostrakoden aus ordovizischen Geschieben II. Geologie und Paläontologie in Westfalen 27:1–273.

Schallreuter, R. 1993b. On *Eopilla ingelorae* Schallreuter gen. et sp. nov. A Stereo-Atlas of Ostracod Shells 20:117–120.

Schallreuter, R. 1994. Schwarze Orthocerenkalkgeschiebe. Archiv für Geschiebekunde 1:491–540.

Schallreuter, R. 1999. Eine neue Ostrakodenfauna aus dem Ordoviz Argentiniens. Greifswalder Geowissenschaftliche Beiträge 6:55–71.

Schallreuter, R., and I. Hinz-Schallreuter. 1998. A Geschiebe from Armorica in the Thuringian Lederschiefer (Ordovician). Archiv für Geschiebekunde 2:323–360.

Schallreuter, R., and L. Koch. 1999. Ostracodes from the Lower Llanvirnian (Ordovician) of Kiesbert (Ebbe anticline, Rhenish Massif). Neues Jahrbuch für Geologie und Paläontologie, Monatshefte 1999:477–489.

Schallreuter, R., and M. Krúta. 2001a. Ostracodes from the Letná Formation (Ordovician) of Blyskava (Bohemia). Sborník Národního muzea v Praze B 56:85–94.

Schallreuter, R., and M. Krúta. 2001b. Ostracodes from the Dobrotivá Formation (Ordovician, Bohemia). Sborník Národního muzea v Praze B 56:95–103.

Schallreuter, R., and D. J. Siveter. 1985. Ostracodes across the Iapetus Ocean. Palaeontology 28:577–598.

Schallreuter, R., J. Verniers, and P. de Geest. 2000. An Ordovician ostracode from Belgium. Neues Jahrbuch für Geologie und Paläontologie, Monatshefte 2000:570–576.

Schallreuter, R., A. V. Kanygin, and I. Hinz-Schallreuter. 2001. Ordovician ostracodes from Novaya Zemlya. Journal of the Czech Geological Society 46:199–212.

Schlotheim, E. F. von. 1820. Die Petrefactenkunde auf ihrem jetzigen Standpunkte durch die Beschreibung seiner Sammlung versteinerter und fossiler Überreste des Thier- und Pflanzenreiches der Vorwelt erläuteri. Becker (Becker'schen buchandlung), Gotha, 437 pp.

Schmidt-Gündel, O. 1994. Die unterordovicizischen Graptolithenfaunen des Bogo- und des Lo-Schiefers (Sr Trøndelag, West-Norwegen). Doctoral thesis, Technical University of Berlin.

Schramm, F. R. 1973. Pseudocoelomates and a nemertine from the Illinois Pennsylvanian. Journal of Paleontology 47:985–989.

Schubert, R. J., and L. Waagen. 1904. Der untersilurischen Phyllopodengattungen *Ribeiria* Sharpe and *Ribeirella* gen. nov. Jahrbuch der Kaiserlich-Königlichen geologischen Reichsanstalt 53:33–50.

Schultze, H.-P. 1996. Conodont histology: An indicator of vertebrate relationship? pp. 275–285 in S. Turner and A. Blieck (eds.), Gross Symposium, Vol. 2. Modern Geology 20(3–4).

Scotese, C. R. 1997. Paleogeographic Atlas. PALEOMAP Progress Report 90-0497. PALEOMAP Project, University of Texas at Arlington.

Scotese, C. R., and W. S. McKerrow. 1990. Revised world maps and introduction; pp. 1–21 in W. S. McKerrow and C. R. Scotese (eds.), Palaeozoic Palaeogeography and Biogeography. The Geological Society, London, Memoir 12.

Scotese, C. R., and W. S. McKerrow. 1991. Ordovician plate tectonic reconstructions; pp. 271–282 in C. R. Barnes and S. H. Williams (eds.), Advances in Ordovician Geology. Geological Survey of Canada, Paper 90-9.

Scrutton, C. T. 1979. Early fossil cnidarians; pp. 161–207 in M. R. House (ed.), The Origin of Major Invertebrate Groups. Academic Press, London.

Scrutton, C. T. 1984. Origins and early evolution of tabulate corals. Palaeontographica Americana 54:110–118.

Scrutton, C. T. 1988. Patterns of extinction and survival in Palaeozoic corals; pp. 65–88 in G. P. Larwood (ed.), Extinction and Survival in the Fossil Record. Systematics Association, Special Volume 42. Clarendon Press, Oxford.

Scrutton, C. T. 1997. The Palaeozoic corals. I. Origins and relationships. Proceedings of the Yorkshire Geological Society 51:177–208.

Scrutton, C. T. 1998. The Palaeozoic corals. II. Structure, variation and palaeoecology. Proceedings of the Yorkshire Geological Society 52:1–57.

Scrutton, C. T., and E. N. K. Clarkson. 1991. A new scleractinian-like coral from the Ordovician of the Southern Uplands, Scotland. Palaeontology 34:179–194.

Seddon, G., and W. C. Sweet. 1971. An ecologic model for conodonts. Journal of Paleontology 45:869–880.

Seilacher, A. 1967. Fossil Behavior. Scientific American 27:72–80.

Seilacher, A. 1970. *Cruziana* stratigraphy of "non-fossiliferous" Palaeozoic sandstones; pp. 447–476 in T. P. Crimes and J. C. Harper (eds.), Trace Fossils. Geological Journal, Special Issue 3. Seel House Press, Liverpool.

Seilacher, A. 1974. Flysch trace fossils: Evolution of behavioural diversity in the deep-sea. Neues Jahrbuch für Geologie und Paläontologie, Monatshefte 1974:233–245.

Seilacher, A. 1977. Evolution of trace fossil communities; pp. 359–376 in A. Hallam (ed.), Patterns of Evolution. Elsevier Science, Amsterdam.

Seilacher, A. 1985. Trilobite paleobiology and substrate relationships. Royal Society of Edinburgh Transactions, Earth Sciences 76:231–237.

Seilacher, A. 1990. Paleozoic trace fossils; pp. 649–670 in R. Said (ed.), The Geology of Egypt. Balkema, Rotterdam.

Seilacher, A. 1992. An updated *Cruziana* stratigraphy of Gondwanan Palaeozoic sandstones; pp. 1565–1580 in M. J. Salem (ed.), The Geology of Libya 5. Elsevier Science, Amsterdam.

Seilacher, A. 1994. How valid is *Cruziana* stratigraphy? Geologische Rundschau 83:752–758.

Seilacher, A. 1999. Biomat-related lifestyle in the Precambrian. Palaios 14:86–93.

Seilacher, A. 2000. Ordovician and Silurian arthrophycid ichnostratigraphy; pp. 237–258 in M. A. Sola and D. Worsley (eds.), Geological Exploration in Murzuk Basin. Elsevier Science, Amsterdam.

Seilacher, A., and S. Alidou. 1988. Ordovician and Silurian trace fossils from Northern Benin (W. Africa). Neues Jahrbuch für Geologie und Paläontologie, Monatshefte 1988:431–439.

Seilacher-Drexler, E., and A. Seilacher. 1999. Undertraces of Sea Pens and Moon Snails and possible fossil counterparts. Neues Jahrbuch für Geologie und Paläontologie, Abhandlungen 214:195–210.

Selden, P. A. 1993. Arthropoda (Aglaspidida, Pycnogonida and Chelicerata); pp. 297–320 in M. J. Benton (ed.), The Fossil Record 2. Chapman and Hall, London.

Selley, R. C. 1970. Ichnology of Paleozoic sandstones in the Southern Desert of Jordan: A study of trace fossils in their sedimentologic context; pp. 477–488 in T. P. Crimes and J. C. Harper (eds.), Trace Fossils. Geological Journal, Special Issue 3. Seel House Press, Liverpool.

Semeniuk, V. 1971. Subaerial leaching in the limestones of the Bowan Park Group (Ordovician) of central western New South Wales. Journal of Sedimentary Petrology 41:939–950.

Šengor, A. M. C., and B. A. Natal'in. 1996. Palaeotectonics of Asia: Fragments of a synthesis; pp. 486–640 in A. Yin and T. M. Harrison (eds.), The Tectonic Evolution of Asia. Cambridge University Press, Cambridge.

Seo, K.-S., and R. L. Ethington. 1993. Conodonts from the Manitou Formation, Colorado, U.S.A. Journal of the Paleontological Society of Korea 9:77–92.

Sepkoski, J. J., Jr. 1975. Stratigraphic biases in the analysis of taxonomic survivorship. Paleobiology 1:343–355.

Sepkoski, J. J., Jr. 1978. A kinetic mode of Phanerozoic taxonomic diversity. I. Analysis of marine orders. Paleobiology 4:223–251.

Sepkoski, J. J., Jr. 1979. A kinetic model of Phanerozoic taxonomic diversity, II. Early Phanerozoic families and multiple equilibria. Paleobiology 5:222–251.

Sepkoski, J. J., Jr. 1981a. A factor analytic description of the Phanerozoic marine fossil record. Paleobiology 7:36–53.

Sepkoski, J. J., Jr. 1981b. The uniqueness of the Cambrian fauna; pp. 203–207 in M. E. Taylor (ed.), Short Papers for the 2nd International Symposium on the Cambrian System. United States Geological Survey, Open-File Report 81–743.

Sepkoski, J. J., Jr. 1982. A Compendium of Fossil Marine Families: Contributions in Biology and Geology. Milwaukee Public Museum, Milwaukee, 125 pp.

Sepkoski, J. J., Jr. 1984. A kinetic model of Phanerozoic taxonomic diversity. Part 3, Post-Paleozoic families and multiple equilibria. Paleobiology 10:246–267.

Sepkoski, J. J., Jr. 1986. Global bioevents and the question of periodicity; pp. 47–61 in O. H. Walliser (ed.), Global Bio-Events. Lecture Notes in Earth Sciences, 8. Springer-Verlag, Berlin.

Sepkoski, J. J., Jr. 1988. Alpha, beta, or gamma: Where does all the diversity go? Paleobiology 14:221–234.

Sepkoski, J. J., Jr. 1991a. A model of onshore-offshore change in faunal diversity. Paleobiology 17:157–176.

Sepkoski, J. J., Jr. 1991b. Diversity in the Phanerozoic oceans: A partisan review; pp. 210–236 in E. C. Dudley (ed.), The Unity of Evolutionary Biology. Dioscorides Press, Portland, Oregon.

Sepkoski, J. J., Jr. 1992a. A Compendium of Fossil Marine Animal Families. Second Edition. Contributions in Biology and Geology. Milwaukee Public Museum, Milwaukee, 156 pp.

Sepkoski, J. J., Jr. 1992b. Phylogenetic and ecologic patterns in the Phanerozoic history of marine biodiversity; pp. 77–100 in N. Eldredge (ed.), Systematics, Ecology, and the Biodiversity Crisis. Columbia University Press, New York.

Sepkoski, J. J., Jr. 1995. The Ordovician Radiations: Diversification and extinction shown by global genus-level taxonomic data; pp. 393–396 in J. D. Cooper, M. L. Droser, and S. C. Finney (eds.), Ordovician Odyssey: Short Papers, 7th International Symposium on the Ordovician System. Book 77, Pacific Section Society for Sedimentary Geology (SEPM), Fullerton, California.

Sepkoski, J. J., Jr. 1996. Patterns of Phanerozoic extinction: A perspective from global databases; pp. 35–51 in O. H. Walliser (ed.), Global Events and Event Stratigraphy in the Phanerozoic. Springer-Verlag, Berlin.

Sepkoski, J. J., Jr. 1997. Biodiversity: Past, present, and future (Presidential Address). Journal of Paleontology 71:533–539.

Sepkoski, J. J., Jr. 1998. Rates of speciation in the fossil record. Philosophical Transactions of the Royal Society of London, B 353:315–326.

Sepkoski, J. J., Jr. 2002. A Compendium of Fossil Marine Animal Genera (D. Jablonski and M. Foote, eds.). Bulletins of American Paleontology 363:1–560.

Sepkoski, J. J., Jr., and C. E. Koch. 1996. Evaluating paleontologic data relating to bio-events; pp. 21–34 in O. H. Walliser (ed.), Global Events and Event Stratigraphy in the Phanerozoic. Springer-Verlag, Berlin.

Sepkoski, J. J., Jr., and A. I. Miller. 1985. Evolutionary faunas and the distribution of Paleozoic benthic communities in space and time; pp. 153–190 in J. W. Valentine (ed.), Phanerozoic Diversity Patterns: Profiles in Macroevolution. AAAS [American Association for the Advancement of Science], Pacific Division, and Princeton University Press, Princeton.

Sepkoski, J. J., Jr., and P. M. Sheehan. 1983. Diversification, faunal change, and community replacement during the Ordovician radiations; pp. 673–718 in M. J. S. Tevesz and P. L. McCall (eds.), Biotic Interactions in Recent and Fossil Benthic Communities. Plenum Press, New York.

Sepkoski, J. J., Jr., R. K. Bambach, D. M. Raup, and J. W. Valentine. 1981. Phanerozoic marine diversity and the fossil record. Nature 293:435–437.

Sepkoski, J. J., Jr., F. K. McKinney, and S. Lidgard. 2000. Competitive displacement between post-Paleozoic cyclostome and cheilostome bryozoans. Paleobiology 26:7–18.

Serpagli, E. 1970. Ordovician conulariids of Sardinia. Bollettino della Società Paleontologica Italiana 8(1969):3–10.

Serpagli, E. 1974. Lower Ordovician conodonts from Precordilleran Argentina (Province of San Juan). Bolletino della Società Paleontologica Italiana 13:17–98.

Servais, T. 1998. An annotated bibliographical review of Ordovician acritarchs. Annales de la Société Géologique de Belgique 120:23–72.

Servais, T., and O. Fatka. 1997. Recognition of the Trans-European-Suture-Zone (TESZ) by the palaeobiogeographical distribution pattern of Early to Middle Ordovician acritarchs. Geological Magazine 134:617–625.

Servais, T., and F. Paris (eds.). 2000. Ordovician Palynology and Palaeobotany. Review of Palaeobotany and Palynology 113:1–212.

Servais, T., and L. Stricanne. 2001. Ordovician phytoplankton (acritarch) diversity: IGCP 410: The Great Ordovician Biodiversity Event. PaleoBios 21:11.

Servais, T., R. Brocke, O. Fatka, A. Le Hérissé, and S. G. Molyneux. 1997. Value and understanding of the term acritarch. Acta Universitatis Carolinae, Geologica 40:631–643.

Sharpe, D. 1853. Description of the new species of Zoophyta and Mollusca, Appendix B; pp. 146–158 in C. Ribeiro, Carboniferous and Silurian formations of the neighbourhood of Bussaco in Portugal. Quarterly Journal of the Geological Society of London, 9.

Sharpe, S. C. F. 1932. Eurypterid trails from the Ordovician. American Journal of Science 24:355–361.

Sheehan, P. M. 1973. Brachiopods from the Jerrestad Mudstone (Early Ashgillian, Ordovician) from a boring in southern Sweden. Geologica et Palaeontologica 7:59–76.

Sheehan, P. M. 1985. Reefs are not so different—they follow the evolutionary pattern of level-bottom communities. Geology 13:46–49.

Sheehan, P. M. 1996. A new look at Ecological Evolutionary Units (EEUs). Palaeogeography, Palaeoclimatology, Palaeoecology 127:21–32.

Sheehan, P. M. 2001a. History of marine biodiversity. Geological Journal 36:231–249.

Sheehan, P. M. 2001b. The Late Ordovician mass extinction. Annual Reviews of Earth and Planetary Sciences 29:331–364.

Sheehan, P. M., and P. J. Coorough. 1990. Brachiopod zoogeography across the Ordovician-Silurian extinction event; pp. 181–190 in W. S. McKerrow and C. R. Scotese (eds.), Palaeozoic Palaeogeography and Biogeography. The Geological Society, London, Memoir 12.

Sheehan, P. M., and D. R. J. Schiefelbein. 1984. The trace fossil *Thalassinoides* from the Upper Ordovician of the eastern Great Basin: Deep burrowing in the early Paleozoic. Journal of Paleontology 58:440–447.

Shen, Y.-B. 1986. *Caryocaris* from the Lower Ordovician of Jiangshan, Zhejiang. Kexue Tongbao 31(11):765–769.

Shergold, J. H. 1975. Late Cambrian and early Ordovician trilobites from the Burke River Structural Belt, western Queensland. Bureau of Mineral Resources, Geology and Geophysics, Australia, Bulletin 153:1–251.

Shergold, J. H. 1991. Late Cambrian and Early Ordovician trilobite faunas of the Pacoota Sandstone, Amadeus Basin, central Australia. Bureau of Mineral Resources, Geology and Geophysics, Australia, Bulletin 237:15–75.

Shergold, J. H., R. Elphinstone, J. R. Laurie, R. S. Nicoll, M. R. Walter, G. C. Young, and W. Zang. 1991. Late Proterozoic and early Palaeozoic palaeontology and biostratigraphy of the Amadeus Basin; pp. 97–111 in R. J. Korsch and J. M. Kennard (eds.), Geological and Geophysical Studies in the Amadeus Basin, Central Australia. Bureau of Mineral Resources, Geology and Geophysics, Australia, Bulletin 236.

Shields, G. A., G. A. F. Carden, J. Veizer, T. Meidla, J.-Y. Rong, and R.-Y. Li. 2003. Sr, C and O isotope geochemistry of Ordovician brachiopods: A major isotopic event around the Middle-Late Ordovician transition. Geochimica et Cosmochimica Acta 67:2005–2025.

Shu, D. 1990. Cambrian and lower Ordovician bradoriida from Zhejiang, Hunan and Shaanxi Provinces. Northwest University Press, Xian, 95 pp. (in Chinese with English abstract).

Shu, D., H. Luo, S. Conway-Morris, X. Zhang, S. Hu, L. Chen, J. Han, M. Zhu, Y. Li, and L. Chen. 1999a. Lower Cambrian vertebrates from south China. Nature 402(6757):42–46.

Shu, D., J. Vannier, H. Luo, L. Chen, X. Zhang, J. Han, and S. Hu. 1999b. The anatomy and lifestyle of *Kunmingella* (Arthropoda, Bradoriida) from the Chengjiang fossil Lagerstätte (Early Cambrian; Southern China). Lethaia 32:279–298.

Shuler, E. W. 1915. A new Ordovician eurypterid. American Journal of Science, ser. 4, 39:551–554.

Shuyskiy, V. P. 1987. Zelenyie vodorosli (Chlorophyta); pp. 38–109 in V. H. Dubatov (ed.), Iskopaemye vodorosli. Akademia Nauk SSSR, Sibirskoe Otdelenie, Trudy Instituta Geologii i geofiziki 674.

Sidaravichiene, N. 1992. Ordovician Ostracods of Lithuania. Litovsk Geological Institute, Vilnius, 252 pp.

Signor, P. W., and G. J. Vermeij. 1994. The plankton and the benthos: Origins and early history of an evolving relationship. Paleobiology 20:297–319.

Simms, M. J., A. S. Gale, P. M. Gilliland, E. P. F. Rose, and G. D. Sevastopulo. 1993. Echinodermata; pp. 491–528 in M. J. Benton (ed.), The Fossil Record 2. Chapman and Hall, London.

Sinclair, G. W. 1940. A discussion of the genus *Metaconularia* with descriptions of new species. Royal Society of Canada, Section 4, Transactions (Series 3) 34:101–121.

Sinclair, G. W. 1941. Notes on *Pseudoconularia* and *P. magnifica* (Spencer). Royal Society of Canada, Section 4, Transactions (Series 3) 35:125–129.

Sinclair, G. W. 1942. The Chazy Conulariida and their congeners. Annals of the Carnegie Museum 29:219–240.

Sinclair, G. W. 1944. Notes on the genera *Archaeoconularia* and *Eoconularia*. Royal Society of Canada, Section 4, Transactions (Series 3) 38:87–95.

Sinclair, G. W. 1946a. Three new conulariids from the Ordovician of Québec. Naturaliste Canadien 73:385–390.

Sinclair, G. W. 1946b. Notes on the nomenclature of *Hyolithes*. Journal of Paleontology 20:72–85.

Sinclair, G. W. 1948. The biology of the Conulariida. Doctoral thesis, McGill University, Montreal.

Sinclair, G. W. 1952. A classification of the Conularida. Fieldiana: Geology 10:135–145.

Sirenko, V. I. 1997. The importance of the development of articulamentum for taxonomy of chitons (Mollusca, Polyplacophora). Ruthenica 7:1–24.

Siveter, D. J., J. K. Ingham, R. B. Rickards, and B. Arnold. 1980. Highest Ordovician trilobites and graptolites from County Cavan, Ireland. Journal of Earth Sciences, Royal Dublin Society 2:193–207.

Siveter, D. J., S. E. Gabbott, R. J. Aldridge, and J. A. Theron. 2001a. The earliest myodocopes: Ostracodes from the Late Ordovician Soom Shale Lagerstätte of South Africa. 14th International Symposium on Ostracoda, Shizuoka, Programs and Abstracts, 42.

Siveter, D. J., M. Williams, and D. Walossek. 2001b. A phosphatocopid crustacean with appendages from the Lower Cambrian. Science 293:356–357.

Skevington, D. 1973. Ordovician graptolites; pp. 27–35 *in* A. Hallam (ed.), Atlas of Palaeobiogeography. Elsevier, Amsterdam.

Skevington, D. 1974. Controls influencing the composition and distribution of Ordovician graptolite faunal provinces. Special Papers in Palaeontology 13:59–73.

Sloss, L. L. 1963. Sequences in the cratonic interior of North America. Geological Society of America, Bulletin 74:93–114.

Smethurst, M. A., A. N. Khramov, and T. H. Torsvik. 1998. The Neoproterozoic and Palaeozoic data for the Siberian Platform: From Rodinia to Pangea. Earth Science Reviews 43:1–24.

Smith, A. B. 1988a. Patterns of diversification and extinction in early Palaeozoic echinoderms. Palaeontology 31:799–828.

Smith, A. B. 1988b. Fossil evidence for the relationships of extant echinoderm classes and their times of divergence; pp. 85–101 *in* C. R. C. Paul and A. B. Smith (eds.), Echinoderm Phylogeny and Evolutionary Biology. Clarendon Press, Oxford.

Smith, A. B. 2001. Large-scale heterogeneity of the fossil record: Implications for Phanerozoic biodiversity studies. Philosophical Transactions of the Royal Society of London, B 356:351–367.

Smith, A. B., and C. R. C. Paul. 1982. Revision of the class Cyclocystoidea (Echinodermata). Philosophical Transactions of the Royal Society of London, B 296(1083):577–684.

Smith, A. B., and J. J. Savill. 2001. *Bromidechinus,* a new Middle Ordovician Echinozoa (Echinodermata), and its bearing on the early history of echinoids. Transactions of the Royal Society of Edinburgh, Earth Sciences 92:137–147.

Smith, A. G., and R. D. Hoare. 1987. Paleozoic Polyplacophora: A checklist and bibliography. Occasional Papers of the California Academy of Sciences 146:1–71.

Smith, A. G., and D. F. Toomey. 1964. Chitons from the Kindblade Formation (Lower Ordovician), Arbuckle Mountains, Southern Oklahoma. Circular of the Oklahoma Geological Survey 66:1–41.

Smith, M. M., and B. K. Hall. 1990. Development and evolutionary origins of vertebrate skeletogenic and ontogenic tissues. Biological Reviews 65:277–373.

Smith, M. M., and I. J. Sansom. 1997. Exoskeletal microremains of an Ordovician fish from the Harding Sandstone of Colorado. Palaeontology 40:645–658.

Smith, M. M., I. J. Sansom, and P. Smith. 1995. Diversity of the dermal skeleton in Ordovician to Silurian vertebrate taxa from North America: History, skeletogenesis and relationships; pp. 65–70 *in* H. Lelièvre, S. Wenz, A. Blieck, and R. Cloutier (eds.), Premiers Vértébrés et Vértébrés inférieurs (VIIIe Congrès International, Paris, 1995). Geobios, Mémoire Spécial 19.

Smith, M. M., I. J. Sansom, and P. Smith. 1996. "Teeth" before armour: The earliest vertebrate mineralized tissues; pp. 303–319 *in* S. Turner and A. Blieck (eds.), Gross Symposium, Vol. 1. Modern Geology 20(3–4).

Smith, M. P. 1991. Early Ordovician conodonts of East and North Greenland. Meddelelser om Grønland, Geoscience 26:1–81.

Smith, M. P., and I. J. Sansom. 1995. The affinity of *Anatolepis* Bockelie and Fortey; pp. 61–63 *in* H. Lelièvre, S. Wenz, A. Blieck, and R. Cloutier (eds.), Premiers Vértébrés et Vértébrés inférieurs (VIIIe Congrès International, Paris, 1995). Geobios, Mémoire Spécial 19.

Smith, M. P., I. J. Sansom, and K. D. Cochrane. 2001. The Cambrian origin of vertebrates; pp. 67–84 *in* P. E. Ahlberg (ed.), Major Events in Early Vertebrate Evolution: Palaeontology, Phylogeny, Genetics and Development. Systematics Association Special Volume 61. Taylor and Francis, London.

Soja, C. M. 1992. Potential contributions of ancient oceanic islands to evolutionary theory. Journal of Geology 100:125–134.

Sokal, R. R., and F. J. Rohlf. 1981. Biometry. Second Edition. W. H. Freeman, San Francisco, 859 pp.

Sokolov, B. S. 1955. Tabulyaty paleozoya evropeyskoy chasti SSSR, Vvedenie: Obshchie voprosy sistematiki i istorii razvitiya tabulyat. Vsesoiuznyi Neftyanoy Nauchno-Issledovatelskii Geologo-Razvedochny Institut (VNIGRI), Trudy, n.s., 85:1–527

Sokolov, B. S. 1962. Podklass Tabulata; pp. 192–265 *in* Yu. A. Orlov (ed.), Osnovy paleontologii. Vol. 2, Gubki,

Arkheotsiaty, Kishechnopolostnye, Chervi. Izdatel'stvo Akademia Nauk SSSR, Moscow.

Sokolov, B. S., and Yu. I. Tesakov. 1963. Tabulyaty paleozoya Sibiri [Paleozoic Tabulata of Siberia]. Akademia Nauk SSSR, Sibirskoe Otdelenie, Institut Geologii i Geofiziki. Izdatel'stvo, Moscow, 188 pp.

Sokolov, B. S., T. N. Alikova, B. M. Keller, O. I. Nikiforova, and A. M. Obut. 1960. Stratigraphy, correlation and paleogeography of the Ordovician deposits of the USSR. Reports of the 21st International Geological Congress, Copenhagen 7:44–57.

Soufiane, A., and A. Achab. 2000a. Chitinozoan zonation of the Late Ordovician and the Early Silurian of the Island of Anticosti, Québec, Canada. Review of Palaeobotany and Palynology 109:85–111.

Soufiane, A., and A. Achab. 2000b. Upper Ordovician and Lower Silurian chitinozoans from central Nevada and Arctic Canada. Review of Palaeobotany and Palynology 113:165–187.

Sowerby, J. 1821. The Mineral Conchology of Great Britain; or Coloured Figures and Descriptions of Those Remains of Testaceous Animals or Shells, Which Have Been Preserved at Various Times, and Depths in the Earth. W. Arding, London, 194 pp.

Spencer, W. K., and C. W. Wright. 1966. Asterozoans; pp. U4–U107 in R. C. Moore (ed.), Treatise on Invertebrate Paleontology. Part U, Echinodermata 3(1). Geological Society of America, New York, and University of Kansas Press, Lawrence.

Spjeldnaes, N. 1967. The palaeoecology of the Ordovician vertebrates of the Harding Formation (Colorado, U.S.A.); pp. 11–20 in Problèmes Actuels de Paléontologie (Evolution des Vertébrés) (Paris, 6–11 juin 1966). Colloques Internationaux du C.N.R.S. [Centre National de la Recherche Scientifique, Paris] 163.

Spjeldnaes, N. 1982. The Ordovician of the districts around Mjøsa. Paleontological Contributions from the University of Oslo 279:148–163.

Spjeldnaes, N., and M. H. Nitecki. 1990a. *Coelosphaeridium*, an Ordovician alga from Norway. Institutt for Geologi, Universitetet i Oslo, Intern Skriftserie 59:1–53. Preprint.

Spjeldnaes, N., and M. H. Nitecki. 1990b. Anatomy and relationship of the Ordovician algal genus *Apidium*. Institutt for Geologi, Universitetet i Oslo, Intern Skriftserie 61:1–37. Preprint.

Springer, D. A., and R. K. Bambach. 1985. Gradient versus cluster analysis of fossil assemblages: A comparison from the Ordovician of southwestern Virginia. Lethaia 18:181–198.

Sprinkle, J. 1971. Stratigraphic distribution of echinoderm plates in the Antelope Valley Limestone of Nevada and Calfiormia. United States Geological Survey, Professional Paper 750:D89–D98.

Sprinkle, J. 1973a. Morphology and Evolution of Blastozoan Echinoderms. Museum of Comparative Zoology, Harvard University, Special Publication, 283 pp.

Sprinkle, J. 1973b. *Tripatocrinus*, a new hybocrinid crinoid based on disarticulated plates from the Antelope Valley Limestone of Nevada and California. Journal of Paleontology 47:861–882.

Sprinkle, J. 1974. New rhombiferan cystoids from the Middle Ordovician of Nevada. Journal of Paleontology 48:1174–1201.

Sprinkle, J. 1980. An overview of the fossil record; pp. 15–26 in T. W. Broadhead and J. A. Waters (eds.), Echinoderms, Notes for a Short Course. University of Tennessee (Knoxville) Department of Geological Sciences, Studies in Geology 3.

Sprinkle, J. (ed.) 1982. Echinoderm Faunas from the Bromide Formation (Middle Ordovician) of Oklahoma. University of Kansas Paleontological Contributions, Monograph 1:1–369.

Sprinkle, J. 1995. Do eocrinoids belong to the Cambrian or to the Paleozoic Evolutionary Fauna? pp. 397–400 in J. D. Cooper, M. L. Droser, and S. C. Finney (eds.), Ordovician Odyssey: Short Papers, 7th International Symposium on the Ordovician System. Book 77, Pacific Section Society for Sedimentary Geology (SEPM), Fullerton, California.

Sprinkle, J., and T. E. Guensburg. 1995. Origin of echinoderms in the Paleozoic Evolutionary Fauna: The role of substrates. Palaios 10:437–453.

Sprinkle, J., and T. E. Guensburg. 1997. Appendix D4: Echinoderm biostratigraphy; pp. 49–50 (and plate 1, chart C) in R. J. Ross, Jr., L. F. Hintze, R. L. Ethington, J. F. Miller, M. E. Taylor, and J. E. Repetski (eds.), The Ibexian, Lowermost Series in the North American Ordovician. United States Geological Survey, Professional Paper 1579–A.

Sprinkle, J., and T. E. Guensburg. 2001. Growing a stalked echinoderm within the Extraxial-Axial Theory; pp. 59–65 in M. Barker (ed.), Echinoderms 2000. Balkema, Lisse, Netherlands.

Sprinkle, J., and T. E. Guensburg. 2003. Major expansion of echinoderms in the early Late Ordovician (Mohawkian, middle Caradoc) and its possible cause; pp. 327–332 in G. L. Albanesi, M. S. Beresi, and S. H. Peralta (eds.), Ordovician from the Andes. Instituto Superior de Correlación Geológica (INSUGEO), Serie Correlación Geológica 17.

Sprinkle, J., and G. P. Wahlman. 1994. New echinoderms from the Early Ordovician of west Texas. Journal of Paleontology 68:324–338.

Sprinkle, J., T. E. Guensburg, and S. V. Rozhnov. 1999. Correlation anomaly shown by Ordovician shelly and trace fossils in Baltic Russia: Redating the Ordovician Radiation; pp. 471–474 in P. Kraft and O. Fatka (eds.), *Quo vadis* Ordovician? Short papers of the 8th International Symposium on the Ordovician System. Acta Universitatis Carolinae, Geologica 43(1–2).

SPSS Inc. 2000. SPSS 10.0 for the Macintosh. Computer program, Chicago.

Stait, B. 1988. Tasmanian nautiloid faunas—biostratigraphy, biogeography, and morphology. Senckenbergiana Lethaea 69:87–107.

Stait, B., and C. Burrett. 1987. Biogeography of Australian and Southeast Asian Ordovician nautiloids; pp. 21–28 in J. Collinson (ed.), Gondwana Six. American Geophysical Union, Washington, D.C.

Stait, B., B. D. Webby, and I. G. Percival. 1985. Late Ordovician nautiloids from central New South Wales, Australia. Alcheringa 9:143–157.

Stanistreet, I. G. 1989. Trace fossil association related to facies of an upper Ordovician low wave energy shoreface and shelf, Oslo-Asker district, Norway. Lethaia 22:345–357.

Stanley, D. C. A., and R. K. Pickerill. 1993. Shallow marine *Paleodictyon* from the Upper Ordovician Georgian Bay Formation of southern Ontario. Atlantic Geology 29:115–119.

Stanley, D. C. A., and R. K. Pickerill. 1998. Systematic ichnology of the Late Ordovician Georgian Bay Formation of Southern Ontario, Eastern Canada. Royal Ontario Museum, Life Sciences Contributions 162:1–56.

Stanley, G. D., Jr. 1986. Chondrophorine hydrozoans as problematic fossils; pp. 68–86 in A. Hoffman and M. H. Nitecki (eds.), Problematic Fossil Taxa. Oxford University Press, New York.

Stanley, S. M. 1968. Post-Paleozoic adaptive radiation of infaunal bivalve molluscs: A consequence of mantle fusion and siphon formation. Journal of Paleontology 42:214–229.

Stanley, S. M. 1972. Functional morphology and evolution of byssally attached bivalve mollusks. Journal of Paleontology 46:165–212.

Stanley, S. M., and L. A. Hardie. 1998. Secular oscillations in the carbonate mineralogy of reef-building and sediment-producing organisms driven by tectonically forced shifts in seawater chemistry. Palaeogeography, Palaeoclimatology, Palaeoecology 144:3–19.

Stanley, S. M., and L. A. Hardie. 1999. Hypercalcification: Paleontology links plate tectonics and geochemistry to sedimentology. GSA Today 9:1–7.

Staplin, F. L. 1961. Reef-controlled distribution of Devonian microplankton in Alberta. Palaeontology 4:392–424.

Staplin, F. L., J. Jansonius, and S. A. J. Pocock. 1965. Evaluation of some acritarchous hystrichosphere genera. Neues Jahrbuch für Geologie und Paläontologie, Abhandlungen 123:167–201.

Starobogatov, Y. I. 1970. On the systematics of the early Paleozoic Monoplacophora. Paleontological Journal 3:6–17 (in Russian).

Stauffer, C. R. 1933. Middle Ordovician Polychaeta from Minnesota. Geological Society of America, Bulletin 44:1173–1218.

Stearn, C. W. 1975. The stromatoporoid animal. Lethaia 8:89–100.

Stearn, C. W. 1980. Classification of the Paleozoic stromatoporoids. Journal of Paleontology 54:881–902.

Stearn, C. W. 1987. Effect of the Frasnian-Famennian extinction event on the stromatoporoids. Geology 15:677–679.

Stearn, C. W., and J. W. Pickett. 1994. The stromatoporoid animal revisited: Building the skeleton. Lethaia 27:1–10.

Stearn, C. W., B. D. Webby, H. Nestor, and C. W. Stock. 1999. Revised classification and terminology of Palaeozoic stromatoporoids. Acta Palaeontologica Polonica 44:1–70.

Steele, H. M., and G. W. Sinclair. 1971. A Middle Ordovician fauna from Braeside, Ottawa Valley, Ontario. Geological Survey of Canada, Bulletin 211:1–97.

Steemans, P. 1999a. Paléodiversification des spores et des cryptospores de l'Ordovicien au Dévonien inférieur. Geobios 32:341–352.

Steemans, P. 1999b. Cryptospores and spores from the Ordovician to the Llandovery: A review; pp. 271–273 in P. Kraft and O. Fatka (eds.), *Quo vadis* Ordovician? Short papers of the 8th International Symposium on the Ordovician System. Acta Universitatis Carolinae, Geologica 43(1–2).

Steemans, P. 2000. Miospore evolution from the Ordovician to the Llandovery. Review of Palaeobotany and Palynology 113:189–196.

Steemans, P. 2001. Ordovician cryptospores from the Oostduinkerke borehole, Brabant Massif, Belgium. Geobios 34:3–12.

Steemans, P., and E. Pereira. 2002. Llandovery miospore biostratigraphy and stratigraphic evolution of the Paraná Basin, Paraguay—palaeogeographic implications. Bulletin de la Société géologique de France 173(5):407–414.

Steemans, P., A. Le Hérissé, and N. Bozdogan. 1996. Ordovician and Silurian cryptospores and miospores from Southeastern Turkey. Review of Palaeobotany and Palynology 93:35–76.

Steemans, P., K. T. Higgs, and C. H. Wellman. 2000. Cryptospores and trilete spores from the Llandovery, Nuayyim-2 Borehole, Saudi Arabia; pp. 92–115 in S. Al-Hajri and B. Owens (eds.), Stratigraphic Palynology of the Palaeozoic of Saudi Arabia. GeoArabia, Bahrain.

Stetson, H. C. 1931. Studies on the morphology of the Heterostraci. Journal of Geology 39:141–154.

Stillman, C. J. 1984. Ordovician volcanicity; pp. 183–194 in D. L. Bruton (ed.), Aspects of the Ordovician System. Palaeontological Contributions from the University of Oslo 295. Universitetsforlaget, Oslo.

Stinchcomb, B. L., and Darrough, G. 1995. Some molluscan Problematica from the Upper Cambrian–Lower Ordovician of the Ozark Uplift. Journal of Paleontology 69:52–65.

Stone, C. G., B. R. Haley, and M. H. Davis. 1994. Guidebook to Paleozoic rocks in the eastern Ouachita Mountains, Arkansas. Geological Society of America, Arkansas Geological Commission GB 94-1:1–46.

Størmer, L. 1951. A new eurypterid from the Ordovician of Montgomeryshire, Wales. Geological Magazine 88:409–422.

Størmer, L. 1955. Merostomata; pp. 4–41 in R. C. Moore (ed.), Treatise on Invertebrate Paleontology. Part P, Arthropoda 2. Geological Society of America, New York, and University of Kansas Press, Lawrence.

Stouge, S. 1984. Conodonts of the Middle Ordovician Table Head Formation, western Newfoundland. Fossils and Strata 16:1–145.

Stouge, S., and G. Bagnoli. 1988. Early Ordovician conodonts from the Cow Head Peninsula, western Newfoundland. Palaeontographia Italica 75:89–179.

Stouge, S., and G. Bagnoli. 1990. Lower Ordovician (Volkhovian-Kundan) conodonts from Hägudden, northern Öland, Sweden. Palaeontographia Italica 77:1–54.

Stover, L. E., H. Brinkhuis, S. P. Damassa, L. de Verteuil, R. J. Helby, E. Monteil, A. D. Partridge, A. J. Powell, J. B. Riding, M. Smelror, and G. L. Williams. 1996. Mesozoic-Tertiary dinoflagellates, acritarchs and prasinophytes; pp. 641–750 in J. Jansonius and D. C. McGregor (eds.), Palynology: Principles and Applications, Vol. 2. American Association of Stratigraphic Palynologists Foundation. Publishers Press, Salt Lake City, Utah.

Strachan, I. 1996. A bibliographic index of British graptolites (Graptoloidea), Part 1. Palaeontographical Society Monograph 150(600):1–40.

Strachan, I. 1997. A bibliographic index of British graptolites (Graptoloidea), Part 2. Palaeontographical Society Monograph 151(603):41–155.

Straelen, V. van, and G. Schmitz. 1934. Crustacea Phyllocarida (= Archaeostraca): Fossilium Catalogus, I. Animalia 64:1–246.

Strand, T. 1933. The Upper Ordovician cephalopods of the Oslo area. Norsk Geologisk Tidsskrift 14:1–118.

Strauss, H. 1997. The isotopic composition of sedimentary sulfur through time. Palaeogeography, Palaeoclimatology, Palaeoecology 132:97–118.

Strother, P. K. 1996. Acritarchs; pp. 81–106 (chapter 5) in J. Jansonius and D. C. McGregor (eds.), Palynology: Principles and Applications, Vol. 1. American Association of Stratigraphic Palynologists Foundation. Publishers Press, Salt Lake City, Utah.

Strother, P. K., and J. H. Beck. 2000. Spore-like microfossils from Middle Cambrian strata: Expanding the meaning of the term cryptospore; pp. 413–424 in M. M. Harley, C. M. Morton, and S. Blackmore (eds.), Pollen and Spores: Morphology and Biology. Royal Botanic Gardens, Kew, England.

Strother, P. K., and A. Traverse. 1979. Plant microfossils from the Llandoverian and Wenlockian rocks of Pennsylvania. Palynology 3:1–21.

Strother, P. K., S. Al-Hajri, and A. Traverse. 1996a. New evidence for land plants from the lower Middle Ordovician of Saudi Arabia. Geology 24:55–59.

Strother, P. K., R. A. MacRae, A. Fricker, R. A. Fensome, and G. L. Williams. 1996b. Phanerozoic phytoplankton diversity is decoupled from marine invertebrate diversity. Ninth International Palynological Congress, Houston, Texas, Program and Abstracts, 152.

Sumrall, C. D., J. Sprinkle, and T. E. Guensburg. 1997. Systematics and paleoecology of Late Cambrian echinoderms from the western United States. Journal of Paleontology 71:1091–1109.

Sumrall, C. D., J. Sprinkle, and T. E. Guensburg. 2001. Comparison of flattened blastozoan echinoderms: Insights from the new Early Ordovician eocrinoid *Haimacystis rozhnovi*. Journal of Paleontology 75:985–992.

Sun, Q.-Y. 1988. Ordovician Ostracoda from Western Hubei. Acta Micropalaeontologica Sinica 5:253–266.

Sutcliffe, O. E., J. A. Dowdswell, R. J. Whittington, J. N. Theron, and J. Craig. 2000. Calibrating the Late Ordovician glaciation and mass extinction by the eccentricity cycles of Earth's orbit. Geology 28:967–970.

Sutton, M. D., D. E. G. Briggs, D. J. Siveter, and D. J Siveter. 2001. An exceptionally preserved vermiform mollusc from the Silurian of England. Nature 410:461–463.

Sweet, W. C. 1958. The Middle Ordovician of the Oslo Region, Norway. Part 10, Nautiloid cephalopods. Norsk Geologisk Tidsskrift 18:1–118.

Sweet, W. C. 1979. Late Ordovician conodonts and biostratigraphy of the Western Midcontinent Province. Brigham Young University Geology Studies 26(3):45–85.

Sweet, W. C. 1984. Graphic correlation of upper Middle and Upper Ordovician rocks, North American Midcontinental Province, USA; pp. 23–35 in D. L. Bruton (ed.), Aspects of the Ordovician System. Paleontological Contributions from the University of Oslo 295. Universitetsforlaget, Oslo.

Sweet, W. C. 1988a. Mohawkian and Cincinnatian chronostratigraphy. Bulletin of the New York State Museum 462:84–90.

Sweet, W. C. 1988b. The Conodonta: Morphology, Taxonomy, Paleoecology, and Evolutionary History of a Long-Extinct Animal Phylum. Clarendon Press, Oxford, 212 pp.

Sweet, W. C. 1995. A conodont-based composite standard for the North American Ordovician: Progress report; pp. 15–20 in J. D. Cooper, M. L. Droser, and S. C. Finney (eds.), Ordovician Odyssey: Short Papers, 7th International Symposium on the Ordovician System. Book 77, Pacific Section Society for Sedimentary Geology (SEPM), Fullerton, California.

Sweet, W. C., and S. M. Bergström. 1984. Conodont provinces and biofacies of the Late Ordovician. Geological Society of America, Special Paper 196:69–87.

Sweet, W. C., and P. C. J. Donoghue. 2001. Conodonts: Past, present, future. Journal of Paleontology 75:1174–1184.

Sweet, W. C., and C. M. Tolbert. 1997. An Ibexian (Lower Ordovician) reference section in the southern Egan Range, Nevada, for a conodont-based chronostratigraphy. United States Geological Survey, Professional Paper 1579-B:51–84.

Syssoiev, V. A. 1957. K morfologii, sistematike i sistematicheskomu polozheniu khiolitiov [To the morphology, systematics and systematic position of the Hyolithoidea]. Doklady, Akademia Nauk SSSR 116:304–307

Sytova, V. A. 1977. On the origin of rugose corals. Mémoirs de Bureau de Recherches Géologique et Minières 89:65–68.

Szaniawski, H. 1970. Jaw apparatuses of the Ordovician and Silurian polychaetes from the Mielnik borehole. Acta Palaeontologica Polonica 15:445–472.

Szaniawski, H. 1982. Chaetognath grasping spines recognized among Cambrian protoconodonts. Journal of Paleontology 56:806–810.

Szaniawski, H. 1987. Preliminary structural comparison of protoconodont, paraconodont, and euconodont element; pp. 35–47 in R. J. Aldridge (ed.), Palaeobiology of Conodonts. Ellis Horwood, Chichester, England.

Szaniawski, H. 1996. New evidence of protoconodont-chaetognath relationship. Sixth European Conodont Symposium (ECOS VI), Instytut Paleobiologii PAN, Warsaw, Abstracts, 56.

Szaniawski, H. 1998. Ancestors of conodonts. Seventh European Conodont Symposium (ECOS VII), Bologna-Modena, Italy, Abstracts, 111.

Sztejn, J. 1985. Ordovician Ostracods in North-Eastern Poland. Biuletyn Panstwowego Instytutu Geologicznego 350:53–89.

Tait, J., V. Bachtadse, W. G. Franke, and H. C. Soffel. 1997. Geodynamic evolution of the European Variscan foldbelt: Palaeomagnetic and geologic constraints. Geologische Rundschau 86:585–598.

Talimaa, V. 2000. Significance of thelodonts (Agnatha) in correlation of the Upper Ordovician to Lower Devonian of the northern part of Eurasia; pp. 69–80 in A. Blieck and S. Turner (eds.), Palaeozoic Vertebrate Biochronology and Global Marine/Non-Marine Correlation—final report of IGCP 328 (1991–1996). Courier Forschungsinstitut Senckenberg 223.

Talyzina, N. M., J. M. Moldowan, A. Johannisson, and F. J. Fago. 2000. Affinities of Early Cambrian acritarchs studied by using microscopy, fluorescence flow cytometry and biomarkers. Review of Palaeobotany and Palynology 108:37–53.

Tappan, H. 1986. Phytoplankton: Below the salt at the global table. Journal of Paleontology 60:545–554.

Tappan, H., and A. R. Loeblich, Jr. 1973. Evolution of the oceanic plankton. Earth-Science Reviews 9:207–240.

Taylor, P. D. 1981. Functional morphology and evolutionary significance of differing modes of tentacle eversion in marine bryozoans; pp. 235–247 in G. P. Larwood and C. Nielsen (eds.), Recent and Fossil Bryozoa. Olsen and Olsen, Fredensborg, Denmark.

Taylor, P. D. 1988. Major radiation of cheilostome bryozoans: Triggered by the evolution of a new larval type? Historical Biology 1:45–64.

Taylor, P. D. 1990a. Preservation of soft-bodied and other organisms by bioimmuration—a review. Palaeontology 33:1–17.

Taylor, P. D. 1990b. Bioimmured ctenostomes from the Jurassic and the origin of the cheilostome Bryozoa. Palaeontology 33:19–34.

Taylor, P. D. 1999. Bryozoa; pp. 623–646 in E. Savazzi (ed.), Functional Morphology of the Invertebrate Skeleton. Wiley, Chichester, England.

Taylor, P. D. 2000. Cyclostome systematics: Phylogeny, suborders and the problem of skeletal organization; pp. 87–103 in A. Herrera Cubilla and J. B. C. Jackson (eds.), Proceedings of the 11th International Bryozoology Association Conference, Smithsonian Tropical Research Institute, Balboa, Republic of Panama.

Taylor, P. D., and J. C. W. Cope. 1987. A trepostome bryozoan from the Lower Arenig of south Wales: Implications of the oldest described bryozoan. Geological Magazine 124:367–371.

Taylor, P. D., and G. B. Curry. 1985. The earliest known fenestrate bryozoan, with a short review of Lower Ordovician Bryozoa. Palaeontology 28:147–158.

Taylor, P. D., and G. P. Larwood. 1990. Major evolutionary radiations in the Bryozoa; pp. 209–233 in P. D. Taylor and G. P. Larwood (eds.), Major Evolutionary Radiations. Systematics Association Special Vol. 42. Clarendon Press, Oxford.

Taylor, P. D., and S. Rozhnov. 1996. A new early cyclostome bryozoan from the Lower Ordovician (Volkhov Stage) of Russia. Paläontologische Zeitschrift 70:171–180.

Taylor, P. D., and M. A. Wilson. 1994. *Corynotrypa* from the Ordovician of North America: Colony form in a primitive stenolaemate bryozoan. Journal of Paleontology 68:241–257.

Taylor, P. D., and M. A. Wilson. 1999. *Dianulites:* An unusual Ordovician bryozoan with a high-magnesium calcite skeleton. Journal of Paleontology 73:38–48.

Taylor, W. A. 2000. Spore wall development in the earliest land plants; pp. 425–434 in M. M. Harley, C. M. Morton, and S. Blackmore (eds.), Pollen and Spores: Morphology and Biology. Royal Botanic Gardens, Kew, England.

Teichert, C. 1967. Major features of cephalopod evolution; pp. 162–210 in C. Teichert and E. L. Yochelson (eds.), Essays in Paleontology and Stratigraphy. R. C. Moore Commemorative Volume (Special Publication 2). University of Kansas, Lawrence.

Teichert, C. 1988. Main features of cephalopod evolution; pp. 11–79 in M. R. Clarke and E. R. Trueman (eds.),

The Mollusca. Vol. 12, Paleontology and Neontology of Cephalopods. Academic Press, New York.

Teichert, C., and B. F. Glenister. 1954. Early Ordovician cephalopod fauna from northwestern Australia. Bulletins of American Paleontology 35:1–112.

Telford, P. G. 1975. Lower and Middle Devonian conodonts from the Broken River Embayment, North Queensland, Australia. Special Papers in Palaeontology 15:1–96.

Termier, G. 1950. Paléontologie marocaine. 1, Invertébrés de l'ère primaire. Fascicule 4, Annélides, Arthropodes, Échinodermes Conularides et Graptolithes. Maroc, Service géologique, Notes et Mémoires 79:1–281.

Termier, G., and H. Termier. 1947. Paléontologie marocaine. 1, Généralites sur les invertébrés fossiles. Maroc, Service géologique, Notes et Mémoires 69:1–391.

Thayer, C. W. 1975. Morphological adaptations of benthic invertebrates to soft substrata. Journal of Marine Research 33:177–189.

Thayer, C. W. 1979. Biological bulldozers and the evolution of marine benthic communities. Science 203:458–461.

Thayer, C. W. 1983. Sediment-mediated biological disturbance and the evolution of marine benthos; pp. 480–625 in M. J. S. Tevesz and P. L. McCall (eds.), Biotic Interactions in Recent and Fossil Benthic Communities. Plenum Press, New York.

Thode, H. G., and J. Monster. 1965. Sulphur isotope geochemistry of petroleum, evaporites and ancient seas; pp. 367–377 in A. Young and J. E. Galley (eds.), Fluids in Subsurface Environments. American Association of Petroleum Geology Memoir 4. Tulsa, Oklahoma.

Thomas, A. T., R. M. Owens, and A. W. A. Rushton. 1984. Trilobites in British stratigraphy. The Geological Society, London, Special Report 16:1–78.

Thorne-Miller, B. 1999. The Living Ocean: Understanding and Protecting Marine Biodiversity. Second Edition. Island Press, Washington, D.C., 214 pp.

Tinn, O., and T. Meidla. 2001. Middle Ordovician ostracods from the Lanna and Holen Limestones, south-central Sweden. GFF [Geologiska Föreningens i Stockhom Förhandlingar] 123:129–136.

Tipper, J. C. 1980. Some distributional models for fossil animals. Paleobiology 6:77–95.

Tjernvik, T. E. 1956. On the Early Ordovician of Sweden: Stratigraphy and fauna. Bulletin of the Geological Institutions of the University of Uppsala 36:107–284.

Tjernvik, T. E., and J. V. Johansson. 1980. Description of the upper portion of the drill-core from Finngrundet in the South Bothnian Bay. Bulletin of the Geological Institutions of the University of Uppsala, n.s., 8:173–204.

Todd, J. A. 2000. The central role of ctenostomes in bryozoan phylogeny; pp. 104–135 in A. Herrera Cubilla and J. B. C. Jackson (eds.), Proceedings of the 11th International Bryozoology Association Conference, Smithsonian Tropical Research Institute, Balboa, Republic of Panama.

Todd, J. A., and H. Hagdorn. 1993. First record of Muschelkalk Bryozoa: The earliest ctenostome body fossils. Sonderbände der Gesellschaft für Naturkunde in Württemberg 2:286–287.

Tollerton, V. P. 1989. Morphology, taxonomy and classification of the order Eurypterida Burmeister, 1843. Journal of Paleontology 63:642–657.

Tollerton, V. P., and E. Landing. 1994. The myth of Ordovician eurypterids in New York State. Geological Society of America, Abstracts with Programs 26(3):76.

Tolmacheva, T. Ju. 2001. Conodont biostratigraphy and diversity in the Lower-Middle Ordovician of Eastern Baltoscandia (St. Petersburg region, Russia) and Kazakhstan. Summary of doctoral dissertation, Department of Earth Sciences—Palaeobiology, Uppsala University.

Tolmacheva, T. Ju., and P. Fedorov. 2001. The Ordovician Billingen/Volkhov boundary interval at Lava River, north-western Russia. Norsk Geologisk Tidsskrift 81:161–168.

Tolmacheva, T. Ju., and A. Löfgren. 2001. Morphology and paleogeography of the Ordovician conodont *Paracordylodus gracilis* Lindström, 1955: Comparison of two populations. Journal of Paleontology 74:1114–1121.

Tolmacheva, T. Ju., T. Danelian, and L. E. Popov. 2001a. Evidence of 15 m.y. of continuous deep-sea biogenic siliceous sedimentation in early Paleozoic oceans. Geology 29:755–758.

Tolmacheva, T. Ju., T. N. Koren, L. E. Holmer, L. E. Popov, and E. Raevskaya. 2001b. The Hunneberg Stage (Ordovician) in the area east of St. Petersburg, north-western Russia. Paläontologische Zeitschrift 74:543–561.

Tongiorgi, M., L. Yin, and A. Di Milia. 1995. Arenigian acritarchs from the Daping section (Yangtze Gorges area, Hubei Province, Southern China) and their palaeogeographic significance. Review of Palaeobotany and Palynology 86:13–48.

Toomey, D. F., and D. LeMone. 1977. Some Ordovician and Silurian algae from selected areas of the southwestern United States; pp. 351–359 in E. Flügel (ed.), Fossil Algae: Recent Results and Developments. Springer-Verlag, Berlin.

Toomey, D. F., and M. H. Nitecki. 1979. Organic buildups in the Lower Ordovician (Canadian) of Texas and Oklahoma. Fieldiana: Geology, n.s., 2:1–181.

Toro, M. A., and H. G. Pérez. 1978. El primer registro de eurypteridos para el Ordovicico de Bolivia. Servicio Geológico de Bolivia, Boletín Serie A 2:13–19.

Torsvik, T. H., and T. B. Andersen. 2002. The Taimyr fold belt, Arctic Siberia: Timing of pre-fold remagnetisation and regional tectonics. Tectonophysics 352:335–348.

Torsvik, T. H., and E. F. Rehnström. 2001. Cambrian palaeomagnetic data from Baltica: Implications for true polar wander and Cambrian palaeogeography. Journal of the Geological Society of London 158:321–329.

Torsvik, T. H., M. A. Smethurst, J. C. Briden, and B. A. Sturt. 1990. A review of Palaeozoic palaeomagnetic data from Europe and their palaeogeographic implications; pp. 25–41 *in* W. S. McKerrow and C. R. Scotese (eds.), Palaeozoic Palaeogeography and Biogeography. The Geological Society, London, Memoir 12.

Torsvik, T. H., M. A. Smethurst, R. Van der Voo, A. Trench, N. Abrahamsen, and E. Halvorsen. 1992. Baltica. A synopsis of Vendian-Permian palaeomagnetic data and their palaeotectonic implications. Earth Science Reviews 33:133–152.

Torsvik, T. H., A. Trench, I. Svensson, and H. J. Walderhaug. 1993. Silurian palaeomagnetic results from Southern Britain: Palaeogeographic significance and major revision of the Apparent Polar Wander Path for Avalonia. Geophysical Journal International 113:651–668.

Tracey, S., J. A. Todd, and D. H. Erwin. 1993. Mollusca: Gastropoda; pp. 131–167 *in* M. J. Benton (ed.), The Fossil Record. Chapman and Hall, London.

Trewin, N. H., and K. J. McNamara. 1995. Arthropods invade the land: Trace fossils and palaeoenvironments of the Tumblagooda Sandstone (?Late Silurian) of Kalbarri, Western Australia. Transactions of the Royal Society of Edinburgh, Earth Sciences 85:177–210.

Tucker, R. D., and W. S. McKerrow. 1995. Early Paleozoic chronology: A review in light of new U-Pb ages from Newfoundland and Britain. Canadian Journal of Earth Sciences 32:368–379.

Tucker, R. D., T. E. Krogh, R. J. Ross, and S. H. Williams. 1990. Time-scale calibration by high precision U-Pb zircon dating of interstratified volcanic ashes in the Ordovician and Lower Silurian stratotypes of Britain. Earth and Planetary Science Letters 100:51–58.

Tucker, R. D., D. C. Bradley, C. A. Ver Straeten, C. A. Harris, A. G. Ebert, and S. R. McCutcheon. 1998. New U-Pb zircon ages and the duration and division of Devonian time. Earth and Planetary Science Letters 158:175–186.

Tuckey, M. E. 1990. Biogeography of Ordovician bryozoans. Palaeogeography, Palaeoclimatology, Palaeoecology 77:91–126.

Tuckey, M. E., and R. L. Anstey. 1992. Late Ordovician extinctions of bryozoans. Lethaia 25:111–117.

Tunnicliff, S. P. 1982. A revision of Late Ordovician bivalves from Pomeroy, Co. Tyrone, Ireland. Palaeontology 25:43–88.

Tunnicliff, S. P. 1987. Caradocian bivalve molluscs from Wales. Palaeontology 30:677–690.

Turner, S. 1991. Monophyly and interrelationships of the Thelodonti; pp. 87–119 *in* M.-M. Chang, Y.-H. Liu, and G.-R. Zhang (eds.), Early Vertebrates and Related Problems of Evolutionary Biology (International Symposium, Beijing, 1987). Science Press, Beijing.

Turner, S., J. J. Kuglitsch, and D. L. Clark. 1999. Llandoverian thelodont scales from the Burnt Bluff Group of Wisconsin and Michigan. Journal of Paleontology 73:667–676.

Ubaghs, G. 1960. Le genre *Lingulocystis* Thoral. Annales de Paléontologie 46:81–116.

Ubaghs, G. 1963. *Rhopalocystis destombesi* n.g., n. sp. Eocrinoïde de l'Ordovicien inférieur (Trémadocien superieur) du Sud marocain. Notes du Service Géologique du Maroc, 23:25–45.

Ubaghs, G. 1966. Ophiocistioids; pp. U174–U188 *in* R. C. Moore (ed.), Treatise on Invertebrate Paleontology. Part U, Echinodermata 3(1). Geological Society of America, New York, and University of Kansas Press, Lawrence.

Ubaghs, G. 1968a [dated 1967]. General characters of Echinodermata; pp. S3–S60 *in* R. C. Moore (ed.), Treatise on Invertebrate Paleontology. Part S, Echinodermata 1(1). Geological Society of America, New York, and University of Kansas Press, Lawrence.

Ubaghs, G. 1968b [dated 1967]. Stylophora; pp. S495–S565 *in* R. C. Moore (ed.), Treatise on Invertebrate Paleontology. Part S, Echinodermata 1(2). Geological Society of America, New York, and University of Kansas Press, Lawrence.

Ubaghs, G. 1969. *Aethocrinus moorei* Ubaghs, n. gen., n. sp., le plus ancien crinoide dicy-clique connu. University of Kansas Paleontological Contributions, Paper 38:1–25.

Ubaghs, G. 1970 [dated 1969]. Les échinodermes carpoïdes de l'Ordovicien Inférieur de la Montagne Noire (France). Cahiers de Paleontologie, Editions du Centre National de la Recherche Scientifique, Paris, 131 pp.

Ubaghs, G. 1979. Trois Mitrata (Echinodermata: Stylophora) nouveaux de l'Ordovicien de Tchecoslovaquie. Paläontologische Zeitschrift 53:98–119.

Ubaghs, G. 1991. Deux Stylophora (Homalozoa, Echinodermata) nouveaux pour l'Ordovicien inférieur de la Montagne Noire (France méridionale). Paläontologische Zeitschrift 65:157–171.

Ubaghs, G. 1994. Échinodermes nouveaux (Stylophora, Eocrinoidea) de l'Ordovicien Inferieur de la Montagne Noire (France). Annales de Paleontologie 80:107–141.

Ubaghs, G. 1999. Échinodermes nouveaux du Cambrien Supérieur de la Montagne Noire (France méridionale). Geobios 31:809–829.

Ulrich, E. O. 1894. The Lower Silurian Lamellibranchiata of Minnesota; pp. 475–628 *in* Vol. 3, Final Report of the Geological and Natural History Survey of Minnesota. [Published and distributed under separate cover prior to publication of the entire Vol. 3 in 1897.]

Ul'st, R. Zh., L. K Gailite, and V. I. Yakovleva. 1982. Ordovik Latvii. Zinatne, Riga, Latvia, 295 pp.

Underhay, N. K., and S. H. Williams. 1995. Lower Ordovician scolecodonts from the Cow Head Group, western Newfoundland. Canadian Journal of Earth Sciences 32:895–901.

Underwood, C. J. 1993. The position of graptolites within Lower Palaeozoic planktic ecosystems. Lethaia 26:189–202.

Ushatinskaya, G. T. 2000 [dated 2001]. Brachiopods; pp. 350–369 in A. Yu. Zhuravlev and R. Riding (eds.), The Ecology of the Cambrian Radiation. Columbia University Press, New York.

Uutela, A., and R. Tynni. 1991. Ordovician acritarchs from the Rapla borehole, Estonia. Geological Survey of Finland Bulletin 353:1–135.

Vaccari, N. E. 1995. Early Ordovician trilobite biogeography of Precordillera and Famatina, western Argentina: Preliminary results; pp. 193–196 in J. D. Cooper, M. L. Droser, and S. C. Finney (eds.), Ordovician Odyssey: Short Papers, 7th International Symposium on the Ordovician System. Book 77, Pacific Section Society for Sedimentary Geology (SEPM), Fullerton, California.

Vaccari, N. E. 2003. Trilobites de la Formación San Juan (Ordovícico Inferior), Precordillera Argentina. Ameghiniana 38:331–348.

Vail, P. R., R. M. Mitchum, Jr., R. G. Todd, J. M. Widmier, S. Thompson, J. B. Sangree, J. N. Bubb, and W. G. Hatlelid. 1977. Seismic stratigraphy and global changes of sea-level; pp. 49–212 in C. E. Payton (ed.), Seismic Stratigraphy—applications to Hydrocarbon Exploration. American Association of Petroleum Geologists Memoir 26.

Valentine, J. W. 1973. Evolutionary Paleoecology of the Marine Biosphere. Prentice-Hall, Englewood Cliffs, New Jersey, 511 pp.

Valentine, J. W., and D. Jablonski. 1983. Larval adaptations and patterns of brachiopod diversity in space and time. Evolution 37:1052–1061.

VandenBerg, A. H. M., and R. A. Cooper. 1992. The Ordovician graptolite sequence of Australasia. Alcheringa 16:33–85.

van den Boogaard, M. 1989. Isolated tubercles of some Palaeoscolecida. Scripta Geologica 90:1–12.

Van Iten, H. 1992a. Morphology and phylogenetic significance of the corners and midlines of the conulariid test. Palaeontology 35:335–358.

Van Iten, H. 1992b. Microstructure of the conulariid test and its implications for conulariid affinities. Palaeontology 35:359–372.

Van Iten, H. 1994. Redescription of *Glyptoconularia gracilis* (Hall), an Ordovician conulariid from North America; pp. 363–366 in E. Landing (ed.), Studies in Stratigraphy and Paleontology in Honor of Donald W. Fisher. New York State Museum/Geological Survey Bulletin 461.

Van Iten, H., R. S. Cox, and R. H. Mapes. 1992. New data on the morphology of *Sphenothallus* Hall: Implications for its affinities. Lethaia 25:135–144.

Van Iten, H., J. Fitzke, and R. S. Cox. 1996. Problematical fossil cnidarians from the upper Ordovician of the north-central USA. Palaeontology 39:1037–1064.

Vannier, J. 1986a. Ostracodes Binodicopa de l'Ordovicien (Arenig-Caradoc) Ibero-Armoricain. Palaeontographica, Abteilung A 193:77–143.

Vannier, J. 1986b. Ostracodes Palaeocopa de l'Ordovicien (Arenig-Caradoc) Ibero-Armoricain. Palaeontographica, Abteilung A 193:145–218.

Vannier, J., and R. Schallreuter. 1983. *Quadritia (Krutatia) tromelini* nov. sp., ostracode du Llandeilo ibéroarmoricain. Intérêt paléogéographique. Geobios 16:583–599.

Vannier, J., P. R. Racheboeuf, E. D. Brussa, M. Williams, A. W. A. Rushton, T. Servais, and D. J. Siveter. 2003. Cosmopolitan arthropod zooplankton in the Ordovician seas. Palaeogeography, Palaeoclimatology, Palaeoecology 195:173–191.

Van Staal, C. R., J. F. Dewey, C. Mac Niocaill, and W. S. McKerrow. 1998. The Cambrian-Silurian tectonic evolution of the northern Appalachians and British Caledonides: History of a complex, west and southwest Pacific-type segment of Iapetus; pp. 199–242 in D. J. Blundell and A. C. Scott (eds.), Lyell: The Past Is the Key to the Present. The Geological Society, London, Special Publication 143.

van Wamel, W. A. 1974. Conodont biostratigraphy of the Upper Cambrian and Lower Ordovician of northwestern Öland, south-eastern Sweden. Utrecht Micropalaeontological Bulletins 10:1–126.

Van Waveren, I. M., and N. H. Marcus. 1993. Morphology of recent copepod egg envelopes from Turkey Point, Gulf of Mexico, and their implications for acritarch affinity. Special Papers in Palaeontology 48:111–124.

Vaslet, D. 1990. Upper Ordovician glacial deposits in Saudi Arabia. Episodes 13:147–161.

Vavrdová, M. 1982. Recycled acritarchs in the uppermost Ordovician of Bohemia. Časopis pro mineralogii a geologii 27:337–345.

Vavrdová, M. 1984. Some plant microfossils of the possible terrestrial origin from the Ordovician of central Bohemia. Věstník Ústředního ústavu geologického 59:165–170.

Vavrdová, M. 1988. Further acritarchs and terrestrial plant remains from the Late Ordovician at Hlásná Třebaň (Czechoslovakia). Časopis pro mineralogii a geologii 33:1–10.

Vavrdová, M. 1989. New acritarchs and miospores from the Late Ordovician of Hlásná Třebaň, Czechoslovakia. Časopis pro mineralogii a geologii 34:403–420.

Vecoli, M. 1996. Stratigraphic significance of acritarchs in Cambro-Ordovician boundary strata, Hassi-R'mel area, Algerian Sahara. Bollettino della Società Paleontologica Italiana 35:3–58.

Vecoli, M. 1999. Cambro-Ordovician palynostratigraphy (acritarchs and prasinophytes) and palaeogeography of the Hassi-R'Mel area and northern Rhadames Basin, North Africa. Palaeontographia Italica 86:1–112.

Vecoli, M. 2000. Palaeoenvironmental interpretation of microphytoplankton diversity trends in the Cambrian-Ordovician of the northern Sahara Platform. Palaeogeography, Palaeoclimatology, Palaeoecology 160:329–346.

Vecoli, M., and J. Samuelsson. 2001. Quantitative evaluation of microplankton palaeobiogeography in the Ordovician–Early Silurian of the northern Trans European Suture Zone: Implications for the timing of the Avalonia-Baltica collision. Review of Palaeobotany and Palynology 115:43–68.

Vecoli, M., M. Tongiorgi, M. Quintavalle, and D. Massa. In press. Palynological contribution to the Cambro-Ordovician stratigraphy of north-western Ghadamis Basin (Libya and Tunisia); in Geology of NW Libya, Elsevier, Amsterdam.

Veizer, J. 1989. Strontium isotopes in seawater through time. Annual Review of Earth and Planetary Sciences 17:141–167.

Veizer, J., P. Fritz, and B. Jones. 1986. Geochemistry of brachiopods: Oxygen and carbon isotopic records of Paleozoic oceans. Geochimica et Cosmochimica Acta 50:1679–1696.

Veizer, J., D. Ala, K. Azmy, P. Bruckschen, D. Buhl, F. Bruhn, G. A. F. Carden, A. Diener, S. Ebneth, Y. Goddéris, T. Jasper, C. Korte, F. Pawellek, O. G. Podlaha, and H. Strauss. 1999. $^{87}Sr/^{86}Sr$, $\delta^{13}C$ and $\delta^{18}O$ evolution of Phanerozoic seawater. Chemical Geology 161:59–88.

Veizer, J., Y. Goddéris, and L. M. François. 2000. Evidence for decoupling of atmospheric CO_2 and global climate during the Phanerozoic eon. Nature 408:698–701.

Vermeij, G. J. 1987. Evolution and Escalation: An Ecological History of Life. Princeton University Press, Princeton, 544 pp.

Vidal, G., and M. Moczydłowska-Vidal. 1997. Biodiversity, speciation, and extinction trends of Proterozoic and Cambrian phytoplankton. Paleobiology 23:230–246.

Viira, V., A. Löfgren, S. Mägi, and J. Wickström. 2001. An Early to Middle Ordovician succession of conodont faunas at Mäekalda, northern Estonia. Geological Magazine 138:699–718.

Vinogradov, A. V. 1996. New fossil freshwater bryozoans from the Asiatic part of Russia and Kazakhstan. Paleontological Journal 30:284–292.

Viskova, L. A., and A. Yu. Ivantsov. 1999. The earliest uncalcified bryozoan from the vicinity of Saint Petersburg. Paleontological Journal 33:25–28.

Vizcaïno, D., and J. J. Álvaro. 2001. The Cambrian and Lower Ordovician of the southern Montagne Noire (Languedoc, France). A synthesis for the beginning of the new century. Annales de la Société Géologique du Nord 8:1–242.

Vizcaïno, D., and B. Lefebvre. 1999. Les échinodermes du Paléozoïque inférieur de Montagne Noire: Biostratigraphie et paléodiversité. Geobios 32:353–364.

Volkheimer, W., A. J. Cuerda, and D. L. Melindi. 1980. Quitinozoos de la Formación Gualcamayo en su localidad tipo al sudoeste de Guandacol, Precordillera de La Rioja, República Argentina. II Congreso Argentino de Paleontología y Bioestratigrafía y I Congreso Latino de Paleontología, Actas 1:23–35.

Volkova, N. A. 1997. Paleogeography of phytoplankton at the Cambrian-Ordovician boundary. Paleontological Journal 31:135–140.

Waddington, J. B. 1980. A soft substrate community with edrioasteroids, from the Verulam Formation (Middle Ordovician) at Gamebridge, Ontario. Canadian Journal of Earth Sciences 17:674–679.

Wade, M. 1988. Nautiloids and their descendants: Cephalopod classification in 1986; pp. 15–25 in D. L. Wolberg (ed.), Contributions to Paleozoic Paleontology and Stratigraphy in Honor of Rousseau H. Flower. New Mexico Bureau of Mines and Mineral Resources, Memoir 44.

Wagner, P. J. 1995a. Testing evolutionary constraint hypotheses with early Paleozoic gastropods. Paleobiology 21:248–272.

Wagner, P. J. 1995b. Diversity patterns among early gastropods—contrasting taxonomic and phylogenetic descriptions. Paleobiology 21:410–439.

Wagner, P. J. 1995c [dated 1996]. Patterns of morphologic diversification during the initial radiation of the "Archaeogastropoda"; pp. 161–170 in J. D. Taylor (ed.), Origin and Evolutionary Radiation of the Mollusca. Oxford University Press, Oxford.

Wahlman, G. P. 1992. Middle and Upper Ordovician symmetrical univalved mollusks (Monoplacophora and Bellerophontina) of the Cincinnati Arch region. United States Geological Survey, Professional Paper 1066-O: 1–203.

Waines, R. H. 1997. The mixopteracean eurypterid Megalograptus—a first occurrence in New York State and earliest occurrence for the genus (Late Medial Ordovician). Geological Society of America, Abstracts with Programs 29(1):87.

Waisfeld, B. G. 1995. Early Ordovician trilobite biofacies in the Argentine Cordillera Oriental, Southwestern Gondwana: Paleoecologic and paleobiogeographic significance; pp. 449–452 in J. D. Cooper, M. L. Droser, and S. C. Finney (eds.), Ordovician Odyssey: Short Papers, 7th International Symposium on the Ordovician System. Book 77, Pacific Section Society for Sedimentary Geology (SEPM), Fullerton, California.

Waisfeld, B. G., and T. M. Sánchez. 1996. "Cambrian Fauna" versus "Palaeozoic Fauna" in the Lower Ordovician of Western Argentina: Provincialism and environment interactions. Geobios 29:401–416.

Waisfeld, B. G., and N. E. Vaccari. 1996. Trilobites de la Formación Suri (Arenigiano) de la Sierra de Famatina. Ameghiniana 33:233.

Waisfeld, B. G., T. M. Sánchez, and M. G. Carrera. 1999. Biodiversification patterns in the early Ordovician of Argentina. Palaios 14:198–214.

Walcott, C. D. 1882. Description of a new genus of the Order Eurypterida from the Utica Slate. American Journal of Science, ser. 3, 23:213–216.

Walcott, C. D. 1892. Preliminary notes on the discovery of a vertebrate fauna in Silurian (Ordovician) strata. Geological Society of America, Bulletin 3:153–172.

Waldron, J. W. F. 1992. The Goldenville-Halifax transition, Mahone Bay, Nova Scotia: Relative sea-level rise in the Meguna source terrane. Canadian Journal of Earth Sciences 29:1091–1105.

Walker, K. R. 1972. Community ecology of the Middle Ordovician Black River Group of New York State. Geological Society of America, Bulletin 83:2499–2524.

Walker, K. R., and L. F. Laporte. 1970. Congruent fossil communities from Ordovician and Devonian carbonates of New York. Journal of Paleontology 44:928–944.

Waller, T. R. 1990. The evolution of ligament systems in the Bivalvia; pp. 49–71 in B. Morton (ed.), The Bivalvia—Proceedings of a Symposium in Honour of Sir Charles Maurice Yonge. Hong Kong University Press, Hong Kong.

Waller, T. R. 1998. Origin of the molluscan class Bivalvia and a phylogeny of major groups; pp. 1–45 in P. A. Johnston and J. W. Haggart (eds.), Bivalves: An Eon of Evolution—Paleobiological Studies Honoring Norman D. Newell. University of Calgary Press, Calgary.

Wallmann, K. 2001. The geological water cycle and the evolution of marine $\delta^{18}O$ values. Geochimica et Cosmochimica Acta 65:2469–2485.

Walossek, D. 1999. On the Cambrian diversity of Crustacea; pp. 3–27 in F. R. Schram and J. Carel von Vaupel Klein (eds.), Crustaceans and the Biodiversity Crisis. Brill, Leiden.

Wang, J.-Q., and M. Zhu. 1996. [An Ordovician vertebrate from Inner Mongolia.] Kexue Tongbao [Chinese Science Bulletin] 42:1187–1189 (in Chinese).

Wang, J.-Q., and M. Zhu. 1997. Discovery of Ordovician vertebrate fossil from Inner Mongolia, China. Kexue Tongbao [Chinese Science Bulletin] 42:1560–1562 (in Chinese).

Wang, X.-F., and X.-H. Chen. 1992. Earliest Ordovician chitinozoans from the eastern Yangtze Gorges. Acta Micropalaeontologica Sinica 9:283–290 (in Chinese with English abstract).

Wang, X.-F., and X.-H. Chen. 1994. Lower Ordovician chitinozoan biostratigraphy and palaeogeography of Upper Yangtze Region. Acta Palaeontologica Sinica 33:720–738 (in Chinese with English abstract).

Wang, X.-F., and X.-H. Chen. 1999. Palaeobiogeography and palaeoclimatology of Ordovician in China. Professional Papers of Stratigraphy and Palaeontology 27:1–27. Geological Publishing House, Beijing.

Wang, X.-F., S.-B. Zhang, and B.-D. Erdtmann. 1994. Ordovician graptolite sequences and palaeogeography of Kalpin, Xingjiang, China; pp. 164–173 in Chen Xu, B.-D. Erdtmann, and Y.-N. Ni (eds.), Graptolites Today. Geological Publishing House, Beijing.

Wang, X.-F., Z.-M. Li, J.-Q. Chen, X.-H. Chen, and W.-B. Su. 1996. Early Ordovician sea-level changes in South China and their worldwide correlation. Journal of China University of Geoscience 7:54–62.

Wang, Y. 1993. Middle Ordovician radiolarians from the Pingliang Formation of Gansu Province, China. Micropaleontology Special Publication 6:98–114.

Wang, Y., J. Li, and R. Wang. 1997. Latest Ordovician cryptospores from southern Xinjiang, China. Review of Palaeobotany and Palynology 99:61–74.

Wang, Z.-H., and S. M. Bergström. 1999. Conodont-graptolite biostratigraphic relations across the base of the Darriwilian Stage (Middle Ordovician) in the Yangtze Platform and the JCY area in Zhejiang, China. Bollettino della Società Paleontologica Italiana 37:187–198.

Wanninger A., B. Ruthensteiner, and G. Haszprunar. 2000. Torsion in *Patella caerulea* (Mollusca, Patellogastropoda): Ontogenetic process, timing, and mechanisms. Invertebrate Biology 119:177–187.

Warn, J. M. 1974. Presumed myzostomid infestation of an Ordovician crinoid. Journal of Paleontology 48:506–513.

Warn, J. M., and H. L. Strimple. 1977. The disparid inadunate superfamilies Homocrinacea and Cincinnaticrinacea (Echinodermata: Crinoidea), Ordovician-Silurian, North America. Bulletins of American Paleontology 72(296):1–138.

Warshauer, S. M., and J. M. Berdan. 1982. Palaeocopid and Podocopid Ostracoda from the Lexington Limestone and Clays Ferry Formation (Middle and Upper Ordovician) of Central Kentucky. United States Geological Survey, Professional Paper 1066H:1–81.

Watkins, R. 1991. Guild structure and tiering in a high-diversity Silurian community, Milwaukee County, Wisconsin. Palaios 6:465–478.

Watkins, R. 1993. The Silurian (Wenlockian) reef fauna of southeastern Wisconsin. Palaios 8:325–338.

Watkins, R., P. J. Coorough, and P. S. Mayer. 2000. The Silurian *Dicoelosia* communities: Temporal stability within an Ecologic Evolutionary Unit. Palaeogeography, Palaeoclimatology, Palaeoecology 162:225–237.

Watson, S. T. 1988. Ordovician conodonts from the Canning basin (W. Australia). Palaeontographica, Abteilung A 203:91–147.

Webb, G. E. 1997. Middle Ordovician *Tetradium* microatolls and a possible bathymetric gradient in tetradiid morphology. Boletín de la Real Sociedad Española de Historia Natural (Sección Geológia) 92:177–186.

Webby, B. D. 1969. Ordovician stromatoporoids from New South Wales. Palaeontology 12:637–662.

Webby, B. D. 1971. The new Ordovician genus *Hillophyllum* and the early history of rugose corals with acanthine septa. Lethaia 4:153–168.

Webby, B. D. 1972. The rugose coral *Palaeophyllum* Billings from the Ordovician of central New South Wales. Proceedings of the Linnean Society of New South Wales 97:150–157.

Webby, B. D. 1974. Upper Ordovician trilobites from central New South Wales. Palaeontology 17:203–252.

Webby, B. D. 1975. Succession of Ordovician coral and stromatoporoid faunas from central-western New South Wales, Australia; pp. 57–68 in B. S. Sokolov (ed.), Drevnie Cnidaria, Vol. 2. Nauka, Novosibirsk.

Webby, B. D. 1976. The Ordovician System in South-Eastern Australia, pp. 417–446 in M. G. Bassett (ed.), The Ordovician System: Proceedings of a Palaeontological Association Symposium, Birmingham [England], September 1974. University of Wales Press and National Museum of Wales, Cardiff.

Webby, B. D. 1977. Upper Ordovician tabulate corals from central-western New South Wales. Proceedings of the Linnean Society of New South Wales 101:167–183.

Webby, B. D. 1978. History of the Ordovician continental platform and shelf margin of Australia. Journal of the Geological Society of Australia 25:41–63.

Webby, B. D. 1980. Biogeography of Ordovician stromatoporoids. Palaeogeography, Palaeoclimatology, Palaeoecology 32:1–19.

Webby, B. D. 1982. *Cliefdenia*, a new stromatolite and associated girvanellid from the Ordovician of New South Wales. Alcheringa 6:185–191.

Webby, B. D. 1983. Lower Ordovician arthropod trace fossils from western New South Wales. Proceedings of the Linnean Society of New South Wales 107:59–74.

Webby, B. D. 1985. Influence of a Tasmanide island-arc on the evolutionary development of Ordovician faunas. New Zealand Geological Survey, Record 9:99–101.

Webby, B. D. 1986. Early stromatoporoids; pp. 148–166 in A. Hoffman and M. H. Nitecki (eds.), Problematic Fossil Taxa. Oxford University Press, New York.

Webby, B. D. 1987. Biogeographic significance of some Ordovician faunas in relation to East Australian Tasmanide suspect terranes; pp. 103–117 in E. C. Leitch and E. Schreibner (eds.), Terrane Accretion and Orogenic Belts. American Geophysical Union, Geodynamics Series 19.

Webby, B. D. 1988. The Ordovician genus *Favistina* Flower and a related colonial coral from New South Wales, Australia. New Mexico Bureau of Mines and Mineral Resources, Memoir 44:139–152.

Webby, B. D. 1990. Comments on a paper supposedly giving the first evidence of aragonitic mineralogy in tetradiid tabulate corals. Paläontologische Zeitschrift 64:379–380.

Webby, B. D. 1992a. Ordovician stromatoporoids from Tasmania. Alcheringa 15:191–227.

Webby, B. D. 1992b. Global biogeography of Ordovician corals and stromatoporoids; pp. 261–276 in B. D. Webby and J. R. Laurie (eds.), Global Perspectives on Ordovician Geology. Balkema, Rotterdam.

Webby, B. D. 1992c. Ordovician island biotas: New South Wales record and global implications. Journal and Proceedings of the Royal Society of New South Wales 125:151–177.

Webby, B. D. 1993. Evolutionary history of Palaeozoic Labechiida (Stromatoporoidea). Association of Australasian Palaeontologists, Memoir 15:57–67.

Webby, B. D. 1994. Evolutionary trends in Ordovician stromatoporoids. Courier Forschungsinstitut Senckenberg 172:373–380.

Webby, B. D. 1995. Towards an Ordovician time scale; pp. 5–9 in J. D. Cooper, M. L. Droser, and S. C. Finney (eds.), Ordovician Odyssey: Short Papers, 7th International Symposium on the Ordovician System. Book 77, Pacific Section Society for Sedimentary Geology (SEPM), Fullerton, California.

Webby, B. D. 1998. Steps toward a global standard for Ordovician stratigraphy. Newsletters on Stratigraphy 36:1–33.

Webby, B. D. 1999. Early to earliest Late Ordovician reef development; pp. 425–428 in P. Kraft and O. Fatka (eds.), *Quo vadis* Ordovician? Short papers of the 8th International Symposium on the Ordovician System. Acta Universitatis Carolinae, Geologica 43(1–2).

Webby, B. D. 2000. In search of triggering mechanisms for the great Ordovician biodiversification event. Palaeontology Down Under 2000, Orange, New South Wales. Geological Society of Australia, Abstracts 61:129–130.

Webby, B. D. 2002. Patterns of Ordovician reef development; pp. 129–179 in W. Kiessling, E. Flügel, and J. Golonka (eds.), Phanerozoic Reef Patterns. SEPM (Society for Sedimentary Geology), Special Publication 72.

Webby, B. D., and W. M. Blom. 1986. The first well-preserved radiolarians from the Ordovician of Australia. Journal of Paleontology 60:145–157.

Webby, B. D., and P. D. Kruse. 1984. The earliest heliolitines: A diverse fauna from the Ordovician of New South Wales. Palaeontographica Americana 54:164–168.

Webby, B. D., and R. S. Nicoll. 1989. Australian Phanerozoic Timescales. 2. Ordovician-Biostratigraphic chart and explanatory notes. Bureau of Mineral Resources, Geology and Geophysics, Australia, Record 1989/32: 1–42.

Webby, B. D., and I. G. Percival. 1983. Ordovician trimerellacean brachiopod shell beds. Lethaia 16:215–232.

Webby, B. D., and V. Semeniuk. 1969. Ordovician halysitid corals from New South Wales. Lethaia 2:345–360.

Webby, B. D., and V. Semeniuk. 1971. The Ordovician tabulate coral genus *Tetradium* Dana from New South Wales. Proceedings of the Linnean Society of New South Wales 95:246–259.

Webby, B. D., and J. Trotter. 1993. Ordovician sponge spicules from New South Wales, Australia. Journal of Paleontology 67:28–41.

Webby, B. D., and Y.-Y. Zhen. 1997. Silurian and Devonian clathrodictyids and other stromatoporoids from the Broken River region, north Queensland. Alcheringa 21:1–56.

Webby, B. D., Q.-Z. Wang, and K. J. Mills. 1988. Upper Cambrian–basal Ordovician trilobites from western New South Wales, Australia. Palaeontology 31:905–938.

Webby, B. D., A. H. M. VandenBerg, R. A. Cooper, I. Stewart, J. H. Shergold, R. S. Nicoll, C. F. Burrett, B. Stait, B. J. Cooper, J. Laurie, and L. Sherwin. 1991. Subdivisions of the Ordovician System in Australia; pp. 47–57 in C. R. Barnes and S. H. Williams, (eds.), Advances in Ordovician Geology. Geological Survey of Canada, Paper 90-9.

Webby, B. D., Y.-Y. Zhen, and I. G. Percival. 1997. Ordovician coral- and sponge-bearing associations: Distribution and significance in volcanic island shelf to slope habitats, Eastern Australia. Boletín de la Real Sociedad Española de Historia Natural (Sección Geológia) 92:163–175.

Webby, B. D., I. G. Percival, G. D. Edgecombe, R. A. Cooper, A. H. M. VandenBerg, J. W. Pickett, J. Pojeta, Jr., G. Playford, T. Winchester-Seeto, G. C. Young, Y.-Y. Zhen, R. S. Nicoll, J. R. P. Ross, and R. Schallreuter. 2000. Ordovician palaeobiogeography of Australasia; pp. 63–126 in A. J. Wright, G. C. Young, J. A. Talent, and J. R. Laurie (eds.), Palaeobiogeography of Australasian Faunas and Floras. Association of Australasian Palaeontologists, Memoir 23.

Welby, C. W. 1961. Occurrence of *Foerstephyllum* in Chazyan rocks of Vermont. Journal of Paleontology 35:391–400.

Welch, J. R. 1976. *Phosphannulus* on Paleozoic crinoid stems. Journal of Paleontology 50:218–225.

Wellman, C. H. 1996. Cryptospores from the type area for the Caradoc Series (Ordovician) in southern Britain. Special Papers in Palaeontology 55:103–136.

Wellman, C. H. In press. Dating the origin of land plants; in P. C. J. Donoghue and M. P. Smith (eds.), Telling the Evolutionary Time: Molecular Clocks and the Fossil Record. Systematics Association Special Volume. Taylor and Francis, London.

Wellman, C. H., and J. Gray. 2000. The microfossil record of early land plants. Philosophical Transactions of the Royal Society of London, B 355:717–732.

Wellman, C. H., K. T. Higgs, and P. Steemans. 2000. Spore assemblages from a Silurian sequence in Borehole Hawiyah-151 from Saudi Arabia; pp. 116–133 in S. Al-Hajri and B. Owens (eds.), Stratigraphic Palynology of the Palaeozoic of Saudi Arabia. GeoArabia, Bahrain.

Wendt, J. 1989. Tetradiidae—first evidence of aragonitic mineralogy in tabulate corals. Paläontologische Zeitschrift 63:177–181.

Wenz, W. 1938–1944. Gastropoda; pp. 1–1639 in O. H. Schindewolf (ed.), Handbuch der Paläozoologie. Gebrüder Borntraeger, Berlin.

Wenz, W. 1940. Ursprung und frühe Stammesgeschichte der Gastropoden. Archiv für Molluskenkunde 72:1–10.

Wenzel, B., C. Lécuyer, and M. Joachimski. 2000. Comparing oxygen isotope records of Silurian calcite and phosphate–$\delta^{18}O$ compositions of brachiopods and conodonts. Geochimica et Cosmochimica Acta 64:1859–1872.

Werle, N. G., T. J. Friest, and R. H. Mapes. 1984. The epizoan *Phosphannulus* on a Pennsylvanian crinoid stem from Texas. Journal of Paleontology 58:1163–1166.

Westermann, G. E. G., and P. Ward. 1980. Septum morphology and bathymetry in cephalopods. Paleobiology 6:48–50.

Westrop, S. R. 1986. Trilobites of the Upper Cambrian Sunwaptan Stage, southern Canadian Rocky Mountains, Alberta. Palaeontographica Canadiana 3:1–179.

Westrop, S. R., and J. M. Adrain. 1998. Trilobite alpha diversity and the reorganization of Ordovician benthic marine communities. Paleobiology 24:1–16.

Westrop, S. R., and M. B. Cuggy. 1999. Comparative paleoecology of Cambrian trilobite extinctions. Journal of Paleontology 73:337–354.

Westrop, S. R., J. V. Tremblay, and E. Landing. 1995. Declining importance of trilobites in Ordovician nearshore paleocommunities: Dilution or displacement? Palaios 10:75–79.

Wetzel, A., and A. Uchman. 1998. Deep-sea benthic food content recorded by ichnofabrics: A conceptual model based on observations from Paleogene Flysch, Carpathians, Poland. Palaios 13:533–546.

Wetzel, A., and A. Uchman. 2001. Sequential colonization of muddy turbidites in the Eocene Beloveža Formation, Carpathians, Poland. Palaeogeography, Palaeoclimatology, Palaeoecology 168:171–186.

Weyer, D. 1973. Über den Ursprung der Calostylidae Zittel 1879 (Anthozoa Rugosa, Ordoviz-Silur). Freiberger Forschungshefte, Leipzig C 282:23–87.

Weyer, D. 1982. Das Rugosa-Genus *Neotryplasma* Kaljo (1957) aus dem Ordoviz der europäischen UdSSR. Freiberger Forschungshefte, Leipzig C 366:89–94.

Weyer, D. 1983. *Lambelasma*-Arten (Anthozoa, Rugosa) aus dem baltoskandischen Mittelordoviz. Freiberger Forschungshefte, Leipzig C 384:7–19.

Weyer, D. 1984. *Lambelasma narvense,* a new rugose coral from the Middle Ordovician of Estonia. Proceedings of the Estonian Academy of Sciences, Geology 33:92–95.

Weyer, D. 1997. *Lambelasma balticum* n. sp. (Anthozoa, Rugosa) aus einem baltoskandischen Oberordoviz-Geschiebe; pp. 43–50 in M. Zwanzig and H. Löser (eds.), Berliner Beiträge zur Geschiebeforschung. Cpress Verlag, Dresden.

Whitfield, R. P. 1894. On new forms of marine algae from the Trenton Limestone, with observations on *Butho-*

graptus laxus Hall. American Museum of Natural History, Bulletin 6 (Art.16):351–358.

Whittard, W. F. 1953. *Palaeoscolex piscatorum* gen. et sp. nov., a worm from the Tremadocian of Shropshire. Quarterly Journal of the Geological Society of London 109:125–135.

Whittington, H. B. 1963. Middle Ordovician trilobites from Lower Head, western Newfoundland. Bulletin of the Museum of Comparative Zoology, Harvard 129:1–118.

Whittington, H. B. 1965. Trilobites of the Ordovician Table Head Formation, western Newfoundland. Bulletin of the Museum of Comparative Zoology, Harvard 132:275–442.

Whittington, H. B., and C. P. Hughes. 1972. Ordovician geography and faunal provinces deduced from trilobite distribution. Philosophical Transactions of the Royal Society of London, B 263:235–278.

Wicander, R., and G. D. Wood. 1997. The use of microphytoplankton and chitinozoans for interpreting transgressive/regressive cycles in the Rapid Member of the Cedar Valley Formation (Middle Devonian), Iowa. Review of Palaeobotany and Palynology 98:125–152.

Wicander, R., G. Playford, and E. B. Robertson. 1999. Stratigraphic and paleogeographic significance of an Upper Ordovician acritarch flora from the Maquoketa Shale, northeastern Missouri, U.S.A. Paleontological Society Memoir 51 (Journal of Paleontology 73[6] supplement):1–38.

Wilde, P. 1991. Oceanography in the Ordovician; pp. 283–298 *in* C. R. Barnes and S. H. Williams (eds.), Advances in Ordovician Geology. Geological Survey of Canada, Paper 90-9.

Wilde, P., and W. B. N. Berry. 1982. Progressive ventilation of the oceans—potential for return to anoxic conditions in the Post Paleozoic; pp. 209–224 *in* S. O. Schlanger and M. B. Cita (eds.), Nature and Origin of Cretaceous Carbon-Rich Facies. Academic Press, New York.

Wilde, P., and W. B. N. Berry. 1984. Destabilisation of the ocean density structure and its significance to marine "extinction" events. Palaeogeography, Palaeoclimatology, Palaeoecology 48:143–162.

Wilde, P., and W. B. N. Berry. 1986. The role of oceanographic factors in the generation of global bio-events; pp. 75–91 *in* O. H. Walliser (ed.), Global Bio-Events. Lecture Notes in Earth Sciences, 8. Springer-Verlag, Berlin and Heidelberg.

Williams, A. 1962. The Barr and Lower Ardmillan Series (Caradoc) of the Girvan District, South-west Ayrshire, with descriptions of the Brachiopoda. The Geological Society, London, Memoir 3:1–267.

Williams, A., and C. H. C. Brunton. 2000. Suborder Orthotetidina; pp. 644–681 *in* R. L. Kaesler (ed.), Treatise on Invertebrate Paleontology. Part H, Brachiopoda, rev., Vol. 3. Geological Society of America, Boulder, Colorado, and University of Kansas Press, Lawrence.

Williams, A., and D. A. T. Harper. 2000a. Suborder Billingsellidina; pp. 690–692 *in* R. L. Kaesler (ed.), Treatise on Invertebrate Paleontology. Part H, Brachiopoda, rev., Vol. 3. Geological Society of America, Boulder, Colorado, and University of Kansas Press, Lawrence.

Williams, A., and D. A. T. Harper. 2000b. Order Protorthida; pp. 709–714 *in* R. L. Kaesler (ed.), Treatise on Invertebrate Paleontology. Part H, Brachiopoda, rev., Vol. 3. Geological Society of America, Boulder, Colorado, and University of Kansas Press, Lawrence.

Williams, A., and D. A. T. Harper. 2000c. Suborder Orthidina; pp. 724–782 *in* R. L. Kaesler (ed.), Treatise on Invertebrate Paleontology. Part H, Brachiopoda, rev., Vol. 3. Geological Society of America, Boulder, Colorado, and University of Kansas Press, Lawrence.

Williams, A., S. J. Carlson, C. H. C. Brunton, L. E. Holmer, and L. E. Popov. 1996. A supra-ordinal classification of the Brachiopoda. Philosophical Transactions of the Royal Society of London, B 351:1171–1193.

Williams, M., and J. Vannier. 1995. Middle Ordovician Aparchitidae and Schmidtellidae: The significance of "featureless" ostracods. Journal of Micropalaeontology 14:7–24.

Williams, M., P. Stone, D. J. Siveter, and P. Taylor. 2001. Upper Ordovician ostracods from the Cautley district, northern England: Baltic and Laurentian affinities. Geological Magazine 138:589–607.

Williams, S. H., and D. L. Bruton. 1983. The Caradoc-Ashgill boundary in the central Oslo region and associated graptolite faunas. Norsk Geologisk Tidsskrift 63:147–191.

Williams, S. H., C. R. Barnes, F. H. C. O'Brien, and W. D. Boyce. 1994. A proposed global stratotype for the second series of the Ordovician System: Cow Head Peninsula, western Newfoundland. Bulletin of Canadian Petroleum Geology 42:219–231.

Williams, S. H., D. A. T. Harper, R. B. Neuman, W. D. Boyce, and C. Mac Niocaill. 1995. Lower Paleozoic fossils from Newfoundland and their importance in understanding the history of the Iapetus Ocean. Geological Association of Canada Special Paper 41:115–126.

Williams, S. H., G. S. Nowlan, C. R. Barnes, and R. S. R. Batten. 1999. The Ledge section at Cow Head, western Newfoundland, as a GSSP candidate for the lower boundary of the second stage of the Ordovician System: New data and discussion of the graptolite, conodont and chitinozoan assemblages. A report to the IUGS/ICS Subcommission on Ordovician Stratigraphy, 30 pp.

Wilson, A. E. 1946. Echinodermata of the Ottawa Formation of the Ottawa-St. Lawrence Lowland. Geological Survey of Canada, Bulletin 4:1–61.

Wilson, A. E. 1948. Miscellaneous classes of fossils, Ottawa Formation, Ottawa-St. Lawrence Valley. Geological Survey of Canada, Bulletin 11:1–25.

Wilson, A. E. 1951. Gastropoda and conularida of the Ottawa Formation of the Ottawa-St. Lawrence Lowland. Geological Survey of Canada, Bulletin 17:1–149.

Wilson, E. O. 1992. The Diversity of Life. Allen Lane, Penguin Press, London, 424 pp.

Wilson, M. A., and T. J. Palmer. 2001a. The Ordovician Bioerosion Revolution. Geological Society of America, Abstracts with Programs 33:248.

Wilson, M. A., and T. J. Palmer. 2001b. Domiciles, not predatory boring: A simpler explanation of the holes in Ordovician shells analyzed by Kaplan and Baumiller, 2000. Palaios 16:524–525.

Wilson, M. A., T. J. Palmer, T. E. Guensburg, C. D. Finton, and L. E. Kaufman. 1992. The development of an Early Ordovician hardground community in response to rapid sea-floor calcite precipitation. Lethaia 25:19–34.

Winchester, J., et al. In press. Palaeozoic amalgamation of central Europe: New results from recent geological and geophysical investigations. Tectonophysics.

Winchester-Seeto, T., C. Foster, and T. O'Leary. 2000. Chitinozoans from the Middle Ordovician (Darriwilian) Goldwyer and Nita formations, Canning Basin (Western Australia). Acta Palaeontologica Polonica 45:271–300.

Withers, T. H. 1921. The "Cirripede" *Plumulites* in the Middle Ordovician rocks of Esthonia. Annual Magazine of Natural History 8:123–127.

Withers, T. H. 1926. Catalogue of the Machaeridia (*Turrilepas* and Its Allies) in the Department of Geology. British Museum (Natural History), London, xv + 99 pp.

Witzke, B. J., and B. J. Bunker. 1996. Relative sea-level changes during Middle Ordovician through Mississippian deposition in the Iowa area, North American craton; pp. 307–320 *in* B. J. Witzke, G. A. Ludvigson, and J. Day (eds.), Paleozoic Sequence Stratigraphy: Views from the North American Craton. Geological Society of America, Special Paper 306.

Won, M.-Z., and W. Iams. 2002. Late Cambrian Radiolarian faunas and biostratigraphy of the Cow Head Group, Western Newfoundland. Journal of Paleontology 76:1–33.

Wood, G. D., and M. A. Miller. 1996. Pre-Carboniferous Chlorophyta: New reports of Hydrodictyaceae, ?Scenedesmaceae and Zygnemataceae. Acta Universitatis Carolinae, Geologica 40:707–717.

Wood, G. D., A. M. Gabriel, and J. C. Lawson. 1996. Palynological techniques—processing and microscopy; pp. 29–50 *in* J. Jansonius and D. C. McGregor (eds.), Palynology: Principles and Applications, Vol. 1. American Association of Stratigraphic Palynologists Foundation. Publishers Press, Salt Lake City, Utah.

Wood, R. 1995. The changing biology of reef-building. Palaios 10:517–529.

Wright, A. D. 1993. Subdivision of the Lower Palaeozoic articulate brachiopod family Triplesiidae. Palaeontology 36:481–493.

Wright, A. D. 2000. Suborder Triplesiidina; pp. 681–689 *in* R. L. Kaesler (ed.), Treatise on Invertebrate Paleontology. Part H, Brachiopoda, rev., Vol. 3. Geological Society of America, Boulder, Colorado, and University of Kansas Press, Lawrence.

Wright, A. D., and A. E. McClean. 1991. Microbrachiopods and the end-Ordovician event. Historical Biology 5:221–254.

Wright, A. D., and M. Rubel. 1996. A review of the morphological features affecting the classification of clitambonitidine brachiopods. Palaeontology 39:53–75.

Wright, A. J., R. A. Cooper, and J. E. Simes. 1994. Cambrian and Ordovician faunas and stratigraphy, Mt. Patriarch, New Zealand. New Zealand Journal of Geology and Geophysics 37:437–476.

Wright, C. A., C. R. Barnes, and S. B. Jacobsen. 2002. The Neodymium isotopic composition of Ordovician conodonts as a seawater proxy: Testing paleogeography. American Geophysical Union, G3 electronic journal, Geochemistry, Geophysics, Geosystems 3(2):10.1029/2001GC000195.

Wright, R. P., and W. C. Meyers. 1981. Organic-walled microplankton in the subsurface Ordovician of northeastern Kansas. Kansas Geological Survey, Subsurface Geology Series 4:1–53.

Wu, H.-J. 1990. Occurrence of Machaeridia in China. Acta Palaeontologica Sinica 29:567–580.

Xu, W. 1995. Acritarchs from the Arenigian Tonggao Formation in the Sandu area, Guizhou Province, with its organic stratigeochemistry. Doctoral thesis, Nanjing Institute of Geology and Palaeontology, Academia Sinica, Nanjing (in Chinese with English abstract).

Yin, L. 1985. Acritarchs; pp. 314–373 *in* J. Chen, Y. Qian, Y. Lin, J. Zhang, Z. Wang, L. Yin, and B.-D. Erdtmann (eds.), Study on Cambrian-Ordovician Boundary and Its Biota in Dayangcha, Hunjiang, Jilin, China. China Prospect Publishing House, Beijing.

Yin, L. 1995. Early Ordovician Acritarchs from Hunjian Region, Jilin and Yichan Region, Hubei, China. Palaeontolgia Sinica, n.s., A 185:1–170.

Yin, T.-H. 1937. Brief description of the Ordovician and Silurian fossils from Shihtien. Geological Society of China, B 16:281–298.

Yochelson, E. L. 1956. Euomphalacea, Trochonemataca, Pseudophoracea, Anomphalacea, Craspedostomataca, and Platyceratacea. [Part] 1, Permian Gastropoda of the southwestern United States. American Museum of Natural History, Bulletin 110:173–275.

Yochelson, E. L. 1961. Notes on the class Conichonchia. Journal of Paleontology 35:162–167.

Yochelson, E. L. 1966. Mattheva, a proposed new class of mollusks. United States Geological Survey, Professional Paper 523-B:1–11.

Yochelson, E. L. 1967. Quo vadis, *Bellerophon*? pp. 141–161 *in* R. C. Moore (ed.), Essays in Paleontology and Stratigraphy. University of Kansas, Lawrence.

Yochelson, E. L. 1968. On the nature of *Polylopia*. United States Geological Survey, Professional Paper 593-F:1–7.

Yochelson, E. L. 1984. Historic and current considerations for revision of Paleozoic gastropod classification. Journal of Paleontology 58:259–269.

Yochelson, E. L. 2000. Concerning the concept of an extinct class of Mollusca: Or what may/may not be a class of mollusks. American Malacological Bulletin 15(2):195–202.

Yochelson, E. L., R. H. Flower, and G. F. Webers. 1973. The bearing of the new Late Cambrian monoplacophoran genus *Knightoconus* upon the origin of the Cephalopoda. Lethaia 6:275–309.

Yoo, E. K. 1994. Early Carboniferous Gastropoda from the Tamworth Belt, New South Wales, Australia. Records of the Australian Museum 46:63–120.

Young, G. A. 1995. A new tetradiid coral from the Late Ordovician of Manitoba. Canadian Journal of Earth Sciences 32:1393–1400.

Young, G. A., and R. J. Elias. 1995. Latest Ordovician to earliest Silurian colonial corals of the east-central United States. Bulletins of American Paleontology 108(347):1–148.

Young, G. C. 1995. The oldest known vertebrates and gnathostomes? New material from the Late Cambrian and Early Ordovician of central Australia. Abstract *in* CAVEPS [Conference on Australasian Vertebrate Evolution, Palaeontology and Systematics] 95, Canberra, April 18–20, 1995, 25.

Young, G. C. 1997. Ordovician microvertebrate remains from the Amadeus Basin, central Australia. Journal of Vertebrate Paleontology 17:1–25.

Young, G. C., and J. R. Laurie (eds.). 1996. An Australian Phanerozoic Timescale. Oxford University Press, Melbourne, 279 pp. and 12 charts.

Young, G. C., V. N. Karatajute-Talimaa, and M. M. Smith. 1996. A possible Late Cambrian vertebrate from Australia. Nature 383(6603):810–812.

Yu, C. C. 1930. The Ordovician cephalopods of central China. Palaeontologia Sinica, B 1:1–71.

Yu, W. 1987. Yangtze micromolluscan fauna in Yangtze Region of China with notes on Precambrian-Cambrian boundary. Stratigraphy and Palaeontology of Systematic Boundaries in China–Precambrian-Cambrian Boundary 1:19–344.

Yuan, W.-W., Z.-Y. Zhou, J.-M. Zhang, Z.-Q. Zhou, X.-W. Sun, and T.-M. Zhou. 2000. Tremadocian trilobite biofacies in western Hunan-Hubei. Journal of Stratigraphy 24:275–282 (in Chinese with English abstract).

Zaslavskaya, N. M. 1982. Chitinozoa; pp. 159–166 *in* B. S. Sokolov (ed.), Ordovician of the Siberian Platform (type section on the Kulyumbe River). Nauka, Moscow (in Russian).

Zaslavskaya, N. M. 1984. Group Chitinozoa: Ordovician chitinozoans of the Siberian Platform; pp. 146–149 *in* Paleontological Atlas. Nauka, Novosibirsk, Siberia (in Russian).

Zaslavskaya, N. M., and A. M. Obut. 1984. Lower Ordovician Chitinozoa of Gorny Altai; pp. 106–114 *in* A. V. Kanygin (ed.), Lower Ordovician Stratigraphy and Fauna of the Altai Mountains. USSR Academy of Sciences, Siberian Branch, 565. Nauka, Moscow (in Russian).

Zaslavskaya, N. M., A. M. Obut, and N. V. Sennikov. 1978. Chitinozoa in the Ordovician and Silurian deposits of Gorny Altai; pp. 42–56 *in* Fauna and biostratigraphy of the Upper Ordovician and Silurian Altai-Sayan Area. USSR Academy of Sciences, Siberian Branch, 405. Nauka, Moscow (in Russian).

Zeng, X.-L, Y.-L. Xu, W. Nie, S.-X. Zhang, Z.-X. Wei, Q.-S. Song, G.-D. Yang, Y.-N. Wang, T. Wu, and Y.-Y. Zhao. 1996. Sequence Stratigraphy and Palaeontology in Northern Tarim Basin, Xinjiang. Geological Publishing House, Beijing, 85 pp.

Zhang, J. 1998. Conodonts from the Guniutan Formation (Llanvirnian) in Hubei and Hunan Provinces, south-central China. Acta Universitatis Stockholmiensis, Stockholm Contributions in Geology 46:1–161.

Zhang, J., C. R. Barnes, and B. J. Cooper. 2000. Middle Ordovician conodonts from the Stokes Siltstone in the Amadeus Basin (Central Australia). Palaeontology Down Under 2000, Orange, New South Wales. Geological Society of Australia, Abstracts 61:137.

Zhang, J.-H. 1995. Ordovician phosphatic inarticulate brachiopods from Cili, Hunan. Acta Palaeontologica Sinica 34:152–170.

Zhang, R.-J. 1984. Early Silurian bivalves and rostroconchs in north-west Hunan, China. Acta Palaeontologica Sinica 23:586–596 (in Chinese with English summary).

Zhang, S., and C. R. Barnes. 2002. Eustatic sea level curve for the Ashgillian-Llandovery derived from conodont community analysis, Anticosti Island, Québec. Palaeogeography, Palaeoclimatology, Palaeoecology 180:5–32.

Zhao, Z.-X., and G.-Z. Zhang. 1992. Ordovician conodonts and stratigraphy in the subsurface of the Tarim Basin; pp. 64–74 *in* X. Tong and D. Liang (eds.), New advances in the study of the petroleum geology of the Tarim Basin. Xinjiang Scientific and Health Publishing House, Urumqi (in Chinese).

Zhao, Z.-X., G.-Z. Zhang, and J. Xiao. 2000. Paleozoic stratigraphy and conodonts in Xinjiang. Petroleum Industry Press, Beijing, 339 pp. (in Chinese).

Zhen, Y.-Y., and I. G. Percival. 2002. Ordovician conodont biogeography—reconsidered. IPC 2002. Geological Society of Australia, Abstracts 68:179–180.

Zhizhina, M. S. 1956. Nekotorye ordovikskie Tabulyaty Vostchnogo Taimyra. Nauchno-Issledovatelskii Institut Geologii Arktiki, Trudy 86(6):91–116.

Zhizhina, M. S. 1966. Tetradiidy Vaigachi i Taimyra. Nauchno-Issledovatelskii Institut Geologii Arktiki, Ucheniye Zapiski, Paleontologiya i Biostratigrafiya 13:5–16.

Zhou, Z.-Y., and W. T. Dean. 1989. Trilobite evidence for Gondwanaland in east Asia during the Ordovician.

Journal of Southeast Asian Earth Sciences 3:131–140.

Zhou, Z.-Y., and R. A. Fortey. 1986. Ordovician trilobites from north and northeastern China. Palaeontographica, Abteilung A 192:157–210.

Zhou, Z.-Y., Z.-Q. Zhou, and J.-L. Zhang. 1989. Ordovician trilobite biofacies of North China Platform and its western marginal area. Acta Palaeontologica Sinica 28:296–313 (in Chinese with English summary).

Zhou, Z.-Y., X. Chen, Z.-H. Wang, Z.-Z. Wang, J. Li, L.-Y. Geng, Z.-J. Fang, X.-D. Qiao, and T.-R. Zhang. 1992. Ordovician of Tarim; pp. 62–139 *in* Z.-Y. Zhou and P.-J. Chen (eds.), Biostratigraphy and Geological Evolution of Tarim. Science Press, Beijing.

Zhou, Z.-Y., T.-Y. Zhang, W.-W. Yuan, and J.-L. Yuan. 1994. Trilobites; pp. 71–82 *in* Sinian to Permian Stratigraphy and Palaeontology of the Tarim Basin, Xinjiang (IV), the Altun Mountains Regions, Southern Xinjiang. China Oil Regional Stratigraphy and Palaeontology Publication Series. Petroleum Industry Press, Beijing, 285 pp. (in Chinese).

Zhou, Z.-Y., W. T. Dean, and H.-L. Luo. 1998a. Early Ordovician trilobites from Dali, west Yunnan, China, and their palaeogeographical significance. Palaeontology 41:429–460.

Zhou, Z.-Y., W. T. Dean, W.-W. Yuan, and T.-R. Zhou. 1998b. Ordovician trilobites from the Dawangou Formation, Kalpin, Xinjiang, North-west China. Palaeontology 41:693–735.

Zhou, Z.-Y., Z.-Q. Zhou, and W.-W. Yuan. 1999. Middle Caradoc trilobite biofacies of western Hubei and Hunan, South China; pp. 385–388 *in* P. Kraft and O. Fatka (eds.), *Quo vadis* Ordovician? Short papers of the 8th International Symposium on the Ordovician System. Acta Universitatis Carolinae, Geologica 43(1–2).

Zhou, Z.-Y., Z.-Q. Zhou, W.-W. Yuan, and T.-M. Zhou. 2000. Late Ordovician trilobite biofacies and palaeogeographical development, western Hubei–Hunan. Journal of Stratigraphy 24:249–263 (in Chinese with English summary).

Zhou, Z.-Y., Z.-Q. Zhou, and W.-W. Yuan. 2001. Llanvirn-early Caradoc biofacies of western Hubei and Hunan, China. Alcheringa 25:69–86.

Zhou, Z.-Y., Z.-Q. Zhou, D. J. Siveter, and W. W. Yuan. 2003. Latest Llanvirn-early Caradoc trilobite biofacies in the northwestern marginal area of the Yangtze Block. Special Papers in Palaeontology 70:281–291.

Zhu, M. 1998. Early Silurian sinacanths (Chondrichthyes) from China. Palaeontology 41:157–171.

Zhu, M.-Y., H. Van Iten, R. S. Cox, Y.-L. Zhao, and B.-D. Erdtmann. 2000. Occurrence of *Byronia* Matthew and *Sphenothallus* Hall in the Lower Cambrian of China. Paläontologische Zeitschrift 74:227–234.

Zhuravlev, A. Yu. 2000 [dated 2001]. Biodiversity and structure during the Neoproterozoic-Ordovician transition; pp. 173–199 *in* A. Yu Zhuravlev and R. Riding (eds.), The Ecology of the Cambrian Radiation. Columbia University Press, New York.

Zhuravlev, A. Yu., and R. Riding. 2000 [dated 2001]. The Ecology of the Cambrian Radiation. Columbia University Press, New York, 525 pp.

Zhuravleva, F. A. 1994. The Order Dissidocerida (Cephalopoda). Paleontological Journal 28:115–148.

CONTRIBUTORS

AÏCHA ACHAB
  Centre Géoscientifique de Québec
  C.P. 7500
  Sainte-Foy, Quebec, G1V 4C7, Canada

JONATHAN M. ADRAIN
  Department of Geoscience
  121 Trowbridge Hall
  University of Iowa
  Iowa City, Iowa 52242, United States

GUILLERMO L. ALBANESI
  CONICET—Museo de Paleontologia
  Universidad Nacional de Córdoba
  Casilla de Correo 1598
  5000 Córdoba, Argentina

ESTHER ASSELIN
  Centre Géoscientifique de Québec
  C.P. 7500
  Sainte-Foy, Quebec, G1V 4C7, Canada

CHRISTOPHER R. BARNES
  School for Earth and Ocean Sciences (SEOS)
    and Centre for Earth and Ocean Research
    (CEOR)
  University of Victoria
  Victoria, British Columbia, V8W 3P6, Canada

MICHAEL G. BASSETT
  Department of Geology
  National Museum and Galleries of Wales
  Cathays Park, Cardiff CF1 3NP, Wales

MATILDE S. BERESI
  CRICYT-IANIGLA
  Avenida, Ruiz Leal s/n
  5500 Mendoza, Argentina

STIG M. BERGSTRÖM
  Department of Geological Sciences
  The Ohio State University
  155 South Oval Mall
  Columbus, Ohio 43210-1397, United States

ALAIN BLIECK
  Sciences de la Terre
  Laboratoire de Paléontologie et Paléogéographie
    du Paléozoïque UMR 8014 du CNRS
  Université des Sciences et Technologies de Lille
  F-59655 Villeneuve d'Ascq Cedex, France

SIMON J. BRADDY
  Department of Earth Sciences
  Wills Memorial Building
  University of Bristol
  Queens Road
  Bristol BS8 1RJ, England

PATRICK J. BRENCHLEY
  Department of Earth Sciences
  University of Liverpool
  Brownlow Street, P.O. Box 147
  Liverpool, L69 3BX, England

MARCELO G. CARRERA
  Cátedra de Estratigrafía y Geología Historíca
  Facultad de Ciencias Exactas, Físicas y Naturales
  Universidad Nacional de Córdoba
  Avenida Vélez Sarsfield 299
  5000 Córdoba, Argentina

CHEN XIAO-HONG
  Center for Stratigraphy and Palaeontology
  Yichang Institute of Geology and Mineral
    Resources
  Yichang, Hubei, China

Contributors

LESLEY CHERNS
Department of Earth Sciences
Cardiff University
P.O. Box 914
Cardiff CF10 3YE, Wales

L. ROBIN M. COCKS
Department of Palaeontology
The Natural History Museum
Cromwell Road
London SW7 5BD, England

ROGER A. COOPER
Institute of Geological and Nuclear Sciences
P.O. Box 30368
Lower Hutt, New Zealand

JOHN C. W. COPE
Department of Earth Sciences
Cardiff University
P.O. Box 914
Cardiff CF10 3YE, Wales

PAUL COPPER
Department of Earth Sciences
Laurentian University
Sudbury, Ontario P3E 2C6, Canada

TANIEL DANELIAN
Laboratoire de Micropaléontologie
CNRS-FRE 2400
Université Pierre et Marie Curie
C. 104, 4, Place Jussieu
75252 Paris Cedex 05, France

MARY L. DROSER
Department of Earth Sciences
University of California
Riverside, California 92521, United States

GREGORY D. EDGECOMBE
Division of Earth and Environmental Sciences
The Australian Museum
6 College Street
Sydney, NSW, 2010, Australia

ROBERT J. ELIAS
Department of Geological Sciences
University of Manitoba
Winnepeg, Manitoba, R3T 2N2 Canada

MATS ERIKSSON
Department of Geology
Lund University
Sölvegatan 13
S-223 62 Lund, Sweden

ANDREJ ERNST
Institut für Geowissenschaften
Universität zu Kiel
Olshausenstrasse 40
D-24118, Kiel, Germany

DAVID H. EVANS
Environmental Impacts Team, English Nature
Northminster House
Peterborough, PE1 1UA, England

RICHARD A. FORTEY
Department of Palaeontology
The Natural History Museum
Cromwell Road
London SW7 5BD, England

ROBERT C. FREY
Bureau of Environmental Health and Toxicity
Ohio Department of Health
246 North High Street
Columbus, Ohio 43266-0588, United States

JIŘÍ FRÝDA
Czech Geological Survey
Klárov 3/131
118 21 Praha 1, Czech Republic

YNGVE GRAHN
Universidade do Estado do Rio de Janeiro—UERJ
Faculdade de Geologia
20559-4001 Rio de Janeiro, Brazil

THOMAS E. GUENSBURG
Physical Science Division
Rock Valley College
Rockford, Illinois 61114, United States

ØYVIND HAMMER
Geological Museum
Postboks 1172
Blindern
N-0318 Oslo, Norway

DAVID A. T. HARPER
Geological Museum
University of Copenhagen
Øster Voldgade 5–7
DK-1350 Copenhagen K, Denmark

OLLE HINTS
Institute of Geology
Tallinn Technical University
7 Estonia puiestee
10143 Tallinn, Estonia

ANETTE E. S. HÖGSTRÖM
Department of Earth Sciences—Palaeobiology
Uppsala University
Norbyvägen 22
SE-752 36 Uppsala, Sweden

LARS E. HOLMER
Department of Earth Sciences—Palaeobiology
Uppsala University
Norbyvägen 22
SE-752 36 Uppsala, Sweden

JISUO JIN
Department of Earth Sciences
University of Western Ontario
London, Ontario N6A 5B7, Canada

DIMITRI KALJO
Institute of Geology
Tallinn Technical University
7 Estonia puiestee
10143, Tallinn, Estonia

ALAN H. KING
English Nature
Roughmoor, Taunton, TA1 5AA, England

PETR KRAFT
Institute of Geology and Palaeontology
Charles University
Albertov 6
128 43, Prague 2, Czech Republic

HUBERT LARDEUX
Géosciences-Rennes
Université de Rennes I
35042 Rennes Cedex, France

JOHN R. LAURIE
Geoscience Australia
G.P.O. Box 378
Canberra, ACT 2601, Australia

OLIVER LEHNERT
Institut für Geologie und Mineralogie
Universität Erlangen
Schlossgarten 5
D-91054 Erlangen, Germany

JUN LI
Nanjing Institute of Geology and
  Palaeontology
Academia Sinica
Chi-Ming-Ssu, Nanjing 210008, China

JÖRG MALETZ
Department of Geology
State University of New York at Buffalo
772 Natural Sciences and Mathematics
  Complex
Buffalo, New York 14260-3050, United States

JOHN M. MALINKY
University of Maryland
Mannheim Campus, Unit 24560
Grenadierstrasse 4
Gebande 485, Box 515 (APO AE 09183)
D-68167 Mannheim, Germany

M. GABRIELA MÁNGANO
INSUGEO
Casilla de Correo 1
4000 San Miguel de Tucumán, Argentina

TIMOTHY MCCORMICK
British Geological Survey
Kingsley Durham Centre
Keyworth, NG12 5GG, England

ARNOLD I. MILLER
Department of Geology
University of Cincinnati
P.O. Box 210013
Cincinnati, Ohio 45221-0013, United States

BJÖRN E. E. NEUMAN
Geological Institute
University of Bergen
Allégatan 41
N-5007 Bergen, Norway

ARNE THORSHØJ NIELSEN
Geological Museum
University of Copenhagen
Øster Voldgade 5–7
DK-1350 Copenhagen K, Denmark

MATTHEW H. NITECKI
　Department of Geology
　Field Museum of Natural History
　Roosevelt Road at Lake Shore Drive
　Chicago, Illinois 60605-2496,
　United States

PAULA J. NOBLE
　Department of Geological Sciences
　University of Nevada
　Reno, Nevada 89957, United States

JAAK NÕLVAK
　Institute of Geology
　Tallinn Technical University
　7 Estonia puiestee
　10143, Tallinn, Estonia

GODFREY S. NOWLAN
　Geological Survey of Canada
　3303 Thirty-third Street NW
　Calgary, Alberta T2L 2A7, Canada

OLGA OBUT
　Institute of Petroleum Geology
　Siberian Branch
　Russian Academy of Sciences
　Novosibirsk 630090, Russia

ALAN W. OWEN
　Division of Earth Sciences
　University of Glasgow
　Gregory Building, Lilybank Gardens
　Glasgow G12 8QQ, Scotland

FLORENTIN PARIS
　Géosciences-Rennes
　UMR 6118 du CNRS
　Université de Rennes I
　35042 Rennes Cedex, France

IAN G. PERCIVAL
　Geological Survey of New South Wales
　P.O. Box 76
　Lidcombe, NSW, 2141, Australia

LEONID E. POPOV
　Department of Geology
　National Museum and Galleries
　　of Wales
　Cathays Park
　Cardiff CF1 3NP, Wales

PATRICK R. RACHEBOEUF
　UFR Sciences et Techniques
　UMR 6538 du CNRS—Domaines Océaniques
　Université de Bretagne Occidentale
　6 Avenue Le Gorgeu, Bôlte Postale 809
　F-29285 Brest Cedex, France

J. KEITH RIGBY
　Department of Geology
　Brigham Young University
　S389 ESC
　Provo, Utah 84602-4606, United States

DAVID M. ROHR
　Department of Geological and Physical Sciences
　Sul Ross State University
　400 North Harrison Street
　Alpine, Texas 79832, United States

RONG JIA-YU
　Nanjing Institute of Geology and Palaeontology
　Academia Sinica
　Chi-Ming-Ssu, Nanjing 210008, China

PETER M. SADLER
　Department of Earth Sciences
　University of California
　Riverside, California 92521, United States

JOAKIM SAMUELSSON
　Department of Earth Sciences—Palaeobiology
　Uppsala University
　Norbyvägen 22
　S-752 36 Uppsala, Sweden

ROGER SCHALLREUTER
　Institut für Geologische Wissenschaften
　Ernst-Morits-Arndt Universität Greifswald
　Friedrich-Ludwig-Jahn-Strasse 17a
　D-17489 Greifswald, Germany

NIKOLAI SENNIKOV
　Institute of Petroleum Geology
　Siberian Branch
　Russian Academy of Sciences
　Novosibirsk 630090, Russia

THOMAS SERVAIS
　Paléontologie—Sciences de la Terre
　UMR 8014 du CNRS
　Université des Sciences et Technologies de Lille
　F-59655 Villeneuve d'Ascq Cedex, France

PETER M. SHEEHAN
Geology Department
Milwaukee Public Museum
Milwaukee, Wisconsin 53233, United States

GRAHAM A. SHIELDS
School of Earth Sciences
James Cook University
Townsville, Qld, 4811, Australia

NILS SPJELDNAES
Institutt for Geofag
Universitetet i Oslo
Postboks 1047
Blindern
N-0316, Oslo, Norway

JAMES SPRINKLE
Department of Geological Sciences
University of Texas
Austin, Texas 78712, United States

PHILIPPE STEEMANS
Paléobotanique, Paléopalynologie et
 Micropaléontologie
Université de Liège
Allée du six Août
B-4000 Liège 1, Belgium

LUDOVIC STRICANNE
Institut und Museum für Geologie und
 Paläontologie Universität Tübingen
Sigwartstrasse 10
D-72076, Tübingen, Germany

LINDSEY TAYLOR
Industrial Minerals
16 Lower Marsh
London SE 7RJ, England

PAUL D. TAYLOR
Department of Palaeontology
The Natural History Museum
Cromwell Road
London, SW7 5BD, England

VICTOR P. TOLLERTON JR.
New York State Museum
The State Education Department
Albany, New York 12230, United States

TROND H. TORSVIK
Geological Survey of Norway
Leif Erikssons vei 39
N-7491 Trondheim, Norway

SUSAN TURNER
Queensland Museum
P.O. Box 3300
South Brisbane, Qld, 4101, Australia

HEYO VAN ITEN
Department of Geology
Hanover College
Hanover, Indiana 47243-0890, United States

MARCO VECOLI
UMR 6538 du CNRS—Domaines
 Océaniques
Université de Bretagne Occidentale
6 Avenue Le Gorgeu,
Bôlte Postale 809
F-29285 Brest Cedex, France

JÁN VEIZER
Ottawa-Carleton Geoscience Centre
University of Ottawa
Ottawa, Ontario, K1N 6N5, Canada

JACQUES VERNIERS
Laboratoire de Paléontologie
Universiteit Gent
Krijgslaan 281 S8
B-9000 Gent, Belgium

ZDENKA VYHLASOVÁ (NEÉ BRABCOVÁ)
Department of Palaeontology
West Bohemian Museum
Kopeckého sady 2,
301 00 Plzen, Czech Republic

BEATRIZ G. WAISFELD
Cátedra de Estratigrafía y Geología Historíca
Facultad de Ciencias Exactas, Físicas y Naturales
Universidad Nacional de Córdoba
Avenida Vélez Sarsfield 299
5000 Córdoba, Argentina

WANG XIAO-FENG
Center for Stratigraphy and Palaeontology
Yichang Institute of Geology and Mineral
 Resources
Yichang, Hubei, China

BARRY D. WEBBY
Centre for Ecostratigraphy and Palaeobiology
Department of Earth and Planetary Sciences
Macquarie University
North Ryde, NSW, 2109, Australia

CHARLES H. WELLMAN
 Centre for Palynology
 Dainton Building
 University of Sheffield
 Brook Hill
 Sheffield S3 7HF, England

STEPHEN R. WESTROP
 School of Geology and Geophysics
 University of Oklahoma
 Museum of Natural History
 2401 Chatauqua Avenue
 Norman, Oklahoma 73019, United States

REED WICANDER
 Department of Geology
 Central Michigan University
 Mount Pleasant, Michigan 48859, United States

MARK A. WILSON
 Department of Geology
 The College of Wooster
 Wooster, Ohio 44691, United States

THERESA WINCHESTER-SEETO
 Centre for Ecostratigraphy and Palaeobiology
 Department of Earth and Planetary Sciences
 Macquarie University
 North Ryde, NSW, 2109, Australia

GRAHAM A. YOUNG
 Manitoba Museum
 190 Rupert Avenue
 Winnepeg, Manitoba R3B 0N2, Canada

JAN A. ZALASIEWICZ
 Department of Geology
 Bennett Building
 University of Leicester
 University Road
 Leicester LE1 7RH, England

ZHEN YONG-YI
 Division of Earth and Environmental Sciences
 The Australian Museum
 6 College Street
 Sydney, NSW, 2010, Australia

ZHOU ZHI-YI
 Nanjing Institute of Geology and Palaeontology
 Academia Sinica
 Chi-Ming-Ssu, Nanjing 210008, China

INDEX

The suffix *t* on a page number indicates a table; *f* indicates a figure.

Acanthodiacrodium, 358, 359
Acanthodiacrodium angustum, 359
Acanthodiacrodium costatum, 359
Acanthodiacrodium uniforme, 359
acanthodians, 328
Acanthodictya, 105
Acanthodii, 328
Acanthodii indet., 333
Acanthodus lineatus, 322
Acerocare Regressive Event, 90
Achinacelloidea, 182
Acidolites, 126, 134t, 142f
Acodus, 322
Acodus deltatus, 314t, 321, 322
acritarchs, 12, 13, 20, 35, 348–360, 352f, 353f; Baltica, 353–354; distribution, 350–351; North Africa, 357–360, 358f; North America, 354–355; previous compilations, 351–352; rock matrix, 349–350; size of, 350; South China, 355–357, 356f
Acrotretida, 158–162, 160f, 163f
Actinocerida, 24, 25, 211, 212
actinocerids, 26, 30–31
"*Actinodonta,*" 200, 201f
Actinodonta, 200
Admixtella, 167
Aegiromena, 164
Aethedionide, 251
Agetolites, 142f, 143
agnathans, 328
agnostids, 242
Ajakmalaisoria, 344
Akelina, 164
algae, 336–347; blue-green, 12, 18, 35, 341–342; brown, 342; cyclocrinitids, 35, 336–337, 339–341, 340f, 344; green, 12, 20, 35, 344–346; red, 12, 35, 342–344
Alleynodictyon, 113, 114
Allodesma, 197
Allotrioceratidae, 211

alpha diversity, 14, 178, 380, 382
alsataspidids, 251
Aluoja substage. *See* Kunda stage
Alwynopora, 149
Alytodonta, 203
Ambitisporites avitus, 361
ambonychiids, 202
Amia, 342
Amphigastropoda, 182–183, 183f, 192f, 194
Amphispongiaceae, 340, 340f
amplexoporines, 150f, 152, 152f
Ampyx, 238f, 251
Ampyxina, 251
Amsassia, 126
Anaconularia, 119, 120, 121f, 122, 123
Ananterodonta, 200, 201f
anaspids, 328, 333
"*Anatolepis,*" 328, 329, 330f, 333, 334
Anatolepis heintzi, 328
Anazyga, 170
Ancistrorhynchoidea, 157
Ancryochitina merga, 308
andinus-evae (A-E) interval, 313, 315, 322–323
Angopora?, 134t, 135f, 137
anomalodesmatans, 200–201, 201f
Anomaloides, 340–341, 340f
Ansella, 324
Ansoporella, 342
Anthaspidellidae, 104, 108, 108f
anthaspidellids, 104
Anticosti Island, 106, 114, 120, 165, 355
Apedolepis, 328
Apedolepis tomlinsonae, 331
Apheathyris, 171
Aphroporella, 344
Apidiaceae, 340, 340f
Apidium, 340, 340f, 344
Apogastropoda, 186f, 191
Arandaspidiformes, 328, 333

Arandaspis, 328
Arandaspis prionotelepis, 330–331
Arborohindia, 105
Arbusculidium filamentosum, 350
Archaeochonetes, 165
Archaeoconularia, 119, 120, 121f, 122, 123
Archaeoconularia fecunda, 121f, 122
Archaeogastropoda, 30, 186–189, 186f, 189f, 190, 191, 192f, 194; slit-bearing, 187–188; without slit, 187, 188–189
Archaeogastropoda sensu stricto, 185, 186
Archaeoscyphia, 18, 104
Archaeotrypa, 148
Archaeotrypa prima, 148
Archaeotrypa secunda, 148
Arctic Ordovician Fauna, nautiloids, 212
Arctohedra, 167
Arenicolites, 373, 374, 377
Arenig (British series); acritarchs, 356f, 358f, 359; bivalve mollusks, 199f, 201f, 203f; brachiopods, 159f, 165f, 166f, 168f–170f, 177f; bryozoans, 149f, 152f; chitinozoans, 44f, 45f, 297f, 301f, 303, 303f, 305f, 309f; conodonts, 44f–45f, 317f, 318f; conulariids, 121f; corals, 130f, 142f; cyclocrinitids, 340f; echinoderms, 268f–271f; graptolites, 42f–43f, 86f, 282f, 290t, 291f; jawed polychaetes, 224f, 227; miospores, 364f; nautiloid cephalopods, 210f; ostracodes, 262f; palaeoscolecidans and chaetognaths, 229f; phyllocarids, 260f; rostroconch mollusks, 205f; stromatoporoids, 113f; trilobites, 243f, 245f, 246f, 248f, 249f; tube-shaped incertae sedis, 216f, 217,

217f; univalve mollusks, 181f; vertebrates, 334
Arenigiphyllum, 343
Arenigiphyllum crustosum, 343
Arenigomya, 201, 201f
Areyongalepis, 328
Areyongalepis oervigi, 331
Argentina; conodonts, 314t, 316t, 321, 322, 324, 325; trilobites, 243, 251
"Aristocystites," 268
?*Aristozoe,* 257
Arkonia, 359
Armoricochitina nigerica, 300
Arthraria, 373
arthropods, 5t, 32, 377; eurypterids, 4t, 5t, 32, 255–257, 256t; ostracodes, 4t, 5t, 32–33, 260–265, 262f; phyllocarid crustaceans, 257–260, 258t, 260f
articulated brachiopods, 4t, 5t, 9, 11, 158, 174, 176
Aryballomorpha, 350, 354
Aryballomorpha grootaertii, 354
Asaphidae, 233, 234f, 252
asaphids, 32, 242, 252
ascocerids, 25, 26
Aseri stage; conodonts and chitinozoans, 44f; graptolites, 42f–43f, 86f
ash fall, 33, 47, 48
Ashgill Lowstand Interval, 92, 93
Ashgill origination/extinction events, 194
Ashgill (British series, informal global stage) 7f, 47; acritarchs, 358f; bivalve mollusks, 199f, 201f, 203f; brachiopods, 159f, 165f, 166f, 168f–170f, 177f; bryozoans, 149f, 152f; chitinozoans, 44f–45f, 297f, 301f, 303f, 305f, 309f; conodonts, 44f–45f; conulariids, 121f; corals, 126, 130f, 142f;

473

Ashgill (cont.)
cyclocrinitids, 340f; echinoderms, 268f–271f; graptolites, 42f–43f, 86f, 282f, 290f, 291f; jawed polychaetes, 224f, 227; miospores, 363, 364f; nautiloid cephalopods, 210f; ostracodes, 262f; palaeoscolecidans and chaetognaths, 229f; phyllocarids, 260f; receptaculitids, 338f; rostroconch mollusks, 205f; stromatoporoids, 113f; trilobites, 243f, 245f, 246f, 248f, 249f; tube-shaped incertae sedis, 216f, 217f; univalve mollusks, 181f; vertebrates, 332–333, 334
Asteriospongia, 106
asterozoans, 33, 268f, 271f
Astraeoconus, 106
Astraeospongium, 106
astrapidiform, 331
astrapids, 331
Astraspidiformes, 333
Astraspis, 328, 331, 333
Astylospongia, 104, 108
Astylostroma micra, 104
Athabascaella, 350, 354
Athabascaella playfordii, 354
Athabascaella rossi, 354
"atheloptic" trilobite assemblage, 26, 32
atheloptic trilobites, 253
Athyridida, 158, 171, 176
Athyridoidea, 157
atmosphere, ocean system, 72–75
Atrypida, 158, 170, 176
Atrypoidea, 157
Attritasporites, 361
Aulacera, 113, 114, 115
Aulaceratidae, 113–116, 113f
Aulacopleura, 251
Auloporina, 125, 141, 142f
auloporinids, 125, 126, 142, 143
Aurelucian stage, graptolites, 42f–43f
Aureotesta, 359
Australasia; corals, 141–144, 142f, 145; graptolites, 42f–43f, 48, 282f, 283, 287f, 288–289, 290t, 291–292, 291f
Australia; conodonts, 316t; ostracodes, 265; trilobites, 239–242, 241f; Whiterock Fauna, 31, 32
Australoharpes, 240
Autolamellibranchiata, 197f, 199
Avalonia terrane, 61, 62, 64; chitinozoans, 304–306, 305f; conulariids, 122; graptolites, 282f, 283–284, 288, 289, 289f, 290t, 291f; ostracodes, 263–264; trilobites, 244–246, 245f, 251; Whiterock Fauna, 32

Babinka, 200, 201f
Baccaoconularia, 120

Bactroceras, 25
Bactroceras latisiphonatum, 25
Bactrotheca, 216f, 217
Baikalides, 64
Bajgolia, 142, 142f, 143
balanced total diversity (BTD), chitinozoans, 296, 297–298, 297f, 301, 301f, 303–309, 303f, 305f, 309f
Baltica, 63; acritarchs, 353–354; chitinozoans, 300–302, 301f; conulariids, 122; graptolites, 282f, 283, 288, 288f, 290t, 291, 292; trilobites, 246–247, 246f
Baltisphaeridium, 355, 359
Baltoceratidae, 211
Baltoniodus, 323, 324
Baltoscandia, 45; chitinozoans, 44f–45f, 300–302, 301f; conodonts, 44f–45f, 314t, 316t, 325; corals, 127, 138–141, 138t, 145; graptolites, 42f–43f; ostracodes, 261–263, 262f; regional subdivisions, 46; rugose corals, 28, 29, 33; sea level changes in, 84–93; stromatoporoid communities, 113; trilobites, 246–247, 246f
"Basal Llanvirn Bio-Event," 193–194
Basilicus, 252
Batanevia, 341
Bathmoceratidae, 211
Bathyuridae, 233
bathyurids, 233, 251
Batostoma, 18, 19
Bellerophon, 183
Bellerophontida, 181, 183
bellerophontids, 30, 179, 181–183
bellerophontiform mollusks, 182, 183
Belodina compressa, 343, 345
Belodina confluens, 345, 346
Belodina monitorensis, 343, 345
Belonechitina, 300
Belonechitina cactacea, 300
Belonechitina robusta, 300
Belonechitina vibrissa, 299
Belubulaspongia, 105
Benambran Orogeny, 144, 241
Bendigonian stage, graptolites, 42f–43f, 46, 241f, 281
benthic realm, 11–12, 13
bentonites. See ash fall; K-bentonite beds
Bergstroemognathus, 322, 323
Bergstroemognathus extensus, 322, 324
Beroun (Czech or Barrandian series), ostracodes, 262f
beta diversity, 14, 178, 382
Beyrichiocopa, 263
beyrichiocopes, 264
Bienvillia, 251
Biernatidae, 160, 161

Big Bentonite Bed (BBB). See Kinnekulle K-bentonite bed
Bighornia, 134t, 135, 135f, 137, 139f
Bija, 343
Billingen stage; conodonts and chitinozoans, 44f; graptolites, 42f–43f, 86f
Billingsaria, 126, 142f
Billingsaria domica, 141
Billingsaria spissa, 126, 141
Billingsella, 167
Billingsellida, 166–167, 176
Billingsellidina, 158
Billingselloidea, 157, 164, 166
Binodicopa, 261, 263
biodiversity, measures of, 52–57, 98
"bioerosion revolution," 19
biofacies patterns, 21–24, 26; trilobites, 249–253; conodonts, 321–326
bioimmurations, 149, 150
biotas, overview, 3–6
bivalve mollusks, 4t, 5t, 10, 30, 196–204
Blackhillsian stage; conodonts and chitinozoans, 44f–45f; sea level curves, 91f
blastozoan eocrinoids, 276
blastozoans, 33, 268f, 270f, 272, 273, 276
blue-green algae, 12, 18, 35, 341–342
Bobinella, 169
Bogutschanophycus, 344
Bohemian Massif, 65; conulariids, 120, 122; nautiloid cephalopods, 26; ostracodes, 33, 262f
Bohemilla, 23
Bohemillidae, 233
bohemillids, 250
Bolindian stage, corals, 142f; graptolites, 42f–43f
Boonderooia, 105
bootstrap tests, 245, 246
bothriocidarids, 271f, 275, 276, 278f
Botomaella, 342
Bowanophyllum, 142
Brachilyrodesma, 199f, 200
brachiopods, 4t, 5t, 9, 10, 11, 29, 157–178; autecology, 173–174; ecologic diversification, 171–177; food resources and size ranges, 172–173; paleoecologic levels, 177; predation, 176; synecology, 174–176; taxonomic diversification, 158–171
Brachiospongia, 106
Brachiospongiidae, 105–106
Brachiospongiodea, 106
Brachyopterus, 256
Brachyopterus stubblefieldi, 257
Bradoriida, 260
bradoriids, 264
Bransonia, 205f, 206

Bransoniidae, 206
Brevitheca, 216f, 217–218
brown algae, 342
bryoniids, 4t, 5t, 31, 220–221, 220f
bryozoans, 4t, 5t, 9–12, 29, 147–156; fossil record, 148–149; phylogeny, 149–151, 149f; taxic diversity patterns, 151–155, 151t, 152f–154f
BTD (balanced total diversity), chitinozoans, 296, 297–298, 297f, 301, 301f, 303–309, 303f, 305f, 309f
Buffalopterus verrucosus, 256
Bulbaspis, 251
Burrellian stage, graptolites, 42f–43f
Byronia, 220
Byronia universalis, 220, 220f
Bythotrepsis, 342

Caenogastropoda, 185, 186, 187, 188f, 190–192, 192f, 194
Calathella, 338, 338f
Calathium, 18
Calceochiton, 180, 181f
Caliculospongia, 104
Callisphenus, 344
Callithamniopsis, 342, 344
Calostylidae, 138
Calostylina, 128
Calostylis, 139, 139f
Calpichitina cf. *C. lata*, 299, 300
Calycocoelia, 104
calymenids, 249
calymenoids, 252
Camarotoechioidea, 157
Cambrian EF, 2f, 9, 10, 11, 12, 124, 159, 171, 173, 175, 176–178, 266, 272, 279, 383–384
Cambrooistodus, 23
Camellaspongia, 104
camerate crinoids, 277
Caradoc (British series, informal global stage), 7f, 10, 19; acritarchs, 358f, 359; bivalve mollusks, 199f, 201f, 203f; brachiopods, 159f, 165f, 166f, 168f–170f, 177f; bryozoans, 149f, 152f; chitinozoans, 44f–45f, 297f, 301f, 303–304, 303f, 305f, 309f; conodonts, 44f–45f; conulariids, 121f; corals, 126, 130f, 140, 142f; cyclocrinitids, 340f; echinoderms, 268f–271f; graptolites, 42f–43f, 86f, 282f, 290t, 291f; jawed polychaetes, 224f, 227; miospores, 363, 364f; nautiloid cephalopods, 210f; ostracodes, 262f; palaeoscolecidans and chaetognaths, 229f; phyllocarids, 260f; receptaculitids, 338f; rostroconch mollusks, 205f; stromatoporoids, 113f;

trilobites, 243f, 245f, 246f, 248f, 249f; tube-shaped incertae sedis, 216f, 217f; univalve mollusks, 181f; vertebrates, 331–332, 334
carbon dioxide, in atmosphere, 73
carbon isotopes: glaciation and, 81, 82f; seawater, 69, 70f, 71, 81
"carbonate inner shelf" facies, infaunal record, 19
Cardiolaria, 200
Cardiolarioidea, 197, 199f
cardiolaroids, 199–200
Carinolithes, 216, 216f, 217
Carolinites, 23, 238f, 250
Carolinites genacinaca, 250
carryover taxa, percentage of, conodonts, 319t, 319f, 320
Caryocaris, 24, 257–260
Caryocaris curvilata, 259
caryocystitid rhombiferans, 270f, 275, 278f
Caryospongia, 104, 108
Castlemainian stage, 42f–43f, 46
Catamarcaia, 197, 203, 203f
Catamarcaiadae, 203
Catazyga, 170
Catenipora, 126, 134t, 135f, 136, 142f, 143
"Catillicephalidae," 238f
Caulerpales, 344, 346
Cautleyan stage, graptolites, 42f–43f
Cavernolithes, 216f, 217
Celidocrania, 162
Celmus, 235, 238f
cemented blastozoans, 273
cephalopods, 4t, 9, 10, 11–12; nautiloid, 24–27, 209–213, 210f
Ceratiocaris, 257
Ceratiocaris primula, 257, 259
Ceratiocarididae, 258
Ceratopyge Regressive Event (CRE), 87, 90, 321, 322, 325
Ceratopygidae, 233, 234f
cerioid forms, corals, 129
Cessipylorum, 100, 100f
Chabakovia, 341
Chaetocladus, 344, 346
chaetognaths, 4t, 5t, 31, 223, 228, 229–230, 229f
Chaetomorpha, 346
Chatsfieldian stage, 47; conodonts and chitinozoans, 44f–45f; sea level curves, 91f
Chazy Group, 120, 126, 343
Chegetella, 260
cheirurids, 242, 249
cheirurines, 253
Chelispongia, 106
Chelodes, 180, 181f
Chelodes? mirabilis, 180
Chelodes whitehousei, 180
Chelsonella, 216, 216f, 217
Cheneyan stage, graptolites, 42f–43f

Chewtonian stage, graptolites, 42f–43f
chiastoclonellids, 104
chiastoclones, 104
Chientangkiangian stage, graptolites, 42f–43f
Chilcaia, 106
Chilidiopsoidea, 157, 166
Chimerolites, 216f, 217–218
China: chitinozoans, 305f, 306–308; conodonts, 316t; graptolites, 42f–43f; ostracodes, 264
chitinozoans, 4t, 5t, 20, 34, 44f–45f, 291f, 294–311, 295f; Avalonia, 304–306, 305f; Baltica, 300–302, 301f; China, 305f, 306–308; East Gondwana, 299; Laurentia, 302–304, 303f; North Gondwana, 296–299, 297f; Siberia, 308; West Gondwana, 299–300
Chlorococcales, 349
chlorophytes, 344–346
Choia, 103
chondrichthyans, 328
Chondrichthyes, 328
Chondrites, 342, 374
Chonetida, 165–166
Chonetoidea, 157, 158
Chonophyllidae, 139
Chosenia, 240
Chuchlinidae, 191
Cincinnatian (North American series), 45; acritarchs, 358f; bivalve mollusks, 199f, 201f, 203f; brachiopods, 159f, 165f, 166f, 168f–170f, 177f; bryozoans, 149f, 152f; chitinozoans, 44f–45f, 297f, 301f, 303f, 305f, 309f; conodonts, 44f–45f; conulariids, 121f; corals, 126–127, 130f, 133, 133f, 134t, 135f, 137, 142f; cyclocrinitids, 340f; echinoderms, 268f–271f; graptolites, 42f–43f; jawed polychaetes, 224f; nautiloid cephalopods, 210f; palaeoscolecidans and chaetognaths, 229f; phyllocarids, 260f; rostroconch mollusks, 205f; sea level curves, 91f; stromatoporoids, 113f; trilobites, 243f, 245f, 246f, 248f, 249f; tube-shaped incertae sedis, 216f, 217f; univalve mollusks, 181f
Circotheca, 216, 216f, 217
Clarkella, 162
Clathrodictyida, 112, 113f, 114, 115, 115f, 116, 117
Clathrodictyidae, 113f, 115
Clathrodictyon, 114
Cliefdenia, 341–342
Cliefdenospongia, 105
Climacoconus, 119–20, 121f, 122, 123
Climacograptus stewarti, 259
Climacograptus zhejiangensis, 259

climate, oceans and, 72–75
Clisospiridae, 189
Clisospiroidea, 189
Clitambonitidina, 158, 167
Clitambonitoidea, 157, 166, 167
Cobcrephora, 179
Coccoseris, 126, 142f
Cocculiniformia, 190
Codiaceae, 344
codiaceans, 12, 35
Coelocerodontus, 229f, 230
Coelosphaeridiaceae, 340, 340f
Coelosphaeridium, 340, 340f, 344
Colaptoconus, 321
Colaptoconus bolites, 322
Colaptoconus floweri, 322
Coleolida, 221
Coleoloids, 219
coleoloids, 4t, 5t, 31, 214, 219
communis-elegans (C-E) interval, 313, 315, 322
Compendium of Fossil Marine Animal Genera (Sepokski), 5, 5t
computer-assisted constrained optimization (CONOP), 6–7, 48–51
Conchiocolites, 219, 221
conjungaspids, 224
Conocardioidea, 206
conocardioids, 204, 206
Conochitina, 300, 306
Conochitina aff. *C. havliceki*, 299
Conochitina baculata, 307
Conochitina brevis, 307
Conochitina cf. *decipiens*, 306
Conochitina cf. *pervulgata*, 306
Conochitina chydaea, 303
Conochitina decipiens, 299–300
Conochitina esthonica, 307
Conochitina havliceki, 307
Conochitina langei, 299, 307
Conochitina parvicolla, 308
Conochitina pirum, 307
Conochitina poumoti, 306, 307
Conochitina raymondi, 307, 308
Conochitina scabra, 302, 306
Conochitina subcylindrica, 299, 303
Conochitina symmetrica, 307
Conodia, 180
conodonts, 4t, 5t, 10, 11, 21–23, 34, 44f–45f, 312–326, 314t–315t
CONOP (Computer-assisted constrained optimization), 6–7, 48–51
Conostichus, 377
Constellatospongia, 106
Contexta, 342, 345
"continental margin" assemblage, corals, 137
continental platforms, Evolutionary Faunas (EFs), 13–14
Contitheca cor, 218
Conularia, 119, 120, 121f, 122, 123
Conulariella, 119, 120, 121f

conulariids, 4t, 5t, 28, 119–123
Conularina, 119, 120, 121f, 122
Coolinia, 166
Copidens, 200, 201f
Coprinisphaera, 370
Corallidomus, 202, 203f
corals, 4t, 5t, 9, 11, 28–29, 124–146; Australasian, 141–144, 142f, 145; Baltoscandian, 127, 138–141, 138t, 145; diversity patterns, 132–133; Laurentian, 133–138, 134t, 135f, 144f, 145; rugose, 28, 29, 33, 125, 127–128, 133–141, 135f, 145; tabulate, 28–29, 124, 125–127, 145; tetradiid, 29, 124, 125, 128–133, 130f, 141, 145
Cordylodus, 23
Cornulitella, 219
Cornulites, 218–219, 221
Cornulitida, 221
Cornulitidae, 218
cornulitids, 4t, 5t, 214, 218
Cornuodus longibasis, 324
cornute stylophorans, 271f, 275, 276, 278f
coronoids, 270f, 275, 278f
corynoidids, 293
Coryphidium, 350, 359
Costiconus, 324
Craniida, 158, 159, 163
craniiformean brachiopods, 29, 158–159, 159f, 162–164, 163f
Craniopsida, 158, 159, 163, 163f
cratons, 72, 73, 74, 79
Cremastoglottos, 23, 250
Crenulites, 134t, 135f, 142, 142f, 143
Crinitella radiata, 346
crinoids, 9, 33, 269f, 273, 277, 278f
crinozoans, 33, 268f, 269f, 272
Crispatella, 216f, 217
Cristallinium, 358
Cristicoma, 162
crustaceans, 5t, 10, 24; phyllocarid, 4t, 5t, 24, 32, 257–260, 258t, 260f
Cruziana, 370, 371, 375
Cruziana furcifera, 376
Cruziana omanica, 377
Cruziana rugosa group, 370, 371, 377
Cruziana semiplicata, 371, 373, 377
Cryptodonta, 204
Cryptolichenaria, 125, 126
Cryptophragmus, 114, 115
cryptospores, 13, 35, 361, 362f
Cryptostomata, 150, 152
Cryptothyrella, 171
Cryptozoon, 341
Ctenoconularia, 119, 120, 121f
Ctenodonta, 197–199, 199f
Ctenodonta nasuta, 197
cyanophytes (blue-green algae), 12, 18, 35 341–342

Index 475

Cyathochitina, 300, 307
Cyathochitina calix, 308
Cyathochitina? cf. *clepsydra,* 306
Cyathochitina hunderumensis, 299
Cyathochitina hyalophrys, 303
Cyathochitina jenkinsi, 300, 307
Cyathochitina protocalix, 307
Cyathophycus, 105
Cyathophylloides, 134, 134t, 135f, 136, 139f, 142, 142f, 143
cybelospine pliomerids, 233
Cyclendoceras, 213
Cycloceras, 213
Cycloconchidae, 200
Cyclocrinaceae, 339, 340f
Cyclocrinales, 337, 339–341, 344
Cyclocrinites, 337, 339–340, 340f, 344, 346
cyclocrinitids, 35, 336–337, 339–341, 340f, 344
Cyclocrinus, 344
Cycloneritimorpha, 191
Cyclopyge, 23, 250
Cyclopygidae, 23–24, 233, 234f, 235t, 236t, 251
cyclopygids, 235, 249
Cyclospira, 170
cyclospirids, 170
Cyclostomata, 150, 151t
cyclostomes, 150f, 152, 152f
Cyclostomiceratidae, 211
Cymatiogalea, 358, 359
Cymatiogalea granulata, 359
Cymbithyris, 167
Cyrtendoceras, 211
Cyrtinoidea, 157
Cyrtoceras, 213
Cyrtodonta, 201f, 202
cyrtodontids, 202
Cyrtodontula, 201f, 202
Cyrtolitida, 182
Cyrtoneritimorpha, 186, 186f, 189f, 191
Cyrtophyllum, 134t, 135, 135f, 142f
cystiphyllids, 142
Cystiphyllina, 128, 141, 142f
Cystistroma, 114, 115, 142–143
Cystoporata, 150, 151t

Dacryoconarida, 221
Dactylofusa, 359
Dalmanellidina, 167
dalmanellidines, 168–169
Dalmanelloidea, 157
dalmanelloids, 168
dalmanitids, 235
Dalmanitoidean Realm, 235, 235f
dalmanitoids, 253
Dapsilodus-Periodon biofacies, 22
Darriwilian-Caradoc boundary, 79, 80
Darriwilian origination/extinction events, 193–194
Darriwilian Stage (ratified global Stage), 7f, 46; acritarchs, 356f, 358f; bivalve mollusks, 199f; 201f, 203f; brachiopods, 16f, 159f, 165f, 166f, 168f–170f, 177f; bryozoans, 149f; chitinozoans, 44f–45f, 297, 297f, 301f, 303f, 304, 305f, 309f; conodonts, 44f–45f, 317f, 318f; conulariids, 121f; corals, 130f, 142f; cyclocrinitids, 340f; echinoderms, 268f–271f; graptolites, 42f–43f, 282f, 290t, 291f; jawed polychaetes, 224f; miospores, 363, 364f; nautiloid cephalopods, 210f; ostracodes, 262f; palaeoscolecidans and chaetognaths, 229f; phyllocarids, 260f; radiolarians, 100f, 101 receptaculitids, 338f; rostroconch mollusks, 205f; stromatoporoids, 113f; trilobites, 243f, 245f, 246f, 248f, 249f; tube-shaped incertae sedis, 216f, 217f; univalve mollusks, 181f; vertebrates, 330–331, 330f, 334
dascyclads, 12, 35
Dasycladales, 344–346
Dasyporella, 344, 345, 346
Dawsonia, 258
Dawsonoceras, 213
Deceptrix, 200
Decipilites, 216f, 218
Degamella, 23, 250
Deike and Millbrig K-bentonite beds, 47, 78
Deiracorallium, 134t, 135, 135f, 136
deltatus-proteus (D-P) interval, 313, 315, 321
demosponges, 102–105
dendroclones, 104
Densigrewingkia, 139, 139f
Denticelox, 202, 203f
Dermatostroma, 114, 115
Desmochitina, 300
Desmochitina complanata, 303
Desmochitina sp. gr. *minor,* 299
Devonian-Carboniferous ("Hangenberg") extinction event, 116
Diagoniella, 105
Dianulites, 149
Diaphorodus, 322
Diasoma, 196
Dicellograptus clingani, 282f, 283
Dichograptidae, 293
Dichograptus maccoyi densus, 281
Dichograptus maccoyi maccoyi, 281
Dicommopalla, 359–360
Dicrodiacrodium, 359
Dictyonellida, 158
Dideroceras, 211
didymograptids, 293
Diexallophasis, 359–360
"differentiation" diversity, 14
Dilytes, 216f, 217
Dimeropygidae, 234f, 235t, 236t, 252
dimeropygids, 233
Dimorphichnus, 371, 375
Dimorphosiphon, 344, 346
Dimorphosiphonoides, 344, 346
dindymenines, 253
dinoflagellates, 348, 350. See also acritarchs
Dinophyceae, 349
Dionide, 251
Dionididae, 233, 234f, 251
Dipleurodonta, 203f
Diplichnites, 375, 378
diplobathrid camerates, 269f, 277, 278f
Diplopodichnus, 378
diploporans, 270f, 273, 275, 278f
Discinidae, 162
Discinoidea, 161
Discoceras, 211
Discosorida, 211, 212
disparity, 156. See also morphological diversity
Dissidocerida, 209, 212
dissidocerids, 30–31
Dithyrocaris longicauda, 259
diversity curves, 8. See also "range-through" diversity; "sampled" diversity
diversity measures, 7–8, 52–57
diversity plateau, 9, 154. See also equilibrium level
Diversoporella, 344
Dobaowanian stage, 43f; chitinozoans, 305f
Docoglossa, 185, 186, 190
Doliporella, 344
Dorsennidium, 355
Dorsennidium hamii, 355
Dorsolinevitus, 216f, 217
Dowlingia, 342
Drepanodus, 322
Drepanodus arcuatus, 324
Drepanoistodus, 322, 324
Drepanoistodus nowlani, 314t
drowning events, 86f–87f, 87–90, 91f, 92–93
Dualites, 142, 142f
Dystactella, 198, 199, 199f
Dystatella, 199f
Dzikodus, 325

early Arenig origination event, 193, 195
early Caradoc origination event, 194
Early Ordovician Epoch, 7f, 11, 107; bivalve mollusks, 198; chitinozoans, 294; echinoderms, 272, 278; graptolites, 282f; nautiloid cephalopods, 211; tube-shaped incertae sedis, 217
Eastonian stage, graptolites, 42f; corals, 142–144, 142f; trilobites, 241f, 241–242
Ecclimadictyon, 114

echinoderms, 4t, 5t, 10, 11, 12, 33, 266–279
Echinognathus clevelandi, 256
echinoids, 10, 271f, 275, 276, 278f
echinozoans, 33, 268f, 271f
ecologic diversity, 2, 13–15, 17–27; of brachiopods, 171–177; of ichnofossils, 371–379. See also biofacies patterns
Ecologic-Evolutionary Subunits (EESs), 313, 315, 320
Ecologic-Evolutionary Units (EEUs), 17, 313
Edenian stage, conodonts and chitinozoans, 44f–45f; corals, 135; sea level curves, 91f
edrioasteroids, 271f, 275, 277, 278f
EFs (Evolutionary Faunas), 2f, 9–10, 13–14, 110, 159, 162, 171, 174–177, 232, 266–267, 272, 276, 279, 380, 383, 384
Eichwaldioidea, 157
Eiffelia, 106
Eione monoliformis, 377
Eisenackitina tongziensis, 307
Eisenackitina uter, 307
Elegantilites, 216f, 217
Elkaniidae, 161
Ellesmerocerida, 11, 24–27, 212
ellesmerocerids, 30, 211
Ellesmoceratidae, 211
Ellipsotaphrus, 250
Elliptoglossa, 160
Encrinurella, 240
Encrinuridae, 233, 234f, 235t, 236t, 252
encrinurids, 32, 242
Endoceras, 213
Endoceratidae, 211
Endocerida, 24, 25, 211, 212
endocerids, 30
end Ordovician mass extinction, 9, 30–31, 33, 144, 233, 236–237, 239, 236t, 247–248; possible causes, 83
Enfieldia, 342
Entactiniidae, 97, 100, 100f
Enteletoidea, 157
enteletoids, 168
Entobia, 370
Eocarcinosoma batrachophthalmus, 257
Eochelodes, 180, 181f
Eoconodontus, 23
Eoconularia, 119, 120, 122
Eoconularia azaisi, 122
Eoconulidae, 160, 161
Eocramatia, 166
Eocramatiidae, 166
eocrinoids, 9, 270f, 276, 277, 278f
Eodinobolus, 142, 162
Eofletcheria, 125–126, 142f
Eogastropoda, 186, 187
Eoharpes, 251

Eoisotelus, 252
Eokosovopeltis-Pliomerina fauna, 240
Eomonoplacophora, 182
Eopilla ingelorae, 265
Eoplacognathus suecicus, 316*t*, 345
Eoplectodonta, 164
Eopteria, 204, 206
Eopteriidae, 206
eopteriids, 204
Eopterioidea, 206
Eosotrematorthis, 167
Eospirfer, 170–171
Eospirfer praecursor, 171
epeiric seas, 72, 73, 74, 110, 123
Ephippelasmatidae, 160, 161, 162
Epiphyton, 18, 341, 342, 343
Epiplastospongia, 105
equilibrium level, 154, 383. *See also* diversity plateau
Eremochitina brevis, 299, 306
Eremos, 260
Ericanthea, 359
Eridostraca, 261, 263
Eriptychiiformes, 333, 334
Eriptychius, 328, 331, 333
Eritropsis, 199*f*, 200
Erraticodon, 324
Erraticodon balticus, 324, 345
esthonioporines, 150*f*, 152
Euconochitina vulgaris, 307
Eumorpholites, 216*f*, 217
Euomphalina, 189
Euomphaloidea, 189, 190
Euomphalomorpha, 30, 186*f*, 189*f*, 190
Eupoikilofusa, 355
Eurymya, 202, 203*f*
eurypterids, 4*t*, 5*t*, 32, 255–257, 256*t*
eustatic oscillations, 73
Evae Drowning Event, 87, 90, 92, 93
Evolutionary Faunas (EFs), 2*f*, 9–10, 13–14, 110, 159, 162, 171, 174–177, 232, 266–267, 272, 276, 279, 380, 383–384
Excultibrachium concinnum, 355
Exoconularia, 119, 120, 121*f*, 122, 123
Exoconularia consobrina, 122
Exoconularia exquisita, 122
Exoconularia pyramidata, 122
extinction (generic- and species-level data), 287*f*, 288*f*, 289*f*, 289, 290*t*, 297*f*, 298, 301*f*, 303*f*, 305*f*, 309*f*, 310; per capita rates of extinction/million years, 57, 152–153, 153*f*, 185, 192*f*, 194, 207*f*, 287*f*, 288*f*, 289*f*, 290*t*, 291, 291*f*, 292*t*, 297*f*, 298, 302, 303*f*, 304, 305*f*, 306, 309*f*, 309, 310, 319*t*, 320, 333*t*, 334; percentage of origination, 57, 207*f*, 287*f*, 288*f*, 289*f*, 290*t*; rates of extinction, 115*f*, 116–117, 130*f*, 132, 139*f*, 140, 143*f*, 144, 183*f*, 183, 185, 188, 189*f*, 190, 192*f*, 194, 224*f*, 225, 241*f*, 319*t*, 320, 333*t*, 338*f*, 340*f*
extinction events, minor intra-Ordovician, 3, 262, 321, 323–324. *See also* Darriwilian origination/extinction events
extinction rate measures, 56, 57, 185, 286, 296

FAD (first appearance datum), 102, 158, 246, 295–296, 298, 300, 307–308
Fahraeusodus, 324
Fallaticella, 262
Fardenia, 166
Fasciocoma, 162
faunal/floral turnover, 57, 241*f*, 286, 287*f*, 288*f*, 289*f*, 290*t*, 315, 318*f*, 333*t*, 338*f*, 340*f*; correlation of origination and extinction rates (r^2 values), 291–292, 292*t*; per capita rate of turnover per million years, 57, 207*f*, 287*f*, 288*f*, 289*f*, 290*t*, 291; percentage of turnover, 57, 207*f*, 287*f*, 288*f*, 289*f*, 290*t*, 291. *See also* extinction (generic- and species-level data); origination (generic- and species-level data); turnover ratio (TR)
Favistina, 127, 128, 134, 142, 142*f*, 143
Favositina, 125, 141
favositinids, 126, 142
Fenestrata, 150, 151*t*
fenestrate bryozoans, 149, 150*f*, 152, 152*f*
Fenestrospongia, 105
Fenhsiangia, 333
Fennian stage, graptolites, 42*f*–43*f*
filter-feeding trilobites, 251
first appearance datum (FAD), 102, 158, 246, 295–296
Fisherites, 338*f*, 339
fletcheriellid gen. indet., 134*t*, 135*f*
Foerstellites, 213
Foerstephyllum, 126, 127, 134*t*, 136, 141, 142*f*
Foerstephyllum wissleri, 127
Foliomena fauna, 171, 176
foraminiferans, 5*t*, 10
Frankea, 359
Furcatoporella, 342

Gacella, 166
Gamachian stage: conodonts and chitinozoans, 44*f*–45*f*; corals, 135, 137; sea level curves, 91*f*
Gamalites, 216*f*, 217
gamma diversity, 14, 15, 52
Gårdlösa Drowning Event, 87

Gasconsia, 163
Gasmascolex, 229
Gastrochaenolites, 374
Gastrochaenolites oelandicus, 374
gastropods, 4*t*, 5*t*, 10, 30, 184–195
Geisonoceras, 213
Geisonocerina, 213
genus turnover rates per lineage million years (Lma), brachiopods, 158, 160*f*, 163*f*, 165*f*, 168*f*
Georginidae, 211
geosynclinal facies, 25
Girvanella, 18, 19, 341–343
Gisbornian stage: corals, 141–142, 142*f*; graptolites, 42*f*–43*f*; trilobites, 240–241, 241*f*
glaciation. *See* Hirnantian glaciation
Gloeocapsomorpha prisca, 4
Glossifungites, 370
glossograptids, 293
Glyptarca, 200, 201*f*
Glyptarcoidea, 200
Glyptoconularia, 119, 120
Glyptoconus, 321
glyptocystitid rhombiferans, 270*f*, 277, 278*f*
Gnathichnus, 370
gnathostomes, 34, 328, 331
Gomphoceras, 213
Gompholites, 217
Gondwana supercontinent, 61, 62–63; conulariids, 122; glaciation and, 81–82, 82*f*; ostracodes, 265
Goniophora, 202, 203*f*
Goniophorina, 202, 203*f*
Gotlandochiton, 180, 181*f*
graptolites, 4*t*, 5*t*, 10, 11, 20, 33–34, 281–293; Australasia, 42*f*–43*f*, 48; integrated zonal framework, 45–46; "running diversity curve," 7
?*Graticula*, 343
green algae, 12, 20, 35, 344–346
Grewingkia, 134*t*, 135, 135*f*, 136, 137, 139*f*, 142, 142*f*
Grimsøya Regressive Event, 89, 90, 92
Gryphochiton, 179
Guttoporella densa, 346

Hadimopanella, 229, 229*f*
Halichondrites, 103
Halimeda, 346
Haljala stage: conodonts and chitinozoans, 44*f*; graptolites, 42*f*–43*f*, 86*f*; ostracodes, 262
halkieriids, 31
Hallopora, 154
halloporines, 150*f*, 152, 152*f*
Halysis, 341, 343
Halysites, 126, 134*t*, 142*f*
Halysitina, 125, 141, 142*f*
halysitinids, 126, 142

"Hangenberg" extinction event, Devonian-Carboniferous boundary, 116
Haplistion, 105
Harju (Baltoscandian series), 45; chitinozoans, 301*f*; conodonts and chitinozoans, 44*f*–45*f*; echinoderms, 268*f*–271*f*; graptolites, 42*f*–43*f*; ostracodes, 262*f*; trilobites, 246*f*; tube-shaped incertae sedis, 216*f*, 217*f*
Harperopsis ohensis, 263
Harpetidae, 233, 234*f*
Hedstroemia, 342, 343
Hedstroemia-Ortonella, 342
Helcionelloida, 182
Helicelasma, 139, 139*f*, 142, 142*f*
Helicotomidae, 190
Heliolitina, 125, 141, 142*f*
heliolitinids, 126, 142, 143
Helminthiodichnites, 375
Helminthochiton, 179, 181*f*
Helminthochiton? aequivoca, 179, 180
Helminthochiton griffithi, 179
Helminthopsis, 376, 377
Helminthopsis abeli, 376
Hemithecella, 180, 181*f*
Herochitina gamachiana, 306
Heteractinida, 103*f*, 106, 108*f*, 109*f*, 110
Heterobranchia, 185, 190, 191
heteroconchs, 200, 201*f*
Hexactinellida, 103*f*, 105–106, 108, 108*f*, 109*f*, 110
Hexitheca, 216*f*, 217
Hibbertia, 251
Hillophyllum, 127, 128, 142, 142*f*
Hindella, 171
Hindia sphaeroidalis, 105
Hintzespongiidae, 105
Hippocardia, 205*f*, 206
Hirnantia fauna, 82, 83, 166, 171, 240
Hirnantian (late Ashgill) extinction, 11, 34, 110, 115–116, 132–133, 140, 145, 153, 169, 183, 212, 262–263, 265, 274–275, 279, 291*f*, 293, 298, 302, 310. *See also* Ashgill origination/extinction events; end Ordovician mass extinction
Hirnantian glaciation, 27, 34, 47, 71, 75, 81, 82, 83, 117–118, 132–133, 140, 145, 208, 293, 298, 310, 341, 363
Hirnantian "Silurian-type" biotas: acritarchs, 35, 359–360; evidence of first vascular plants, 36, 363, 364*f*; vertebrates, 35, 334, 330*f*
Hirnantian stage, 7*f*, 47, 72, 90, 133, 140–141, 206; carbon and oxygen isotopes in, 81, 82; graptolites, 42*f*–43*f*
Hirsutodontus, 23

Index 477

Hiscobeccus, 173
Histiodella, 323, 324
Histiodella altifrons, 324
Histiodella sinuosa, 316t
holdover taxa, percentage of, conodonts, 319t, 319f, 320
Holorhynchus, 163
holothurians, 271f, 275, 276
Homalonotidae, 235t, 236t
Homalozoa, 267
homalozoans, 33, 268f, 271f, 272, 273
Homoctenida, 221
homoiosteleans, 267, 278f
homosteleans, 267
"*Hubeipora*," 149
Hunderum substage. *See* Kunda stage
Hunneberg stage: conodonts and chitinozoans, 44f; graptolites, 42f–43f, 86f; sea level changes, 87
Husbergøya Drowning, 89, 90
Hyattidina, 171
Hyattidina? sulcata, 171
Hydrodictyacea, 349
Hymenocaris, 258
Hyolitha, 214
Hyolithes, 216f, 217
Hyolithes kotoi, 218
Hyolithida, 214, 217f
hyoliths, 4t, 5t, 9, 31, 214–218, 216f
Hyolithus pennatuloides, 216
Hypermecaspis, 238f, 251
Hyperobolus-Talasotreta association, 161
Hysterolenus fauna, 240, 248
"hystrichospheres," 348
hystricurids, 233, 242
hystricurines, 32

Iapetognathus-Cordylodus fauna, 313
Iapetus Ocean, 2, 15–16, 66, 74, 78, 110, 313
Ibex Fauna (trilobite cohort), 232, 233, 234f, 242, 252
Ibexian (North American series), 45; acritarchs, 356f, 358f; bivalve mollusks, 199f, 201f, 203f; brachiopods, 159f, 165f, 166f, 168f–170f, 177f; bryozoans, 149f, 152f; chitinozoans, 44f–45f, 297f, 301f, 303f, 305f, 309f; conodonts, 44f–45f, 317f, 318f; conulariids, 121f; echinoderms, 268f–271f; graptolites, 42f–43f; jawed polychaetes, 224f; nautiloid cephalopods, 210f; palaeoscolecidans and chaetognaths, 229f; phyllocarids, 260f; rostroconch mollusks, 205f; sea level curves, 91f; stromatoporoids, 113f; trilobites, 243f, 245f, 246f, 248f, 249f; tube-shaped incertae sedis, 216f, 217f; univalve mollusks, 181f
Ichangian stage, graptolites, 42f–43f
ichnofaunas, 372f; continental environments, 378; deep marine environments, 374–376; marginal marine environments, 377–378; shallow marine carbonate environments, 373–374; shallow marine siliciclastic environments, 371, 373; volcanic environments, 376
ichnology, 369–79
"ichthyoliths," 328
Idavere substage. *See* Haljala stage
Idiospira, 170
Illaenidae, 233, 234f, 235t, 236, 236t
illaenids, 242, 249
Illaenopsis, 253
Illaenus, 235, 238f
Imperfectotriletes, 361, 363
Inaniguttidae, 97, 100, 100f, 101
inarticulated brachiopods, 4t, 5t, 9, 10, 11, 158
Incaia fauna, 240
Inmostia, 342
Intermurella, 344, 345
"inventory" diversity, 14
Ischadites, 338f, 339
Ischaditidae, 338f, 339
Ischyrinia, 205f, 206
Ischyrinioids, 206
island arcs, 15–16, 61, 66–67; as biotal centers of origin and refugia, 15, 16, 108–109, 114, 164
Isocolidae, 234f, 235t, 236t
isocolids, 249
isograptids, 293
Isograptus fauna, 323
Isograptus manubriatus, 281
Isograptus victoriae, 281
isophorid edrioasteroids, 271f
Isotelus, 252, 253
isotopes: carbon, 69, 70f, 71, 81; glaciation and, 81, 82f; neodymium, 69, 71, 75f; Ordovician seawater, 68–71; oxygen, 69, 70f, 71, 81, 82; strontium, 68, 70f, 71, 79; sulfur, 71
Ivoechiton, 180, 181f

jawed polychaetes, 223–225, 224f
jawed vertebrates, 328
jawless vertebrates, 328
Jawonya, 106
Jbel Gaiz succession, 377
Jõhvi substage. *See* Haljala stage
Juab/Kanosh shift, 92
Juanognathus, 322
Juanognathus jaanussoni, 324
Juanognathus variabilis, 322, 324
Jumudontus gananda, 324

K-bentonite beds (or ash falls). *See* Deike and Millbrig K-bentonite beds; Kinnekulle K-bentonite bed; trans-Iapetus ash marker
Kaimenella Märss, 229
Kalochitina multispinata, 300
Kazakhstanelia, 344
Keelophyllum, 134t
Keila Drowning, 88, 92, 93
Keila stage, 47: conodonts and chitinozoans, 44f; graptolites, 42f–43f, 86f
Kellerella, 171
Kellerella ditissima, 171
"Kellerwassser" faunal crisis, end Frasnian, 117
"Kelly Creek Eustatic Event," 321
keratose demosponges, 102
Kiaerograptus Drowning Event, 87, 90, 93
Kilbuchophyllida, 125, 145
Kindbladochiton, 180, 181f
Kinnekulle K-bentonite bed, 33, 47, 261–262
Kinwowia, 342
Kometia, 106
Komstad Regressive Event, 87
Koraipsis, 240
Kozhuchinella, 167
"kukersite" oil shales, 4
Kukruse stage: conodonts and chitinozoans, 44f; graptolites, 42f–43f, 86f; ostracodes, 262
Kullervo, 167
Kunda stage, 47; chitinozoans, 301f; conodonts and chitinozoans, 44f–45f; echinoderms, 268f–271f; graptolites, 42f–43f, 86f; ostracodes, 262f; sea level changes, 87; trilobites, 246f; tube-shaped incertae sedis, 216f, 217f

Labechia, 114
Labechiella, 114
Labechiida, 112, 113f, 114, 115, 115f, 116, 117
Labechiidae, 113f, 115
Labyrinthites, 126, 134t
LAD (last appearance datum), 102, 158, 246, 295–296
Ladogella, 358
Lady Burn Starfish Beds, 277, 278f, 333
laevis-norrlandicus (L-N) interval, 313, 315, 323–324
Laevolancis, 363
Lagenochitina, 44f, 300, 301
Lagenochitina aff. *deunffi*, 308
Lagenochitina baltica, 303
Lagenochitina cf. *obelgis*, 307
Lagenochitina dalbyensis, 298
Lagenochitina destombesi, 300, 306
Lagenochitina deunffi, 308
Lagenochitina esthonica, 300
Lagenochitina langei, 299, 300
Lagenochitina obelgis, 307
Lagenochitina prussica, 300
Lagenochitina stentor, 306, 308

Lambelasma, 127, 128, 139, 139f
Lambelasmatidae, 138, 139
Lambeophyllum, 127, 128, 133
Lambeophyllum cf. *L. profundum*, 127
Lancefieldian stage, graptolites, 42f–43f
land plants, 362–365
Langenochitina obeligis, 299
Langevoja substage. *See* Volkhov stage
Langøyene Drowning Event, 89, 90
Lasnamägi stage: conodonts and chitinozoans, 44f; graptolites, 42f–43f, 86f
Last appearance datum (LAD), 102, 158, 246, 295–296
"Late Arenig-Early Llanvirn Lowstand Interval," 92, 93
late Ashgill. *See* Hirnantian stage
"Late Llanvirn-Caradoc Highstand Interval," 92, 93
Late Ordovician Epoch, 12, 14; Baltoscandia facies belts, 84–85, 85f; bivalve mollusks, 198; chitinozoans, 297, 298; corals, 138; echinoderms, 272; glaciation, 82; graptolites, 282f; nautiloid cephalopods, 212; sponges, 107–108; vertebrates, 334
"Late Tremadocian-Early Arenig Lowstand Interval," 90, 93
Latorp "superstage": chitinozoans, 301f; echinoderms, 268f–271f; ostracodes, 262f; trilobites, 246f; tube-shaped incertae sedis, 216f, 217f. *See also* Billingen stage; Hunneberg stage
Laufeldochitina aff. *stentor*, 308
Laufeldochitina stentor, 298
Laurentia, 14, 64; chitinozoans, 302–304, 303f; conulariids, 120; corals, 133–138, 134t, 135f, 144f, 145; ostracodes, 264; vertebrates, 334; Whiterock Fauna in, 237, 239
Lazarus taxa, 29, 151, 154, 154f, 156, 164, 247, 320
Lebesconteia, 257
Leiofusa, 355
Lenargyrion, 229
Lenodus, 324, 325
Lenodus variabilis, 316t, 324
Lenodus variabilis-Eoplacognathus suecicus transgression, 104
Leolasma, 128, 139f
Leolites, 216, 216f, 217
leperditiocopes, 264
Lepidocoleidae, 226
Lepidocoleomorpha, 226
Lepidocoleus jamesi, 227
Lepidocoleus strictus, 227
Lepidocoleus suecicus, 226f, 228
Lepidocyclus, 173

Lepidolites, 339
lepidopleurid, 179
lepidotrichia, 328
Leprotolypa, 359–360
Leptembolon, 161
Leptopoterion, 339
Leurocycloceras, 213
level-bottom communities, Ordovician radiation, 17–18, 174–177
Lewinia, 105
Lichenaria, 18, 125, 126, 127, 133
Lichenaria cloudi, 125
Lichenariina, 125
lichenariinids, 142
Lichidae, 234*f,* 235*t,* 236*t,* 252
Liliosphaeridium, 359
Liliosphaeridium pennatum, 359
Linearis Drownings, 89, 92, 93
Lingulida, 158, 159, 160, 160*f,* 163*f*
Linguliformea, 159–162, 163*f*
linguliformean brachiopods, 29, 158–162, 159*f,* 160*f,* 173, 175
Lissocoelia, 104
Lithistida, 18, 103–104, 106, 107, 108
lithofacies, 84–85
Litoceras, 211
Lituitidae, 211, 212
Llandeilian stage, graptolites, 42*f*–43*f,* 86*f*
Llandovery Series, miospores, 363, 364*f*
Llanvirn (British series), 47; acritarchs, 356*f,* 358*f,* 359; "Basal Llanvirn Bio-Event," 193–194; bivalve mollusks, 199*f,* 201*f,* 203*f;* brachiopods, 159*f;* bryozoans, 149*f,* 152*f;* chitinozoans, 44*f*–45*f,* 297*f,* 301*f,* 303*f,* 305*f,* 309*f;* conodonts, 44*f*–45*f,* 317*f,* 318*f;* conulariids, 121*f;* corals, 130*f,* 142*f;* cyclocrinitids, 340*f;* echinoderms, 268*f*–271*f;* graptolites, 42*f*–43*f,* 86*f,* 282*f;* jawed polychaetes, 224*f,* 227; miospores, 364*f;* nautiloid cephalopods, 210*f;* ostracodes, 262*f;* palaeoscolecidans and chaetognaths, 229*f;* phyllocarids, 260*f;* receptaculitids, 338*f;* rostroconch mollusks, 205*f;* sea level changes, 88; stromatoporoids, 113*f;* trilobites, 243*f,* 245*f,* 246*f,* 248*f,* 249*f;* tube-shaped incertae sedis, 216*f,* 217*f;* univalve mollusks, 181*f*
Lma (genus turnover rates per lineage million years, brachiopods), 158, 160*f,* 163*f,* 165*f,* 168*f*
Lobocorallium, 134*t,* 135, 135*f,* 136
longicones, 24–25, 25
Lophosphaeridium, 355

lower Middle Ordovician Stage (unnamed), 7*f,* 46
Lower Ordovician Series: acritarchs, 356*f,* 358*f;* Baltoscandian sea level changes, 85, 86*f,* 87; bivalve mollusks, 199*f,* 201*f,* 203*f;* brachiopods, 159*f,* 160*f,* 163*f,* 165*f,* 166*f,* 168*f*–170*f,* 177*f;* bryozoans, 149*f,* 152*f;* chitinozoans, 44*f*–45*f,* 297*f,* 301*f,* 303*f,* 305*f,* 309*f;* conodonts, 44*f*–45*f,* 314*t*–315*t,* 317*f,* 318*f,* 321–323; conulariids, 119, 120, 121*f;* echinoderms, 268*f*–271*f;* graptolites of Australasia, 42*f*–43*f;* jawed polychaetes, 224*f;* nautiloid cephalopods, 210*f;* North American sea level changes correlated with Baltoscandian, 90, 91*f,* 92; ostracodes, 262*f;* palaeoscolecidans and chaetognaths, 229*f;* phyllocarids, 260*f;* radiolarians, 99, 100*f;* receptaculitids, 338*f;* rostroconch mollusks, 205*f;* sea level changes, 85, 86*f,* 87, 90, 92; stromatoporoids, 113*f;* trilobites, 243*f,* 245*f,* 246*f,* 248*f,* 249*f;* tube-shaped incertae sedis, 216*f,* 217*f;* univalve mollusks, 181*f;* vertebrates, 329–330
Lowvillia, 344, 346
Loxonematoidea, 186, 189*f,* 193
Lua, 350, 354
Lua erdaopuziana, 354
Ludvigsenites, 260
Lunoprionella, 224
Lyopora, 126, 142*f*
Lyrodesma, 199*f,* 200

Macerodus dianae, 315*t,* 322
machaeridians, 4*t,* 5*t,* 31, 223, 225–228, 226*f*
Macluritoidea, 30, 189, 189*f,* 190–191, 192*f*
Macquarie Arc, 15, 16, 141, 142
Madaraspis, 238*f,* 251
magnetic reversals, 78–79, 78*f*
Malacostraca, 257
Mamelohindia, 105
Manespira, 170
Manipora, 134*t,* 135, 135*f*
Manitobia, 342
manitouensis-deltifer (M-D) interval, 313, 315, 321
mantle superplume, 72, 74, 77–80
Maquoketa Group, 110, 331
Margaritichnus reptilis, 377
marine environments, ichnofaunas: deep marine, 374–376; marginal marine, 377–378; shallow carbonate, 373–374; shallow siliciclastic, 371, 373
Mastopora, 339, 340*f,* 344
Mastopora parva, 339

Matapedia Group, 375
Matthevia, 180
Maysvillian stage: conodonts and chitinozoans, 44*f*–45*f;* conulariids, 120; corals, 135; sea level curves, 91*f*
mean standing diversity (MSD), 52–57. *See also* balanced total diversity; normalized diversity; taxa (genera/species) per million years
Mediolites, 216*f,* 218
megalograptids, 255
Megalograptus, 256, 257
Megalograptus ohioensis, 256
Megamorina, 103*f,* 105, 107, 108*f,* 109*f*
Megistapidine Realm, 235, 235*f*
megistaspidine asaphids, 235
Mermia, 370
Mesonomia, 169
Metaconularia, 119, 120, 121*f,* 122, 123
Michelinoceras, 213
Michelinoceratida, 24
Microparia, 250
micro-atolls, 143
microbrachiopod assemblages, 160
microphytoplankton, acritarchs, 12, 13, 20, 35, 348–360
Microzarkodina, 314*t,* 324
Microzarkodina flabellum, 316*t,* 324
"Mid Arenig Highstand Interval," 92, 93
mid Caradoc, 11, 12, 19
mid Caradoc peak, 12
Mid Cretaceous Superchron, 78
Mid Ordovician Epoch, 11; Baltoscandia facies belts, 84–85, 85*f;* bivalve mollusks, 198; chitinozoans, 294; echinoderms, 272, 277; food resources in, 172; graptolites, 42*f*–43*f;* nautiloid cephalopods, 211; ostracodes, 262; sponges, 107; superplume, 78–79; use of term, 6
mid Tremadocian highstand, 90
Middle Ordovician Series: acritarchs, 356*f,* 358*f;* Baltoscandian sea level changes, 86*f,* 87–88; bivalve mollusks, 198, 199*f,* 201*f,* 203*f,* 204; brachiopods, 159*f,* 160*f,* 163*f,* 165*f,* 166*f,* 168*f*–170*f,* 177*f;* bryozoans, 149*f,* 152*f;* chitinozoans, 44*f*–45*f,* 297*f,* 301*f,* 303*f,* 305*f,* 309*f;* conodonts, 44*f*–45*f,* 316*t,* 317*f,* 318*f,* 323–325; conulariids, 119, 120, 121*f;* corals, 130*f,* 142*f;* cyclocrinitids, 340*f;* echinoderms, 268*f*–271*f;* graptolites, 42*f*–43*f,* 282*f;* jawed polychaetes, 224*f;* miospores, 364*f;* nautiloid cephalopods, 210*f;* North American sea level changes correlated with Balto-

scandian, 91*f,* 92; ostracodes, 262*f;* palaeoscolecidans and chaetognaths, 229*f;* phyllocarids, 260*f;* receptaculitids, 338*f;* rostroconch mollusks, 205*f;* stromatoporoids, 113*f;* trilobites, 243*f,* 245*f,* 246*f,* 248*f,* 249*f;* tube-shaped incertae sedis, 216*f,* 217*f;* univalve mollusks, 181*f;* use of term, 6; vertebrates, 330–331, 330*f*
Migneintian stage, graptolites, 42*f*–43*f*
Milaculum, 229, 328
Mimospira, 189
Mimospirina, 30, 186*f,* 189–190, 189*f,* 192*f*
Miospirina, 194
miospore, 35–36
mitrate stylophorans, 271*f,* 278*f*
Mixosiphonoceras, 213
Mochtyella, 224
Modern EF, 2*f,* 9, 10, 11, 12, 13, 266, 272, 380, 383–384
modiolopsids, 202
Mohawkian (North American series), 18, 45; acritarchs, 358*f;* bivalve mollusks, 199*f,* 201*f,* 203*f;* brachiopods, 159*f,* 165*f,* 166*f,* 168*f,* 169*f;* bryozoans, 149*f,* 152*f;* chitinozoans, 44*f*–45*f,* 297, 297*f,* 301*f,* 303*f,* 305*f,* 309*f;* conodonts, 44*f*–45*f;* conulariids, 121*f;* corals, 130*f,* 133, 142*f;* cyclocrinitids, 340*f;* echinoderms, 268*f*–271*f,* 274; graptolites, 42*f*–43*f;* jawed polychaetes, 224*f;* palaeoscolecidans and chaetognaths, 229*f;* phyllocarids, 260*f;* rostroconch mollusks, 205*f;* sea level curves, 91*f;* stromatoporoids, 113*f;* trilobites, 243*f,* 245*f,* 246*f,* 248*f,* 249*f;* tube-shaped incertae sedis, 216*f,* 217*f;* univalve mollusks, 181*f*
mollusks, 4*t,* 5*t,* 10, 30; bellerophontiform, 182, 183; bivalves, 4*t,* 5*t,* 10, 30, 196–204; rostroconchs, 4*t,* 5*t,* 10, 11, 30, 204–206, 205*f,* 207–208, 207*f;* univalves, 4*t,* 5*t,* 179–183
Monaxonida, 102–103, 103*f,* 106, 108*f,* 109*f*
Mongolepida?, 328
mongolepids, 328
Mongolitubulus, 334
Moniliporella, 342, 345, 346
monobathrid camerates, 278*f*
Monocraterion, 373, 377
Monomorphichnus, 371, 375
Monoplacophora, 181, 182, 186
monoplacophorans, 9, 30
Moridunian stage, graptolites, 42*f*–43*f*
morphological diversity, 2, 36, 387

Moyeria, 361
MSD (mean standing diversity), 52–54. *See also* balanced total diversity; normalized diversity; taxa (genera/species) per million years
Multiplicisphaeridium, 355, 359
Multiplicisphaeridium irregulare, 355
Multivasculatus, 105
Murchisonia, 187
Murchisonioidea, 187
Myodakryotus, 204
Mytocaris, 257, 258
Myzostomites, 374

Nabala stage: conodonts and chitinozoans, 44f; graptolites, 42f–43f, 86f
Nakkholmen Drowning Event, 92, 93
Nanopsis nanella, 261
Nasutimena, 173
Naticopsidae, 191
nautiloid cephalopods, 4t, 5t, 24–27, 30–31, 209–213, 210f
Neichianshanian stage, graptolites, 42f–43f
nematomorphs, 228
Neocramatia, 166
neodymium isotopes, seawater, 69, 71, 75f
Neoloricata, 179
Neomphalidae, 190
Neomphalus, 190
Neopilinidae, 182
Nephrotheca, 216f, 217
Neritimorpha, 185, 186, 186f, 189f, 190, 191, 192f, 194
Nervolites, 216f, 217
Neseuretus biofacies, 252
New South Wales, corals, 126, 127, 130, 143
New South Wales island arc, 25. *See also* Macquarie Arc
New Zealand: corals, 141; trilobites, 239–242, 241f
Newfoundland, conodonts, 314t, 321
Nexospongia, 105
Nexospongiidae, 105
Nidulites, 339, 340f
Nikolaispira, 171
Nikolaispira rasilis, 171
nileids, 242
Nobiliasaphus, 252
nonarticulated brachiopods, 158
Noradonta, 200
normalized diversity (Cooper's best approximation to true MSD), 8, 53f, 54, 54f, 99t, 103f, 115f, 130f, 132, 139f, 143f, 143–144, 149f, 151, 152, 183f, 183, 185, 189f, 190, 192f, 207f, 210f, 224f, 225, 241f, 243f, 244, 245f, 248f,
286, 287f, 288f, 289f, 290t, 291, 291f, 315, 318f, 319t, 320, 338f, 340f; corrected first and last appearance data (FADs and LADs) approximates normalized diversity, brachiopods and trilobites, 158, 164, 165f, 168f, 170f, 246, 246f; diversity curve shown as logarithmic plot, 152, 152f. *See also* balanced total diversity
North Africa, acritarchs, 357–360, 358f
North America: acritarchs, 354–355; Baltoscandian sea level curves correlated with, 90, 91f, 92; chitinozoans, 44f–45f; conodonts, 44f–45f, 314t–316t; conulariids, 120; graptolites, 42f–43f
North China: conodonts, 325; corals, 131
North Gondwana: chitinozoans, 44f–45f, 46, 296–299, 297f; conodonts, 44f–45f, 46
Nothozoe, 257, 258
Notonychia, 201f, 202
Novantiella, 344
Novocrania, 159
nuculoids, 197, 198, 199f
Nuia, 342, 344
numbers of taxa per sample, radiolarians, 98, 98t
Numericoma, 162
Nyctopora, 126, 134t, 135f, 142, 142f

Oandu stage: conodonts and chitinozoans, 44f; graptolites, 42f–43f, 86f
obolids, 161
Obruchevella, 341
ocean–atmosphere system, 72–75
oceans: climate and, 72–75; glaciation and, 81; isotope chemistry of seawater, 68–71
odontodes, 328
odontodium, 328
Odontopleuridae, 233, 234f, 235t, 236t, 238f, 252
Oeland (Baltoscandian series), 45, 47; conodonts and chitinozoans, 44f–45f; graptolites, 42f–43f
Oelandodus, 322
Oenonites, 224
Oepikina, 173
Oepikodus communis, 314t, 322
Oepikodus evae, 23, 314t, 322
Oepikodus evae transgression, 104
Oepikodus intermedius, 314t, 324
ogygiocarinine asaphids, 235
Oldhamia, 342, 375
olenid trilobites, 250
Olenidae, 233, 234f
Omphalotrochidae, 190
Oncocerida, 211, 212
Onniella meeki, 374
Onychochilidae, 189
Onychoplecia, 166
Onychopterella, 256
Onychopterella augusti, 257
Ooidium, 358
Oopsites, 250
"Open-Sea Realm" (OSR), conodonts, 22
ophiocistioids, 271f, 273, 275, 276
Opipeuter, 23, 238f, 250
Orbipora, 149
Orbiporidae, 149
Orchocladina, 103f, 104, 106, 107, 108, 109f, 110
ORDOCON ("Ordovician Conulariid Species and Genera"), 119
Ordovician biodiversity: continental platforms, 13–14; diversity measures, 7–8; earlier work, 9–27; floral patterns, 12–13; fossil groups described, 4–6, 4t, 5t, 8–9; geotectonic settings, 15–16; overview of, 2f, 10–12; sampling biases, 16–17; sea level changes, 84–93; stratigraphy, 41–45; timescale and time slices, 6–7, 7t, 10, 41, 46–47, 50, 50f
Ordovician "bioerosion revolution," 19, 374
Ordovician Period: extinction episodes, 3. *See also* Ordovician biodiversity; Ordovician timescale
Ordovician radiation, 2, 13, 171, 174, 380–387; in faunal communities, 19, 29; level-bottom communities, 17–18; pelagic organisms, 19–27; reef communities, 18, 19
Ordovician System, 41, 43
Ordovician timescale, 6–7, 7t
Ordovicidium, 359
origination (generic- and species-level data), 241f, 262, 287f, 288f, 289f, 290t, 289, 287f, 288f, 289f, 290t, 289, 297f, 298, 301f, 302, 303f, 305f, 309f, 310; per capita rate of origination/million years, 57, 152–153, 153f, 185, 192f, 192–193, 207f, 287f, 288f, 289f, 290t, 291, 291f, 292t, 297f, 298, 301f, 302, 303f, 304, 305f, 309f, 309, 310, 319t, 320, 333t, 334; percentage of origination, 57, 207f, 287f, 288f, 289f, 290t; rates of extinction 115f, 116–117, 130f, 132, 139f, 140, 143f, 183f, 183, 185, 189f, 190, 192–194, 192f, 224f, 225, 319t, 320, 333t, 338f, 340f
origination rate measures, 56, 57, 185, 286, 296
Oriostomatidae, 191
Orthida, 158, 167–169, 168f
orthide brachiopods, 173, 176
Orthidina, 167
Orthisocrania, 163
Orthoceras, 211
Orthocerida, 209, 211, 212
orthocerids, 25, 26, 27, 30–31
Orthogastropoda, 186, 187
Orthoidea, 157, 167
Orthonychia, 186f
Orthonychiidae, 191
Orthosphaeridium, 355, 359
Orthosphaeridium insculptum, 355
Orthosphaeridium rectangulare, 355
Orthotetida, 158, 165, 166, 176
Orthothecida, 214, 217f
Ortonella, 201f, 341
OSR ("Open-Sea Realm"), conodonts, 22
Osteostraci?, 328
ostracodes, 4t, 5t, 9, 32–33, 260–265, 262f
Ovatoconcha, 198
Oxoplecia, 166
oxygen isotopes: glaciation and, 81, 82; seawater, 69, 70f, 71, 81

Pachystylostroma, 114
Padunoceratidae, 211
Paenetradium, 130, 132
Pakerort stage: conodonts and chitinozoans, 44f; graptolites, 42f–43f, 86f
Palaeocopa, 261, 263
Palaeomanon, 104
Palaeophycus, 371, 373, 375
Palaeophycus tubularis, 376
Palaeophyllum, 128, 134t, 135f, 136, 139, 139f, 142, 142f, 143
Palaeoporella, 344, 346
Palaeopteria, 202, 203f
Palaeosabella, 374
Palaeosaucus, 105
Palaeoscenididae, 100, 100f, 101
Palaeoscolecida, 229
palaeoscolecidans, 4t, 5t, 31, 223, 228–229, 229f
?*Palaeoscolex*, 229, 229f
Palaeostrophia, 169
Paleoalveolites, 129, 130f, 131, 132, 141, 142f
Paleoalveolites carterensis, 131
?*Paleoalveolites explanatus*, 131
Paleoalveolites tasmaniense, 131
Paleoalveolitidae, 125, 129
paleoclimate modeling, 73
Paleodictyon, 375
Paleofavosites, 126, 134t, 135f, 137, 142f
Paleoloricata, 179
Paleozoic EF, 2f, 9–10, 11, 12, 13, 75, 79, 80, 162, 171, 174–177, 232, 266–267, 272, 276, 279, 380, 383–384
Paliphyllidae, 139

Paliphyllum, 128, 134t, 139f
Palmatohindia, 105
Palmatoporella lata, 346
Palmatoporella stena, 346
Palmichnium, 377
Paltodus, 322
Paltodus deltifer, 313, 314t, 321
palynomorphs, 361
Panitheca, 216f, 217
Paoshanella, 219
parablastoids, 270f, 278f
Parabolinella, 251
Parachaetetes, 342
Parachaetetes cf. *compacta*, 342
Paracordylodus, 322
Paracordylodus gracilis, 314t, 322
Paracraniops, 163
paracrinoids, 270f, 273, 278f
Paragastropoda, 189
Parallelodus, 202, 203f
Parapaltodus simplicissimus, 324
Parapanderodus, 321
Parataenidium monoliformis, 377
Paratetradium, 129, 130f, 131, 132, 134t
Paratetradium clarkei, 131
Paroistodus, 322, 324
Paroistodus originalis, 324
Paroistodus proteus, 23, 321
particle-feeding trilobites, 252
Pasceolaceae, 340, 340f
Pasceolus, 340, 340f
Patellogastropoda, 185, 186, 186f, 190
Paterinida, 159, 163f
Paterula, 160
Pattersonia, 106
Pattersoniidae, 106
Pauxillites, 216f, 217
pelagic bohemillids, 250
pelagic organisms, 19–27
pelagic realm, 11, 19–27
pelagic telephinids, 235
pelagic trilobites, 250
Pelagiellida, 182
Pelicaspongiidae, 106
Penelope effect, 320
Pentamerida, 158, 169, 170f, 176
Pentameridina, 169
Pentameroidea, 157
Peri-Gondwana, ostracodes, 265
Periodon, 322, 324
Periodon-Cordylodus? biofacies, 324
Periodon flabellum, 324
Periodon gladysi, 324
Periodon-Texania biofacies, 323
Peritritoechia, 167
Peruneloidea, 191–192
Perunica terrane, 61, 62, 65–66; conulariids, 122; ostracodes, 263
Petalichnus, 377
Peteinosphaeridium, 355, 359
Peteinosphaeridium velatum, 359
Petrocrania, 162, 163
Petrophyton, 342

Petrophyton kiaeri, 343
Petroxestes, 374
Phacopina, 252
phacopoids, 32, 252
phaenophytes, 342
Phakelodus, 230
Phakelodus elongatus, 230
Phakelodus tenuis, 230
Pharcidoconcha, 201f, 202
Phenacoon, 358
Phialaspongia, 104
Phorocephala, 23, 250
Phosphannulus, 220
"*Phosphannulus*-type" bryoniids, 220, 221
Phosphatocopida, 260
Phragmodus undatus, 345, 346
Phycodes, 371, 373, 376
Phylactolaemata, 149
phyllocarid crustaceans, 4t, 5t, 24, 32, 257–260, 258t, 260f
phytoplankton, 12–13
"*Phytopsis*," 131, 132, 130f, 142f
"*Phytopsis*" *cellulosum*, 131
"*Phytopsis*" *variabile*, 142
Phytopsis cellulosum (tetradiid coral), 129
Phytopsis Hall 1847, 129
Phytopsis tubulosum (wormlike trace fossil), 129
pilekiids, 242
Piloceratidae, 211
Pinnocaris, 205f, 206
Pircanchaspis, 333
Pircanchaspis rinconensis, 330
Pirgu stage: conodonts and chitinozoans, 44f; graptolites, 42f–43f, 86f; ostracodes, 262
Placodermi?, 328
Plaesiomys, 173
Plagiothyridae, 191
planktic graptolites, 20
plankton, 19–20
Planolites, 371, 374, 375, 377
Planolites beverleyensis, 376
Planolites montanus, 376
Plasmopora, 134t, 137, 142f
Plasmoporella, 134t, 142f, 143
Plasmuscolex, 229, 229f
Platyceratidae, 191
Platycopa sensu lato, 261
Plectambonitoidea, 157, 164
Plectolites, 211
Plectorthoidea, 157, 167
plectorthoids, 167
Plectostrophia, 169
Pleurotomarioidea, 187
Plexa, 342, 345
Plexodictyon, 114
pliomerids, 242
Plumulites, 226f, 227, 228
Plumulites gummunsoensis, 227
Plumulites primus, 227
Plumulitidae, 226
Poidilofusa, 359
Pojetaia, 198
"*Polonodus*" *corbatoi*, 322

polychaetes, 5t; jawed, 223–225, 224f
Polycostalis, 264
Polycostatus oneotensis, 322
Polygonium, 355
Polylasma, 162
Polylopia, 219
polyplacophorans, 4t, 5t, 10, 30, 179–181, 181f
Polytoechioidea, 157, 166, 167
Pontides terrane, 61, 62f, 63f, 66; conulariids, 122
Porambonitoidea, 157
Porites, 143
Porkuni stage: algae, 345; conodonts and chitinozoans, 44f; graptolites, 42f–43f, 86f
Porophoraspis crenulata, 330–331
Praeaechmina, 264
Praoistodus parallelus, 322
prasinophytes, 12
Preacanthochiton, 180, 181f
Precordillera terrane, 61, 66, 78, 243, 314t, 316t, 321, 322, 324
predators, trilobites, 252–253
priapulids, 228
Pricyclopyge, 23
pricyclopygine, 250
Primicorallina, 344, 346
Primitophyllum, 127, 128, 139, 139f
Prioniodus, 322
Prioniodus adami, 314t, 321
Prioniodus elegans, 23, 314t, 322
Prioniodus-Texania biofacies, 322
Prisochiton, 180, 181f
Proauloporа, 341
Productida, 158
proetids, 32
Prolobella, 204
Propora, 134t, 137, 142f
Protambonites, 167
Proterocameroceratidae, 211
Protoceratoikiscum, 100, 100f
protoconodonts, 230
Protocycloceratidae, 211
Protokionoceras, 213
Protopanderodus, 322, 324
Protopanderodus gradatus, 322, 324
Protopanderodus leonardii, 322
Protoprioniodus, 323
Protoprioniodus aranda, 324
Protorthida, 158, 167
Protosagitta, 229
"*Protoscolex*," 229, 229f
Protospongia, 105
Protospongiidae, 105
Protovirgularia, 373, 377
Protrochiscolithus, 134t, 135, 135f
Protyria, 139, 139f
Proventocitum, 100f, 101
Pseudarca, 200
Pseudobelodina dispansa, 343, 345
Pseudochaetetes, 342
Pseudoconularia, 119, 120, 121f, 122, 123
Pseudocrania, 162

Pseudolancicula, 106
Pseudopholidops, 162, 163
Pseudorthocerida, 209, 211, 212
Pseudosagitta maxima, 230
Pseudostylodictyon, 114
Psiloconcha, 198, 199f
Psilonichnus, 370
Pteracontiodus, 324
Pteracontiodus cryptodens, 316t, 323
pteraspidomorphs, 328, 329, 331
pterioideans, 202
pteriomorphians, 201–204, 201f, 203f
Pterygometopidae, 234f, 235t, 236t
ptilodictyines, 150f, 152, 152f
"ptychoparioids," 252
Pulchrilamina, 18, 114
Pulchrilaminidae, 112, 113f, 114
Pumilocaris, 258
Pycnaspis, 328
Pycnostylus, 134t, 137
Pygodus, 325
Pygodus anserinus, 343, 345
Pyrrhophyta, 348
Pyruspongia, 106
Pyruspongiidae, 106

quadripartite fission, corals, 129
Quadrotheca, 216, 216f, 217
Quasiaulacera, 116
Quepora, 126, 142f

radiolarians, 4t, 5t, 21, 28, 97–101, 100f
Radiostroma, 114
Railites, 218
Railtonella, 240
Rakvere stage: conodonts and chitinozoans, 44f; graptolites, 42f–43f, 86f
"range-through" diversity, 8, 113f, 115f, 116, 241f, 241, 244–245, 245f, 247, 248f
Rangerian stage, 46; conodonts and chitinozoans, 44f–45f; sea level curves, 91f
Raphiophorid biofacies, 251
Raphiophoridae, 233, 234f, 235t, 236, 236t
raphiophorids, 32, 242, 249, 253
Rawtheyan stage, 90; graptolites, 42f–43f
Raymondaspis, 235
Receptaculites, 337
Receptaculitida, 337–339
Receptaculitidae, 338f, 339
receptaculitids, 4t, 5t, 18, 35, 336–337, 338f
Recilites, 216, 216f, 217
red algae, 12, 35, 342–344
Red River-Stony Mountain Province, 110; corals, 133–134, 133f, 135–136
reedocalymenine calymenids, 235
Reedocalymenine Realm, 235, 235f, 236

reef communities, 18, 19
reef mounds, 104, 107
reefs, 112–118
relative reversal of magnetic field, 78f
remopleuridids, 251
remopleuridioids, 250
Renalcis, 18, 341, 342
Reutterodus, 322
Reutterodus andinus, 314t, 322–324
Rhabdelasma, 142, 142f
Rhabdinopora flabelliformis series, 46, 282f, 283
Rhabdinopora praeparabola, 282f, 283
Rhabdinopora socialis, 87
Rhabdochitina, 307
Rhabdochitina turgida, 307
Rhabdochitina usitata, 307
Rhabdopleurella, 345
Rhabdoporella, 344, 346
Rhabdotetradium, 129, 130, 130f, 131, 132, 134t, 135f, 141, 142f, 143
Rhabdotetradium apertum, 129, 132
Rhabdotetradium cf. *syringoporoides*, 131
Rhabdotetradium cribriforme, 130, 132, 142, 143
Rhabdotetradium cylindricum, 130, 131
Rhabdotetradium duplex, 142
Rhabdotetradium floriforme, 129
Rhabdotetradium frutex, 130
Rhabdotetradium nobile, 129, 130
Rhabdotetradium nobile-cribriforme group, 130, 132
Rhabdotetradium quadratum, 132
Rhabdotetradium subapertum, 132
Rhabdotetradium syringoporoides, 130, 131
Rhabdotetradium? tessulatiformis, 132
Rhabdotetradium? tessulatus, 132
Rhegmaphyllum, 134t, 137
Rhinopterocaris, 258
rhipidocystids, 270f, 273, 278f
Rhizomorina, 103f, 105, 107, 109f, 110
rhodophytes, 342–344
rhombiferans, 270f, 275, 277, 278f
Rhopalocoelia, 104
Rhynchonellata, 167–171
Rhynchonellida, 158, 169–170, 176
Rhynchonelliformea, 157, 164–171
rhynchonelliformean brachiopods, 29, 158, 159f, 162, 175, 176, 177f
rhyniophytes, 36
Ribeiria, 205f, 206
Ribeiriidae, 205
ribeirioids, 204, 205–206

Ribeiroida, 205
Richmond Province, 110; corals, 133, 135, 136–137
Richmondian stage; conodonts and chitinozoans, 44f–45f; corals, 135; sea level curves, 91f
richthofenioids, 157
Robergia, 251
Robustum, 180, 181f
Rodingan Movement, 144
Rodinia supercontinent, 73
Ropalonaria, 374
Rossodus, 324
Rossodus barnesi, 322
Rossodus manitouensis, 314t, 321, 322
rostroconch mollusks, 4t, 5t, 10, 11, 30, 204–206, 205f, 207–208, 207f
Rostroconchia, 204–206, 205f
Rothpletzella, 342
Rowellella, 160
Rozmanospira, 170
rugose corals, 28, 33, 125, 127–128, 133, 135f, 145; Baltoscandian, 138–141, 138f, 139f; Laurentian, 133–138, 133f, 134t, 135f
Rugospongia, 105
Rusophycus, 371, 373, 377
Rusophycus latus, 373, 377

Sacabambaspis, 328, 331, 334
Sacabambaspis janvieri, 331
Saccocaris, 257, 258
Saccospongia, 105
Saccospongiidae, 105
Saffordophyllum, 126, 134t, 135, 135f
Sagavia, 250
Sagenachitina oblonga, 307
Saka substage. See Volkhov stage
Salopiella, 219
Salvadorea, 134t, 135, 135f
"sampled" diversity, 8, 241f, 241, 244–245, 245f, 247, 248f
sampling biases, 15–17
Sandivia, 331
Sandivia augusta, 332
Sandivia melnikovi, 332
Sarcinulina, 125, 141, 142f
sarcinulinids, 126, 137, 141
Scaphelasmatidae, 160, 161
Scaphopoda, 196
scavengers, trilobites, 252–253
Scenedesmaceae, 349
sclerite-bearing machaeridians, 223, 225–228, 226f
sclerites, 31
sclerosponges, 106
scolecodonts, 4t, 5t, 31, 223–225, 224f
Scolopodus, 322
Scolopodus krummi, 324
Scoyenia, 370
scyphozoans, 119

sea level changes, 84–93, 86f–87f, 91f; depositional model, 84–85, 85f; Lower Ordovician, 85, 87, 90, 92; Middle Ordovician, 86f, 87–88, 92; Upper Ordovician, 88–90, 92–93
sea level curves, 27
sea level oscillations, 73
seawater: carbon isotopes, 69, 70f, 71, 81; glaciation and, 81; isotope chemistry, 68–71; neodymium isotopes, 69, 71, 75f; oxygen isotopes, 69, 70f, 71, 81, 82; strontium isotopes, 68, 70f, 71, 79; sulfur isotopes, 71
Secuicollactinae, 100, 100f
Seleneceme, 251
"Selenimorpha," 30, 187–188, 188f, 189f, 192f, 194
Selenoides, 338f, 339
Selenopeltis Province, 235
Semibolbina, 263
Semicorallidomus, 202, 203f
Septadella, 260
septatrypinids, 170
Serpulospira 186f
Setula, 341
"Shallow-Sea Realm" (SSR), conodonts, 22
Shuguria, 341
Shumardiidae, 234f, 238f
shumardiids, 242, 253
Siberian Platform, 18; chitinozoans, 308; corals, 129, 132; ostracodes, 264
Sigmagraptidae, 293
sigmagraptids, 293
Silicuncudus, 106
Sinacanthus, 331
Sinodictyon, 114
sinograptids, 293
Sinuatoporella, 344
sinuosa-suecicus (S-S) interval, 313, 315, 324–325
Siphonotretida, 158, 159, 160, 160f, 163f
Skenidioidea, 157, 167
Skiichthys, 328, 331, 333
Skolithos, 370, 373, 377
Skullrockian stage: conodonts and chitinozoans, 44f–45f; sea level curves, 91f
slit-bearing Archaeogastropoda, 187–188
"snowshoe morphologies," 174, 226
solemyoids, 198–199, 199f
Solenocaris, 180, 181f
Solenopora, 342, 343
Solenopora compacta, 343
Solenopora gotlandica, 343
Solenoporaceae, 342
"solenoporaceans," 343–344
solenoporans, 12, 35, 342
Solenotheca, 216, 216f, 217
solutans, 267

Songxites, 253
South America: ostracodes, 265; phyllocarids, 259–260, 260f; trilobites, 242–244, 243f; Whiterock Fauna, 32
South China, 14, 61, 64; acritarchs, 355–357, 356f; trilobites, 247–249, 248f, 249f; Whiterock Fauna, 32
Southern Hemisphere, terranes, 62f, 63f, 65f
Sowerbyellidae, 165
Spannslokket Drowning, 89, 90, 92, 93
Sphaerellarians, 100
Sphaerocladina, 103f, 104, 107, 109, 109f, 110
Sphaerocodium, 342
sphenothallids, 4t, 5t, 31, 219–220
Sphenothallus, 219–220
sphinctozoans, 103f, 106, 107, 108, 109, 109f
Spinachitina, 44f, 300
Spiriferida, 158, 170–171, 176
spirigerinids, 170
sponge-microbial facies, 104
sponges, 4t, 5t, 10, 18, 28, 102–110, 108f; demosponges, 102–105; diversification pattern, 103f, 106–109, 106f; diversity curve, 103f; heteractinids, 103, 106; hexactinellids, 103f, 105–106; monaxonids, 102–103, 103f; sphinctozoans, 103f, 106, 107; stromatoporoids, 4t, 5t, 28, 112–118
Spyoceras, 213
SSR ("Shallow-Sea Realm"), conodonts, 22
Stairsian stage: conodonts and chitinozoans, 44f–45f; sea level curves, 91f
standing diversity, comparison with corrected diversity, brachiopods,166f, 169f
Stauridae, 139
Stauriina, 128, 141, 142f
stauriinids, 142
stauromedusans, 119
Stellechinatum, 359
Stelliferidium, 358, 359
Stelliferidium stelligerum, 359
Stelliferidium striatulum, 359
Stelodictyon, 114
Stelterella, 216f, 217
stenolaemate bryozoans, 9, 150, 150f
Stolodus stola, 324
Stonehenge Drowning Event, 93
stratigraphy, 41–45; conodont and chitinozoan zonal sequences, 44f–45f; correlation of data and calibration, 48–51; graptolite zonal sequences, 42f–43f; isotopic chemistry of seawater, 68–71; stratigraphic ties with

sea level curves, 91f; timescale nomenclature, 6–7
Stratodictyon, 114, 115
Streffordian stage, graptolites, 42f–43f
Streptelasma, 127, 128, 134t, 135f, 137, 139f, 140, 142, 142f
Streptelasma cf. *S. expansum*, 127
Streptelasmatidae, 138, 139
Streptelasmatina, 128, 141, 142f
streptelasmatinids, 128, 142
Streptosolen, 104
Streptosolenidae, 104, 108, 108f
Striatoteca, 359
Striatotheca, 350
Stricklandioidea, 157
Striola, 344
Stroinolepis maenniki, 332
Stromatoceriidae, 113f, 114
Stromatocerium, 114, 115, 127–128
stromatoporoids, 4t, 5t, 28, 112–118, 127–128, 142; distribution of, 114–116, 115f; diversity patterns, 115f, 116–118
Strombodes, 139, 139f
strontium isotopes, seawater, 68, 70f, 71, 79
Strophomenata, 164–167
Strophomenida, 158, 164, 165f, 166f
strophomenides, 174
Strophomenoidea, 157, 164, 165f, 166f
Stultodontus, 321
Styginidae, 234f, 235t, 236t, 238f
styginids, 242
stylonurid eurypterids, 255
stylophorans, 271f, 275, 276, 277, 278f
Stylostroma, 114
Subtifloria, 341
?*Subtifloria*, 342
Subulites, 192
Subulitoidea, 186, 188f, 189f, 191–192, 193
Suecoceras, 211
Sulcatospira, 170
sulfur isotopes, seawater, 71
superplumes, 77–80; mantle superplume, 72, 74, 77
suspension feeders, 9
suspension-feeding trilobites, 32, 251–252
Syltrochos, 104
Symphysops, 250
Syntrophiidina, 169
syringoporoid-like auloporinids, 126

tabulate corals, 28–29, 124, 125–127, 145
Taconic Orogeny, 14, 75, 79, 110, 274
Taeniidium, 375
Taklamakania, 251
Tanuchitina anticostiensis, 300
Tanuchitina bergstroemi, 303
Taoqupognathus blandus, 345, 346
Taplowia, 105
Tarphyceratidae, 211
Tarphycerida, 24, 25, 209, 211, 212
tarphycerids, 30
tasmanitids, 4
Tasmanoconularia tuberosa, 119
taxa (genera /species) per million years, 53f, 53, 54f, 115f, 116, 139f, 183, 183f, 185, 189f, 190, 192f, 207f, 224f, 225, 286, 287f, 288f, 289f, 290t
Technophorus, 205f, 206
tectonic events, 72–75, 110. *See also* volcanism
Teganiidae, 106
Teichichnus, 371
Telephina, 23, 238f, 250
Telephinidae, 23, 234f, 250
telephinids, 235, 242
Tentaculita, 221
Tentaculites, 221, 222
Tentaculites anglicus, 222
Tentaculites lowndoni, 222
Tentaculites richmondensis, 222
Tentaculites sterlingensis, 222
Tentaculitida, 221
tentaculitids, 214, 221–222
Tentaculitoidea, 221
Teredolites, 370
Tergomya, 181
Teridontus, 23
terranes: described, 61–67; paleographic reconstruction of, 74f
"*Tesakoviaspis*," 328, 332
tetradiid corals, 29, 124, 125, 128–133, 130f, 141, 145
Tetradiida, 134t, 141, 142f; taxonomic relationships, 128–131
Tetradiidae, 128, 129
Tetradium, 129, 130, 130f, 131, 134t, 135f, 142f
Tetradium aff. *fibratum*, 131
Tetradium cribriforme, 130
Tetradium, 129
Tetradium marylandicus, 131
Tetradium (P.) *huronensis*, 130
Tetraphalerella, 173
Tettragonis, 338f, 339
Texania, 323
Texturata, 342, 345
Thalassinoides, 19, 373
thallophytes, 12
Thamnobeatricea, 113, 114
Thelodonti, 328
thelodonts, 34, 328, 331, 332
Thuringia, ostracodes, 265
Tilasia, 161
time slices, 7, 7t, 10, 41, 46–47, 50, 50f
timescale, 6–7, 48
Timofeevia, 358
Titerina, 229f, 230

Titerina rokycanensis, 230
Tollina, 134t, 135f
Tolmachovia, 205f
tommotiids, 31
Toquimiella, 106
Torynelasmatidae, 160, 161, 162
total diversity, 53f, 53, 54f, 56, 99t, 135f, 183f, 185, 189f, 192f, 207f, 285–286, 287f, 288f, 289f, 290t, 315, 318f, 319, 319t, 320, 333t
TR (turnover ratio), chitinozoans, 296 297f, 298, 300, 301f, 302, 303, 304, 305f, 309, 309f
Trabeculites, 134t, 135, 135f
trace fossils, 36; significance of, 370–371. *See also* ichnofaunas
trachelocrinid eocrinoids, 277
trackways, eurypterids, 257
trans-Iapetus ash marker, 47, 78
Transgressive System Track (TST), 307
Trapezognathus, 323
Trapezotheca, 216f, 217
Tremadoc (British series): acritarchs, 356f, 358f; bivalve mollusks, 199f, 201f, 203f; brachiopods, 159f, 165f, 166f, 168f–170f, 177f; chitinozoans, 297f, 301f, 303f, 305f, 309f; conodonts, 317f, 318f; conulariids, 121f; echinoderms, 268f–271f; graptolites, 282f; jawed polychaetes, 224f; nautiloid cephalopods, 210f; ostracodes, 262f; palaeoscolecidans and chaetognaths, 229f; phyllocarids, 260f; receptaculitids, 338f; rostroconch mollusks, 205f; stromatoporoids, 113f; trilobites, 243f, 245f, 246f, 248f, 249f; tube-shaped incertae sedis, 216f, 217f; univalve mollusks, 181f
Tremadocian highstand, 90
Tremadocian Stage (ratified global Stage), 7f, 45, 46; acritarchs, 356f, 358f; bivalve mollusks, 199f, 201f, 203f; brachiopods, 159f, 165f, 166f, 168f–170f, 177f; bryozoans, 149f, 152f; chitinozoans, 44f–45f, 297f, 301f, 303f, 305f, 309f; conodonts, 44f–45f, 317f, 318f; conulariids, 121f; echinoderms, 268f–271f; graptolites, 42f–43f, 86f, 282f, 290f, 291f; jawed polychaetes, 224f, 227; nautiloid cephalopods, 210f; ostracodes, 262f; palaeoscolecidans and chaetognaths, 229f; phyllocarids, 260f; receptaculitids, 338f; rostroconch mollusks, 205f; stromatoporoids, 113f; trilobites, 243f, 245f, 246f, 248f, 249f; tube-shaped incertae sedis, 216f, 217f; univalve

mollusks, 181f; vertebrates, 329–330, 330f, 334
Trematidae, 161–162
Trepostomata, 150, 152
Trichichnus, 373
Trichophycus, 373, 377
Tricranocladina, 103f, 105, 107, 109, 109f, 110
trigonioids, 199f, 200
Trigonocarys, 257–258
trilobites, 4t, 5t, 9, 10, 11, 17, 18, 23–24, 26, 31–32, 231–253; adaptive deployment, 249–253; atheloptic, 253; Australia and New Zealand, 239–242, 241f; Avalonia, 244–246, 245f, 251; Baltica, 246–247, 246f; filter-feeding, 251; global patterns of, 231–233, 234f, 235–237, 235f, 235t, 236t, 238f–239f, 239; olenid biofacies, 250–251; particle feeders, 252; pelagic biofacies, 250; predators/scavengers, 252–253; South America, 242–244, 243f; South China, 247–249, 248f, 249f; suspension-feeding, 32, 251–252
Trimerellida, 158, 159, 162, 163, 163f
trimerellides, 163
Trinucleidae, 233, 234f, 235t, 236t, 251
trinucleids, 32, 235, 242, 249
Triplesia, 166
Triplesiidae, 158, 166
Tripodus laevis, 316t, 323
Tritoechia, 162, 167
Trocholitidae, 211, 212
"Trochomorpha," 30, 187, 188–189, 188f, 189f, 192f, 194
Tromelinodonta, 199f, 200
Tropodus, 322, 323
Tropodus australis, 322, 324
"true" mean standing diversity (MSD), 8, 285
Trunculumarium, 358
Tryblidiida, 181, 182–183, 186
tryblidiids, 30, 179, 181–183
Trypanites, 370, 374
Tryplasma, 134t, 135f, 136, 137, 139, 139f
Tryplasmatidae, 138–139
TST (Transgressive System Track), 307
Tuarangia, 196, 198
tube-shaped incertae sedis, 214–222
Tubomorphophyton, 342
Tubophyllum, 341
Tulean stage: conodonts and chitinozoans, 44f–45f; sea level curves, 91f
Turinian stage: conodonts and chitinozoans, 44f–45f; sea level curves, 91f

turnover ratio (TR), chitinozoans, 296, 287f, 298, 300, 301f, 302, 303f, 304, 305f, 309, 309f
Turrilepadidae, 226
Turrilepadomorpha, 226

udotaceans, 12, 35
Udoteaceae, 346
Uhaku stage: conodonts and chitinozoans, 44f; graptolites, 42f–43f, 86f
Ullernelasma, 139, 139f
Ulotrichales, 344
Ulvophyceae, 344
Undulograptus austrodentatus, 282f, 283
Undulograptus dentatus, 282f, 283
Uniconus, 222
univalve mollusks, 4t, 5t, 179–183
upper Lower Ordovician Stage (unnamed), 7f, 46
upper Middle Ordovician, radiolarians, 99, 100, 100f, 101. *See also* Darriwilian Stage
Upper Ordovician Series, 47; acritarchs, 358f; Baltoscandian sea level changes, 86f, 88–90; bivalve mollusks, 198, 199f, 201f, 203f, 204; brachiopods, 159f, 160f, 163f, 165f, 166f, 168f–170f; bryozoans, 149f, 152f; chitinozoans, 44f–45f, 297f; conodonts, 44f–45f; conulariids, 119, 120, 121f; corals, 130f, 142f; cyclocrinitids, 340f; echinoderms, 268f–271f; graptolites of Australasia, 42f–43f; jawed polychaetes, 224f; miospores, 364f; nautiloid cephalopods, 210f; North American sea level changes correlated with Baltoscandian, 91f, 92–93; ostracodes, 262f; palaeoscolecidans and chaetognaths, 229f; phyllocarids, 260f; radiolarians, 99, 100–101, 100f; receptaculitids, 338f; rostroconch mollusks, 205f; stromatoporoids, 113f; trilobites, 243f, 245f, 246f, 248f, 249f; tube-shaped incertae sedis, 216f, 217f; univalve mollusks, 181f; vertebrates, 330f, 331–333
Uralella, 344
Uralichas, 252
Utahphospha, 229

Vääna substage. *See* Volkhov stage
Valaste substage. *See* Kunda stage
Valyalepis, 331
Vanuxemia, 201f, 202
Varangu stage: conodonts and chitinozoans, 44f; graptolites, 42f–43f, 86f
Variabiloconus bassleri, 322
Variabiloconus transiapeticus, 321
Vavrdovella, 359
Vellamo, 167
Vermiporella, 344, 345, 346
vertebrates, 4t, 5t, 327–334, 333t
Veryhachium, 355
Vietnamostroma, 114
Villosacapsula, 359
Villosacapsula setosapellicula, 355
Villosoporella, 342, 345
Virgatasporites, 361
Virgoceras, 213
Viru (Baltoscandian series), 45, 47; chitinozoans, 44f–45f, 301f; conodonts, 44f–45f; echinoderms, 268f–271f; graptolites, 42f–43f; ostracodes, 262f; trilobites, 246f; tube-shaped incertae sedis, 216f, 217f
Vltaviellidae, 191
Vogtlandia, 359
volcanism, 72, 75, 78–79, 110; mantle superplume, 72, 74, 77–80
Volkhov stage, 46, 87; chitinozoans, 44f–45f, 301f; conodonts, 44f–45f; echinoderms, 268f–271f; graptolites, 42f–43f, 86f; ostracodes, 262f; trilobites, 246f; tube-shaped incertae sedis, 216f, 217f
Vormsi stage: conodonts and chitinozoans, 44f; graptolites, 42f–43f, 86f
Vuelta de Las Tolas Member, 376
Vulcanisphaera, 358

Wagga Sea (marginal sea), 25
Wagima, 106
Walliserodus, 324
Warrigalia, 105
Webbyspira, 170
Weibeia, 171
West Gondwana, 62; chitinozoans, 299–300
Westonia, 342
Wetherdella, 342
Whiteavesia, 203f, 342
Whiterock Fauna (trilobite cohort), 13, 31, 32, 232, 233, 234f, 235t, 236, 237, 239, 253
Whiterockian (North American series), 17, 45, 46; acritarchs, 356f, 358f; bivalve mollusks, 199f, 201f, 203f; brachiopods, 159f, 165fl, 166f, 168f–170f, 177f; bryozoans, 149f, 152f; chitinozoans, 44f–45f, 297f, 301f, 303f, 305f, 309f; conodonts, 44f–45f, 317f, 318f; conulariids, 121f; corals, 130f, 142f; cyclocrinitids, 340f; echinoderms, 268f–271f; graptolites, 42f–43f; jawed polychaetes, 224f; nautiloid cephalopods, 210f; palaeoscolecidans and chaetognaths, 229f; phyllocarids, 260f; rostroconch mollusks, 205f; sea level curves, 91f, 92; stromatoporoids, 113f; trilobites, 243f, 245f, 246f, 248f, 249f; tube-shaped incertae sedis, 216f, 217f; univalve mollusks, 181f
Whitfieldella, 171
Whitlandian stage, 46; graptolites, 42f–43f
Winnepegia, 342; worms, 5t
Wutinoceratidae, 211

xanioprionids, 224
Xenorthis, 167

Yangtze Platform, acritarchs, 355–357
Yaoxianognathus lijiapoensis, 343, 345
Yaoxianognathus? tunguskaensis, 346
Yapeenian stage, 46; graptolites, 42f–43f
"Yichangopora," 149

Zadimerodia, 200
Zangerlispongia, 106
zhanatellids, 161
Zhexiella, 264
Zondarella, 114
Zoophycos, 370
Zuegelepis, 331
Zygnemataceae, 349
Zygnematales, 349
Zygospira, 170